ANCIENT SCIENCE THROUGH THE GOLDEN AGE OF GREECE

George Sarton

DOVER PUBLICATIONS, INC.
New York

Published in Canada by General Publishing Company, Ltd., 30 Lesmill Road, Don Mills, Toronto, Ontario.
Published in the United Kingdom by Constable and Company, Ltd., 3 The Lanchesters, 162–164 Fulham Palace Road, London W6 9ER.

This Dover edition, first published in 1993, is an unabridged and unaltered republication of the W. W. Norton & Co. 1970 edition of the work first published by The Harvard University Press, Cambridge, Massachusetts, in 1952 under the title *A History of Science, Volume I: Ancient Science Through the Golden Age of Greece.*

Manufactured in the United States of America
Dover Publications, Inc., 31 East 2nd Street, Mineola, N.Y. 11501

Library of Congress Cataloging-in-Publication Data

Sarton, George, 1884–1956.
Ancient science through the golden age of Greece / George Sarton. — Dover ed.
p. cm.
Originally published: Cambridge : Harvard University Press, 1952–59.
Includes bibliographical references and index.
ISBN 0-486-27495-0 (pbk.)
1. Science—History. 2. Science, Ancient. 3. Civilization, Ancient. I. Title.
Q125.S238 1993
509.3—dc20 92-38351
CIP
AC

This book is gratefully dedicated
to my colleague and friend
Werner Jaeger

PREFACE

Many years ago, soon after the publication of volume 1 of my *Introduction*, I met one of my old students as I was crossing the Yard, and invited him to have a cup of coffee with me in a cafeteria of Harvard Square. After some hesitation, he told me, "I bought a copy of your *Introduction* and was never so disappointed in my life. I remembered your lectures, which were vivid and colorful, and I hoped to find reflections of them in your big volume, but instead I found nothing but dry statements, which discouraged me." I tried to explain to him the purpose of my *Introduction*, which was severe and uncompromising; a great part of it was not meant to be read at all but to be consulted, and I finally said, "I may be able perhaps to write a book that pleases you more."

Ever since, I have often been thinking of this book, which reproduces not the letter but the spirit of my lectures. It was written primarily for my old students and for historians of science, all of whom have been my companions as readers of *Isis* and *Osiris*, and many of whom have worked with me or helped me in various ways. It was written also for educated people in general, but not for philologists.

This requires a word of explanation. I am not hostile to philologists and am in some respects one of them, though they would probably repudiate me. Nature is full of wonderful things — shells, flowers, birds, stars — that one never tires of observing, but the most wonderful things of all to my mind are the words of men, not the vain multiplicity of words that flow out of a garrulous mouth, but the skilful and loving choice of them that falls from wise and sensitive lips. Nothing is more moving than the contemplation of the means found by men to express their thoughts and feelings, and the comparison of the divers means used by them from time to time and from place to place. The words and phrases used by men and women throughout the ages are the loveliest flowers of humanity. There is so much virtue in each word; indeed, the whole past from the time when the word was coined is crystallized in it; it represents not only clear ideas, but endless ambiguities; each word is a treasure house of realities and illusions, of truths and enigmas. That is why I so often pause in my thought, speech, or writing and wonder what this or that word really signifies. Such preoccupation will frequently obtrude itself in my book, especially in the footnotes, which indifferent readers can easily skip if they wish.

And yet my scientific studies have been too deep and too long to make me feel at ease with philologists, or they with me. As far as I can judge, my interest in languages is more genuine than the interest of the average philologist in science. My main regret as a teacher of ancient science is that my large audiences hardly ever included students of classical philology, and yet my course might have been a revelation to them; the probable reason for their absence was that their faculty advisers were not concerned about science, nor even about the history of science. Too bad!

The book is not written for classical philologists, but rather for students of science whose knowledge of antiquity is rudimentary, who may never have studied Greek, or whose knowledge of it was too shallow to endure. Therefore, my Greek quotations are restricted to the minimum and are always translated, and I explain many things that every philologist already knows. On the other hand, I explain scientific matters as much as can be done briefly; complete scientific explanations are out of the question, for one cannot teach science and the history of science at the same time.

My teaching of the history of science was divided into four courses, dealing respectively with antiquity, the Middle Ages, the fifteenth to the seventeenth centuries, and the eighteenth century until now. Each of these courses extended to some 35 lectures, and its publication will require two volumes. This is thus the first of eight volumes. Each volume is complete in itself. The present one explains the development of science from the beginnings to the end of the Hellenic period.

As it took me two years to complete the full cycle of my lectures, I did not return to a definite subject, say Empedocles or Eudoxos, in a shorter interval of time. Now two years is for a wide-awake scholar a pretty long period. Many things may and do happen; memoirs and books are published that throw a new light on the subject; the very advance of science obliges one to reconsider old ideas; above all, I was changing. As a result of all this, I have never given twice the same lecture, and no lecture was ever fixed; they remained in a state of fluidity until now when the necessity of writing and printing freezes them. That freezing is very uncongenial to me, but it cannot be helped. I hope that some of my readers, at least, will unfreeze the printed words, and give a new life to them by their own critical attention.

The history of science is an immense field which it would be impossible to cover completely in a hundred or a thousand lectures, and I preferred to deal with a few selected subjects as well as possible rather than try the impossible. There is no space and time to say everything, but the selection of items is more careful and richer in this book than it could be in the spoken lectures.

For each selected topic, say Homer, it is impossible to state all the facts, nor is it necessary to do so. A few elementary things must be repeated, yet space must be kept for things that are less hackneyed and withal more important. In this I have been helped by my faith in the reader, who need not be told everything but requires only a few hints.

It is the eternal conflict between knowledge and wisdom. The known facts, the technical details, are fundamental but insufficient. They must be simplified, symbolized, and informed with a deeper understanding of the problems involved.

As I grew older my lectures became simpler; I tried to say fewer things and to say them better, with more humanity. This book continues in a different way the same evolution, but it is not yet as simple as I would have liked to make it.

Some technical questions of great difficulty have been left out, because the explanation of them to nonspecialists would have required considerable space and, what is worse, would have sidetracked their attention and diverted it from things of greater importance. The conflict between technique and wisdom existed in the past even as it does now, and there were then as now plenty of dunces who fussed about trifles and overlooked the essential.

The ability of nonintelligent people to understand the most complicated mechanisms and to use them has always been to me a cause of astonishment; their inability to understand simple questions is even more astonishing. The general acceptance of simple ideas is difficult and rare, and yet it is only when simple, fundamental, ideas have been accepted that further progress becomes possible on a higher level.

Erudition without pedantry is as rare as wisdom itself.

The understanding of ancient science has often been spoiled by two unpardonable omissions. The first concerns Oriental science. It is childish to assume that science began in Greece; the Greek "miracle" was prepared by millennia of work in Egypt, Mesopotamia and possibly in other regions. Greek science was less an invention than a revival.

The second concerns the superstitious background not only of Oriental science but of Greek science itself. It was bad enough to hide the Oriental origins without which the Hellenic accomplishment would have been impossible; some historians aggravated that blunder by hiding also the endless superstitions that hindered that accomplishment and might have nullified it. Hellenic science is a victory of rationalism, which appears greater, not smaller, when one is made to realize that it had to be won in spite of the irrational beliefs of the Greek people; all in all, it was a triumph of reason in the face of unreason. Some knowledge of Greek superstitions is needed not only for a proper appreciation of that triumph but also for the justification of occasional failures, such as the many Platonic aberrations.

If a history of ancient science is written without giving the reader a sufficient knowledge of these two groups of facts — Oriental science on the one hand and Greek occultism on the other — the history is not only incomplete but falsified.

My account is based as far as possible on the sources; I have always tried to go down to bedrock. Our documentation is often very imperfect. For example, primitive men applied a large amount of knowledge before they were conscious of having any; if they were not conscious of it, how could we be?

On the other hand, the documentation concerning Egyptian and Mesopotamian science is often very precise, more precise than that concerning Greek science. Indeed, Egyptologists and Assyriologists are priviledged to handle original documents, while Hellenists must generally be satisfied with fragments, with indirect quotations and opinions, with copies of copies many times removed from the originals. Sometimes a fair text has come down to us, say the *Iliad*, but the author has remained practically unknown; at other times, the author, say Thales or Epicuros, has been made familiar to us by various anecdotes, but the bulk of his writings has disappeared.

The historian must do his best within the limitations of each case. The "sources" are of different value. There is no harm in using poor ones, *faute de mieux*, as long as one remembers their nature and does not confuse *n*th-hand copies with originals, rumors with certainties. Indeed, there can hardly be real certainty in our knowledge of the past, but this does not weaken our responsibility.

The major part of this book is of necessity devoted to Greece, to a new or less known aspect of the glory that was Greece. Her early men of science were comparable in greatness to the great architects and sculptors, or to the poets and other men of letters. Scientific achievements seem evanescent, because the very progress of science causes their supersedure; yet some of them are of so fundamental a nature that they are immortal in a deeper way. Some of the conclusions reached by Eudoxos and Aristotle are still essential parts of the knowledge current today. Moreover, from the humanistic point of view every human achievement is unforgettable and immortal in its essence, even if it is replaced by a "better" one.

Greek culture is pleasant to contemplate because of its great simplicity and naturalness, and because of the absence of gadgets, each of which is sooner or later a cause of servitude.

The rationalism of the creative minds was tempered by abundant fantasies, and the supreme beauty of the monuments was probably spoiled by the circumambient vanities and ugliness; in a few cases, the Greeks came as close to perfection as it was possible to do, yet they were human and imperfect.

The most astounding feature of Greek science is to find in it so many adumbrations of our own ideas. To be ahead of other people, a thousand years ahead of them — that is genius indeed. The Greek genius appears as brilliantly in science as it does in art or literature; if we fail to appreciate the scientific aspect of it, we cannot say that we have really grasped it.

It is not enough to underline cultural anticipations; we must also recall everything in the present that may help us to understand the past, and everything in the past that may help us to understand the present, our own selves. For the artist, indeed, and for the philosopher, who are used to contemplating everything *sub specie aeternitatis*, there is no past, no future, but only an eternal present. Homer and Shakespeare are as alive today as they ever were; from the time of their first appearance they have always been on hand; it is we who were not.

Our account of the past is limited in many ways. One of the necessary limitations is that we must restrict ourselves to our own ancestry. Early Hindu science and Chinese science are generally left out, not because they lack importance, but simply because they lack signification for us Western readers. Our thinking has been deeply influenced by Hebrew and Greek thoughts, hardly any by Hindu or Chinese ones, and whatever influences came from southern and eastern Asia reached us in a roundabout way.

Our own culture, of Greek and Hebraic origin, is the one that interests us the most, if not exclusively. We do not say that it is the best culture, but simply that it is ours. To claim that it is of necessity superior would be wrong and evil. That attitude is the main source of international trouble in the world. If I were superior to my neighbors, it would be not for me but only for them to say so. If I should make a claim of superiority that they could not or would not confirm, that could only prepare enmity between us. The same is true in a more complex and deeper way whenever nations are compared. Each nation prefers its own usages.

My main interest, almost the only one, is the love of truth, whether pleasant or not, whether useful or not. Truth is self-sufficient, and there is nothing to which it can be subordinated without loss. When truth is made subservient to anything else, however great (say religion), it becomes impure and sordid.

My purpose is to explain the development not of any one science, but of ancient science in its wholeness. We shall consider problems of mathematics, of astronomy, of physics, of biology, but always in their mutual relation and with as good a comprehension as possible of their background. Our main interest is ancient culture, the whole of it, but is focused, as it should be, upon ancient science, ancient wisdom. Wisdom is not mathematical, nor astronomical, nor zoölogical; when it talks too much of any one thing it ceases to be itself. There are wise physicists, but wisdom is not physical; there are wise physicians, but wisdom is not medical.

The main misunderstandings concerning the history of science are due to historians of medicine who have the notion that medicine is the center of science. That misunderstanding was magnified by a great scholar, Karl Sudhoff, who was primarily a historian of medicine, a very distinguished one, but whose scientific (nonmedical) knowledge was insufficient.[1] Anyone who has a good scientific and philosophic mind realizes that there is a general hierarchy in the growth of knowledge: the simplest and most fundamental ideas are mathematical; if to space and number one adds the concept of time, one enters the mechanical field; other assumptions introduce us into the fields of astronomy, physics, chemistry. Or one might consider the earth in its past and present, and begin geographic and geologic studies; one might investigate seismologic problems, start the study of mineralogy and crystallography.

Thus far, all our thinking has concerned lifeless matter only. Add the idea of life and we introduce biology and all its branches, botany, zoölogy, paleontology, anatomy, physiology. One might then move up to a higher level still, and consider man, the spirit of man. This brings in the humanities and the social sciences.

All the branches of knowledge that have been enumerated may be and are applied to human needs of various kinds, and this introduces various applications, such as technology, medicine, education. It is true that in practice the applications have often preceded their own principles; early people were obliged to practice obstetrics and surgery long before they paid attention to anatomy or embryology. The order that was described above is the logical, not by any means the historical, order. The physicians preceded the physicists and chemists, yet it is the latter who gave tools to the former and not vice versa. We must see things in their proper perspective. The historical order is very interesting, but accidental and capricious; if we would understand the growth of knowledge, we cannot be satisfied with accidents, we must explain how knowledge was gradually built up. This does not mean that we should explain the history of mathematics first, then the history of mechanics, and so on. That method would be certainly wrong; we must proceed

[1] G. Sarton, "Acta atque agenda," *Arch. internat. d'histoire des sciences* 30, 322–356 (1951). Sudhoff was the founder of the *Mitteilungen zur Geschichte der Medizin und der Naturwissenschaften* (40 vols.; 1902–1942), which, as the title indicated, was devoted to the history first, of medicine, and second, of science.

from one chronologic level up to the following one, but on each level we must pay attention to mathematical ideas, then to physical ones, and so forth.

The problems of health vs. disease, of life vs. death are so important to the average man that he may be excused if he is led to believe that medicine is the hub of science. The philosopher and the mathematician are willing enough to concede the practical importance of those problems, but not their spiritual hegemony. They are deeply interested in other problems concerning the nature of God and of ourselves, the implications of number and continuity, space and time; the problems of life, but not simply of our own lives; the problems of equilibrium, not only those of our own health.

Medicine began very early, but it is not certain that it began before mathematics and astronomy. As a child I was thinking of numbers and shapes long before any medical idea ever crossed my mind. If I had been ailing or crippled, however, my scale of values and my perspective might have been different.

Men understand the world in different ways. The main difference lies in this, that some men are more abstract-minded, and they naturally think first of unity and of God, of wholeness, of infinity and other such concepts, while the minds of other men are concrete and they cogitate about health and disease, profit and loss. They invent gadgets and remedies; they are less interested in knowing anything than in applying whatever knowledge they may already have to practical problems; they try to make things work and pay, to heal and teach. The first are called dreamers (if worse names are not given to them); the second kind are recognized as practical and useful. History has often proved the shortsightedness of the practical men and vindicated the "lazy" dreamers; it has also proved that the dreamers are often mistaken.

The historian of science deals with both kinds with equal love, for both are needed; yet he is not willing to subordinate principles to applications, nor to sacrifice the so-called dreamers to the engineers, the teachers, or the healers.

This history of ancient culture, focused upon science, is of necessity a form of social history, for what is "culture" but a social phenomenon? We try to see the development of science and wisdom in its social background, because it can have no reality outside of it. Science could not develop in a social vacuum, and therefore every history of science, even of the most abstract one, mathematics, includes a number of social events. Mathematicians are men, subject to every kind of human fancy and frailty; their work may be and often is dominated by all kinds of psychologic deviations and social vicissitudes.

The psychologic reactions of individuals are innumerable and the social vicissitudes are caused by the endless and unpredictable conflicts of those reactions; the historian cannot possibly tell the whole story and the best that he can do is to select the conflicts that happen to be the most significant.

Under the influence of dialectical materialism, there has spread a belief that the history of science ought to be explained primarily, if not exclusively, in social and economic terms. That seems to me to be all wrong. Let me introduce a new dichotomy. There are two kinds of people in the world, whom we might dub the jobholders and the enthusiasts. The term jobholder is not derogatory; there are good jobholders and bad ones, and they are found at every social level from top to bottom. The majority of kings and emperors were jobholders and so were many of the popes. All these men were accomplishing duties connected with the jobs entrusted to them; they might hold, and often did hold, in succession different jobs, sometimes very different ones. The enthusiasts, on the contrary, are men who are anxious to do their own self-appointed tasks and can hardly do anything else. The term is not necessarily approbative; there are bad enthusiasts as well as good ones; some of them follow a mirage, they delude themselves as well as their neighbors; others are real creators. Indeed, most of the creators in the field of art and religion, and many of them in the field of science, were enthusiasts.

Now economic conditions may deeply affect the jobs and the jobholders, but they make little impression upon the enthusiasts. The latter must not be denied the primary needs of life, they must live, but as soon as those needs are satisfied in the humblest manner, the real enthusiasts bother about nothing but their work or their mission.

It is really the jobholders who keep things going with enough continuity and smoothness; they are the builders of usages and customs, the defenders of morality and justice. It is they who do all the routine work without which everything would soon degenerate into chaos, yet, by and large it is the enthusiasts who are the poets, the artists, the saints, the men of science, the inventors, the discoverers. They are the main instruments of change and progress; they are the real creators and troublemakers. The enthusiasts are the salt of the earth, but man cannot live by salt alone.

In this book pains have been taken to evoke the social background of living science, but no attempt has been made to explain the growth of science in terms of "diamat" jargon, because such an explanation would apply only, at best, to the jobholders, hardly to the enthusiasts, the crazy individuals, like Socrates, whom the menace of death could not swerve from their chosen path.

This book tries to show the growth of the human spirit in its natural background. The spirit is always influenced by the background, but its originality and integrity are in itself. A cabbage may grow better or worse in this or that field, but its cabbageness is in itself and nowhere else; if this is true of a humble cabbage, it is even more true of a man of genius. The ideas of men, however, are never completely independent and original; they hold together and form chains, the golden chains that we call traditions. Those chains are infinitely precious, but sometimes they may become embarrassing and dangerous; at their best they are light golden chains to which it is a joy and a pride to hold on; sometimes they become heavy like iron shackles and there is no way of escape but in breaking them. That has happened often, and we shall tell the story (it must be told) whenever it occurred. Such stories are part of the history of thought, but they are also essential parts of social history.

My insistence upon the necessity of referring, however briefly, to the old superstitions is a proof of my social concern. Science never developed in a social vacuum, and in the case of each individual it never developed in a psychologic vacuum. Every man of science was a man of his time and place, of his family and people, of his group and church; he was always obliged to fight his own passions and prepossessions as well as to assail the superstitions that clustered around him and threatened to choke out the novelties. It is just as foolish to deny the existence of those superstitions as it is to ignore contagious diseases; one must throw light upon them, describe them, and fight them. The growth of science implies at every step the fight against errors and prejudices; the discoveries are largely individual, but the fight is always collective.

Every good historian of science, not to mention every historian of medicine, is of necessity a historian of society, a social historian. How else could it be? The Russians' claim that their histories of science, or the histories inspired by them, are the first social histories is just moonshine. Like all fanatics, they are not interested so much in truth as in their "truth," which is incomplete, lopsided, and, *ipso facto*, false.

The history of science should not be used as an instrument to defend any kind of social or philosophic theory; it should be used only for its own purpose, to illustrate impartially the working of reason against unreason, the gradual unfolding of truth, in all its forms, whether pleasant or unpleasant, useful or useless, welcome or unwelcome.

At the moment of concluding a work that has occupied my mind for so many years, I wish to express my gratefulness to all the men whose own activities have made mine possible. My main debt is to nine scholars, three of them French, two German, two Belgian, one English, and one Danish — all dead. The earliest of these debts is to the brothers Croiset whose *Histoire de la littérature grecque* I bought and read when I was in the "rhétorique" of the Athénée of Chimay (=senior class in high school). These were the first important books that I bought (five big volumes); to them I owe my Greek initiation; I have treasured them ever since and I often consult them, for in addition to giving me the "first help" which I may need they evoke my youthful enthusiasm. Some of these volumes were written by Alfred, others by Maurice, but I was never able to distinguish them, and I thought of them both under a single name, Croiset. I am fully aware that much has been done since their time,[2] that much is known today that they did not know — many other books have taught me that — yet the criticism of many scholars, who were more learned than the Croisets but less sensitive, has not shaken my gratitude. It is they who awakened my admiration for the Greek genius.

At the University of Ghent, I worked for a time under Bidez — unfortunately, a very short time, because I soon abandoned the "faculté de philosophie et lettres" to begin scientific studies. Joseph Bidez influenced me, not so much then but later when I was separated from him by the Atlantic Ocean and more so by endless investigations that would have seemed irrelevant to him. It was he who introduced me (impersonally) to Franz Cumont and to Wilamowitz-Moellendorff. Bidez used the latter's *Griechiches Lesebuch* in his teaching and thus it happened that the first Greek scientific text that I read (very haltingly) was Hippocrates' treatise on the *Sacred disease*. My youthful impression of Greek science is as unforgettable and irrevocable as my first visions of the sea, of the high Alps, or of the desert.

Toward the end of my long scientific studies (during which time I had stopped completely my Greek studies and almost forgotten the language), Paul Tannery

[2] I used the second revised edition of vols. 1–4 (1896, 1898, 1899, 1900) and the first of vol. 5 (1899).

brought me back from science to the humanities. Thanks to his posthumous help, I learned to know of many other scholars, chiefly Diels and Heiberg. Later still, when I had moved to America and the English language was more familiar to me, I began to use more often the works of Thomas Little Heath.

Of these nine men,[3] I knew in the flesh only one, Bidez, and had some correspondence only with four, Bidez, Cumont, Heiberg and Heath. My debt to Tannery, the greatest of all, was partly repaid in an article on Paul, Jules, and Marie Tannery, *Isis* 38, 33–51 (1948) and in volume 4 of *Osiris*, dedicated to Paul and Marie. Volumes 2 and 6 of *Osiris* were dedicated respectively to Sir Thomas Heath and to Joseph Bidez. A biography of Heiberg appeared in *Isis* 11, 367–374 (1928). Cumont wrote a paper for *Isis* 26, 8–12 (1936), and many of his works, especially the catalogues of Greek astrologic and alchemical manuscripts that he inspired, were reviewed by me as they appeared.

It is better not to try to enumerate the Hellenists and men of science, now living in many countries, who have helped me in various ways, for my list would be incomplete and invidious. Whenever they greet me, I am happy to see them; whenever they write, I am grateful; and when I am writing to them, I am gladly aware of our common interests and our mutual debts. I do not always express my thanks but my heart is full of them. Above all, I share their delight in contemplating the greatest and purest achievements of mankind.

Cambridge, Massachusetts GEORGE SARTON
April 18, 1951

NOTES ON THE USE OF THIS BOOK

The following notes will help readers to make the best use of what I have to offer to them.

1. *Precautions and indecisions.* When we deal with ancient times our knowledge can never be certain, and the author is sorely tempted to recall his own uncertainty and indecision apropos of almost every statement. Yet, if he were to repeat continuously such phrases as "to the best of my knowledge," "as far as one has been able to ascertain," or simply "perhaps," the reader would lose patience. I have generally suppressed all these qualifications, though there are a few cases when I lacked the courage to take them off. The reader is here told once for all that all that I write is "to the best of my knowledge," and that whatever the results of my efforts may be, I am doing my best all the time, neither more nor less.

The same remark applies to dates. Should we say that Socrates was born "in 469 or 470" or "about 469," or give but one of these dates and let it go at that? I have tried to simplify my accounts but have not always been consistent. Some times I have been more definite than the available evidence warranted. Long discussions concerning very close dates seem to be nothing but futile pedantry. What difference can it make to anybody whether Socrates' birth year is 469 or 470 (it was the year 470–69).

2. *Chronology.* The foregoing paragraph does not mean that I do not attach

[3] A list of them, in order of death years, may be convenient: Paul Tannery (1843–1904); Hermann Diels (1848–1922); Alfred Croiset (1845–1923); Johan Ludvig Heiberg (1854–1928); Ulrich von Wilamowitz-Moellendorff (1848–1931), Maurice Croiset (1846–1935), Sir Thomas Little Heath (1861–1940), Joseph Bidez (1867–1945), Franz Cumont (1868–1947).

importance to dates. Dates are very important. A correct chronology is the skeleton of historiography. One could not take too many pains to set it right.

For Egyptian and Mesopotamian matters the best way of dating events is by the rule of such or such a king, or if that is not possible by dynasty. My way of indicating this is *x*th Dynasty (*y*–*z*), *y* and *z* being dates B.C. This equation is not always accurate; the original dating is the dynastic one, the second dating being added for the reader's convenience. Its validity may be questioned by some scholars, but it is not possible to reconsider the general problem of Egyptian (or Mesopotamian) chronology at every step. The reader is warned that the first dating may be uncertain, and that the second, which seems to be more precise, is in reality less so for it has the same uncertainty as the first plus new ones.

When referring to millennia, I generally write out third millennium, second millennium, first millennium, but without adding B.C. For dates in centuries or years, B.C. is generally left out, unless there is a danger of ambiguity. For instance, it is enough to say that Aristotle died in 322 (nobody would think that A.D. 322 is meant), while in the case of Vergil it is better to say that he died in 19 B.C. (he might have lived until A.D. 19). When two or more dates are given, no ambiguity is possible. For example, Tissaphernēs was satrap of western Anatolia from 413 to 408 and from 401 to his political murder in 395; that can only be B.C.

After the name of an author, say Diogenes Laërtios, there may be two kinds of indication: thus, x, 16–21 refers to chapters 16 to 21 of Book x of his *Lives of the philosophers*; III–1 means two things, first, that he flourished in the first half of the third century after Christ, and second, that a special section is devoted to him in my *Introduction*. That section occurs in vol. 1, p. 318, but those details are not added, being superfluous. There can be no confusion between those kinds of indication. In the second kind B.C. is always added if needed: for example, Hippocrates of Chios (V B.C.).

3. *Geographic names.* Geographic precision is just as necessary as chronologic precision. We should be able to locate each event in place and time. Therefore, pains have been taken to find whence every important man came and where he flourished. Strictly speaking, the terms used should be the ancient ones referring to ancient conditions. For example, in describing the voyage of a man sailing from Greece to the eastern coast of Thracia or to the northern coast of Paphlagonia one should say that he passed through the Hellēspontos, crossed the Propontis, sailed along the Bosporos, and thus reached the Pontos Euxinos. Such language would be correct but puzzling to men of science (not philologists). Therefore, I would rather say that the man sailed through the Dardanelles, the Sea of Marmara and the Bosporus, finally reaching the Black Sea. The things are the same; it is only the names that have changed. It is always better to be clear than pedantic, but I have not always been consistent.

4. *Bibliography.* Bibliographic references have been restricted to a minimum. In the case of an important text, the first Greek edition is mentioned, also the best and handiest ones, and finally the translation into English, or, if such does not exist, the translation into any other language of international currency.

References to my *Introduction* are not always given but are always implied, and readers are warned once for all that information concerning, say, Aristotle, is found not only in vol. 1 of the *Introduction* but also in vol. 2 and 3. It is good practice to consult first the index to vol. 3. No references are needed for statements that have become commonplace, but they are always given for novelties.

See the general bibliography.

5. *Quotations.* Quotations are always in English translation. As the Loeb Classical Library editions, which include an English translation opposite the Greek text, are especially convenient to English readers, reference has been made to them whenever possible. My quotations are not very numerous (that is, it would have been tempting to multiply them), but they have sometimes been extended beyond the immediate need, in order that the reader may sense the context. As abrupt quotations may be misleading and dangerous, it is better to avoid them.

6. *Transcription of Greek words in the English alphabet.* This is a moot question that has vexed my mind for half a century, and that cannot be answered to everybody's satisfaction, not even to the author's. Since the printing of Greek has become too onerous, the transcriptions had to be more accurate than in my *Introduction*, where the Greek forms are always given.

The diphthongs are written as in Greek with the same vowels (e.g., *ai*, not *ae; ei*, not *i; oi*, not *oe*), except *ou*, which is written *u* to conform with English pronunciation. The *omicron* is always replaced by an *o*, and hence Greek names are not Latinized but preserve their Greek look and sound. Our transcription has the advantage of distinguishing Greek writers like Celsos and Sallustios from Latin writers like Celsus and Sallustius. There is really no reason for giving a Latin ending to a Greek name, when one is writing not in Latin but in English. Hence, we write Epicuros, not Epicurus (the two *u*'s of that Latin word represent different Greek vowels!). When two *gammas* follow each other, they are transliterated *ng* to conform with pronunciation, e.g. *angelos, lyngurion.* In names ending in *on* we keep the final *n* instead of dropping it Latinwise. Thus, we write Heron, not Hero, but we found it impossible to write Platon. Old habits have probably introduced other inconsistencies, e.g., Achilles instead of Achilleys.

We indicate the differences between the short vowels *epsilon* and *omicron* and the long ones *ēta* and *ōmega* as we have just done in their names, but we had to abandon the idea of adding the accents because that would have given to many transcriptions such an outlandish look that the non-Greek reader would have been put off instead of being helped. As to the Greek reader, he does not need those indications; he knows how each word is accented, or, if not, he can easily find it in a dictionary or in my *Introduction.*

There remain inconsistencies in our transliteration because we prefer to be inconsistent rather than pedantic and do not wish to disturb our readers more than we can help. We hope that they will appreciate the situation and not judge us too severely. They should realize that English usage is full of inconsistencies, e.g., one writes habitually Aristarch*us* of Sam*os* and Eudox*us* of Cnid*os*. Old Greek names are transliterated in the Latin way, but Byzantine names otherwise (Psellos, Moschopulos); as to modern Greek names, one is obliged to respect the decisions of their bearers (Eleutheroudakis, Venizelos).

7. *Use of capitals.* We have tried to restrict capitals to proper words, and to use them sparingly for common words. There may be doubtful cases. For example, Earth, Moon, Sun are written with a capital when the celestial bodies are meant and not the common earth, the common sun, and moonshine.

CONTENTS

PART ONE

ORIENTAL AND GREEK ORIGINS

CONTENTS

EPILOGUE

PART ONE

ORIENTAL AND GREEK ORIGINS

I

THE DAWN OF SCIENCE

When did science begin? Where did it begin? It began whenever and wherever men tried to solve the innumerable problems of life. The first solutions were mere expedients, but that must do for a beginning. Gradually the expedients would be compared, generalized, rationalized, simplified, interrelated, integrated; the texture of science would be slowly woven. The first solutions were petty and awkward but what of it? A *Sequoia gigantea* two inches high may not be very conspicuous, but it is a *Sequoia* all the same. It might be claimed that one cannot speak of science at all as long as a certain degree of abstraction has not been reached, but who will measure that degree? When the first mathematician reognized that there was something in common between three palm trees and three donkeys, how abstract was his thought? Or when primitive theologians conceived the invisible presence of a supreme being and thus seemed to reach an incredible degree of abstraction, was their idea really abstract, or was it concrete? Did they postulate God or did they see Him? Were the earliest expedients nothing but expedients or did they include reasonings, religious or artistic cravings? Were they rational or irrational? Was early science wholly practical and mercenary? Was it pure science, such as it was, or a mixture of science with art, religion, or magic?

Such queries are futile, because they lack determination and the answers cannot be verified. It is better to leave out for the nonce the consideration of science as science, and to consider only definite problems and their solutions. The problems can be imagined, because we know the needs of man; he must be able to feed himself and his family, to find a shelter against the inclemencies of the weather, the attacks of wild beast or fellow men, and so on. Our imaginations are not arbitrary, for they are guided by a large number of observed facts. To begin with, archaeologic investigations reveal monuments which help us to realize the kind of objects and tools that our forefathers created and even to understand their methods of using them, and to guess their intentions. The study of languages brings to light ancient words which are like fossil witnesses of early objects or early ideas. Anthropologists have made us familiar with the manners and customs of primitive men who were living under their own eyes. Finally, psychologists have analyzed the reactions of children or of undeveloped minds in the face of the very problems that primitive men had to solve. The amount of information thus obtained from several directions is so large that a scholar's life is too short to encompass it. There is no place here for a review of it, however brief, but only for a few hints.

In order to simplify our task a little, let us assume that the primitive men we are dealing with have already solved some of the most urgent problems, for otherwise their very existence would have remained precarious, not to speak of their progress, material or spiritual. Let us assume that they have discovered how to make a fire and have learned the rudiments of husbandry. They are already — that is, some of them are — learned people and technicians, and they may already be speaking of the good old days when life was more dangerous but simpler and a man did not have to remember so many things. I say "speaking," for by this time they have certainly developed a language, though they are still unable to write it; indeed, they are still unconscious of the possibility of doing so. At this stage, and for a long time to come, writing is neither essential nor necessary. Our own culture is so closely dependent on writing that it requires some effort to imagine one independent of it. Man can go very far without writing,[1] but not without language. Language is the bedrock upon which any culture is built. In the course of time it becomes the richest treasure house of that culture.

One of the greatest mysteries of life is that the languages of even the most primitive peoples, languages that have never been reduced to writing (except by anthropologists), are extremely complex. How did those languages develop as they did? The development was very largely unconscious and casual.

Our reference to investigations made today by field anthropologists is sufficient warning that when we speak of the dawn of science or of any prehistoric period we are not thinking in terms of a chronologic scale of universal application. There is no such scale. The dawn of science occurred ten thousand years ago or more in certain parts of the world; it can still be witnessed in other parts today; and irrespective of place we can observe it to some extent in the mind of any child.

EARLY TECHNICAL PROBLEMS

Let us consider rapidly the multitude of technical problems that early men had to solve if they wished to survive, and, later, to improve their condition and to lighten the burden of life. They had to invent the making of fire and experiment with it in various ways. Not only the husbandman but also the nomad needed many tools, for cutting and carving, flaying, abrading, smoothing, crushing, for the making of holes, for grasping and joining. Each tool was a separate invention, or rather the opening up of a new series of inventions, for each was susceptible of improvements which would be introduced one by one. In early times there was already room for key inventions, which might be applied to an endless group of separate problems and which ushered in unlimited possibilities. For example, there was the general problem of how to devise a handle and how to attach it firmly to a given tool. Many different solutions were found for that problem, one of the most ingenious being that of the Eskimos and Northern Indians, namely, the use of babiche (strings or thongs of rawhide) by means of which the tool and handle are bound together; as the hide dries it shrinks almost to half its length and the two objects are inseparable. A tighter fit could hardly be obtained otherwise.

The husbandman had to discover the useful plants one by one — plants to use as food, or as drugs, or for other domestic purposes — and this implied innumerable

[1] Witness the Incas of Peru, whose civilization was very complex and advanced. They had an elaborate language but no system of writing, [*Isis* 6, 219 (1923–24)].

experiments. It was not enough for him to discover a plant; he had to select among infinite variations the best modalities of its use. He had to capture animals and to domesticate the very few that were domesticable,[2] to build houses and granaries, to make receptacles of various kinds. There must have been somewhere a first potter, but the potter's art involved the conscious or unconscious coöperation of thousands of people. Heavy loads had to be lifted and transported, sometimes to great distances. How could that be done? Well, it had to be done and it was done. Ingenious people invented the lever, the simple pulley, the use of rollers, and later, much later, that of wheels.[3] A potter of genius applied the wheel to his own art. How could a man cover his body to protect it from the cold or the rain or the burning sun? The use of hides was one solution, the use of leaves or bark another, but nothing equaled the materials obtained by the weaving of certain fibres. When this idea occurred to a great inventor, the textile industry was born.[4] The earliest tools were made of stone or bones; when the practical value of metals was finally realized it became worth while to dig for their ores and to smelt them, to combine them in various ways; this was the beginning of mining and metallurgy. Each of the sentences of this paragraph could easily be expanded into a treatise.

In order to illustrate the almost uncanny ingenuity of "primitive" people, it may suffice to display the three following examples, taken in three parts of the world very distant from each other. The Australian boomerang is so well known that it hardly requires discussion; it is a missile weapon the curved shape of which is so cunningly devised that the weapon when thrown describes extraordinary curves and may even return to the sender. The South American[5] tipiti is an elastic plaited cylinder of jacitara-palm bark which is used to express the juice of the cassava (or manioc); as the cylinder is lengthened, by the weight of a stone or otherwise, the internal pressure increases and the juice flows out. This invention is admirable in its simplicity and effectiveness, but what is more astonishing is that the Indians were able to discover the great nutritive value of cassava. The juice contains a deadly substance (hydrocyanic acid) which must be removed by cooking; otherwise, the consumer would be killed instead of nourished. How did the Indians find the treasure which could be enjoyed only after the poison spoiling it had been removed? My third example is the *li* 鬲, a tripod used in China in prehistoric

[2] William Henry Hudson remarked, "It is sad to reflect that all our domestic animals have descended to us from those ancient times which we are accustomed to regard as dark or barbarous, while the effect of our modern so-called humane civilization has been purely destructive to animal life. Not one type do we rescue from the carnage going on at an ever-increasing rate over all the globe." *The naturalist in La Plata* (London: Chapman and Hall, 1892), p. 233. The only animal domesticated in historic times is the ostrich [*Isis* *10*, 278 (1928)]; this was a poor achievement, which was justified only because some women and generals wanted feathers for their hats.

[3] Wheels remained unknown in the Americas; see *Isis* 9, 139 (1927).

[4] The finest weaving, that of silk, was invented by the Chinese in times immemorial. Consider what the invention implied – the domestication of an insect, the "education" of silkworms, the cultivation of the white mulberry, the whole of sericulture! The Chinese ascribe the first idea of sericulture and silk weaving to Hsi-ling Shih 西陵氏, the lady of Hsi-ling (in Hupeh), consort of the mythical Yellow Emperor, Huang Ti 黃帝, supposed to have ruled from 2698 to 2598 B.C. It must be added that the earliest specimens of silk that have come to us date only from the Han dynasty.

[5] It is often called Brazilian, but is used also in other parts of South America than Brazil. See a map of its distribution in Albert Métraux, *La civilisation matérielle des tribus Tupi-Guarani* (Paris, 1928) [*Isis* *13*, 246 (1929–30)], p. 114. See also Victor W. von Hagen, "The bitter cassava eaters," *Natural History* (New York, March 1949), with many illustrations.

times.[6] It is a three-legged cooking pot, the legs of which are shaped like cows' udders; various foods may be cooked in each leg with a single fire burning in the middle.

These examples might easily be multiplied. Selected as they have been in three corners of the world as remote from one another as could be, they illustrate the wide distribution of genius. We well know that whatever amount of civilization we enjoy today is the gift of many nations; we do not know so well that the same was already true thousands of years ago. Prehistorians have proved beyond doubt the existence of sophisticated cultures at very early times in many places. This does not disprove the monogenesis of mankind. It is highly probable that the new species *Homo sapiens* originated in a single place, but so long ago that by the time at which the earliest observable cultures flourished man had already invaded a good part of the world.

PREHISTORIC TRAVEL AND TRADE

Travel was much slower and more difficult in the past than it is now, and one might be tempted to conclude that primitive man moved very little and did not rove far away from his hiding place. That conclusion would be wrong. To begin with, we may observe that the speed of communication did not increase materially until the steam age, a century ago. Primitive people could move as fast as Napoleon's soldiers; sometimes they moved much faster. It is now generally agreed that there was considerable travel, individual and tribal (migrations), in the earliest days that scientific research can reach. For example, the Americas were discovered and colonized thousands of years ago by people coming over from Siberia and crossing the Bering Strait region; every American Indian is ultimately of Asiatic origin. The migrations were probably more frequent and more abundant in the oldest prehistoric periods before the invention of the agricultural arts, for as soon as people mastered those arts they became naturally more sedentary and more timid.

The passage from nomadic to settled life was perhaps the most pregnant step up in the whole history of mankind. That passage was far more important than the ones from stone to bronze or from bronze to iron; it might be called the passage from food gathering to food producing. Man could not settle down for life in any one place until he was secure from enemies, and this implied association with other men and some kind of government, nor until he was secure from want, and this implied the possibility of obtaining in the neighborhood enough food for himself, his family, and his beasts; it implied the arts and the folklore of agriculture. It has been remarked above that the development of mankind does not synchronize everywhere. Some people are more advanced than others, nor do they all pass through the same stages. The passage from nomadic to settled life occurred many millennia ago in some places, yet it has not been completed today by the Arab Bedouins. Man always was the child of circumstances, and since his environment varied enormously from place to place, he was bound to develop differently in different regions.

Men who had learned to cultivate the land were gradually blessed (and cursed)

[6] Yang Shao culture, so called after the place Yang Shao Tsun in Honan; latest stone age. See J. Gunnar Anderson, *Children of the yellow earth* (London: Kegan Paul, 1934), pp. 221, 330 [*Isis* 23, 274 (1935)].

with the ownership of more and more things and bound to the soil by more and more ties. As to their nomadic brethren, roving in search of better hunting or fishing, they might come back periodically to the same grounds, but there was nothing save habit and incipient domestication to oblige them to do so. The real nomads kept moving on without retracing their steps and were likely to cover immense distances.

The distinction between settled people, seminomads, and nomads is generally made with regard to people moving on land, but it applies equally well to those moving on water. No savages have ever been found near water who were not able to navigate it, but some of them were more settled than others, and some were regular sea rovers. The canoe is probably one of the oldest inventions of man, older than the bow; in favored places, where canoes were especially needed and materials for making them were handy, they were invented perhaps as early as thirty thousand years ago. Seaworthy ships came later, yet so early that deep-sea navigation reached a climax many millennia ago. According to the Norwegian archaeologist Anton Wilhelm Brøgger,[7] there was a golden age of oceanic navigation during the period roughly defined as 3000 to 1500 B.C., that is, before the days of Phoenician navigation. This is an archaeological interpolation, but its plausibility is confirmed from many sides. Sailing appealed to early men as it does to the young and strong of every time, and there are few fields wherein their inventiveness appeared more brilliantly. In this field, as in every other, it was not a matter of one invention but of a thousand, and the complete story would be endless. Among the masterpieces of primitive technology we may mention the wooden outrigger canoe of the South Seas, the Irish curragh (or coracle), the Eskimo dory-shaped umiak and their watertight kayak.

The early inhabitants of the northwestern European shores were not afraid of exploring the foggy and tempestuous Atlantic, and the South Sea islanders navigated the Pacific in every direction. For example, Polynesians did not hesitate to sail their canoes from Tahiti to Hawaii, a distance of 2400 nautical miles.

As to primitive commerce, there are many witnesses to it, one of the clearest being the relics of the amber trade. The best-known kind of amber (succinite) is a natural product of the Baltic shores, but pieces of it have been found in prehistoric tombs scattered in so many places that it has been possible to draw maps of the prehistoric amber routes.[8] As amber was very valuable and easy to transport, Scandinavians were able to obtain in exchange for it many goods of the southern regions, which had been favored by nature and were more advanced. Trade, then as now, was one of the main occasions of intercourse, one of the vehicles of civilization.

In the Stone Age the special value of flints for tools was soon realized, and good flints, breaking with sharp edges, were not found everywhere. The existence of flint quarries and of an international flint trade has been proved repeatedly. Alluvial gold must have been observed and collected very early and used for ornaments.

[7] In a lecture delivered at the Second International Congress of Prehistoric and Protohistoric Sciences, Oslo, 1936, and referred to by Vilhjalmur Stefansson in his *Ultima Thule* (New York: Macmillan, 1940), p. 31, and *Greenland* (New York: Doubleday, 1942), p. 26 [*Isis 34*, 379 (1942–43)].

[8] J. M. de Navarro, "Prehistoric routes between Northern Europe and Italy defined by the amber trade," *Geographical J.* 66, 481–507 (1925); maps, referring to the Bronze and Early Iron Ages.

The first ores to be exploited were probably sulfides of copper and antimony, both of which are very easily reducible, and thus copper and antimony were discovered. When grains of cassiterite were reduced, tin was obtained, and one of the first metallurgic geniuses had the idea of alloying a little tin with copper and thus obtaining a new metal, bronze, much harder and more serviceable than copper. Wherever that discovery was made or introduced, the Stone Age was followed by a Bronze Age. Later, other inventors found means of reducing the most fusible of the iron ores and the Iron Age began.[9]

It is not necessary to insist upon these momentous facts, with which the reader is presumably acquainted, but it is well to repeat a double warning. First, the Stone Age (or Ages), the Bronze Age, the Iron Age did not synchronize in every country; they might begin earlier and last longer in one region than in another. In the Americas, the Stone Age lasted until the European conquest. Second, they were never sharply separated from one another. Stone tools continued to be used in the Bronze Age and bronze tools in the Iron Age. Sometimes the use of old-fashioned materials was continued for religious or ceremonial purposes, for example, stone knives for circumcision in Egypt and Palestine,[10] and jade implements in China. Social inertia often sufficed to perpetuate old usages and prevent the substitution of new tools for old ones. Thus one of Mariette's[11] foremen was still shaving his head with a flint razor. Indeed, prehistoric tools are still in use today. Women may still be seen in various parts of Europe (Scottish Highlands, Pyrenees, etc.) spinning with a hand spindle loaded with a stone whorl.[12]

The decorative arts, not only ancient and medieval, but even the present ones, ring the changes on many prehistoric motives. We might say that there are as many prehistoric relics embedded in the language of forms as there are in the language of words; it is one of the delights of art historians as well as of philologists to detect these immortal witnesses of the distant past.

PREHISTORIC MEDICINE

We have already referred to the prehistoric knowledge of herbs and other drugs, knowledge distilled from immemorial empiricism, trial and error doggedly continued for hundreds and thousands of years. It is impossible for us to understand how such vague and casual experiments could be repeated long enough, their results taken note of and transmitted from generation to generation, but the fact is there: our prehistoric ancestors, like the primitive people who can still be observed, had managed to try many plants and other objects and to classify them

[9] Some idea of the earliest iron metallurgy may be obtained from E. Wyndham Hulme, "Prehistoric and primitive iron smelting," *Trans. Newcomen Soc. 18*, 181–192 (1937–38); *21*, 23–30 (1940–41). The best book on early metallurgy now available is R. J. Forbes, *Metallurgy in antiquity* (Leiden: Brill, 1950).

[10] According to the interpretation by W. Max Müller, in his *Egyptological researches. Results of a journey in 1904* (Washington: Carnegie Institution, 1906), p. 61, pl. 106, of a monument (Fig. 10) in the necropolis of Ṣaqqāra of the beginning of the Sixth Dynasty (*c.* 2625–2475). It is true that Jean Capart, *Une rue de tombeaux à Saqqarah* (2 vols., Brussels, 1907), vol. 1, p. 51; vol. 2, pl. lxvi, does not accept that interpretation unreservedly. At any rate, stone knives are mentioned in Exodus 4:25 and Joshua 5:2 (the translation of *ḥarbot ẓurim* in the Authorized Version by "sharp knives" is wrong; the correct meaning is "flint knives").

[11] Auguste Edouard Mariette (1821–1881), French Egyptologist.

[12] A perforated disk of stone (or terra cotta) slipped along a spindle, acting by its weight as a flywheel, causing the spindle to rotate more steadily.

in various groups according to their utility or danger.[13] Shepherds must have learned simple ways of setting broken or dislocated bones. Midwifery was necessarily practiced and the more intelligent midwives improved their methods and taught them to younger helpers. In all such cases the best as well as the hardest of teachers was always close at hand: necessity. If a man had his arm mauled by a wild beast or a falling rock, if he broke his leg, if a woman experienced unusual trouble in her travail, something had to be done, quickly. Other pathologic disturbances called for immediate solutions. Healing was probably one of the earliest vocations and professions. Sometimes the healer succeeded, and his successes were more likely to be remembered than his failures; he became famous and was imitated. Prehistoric medicine may be understood by comparison with the practice, half empirical, half magical, of primitive medicine men or shamans. It is possible that the extraordinary success of some of those shamans was due to their mediumistic power or to the popular belief in such a power. We may assume that faith healing began, at least in some places, at the very dawn of civilization.

All this is necessarily conjectural, but in at least one case we have direct and abundant evidences of a peculiarly daring kind of procedure. Many of the prehistoric skulls that have come down to us show signs of trepanation. The reader will ask: "How do you know that the operation was made on living men, not simply, for some ritual purpose, on empty skulls?" We know it well, because the hole made in the skull of a living man tended to heal itself and the growth of new bone can be recognized without ambiguity.[14] Why was the skull perforated? That question cannot be answered. It is possible that the surgeon tried to relieve unendurable pressure due to concussion of the brain. One may also ask, How was it done? Some kinds of drills were already used by paleolithic craftsmen; witness the existence of perforated stones and of actual drills in ancient sites.[15] The perforation of a stone

[13] This process of discovery and selection is the more mysterious, because (like the creation of language) it is largely unconscious. The following remarks, taken from Carl Binger, *The doctor's job* (New York: Norton, 1945), p. 153, will fascinate the reader as they did me. "Dr. Curt Richter of the Johns Hopkins Medical School, whose ingenious and important experiments on white rats I shall presently refer to, tells the story of a three-and-a-half-year-old boy who was admitted to the Johns Hopkins Hospital with a tumor of the adrenal gland – a fatal disease. The child had the habit of eating salt by the handful. He took to it as another child might to sugar or to jam. When he entered the hospital his salt-eating habit was stopped and he was put on the regular hospital diet. Unfortunately, he died soon thereafter. Now it appears that this child had discovered independently what it has taken experimental scientists many years to find out – that patients suffering from lesions of the adrenal glands are greatly benefited by the addition of large quantities of common salt to the diet.

Dr. Richter's white rats are also gifted scientists. He has shown that on a mixed standard diet of carbohydrates, proteins and fats, plus minerals and vitamins, the rats will maintain a predictable rate of growth and weight increase. Now if he offers his rats the ingredients of the diet unmixed they will still select just what they need to continue their growth and development at the usual rate. But even more remarkable is the fact that whereas a normal rat will consume relatively little salt a rat whose adrenal glands have been surgically removed will quickly and automatically increase its salt intake sufficiently to survive, whereas cagemates, similarly operated upon, will die when allowed only the normal salt ingredient in their diets. Rats deprived of their parathyroid glands will eat enough calcium to keep themselves alive and free from tetany. If the rats could consult medical literature they would find that calcium is given to babies with tetany as it is to adults whose parathyroid glands have been removed during an operation for goiter. Rats fed on thyroid extract develop a morbid appetite for weak solutions of iodine, the standard medicine given to patients whose thyroid glands are overactive."

[14] There is a considerable literature on the subject. See, for example, Stéphen-Chauvet, *La médecine chez les peuples primitifs* (Paris: Librairie Maloine, 1936); Henry E. Sigerist, *History of medicine* (New York: Oxford University Press, 1951), vol. 1 [*Isis* 42, 278–281 (1951)]. Sigerist's volume had not yet appeared when I wrote this chapter.

[15] Franz M. Feldhaus, *Die Technik* (Leipzig, 1914), p. 115.

with a stone drill must have been a very long task; trepanning must have been relatively easy, at least for the surgeon; not so easy for the patient.[16]

PREHISTORIC MATHEMATICS

The transition from empiricism to rational knowledge was necessarily very slow in medicine, because the number of independent variables was very great and each ailment might vary considerably from one individual to another. Let us pass to another field, mathematics, where some humble kind of rationalization was possible, and abstraction natural, at an early stage. One of the fundamental ideas of mathematics is the idea of number, which, in its simplest form, may have occurred to very early men. The first mathematician — a great unknown genius — was perhaps the man who adumbrated that idea.

How did that happen? We can only guess, but our guesses are neither arbitrary nor futile. The first theologian adumbrated the idea of oneness or wholeness, one cause, one world, one self, one God. The idea of twoness or duality must have occurred almost as early, for there are many obvious pairs in nature. We have two eyes, two nostrils, two ears, two hands, two feet; women have two breasts. The hands were particularly instructive, for man must have used them unequally from the very beginning. The simplest acts, such as eating, drinking, using tools, loving, or fighting, imply different tasks for each hand. The two hands revealed the right and left sides of everything, not a simple duality but a polarity wherein one side is different from, and preferable to, the other. Above all, dominating everything, is the polarity of sex. Not only all men, but also every animal that they could observe, was either male or female. That was not only obvious, but imperative, obsessing, unavoidable. Moreover, every quality appears necessarily under a dual aspect; things are hot or cold, dry or moist, large or small, pleasant or unpleasant, good or evil.

Larger groups, though less universal, were noticeable enough. A father and mother with their first baby — that is a trinity. On a river there are two directions, up and down, but to the man standing in a plain there are many more. Let him stand with outstretched arms; there are revealed to his mind four privileged positions — straight ahead, backward, and the directions of either arm. His language will soon express this with four significant words, such as front, back, right, left. If his hands were stretched out, the right toward the place of the rising sun and the left toward that of the setting one, the idea of four cardinal points was emerging. To those four elements might be added a fifth, the center, the very place where he stood, or two others, the sky above and the earth below. Hence arise the categories of fiveness, sixness, sevenness. The first of these categories was strongly reinforced by the existence of five fingers. When counting things on one hand or one foot, it was natural enough to group them in fives and to speak of so many "hands." Larger groups, such as ten or twenty, were almost equally natural, but a little more difficult to recognize.

Most people, almost all of them, took these categories for granted and did not give a thought to them, but if there was a born mathematician among them — and why should there not have been? — he must have realized the existence of

[16] Though means of intoxicating or benumbing him may already have been available. Such means were used very early in many parts of the world.

numbers, abstract numbers, independent of the objects counted. The fiveness of hand or of foot or of Cassiopeia, he must have thought, is essentially the same thing. As to the theologians or the cosmologists, they may have been hypnotized by the number one, generative of all others, or by two, expressing the universal polarity, or even by three, the mystic triangle. Dualism, such as was elaborated in the Zoroastrian religion, is rooted in the deepest recesses of the human conscience.

These numerical categories were the seeds of arithmetic, that is, of pure science, but also of number mysticism, or pure nonsense. The two roots grew exuberantly. Let us consider the situation in China; we can do this without abandoning the prehistoric level, for the numerical groupings of which the Chinese mind is so fond are immemorial, and if we could trace them to their origins they would very probably take us back to the most remote antiquity. Chinese ideology is dominated by the universal polarity of *yang* 陽 and *yin* 陰, the male and female, positive and negative, principles of life. *Yang* is male, light, hot, active, it is heaven, the sun, rocks and mountains, goodness . . .; *yin* is female, dark, cold, passive, it is the earth, the moon, water, trouble and evil . . . (It is clear enough that the earliest Chinese cosmologists were men, not women!) Every example of duality can be expressed in terms of *yang* and *yin*. The sexual origin of every form of life, the fact that each child needs two parents, is extended to the whole universe. What is most curious is that that sexual cosmology received very early a mathematical interpretation. Not only is negative opposed to positive (a fundamental distinction to be developed later on in geometry as well as in arithmetic), but *yang* is represented by a solid line and *yin* by a broken one. Take these lines three by three, and eight combinations, the eight diagrams *pa kua* 八卦, are possible, neither more nor less (Fig. 1). The discovery of that mystery was ascribed to the legendary founder of Chinese culture, Fu hsi 伏羲, the first emperor, supposed to have ruled from 2953 to 2838 B.C. Such an ascription is simply a patent of hoary antiquity. If the *yang* and *yin* lines are combined six by six there are 64 possible hexagrams, each of which was given a definite meaning; this process might be continued and was indeed continued (the mathematical mind at work!), but we need not worry about that. It is interesting to realize that those early Chinese savants and mystics were playing, without being aware of it, with combinatorial analysis. It would be foolish to expect that they should have realized the mathematical implications of their thinking at that early stage, but their instinctive tendency in that direction is confirmed by their invention of a sexagenary period (a "cycle of Cathay") obtained by combining, two by two, the twelve earthly branches (*shih êrh ti chih* 十二地支) with the ten heavenly stems (*shih t'ien kan* 十天干).[17] As $12 \times 5 = 10 \times 6 = 60$, sixty different combinations are possible (Fig. 2). This discovery is ascribed to another mythical emperor, Huang Ti 黃帝, who ruled from 2698 to 2598. At first, it was applied only to days and hours; the application to years occurred only later under the Han dynasty (let us say, about the time of Christ), but we are concerned here only with the fundamental idea of a sexagenary cycle, not with its applications.[18]

[17] The Chinese name of the sexagenary cycle, *chia tzŭ* 甲子, is made up of the names of the first stem *chia* and of the first branch *tzŭ*. The names of the twelve branches are names of animals (as for the zodiac); *tzŭ* is the rat.

[18] It is interesting to compare the Chinese calendar with the Maya, for each was as independent of the other as if they had developed on different planets. The Mayas enmeshed a civil year (*haab*) of 365 days with a sacred year (*tzolkin*) of 260

Fig. 1. Symbols of *yang* (white, male) and *yin* (dark, female) in the center and the eight diagrams around.

The average Chinese did not indulge in such speculations; he accepted the *pa kua* and the *chia tzŭ* as naturally as the seasons or the phases of the moon, yet the habit of numerical categories was deeply ingrained in his mind. Some such desire of grouping things by twos or threes and so forth exists in every mind (it expresses an instinctive need of order and symmetry, fundamental to science as well as to art), but the Chinese allowed it to expand more freely than any other people. Thus a large collection of groupings are as familiar to them, as, let us say, the four cardinal points to us, groupings by twos, threes, fours, fives,[19] sixes, sevens, eights, nines, tens, twelves, thirteens, seventeens, eighteens, twenty-fours, twenty-eights, thirty-twos, seventy-twos, hundreds. William Frederick Mayers[20] listed 317 such groupings, and I am confident that his list could be extended. Of course, many of these groups are of late origin, others will be added in the future, but the primary idea is almost as old as Chinese culture.

We have come very close to mathematics and then drifted away. This must have happened many times in the past; it continues to happen within our experience. Any scientific idea may be, and often is, perverted; that cannot be helped. It is like a tool that may be used for good purposes or for evil ones.

To return from fancy to reality, the progress of arithmetic was probably due to the fact that people could not stop at small and familiar categories, but were obliged to count things and faced relatively large numbers at a very early stage. A chieftain, wanting, as was natural enough, to assess his resources, would ask himself how many men he could depend upon, how many horses, sheep, and goats. In short, he would need a census, and, even if his tribe were small, that census would quickly lead to numbers too large to be counted on one's fingers. How would he do it? In his delightful account of how the rajah of Lombok took the census,[21] Wallace gives us the diplomatic side of the story and stops at the point where the mathematical difficulties begin; those difficulties could not be evaded. The tangible result of the rajah's survey was many bundles of needles. How did he count the needles? Now grouping is the very basis of counting. Every language betrays the presence of what mathematicians call a number base, which was often five (among many American tribes), sometimes twenty (among the Mayas), but more often

days; this implied a great year or a "year bundle" as they called it (*xiuhmolpilli*) of 18,980 days (=52 haab=73 tzolkin). For details, see Silvanus Griswold Morley (1883–1948), *The ancient Maya* (Stanford: Stanford University Press, 1946), pp. 265–274 [*Isis* 37, 245 (1947); 39, 241 (1948)].

[19] For the fives, see synoptic table in *Isis* 22, 270 (1934–35).

[20] W. F. Mayers, *Chinese reader's manual* (Shanghai, 1874).

[21] Alfred Russel Wallace, *The Malay archipelago* (London, 1869), chap. 12. Lombok is one of the smaller islands lying between Java and Australia; its western coast faces Bali.

ten.[22] These bases were more popular than others, because almost every primitive man used the same counting machine, to wit, his fingers or toes. He might stop at one hand (or foot) in which case his base was five, or use both hands (or both feet) when the base was ten, or use the whole outfit, when the base was twenty.[23] *In medio virtus!* The peoples whose culture patterns were destined to dominate all others agreed unconsciously on the use of ten. How do we know the number bases of primitive people? We can easily deduce it from their language, even as our own decimal base is clearly represented by our number words. Indeed, it was partly because of the words themselves that a base was needed and was instinctively created. The base makes it possible to use periodically the same few words, with slight modifications, if any; without it an infinity of words would be required.[24]

The spontaneous agreement of the leading nations on the decimal base is wonderful, but not more so after all than the marvelous symmetry of each language. These things pass our understanding. How is it possible to account for the unconscious development of such complex and symmetric structures, not in one place, but wherever men flourished? Each language evidences not a perfect symmetry, like that of a geometric drawing, but one that is imperfect in many ways, like that of a tree or of a beautiful body — a living symmetry.

How were the returns of a primitive census enumerated? Let us assume that each item to be counted was represented by a twig[25] and that the base was decimal.

[22] There were still other bases, for which see Levi Leonard Conant, *The number concept* (New York, 1896). For the decimal ones, see G. Sarton, "Decimal systems early and late," *Osiris* 9, 581–601 (1950).

[23] Counting on the feet was natural enough in warm countries where people remain barefooted. In many languages, for example, Greek, Latin, and Arabic, the same words are used for fingers and toes; if more precision is needed, the latter are called fingers of the foot.

[24] Consider our own language. To count up to a hundred we need nineteen words: one, two, . . ., ten; twenty, . . ., ninety; hundred; but we must remember a few modifications of them for the second decade, as eleven (for one ten), twelve, thirteen, . . ., nineteen. To count up to 999,999 we need only one more word, thousand.

[25] One may see in the National Museum, Washington, D. C., five bundles of reeds which constituted a census made by the Comanche Indians (originally in western Wyoming, later ranging widely between Kansas and northern Mexico). These bundles indicate respectively the number of women in the village, the number of young men, the number of warriors, the number of children, and the number of lodges. They were collected by Edward Palmer in the 1880's (letter from Alexander Wetmore, Washington, D. C., 20 June 1944).

1.	甲子	11.	甲戌	21.	甲申	31.	甲午	41.	甲辰	51.	甲寅
2.	乙丑	12.	乙亥	22.	乙酉	32.	乙未	42.	乙巳	52.	乙卯
3.	丙寅	13.	丙子	23.	丙戌	33.	丙申	43.	丙午	53.	丙辰
4.	丁卯	14.	丁丑	24.	丁亥	34.	丁酉	44.	丁未	54.	丁巳
5.	戊辰	15.	戊寅	25.	戊子	35.	戊戌	45.	戊申	55.	戊午
6.	己巳	16.	己卯	26.	己丑	36.	己亥	46.	己酉	56.	己未
7.	庚午	17.	庚辰	27.	庚寅	37.	庚子	47.	庚戌	57.	庚申
8.	辛未	18.	辛巳	28.	辛卯	38.	辛丑	48.	辛亥	58.	辛酉
9.	壬申	19.	壬午	29.	壬辰	39.	壬寅	49.	壬子	59.	壬戌
10.	癸酉	20.	癸未	30.	癸巳	40.	癸卯	50.	癸丑	60.	癸亥

Fig. 2. Sexagenary cycle ("a cycle of Cathay"). The ten symbols of each first column are alike; they are ten celestial stems. The twelve earthly branches are written in the second columns, from 1 to 12, 13 to 24, 25 to 36, 37 to 48, 49 to 60. Each group of two characters is different from every other. [Herbert A. Giles, *Chinese-English dictionary* (Shanghai, ed. 2, 1912), vol. 1, p. 32.]

One made bundles of ten twigs each, then the total number of twigs was ten times that of the bundles. If there were too many bundles, it may have occurred to the computer to consider each bundle as a single twig, a kind of superior twig, and make new bundles of ten bundles each. If the computer had done that much and was mathematically minded, there was nothing to stop him from repeating that operation as often as necessary. After having recognized tens, he might recognize hundreds, thousands, tens of thousands, and so on, creating new words, as well as new symbols if he had already reached that particular stage. Please note that the number of new words (or symbols) which are needed decreases rapidly. It took probably a long time before the word million was actually needed and we are but just beginning to use with any frequency the word billion.[26]

What we call the fundamental operations (addition, subtraction, multiplication, division) emerged naturally, if not explicitly, from the very process of enumerating collections and sharing them. The idea of subtraction arose also from the fact that when numbers are a little smaller than round numbers it is easier to approach them from above than from below, to say, for example, that there are 2 less than 20 rather than 18, 100 less 1 rather than 99, 10,000 less 300 rather than 9700.[27]

We have assumed thus far that early accounts were made by means of twigs or other objects, say pebbles (*calculi* in Latin, hence our words calculus, calculation, etc.); it might be done and was done as well by means of knots in strings or notches in tallies. The same periodicities would naturally reappear. Anyone having in his mind, however unconsciously, the decimal rhythm, would carve a longer notch for a ten, and one still longer for a hundred; numbers approaching the longer notches would be grasped more easily by retrogression from those notches, that is, subtractively.

The concepts of rhythm and pattern awakened by the necessity of counting reappeared more tangibly in the process of ornamentation. The simplest measurements, such as were needed to build an altar or a house, may have inspired the earliest geometric ideas, but the love of beauty innate in most men was probably the real cradle of geometry, for in order to decorate pleasantly various objects or one's body, not only were some measurements required but a whole gamut of them, plus as many symmetric and periodic combinations of decorative elements as fancy would suggest. Mother Nature was the best art teacher; the infinite patterns evidenced in natural objects, such as trees, leaves, flowers, birds, snakes, were a continual source of inspiration to the men who had in them the love of beauty. Some of the paleolithic drawings that have come down to us were made by genuine artists. The decorations of ceramics and textiles that may be viewed in anthropologic museums reveal an astonishing amount of imagination and sensitiveness. Not only were the craftsmen able to create patterns of great complexity, but they rang the changes on them with virtuosity and were subtle enough to realize the value of little deviations. Any such composition implied the solution, however crude, of many geometric problems.

It was easy enough to measure a distance and to divide it — let us say by means of a string which could be folded twice or more often — but a more difficult prob-

[26] Its meaning is not yet agreed upon; for the English — in this, more logical than we are — it is 10^{12}; for us, 10^9.

[27] Witness the words (popular creations!) *duodeviginti* and *undecentum* in Latin and *triacosiōn apodeonta myria* in Greek; these words mean 18, 99, 9700.

lem arose when early "scientists" tried to estimate the relative distances of the stars of a familiar constellation, or the change of distance of a moving star (planet) approaching a fixed one (that is, one moving regularly with all the others), or the changes of distances between the Moon and the constellations through which it never ceased to proceed. They may have tried to measure such a distance with a string, but if so, they must have promptly noticed that the length to be measured diminished when the string was brought closer to the eyes. It finally dawned upon the mind of a prehistoric Newton that astronomic distances were not linear but angular; the idea of angle was a geometric and astronomic invention of fundamental importance.

It did not suffice to make measurements; such measurements had to be expressed and that expression implied a choice of units. Nor did it suffice to choose units; it was necessary to keep them. The preservation of standard units was perhaps one of the first steps in scientific organization, though this was naturally as unconscious as were all the other early steps. It would seem that almost every nation agreed upon selecting as units parts of the adult body (cubit,[28] foot, span, etc.). Our earliest ancestors realized as naturally as we do that many units are needed, small ones for small distances, larger ones for longer distances, and so on, but they did not try to establish fixed relations between those units. We should not blame them, bearing humbly in mind that highly civilized peoples of our own days have not yet understood that need.

PREHISTORIC ASTRONOMY

We have already spoken of the stars. It was impossible for any reflective man to observe them night after night without asking himself a number of questions, which were primarily scientific questions. Early people, especially those encouraged by a hot climate to spend nights out of doors, cannot fail to have observed the changing places throughout the year of sunrise and sunset, the phases of the Moon, the regular motion of the Moon to the left[29] among the stars at different altitudes of culmination but with about the same speed, the seasonal appearance and disappearance of some constellations, the more complex motions of the morning and evening stars[30] and of other planets. They were aware in many ways of the march of time, for they could not help recognizing the ever-recurring periods of day and night, of moon phases, of meteorologic seasons, and years. They made for themselves calendars wherein those events were foretold on the basis of past experience, calendars based on meteorologic events, on the Moon cycle, or the Sun cycle, or on many of these events combined. Those calendars might be gradually improved as the observations from which they were derived were repeated and refined.

We need not continue this enumeration. It is certain that at least a few privileged peoples, favored by better opportunities of climate or site or by greater intelligence, had already accumulated a large amount of knowledge before the invention of

[28] The Latin word *cubitum* means elbow, but also the distance from the elbow to the end of the middle finger.

[29] In the northern hemisphere.

[30] Lucifer, *Heōsphoros*, or *Phōsphoros*, and Hesperus or *Hesperos*; the identity of the two stars may have been recognized very early, but we cannot tell how early. Both are identical with the planet Venus, *Aphroditēs astēr*. In low latitudes (that is, in the subtropical countries where a higher culture began) it was possible to observe another couple, the morning star Apollo (*Apollōn*) and the evening star Mercurius (*Hermēs*), both identical with the planet Mercury. One could not fail to see Mercury even in latitudes as high as 50°.

writing. Prehistoric knowledge was so vast and varied in some parts of the earth that a complete catalogue of it, if it were possible to reconstruct it, would occupy considerable space.

PURE SCIENCES

Some readers will object that whatever knowledge there was, was purely practical, empirical, too raw and rough to deserve the name of science. Why should we not call it science? It was a very poor science, very imperfect, yet perfectible; our science is decidedly deeper and richer, yet the same general description applies to it — it is very imperfect yet perfectible. Or one might say, There was no pure science. Why not again? How pure must science be to be called pure? If pure science is disinterested science, knowledge obtained for its own sake without thought of immediate use, surely the early astronomers were or might be as pure as our own. It is possible that astrologic fancies had already developed, but it is equally possible that they had not, for that would have implied a degree of sophistication which those astronomers had not yet reached. Their main reason for observing the strange behavior of certain planets may have been simply curiosity.

Curiosity, one of the deepest of human traits, indeed far more ancient than mankind itself, was perhaps the mainspring of scientific knowledge in the past as it still is today. Necessity has been called the mother of invention, of technology, but curiosity was the mother of science. The motives of primitive scientists (as opposed to those of primitive technicians and shamans) were perhaps not very different from those of our contemporaries; they varied considerably from man to man and time to time and then as now covered the whole gamut from complete selflessness, reckless curiosity, and spirit of adventure down to personal ambition, vainglory, covetousness.

If research had not been inspired and informed from the beginning by a certain amount of disinterestedness and adventurousness, and by what its enemies would later call indiscretion and impiety, the progress of science would have been considerably slower than it was. The amount of knowledge attained by some primitive men can be deduced from anthropologic records and also from the amount observable in the most ancient civilizations. When man appears on the scene of history, we find him already a master of many arts, expert in many crafts, as full of lore as of cunning.

Then as now the true scientist, even as the true artist, was likely to be or to seem a bit queer and secretive; it is highly probable that his more practical neighbors already made jokes about his absent-mindedness. Of course, he was not more absent-minded than they were, but their minds were focused on different interests. He was engrossed in his own reflections; his motives being less tangible, his life seemed mysterious. Sometimes he may have wished for praise and recognition, or he may already have discovered that such praise was futile and that it was better not to try for it. If he were selfish and jealous, the primitive inventor might prefer to keep his new idea — say a better hook, or a better ax, or better materials for the making of either — to himself and his family. In almost every case the scientist or the inventor tended to be reticent. The growth of science was always entangled in psychologic and social accidents.

Not only was the development of primitive invention somewhat confidential and secret, it was also of necessity antagonistic to the regular habits and traditions that it tended to subvert. Every invention, however useful it may turn out to be (and it cannot be useful before it is used), is disturbing, and the more pregnant it is, the more disturbing. There were vested interests in prehistoric times as well as now, though they could not be described in exactly the same way and were perhaps less blatant. There was, then as now, a strong inertia impeding progress, the inertia of habit and complacency, distrust and contempt of everything novel or foreign. That inertia, however, was not simply a hindrance, but a necessity, like a flywheel or a brake, to steady and warrant mankind's invasion of the unknown. Men's resistance to new tools or newfangled ideas was useful, because novelties should be thoroughly tested before being adopted. Every accepted tool was the fruit of a very long process of trial and error, of a very long tussle between inventors, innovators, reformers at one end and conservatives at the other. The latter were far more numerous; the former were more enthusiastic and aggressive.

DIFFUSION AND CONVERGENCE

Some anthropologists (the "diffusionists") seem to believe that each invention was made in but one place, and that this sufficed to spread it elsewhere, if it was worth while. Sir Grafton Elliot Smith (1871–1937) and William James Perry, arguing in that way, would have us consider Egypt the cradle of civilization. A generalization of such boldness is not susceptible of proof, and the history of science tends to disprove it. Simultaneous discoveries, that is, identical or similar discoveries made at about the same time by different people in separate places, are not uncommon in modern times, and their circumstances have been investigated. They are generally explained by a common ancestry of problems or instruments; the inventors were trying to solve the same problems, and drew their information from the same sources and their inspiration from similar needs; the simultaneity (or quasi simultaneity) of their triumphs is accounted for by the simultaneity of their needs. "The idea was in the air," as we say. Moreover, each problem, as soon as it is solved, creates new problems; each discovery entails a logical sequence of other discoveries. Why should it not have been the same in prehistoric times? The only difference in this respect between the distant past and the present is that, everything being much slower in the past than now, the simultaneities would be counted in centuries instead of in years or months as they are now.

The most impressive example of convergence (as opposed to imitation) is the independent invention of a decimal system of numeration in distant parts of the world, its almost unanimous (yet unconscious) acceptance by the very nations whose cultures became the dominating ones. That is one of the miracles of the dawn of science. The anatomic explanation given above is convincing enough as far as it goes, but it is far from complete. Why did men unite on ten, rather than on five or twenty?

The theory of convergent evolution, or convergence (as the anthropologists call it), does not deny the frequent occurence of borrowings and imitation between one people and another; it claims that similarities between different cultures are not necessarily the result of imitation but may be and often are due to independent inventions. Even when a people borrows a cultural trait, a tool, a word, or an idea from another, the imitation is more often active than passive. Indeed, the tool or the idea must be acceptable to the new people; if not immediately acceptable, it must be put in an acceptable shape; and even when acceptable, it must still be accepted and that may involve as long and painful a struggle as was needed for the acceptance of the original invention. The cultural trait is not really a trait of the new people until it has been thoroughly understood (or misunderstood!), liked, assimilated. Its introduction is never a process of simple addition, but one of biological intussusception, re-creation. In order to use metal tools or weapons instead of stone ones, men had to discard old notions and become — to use modern slang — metal conscious. That sort of thing did not happen in a day, nor in a year, nor perhaps in a century.

Even if mankind originated in but a single place, so many millennia elapsed between its emergence and the dawn of culture that men had had innumerable opportunities of spreading in many directions as fate and circumstances pushed them hither and thither. Though modified by climatic and geographic conditions, the problems that they had to solve were essentially the same. Is it surprising, then, that they hit upon the same or similar solutions? Were they not essentially the same people? Sometimes they would find a solution unaided by others; sometimes another solution having reached their eyes or ears would be accepted by them, stolen, or reinvented. The borrowing could be interpreted in various ways, and it would vary considerably from all to almost nothing, or from servile imitation to the taking of the least hint.

Each settlement had its men of genius, its dullards, and its great majority of "average" people. The average varied from settlement to settlement not only for hereditary reasons, but also because the climatic and geographic conditions (including the availability of definite plants and animals) were more favorable in some places than in others. There was from the beginning a great variety of men and women as well as a great variety of opportunities. People who had settled at the edge of a lake or of the sea had different opportunities from their distant cousins who had found a refuge in mountain caves or in desert oases. Each gift of nature created distinctive needs. Some of those needs vanished in the course of time, and this accounts for the "lost arts." Primitive man could do a great many things that we would be incapable of, and he managed to survive in the midst of dangers that we could not face any more.

Even as some people excelled among other people, some communities excelled among other communities, and were able to accomplish certain tasks that those others did not even think of, and thus to help mankind rise a step higher. The next step was made possible by another community, at another time, in another place. Thus it was at the beginning and thus it has always been. The student of human evolution cannot escape feeling that mankind is working in shifts. There is no privileged "race" or community in any absolute way, but for each task and for each time some people or some nations may excel all others.

The dawn of science did not break out everywhere with the same beauty and the same hopefulness. There were probably precocious peoples, as there are precocious children, who began very early but did not go very far. We shall concern ourselves in the following chapters with the ancient peoples whose cultural dawn was only the prelude to the greatest achievements of the third and second millennia before Christ.[31]

[31] No attempt has been made to discuss the mixed origins of science and magic, and we might add of religion and art, because a sufficient account of the facts involved would cover too much space. The reader will find an excellent account of those moot questions in "Magic, science and religion," by the late Bronislaw Malinowski, in Joseph Needham, ed., *Science, religion and reality* (New York, 1928), pp. 19–84 [*Isis 36*, 50 (1946)], with bibliography. See also M. F. Ashley Montagu, "Bronislaw Malinowski, 1884–1942," *Isis 34*, 146–150 (1942).

II

EGYPT

The outstanding cultural patterns coalesced in the valleys of great rivers in northern subtropical regions. It is clear that a culture of great complexity could develop only where a sufficient number of people were able to come together in relative peace and comfort, share their many tasks and the fruits thereof, and stimulate one another. Those rivers are the Nile, the Euphrates and the Tigris, the Indus and the Ganges, the Hwang Ho and the Yangtze, and perhaps also the Menam and the Mekong.[1] They are all of considerable length (the shortest, the Menam, being about 750 miles long, while the longest, the Nile and the Yangtze, measure respectively 3473 and 3200 miles) and they drain and irrigate enormous territories. That coincidence is not accidental. The rivers that were carrying to the sea not only water but men, goods, ideas, had to be pretty large to provide a sufficient concentration and competition in the lower reaches. Any culture, even the least developed, is so complex that it cannot be created by small groups, but only by relatively large ones — thousands or millions of men. To appreciate the immensity of the tasks that had to be accomplished, one need think of only one element, language, the perfection of which implied a multitudinous, anonymous, unconscious fermentation of unimaginable intricacy.

As we are primarily concerned with the origins of our own culture, we shall consider only in this and the following chapter the two civilizations of the Ancient Near East, for these two influenced the Mediterranean world most deeply. Indeed, these two civilizations were closest to the Mediterranean, though neither was a complete part of it. This is obvious enough for Mesopotamia; the upper Euphrates came very near the Mediterranean Sea, yet the outlets of both that river and the Tigris were in the Persian Gulf. The Nile – the only one of the great rivers named above that flows northward – poured its waters into the Mediterranean, yet the Old Egyptian culture did not grow near the sea but some distance away from it and the sea of the Egyptians was not the Mediterranean but the Nile itself. Egypt was like "a long river oasis in the middle of the desert." [2]

The periodic overflow of the Nile fertilized the narrow valley and helped to produce abundant crops. The dry sterile climate was tempered by those inundations and Egypt was favored above all the nations of the Mediterranean world. It is of course impossible to say when Egyptian culture began and to decide whether it is anterior to Mesopotamian and Chinese cultures or not. These questions of

[1] The lower courses of the two last named are distinctly tropical, and so is the Ganges estuary.

[2] *Osiris* 2, 410 (1936).

priority are not sufficiently relevant to our purpose to be discussed here. Indeed, we shall not describe the conditions obtaining in prehistoric Egypt [3]; it will suffice to say that its prehistoric culture was that of the late Stone Age and that the early Egyptians had already developed many agricultural arts; they cultivated barley, spelt (a kind of wheat),[4] and flax, wove linen, had a yearly calendar. When the curtain of history rises on the First Dynasty, the cultural achievements that we are able to witness are not by any means a beginning; they are rather a climax, the existence of which would have been impossible but for a preparation lasting many thousands of years.

The oldest historical period of Egypt, called the Old Kingdom, is a succession of six dynasties (First to Sixth) which lasted from *c.* 3400 to *c.* 2475 B.C., or almost a thousand years.[5] The first half of that period is not well known, and when we speak of the Old Kingdom we are thinking primarily of the second half, the Pyramid Age (Third to Sixth Dynasties, *c.* 2980 to *c.* 2475, half a millennium). The Pyramid Age is immortalized by a number of inscriptions and a few other writings, above all by prodigious monuments.

THE INVENTION OF WRITING

The greatest achievement of the early Egyptians was the invention of writing. Whether they were the first to make that invention, or were anticipated by the Sumerians or the Chinese, is a moot question. At any rate, they made it independently. One must bear in mind that such an invention, wherever it was made, cannot be very conveniently marked off on a time scale, because it was not made in one step nor at one definite time. As far as Egypt is concerned, it was begun during the prehistoric age and may have been brought to a fair stage of completion before the end of that age. The earliest writings that have come down to us are of the Old Kingdom.

We may assume that the Egyptians began by using pictograms (images) representing things or ideas rather than words. In the course of time such images would be gradually conventionalized, simplified, and standardized, and they would finally come to be associated with spoken words. Then each image would represent not simply an idea but a definite word of the Egyptian language. Later still the original idea might be forgotten, and the image would retain only a phonetic value. The scribes, having at their disposal a sufficient number of such phonemes, might use them, and did actually use them, to write words containing the same sounds, especially proper names or abstract words, which were not susceptible of pictographic representation. The Egyptians went even a step further; in the course of long usage certain symbols were used to represent only the consonantal beginnings of such phonemes. During the Old Kingdom, they had thus obtained a group of twenty-four alphabetic signs, which was not increased afterward (Fig. 3).

May we then say that the Egyptians invented the alphabet? No, they invented alphabetic symbols, but did not grasp their full implications, for they continued

[3] There was no glacial period in Egypt, and hence no interruption of its prehistoric development. This gave Egypt a tremendous but incalculable advantage over other countries.

[4] *Isis* 37, 96 (1947).

[5] I use throughout the "short" chronology, according to which the first king of the First Dynasty, Menes, began to rule *c.* 3400. Other chronologists would place him much earlier, the extreme dating being that of Champollion-Figeac, 5867! For an explanation and justification of the "short" chronology, see James Henry Breasted, *Ancient records of Egypt* (Chicago, 1906), vol. 1, pp. 25–48. One should always mention the dynasty; I have done so.

THE ALPHABET

SIGN	TRANS-LITERATION	OBJECT DEPICTED	APPROXIMATE SOUND-VALUE	REMARKS
𓄿	*ꜣ*	Egyptian vulture	the glottal stop heard at the commencement of German words beginning with a vowel, ex. *der Adler*.	corresponds to Hebrew א *'āleph* and to Arabic أ *'elif hemzatum*.
𓇋	*i͗*	flowering reed	usually consonantal *y*; at the beginning of words sometimes identical with *ꜣ*.	corresponds to Hebrew י *yōdh*, Arabic ى *yā*.
(1) 𓇌 (2) 𓏭	*y*	(1) two reed-flowers (2) oblique strokes	*y*	used under specific conditions in the last syllable of words
𓂝	*ꜥ*	forearm	a guttural sound unknown to English	corresponds to Hebrew ע *ʿayin*, Arabic ع *ʿain*.
𓅱	*w*	quail chick	*w*	
𓃀	*b*	foot (position of foot)	*b*	
𓊪	*p*	stool	*p*	
𓆑	*f*	horned viper	*f*	
𓅓	*m*	eagle owl	*m*	
𓈖	*n*	water	*n*	corresponds to Hebrew נ *nūn*, but also to Hebrew ל *lāmedh*.
𓂋	*r*	mouth	*r*	corresponds to Hebrew ר *rōsh*, more rarely to Hebrew ל *lāmedh*.
𓉔	*h*	courtyard	*h* as in English	corresponds to Hebrew ה *hē*, Arabic ه *hā*.
𓎛	*ḥ*	twisted hank of flax	emphatic *h*	corresponds to Arabic ح *ḥā*.
𓐍	*ḫ*	placenta (?)	like *ch* in Scotch *loch*	corresponds to Arabic خ *ḫā*.
𓄡	*ẖ*	animal's belly with teats	perhaps like *ch* in German *ich*	interchanging early with 𓈙 *š*, later with 𓐍 *ḫ*, in certain words.
(1) 𓊃 (2) 𓋴	*s*	(1) bolt (2) folded cloth	*s*	originally two separate sounds: (1) *z*, much like our *z*; (2) *ś*, emphatic *s*.
𓈙	*š*	pool	*sh*	early hardly different from 𓄡 *ẖ*.
𓈎	*ḳ*	hill-slope	backward *k*; rather like our *q* in *queen*	corresponds to Hebrew ק *qōph*, Arabic ق *ḳāf*.
𓎡	*k*	basket with handle	*k*	corresponds to Hebrew כ *kaph*, Arabic ك *kāf*.
𓎼	*g*	stand for jar	hard *g*	
𓏏	*t*	loaf	*t*	
𓍿	*ṯ*	tethering rope	originally *tsh* (*č*)	during Middle Kingdom persists in some words, in others is replaced by 𓏏 *t*.
𓂧	*d*	hand	*d*	
𓆓	*ḏ*	snake	originally *dj* and also a dull emphatic *s* (Hebrew צ)	during Middle Kingdom persists in some words, in others is replaced by 𓂧 *d*.

Fig. 3. The Egyptian alphabet. [Borrowed with permission from Alan H. Gardiner, *Egyptian grammar* (Oxford: Clarendon Press, 1927), p. 27.]

to use all kinds of other complicated symbols — hieroglyphics [6] — together with the twenty-four "letters" which they had succeeded in abstracting from their language. That stopping short in sight of the goal may seem strange, but in the history of science it is the rule rather than the exception. The great inventions were seldom completed by the great inventors; other men — often smaller men but more practical or more radical — were needed to realize the full value of the invention and to exploit it ruthlessly. The Faradays and the Maxwells sow the seeds, the Edisons and the Marconis pluck the fruits. The Egyptians were so accustomed to their hieroglyphics that they would not forsake them and carried them along for thousands of years, together with the alphabetic signs which they had invented but did not use properly.[7] The invention was brought to a higher stage of perfec-

[6] From *hieros*, sacred, and *glyphein*, to engrave.

[7] It should be borne in mind that hieroglyphics or other conventional signs are easier to read, if one knows them, than alphabetic writing, and therefore such signs are introduced into every language, especially into the scientific language. Consider the signs used to convey astronomical, chemical, mathematical meanings, or more homely ones

tion by the Phoenicians, who created the first Semitic alphabet (purely consonantal); it was completed by the Greeks, who added vowels. The whole development lasted two or three thousand years, if not longer.

How did the Egyptians finally write down a word of their language? Most of the hieroglyphics contain two kinds of signs, "phonetic" and "determinative." The first indicate the sound, the latter the idea, the class to which the word would belong in any classification according to meaning. The phonetic signs may be simply alphabetic (consonantal) or they may represent combinations of consonants, such as *mr, tm, nfr.* The combination of the two kinds of signs completes the identification of a word, and facilitates its recognition and one's remembrance of it, among thousands of others. The Egyptian writing, born out of inherent compromise, is cumbersome enough and often redundant, but English-speaking people should not judge it too harshly, for their own perversion of the alphabet due to similar compromises is just as shocking. They have inherited a marvelous instrument but have failed to use it consistently and unambiguously for the spelling of their own language.

Any Chinese or Sinologist reading my brief description of hieroglyphics will say to himself that it applies very well to Chinese characters. The Egyptians and the Chinese, working independently at two ends of the world, created two vast collections of word symbols. It is very interesting to compare the fruits of those gigantic experiments. They started with pictograms as everybody would; moreover, the early Chinese and Egyptian pictograms of the same objects — sun, moon, mountains, water, rain, man, bird — were often analogous. As the two kinds of word symbols were standardized and simplified, and became more and more numerous, both peoples reached the same general conclusion — that each word should contain a phonetic element (sound sign) and a determinative one (sense sign). The Chinese did this very consistently. About 80 percent of their characters are made up of two parts, one of which is a clue to the sound, while the other (one of 214 "classifiers") is a clue to the meaning; generally speaking, the pronunciation of the classifier and the meaning of the phonetic element are disregarded.

Thus far the Chinese and Egyptian achievements are very much alike; there are fundamental differences between them, however — and what else could we expect, considering that the two nations were very unlike and had been submitted for thousands of years to very different physical and psychologic environments? In Egyptian writing the vowels are omitted and in speech they are frequently changed either to obey grammatical inflections or to indicate variations of meaning; in Chinese, on the contrary, the vowels belong to the root, have a semantic value, and are constant. The study of the meanings of Chinese words cannot be separated from the study of their sounds. One can see how alphabetic signs could eventually emerge from the Egyptian habit of script; they could not have emerged from the Chinese one.[6] The Chinese word is always concentrated in a single character, more or less complex, yet meant to occupy the same space as any other character; the

such as $ for dollar, or the ampersand, &. The weakness of all such signs is that one cannot understand them at all unless one is already familiar with them, whereas everybody can read such words as "Venus," "ascending node," or "antimony," and look them up in the dictionary, if necessary.

[6] For further discussion and exemplification, see Won Kenn (=Huang Chüan-shêng), *Origine et évolution de l'écriture hiéroglyphique et de l'écriture chinoise* (Lyons: Bosc Frères and Riou, 1939).

Egyptian word is more like a word in any syllabic script, it may cover more or less space.

The early students of Chinese and Egyptian were far more impressed by the resemblances between the two scripts than by their divergencies. Having more enthusiasm than knowledge, they jumped eagerly to the conclusions that appealed to their sensibility. In 1759 the French Sinologist Joseph de Guignes wrote a memoir in which he claimed that the Chinese characters were derived from the Egyptian, and that China was originally an Egyptian colony! [9] This started a controversy that we have no time to analyze. But a century ago Samuel Birch (1813–1885) was still approaching the study of the hieroglyphics from the Chinese point of view.[10] Birch was not by any means a dilettante, but a man of incredible zeal, the author of the first Egyptian dictionary in alphabetic order (1867).

In the meanwhile the consonantal nature of the Egyptian script had started another polemic. Indeed, alphabets restricted to consonants are a feature common to every Semitic language. Should we not then consider Egyptian a member of the Semitic family? This controversy is far more serious than the Chinese-Egyptian one. The Chinese-Egyptian similarities are due to the sameness of the tasks to which the Chinese and the Egyptians had addressed themselves and the essential identity of their own natures. The Egyptian-Semitic similarities were due to definite contacts and borrowings. This cannot be denied and the discussion turns around the quantity of the borrowing rather than its reality. Many distinguished Egyptologists concluded that the Egyptian and Semitic languages were closely related, and one of them, the Italian Simeone Levi, published a Coptic-Hebrew-hieroglyphic dictionary wherein the many affinities that he had (or thought he had) discovered between Egyptian and Hebrew were brought together.[11] Not only are there parallelisms in words and word building but the formation of pronouns and numbers is similar. Yet for all that the differences between Egyptian and the Semitic family are far greater than those obtaining between the different members of that family.

Consider the Egyptian number words. The words for 1, 2, 3, 4, 5, 10 are African, those for 6, 7, 8, 9 are Semitic. What does that mean? It means that the original linguistic stock was African (Hamitic), for surely the words for 1, 2, 3, 4, 5 are among the first to be needed and coined in any language; it also means (see previous chapter) that the number base of the early Egyptians was five. Later contacts with Semitic peoples in the South and East introduced Semitic features into their language plus the base ten. As the Egyptians became more powerful (during the Eighteenth to Twentieth Dynasties, from the end of the sixteenth century to the twelfth, Egypt controlled a world empire), they influenced the Semitic peoples of the Near East. Many traces of Egyptian influence can be detected in the form and contents of the Hebrew Bible.[12] Those exchanges of influence are of deep concern to the historian of humanity. They show that Egypt was after all an intrinsic part

[9] Joseph de Guignes (1721–1800), *Mémoire dans lequel on prouve que les Chinois sont une colonie égyptienne* (Paris, 1759; 59 p., 1 pl.).

[10] Sir E. A. Wallis Budge, *Egyptian dictionary* (London, 1920), p. xiv.

[11] Simeone Levi, *Vocabolario geroglifico-copto-ebraico* (10 parts in 3 vols.; Turin, 1887–1894).

[12] Just as the Semitic elements of Egyptian have been exaggerated by some scholars, so the Egyptian elements in the Old Testament have been exaggerated by others, for example Abraham Shalom Yahuda, *The language of the Pentateuch in its relation to Egypt* (London: Oxford University Press, 1933).

of the Mediterranean world. Though Egyptian wisdom reached us largely through Semitic channels, Egyptian manners and arts came to us also via Crete and other islands.[13]

THE INVENTION OF PAPYRUS

The invention of writing was given its full social value by another invention, that of a suitable material to write upon, easily available and not too expensive. It is clear that as long as writing was restricted to inscriptions on stone (as was apparently the case in Greece for centuries), its scope was limited to records considered to be of outstanding importance. Literary productions were too long to be chiseled on stone or metal; a cheaper material was needed for their nonoral preservation.

The early Egyptians solved that fundamental problem in a magnificient way by the invention of papyrus. This was a remarkably good writing material, made of the pith of the stem of a tall sedge (*Cyperus papyrus*), then abundant in the Delta marshes.[14] The pith was cut into longitudinal strips, and these strips were arranged crosswise in two or three layers, soaked, pressed, and burnished. The cost could not be very high for there was no end of reeds in the marshes; it sufficed to collect them and the fabrication was simple enough.

Every invention calls for complementary inventions. It is not enough to have something handy to write upon; one must have tools for writing with. The Egyptians used various kinds of pigment (or ink) and spread it on the papyrus with a fine brush made of a thin rush (*Juncus maritimus*)[15] found in the same marshes as the sedge.

The tremendous importance of the invention of papyrus is immortalized by two words common in many languages, paper and bible. The first word is somewhat misleading, for our paper, made of pulp, is a Chinese invention essentially different from the Egyptian one. The Greeks called papyrus *byblos* and a strip of it *byblion* or *biblion*; later they used the word for a whole book (compare the similar evolution of the Latin word *liber*). It is possible but not certain that the word *byblos* was itself derived from the name of a busy marketplace and harbor north of Beirūt (Byblos = Jubayl), for the international trade in papyrus was largely controlled by Phoenicians. Indeed, objects have often been named from their best-known place of importation, rather than from the place of origin, which might be, and often was, unknown (India ink, Arabic numerals, etc.).

The superiority of papyrus over other writing grounds that the Egyptians used at one time or another (such as bones, clay, ivory, leather, linen) was clear enough; there is one aspect of it that may not have been immediately obvious, though it is perhaps to our mind the most important. Accounts written on pieces of bone, leather, or what not were bound to remain *disjecta membra*, which it would have been almost hopeless to keep together for centuries. The ingenious creators of papyrus, after having fabricated isolated sheets, discovered that many sheets, in

[13] J. D. S. Pendlebury, *Aegyptiaca. A catalogue of Egyptian objects in the Aegean area* (Cambridge: The University Press, 1930) [*Isis 18*, 379 (1932–33)].

[14] It is no longer found in those marshes but still flourishes in the Sudan. Would its disappearance from the Delta be due to exhaustive use of it in ancient and medieval times? According to Pliny, who gives much information on papyrus (*Natural history*, XIII, *21–27*), there was such a scarcity of it under Tiberius (emperor, 14–37), that senators were obliged to regulate its distribution. Thus paper rationing is not a novelty of our own day!

[15] A reed was not used until much later (Greco-Roman times); it is still used by natives today (qalam).

fact almost any number of them, could be glued together, each one to the edge of the preceding one, and thus it was possible to make a roll (*volumen*, hence our word volume) to contain a text of any length and preserve it completely in its proper sequence. The width of a roll varied from 3 to 18½ in.; the length depended naturally on the text it included; the longest papyrus is the Papyrus Harris No. 1 (British Museum No. 9999), measuring 133 feet by 16½ in. Thanks to the invention of the roll, many ancient texts have come down to us in their entirety.

The makers of papyrus supplied the ancient Western world with an excellent, attractive, and cheap[16] vehicle for the diffusion of its main cultural achievements. Most of the rolls that we now have were found in tombs. The preservation of papyrus, which would have been impossible under those conditions in most climates and highly problematic in many, was guaranteed by the dry weather of Egypt. A great part of ancient literature was thus safeguarded by the miraculous coincidence of a great invention with an extraordinary climate. Without nature's help, man's efforts would have been wasted.[17] Though we are concerned now mainly with ancient Egypt, the literary remains of which have been preserved almost exclusively by means of papyrus, it may be mentioned that we owe to the same material the preservation of a vast number of other documents, Biblical, Greek, and Roman. The accumulated knowledge at the disposal of the Romans would have been considerably less without papyrus, and the course of intellectual history very different.

To be sure, other materials for writing might have been invented, but the only ones that proved of comparable value,[18] parchment and paper, did not become available until much later. If the story connecting the invention of the former with the library of Pergamon is correct, its invention dates only from the second century B.C. Paper was invented in China at the beginning of the second century after Christ. Thus both parchment and paper are definitely posterior to Pharaonic Egypt; we may say that even the oldest of both was invented more than twenty-seven centuries later than papyrus! That is, during that very long period of time, papyrus was not only the best but, with the exception of clay tablets, the only suitable material for the diffusion of culture.

In fact, papyrus was so good that it continued to be used until the eleventh century,[19] though Chinese paper was known in Egypt *c.* A.D. 800 and manufactured there a century later. Parchment (or vellum) was an excellent material, but far more expensive, prohibitively so for the simpler purposes of life.

As long as writing was needed only for monumental purposes, it was done very

[16] Relatively cheap. Papyrus was never as cheap or abundant as handmade paper was, not to mention the paper of today which is so cheap that it is continually wasted for vain and vile purposes. Papyrus was always a de luxe material. We know but little about its early production, but for later times see Naphtali Lewis, *L'industrie du papyrus dans l'Egypte greco-romaine* (200 pp.; Paris: Rodstein, 1934) [*Isis* 35, 245 (1944)].

[17] A good example of this is the use of palm leaves for writing in Ceylon and India. They used the leaves of talipot (*Corypha umbraculifera*), which grows in Ceylon and Malabar, and produced a kind of papyrus, in narrow strips, called olla. Unfortunately, the climate of India was not as favorable for the preservation of olla documents as that of Egypt was for papyrus.

[18] The clay tablets used in Mesopotamia were excellent from the point of view of preservation of separate items, but did not make possible the invention of anything comparable to a roll, and hence the integrity of long documents was jeopardized.

[19] Papal bulls were published on papyrus until 1022. *Pontificum Romanorum Diplomata papyracea quae supersunt in tabulariis Hispaniae, Italiae, Germaniae, phototypice expressa jussu Pii PP.XI consilio et opera procuratorum Bibliothecae Apostolicae Vaticanae* (18 pp., 15 facsimiles on 43 pls.; Rome, 1929).

slowly. Inscriptions, especially on hard stone like granite, were exceedingly difficult to carve. That difficulty was not a serious obstacle, for inscriptions, even the longest, are relatively short. From the artistic point of view it was a blessing. The artisan was put on his mettle, did his best, and often surpassed himself. Some of the monumental hieroglyphics, chiseled in the hard stone, inlaid, or simply painted, are among the treasures of Egyptian art. However, when scribes began to write on papyrus, they had to proceed much faster and the old hieroglyphics became unmanageable. Thus there gradually developed a new, easier, script, a cursive or running hand, called hieratic (*c.* 1900 B.C.). Later still (*c.* 400 B.C.), as writing was popularized, even the hieratic script was too slow, and it was replaced by a kind of shorthand called enchorial or demotic (Fig. 4).[20] Of course every script has been submitted to a similar evolution, but the range of that evolution was greater for Egyptian than for any other script because the hieroglyphics were the most elaborate symbols ever invented. The only characters comparable to them are the Chinese, but these were much simpler and also less beautiful. In the course of time Chinese calligraphy achieved a remarkable beauty of its own, but always more abstract than the beauty of hieroglyphics.

[20] *Hieraticos* = priestly (because the scribes were generally clerics); *enchōrios* = rural, *dēmoticos* = popular.

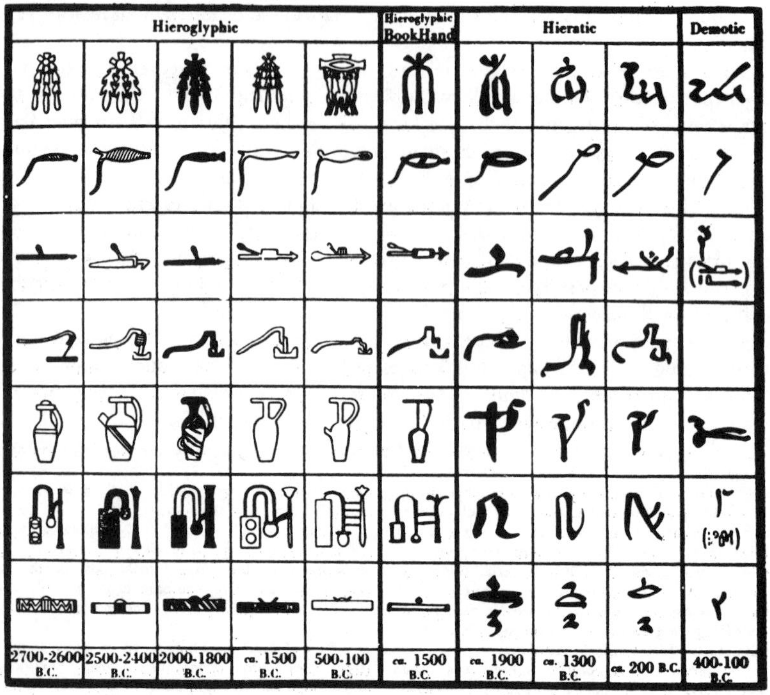

Fig. 4. The passage from hieroglyphics to demotic script. [Borrowed with permission from George Steindorff and Keith C. Seele, *When Egypt ruled the East* (Chicago: University of Chicago Press, 1942), p. 123.]

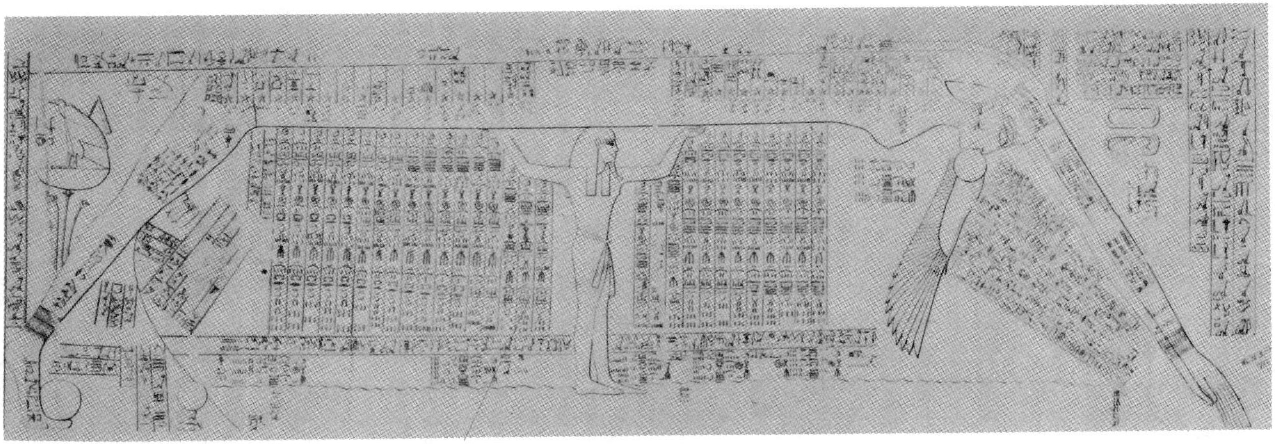

Fig. 5. Nut and Shu. A colossal figure of Nut, goddess of the sky, supported by Shu, god of the air, in the cenotaph of Seti I (1313–1292, Nineteenth Dynasty) in Abydos. Nut gives birth every day to the Sun and the stars. On her body are given the names of the decans and underneath her, and also on her arms and legs, are tabulated the days and months upon which a morning, a midnight, or an evening rising of the corresponding constellation occurred. [From H. Francfort, *The cenotaph of Seti I at Abydos* (2 vols.; London: Egypt Exploration Society, Memoir 39, 1933), vol. 1, pp. 27, 72–75; vol. 2, pl. lxxxi.]

A similar allegory may be seen in the tomb of Ramses IV (1167–1161, Twentieth Dynasty) in Thebes. See diagram and commentary in Heinrich Brugsch, *Astronomische und astrologische Inschriften altaegyptischer Denkmaeler* (Leipzig, 1883), p. 174.

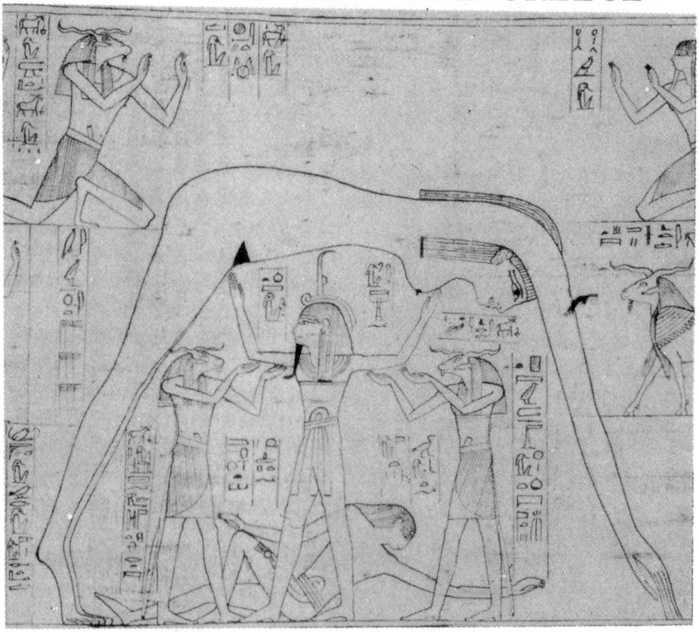

Fig 6. Nut and Shu. The goddess of the sky, Nut, is represented surrounding heaven and supporting herself on her hands and feet. The earth god, Qeb, is stretched on the ground. The god Shu stands in the middle after having lifted up Nut on his two hands. Sheet 87 of the Greenfield papyrus in the British Museum, the longest papyrus of the Theban recension of the *Book of the Dead* (before being divided into 96 sections, the roll was nearly 123 ft long and 1 ft 6½ in. high). [Reproduced with the Trustees' permission from E. A. Wallis Budge, *The Greenfield papyrus. The funerary papyrus of Princess Nesitanebtashu, priestess of Amen-Ra at Thebes c. 970* B.C. (British Museum, 1912), pl. 106.]

ASTRONOMY

The Egyptians' familiarity with the stars dates back to the most remote prehistoric age. This is no wonder, for their cloudless atmosphere and the pleasant freshness of their climate at night invited men to contemplate the rotation of the sky. They could not help noticing that the stars were unequally distributed and formed groups (or constellations) of recognizable shape. One of their early mythologic fancies was to conceive the whole heaven as surrounded by the body of a goddess (Nut), who is supporting herself on her hands and feet (Figs. 5 and 6). That vast conception gave them the habit of sweeping the whole sky with their eyes and enabled them to recognize constellations of gigantic size as compared with ours. The longest of them, the man Nekht, took almost 6 hours to pass across the meridian. For the sake of easy reference they divided a wide belt along the equator into 36 parts, each part including the most conspicuous stars and constellations (or parts of constellations) the rising of which could be observed during each successive 10-day period or decad (*hē decas*); hence each such group of stars was called a *decan* (*ho decanos*). We have early tables of the decans, listing the stars characteristic of each.[21]

The most important event in the life of Egypt was the annual overflow of the Nile, upon which depended the farmer's prosperity (or, if it were to fail, his misery). That event coincided or almost coincided (for its regularity was imperfect) with the heliacal rising of the brightest star in the sky, Sothis.[22]

The Egyptians had first tried to take account of the passing of time by means of the Moon, but, fortunately for them, they discovered the ambiguities of such a method before being committed to it by religious ceremonies and could thus easily overthrow it in favor of a solar calendar. Their year was at first divided into twelve months of three decads each (corresponding to the 36 decans), but they soon added a holiday season of five days (*hai epagomenai* sc. *hēmerai*). The civil or calendar year began with the first day of the month Thot; the Sothic or astronomic year began with the heliacal rising of Sothis. The continued observation of that rising, year after year, must have given their astronomers considerable perplexity. Indeed, their civil year had 365 days, while the heliacal rising of Sothis recurred after a somewhat longer interval of about 365¼ days. After four years (*tetraetēris*) there was a difference of one day; Sothis did not reappear on the first day of the new civil year but a day later; after forty years, the difference was 10 days. It seems easy enough to conclude, and the ancients did so, that after 1460 years the Sothic cycle would be completed (for $365 \times 4 = 1460$).

However, Carl Schoch[23] has shown that the length of the Sothic cycle was 1456 rather than 1460 years; he took into consideration the secular acceleration of the sun, the large proper motion of Sirius, and improved values of the *arcus visionis*. The following table, based on Schoch's discussion, shows that the Julian-calendar date corresponding to the first day of Thot, the civil New Year, changed from 16 July to 19 July at the beginning of the four Sothic cycles of Egyptian history; the heliacal rising of Sothis, on the corresponding July dates, fell on 1 Thot during the four *tetraeterides* indicated in the second column of the table.

Sothic cycle	First tetraeteris of the cycle	Civil New Year, first day of Thot (extrapolated Julian calendar)	Heliacal rising of Sothis
1	4229–4226 B.C.	16 July	16 July
2	2773–2770 B.C.	17 July	17 July
3	1317–1314 B.C.	18 July	18 July
4	140– 143 A.D.	19 July	19 July

The Sothic (or Julian) year of 365¼ days was introduced in Rome in 45 B.C. by Julius Caesar, with the technical assistance of an Egyptian Greek, Sōsigenēs. The beginning of the new Sothic cycle (the fourth in the list above), that is, the coin-

[21] Alexander Pogo, "Three unpublished calendars from Asyūṭ," *Osiris 1*, 500–509 (1936); 10 pls., 3 figs., 1 table.

[22] Sothis = Sirius = *cyōn* = Dog Star. The "dog days" or "canicular days" refer to the hot period beginning with Sothis' heliacal rising (that is, the first observable rising of Sothis at dawn). The date of that rising varies with the latitude and changes slowly in the course of time. It was 19 July in Roman times, and is now 21 July Julian (= 3 August Gregorian) for Memphis. It is not clear to me how well the heliacal rising could be observed, for this implies the ability to distinguish a star when its elongation from the sun is less than, say, 1°.

[23] Carl Schoch, "Die Länge der Sothisperiode beträgt 1456 Jahre," *Astron. Abhandl.*, *Ergänzungshefte Astron. Nachr.* 8, no. 2, B9–B10 (1930).

cidence of the first day of Thot with the heliacal rising of Sothis, was actually observed in Egypt in A.D. 140–143. Calculating backward from that date and assuming wrongly that the Sothic cycle equaled 1460 years and was constant, Breasted determined what he called "the oldest fixed date" in history, the era of the Sothic calendar, the year 4241 B.C.[24] After taking Schoch's corrections into account we conclude that the "oldest fixed date" was not 4241 but 4229–4226. In any case, we should bear in mind that that date is the result of backward extrapolation, and not attach too much importance to it.

The astronomic ability of the early Egyptians is proved not only by their calendar, tables of star culminations, and tables of star risings, but also by some of their instruments, such as ingenious sundials or the combination of a plumb line with a forked rod that enabled them to determine the azimuth of a star. Specimens of such instruments are preserved in the Cairo and Berlin museums, and accurate replicas of them may be examined in many Egyptologic or astronomic collections.[25]

ARCHITECTURE AND ENGINEERING[26]

The pyramids are so well known to everybody that they do not require any description. The average reader, however, thinks only of the three Pyramids of Gīza, which are the largest but not by any means the only ones, nor the earliest. The oldest one, built for King Zoser of the Third Dynasty (in the thirtieth century), is the so-called step pyramid (al-haram al-mudarraj) of Saqqarā (near the old capital, Memphis, south of Cairo); it is about 200 ft high. The Great Pyramid, the largest of the three in Gīza, was built a century later for Khufu (Cheops) of the Fourth Dynasty; this was the largest building of ancient times and one of the largest ever built by man. Each side measures about 775 ft, and when it was intact the monument was 480 ft high. The pyramids, erected to house and protect royal tombs, are structures of limestone blocks, built solid except for the funereal chamber and devious passages leading to it.

The construction of such immense buildings forty-nine centuries ago raises a number of technical problems, many of which have not yet been solved. How Cheops' architects could devise and his subjects construct such edifices still baffles the imagination. Their mechanical equipment, however advanced it might be as compared with that of illiterate savages, was vastly inferior to ours. The Great Pyramids are so wonderful that some of the scholars who tried to penetrate their secrets became the victims of a mild form of insanity and ascribed to the ancient builders occult and metaphysical intentions and an esoteric knowledge the possession of which would have been even more marvelous than the mechanical and engineering ability that they certainly possessed. But the pyramids were built, there they stand in the desert, the most massive facts of antiquity, the best witnesses of

[24] Breasted, *Ancient records of Egypt* (vol. 1, p. 30).

[25] They have been elaborately discussed by Ludwig Borchardt, *Altägyptische Zeitmessung* (folio, 70 pp., 18 pls., 25 figs.; Berlin, 1920) [*Isis 4*, 612 (1921–22)].

[26] Henry Honeychurch Gorringe, *Egyptian obelisks* (folio, 197 pp., 51 pls.; New York, 1882); Edward Bell, *The architecture of ancient Egypt* (280 pp., 1 map; London, 1915); Reginald Engelbach, *The problem of the obelisks. From a study of the unfinished obelisk at Aswān* (134 pp., 44 figs.; London, 1923), valuable for technical details, but inferior for historical matters; Somers Clarke and R. Engelbach, *Ancient Egyptian masonry. The building craft* (258 pp., 269 ills.; London, 1930); Alfred Lucas, *Ancient Egyptian materials and industries* (460 pp.; rev. ed., London, 1934); Flinders Petrie, *Wisdom of the Egyptians* (162 pp., 128 figs.; London: Quaritch, 1940) [*Isis 34*, 261 (1942–43)].

their builders to this day, and they will probably outlast most of the buildings of which modern man is so proud.

The achievements of the pyramid builders have been pooh-poohed by people who say: "The Egyptians used many thousands of men working for long periods of time. They replaced machine power by manpower in unlimited quantities." Of course, they used large numbers of men, but this does not solve the main architectural and technical riddles, and it introduces new ones — human riddles — of almost equal difficulty. It is easy to speak of harnessing 30,000 men to a task and making them work together, but exactly how was it done? The number of people that can be usefully associated for a definite task in a limited space is limited, but assuming that a very large number, say tens of thousands, could be used at one and the same time, the direction of their efforts required considerable skill and forethought, and the satisfaction of their hunger and other needs required administrative experience and a commissariat technique of great complexity. Whether the power needed for a task is supplied by a dynamo or by an army of men, the planning and execution of the task imply knowledge, intelligence, and adaptability.

It is impossible to review here all the problems involved in Egyptian architecture, for they are legion. Let us consider a special case, the erection of granite obelisks.[27] To see the pyramids one must needs go to Egypt, but obelisks are available in many European countries and even in New York. How were they created? All the granite obelisks were quarried in Aswān, just below the First Cataract.[28] The very quarries (maḥājir) from which they were taken can be examined today and are indeed a great attraction to tourists, especially because one can see *in situ* a gigantic obelisk that had to be abandoned when fissures developed in its mass. If it had been possible to extract and to erect it, that obelisk would have been the largest of all, reaching a height of 137 ft and a weight of 1,168 tons. That abandoned obelisk has made it possible to figure out how the ancient engineers proceeded to remove the top layers of the granite stratum, to set out the mass of stone to be detached, and finally to separate it all around from its matrix. Reginald Engelbach, taking advantage of all the evidence available in Aswān and elsewhere, has explained these matters to us, as well as the transportation of the detached obelisk on sleds to the Nile, its embarkation on a ship, later its disembarkation, its transport to the point of erection, and finally its erection. In spite of his mechanical and archaeologic experience, Engelbach has not been able to explain everything. For example, what kind of tools did the Egyptians use to cut the very hard rock? They probably used dolerite balls (many of which are found *in situ*) to bash it out, rather than cut it out, but they needed other tools — metal tools probably, but what? How were the intricate and long hieroglyphic inscriptions engraved into the hard granite?[29]

The sophistication of the Egyptian architect is proved by the existence of a definite entasis in the Paris obelisk.[30] The final erection of an obelisk was an ex-

[27] In order to consider the obelisks we have to make a big jump, from the so-called Old Kingdom into the New one. The great pyramids date from the Fourth Dynasty (2900 to 2750), the Obelisk Age is that of the Eighteenth and Nineteenth Dynasties (1580 to 1205); mean interval between the two ages, fourteen centuries!

[28] That is, 7°27′ south of the Mediterranean (Damietta mouth). Aswān is about half a degree north of the Tropic of Cancer. Aswān = Syene of the Greeks.

[29] Some Egyptian tools are discussed by Clarke and Engelbach, *Ancient Egyptian masonry*, p. 224; 3 pls.

[30] The Greeks called by the name *entasis* the swelling in the middle of a column necessary to correct an illusion of concavity (*Vitruvius*, III, 3, 13). A convexity was intentionally left on the front face of the Paris obelisk. That obelisk dates from the Nineteenth Dynasty (1350–1205).

Fig. 7. Statue of Senmut, architect of queen Hatshepsut (1495–1475), holding in his lap her eldest daughter Nefrure whom he reared (Cairo Museum). It is 60 cm high. For Senmut see J. H. Breasted, *Ancient Records of Egypt* (Chicago: University of Chicago Press, 1906), vol. 2, secs. 345–368.

tremely delicate job, on which the architect risked his reputation and possibly his life. If the obelisk did not fall gently enough [31] it would break, and years of labor would be lost, or if it was not properly adjusted to its base, the damage was irreparable and the architectural effect spoiled.[32] The task was complex and so full of hidden difficulties that one cannot help wondering whether the Egyptians did not experiment with scale models in order to determine the weight and balancing points of the obelisks, rehearse the erection process, and thus escape fatal disappointments.[33] In any case, the Egyptian architects and thei. royal masters were fully conscious of their achievements and proudly recorded them. Half a dozen of the obelisk architects are personally known to us, because they were rewarded with tombs in the Theban necropolis and with statues in the temple. The inscriptions on the tombs and statues relate the erection of obelisks, but unfortunately do not explain how it was done. Maybe the explanation would have taken too much space and would have been of no interest except to other architects, who did not need it (or who needed technical details rather than generalities). In the same way, when we place an inscription upon a bridge we do not try to explain, even in the briefest manner, how the bridge was built.

Let me evoke two of those archiects. The first is Senmut, chief architect of Queen

[31] It seems certain that an obelisk was not erected from its position on the ground to a position perpendicular to it; that would have been impracticable. The obelisk was pulled up a long sloping embankment until it was at a height well above that of its balancing point, or center of gravity; than the earth was cut from below it carefully until the obelisk settled down onto the pedestal with its edge 'n the pedestal notch, leaning against the embankment. From this position it was pulled upright. For details and drawings, see Engelbach, *The problem of the obelisks*, pp. 66–84.

[32] The obelisk of Queen Hatshepsut (1495–1475) at Karnak came on to its 'pedestal askew, but the irregularity is too small to be unpleasant.

[33] Modern architects, beginning with Fontana, do use scale models.

Fig. 8. Reërection of an Egyptian obelisk in the Vatican, Rome, in 1586 by Domenico Fontana. [From G. Sarton, *Agrippa, Fontana and Pigafetta*, *Arch. internat. d'histoire des sciences 28*, 827–854 (Paris, 1949), with 14 figures.]

Hatshepsut (1495–1475), builder of her obelisks and of the great temple of Deir al-Baḥarī. In his statue, he is represented holding her eldest daughter, Nefrure, whose tutor he was (Fig. 7). The second is Beknekhonsu, living a century later, creator of the Paris obelisk and perhaps inventor of the entasis; his statue, bearing a long autobiographic inscription, is now in the Glyptothek of Munich.[34]

Many of the obelisks were moved from Egypt and taken to Rome,[35] Constantinople, later to Paris, London, and other cities, and even across the Atlantic to New York. The Romans, who were connoisseurs of engineering difficulties, were the leaders in the obelisk exodus. The largest obelisk to be seen anywhere today is the one standing in front of S. Giovanni in Laterano. It was begun by Thutmosis III and completed by Thutmosis IV (1420–1411) for the temple of Karnak. It was transported to Alexandria in A.D. 330 by order of Constantine the Great, who wanted it to embellish Constantinople, but in 357 his son Constantius II took it to the Circus Maximus in Rome. In 1587, it was discovered there, broken in three pieces; in the following year it was set up in its present location by Domenico Fontana. The same Fontana achieved greater fame by the erection of another obelisk, the Vatican one, smaller but unbroken. That obelisk had not been finished by the Egyptians, for it bears no hieroglyphic inscription (hence we do not know its early history). It was brought from Heliopolis by order of Caligula (emperor, 37–41) and set up in the circus later called the Circus of Nero. Pope Sixtus V ordered its transportation to the Piazza di San Pietro, which was done under Fontana's direction in 1586 (Fig. 8). The event attracted considerable attention and was discussed in detail by Fontana himself in a remarkable book.[36]

The Paris obelisk was taken from Luxor and moved to its present location by the naval engineer J. B. A. Lebas in 1836. The New York and London obelisks were originally standing together in Heliopolis where they had been erected by Thutmosis III (1501–1448). Both had been moved by the Romans *c.* 22 B.C. to Alexandria. 'Abd al-Laṭīf (XIII–1), writing at the beginning of the thirteenth century, saw them both standing; Pierre Belon (1517–1564), who visited Alexandria about the middle of the sixteenth century, saw only one. In the meantime, one of them had fallen; happily, the piles of sand that had accumulated around it throughout the centuries had broken its fall and it had remained whole. That one was erected on the London embankment in 1878; the standing one was taken down and reërected in Central Park, New York, in 1881. The engineer responsible for the transportation to America and erection in New York was the Barbados-born Henry Honeychurch Gorringe (1841–1885), Lt. Com., U.S. Navy, who published an excellent account of the achievement, together with information on all the other obelisks. This is still the standard work on the subject.

It has already been mentioned that the abandoned Aswān obelisk would have weighed 1,168 tons. The other obelisks named above (I name them again in order of size) — Lateran, Vatican, Paris, New York, London — weigh, respectively, 455,

[34] The translation of that moving inscription may be read in Breasted, *Ancient records of Egypt*, vol. 3, pp. 561–568.

[35] There are a dozen obelisks in the public squares of Rome.

[36] Domenico Fontana (1543–1607), *Della trasportatione dell' obelisco vaticano* (Rome, 1590). Fontana was the main architect and collaborator of Sixtus V (pope, 1585–1590) in the creation of "Sixtine Rome." See G. Sarton, "Agrippa, Fontana and Pigafetta. The erection of the Vatican obelisk 1586," *Arch. internat. d'histoire des sci. 28*, 827–854 (1949), 14 figs.

331, 227, 193, 187 tons.[37] The ancient Egyptians were prepared to handle obelisks much larger than those familiar to us in the West; the Aswān obelisk would have been almost six times as heavy as the London one. The erections directed by Fontana in 1586 and by Gorringe in 1881 were talked about as nine-day wonders, and yet these men were but repeating a part of the work that their Egyptian forerunners had done thousands of years earlier.

The boastful accounts of modern engineers [38] who have at their disposal mechanical means of incredible power (the fruits of centuries of accumulated efforts) constitute the best proof of the genius of the Egyptian engineers, who were able to accomplish similar deeds without such means. From that point of view, modern Egyptians should not regret that so many obelisks were taken away from their native country. Each one of the exiled obelisks is an almost imperishable monument to the glory of ancient Egypt.

MATHEMATICS [39]

The architectural and engineering events of Egypt imply a good deal of arithmetic and geometric knowledge. To begin with, simple means of keeping complicated accounts were indispensable. Such needs were satisfied early. There is a royal mace in the Ashmolean Museum, Oxford, dating from King Nar-Mer, before the First Dynasty (before 3400 B.C.); it records the taking of 120,000 prisoners, 400,000 oxen and 1,422,000 goats.[40] These are large numbers; they are written somewhat in the Roman manner, there being symbols for each decimal multiple (up to a million) which are repeated as often as necessary.[41] In general, the largest units were listed first, then the others in order of importance, but that was not essential; they might be grouped in any order that would be pleasing to the eye. Later a simplified method was used and one wrote 100,000 × 101 instead of 10,100,000.[42]

As to geometry, the need of it was obvious, even for the building of monuments as simple in outward appearance as the pyramids, and these take us back to the thirtieth century. The pyramid builders were obliged to cut limestone blocks exactly before bringing them to their proper position; the largest blocks were those placed in a complex arrangement above the royal chamber with the objective of diverting

[37] The weights are quoted from Engelbach, *The problem of the obelisks*, p. 30. Engelbach's tons are long tons (= 2,240 lb avoirdupois). In short tons (= 2,000 lb), the weights would be 1308, 510, 371, 254, 216, 209.

[38] To those already mentioned one might add A. Richard de Montferrand, *Plans et détails du monument consacré à la mémoire de l'empereur Alexandre* (elephant folio; Paris, 1836); copy in Harvard Library. The Leningrad column is a granite monolith, 12 ft in diameter and 84 ft long; the whole monument is 154 ft high. The Russian undertaking is more directly comparable to the Egyptian one, for the Russians did all the work, beginning with the quarrying of the granite in Finland. Montferrand's original idea was to create an obelisk but the emperor preferred a column.

[39] T. Eric Peet, *The Rhind mathematical papyrus* (folio, 136 pp., 24 pls.; Liverpool University Press, 1923 [*Isis 6*, 553–557 (1924–25)]; Arnold Buffum Chace, Ludlow Bull, Henry Parker Manning, and Raymond Clare Archibald, *The Rhind mathematical papyrus* (2 vols.; Oberlin, Ohio, 1927–1929) [*Isis 14*, 251–253 (1930)]; W. W. Struve, *Mathematischer Papyrus des Staatlichen Museums der Schönen Künste in Moskau* (210 pp., 10 pls.; Berlin, 1930) [*Isis 16*, 148–155 (1931)]; Otto Neugebauer, *Vorlesungen über Geschichte der antiken mathematischen Wissenschaften. 1. Band, Vorgriechische Mathematik* (Berlin: Springer, 1934) [*Isis 24*, 151–153 (1935–36)].

[40] James Edward Quibell, *Hierakonpolis* (London, 1900), p. 9, pl. xxviB.

[41] Just as the Romans would write MMCCCIIII for 2304.

[42] Alan H. Gardiner, *Egyptian grammar* (Oxford, 1927), p. 191, gives two examples, one of the Middle Kingdom (2160–1788), the other of the time of Ramses III (1198–1167).

the pressure from its ceiling; there are fifty-six such roofing beams over the chamber of the Great Pyramid, the average weight of which is 54 tons. The accuracy obtained in the building of that pyramid (Cheops, Fourth Dynasty) is almost incredible. According to Flinders Petrie,

> the mean error of length of the sides, 755 ft long, is 1 in 4,000, an amount which would be produced by a difference of 15°C in the temperature of copper measuring bars. The error of squareness is 1′12″. The error of leveling averages 5 in. between the different sides, or 12″. On shorter lengths of 50 ft the differences are only 0.02 in.
>
> The accuracy of three granite sarcophagi of Senusert II, Twelfth Dynasty, averages 0.004 in. from a straight line in some parts, 0.007 in. in others. The curvature of the planes of the sides is only 0.005 in. on one, 0.002 in. on another face. The mean error of the proportions of the different dimensions in even numbers of palms is 0.028 in. This is more like the work of opticians than of masons.[43]

The cutting of stones that were meant to fit nicely together implied some knowledge of stereometry (we shall see presently that the Egyptians had gone remarkably far in that field); one might claim that it also involved some knowledge of descriptive geometry and stereotomy. It was not enough to solve such problems in a general way, for the stonecutter must be shown very clearly how the limestone blocks should be cut. That knowledge, however, remained empirical and probably unformulated.[44]

Though we may be certain that the pyramid builders had already a fair mathematical equipment, without which the scientific part of their task could not have been accomplished, we have no mathematical texts of the Old Kingdom, nor any prior to the Twelfth Dynasty (2000–1788). The two most important texts have come down to us in somewhat later editions, but they go back very probably to that same dynasty.

Archibald[45] has listed some thirty-six original documents concerning Egyptian mathematics; they are written in Egyptian, Coptic, and Greek, and range in date from *c.* 3500 B.C. to *c.* A.D. 1000 (forty-five centuries); the documents prior to 1000 B.C. are only sixteen in number and two of them are so much longer and more complete that they overshadow all the others.

Let us consider them more closely. They are two collections of mathematical problems — we might call them treatises — the oldest mathematical treatises in existence. They are in the form of papyrus rolls, called respectively (after the names of former owners), the Golenishchev papyrus (in Moscow) and the Rhind papyrus (in London).[46] The Golenishchev is the older, dating from the Thirteenth Dynasty (beginning in 1788), but reflecting manners of the preceding dynasty; the Rhind papyrus dates from the Hycsos time (say the seventeenth century), but it professes to be a copy of an older document of the Twelfth Dynasty. Thus these two venerable treatises, though different in age, may be said to represent the same time, the Twelfth Dynasty (2000–1788), or roughly the nineteenth century.

[43] Petrie, *Wisdom of the Egyptians*, p. 89.

[44] Marcelle Baud, *Les dessins ébauchés de la nécropole thébaine au temps du Nouvel Empire* (folio, 272 pp., 33 pls.; Cairo: Institut français d'Archéologie Orientale, 1935) [*Isis* 33, 71–73 (1941–42)].

[45] Chace, Bull, Manning, and Archibald, *The Rhind mathematical papyrus*, vol. 2, pp. 192–193.

[46] The Rhind papyrus consists really of two papyrus rolls (British Museum Nos. 10057 and 10058), but a fragment connecting these two rolls was discovered in the New York Historical Society, New York. The two British Museum rolls and the New York fragment constituted a single roll or a single treatise.

The period extending from the twentieth to the seventeenth century (four centuries) marked the scientific climax of Egypt, while the period immediately following, say from the sixteenth to the twelfth century, marked its political climax, Egypt being then the head of a world empire. Note that the intellectual climax preceded the political one, instead of coinciding with it or following it as we might expect.

Strangely enough, those two extraordinary papyri have the same length (544 cm), but while the Rhind one is of full width (33 cm) the Golenishchev one is a sort of pocket edition only one quarter of that width (8 cm). Though the latter is ostensibly the earlier, it is convenient to speak first of the Rhind papyrus.

The immense constructions undertaken during the Pyramid Age necessitated the activities of clerks who preserved and expanded the traditions in the form of methods and recipes, problems, accounts and tables, and what was the equivalent of our blueprints. We must assume the preservation of such traditions, gradually enriched, until the end of Egyptian splendor. For example, the erection of so many obelisks during the Eighteenth and Nineteenth Dynasties suggests that the results of many experiments, and the methods gradually developed by trial and error, were transmitted by each architect to his apprentices, and from court to court. It is probable that the priests, who were the only educated people, or at any rate the best-educated ones, were the guardians of such scientific traditions, or that they helped to preserve them. The Rhind papyrus was actually written by a responsible scribe who names himself in the introductory paragraph.

> Rules for enquiring into nature, and for knowing all that exists, [every] mystery, ... every secret. Behold this roll was written in Year 33, month 4 of the inundation season, ... [under the majesty of the King of Upper] and Lower Egypt Aauserrē', endowed with life, in the likeness of a writing of antiquity made in the time of the King of Upper and Lower Egypt Nemarē'. It was the scribe Aḥmōse who wrote this copy.[47]

This statement suggests that Aḥmōse realized the great importance of his mission. He was actually writing a treatise, that is, a systematic account of available knowledge in his field. To be sure, his treatise is not by any means as systematic as one written today, but as much method as it does contain is tremendously impressive. Think of it; here is a man, Aḥmōse, who lived almost as many centuries before Christ as we do after Christ, undertaking to set forth the main problems of arithmetic and geometry as they appeared to his contemporaries.

We have two excellent editions of it in English, Peet's and Chace's, either or both of which may be found in almost every reference library. Chace's edition, published six years later than Peet's, is far more instructive, since it enables one to pass gradually from the original hieroglyphics to the free English version.

Before describing the contents of the Rhind papyrus, it is necessary to explain the Egyptian idea of a fraction. For some strange reason the only fractions acceptable to them were those of the type $1/n$ (the nth part); they wrote it "part 125," meaning 1/125. They also used two "complementary" fractions, 2/3 and 3/4 (each expressing the remainder when "part three" or "part four" is taken away). Their use of the second one — "three parts" — is rare, but that of the first — "two parts"

[47] Peet, *The Rhind mathematical papyrus*, p. 33.

(meaning two thirds) — is very common. The fraction 2/3 was represented by a separate symbol, which occurs frequently in the mathematical texts.

The Rhind papyrus begins with a table of the resolution of fractions of the type $2/(2n + 1)$, wherein n is given every integral value from 2 to 50, into sums of fractions with numerator one:

$$\begin{aligned} 2/5 &= 1/3 + 1/15, \\ 2/7 &= 1/4 + 1/28, \\ 2/9 &= 1/6 + 1/18, \\ &\dots\dots\dots \\ 2/99 &= 1/66 + 1/198, \\ 2/101 &= 1/101 + 1/202 + 1/303 + 1/606. \end{aligned}$$

The very publication of this table at the beginning of the book is typical of its semitheoretical, semipractical nature. The scribe or his unknown predecessor had already reached experimentally a certain amount of abstraction and found it advantageous to put it forward.

Then follow forty arithmetic problems (see problem 4 in Fig. 9) concerning the division of 1, 2, . . . , 9 by 10, the multiplication of fractions, problems in completion (complete 2/3 1/30 to 1; the correct answer is 1/5 1/10), quantity problems (a quantity and its 1/7 added together become 19, what is the quantity?; the answer is 16 1/2 1/8), division by fractions, division of the measure hekat, division of loaves in arithmetic progression (see the example given below). These problems lead to equations of the first degree with one unknown quantity. Of course, there are no equations in the papyrus, but we notice symbols denoting addition and subtraction, and even one representing the unknown quantity. A problem in a Berlin papyrus (No. 6619) of Kahun (Twelfth Dynasty) leads to two equations, one of them quadratic, with two unknown quantities.[48] In modern notation,

$$x^2 + y^2 = 100,$$
$$y = \frac{3}{4}x.$$

The answer, correctly given, is $x = 8$, $y = 6$. Then $8^2 + 6^2 = 100$, or $4^2 + 3^2 = 5^2$, and we recognize the numbers involved in the Pythagorean theorem, to which we shall come back in a moment.

Here is the final arithmetic problem as translated by Chace: [49]

Problem 40.

Divide 100 loaves among 5 men in such a way that the shares received shall be in arithmetic progression and that 1/7 of the sum of the largest three shares shall be equal to the sum of the smallest two. What is the difference of the shares?

Do it thus: Make the difference of the shares 5½. Then the amounts that the 5 men receive will be

23 17½ 12 6½ 1, total 60

As many times as is necessary to multiply 60 to make 100, so many times must these terms be multiplied to make the true series.

1	60
⅔	40

The total, 1⅔, times 60 makes 100.

Multiply by 1⅔

	23 it becomes	38⅓
	17½ " "	29⅙
	12 " "	20
	6½ " "	10⅔ ⅙
	1 " "	1⅔
total	60 " "	100.

[48] Moritz Cantor, *Vorlesungen zur Geschichte der Mathematik* (Leipzig, ed. 3, 1907), vol. 1, p. 95.

[49] Chace, Bull, Manning, and Archibald, *The Rhind mathematical papyrus*, vol. 2, p. 84.

Fig. 9. The Rhind papyrus, problem 4 (partly in the British Museum and partly in the New York Historical Society). The top part reproduces the original hieratic script; below is a transcription in hieroglyphic with a literal transliteration in our own alphabet. A free translation reads:

Divide 7 loaves among 10 men
Each man receives 2/3 1/30
Proof. Multiply 2/3 1/30 by 10, the result is 7

Do it thus:	1	2/3 1/30
	2	1 1/3 1/15
	4	2 2/3 1/10 1/30
	8	5 1/2 1/10

Total 7 loaves, which is correct.
[Reproduced with kind permission from the edition by A. B. Chace, *The Rhind mathematical papyrus* (Oberlin, 1927–1929), vol. 1, p. 61; vol. 2, p. 36.]

Problems 41 to 60 deal with the determination of areas and volumes, and problems 61 to 84 are miscellaneous. The area of a triangle was obtained by multiplying its base by half of its side; this was correct only in the case of narrow triangles. The volume of a cylindrical granary of diameter d and height h is said to be $(d-1/9d)^2h$. This is a remarkably close approximation for the area of the circle – $0.7902\ d^2$ instead of $0.7854\ d^2$, as if π equaled 3.16 instead of 3.14.

There is no reason to believe that the Egyptians knew the Pythagorean theorem, except the indirect one suggested above apropos of the Berlin papyrus. They might have obtained an empirical knowledge of it in many ways, but the matter is very uncertain. The fact that that knowledge was relatively easy to obtain and that they overcame greater difficulties is no valid argument. It is one of the commonplaces of the history of science that problems have not always been solved, either by one nation or by all of them collectively, in order of increasing difficulty.

The allusion of Democritos of Abdera (V B.C.) to the wise *harpedonaptai*, the rope stretchers or rope fasteners, of Egypt has been wrongly interpreted. According to Democritos,[50] no one of his time had surpassed him in constructing figures from lines and in proving their properties, not even the rope stretchers of Egypt. It has been assumed without further proof that the rope stretchers were able to draw right angles by using ropes divided by knots in the ratios 3:4:5. It is more probable that the function of the rope stretchers was astronomical rather than mathematical. "Stretching the cord" was one of the initial ceremonies at the foundation of a temple. The cord had to be stretched in the direction of the meridian in order to orient the temple properly.[51] It is not impossible that the rope stretchers were able also to draw a perpendicular to the meridian and they may have done it with a rope divided in segments of 3, 4, 5 units, but that is guesswork, as are all the theories ascribing the discovery of the Pythagorean theorem to Hindus or Chinese.

There are only twenty-five problems in the Golenishchev papyrus, but one of

[50] As quoted in the John Potter, ed., *Miscellanies [Strōmateis] of Clement of Alexandria* (Oxford, 1715), vol. 1, p. 357. Clement died *c.* 590 years after Democritos.

[51] Peet, *The Rhind mathematical papyrus*, p. 32.

them is breath-taking.[52] It seems to prove that the Egyptians knew how to determine the volume of the frustum of a square pyramid, and their solution was essentially the same as ours, represented by the formula

$$V = (h/3)\ (a^2 + ab + b^2),$$

where h is the height of the frustum and a and b are the sides of its base and its top.

That solution may be called the masterpiece of Egyptian geometry. It is typical of Egyptian precocity and of the limitations of their genius that that solution was found by them perhaps as early as the nineteenth century, if not earlier, and that they never found anything better though they continued to work for three more millennia.

TECHNOLOGY[53]

The most important technical achievement from the point of view of its cultural implications was the manufacture of papyrus, which has been discussed above. Let us say a few words about two other departures, each of which opened up infinite possibilities — the making of glass and the weaving of textiles.

It is impossible to say when glass was first made deliberately (there are a few predynastic specimens), but by the beginning of the Eighteenth Dynasty (*c.* 1580) it was already produced on a large scale, and by the middle of that dynasty (*c.* 1465) the technique had reached a high standard of excellence.[54] Glass is obtained by fusing together silica (sand) with alkali; the alkali found in Egyptian examples is very largely soda, potash being present only in very small proportion. This shows that their alkali was obtained mainly from natron (a native sodium carbonate) and not from the leaching of plant ashes; remains of glass factories have been excavated in the Wādī Natrūn.[55] The Egyptians made glazes of many kinds, notably to coat earthen vessels, and glasses of many colors — amethyst, black, blue, green, red, white, yellow. This means that they had discovered that the addition of certain metals or earths to the basic materials (quartz sand and natron) produced desirable effects. It would be highly misleading to call such empirical knowledge by the name of chemistry, or to say, for example, that they knew cobalt because cobalt has been found in ancient glass (even as early as the Eighteenth Dynasty). Yet the presence of that cobalt is significant, because cobalt compounds do not occur in Egypt and had to be imported from other regions (Persia, Caucasus). This implied that the Egyptian glassmakers were already sophisticated enough to hunt abroad for sundry ingredients in order to obtain new colors, in this case a dark blue one.

They fashioned glass into beads, mosaic, and vases. The last were formed on a sandy clay core. Blown glass was not known until considerably later, in Roman times.

Some fabrics were woven in prehistoric days. The Egyptian methods of spin-

[52] Struve, *Mathematischer Papyrus*, No. 14, p. 134–145.

[53] See Lucas, *Ancient Egyptian materials and industries*.

[54] *Ibid.*, p. 116.

[55] The Wādī Naṭrūn, in the Libyan desert between Alexandria and Cairo, is so called because of the vast amount of naṭrūn (natron) that it contains. That rich source of salt and soda is exploited to this day.

ning and weaving can be understood from a model[56] of the Eleventh Dynasty (2160–2000) and from wall paintings of the Twelfth and later dynasties. Some linen found in royal tombs is so finely woven that one can hardly distinguish it from silk with the naked eye, and it was translucent. Even if we did not have actual specimens of such linen (of the Old Kingdom!) we could deduce its existence from ancient paintings showing a woman's limbs through the fabric she is wearing. The painter reproduced exactly what he saw.[57]

METALLURGY AND MINING

The value of hard metals for technical purposes is one of the fundamental discoveries of mankind. That discovery was made independently in many places. Wherever it was made it created or prepared an industrial revolution. We think of the metal ages as succeeding the stone ages, and ancient Egypt impresses us as a kind of triumphal stone culture, because the metal tools have vanished, while the stone monuments still dominate the Nile valley. As a matter of fact, it was probably metal chisels that made those monuments possible, or at least increased their number. Metal instruments changed not only the mason's craft, but many other crafts; metal weapons modified profoundly the political equilibrium.

How were the first metals discovered? This is not an Egyptian problem, but a problem of prehistory in general. The discovery was probably accidental, and it may have occurred in more than one way. There was plenty of copper ore in the Sinai peninsula; a native or an Egyptian visitor banking his campfire with pieces of that ore might have reduced some of it, and shining bits of copper would be found next morning in the embers. Egyptian women of the earliest age known to us (the Badarian) used malachite as an eye paint. Malachite is a copper ore (green basic carbonate of copper); if a piece of malachite fell into a charcoal fire it might be reduced and a bead of copper would appear. If the man in the first case, or the woman in the second, was intelligent enough to learn anything from a casual and irrelevant experience (very few people are, but there were such in every time), he would repeat and vary the experiment, obtain more copper, learn to hammer or to cast it into a desired shape, make a tool of a new kind, use that tool . . . As always, there is not one discovery to be considered, but a chain of them, so long a chain that no single man, yea, no single people, would be able to forge it alone; each would have followers, and each of the followers more followers. By the time the Pyramids were built the Copper Age was already well advanced.

Ores are seldom restricted to one metal. The early metallurgists were bound to obtain impure metals, that is, a mixture of one main metal, copper, with others. They may have noticed the superior value of some alloys and, at a later stage, prepared similar alloys by the mixture of different ores. That is, they may have noticed that a better kind of metal was obtained when various ores were melted together. Later still, much later, they may have prepared definite alloys by mix-

[56] Small models of every kind of object and representing many kinds of activities were placed in the tombs. This particular one, showing women engaged in spinning and weaving, was found in Thebes and is now in the Cairo Museum.

[57] One example, quoted because the author is very familiar with it, though there are many others, is a wall painting in the tomb of Nefretere, queen of Ramses II (1292–1225), showing Isis conducting her to her tomb. There is an excellent reproduction in Nina de Garis Davies, *Ancient Egyptian paintings selected, copied and described* (2 vols., 91 pls.; Chicago: University of Chicago Press, 1936).

ing metals in a constant proportion. This paragraph summarizes many millennia of metallurgic experience.

The best-known alloy of antiquity was bronze (that is, copper with tin); it was probably unintentional before the Eighteenth Dynasty (1580–1350). Specimens of copper older than that dynasty contain varying amounts of tin, arsenic, manganese, or bismuth. The invention of bronze, that is, the deliberate mixing of copper with a definite amount of tin (2 to 16 percent in ancient times; 9 or 10 percent now), was a step up almost as important as the discovery of copper itself; it marked the beginning of a new age. Bronze is stronger and harder than copper, particularly when hammered; [58] its melting point is lower than that of copper, and casting is easier in various ways; melted bronze does not contract like melted copper and does not absorb gases as readily. Much use was made of bronze during the Eighteenth Dynasty and later.

Where did the Egyptians get the tin? It is possible that it was already imported into Egypt before the end of the Old Kingdom.[59] Tin was brought from some of the islands, from Byblos, possibly even from Central Europe. The most obvious source was Byblos, in the vicinity of which ores of copper and ores of tin were obtainable together. It is thus possible that the mixture of those ores occurred very early in that city, accidentally to begin with, then more and more deliberately.

After having used up ores that were close to the surface, if those ores proved to be especially valuable, if there was a demand for them, the natives must have learned to dig for them, to dig deeper and deeper. The mines of Sinai were already exploited during the Old Kingdom; their exploitation was reorganized during the Twelfth Dynasty, in the time of Sesostris I (1980–1935), and much developed under Amenemhēt III (1849–1801), who dug wells and cisterns, and built barracks for the workmen, houses for the overseers, and fortifications to keep the Beduins out. At Sarābīt al-Khādim (Sinai) he excavated a large cistern in the rocks; the mines were administered very methodically. Ruins of that mining settlement of almost thirty-eight centuries ago can be seen today.[60]

The Egyptians made occasional use of meteoric iron, but their main metals were copper and bronze. The metallurgy of iron is far more difficult than that of copper; it was initiated and developed in Western Asia and not introduced into Egypt until very late (in Naucratis, sixth century B.C.). It is possible that Asiatic blacksmiths came to Egypt before that time and that would explain the presence of a few iron tools, more or less carburized and annealed, dating from 1200 B.C. and later.

[58] This is true only if the amount of tin is small, say 4 percent; if it is a bit larger, say 5 percent, the alloy becomes brittle when hammered, unless frequently annealed during the process; see Lucas, *Ancient Egyptian materials and industries*, p. 174. This is quoted to illustrate the great complexity of metallurgic problems. There were probably some great metallurgic artists in antiquity, while lesser artists must have been perplexed by mysterious failures.

[59] Tin was used in Egypt apart from bronze; on the other hand, bronze may have been made before tin or tin ore were recognized as such. For the antiquity of tin in Egypt, see W. Max Müller, *Egyptological researches* (Washington, 1906), vol. 1, pp. 5–8, pl. 1; G. A. Wainwright, "Early tin in the Aegean," *Antiquity 18*, 57–64, 100–102 (1944); and as always Lucas, *Ancient Egyptian materials and industries*.

[60] J. H. Breasted, *History of Egypt* (New York, 1909), p. 190, fig. 85.

In order to increase the temperature of their metallurgic furnaces, the Egyptians used blowpipes as early as the Fifth Dynasty, and bellows in the Eighteenth Dynasty and later.

MEDICINE[61]

It is not necessary to emphasize the antiquity of Egyptian medicine; in every culture medicine develops very early, for the need of it is too universal and too pressing ever to be overlooked. We may be sure that some kind of medicine was already practiced in Egypt in the earliest prehistoric days, many millennia before Christ. To quote an example, the use of malachite as an eye paint and an eye salve goes back to the Badarian age; the use of galena for similar purposes was introduced later, though still in predynastic times. Circumcision is a rite of immemorial age; bodies exhumed from prehistoric graves (as early as, say, 4000 B.C.) show traces of it. A very clear representation of the operation was sculptured on the wall of a tomb of the Sixth Dynasty (*c.* 2625–2475); see Fig. 10.

[61] See J. H. Breasted, *The Edwin Smith surgical papyrus* (2 vols, Chicago, 1930) [*Isis 15*, 355–367 (1931)]; B. Ebbell, *The papyrus Ebers* (136 p.; Copenhagen: Levin and Munksgaard, 1937) [*Isis 28*, 126–131 (1938)].

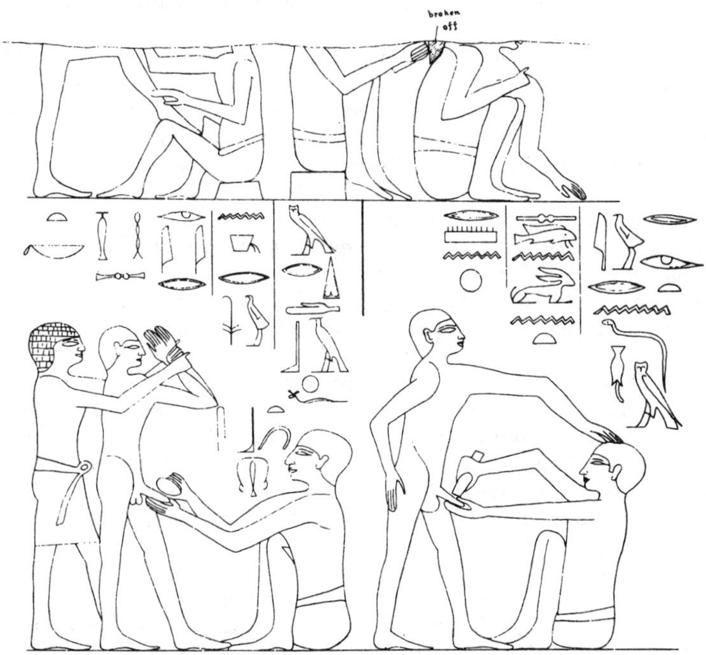

Fig. 10. The earliest representation of a surgical operation: circumcision with a stone knife. Saqqâra, beginning of the Sixth Dynasty (say the end of the twenty-seventh century). [From the drawing by W. Max Müller, *Egyptological researches* (Washington, 1906), vol. 1, pl. 106. Courtesy of Carnegie Institution of Washington.]

The earliest physician whose name has been recorded, Imhotep,[62] was the wazīr of Zoser, founder of the Third Dynasty, in the thirtieth century. Imhotep was a learned man, astronomer, physician, architect (he may have been the builder of the first pyramid, the step pyramid of Ṣaqqāra). In later times he was worshiped as a hero, as a blameless physician, and later still as the god of medicine, the prototype of Asclepios (even as the learned God Thoth was the prototype of Hermes and Mercury). We know precious little about Imhotep's medical knowledge but his apotheosis is significant and we may well take him at the Egyptian valuation as the first great man in medicine. The people who speak of Hippocrates as the father of medicine should bear in mind that Hippocrates comes about half way between Imhotep and us. That would improve their perspective of ancient science.

Not only were there many physicians in the Pyramid Age, but there were very specialized ones. The skill of an early dentist is beautifully illustrated by a mandible found in a tomb of the Fourth Dynasty (2900–2750), in which an alveolar process was pierced to drain an abscess under the first molar. From the tombstone of Iry, chief physician to a pharaoh of the Sixth Dynasty (2625–2475), we learn that he was also "palace eye physician" and "palace stomach-bowel physician" and bore the titles "one understanding the internal fluids" and "guardian of the anus."[63]

The medical papyri that have come to us, seven or more, are relatively late. They date from the Twelfth Dynasty to the Twentieth (2000 to 1090), but most of them reflect professedly earlier knowledge, going back to the Old Kingdom, as far back as the Fourth Dynasty. The two earliest papyri, the Kahun and the Gardiner fragments (*c.* 2000), deal with diseases of women, children, and cattle. The two most important ones, the so-called Smith and Ebers papyri, date from the seventeenth and sixteenth centuries. The Smith one is of the same age as the Rhind mathematical papyrus. Roughly speaking, we may say that the outstanding mathematical and medical treatises that have come to us are of the same general period, the end of the Middle Kingdom and the beginning of the New Kingdom, just prior to the imperial age, when Egypt dominated the world.

Let us consider more carefully the two outstanding papyri, the Smith and the Ebers, both of which are much larger than any others. On the basis of the figures given by Sarton,[64] the seven medical papyri listed by him include 3746 lines; the Smith has 469 lines and the Ebers 2289, so that together they have 2758 lines, which is almost 74 percent of the total. As all the manuscripts are ultimately derived from similar Old Kingdom sources, we may safely assume that the study of the Ebers and the Smith papyri will give us a fair knowledge of ancient Egyptian medicine.

We shall begin with the younger one, the Ebers papyrus, because it is by far the largest (almost five times as large as the Smith) and was the best known until very recent times. The difference in age is small anyhow, about a century, and negligible if one bears in mind that both texts represent older traditions. We are sure that the Ebers papyrus was written somewhat later than the Smith one,

[62] Jamieson B. Hurry, *Imhotep, the vizier and physician of King Zoser and afterward the Egyptian god of medicine* (ed. 2, 228 pp., 26 figs.; London, 1928) [*Isis 13*, 373–75 (1930)].

[63] Hermann Junker, "Die Stele des Hofarztes Irj," *Z. aegyptische Sprache 63*, 53–70 (1927) [*Isis 15*, 359 (1931)].

[64] G. Sarton, *Isis 15*, 357 (1931).

but it would be unwise to conclude that the contents of the former are of later date than the contents of the latter.

The Ebers papyrus is a roll 20.23 m long and 30 cm high; the text is distributed in 108 columns of 20 to 22 lines each. It contains 877 recipes concerning a great variety of diseases or symptoms. Spells are recommended only in twelve cases, and in other cases the therapeutics does not seem irrational, though we are seldom able to understand either the trouble or the remedy. The contents are arranged in the following order:

Recitals before medical treatment, to increase the virtue of the remedy. Internal medical diseases. Diseases of the eye. Diseases of the skin (with an appendix of sundries). Diseases of the extremities. Miscellanea (especially diseases of the head, for example, of the tongue, teeth, nose, and ears, and cosmetics). Diseases of women (and matters concerning housekeeping). Information of an anatomic, physiologic, and pathologic nature, and explanations of words. Surgical diseases.[65]

That order is open to many objections, but the author's intention is clear enough. He wanted to put together as well as possible all the information that a physician might need; he wrote a medical treatise, one of the earliest ever written (thirty-six centuries ago!).

The Smith papyrus is much shorter. It is 33 cm high and was probably 5 m long, but the beginning has been lost and it now measures 4.70 m. It is a copy of a much older text, dating back to the Pyramid Age, perhaps even early in that age, let us say the thirtieth century. After it had circulated for some generations it was found that its terms were antiquated.

Toward the end of the Old Kingdom, say in the twenty-sixth century, a learned physician had the idea of rejuvenating it by the addition of glosses (69 in all), explaining obsolete terms and discussing dubious matters. (N.B. the Papyrus Ebers has also some glosses, 26 in all, but they have been badly messed up.) These glosses constitute the most valuable part of the papyrus.[66]

The text as we have it now comprises two very distinct parts — 17 columns (377 lines) on the front and 4½ columns (92 lines) on the back. The latter part contains only recipes and incantations and need not detain us. The main part is a surgical treatise, informed by a scientific spirit far superior to that of the Ebers papyrus.

To be sure, the field of surgery is much less likely than that of internal medicine to be contaminated by irrational ideas, for in most surgical cases dealt with by ancient physicians the cause of the injury was too obvious to require the insertion of magical antecedents. On the contrary, an internal disease is always mysterious and likely to breed superstitious ideas in the patient's mind, even in the physician's mind. The Smith papyrus consists not of recipes but of definite cases. It was planned to deal with the ailments in the order of the bodily parts from head to foot, but unfortunately it stops a little below the shoulders, whether because the scribe was interrupted or because the end of the manuscript got lost, we do not know. That order — *eis podas ec cephalēs, a capite ad calces* — remained the standard one throughout the Middle Ages, but it was so natural, as a first approximation, that we should not assume it was determined by the Egyptian example.

[65] As quoted in Ebbell, p. 27.

[66] *Isis 15*, 359 (1931).

The forty-eight cases dealt with in the papyrus, as it has come to us, are classified as follows:

The discussion begins with the head and skull, proceeding thence downward by way of the nose, face and ears, to the neck, clavicle, humerus, thorax, shoulders and spinal column, where the text is discontinued, leaving the document incomplete. Without any external indication of the arrangement of the text, the content of the treatise is nevertheless carefully disposed in groups of cases, each group being concerned with a certain region.

These groups are as follows:

A. Head (27 cases, the first incomplete):
   Skull, overlying soft tissue and brain, Cases 1–10.
   Nose, Cases 11–14.
   Maxillary region, Cases 15–17.
   Temporal region, Cases 18–22.
   Ears, mandible, lips and chin, Cases 23–27.
B. Throat and neck (cervical vertebrae), Cases 28–33.
C. Clavicle, Cases 34–35.
D. Humerus, Cases 36–38
E. Sternum, overlying soft tissue, and true ribs, Cases 39–46.
F. Shoulders, Case 47.
G. Spinal Column, Case 48.[67]

The incompleteness of Case 48 confirms our suspicion that the rest of the treatise is lost. The discussion of each case is done systematically in the following way:

1. Title.
2. Examination.
3. Diagnosis.
4. Treatment (unless a fatal case, considered untreatable).
5. Glosses (a little dictionary of obscure terms, if any, employed in the discussion of the case).[68]

The title of Case 4 reads, "Instructions concerning a gaping wound in his head, penetrating to the bone, and splitting his skull"; that of Case 6, "Instructions concerning a gaping wound in his head, penetrating to the bone, smashing his skull, and rending open the brain of his skull."

The examination regularly begins thus: "If thou examinest a man having . . ." The form adopted is that of a teacher instructing a pupil that he shall do so and so. The methods of observation expressly stipulated or implied are answers elicited from the patient, ocular, olfactory, and tactile observations, movements of parts of the body by the patient as directed by the surgeon. Strange to say, eight out of eleven surgical operations are classified with the examination rather than with the treatment. This would suggest that the surgical work was considered a preparation to the medical treatment, independent of it.

The diagnosis is always introduced by the words: "Thou should say concerning him [the patient] . . ." and ends with one of three statements:

1. An ailment which I will treat.
2. An ailment with which I will contend.
3. An ailment not to be treated.

Three diagnoses consist of this final hopeless verdict and nothing more; but in forty-nine diagnoses in our treatise the three verdicts are preceded by other observations on the case. In thirty-six of these forty-nine diagnoses the other observations are nothing more than a repetition of the title of the case, or of observations already made in the examination; but in the remaining thirteen, the diagnosis adds one or more *conclusions based on the facts determined in the examination*. These are the earliest surviving examples of observation and conclusion, the oldest known evidences of an inductive process in the history of the human mind.[69]

Parallel with the systematic use of these three verdicts is a similar series of temporal clauses bearing more directly on the condition of the patient although not so regularly

[67] Breasted, *The Edwin Smith surgical papyrus*, vol. 1, p. 33.

[68] *Ibid.*, p. 36.

[69] *Ibid.*, p. 7.

employed, and placed at the end of the treatment. These read:

A. "Until he recovers."

B. "Until the period of his injury passes by."

C. "Until thou knowest that he has reached a decisive point." [70]

The matter-of-factness and soberness of those early medical texts is very impressive. The doctor who wrote them down was not only an experienced man but a wise one, whose general point of view sometimes adumbrates that of the Hippocratic writings. For example, he recommends an expectant attitude, trusting in the healing power of nature, or he recommends waiting "until thou knowest that he [the patient] has reached a decisive point"; this reminds us of the Hippocratic notion of crisis.

There is no reason to believe that the ancient Egyptians had studied anatomy by means of deliberate dissections, but they had taken advantage of the accidental experiments falling under their eyes and had accumulated much knowledge. Of course, the mummification of dead bodies of men and animals, which had been practiced from time immemorial, might have taught them many things, but I am rather skeptical about that; the embalmers were too much concerned about their own difficult art to pay attention to irrelevant anatomic details. It is possible that the practice of mummification made it easier later, much later, in Ptolemaic times, for Greek scientists to undertake systematic dissections, but that is another story. As far as ancient Egypt is concerned there is no evidence of the influence of mummification on anatomic knowledge.

The author whose work is recorded in the Smith papyrus had meditated on anatomic and physiologic questions. He was aware of the importance of the pulse, and of a connection between pulse and heart. He had some vague idea of a cardiac system, though not of course of a circulation, which nobody clearly understood before Harvey. His knowledge of the vascular system was made hopelessly difficult by his inability to distinguish between blood vessels, tendons, and nerves. Yet consider these astounding observations of the brain (Fig. 11):

If thou examinest a man having a gaping wound in his head, penetrating to the bone, smashing his skull, and rending open the brain of his skull, thou shouldst palpate his wound. Shouldst thou find that smash which is in his skull like those corrugations which form in molten copper, and something therein throbbing and fluttering under thy fingers, like the weak place of an infant's crown before it becomes whole – when it has happened there is no throbbing and fluttering under thy fingers until the brain of his [the patient's] skull is rent open – and he discharges blood from both his nostrils, and he suffers with stiffness in his neck.[71]

He had observed the meninges, the cerebrospinal fluid, and the convolutions of the brain (compared in the previous quotation to the rippling surface of metallic slag). Moreover, he had realized that the brain was the seat of the control of the body, and that special kinds of control were localized in special parts of the brain. For further details I must refer to Breasted's masterly edition or to my long review of it.[72]

To conclude, the Smith papyrus, and to a lesser extent the Ebers one, give us a very favorable idea of the medicine, anatomy, and physiology of the Egyptians,

[70] *Ibid.*, p. 47.

[71] *Ibid.*, p. 165, Case 6.

[72] G. Sarton, *Isis 15*, 366 (1931); see Case 31.

and of the scientific outlook that they had obtained at least two thousand years before Hippocrates.

EGYPTIAN "SCIENCE"

The preceding accounts of Egyptian engineering, mathematics, and medicine, brief as they are, are sufficient, I believe, to answer a query that the reader is bound to make (as I am keenly aware because of my experience as a teacher). Can we speak of Egyptian "science," or is all that simply empiricism and folklore?

What is science? May we not say that whenever the attempt to solve a prob-

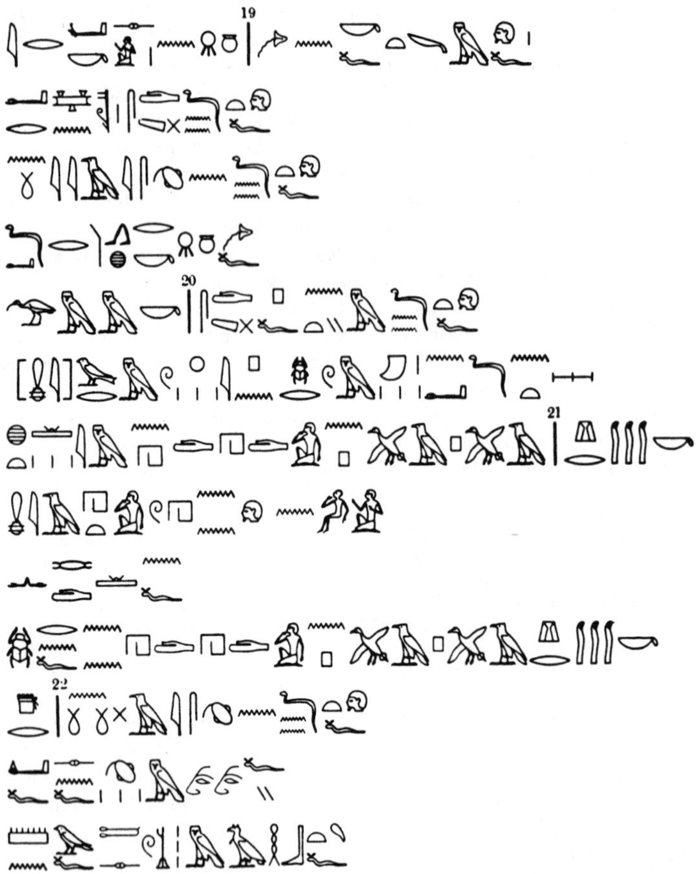

Fig. 11. The Smith papyrus, case 6, translated in the text. This is not the original hieratic text, but the hieroglyphic transcription, reproduced with kind permission from James Henry Breasted, *The Edwin Smith papyrus* (Chicago: University of Chicago Press, 1930) [*Isis 15*, 355–367 (1931)]. For the original hieratic, see the same work, vol. 2, pl. II.

lem is made methodically, according to a predetermined order or plan, we are witnessing a scientific procedure, we are witnessing the very growth of science? To be sure, early methods seem childish and weak as compared with ours, but will the scientists of the year 5000 think as favorably of our methods as we do ourselves? A beginning had to be made and not only did the Egyptians make it, but they proceeded remarkably far along the road that we are still following. For example, do not the tables of the Rhind papyrus represent a deliberate attempt to solve problems in a general way and by anticipation? Such tables are the true ancestors of all the mathematical tables — their name is legion — of which we are so proud today. It is probable that other tables were compiled by the scribes in charge of the accounts and measurements entailed by the gigantic constructions. It is not surprising that such documents have failed to reach us, for they would not be preserved in tombs for eternity, but would be used by living men and worn out of existence. And the classification of cases in the Smith papyrus, the method followed in the discussion of each case — is not that science?

Some readers having at the back of their minds the prejudice that science is a Greek invention (have not scholars repeated that for centuries?) will insist and say, "That may be science, but not pure sicence." Why not? At the end of his admirable investigation of the Smith papyrus, Breasted concluded:

> Indeed these two men, the surgeon who was the original author of the treatise, and his later successor, who wrote the glosses forming the ancient commentary, both living in the first half of the third thousand years B.C., were the earliest known natural scientists. In the long course of human development they are the first men whom we can see confronting a great body of observable phenomena, which they collected and stated, sometimes out of interest in the rescue of the patient, sometimes out of pure interest in scientific truth, as inductive conclusions which they drew from observed fact.[73]

I am sure that not only the Egyptians who had reached the stage of composing mathematical and medical treatises, but simpler men, living perhaps thousands of years earlier, were already pure scientists, that is, men moved by such an intense curiosity that the practical results and immediate fruits of their research became of secondary importance to them. As to Ahmōse and the unknown author of the Smith papyrus, I am sure that no men of science can read their books today without emotion, for they cannot help recognizing in them some of their own intellectual traits.

If disinterestedness is the criterion of pure science, we may say that science was never completely pure or completely impure. The circumstances of life and the inexorable drift of their own endeavors obliged the Egyptians to solve many technical problems; the exploration of those problems created scientific interests extending beyond the immediate solution of particular cases. The development of Egyptian science was simply an anticipation of the development of science in general.

There can be no doubt about the efflorescence of a scientific spirit in Egypt before the middle of the second millennium, but alas! its development was arrested and it gradually died down. What were the causes of its fall and decadence? Similar questions have been asked about China, Greece, Rome, Islām, and have

[73] Breasted, *The Edwin Smith surgical papyrus*, vol. 1, p. 12.

never been answered completely. At first the growth of Egyptian science, and later its very life, were stopped by the combination of political with religious obscurantism. The science and wisdom of the Egyptians were blighted, but their efforts were eventually continued by other nations. This has happened again and again in the past and even within our own experience; it may happen again in the future, but obscurantism, however well organized, can never be universal and perpetual.

ART AND LITERATURE

Though we are primarily concerned with science, it is necessary to say a few words about Egyptian art and letters because the general reader is not so well acquainted with them as he may be with art and letters of later times. If he lives near one of the great museums he may have some familiarity with Egyptian art, but even so, prejudices may have prevented him from seeing it well. I have heard educated people remark that everything in Egyptian art was stereotyped and static, that the representation of human figures was dominated by the law of frontality, and so forth. The fact is that much in Egyptian art, even of the Old Kingdom, is tremendously active and sensitive, and that that art, far from being fixed, did evolve considerably during its long existence. Moreover, it is extremely complex, for it includes such gigantic monuments as the Pyramids, the Sphinx, the Colossi of Memnon, the Temples; conventional statues of kings, stiffened by ritual and symbolism; and many other statues – even of kings and queens – full of individuality and expressing many peculiarities, many moods, many graces. To mention only the most popular examples, think of the bust of Prince Ankhhaef (Fourth Dynasty) in Boston, the Shaykh al-balad (Fifth Dynasty) in Cairo, the squatting clerk (Fifth Dynasty) in the Louvre ("Le scribe accroupi"), the head of Queen Nefertete (Eighteenth Dynasty)[74] in Berlin. We owe to Egypt some of the most individualized and the most moving portraits of ancient times. Descriptions of such matters are futile. Take an album of Egyptian art and examine it leisurely with an open mind.

Art cannot be dissociated from literature, for in Egypt (as in the Christian Middle Ages) it was the literature of the illiterate. Of course the vast majority of the people were illiterate, for the various forms of script were so forbidding that only a few people in a thousand could read them at all. We find in the Egyptian tombs immense collections of the objects that living people used (small replicas of them were placed in the tombs for use in the hereafter; see Fig. 12); moreover, bas-reliefs and paintings describe most of their occupations. Such delineations are far more effective than verbal explanations. We can see the *fallāhīn* of the Pyramid Age plowing and sowing, harvesting, threshing, weaving; we can see the carpenters, the potters, the bakers, the smiths, the charioteers, the boatmen and sailors, the scribes, and also the jugglers and acrobats, the wrestlers, the dancing girls and musicians, the women on their way to market; we witness hunting scenes in the papyrus marshes (Fig. 13) or in the desert; we become acquainted not only with people but with their animal companions,

[74] She was the wife of Ikhnaton (1375–1358). There are many portraits of her.

Fig. 12. Hippopotamus in blue faience, Seventeenth Dynasty (seventeenth or sixteenth century), to illustrate Egyptian freedom from the law of frontality; such freedom was not by any means exceptional with the Egyptians. [Courtesy of the British Museum.]

cows and calves, donkeys, rams, dogs, cats, horses,[75] and also poultry, rabbits, geese and ducks, owls and cranes, mice, gazelles and oryxes, ibexes, ichneumons, leopards, crocodiles, hippopotami, giraffes, elephants; we visit the gardens and the fields, the villas of the noblemen with all their appurtenances; we watch the chariots and the ships. There was everywhere a great love of beauty, which is immortalized in the models, the reliefs, the paintings and drawings, and in innumerable details of the objects that have come down to us in abundance. In short, we have no difficulty in visualizing the life of ancient Egypt in all its infinite variety and we have a better knowledge of it than we have of periods much nearer to our own. We certainly know the Egyptians of the Pyramid Age far more intimately than the Greeks of the Homeric Age; in the second case, it is true, we have the *Iliad* and the *Odyssey*, but we lack the wealth of illustrations that enable us to evoke the life that was lived two thousand years earlier.

Egyptian literature is not at all on the level of Egyptian art, either in quality or in quantity, but it is original,[76] meaningful, and impressive. We know it very imperfectly, for the written documents preserved only a part of it to begin with, and the majority of those documents have been lost. Only those enclosed in tombs had a chance of survival. Little has come down to us from the Old Kingdom beyond the so-called Pyramid Texts, which are hardly more than magical incantations. From the period following the Sixth Dynasty, however, we have a fair collection of literary efforts, varied and conscious. By the time of the Twelfth Dynasty (2000–1788), an "author" was already complaining of the difficulty of saying anything new! We have the heterogenous collection united under the misleading name of the *Book of the Dead*,[77] the book of what is in the nether world (Am Duat), rituals, litanies, moving hymns, royal and private letters, historical

[75] The horses appeared only in the New Kingdom (begun in 1580); they were brought from Asia. As to camels, now ubiquitous, they were very rare and did not come into general use until Greco-Roman times. Joseph P. Free, "Abraham's camels," *J. Near Eastern Studies* 3, 187–193 (1944) [*Isis* 36, 40 (1946)].

[76] It was not necessarily original, for Egypt was submitted early to outside influences; yet it remained *sui generis*.

[77] Most of the papyri are of the New Kingdom or later, but many of the chapters of the *Book of the Dead* were composed in the Middle Kingdom and some even in the Old Kingdom; the so-called Pyramid Texts can be traced back to the Fourth Dynasty and even to the First. The God Thoth, father of arts and letters, personification of justice, "recording angel," was considered to be the author.

Fig. 13. View of a papyrus marsh along the Nile. The flowers and vertical lines represent the papyrus thicket. Observe the men in a reed boat, hippopotami, birds, fishes, mongoose (right of center). This is one of the many bas-reliefs concerning fishing and hunting in the marshes, in the mastaba of Mereruka. [From the Sakkarah Expedition, Prentice Duell, field director, *The mastaba of Mereruka* (2 vols., folio; Chicago: Oriental Institute, University of Chicago Press, 1938), pl. 19; reproduced with kind permission of the Oriental Institute; see also pls. 9–13, 15–21, showing other hunting and fishing scenes in the marshes.] The mastaba is a monument of the Old Kingdom, Sixth Dynasty (2625–2475).

records, laws and treaties, touching stories such as the tale of Sinuhe,[78] and other stories anticipating *The Arabian Nights*, collections of maxims for the edification of young princes (prototypes of the medieval Regimina principum), lamentations and books of wisdom suggesting comparison with similiar books in the Old Testament. That literature is often bombastic, and hackneyed metaphors create an impression of monotony; on the other hand, it is enlivened by directness, picturesqueness, and humor. When attempting to judge it we should bear in mind the possibility of misunderstanding it or at least of failing to appreciate it completely because of our insufficient knowledge of the language and of the people who spoke it; we should remember also that, such as it is, it is stretched out during a long period, two millennia, which completely antedates the whole of the Greek and Hebrew literatures.[79]

[78] Alan H. Gardiner's favorite! See his article in S. R. K. Glanville, ed., *The legacy of Egypt* (Oxford: Clarendon Press, 1942), pp. 74–75.

[79] For general orientation on Egyptian art, the best way is to consult an album of pictures. There are many such albums.

For Egyptian literature, see Adolf Erman, *The literature of the ancient Egyptians*, translated into English by Aylward M. Blackman (336 p.; London, 1927). The German original edition appeared in Leipzig in 1923.

Max Pieper, *Die ägyptische Literatur* (Potsdam, 1928).

T. Eric Peet, *Comparative study of the litera-*

THE DAWN OF CONSCIENCE [80]

There is no point in trying to explain the very complicated religion of the Egyptians, for this would illustrate their mythopoeic imagination rather than their scientific ability. However, the growth of science postulates a sufficient development of moral and social ideals. We may ask ourselves, why did it grow so early in the land of Egypt? The answer involves many factors, some of which are beyond our ken; it must suffice here to discuss briefly the political and religious ones.

No culture can be built in one day or in a single century; its elaboration implies the persistence of convergent efforts for a long time, and this is hardly possible without a sufficient amount of political concentration and stability. Such conditions obtained very early in the Nile Valley and help to account for what might be called the Egyptian miracle.

Some kind of political unity was already achieved in prehistoric times (say by 4000 or before), but it did not yet involve the whole of Egypt. There were two kingdoms, that of Lower Egypt (the Delta) and that of Upper Egypt, a long ribbon stretching from Memphis (Cairo) to the First Cataract (Aswān, ancient Syene, 24°5′ N). The dynastic periods began when King Menes united the two kingdoms, wore the double crown, and styled himself "King of Upper and Lower Egypt" or "Lord of Both Lands." That union did not last forever, but it lasted during the first six dynasties (the Old Kingdom), or from *c.* 3400 to 2475, that is, almost a millennium, enough time to crystallize moral ideas and moral habits. For the sake of readers who insist on thinking of ancient Egypt as a monotonous sameness, let me recall that there were three periods of stability:

Old Kingdom	Dyn. I–VI	3400–2475
Middle Kingdom	Dyn. XI–XII	2160–1788
New Kingdom	Dyn. XVIII–XX	1580–1090

These periods lasted, respectively, 925, 372, and 490 years, and were separated by two periods of anarchy or at least of instability lasting 315 and 208 years. Happily for the Egyptians, the periods of stability were long enough, especially the first and fundamental one, to establish their institutions and enable traditions to take root. In order to appreciate the length of those periods, let us express them in terms of American history. If we consider the length of that history from the Revolution, 1775, until 1950 (175 years) as one unit; then the Old, Middle, and New Kingdoms lasted, respectively, 5.3, 2, and 3 units, and the twenty-six dynasties of ancient Egypt (3400 to 525 = 2875 years) lasted 16.4 units. While the periods of stability were long enough to give to the whole of Egyptian culture a certain unity, various upheavals and interruptions, changes of political gear and religious mood, prevented excessive uniformity. The simplest way of measuring the evolution is to consider good series of works of art in their chronological sequence; if one is sufficiently sensitive to them one thus obtains

tures of Egypt, Palestine and Mesopotamia (142 p.; London: Oxford University Press, 1931) [*Isis 21*, 305–316 (1934)].

Josephine Mayer and Tom Prideaux, *Never to die. The Egyptians in their own words* (New York: Viking, 1938); popular.

Brief surveys have been given by Alan H. Gardiner in his *Egyptian grammar* (Oxford, 1927), pp. 17–24, and in Glanville, ed., *The legacy of Egypt*, pp. 53–79.

[80] See J. H. Breasted, *The dawn of conscience* (450 p., 19 fig., New York: Scribner, 1933) [*Isis 21*, 305–316 (1934)].

immediately an intuitive understanding of the development up and down of the Egyptian genius.

During the Old Kingdom the Egyptians were already discussing the problem of right and wrong; witness the so-called Memphite drama, which we know only through a late Ethiopian copy (Dyn. XXV, 712–663) but the contents of which are of great antiquity. The Proverbs of Ptathotep, which can be traced back to the Vth Dynasty, evidence the progress of moral fermentation, what might be called the birth and growth of human conscience. Here is an example: [81]

> Be not arrogant because of thy knowledge, and be not puffed up for that thou art a learned man. Take counsel with the ignorant as with the learned, for the limits of art cannot be reached, and no artist is perfect in his excellence. Goodly discourse is more hidden than the precious green-stone, and yet it is found with slave-girls over the millstones.

This is something different from art and science, even from religion, something the absence of which would make any lasting culture impossible. Meanwhile Egyptian religion was developing into two main directions leading respectively to Heaven and to Hell — on one hand, a solar cult with the conception of an empyrean realm of the dead; on the other hand, the Osiridian cycle of myths, suggested by the miraculous fertility of plants, animals, and men, together with the conception of underground mysteries. Those fables can be followed (with difficulty) in the Pyramid Texts and the Coffin Texts, but in the latter we find occasional utterances that adumbrate the idea of human brotherhood. Says Re, the Sun-god,

> I have made the four winds that every man might breathe thereof like his brother during his time.
>
> I have made the great waters that the pauper like the lord might have use of them.
>
> I have made every man like his brother, and I have forbidden that they do evil, (but) it was their hearts which undid that which I had said.[82]

To be sure, those hoary texts, the Coffin Texts and the Book of the Dead, are full of magic and nonsense, but the seeds of morality that they contain vindicate and redeem them. The dawn of morality was as important as the dawn of science. The Book of the Dead explains the idea of a moral judgment and its illustrations give it a very concrete form. We witness the actual weighing of the man's heart in the temple of Osiris (Fig. 14).[83]

This moral and religious fermentation reached a climax toward the end of the XVIIIth Dynasty. That dynasty was an age of great power; Egypt dominated the Western World. Political imperialism suggested a kind of religious imperialism. There was but one Pharaoh, there should be but one god. The last king of that dynasty, Amenhotep IV (*c.* 1375–1350), tried to establish a new monotheistic religion and as a symbol of his own conversion changed his own name to Ikhnaton. His fervor is illustrated by hymns, the most remarkable of which is the "Adoration of the Disk [the Disk of the Sun, Aton, name of the one God] by King Ikhnaton and Queen Nefertete." [84] According to Breasted, that hymn is the

[81] Peet, *Comparative study of literatures*, p. 101.

[82] Breasted, *The dawn of conscience*, p. 221.

[83] Incidentally, this shows that the ancient Egyptians were familiar with the use of scales of a relatively elaborate kind.

[84] Translations of it may be read in Peet, *Comparative study of literatures*, pp. 78–81, or Breasted, *The dawn of conscience*, pp. 281–286.

earliest truly monotheistic hymn in the world literature; parts of it suggest comparison with the 104th Psalm.

In order to consecrate his reformation of the religion of his fathers, Ikhnaton moved his capital from the priest-ridden Thebes to a new site, Tell al-'Amārna.[85] Many literary and artistic treasures have been found in the ruins of that place, as well as a part of his political correspondence with the kings of Western Asia, written on clay tablets in cuneiform characters (more about that anon).

Ikhnaton was a powerful king, but no ruler can rule alone, and the greater his empire the more assistants he needs; in the long run those assistants are bound to limit and perhaps to control his power. The Egyptian empire (not unlike almost every other empire) rested on three pillars, the King, the Clergy, and the Army. Ikhnaton's bold reform, a kind of Reformation occurring twenty-nine centuries before the European one, was premature. Moreover, the Empire had passed its climax and was beginning to slip out of the Pharaoh's hands. The

[85] Not far from Mallawi, about halfway between Memphis and Thebes (between Cairo and Luxor).

Fig. 14. Papyrus of the lady Anhai (British Museum, papyrus No. 10472), *Book of the Dead*, chap. cxxv. [Reproduced with kind permission from E. A. Wallis Budge, *The Book of the Dead. Facsimiles of the papyri of Hunefer, Anhai, Kerasher and Netchemet* (folio; London, 1899), pl. 4 of Anhai.]

The lady Anhai was a priestess in the college of Amon-Rē in Thebes, at the time of the Twentieth or Twenty-first Dynasty (*c.* 1200–945). The scene represented is the weighing of the conscience (*psychostasia*). At the top left are the gods seated near tables covered with offerings. The weighing is taking place below them. The jackal-headed Anubis is weighing the heart of Anhai (right-hand scale) against a little figure of Maat, goddess of truth; Anubis knows that the weights are equal when the pointer of the balance is parallel to a plumb line or to the vertical stand supporting the balance. On the extreme left are Maat and, below her, the ibis-headed Thoth, god of knowledge and justice, who records the result of the trial. The larger figures on the right are the falcon-headed Horus, leading Anhai into the presence of Osiris (not included in this plate). At the extreme right, the goddess Maat is embraced by the goddess Amentet.

monotheistic cult of Aton was rejected by the priests. After Ikhnaton's death they reëstablished the old mythology, resumed their power, and discouraged new adventures. Religion and science were fossilized; further progress became more difficult, if not impossible. Ikhnaton's failure was sealed when his second successor and son-in-law, Tutankhamon, abandoned Tell el-'Amārna and moved his capital back to Thebes.[86]

Ikhnaton's folly or genius being repudiated, a chapter in the history of mankind was closed or seemed closed. In spite of their immense power and their mystical hold on the people, the priests could not eradicate the monotheistic ideal. Ideas can never be completely eradicated; they are bound to crop out again and again. Ikhnaton's noble vision reappeared three and a half centuries after his death in the Wisdom of Amenemope (or Amenophis)[87] and later still in the Proverbs of Solomon.

One does not know what to admire most in Egyptian achievements, especially those of the third and second millennia — the glory of art, the beginnings of mathematics and medicine, the variety and perfection of their techniques, the dawn of conscience. We should bear in mind that the scientific achievements, which are our main concern, were of necessity the least developed, while the artistic achievements and even the religious ones could attain a climax, comparable to the climaxes of later ages. Ikhnaton could reach as near to God as we can, and the artists of the Old Kingdom could come as close to beauty as the artists of any time. On the other hand, the Egyptian mathematicians and physicians were standing near the bottom of a ladder which we are still climbing. Their position was necessarily low, and if ours is a bit higher, it is partly to their efforts that we owe it. They were our first guides and our first teachers.

[86] Tutankhamon became the best-known Egyptian Pharaoh when the Earl of Carnarvon and Howard Carter discovered his unviolated tomb in Thebes, in 1922. The extraordinary treasures revealed in that tomb (now in the Cairo museum) made a tremendous sensation. Howard Carter, *The tomb of Tut.ankh.Amen* (3 vols., ill.; London, 1923–1933).

[87] Papyrus B.M. 10474. Sir E. A. Wallis Budge, *Facsimiles of Egyptian hieratic papyri in the British Museum* (second series, pls. I–XIV; London, 1923); *The teaching of Amen-em-apt, son of Kanekht*, hieroglyphic text and English version (London, 1924). A better English translation is by F. Ll. Griffith, *J. Egyptian archaeology 12*, 191–231 (1926); for detailed comparison with the Book of Proverbs, see D. C. Simpson, *Ibid.*, pp. 232–239.

III

MESOPOTAMIA

GEOGRAPHIC AND HISTORICAL BACKGROUND

There are many points of similarity between Mesopotamia and Egypt, and a few of them will be indicated presently, because this will help the reader to understand both cultures more clearly. To begin with, the background of Egyptian history was relatively simple, the Delta and the narrow Nile valley. Yet that simplicity should not be exaggerated.

There was but one river in Egypt as against two in Mesopotamia, but in both cases there were two seas. In the case of Egypt, there are the Mediterranean in the north and the Red Sea in the east, both seas playing a tremendous part in her history. In the case of Mesopotamia, there are the Persian Gulf in the southeast and the Mediterranean in the west. Most of the historical events happened in the two valleys and the plain stretching between them,[1] the plain of Shinar often mentioned in the Bible; yet in order to grasp the sequence of those events one must take into account the mountainous country east of the Tigris and the country extending along the Eastern Mediterranean coastline. The two Mesopotamian seas are connected by a semicircular area of fertile land which Breasted has very aptly called the "Fertile Crescent." As the map (Fig. 15) shows, that Crescent, which unites the Mediterranean Sea to the Persian Gulf, faces and surrounds the Syrian desert, which might be likened to another sea, albeit a dry one. One does not settle in the desert but one may cross it in various directions.

For a complete recital of ancient Mesopotamia history one needs the whole Crescent as the geographical background, but for the earliest times it suffices to consider the region bordering on the Persian Gulf and the lower courses of the Euphrates and the Tigris, chiefly the former. In those days the Gulf was somewhat longer than it is now and the two rivers reached it separately; since that time it has been gradually shortened by sedimentation. The main difference between the two countries is that Mesopotamia has two rivers for one in Egypt; the courses of those two rivers are very capricious; between them they embrace the Mesopotamian plain, but the Euphrates faces the Syrian desert, while the Tigris

[1] It is for that reason that we give this chapter a purely geographical title, "Mesopotamia," instead of one like "Babylonia and Assyria," which is correct only for certain periods. However, the term Babylonia is often used in a more general way without chronological restrictions. Thus, one speaks of "Babylonian mathematics," meaning Sumerian mathematics as well as Babylonian *stricto sensu*. There is no harm in that, provided one is careful. No term is completely satisfactory or can remain so very long, for the region of applicability of geographic and historical terms varies from time to time.

valley is dominated in the east by the Persian mountains; both rivers originate from the highlands of Cappadocia and Armenia.

Aside from the dissymmetry of the rivers, there is a curious symmetry between the two countries. Both extend between two seas, and those seas are the same seas, the Mediterranean and the Arabian ones; the two countries are separated by the Syrian desert, or perhaps we should say they are united by the intermediate desert as well as by their common seas.

The earliest records of Mesopotamian civilization come from the country of

Fig. 15. Sketch map of the Near and Middle East in ancient times. What Breasted called the Fertile Crescent is the zone following the Mediterranean coast of Phoenicia (Lebanon and Syria), rejoining the middle course of the Euphrates, sweeping all along Mesopotamia (the country between the Two Rivers) to the Persian Gulf. It lies south of the highland zone of Anatolia and surrounds the Syrian desert. It has the general shape of a crescent and includes all the fertile lands of that territory. The main point is that the Fertile Crescent connected the Arabian Sea with the Mediterranean; it joined Mesopotamia (Persia, India, etc.) on the one hand with Egypt, the Phoenician and Aegean worlds on the other. (Drawn by Erwin Raisz.)

Sumer between the two rivers close to the Gulf's head, but it must have involved other people than those who were settled in that plain. One can never be sure of how and where a civilization began because the earliest documents available to us never represent the beginning, but a stage which is somewhat later and may be considerably later. Did the Mesopotamian culture begin in Sumer? or was it brought there from higher lands up the rivers or to the east of them?

When a new culture develops in a geographical background like the Mesopotamian one, we must fully expect a three-cornered conflict between the bearers of that culture settled in towns, the nomadic people moving across the desert and along the edge of cultivated lands, and finally the mountaineers, who are used to a harder life than the people living in the plains and never cease to covet the latter's ease and chattels. The relations of the Sumerians to the two other groups are but imperfectly known to us. In some of the earliest texts they speak with contempt of the nomads as "the people who do not know houses and who do not cultivate wheat."[2] It is clear that those early Sumerians did not view themselves as cultural upstarts but were already reminiscent of a past too deep to be fathomed. Long before 3000 B.C. they had already reclaimed the marshes in the lowlands near the Gulf and along the lower Euphrates. They had learned to drain the land and at the same time to irrigate it by means of canals, the traces of which can still be seen from airplanes. They cultivated barley and spelt (as the Egyptians did) and had domesticated cattle, goats, and sheep; they used oxen or donkeys to draw wheeled chariots. No stones being available to them, they built houses of mud bricks dried in the sun (adobe).

The Sumerians were very different from the Semitic people,[3] living higher up between the rivers; at any rate their language was not a Semitic language; neither was it Aryan. It is possible that they originated in the Elamite plateau east of the Tigris. Their origin in higher lands is inferred from the fact that they used the same word for mountain and country, and from similar facts equally suggestive yet unconvincing. We need not concern ourselves with the origins of the Sumerians and with their culture prior to their settlement in Sumer. As soon as we hear of them we find them flourishing in a Copper-Age pattern, and, as we shall see presently, they were astonishingly advanced in many respects.

They were conscious of the great antiquity of their culture, and like other nations (for example, the Chinese and the Japanese) they had rationalized their beliefs by the fabrication of a long mythological history (this was done by them as early as 2000 B.C. if not earlier). One of their sacred legends relates to a flood, which may have been a real flood or tidal wave in the Gulf region and may be identical with the Biblical flood. They postulated the existence of a number of antediluvian kings, each of whom ruled for many thousands of years. With the dynasties that held sway after the flood we are on surer ground, and

[2] Edward Chiera (1885–1933), *They wrote on clay*, ed. by George G. Cameron (Chicago: University of Chicago Press, 1938), p. 51. Here is as fine an example of cultural lag as may be found anywhere. Sumerians of 3000 B.C. spoke of nomads as people behind the times, yet similar nomads (Beduin Arabs) are still living in that vicinity today, fifty centuries later!

[3] It is better to leave racial considerations out, for we cannot obtain certain knowledge about the races of the Ancient East. One thing is clear – that by 2000 B.C., if not long before, those races had already been subjected to considerable mixture. One should always hesitate to conclude from language to race. It is easy enough for men, and especially for children, to learn a new language, but they cannot change their chromosomes. Reference to Semitic people in what follows should always be understood to mean people speaking Semitic languages – nothing more.

archaeological discoveries have substantiated the reality of one after another. Sir Charles Leonard Woolley's excavation of Ur, the Biblical "Ur of the Chaldees," birthplace of Abraham, aroused the world's attention. The first dynasty of Ur became tangible. Sumerian cities had time to develop not only in Ur, but also in Kish, Erech, Nippur, Larsa, Eridu, Lagash, Umma, Tello, and elsewhere. Our knowledge of those cities is not legendary or fanciful, but is based on scientific excavations; each of those places is now known with a fair amount of detail. The archaeological discoveries harmonize with the information culled from Sumerian or later texts.

In the meanwhile Semitic people had developed their own culture higher up between the rivers, in the region called Accad. Led by their king Sharrukīn (Sargon, 2637–2582), the Accadians subjugated the Sumerians and created the united kingdom of Sumer and Accad. The Sumerian culture was far superior to the Accadian and continued to be for millennia the dominating one; the Sumerians conquered their conquerors.[4]

Sargon's followers lacked his strength and the southern country soon regained its independence from the northern one, yet Sumer and Accad remained united. The Accadian dynasty was followed by many others, and the kings styling themselves "Kings of Sumer and Accad" were drawn from both races.

A new climax occurred when the sixth king of the Amurru[5] dynasty, Hammurabi (1728–1686) became the supreme ruler of Mesopotamia. His capital was Babylon and he gave so much luster to it that the whole country was later called Babylonia and the name of Sumer was almost forgotten. When one speaks of Babylonian culture one thinks primarily of the time of Hammurabi, which was its golden age. Moreover, we know that illustrious ruler very well, not only because of his code but also because of other inscriptions and of his correspondence; fifty-five of his own letters have come down to us.[6] The Babylonians used the Accadian or Babylonian, a Semitic, language, but had not forgotten the Sumerian. Indeed, Sumerian was to them a kind of sacred language, which educated people must know, just as we must know Greek and Latin (or rather much more, for alas! we have ceased to feel that obligation).

The Babylonian peace established by Hammurabi did not last very long, for the fight between the peoples of the plains and those of the mountains continued. His power was broken by Easterners who brought horses with them. There follows a period of chaos and stagnation which lasted until the consolidation of the Assyrian empire in the seventh century. The name of Assyria then replaced that of Babylonia. Because of the accident that Assyrian documents were the first to be investigated, the scholars engaged in the study of Mesopotamian antiquities are called to this day "Assyriologists," in spite of the fact that many of

[4] Just as the Greeks twenty-five centuries later conquered their Roman conquerors. Remember Horace's lines (*Epistolae* II, 1, 156):

Graecia capta ferum victorem cepit et artes
Intulit agresti Latio. . .

[5] That is, the Amorites of the Old Testament, a Semitic tribe of northern Syria. Their intervention introduced the Mediterranean shores into Mesopotamian history. The dating of Hammurabi is very controversial; the one mentioned in the text is given by Theophile J. Meek in James B. Pritchard, *Ancient Near Eastern texts* (Princeton: Princeton University Press, 1950), p. 163 [*Isis 42*, 75 (1951)].

[6] Leonard W. King, *The letters and inscriptions of Khammurabi, king of Babylon, about* B.C. *2200* (3 vols.; London, 1898–1900). English translations are in vol. 3.

them restrict themselves to pre-Assyrian times and that the dominating culture remained the Sumerian one.

To be sure, the original Sumerian culture was modified in many ways by the Babylonian and later by the Assyrian invaders. Not only that, but during the second millennium, if not before, Egyptian influences spread to Mesopotamia through the western side of the Fertile Crescent; that cultural invasion was especially vigorous during the period when the Egyptians dominated the Near East (sixteenth to twelfth centuries). As far as we, modern onlookers, are concerned, the Egyptian pattern was long more obvious and more familiar than the Mesopotamian one, so much so that for a long time we thought only or chiefly of ancient Egypt. The immense stone monuments of Egypt could never be overlooked, while the adobe cities of Mesopotamia disappeared one after another almost completely (dust unto dust) leaving underground ruins that could not be interpreted without difficult research. Moreover, Egyptian archaeology began half a century earlier than the Mesopotamian one.

The Tell al-ʿAmārna tablets — cuneiform tablets in the Babylonian language discovered in the Nile valley — revealed in detail the relations established about the middle of the second millennium between Egypt and the nations of Western Asia. They prove that the Babylonian language had become at that time the language of international diplomacy. This was not due to power, for the Egyptians were then stronger than the Babylonians, but to tradition (much as the French language continued to be the language of diplomacy long after French hegemony had ceased to be a reality).

The kings of Mesopotamia were also involved in many dealings and struggles with their northwestern neighbors, inhabiting the mountain lands of Anatolia and Armenia — the Hurrians who had moved westward from the Lake Van region and had eventually come together with the Hittites under the kings of Mitanni. The Hurrians overran the territory of the Hittites, even the latter's capital, Boghāzköy (90 miles east of Ankara), and they moved southward along the Syrian coast as far as the land of Edom, south of the Dead Sea. Traces of their settlements have been found in Rās Shamrā, Jerusalem, and farther south. They were possibly involved with the mysterious Hycsos, who invaded Egypt in the period 1788–1580. The kings of Mitanni had Indo-Iranian origins and swore by Indra, Mithra, and other such gods. The Hittites, as far as we can judge from their language, had also some Indo-Iranian connections. The main Hurrian achievement was the introduction of horse-drawn war chariots, possibly from India.

The story which we have been obliged to tell so rapidly is tantalizing, for it evokes all kinds of cultural contacts between the Mesopotamians, the Egyptians, the Syrians, and many other nations of Western Asia on the one hand and the nations of Iran and India on the other. On account of their position at the head of the Persian Gulf it is highly probable that the early Sumerians were in touch with India. Further study of the prehistoric civilization of the Indus valley (Mohenjo-daro, Harappa) and decipherment of their writing may substantiate that claim, which at present rests only on similarities between Sumerian and Hindu seals.[7]

[7] G. Sarton, "A Hindu decimal ruler of the third millennium," *Isis 25*, 323–326 (1936), *26*, 304–305 (1936).

In spite of all those outside influences, of which the Egyptian was the greatest, the Mesopotamian culture preserved its high originality for a very long period of time, say three millennia. That civilization, let me repeat once more, had been so deeply marked by the Sumerian pioneers that it remained until the end Sumerian, as our civilization is Greco-Latin, or as the Japanese one is Chinese.

For general reference, see Leonard William King, *History of Sumer and Akkad from prehistoric times to the foundation of the Babylonian monarchy* (404 pp., 34 pls., 69 figs., 12 maps; London, 1910); *History of Babylon from the foundation of the monarchy to the Persian conquest* (364 pp., 32 pls., 72 figs., 18 maps; London, 1915); Bruno Meissner, *Babylonien und Assyrien* (2 vols.; Heidelberg, 1920–1925) [*Isis 8*, 195–198 (1926)]; Georges Contenau, *Manuel d'archéologie orientale* (3 vols., Paris, 1927–1931) [*Isis 20*, 474–478 (1933–34)].

INVENTION OF WRITING

It has already been observed that two essentially different languages were used in Mesopotamia — the Sumerian and later the Accadian. Sumerian is neither Semitic nor Aryan, but an agglutinative language inviting comparison with Mongolian, Japanese, or Chinese [8] yet distinct from them and from any other Asiatic language. Accadian, on the contrary, is definitely a Semitic language, coming close to Hebrew — so close, indeed, that Accadian readings have enabled us to understand Biblical words more clearly. It is known in various dialectal forms which are called Babylonian, Assyrian, Chaldean, but this is the affair of philologists. We are primarily concerned with the fact that there was in Mesopotamia as well as in Egypt a conflict between two languages one of which was Semitic. This comparison — like every comparison with Egypt — does not go very far. The linguistic situation was very different in the two countries. In Egypt the conflict was soon ended by assimilation, the earliest records already revealing the existence of a single language, partly Hamitic, partly Semitic. In Mesopotamia, the Sumerian language was commonly used until the end of the third millennium, then displaced gradually by various Eastern Semitic languages, closely related, Accadian, Babylonian, Assyrian, Chaldean. The Sumerian language was completely free from Semitic features, but the Semitic dialects preserved many Sumerian elements.

Now all those languages were written in a special script which was called cuneiform because it is composed of wedge-shaped signs (*cuneus* = wedge). That script was invented by the Sumerians. Was that invention independent of the Egyptian one? Before trying to answer the question, it should be made clear that the transmission of an invention can be understood in two very different ways, according to whether one thinks of the invention in general or of its technical aspect. The general idea in this case is that the spoken language can be exactly represented, standardized, and perpetuated by means of written signs. That idea has occurred independently to many nations. In its earliest stage it is natural and simple enough. Pictographic symbols may easily serve as reminders of ideas or facts. Such symbols have been used by American Indians, Hindus, Chinese, Sumerians, Egyptians, and others. We still use some; the skull and

[8] The possible relation with Chinese has been patiently explored by C. J. Ball, *Chinese and Sumerian* (quarto, 192 pp.; London, 1913). Various other attempts have been made to connect Sumerian with Chinese antiquities, but none is convincing.

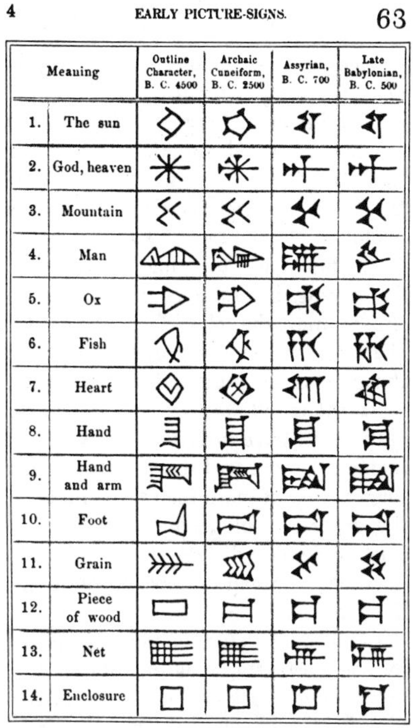

4
EARLY PICTURE-SIGNS.
Meaning
Outline Character, B. C. 4500
Archaic Cuneiform, B. C. 2500
Assyrian, B. C. 700
Late Babylonian, B. C. 500
1. The sun
2. God, heaven
3. Mountain
4. Man
5. Ox
6. Fish
7. Heart
8. Hand
9. Hand and arm
10. Foot
11. Grain
12. Piece of wood
13. Net
14. Enclosure

Fig. 16. Development of cuneiform. [From Leonard William King, *The Assyrian language* (London, 1901), p. 4.] Note that the resemblance of signs to things is greater when the characters are looked at from the right (for example, No. 3).

bones of our drug bottles requires no interpretation. However, intelligent people must have realized sooner or later that that method was seldom free from ambiguity, and that its scope was narrowly limited. It does not permit the graphical representation of abstractions, or feelings, or proper names (the names of individuals or of places). As to the technical implementation of the idea, the Egyptian and Sumerian methods are so different that we may be certain neither nation influenced the other.

The Sumerians (or some unknown predecessors of theirs) did not start their experiments with cuneiform symbols. Like the Chinese and the Egyptians they began with pictographs, some of which have been preserved (Fig. 16). Later they used the so-called line characters derived from the early pictures. That was natural enough as long as writing remained an exceptional performance, and the characters might be inscribed, let us say, on hard rocky surfaces. When writing became more popular and more frequent, the problem necessarily arose of finding a suitable material to write upon. The Egyptians, we may recall, found an admirable one — papyrus. The Sumerians, taking advantage of an inexhaustible supply of clay in Lower Mesopotamia, invented the use of clay tablets for writing purposes. They found that it was possible to make quickly long series of marks on soft fresh clay with a reed, and that the marks were fixed as soon as the clay was dry and remained clearly visible for an indefinite length of time; the process could be improved by baking the tablets. However, the scribe writing on clay had not by any means the same freedom as his Egyptian colleague writing on glossy papyrus. The latter was like a painter or draftsman; the former could only make two or three kinds of signs or wedges. The cuneiform script was an unavoidable consequence of the choice of clay as writing material.

The Sumerian script was based on the use of some 350 syllabic signs; it never reached a truly alphabetic stage, even in a limited way, as was the case for Egyptian. The Semitic followers of the Sumerians used the same script, adapting it to their own language, and sometimes preserving Sumerian words as ideograms. The evolution of cuneiform is comparable to that of Chinese and Egyptian in two ways. In the first place, the same needs introduced the addition of phonetic complements (suggesting the pronunciation) and of determinatives never pronounced (suggesting the meaning, the "class"). In the second place, as the speed of writing increased, the characters were necessarily simplified; various forms of cursive or shorthand changed the appearance of the script profoundly.[9]

To the uninitiated, cuneiform writing seems very clumsy and very hard to read, yet it must have merits of its own, for in spite of many political vicissitudes it remained the standard script in Mesopotamia, almost until the time of Christ, that is, for a period of more than three millennia. It was used by different nations, and for the expression of languages as unrelated as Sumerian on the one hand and the Eastern Semitic dialects on the other. Nor was it restricted to the peoples of Mesopotamia. It spread to the countries east of the Tigris and to those north and west of the two rivers.

Let us quote a few examples. The largest of the 'Amārna tablets is a letter written by one Tushratta, King of Mitanni, to Amenhotep III (1411–1375); that letter is not written in Babylonian but in Hurrian; incidentally, it is the longest Hurrian text thus far known to us. Many thousands of cuneiform tablets have been found in Boghāzköy and other Anatolian sites. The most ancient of those tablets were written in Accadian (or Babylonian), but later ones (say by 1400) were written by those Anatolians in their own language, Hittite. From Boghāzköy have come syllabaries or dictionaries, giving in parallel columns equivalent words in Hittite, Sumerian, and Accadian; some tablets (relatively few) contain Hurrian texts, but the majority are Hittite. The Hittite influence was recognized as far as Egypt; witness a treaty made between one of their kings and Ramses II (1292–1225). Two tablets have come down to us, one bearing the original Babylonian text of that treaty, the other a translation of it in hieroglyphics. The most interesting Hittite text thus far discovered is a treatise on horse training of the fourteenth century; we shall come back to that presently.[10]

The outstanding merit of cuneiform writing was its peculiar adaptation to clay; hence, wherever clay tablets were used cuneiform went with them. Such was the case in Anatolia, and also in Elam east of the lower Tigris, where cuneiform was the standard script from very early times on. The inertia of tradition preserved the use of cuneiform script, even in the relatively exceptional cases when writing was

[9] Compare the differences between our printing and various forms of calligraphy, abbreviations, and shorthand.

[10] Hittite is closely related to the Indo-European languages, being derived together with them from a common source. On the contrary, Hurrian has no genetic connection with those languages, nor with Egypitan or Sumerian. Edgar H. Sturtevant, *Comparative grammar of the Hittite language* (Philadelphia: Linguistic Society of America, University of Pennsylvania, 1933). E. A. Speiser, *Introduction to Hurrian* (New Haven: American Schools of Oriental Research, 1941). Many specimens of Hittite literature have been translated by Albrecht Goetze for James B. Pritchard, *Ancient Near Eastern texts* (Princeton: Princeton University Press, 1950), p. 503 [*Isis* 42, 75 (1951)].

done on other materials than clay, for example, in monumental rock inscriptions or on standard weights. The Achaimenian inscriptions, thanks to which the cuneiform enigma was penetrated, were written in three columns and represented three separate languages — Old Persian, Babylonian and Elamite — but a single script.[11]

To return to earlier times and to conclude, before the end of the fifteenth century Babylonian had become the language, and cuneiform the script, of diplomacy. The language was popular but the script much more so. It was used to transcribe not only the Babylonian language of the day, but the old Sumerian, and the dialects of a number of foreign nations, Elamite, Hittite, Hurrian, Phoenician, and others. Cuneiform tablets bearing texts in one or the other of those languages were scattered over Western Asia.

Any one remembering that that part of the world was the cradle of some of the most precious features of our own civilization, our cradle, cannot help being deeply moved when he observes the intense miscegenation that had already taken place there before the year 1000 B.C. (indeed long before that date), the multiplicity of languages together with the unity of script.

ARCHIVES AND SCHOOLS. THE BIRTH OF PHILOLOGY

Cuneiform inscriptions on stone or other materials than clay are relatively rare; the great mass of the cuneiform texts were preserved on clay tablets. We have already indicated that the popularity of that writing material, clay, determined the popularity of the cuneiform script. It is worth while to consider the tablets themselves a little more carefully. Clay was abundant and cheap, the making of tablets exceedingly simple, much more so than that of papyrus. Moreover, the clay tablets left to themselves were practically indestructible even when unbaked. The inviolability of certain documents was secured by placing them in clay envelopes; clay shrinks considerably in drying and the document could not be extracted from the envelope without breaking the latter; nor could a fresh envelope be placed on a preshrunk tablet.[12] Note that the durability of papyrus was due not so much to the material itself as to the dry climate of Egypt; if papyrus had been employed in Mesopotamia, nothing would have remained of it. Large numbers of tablets were used for the preservation of documents of every kind, public and private. Many thousands of them prior to *c.* 1500 are available in our museums, and the number of later ones that have been rescued is so large that a long time will elapse before they are all interpreted.

Clay did not lend itself to calligraphy as papyrus did, and cuneiform script never became a definite branch of art as did the writing of hieroglyphics. What is worse, clay dries out quickly and a tablet had to be written all at one time;[13] the majority of tablets were relatively small. Longer texts, such as annals, might be written on the outside surfaces of hollow polyhedra of clay (cylinders or prisms with hexagonal, heptagonal, or octogonal bases), but the usual method was to write them on many tablets.

[11] The largest and most famous of those polyglot inscriptions is that of Behistūn (or Bīsutūn, near Kirmānshāh on the road from Baghdād to Hamadān) wherein Darius the Great related his victories, in 516 B.C. It was that inscription which gave Sir Henry Rawlinson in 1847 the key to the decipherment of Babylonian and led to the foundation of "Assyriology" as a science (1857).

[12] For further discussion see the excellent semipopulár book of Edward Chiera, *They wrote on clay*, chap. 6.

[13] Unless the scribe placed wet towels over the half-written tablet, as a sculptor does for an unfinished clay model.

Both the Egyptians and the Sumerians invented writing and they developed their invention and utilized it on a large scale. The former, being favored by a more convenient material, made an additional invention, the roll or the book, thanks to which a text, however long, could be preserved in its entirety. The Sumerians were not so fortunate. A few long texts they wrote on large polyhedra or on a huge block of stone (like the code of Hammurabi), but it is clear that even in those cases they did not provide the equivalent of a book. In the majority of cases, a long text was written on as many independent tablets as were needed. In order to insure their proper sequence the scribes would write at the bottom of each tablet "table *x* of series *y*" and add the opening line of the following tablet, yet that did not suffice to preserve the integrity of the text. Papyrus rolls have generally been found complete,[14] but the tablets constituting a text have almost never reached us in due order. The tablets have been shuffled and reshuffled, some have been lost or widely separated from the others;[15] and the reconstruction of the text is like the solving of a jigsaw puzzle of fantastic complexity.

Their failure to invent the book may have caused Sumerians to develop more rapidly the creation of archives and libraries. We must assume that there existed collections of papyrus rolls in Egyptian temples and palaces, but the keeping of tablets in good order was even more urgent than the collection of whole books; hence it is highly probable that archives and libraries existed very early in Mesopotamia. To put it briefly (too briefly), we might say that the Egyptians invented books, while the Sumerians invented record offices!

An enormous "library" was excavated by American archaeologists at Nippur, and many thousands of tablets found there are now in the museums of Constantinople and Philadelphia. The majority of the tablets were unbaked and therefore less well preserved than baked ones would be and more difficult to decipher, yet they have finally revealed to us a number of literary and scientific texts which, because of their great antiquity, are of unsurpassed interest. Nippur was one of the most famous centers of Sumerian religion and its temple dedicated to the great god Enlil[16] was a conservatory of early traditions. The tablets of the library seem to have been generally arranged on shelves made of clay and about 18 in. wide. Connected with the temple was not only that library or record office, but also a school,[17] and many model texts prepared by the teachers and exercises written by the students have been discovered among its ruins. This enables us to understand how the cunei-

[14] Sometimes the head or the ending might be missing, or even a middle part; but in any case the roll would preserve a relatively long sequence of the original text.

[15] They were separated at first because the places where they had been deposited were burned or fell to pieces in the way adobe houses do; further separations were due to rebuilding, and to furtive or scientific excavations, to sales, and so on. A good many of the tablets in our museums have been bought from dealers who obtained them from Arab diggers hiding their sources of supply. Thus some tablet of a text may be in a Russian museum, while others of the same text are in an American collection. Even individual tablets might be broken and the fragments dispersed. A medical text studied by Edward Chiera was based on a broken tablet, a part of which was in Philadelphia, the rest of it in Constantinople! See Chiera, *They wrote on clay*, p. 117.

[16] Enlil, the god of air and earth, became the supreme god of the Sumerians. Under Babylonian domination he became Marduk (or Bel). Bel is the Semitic name of Enlil. The gods change with the people. Compare the transformation of Zeus and Aphrodite into Jupiter and Venus.

[17] This was natural enough. A temple needs priests and scribes for its rites, traditions, and business, and these must be trained; the logical place of training is in the temple itself or close to it, the men in office being the best teachers of their successors. The same conditions produced everywhere similar results, e.g., schools in Egyptian and Buddhist temples, medieval cathedral schools.

form script and Sumerian formulas were taught to the young. An actual schoolhouse of Hammurabi's time has been unearthed which was claimed to be the earliest in existence. This may be true if we take "schoolhouse" in a technical sense, a house definitely meant for teaching purposes, but we may be certain that there were already schools before Hammurabi (in Egypt as well as in Sumer), though even if they were excavated there might be nothing to prove their reality. Any room might be used for a schoolroom and the children might even be taught in the open air; all that was needed was a few tablets illustrating the signs, words, or formulas to be copied and memorized, a lump of fresh clay, and a bundle of reeds.

The evocation of schools and libraries suggests that the invention of writing had another purpose than the preservation of records, a deeper purpose, which escaped the attention of the average scribe but must have exercised the minds of the early "philologists." That purpose was the preservation, correction, and standardization of the language itself. As long as the language remains unwritten, it is bound to fluctuate rather rapidly, and perhaps too rapidly. Writing helps to fix it. The invention of writing must be conceived as a very long process. The fundamental idea is simple enough, but, however great the intelligence of the early "philologists" who tried to realize it, they could not possibly imagine at once all the difficulties nor the means of overcoming them. The very process of reducing a language to writing introduces philological problems and may awaken a kind of philological consciousness in the mind of a few men of genius. Those early grammarians, who may have been also the early schoolmasters (for teaching a subject has always been one of the best ways of mastering it), compiled lists of classified words which are the prototypes of our dictionaries. Such lists have been found in the Sumerian site of Erech (Warka), dating from before 3000 B.C. The Semitic invaders compiled more elaborate lists containing the Sumerian words and their Accadian equivalents, and made investigations on the morphology and syntax of those languages. We have already referred to Hittite glossaries, which continued the same tendencies in a neighboring country. The fact that the Accadian, Babylonian, or Hittite grammarians were using simultaneously two or more languages, the structure of which was absolutely different, must have quickened their philologic sensitiveness.[18]

In spite of many statements to the contrary, we must say that philology is not one of the latest sciences, but rather one of the earliest. How else could it be? No scientific work of any kind could ever be published without a linguistic tool of sufficient exactitude; the common people created the language, but philologists were needed almost from the beginning in order to standardize it, to refine it, and to increase its precision. It is possible that one of the differences between the peoples who gradually developed a high civilization and those who did not lies in the fact that the first were not satisfied very long with a traditional and unconscious language, but were eager to analyze it and to use it deliberately and exactly. Philologic consciousness was a part, an essential part, of scientific curiosity; in some nations this curiosity and that consciousness were more highly developed than in others. Those nations were our spiritual ancestors.

[18] The Egyptians lacked that advantage, yet their own language had evolved sufficiently by the end of the Old Kingdom (say by the twenty-sixth century) to require philological commentaries, and we do find such glosses in the Smith surgical papyrus; *Isis 15*, 359 (1931).

BABYLONIAN SCIENCE

Having obtained some idea of the instruments, physical (tablets) and mental (philology), let us see how they were applied to the understanding of the world and the enrichment of knowledge. Everything considered, the best expression to designate that body of knowledge is "Babylonian science," for the bulk of our information comes from Babylonian tablets. These tablets reflect Sumerian knowledge as explained and transformed by Accadian (Babylonian) scribes. We might call that science Mesopotamian, or speak of the science of Sumer and Accad, but that would be cumbrous and on the whole less evocative than to call it "Babylonian." The essential point is to bear always in mind the Sumerian origin and coloring of that science.

The scientific tablets are generally undated and undatable, except when their exact provenance is known, as when they were found by scientific excavators at a definite level. Unfortunately, a great many of the tablets available to scholars have been obtained through clandestine diggings. In the case of astronomical tablets the dating of the original text (not necessarily of the tablet) may sometimes be determined from internal evidence. With regard to mathematics, there is only a little fragment of a Sumerian text; the majority of the problems are Old Babylonian,[19] the rest are Seleucid (that is, of the last three centuries B.C.).

Many misunderstandings have been caused by careless scholars[20] dealing in the same chapters or even in the same paragraphs, with Old Babylonian texts, definitely pre-Hellenic, and with Seleucid ones, post-Hellenic. Let us repeat once more that the whole of Greek science (as opposed to Hellenistic and Roman science) was developed in a period of time that was not only preceded but followed by Mesopotamian (and Egyptian) activities. If we replaced time by space, we might visualize Greek science as a small island surrounded by an Oriental sea. Our readers will be protected against such grave misunderstandings, for the Seleucid tablets, belonging to the Hellenistic age, will not be discussed at all, not only in this chapter, but not even in this volume. With the exception of occasional brief references to later tablets, all the documents examined in this chapter represent the old Sumerian-Babylonian culture, considerably older than the beginning of Greek science.

MATHEMATICS[21]

The number of mathematical clay tablets that have thus far been deciphered is not very large — some sixty — to which must be added some two hundred containing tables. Moreover, the majority (say two-thirds) of those tablets are of very late date (Seleucid). Hence, we have less than a hundred tablets to represent ancient Babylonian mathematics. Almost all these tablets, having come to us from clandestine excavations, are undatable except in a very indirect and imperfect manner. Moreover, we have no treatise or textbook comparable to the Rhind papyrus. This

[19] That is, the most ancient are hardly anterior to Hammurabi, the bulk dates probably from the second third of the second millennium.

[20] Reference is made here not to Assyriologists, but to historians of science and culture.

[21] R. C. Archibald, *Bibliography of Egyptian and Babylonian mathematics* (2 parts; Oberlin, Ohio, 1927–1929) [*Isis 14*, 251–255 (1930)]. Otto Neugebauer, *Vorlesungen über Geschichte der antiken Wissenschaften* (vol. 1; Berlin, 1934) [*Isis 24*, 151–153 (1935)]; *Mathematische Keilschrift-Texte* (3 vols.; Berlin, 1935–1937) [*Isis 26*, 63–81 (1936), *28*, 490–491 (1938)]. François Thureau-Dangin, *Textes mathematiques babyloniens* (Leiden: E. J. Brill, 1938) [*Isis 31*, 405–425 (1939–40)].

may be due to the fact, already explained, that publication on clay tablets discouraged the composition of long texts, while the papyrus roll tended to encourage them. Or, if textbooks were written,[22] they have not come to us. Not only were tablets forming a series dispersed but even single tablets were sometimes broken into fragments. The student of Babylonian mathematics is thus much less fortunate than his colleague investigating Egyptian mathematics.

Sumerian numeration was at first a strange mixture of decimal and sexagesimal ideas. It would seem that their earliest mathematicians started with a decimal base but soon afterwards realized that a sexagesimal base would be better.[23] That change of mind, which must have been deliberate, is in itself very remarkable. The sexagesimal system was not pure, the successive orders being obtained by the alternate use of the factors 10 and 6, thus: 1, 10, 60, 600, 3600, 36000, etc. (Fig. 17). The variety of numerical symbols being restricted by the cuneiform script, they had only two separate elementary signs for numerals: ▽ for 1 and ∢ for 10, but the first sign was used not only for 1 but also for 60 and for every power of 60, and the second was used not only for 10 but also for ten times any power of 60. Thus we may write ▽ $= 60^n$ and ∢ $= 10 \times 60^n$, where n is any positive or negative integer or zero. The system of numeration was thus mainly sexagesimal, for the symbol for 10 was subordinate and there was no symbol for 100, 1000, . . . One hundred would be written 1,40; one thousand, 16,40.[24]

The absolute value of a number could be determined only from the context. The Sumerians had discovered the principle of position and hence, if in a given number the absolute value of one place was known, the value of other places could be deduced. However, they had no medial zero until late (Seleucid) times; the absence of units of a certain order was indicated by an empty space and this was unclear and precarious. Those ambiguities increase materially the difficulty of deciphering mathematical tablets.

A number like *abcdef* (without empty spaces) is to be interpreted as $a(60)^n + b(60)^{n-1} + c(60)^{n-2} + d(60)^{n-3} + e(60)^{n-4} + f(60)^{n-5}$, wherein n may be null or have any positive or negative integral value. In general, the questions dealt with, or the sequence of operations, would suppress or reduce the ambiguities. The large

[22] Statements made by Hypsiclēs (II-1 B.C.) and by Geminos (I-1 B.C.), quoted by Neugebauer, *Mathematische Keilschrift-Texte*, vol. 3, p. 76, refer probably to late, post-Hellenic, textbooks. We are thinking of pre-Hellenic Babylonian textbooks and there is no evidence that such existed.

[23] The coexistence of sexagesimal ideas in China and Mesopotamia is very striking (see pp. 11–13). That is too slender a basis to conclude that one of these two cultures was influenced by the other; yet it is more convincing to me than linguistic analogies. Sixty is a very large number to agree upon. Its use as a number base or as a cycle implies a high degree of sophistication.

[24] For the printer's and the reader's convenience, in our examples of Babylonian (sexagesimal) numbers we will separate each sexagesimal power from the preceding one by a comma, and the negative powers from the positive ones by a semicolon; also we will use zeros, though the Babylonians did not use them. Thus 11,7,42;0,6 means $(60^2 \times 11) + (60 \times 7) + 42 + (60^{-2} \times 6) = 40{,}062.00166$.

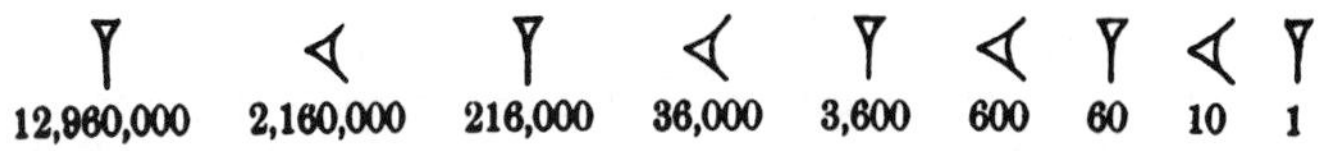

Fig. 17. Sumerian numerals. [From H. V. Hilprecht, *The Babylonian expedition of the University of Pennsylvania. Series A, Cuneiform texts* (Philadelphia, 1906), vol. 20, part 1, p. 26.]

size of the base 60 also helped to restrict the reader's choice, for there would be such a tremendous difference between, let us say, a length of 7 cubits and one of 420 cubits or 25,200 cubits that one or the other was indubitably meant.

Imperfect as it was, the Sumerian system implied a degree of arithmetical abstraction that is astounding. It is impossible to reconstruct the genesis of their discovery. Were they calculators of genius who devised such a system out of long experience, or did the system stimulate their efforts in the direction of computations of increasing complexity and of algebraic experiments? Perhaps it worked both ways, as it always does in the development of science: new abstractions suggest new experiments, and vice versa.

The oldest Sumerian tablets contain all kinds of numerical tables: tables of multiplication, tables of squares and cubes which being inverted gave tables of square roots and cube roots, tables of reciprocals. If one reads such a table consecutively there is little room for ambiguity. For example,

The square of 1 is 1,
The square of 2 is 4,
The square of 3 is 9,
. . . .
The square of 8 is 1, 4 (meaning $60+4$),
The square of 60 is 60 (meaning 60^2).

That is easy enough. But what happened to the computers who consulted a single item of the table? They had to be careful, that is all, and not consider one item without the neighboring ones. They might read "The square of 59 is 58,1"; this must mean $(60\times58)+1$, for the square of 59 must be just a little smaller than the square of 60. "The cube of 59 is 57,2,59"; this cannot but mean $(60^2\times57)+(60\times2)+59$.

The table of reciprocals, which are numerous and extensive, are specially interesting. Having discovered the use of fractions built on the same pattern as the integers, they had thus by a precocious stroke of genius suppressed most of the fractions. Sexagesimal fractions, they had realized, were only a kind of sexagesimal integers not essentially different from them (just as decimal fractions are simply a kind of decimal integers, though there are educated and intelligent people now living who cannot yet see that!). However, sexagesimals did not suppress every fraction. What about such fractions as ½, ⅔, ⅗, not to speak of more complex ones? The circumstances of life would unavoidably introduce nonsexagesimal fractions. What was one to do with them? One might reduce them to sexagesimals but that was not always possible. Giving us another proof of their genius for arithmetical creation, the Sumerians replaced the consideration of fractions by that of reciprocals; or, to put it otherwise, reciprocals enabled them to replace every division by a multiplication. The third of sixty is twenty; they said the reciprocal of 3 is 20; to divide by three (to take one-third) can be replaced by multiplication by twenty. The base 60, having an extraordinarily large number of factors (2, 3, 4, 5, 6, 10, 12, 15, 20, 30), lent itself so well to reciprocal calculation that one cannot help wondering once more whether the Sumerians did not use that base for the very reason that it has so large a number of factors. Their use of reciprocals was so habitual to them that they sometimes complicated their reckoning needlessly because of it. The third of 6 cubits they would say is $6\times20=120=2$ cubits. Or, having to find the square of 12, they took the reciprocal of 12, which is 5; squared 5, which is 25, and took the reciprocal of 25, which is 2,24; the final result is correct but might have been obtained more easily. This is a well-known mathematical foible; its existence gives us an additional proof that the Sumerians were real mathematicians; they were carried away by their abstractions to such an extent that they sometimes forgot simpler methods.

The example just quoted [25] involved very small numbers, but the Sumerians extended their tables of reciprocals to very large ones, up to the order of 60^{19}.

Among the powers of 60, one occurs very frequently in the old tables, namely, $60^4 = 12{,}960{,}000$. Now this is the geometric number of Plato,[26] and 12,960,000 days = 36,000 years of 360 days, the "great Platonic year" (the duration of a Babylonian cycle). A man's life of 100 years [27] contains 36,000 days, as many days as there are years in the "great year." The "geometric number," that is, a number measuring or governing the earth and life on earth, was thus clearly of Babylonian origin.[28]

Not only did the Sumerians use a positional notation (though without zero) and extend it to submultiples of the base as well as to multiples, but their number system was closely connected with the subdivision of weights and measures. That is, they had devised a complete sexagesimal system before 2000 B.C.; in order to appreciate their genius it will suffice to recall that the extension of the same ideas to the decimal system was only conceived in 1585 (by the Fleming Simon Stevin),[29] that its implementation was begun only during the French Revolution and is not yet completed today. The old Sumerians were more consistent than are many of our own contemporaries who persist in defending English metrology in a decimal world. Having realized that, it becomes a little difficult to consider the former as primitives or the latter as truly civilized.

How can one account for the sexagesimal basis and Sumerian precocity? One way of explaining the matter, as far as it can be explained at all, is to say that the Sumerian metrology and the Sumerian number system harmonize so well because they grew together. It is hard to believe that the Sumerians would have selected the base 60 on purely mathematical grounds; it is easier to assume that their metrological practice suggested that base. Indeed, when one measures things one cannot help coming across many parts of the chosen standard; fractions come in willy-nilly and one may be thus led to take as standard (of length, weight, and number) a unit accommodating as many fractions as possible. The natural relation between fractions and metrology is illustrated by the Roman system; the *as* or *libra* divided into twelve *unciae* suggested the fractions most commonly used by the Romans. That was very neat, the only trouble being that the *as* introduced a duodecimal system in a decimal numeration. The Sumerians' natural genius precluded that fundamental blunder; they used sexagesimal fractions and a sexagesimal metrology together with a sexagesimal system of integers.

[25] It is one actually found on an old Babylonian tablet; see Thureau-Dangin, *Textes mathematiques babyloniens*, p. 18.

[26] *Republic*, VIII, 546 B–D.

[27] *Ibid.*, X, 615 B.

[28] For further discussion of this topic see Hermann Vollrat Hilprecht, *Mathematical, metrological and chronological tablets from the Temple library at Nippur* (Philadelphia, 1906), pp. 29–34; Sir Thomas Heath, *History of Greek mathematics* (Oxford, 1921), vol. 1, pp. 305–308 [*Isis 4*, 532 (1922)].

[29] G. Sarton, "Simon Stevin of Bruges, 1548–1620," *Isis 21*, 241–303 (1934); "The first explanation of decimal fractions and measures, 1585," *Isis 23*, 153–244 (1935).

The sexagesimal base was strangely reinforced in the course of time by the occurrence of another unit six times as large. The Sumerians thought at first (like the earliest Egyptians) that there were 360 days in the year.[30] They began by dividing the days into six watches (three day watches and three night ones, which were naturally of varying length),[31] but they soon realized the impracticality of unequal time periods for astronomical work and divided the whole day (day and night, *nychthēmeron*) into 12 equal hours of 30 *gesh* each.[32] That is, their astronomical day was divided into 360 equal parts. There were thus 360 days in the year and 360 *gesh* in the day; the same division into 360 parts was applied later to parallels, and later still (in Achaemenian times, *c.* 558–330) to the ecliptic (zodiac, dodecatemories).[33] We divide the circle into 360 degrees to this day, and subdivide the degrees on a sexagesimal basis, because of the Sumerian mathematicians who flourished more than two millennia before Christ.[34]

The reader has already observed that there are three confluent sources of Babylonian mathematics – arithmetic, metrology, and astronomy. We shall come back to the last-named one presently. Metrology is the daughter of business; selling and buying imply the existence of unit prices and the practice of measuring and weighing. Innumerable tablets are simply business documents, the mathematical structure of which is sometimes very instructive. In a Louvre tablet (AO 6770) of *c.* 2000 B.C. there is a problem [35] to find how long it would take for a sum of money to double itself at compound interest, interest being computed at 20 percent. The problem as we would put it is to find x in the equation $(1 + 0;12)^x = 2$. The correct result, 3;48 (3 years and 4/5), was duly obtained by the Sumerian computer! If he thus succeeded in solving an exponential equation, we shall not be surprised to hear that he was able to solve other equations. He certainly did solve linear equations, simultaneous linear equations with many unknown quantities, and quadratic and cubic equations. For the quadratic he seems to have known a formula comparable to our own. Neugebauer has suggested that even some cubics were reduced to a normal form, and that a table [36] gave the values of $n^2 + n^3$ for such a purpose. This may be going a little too far. From the examples that have come to us, we can only conclude that the Sumerian computer was able to solve some cubic equations. But even if he had done nothing but solve quadratic equations as generally as he did, as well as systems of two quadratics with two unknowns, we would already have sufficient reason for admiring him.

[30] We must remember that the passage from 60 to 360 was not an unnatural one for Sumerians. It would seem that at first, at least, they passed from one sexagesimal order to the next one in two steps; that is, they did not multiply by 60, but by 10 then by 6 (see above).

[31] The use of unequal divisions of the day was almost universal in ancient times and it continued in some parts of Europe as late as the eighteenth century. The Egyptians divided the day and the night each into 12 hours, the Greeks and the Romans did the same; those hours were of varying length like the watches. As to the latter, we find them in the Bible, *ashmūrāh* in Exodus 14:14 and *phylacē* in Matthew 14:25. The Jews divided the night into three watches, the Romans into four, the guards being changed at the end of each watch.

[32] Each *gesh* was thus equal to four minutes of our time.

[33] The earliest Greek work in which the division of the ecliptic into 360° occurs is the one ascribed to Hypsicles (II–1 B.C.).

[34] François Thureau-Dangin, "Sketch of a history of the sexagesimal system," *Osiris* 7, 95–141 (1939). Solomon Gandz, "Egyptian and Babylonian mathematics," in M. F. Ashley Montagu, ed., *Studies and essays in the history of science and learning offered in homage to George Sarton on the occasion of his sixtieth birthday* (New York: Schuman, 1944), pp. 449–462 [*Isis* 38, 127 (1947)].

[35] I quote it from Archibald's analysis of Neugebauer's publication, *Isis* 26, 71 (1936), 28, 491 (1938), where more details and references to the original tablet may be found.

[36] Berlin VAT 8492.

In spite of the fact that he had no equations and no symbolism of any kind [37] (not even a symbol for the unknown quantity), his algebraic ingenuity was such that he was able to do the equivalent of many processes familiar to us, such as reduction of similar terms, elimination of one unknown quantity by substitution, introduction of an auxiliary unknown quantity. Moreover, in spite of the complete absence of algebraic symbolism he was aware of the identity which we express $(a + b)^2 = a^2 + 2ab + b^2$ and he had an algebraic means of finding successive approximations of the square root of a number.[38] These achievements are almost uncanny, and the only (very incomplete) explanation that I can offer is that his abstract reckonings and tables had given his mind a kind of algebraic coloring and motivation.

Finally, it is clear that the Sumerians were not afraid of handling negative numbers;[39] this may seem a small matter, yet the concept of negative quantity did not penetrate Western minds until the time of Leonardo of Pisa (XIII-1) and its proper development required many more centuries.

It is not necessary to continue this enumeration; the algebraic achievements of the Sumerian people of 4,000 years ago are more than sufficient to daze the youthful mathematicians of today. The average philologist would be utterly unable to understand Sumerian mathematics, yet he will complacently repeat that there was no real mathematics before the Greeks! It is quite clear to us that the old Sumerians had as much natural genius for algebra as the Greeks for geometry.

The Babylonians of the period 2200–2000 knew how to measure the area of rectangles and of right and isosceles triangles; they had some knowledge of the Pythagorean theorem,[40] and were aware that the angle in a semicircle is a right angle; they could measure the volume of a rectangular parallelepiped, of a right circular cylinder, of the frustum of a cone or of a square pyramid. Their solution of the last-named problem (volume of the frustum of a square pyramid) was a little different from the Egyptian solution. It might be represented by the formula

$$V = h\left[\left(\frac{a+b}{2}\right)^2 + \frac{1}{3}\left(\frac{a-b}{2}\right)^2\right].$$

The Egyptian solution given above (p. 40) is simpler, but the two solutions are equivalent. It is interesting to note that when the Hellenistic mathematician, Heron of Alexandria, tackled the same problem almost two millennia later his solution was like the Babylonian one.[41]

With regard to circular measurements the Babylonian mathematicians were definitely inferior to their Egyptian contemporaries. The best way to compare the two methods is to calculate the value of π corresponding to each. While the

[37] Bear in mind that the development of symbolic algebra hardly began before the sixteenth century, more than three millennia later!

[38] Essentially the same as the Archimedian-Heronian method. If a is an approximate root of A, and $A - a^2 = b$, then better approximations are $a_1 = a \pm b/2a$, $a_2 = a_1 \pm b_1/2a_1$, $\cdots$

[39] R. C. Archibald, *Isis 26*, 76 (1936); yet see Thureau-Dangin, *Textes mathematiques babyloniens*, p. xxxiv.

[40] Archibald is certain of this and quotes examples tending to prove it, *Isis 26*, 79 (1936).

[41] Heron, *Opera* (Leipzig, 1914), vol. 5, pp. 30–35. Heron's date is uncertain; in my *Introduction*, I placed him tentatively in (I-1 B.C.). We know better now; he flourished between A.D. 62 and 150. *Isis 30*, 140 (1939); *32*, 263–266 (1947–49); *39*, 243 (1948).

Egyptian method was equivalent to taking $\pi = 3.16$ (real value, 3.14), the Babylonian one was equivalent to taking $\pi = 3$.[42]

How did the Babylonian achievements affect other nations? Their algebraic ingenuity was largely forgotten but it reappeared in Archimedes (III–2 B.C.), Heron (I–2), and more completely in Diophantos (III–2), then disappeared again for many centuries until the Arabic speaking people resurrected it (the name of the science algebra is of Arabic origin). The Arabic invention was not appreciated in the West except by a very few men, and the use of symbols continued to be small and erratic until the sixteenth and seventeenth centuries. The history of algebra is very puzzling, because much of its development was underground and secret. It was only with the beginning of its symbolic phase that progress could become steady and rapid. That final progress is easy enough to understand, but the achievements of the mathematicians who were groping in the darkness of the presymbolic age are astounding.

The Sumerians and their Babylonian successors left three legacies the importance of which cannot be exaggerated.

(1) The position concept in numeration. This was imperfect because of the absence of zero (until Seleucid times) and because the absolute value of the numbers quoted was often ambiguous. That concept was lost until its very slow revival in connection with the use of Hindu-Arabic numerals.

(2) The extension of the numerical scale to submultiples of the unit as well as to multiples. This was lost, and was not revived until 1585 with reference to decimal numbers.

(3) The use of the same base for numbers and for metrology. This was lost and not revived until the foundation of the metric system in 1795.

These three gifts were perhaps too great to be appreciated by posterity except after an interval of many millennia. Strangely enough, another gift, far less precious — the sexagesimal idea — was accepted far more readily and its acceptance delayed for centuries the reception and development of the decimal system. It is still weighing upon us today. Of course, that is not the Babylonian's fault; the tradition was capricious and defective, as it often is.

ASTRONOMY

In spite of the fact that their astronomical achievements were far inferior to their mathematical ones, the ancient Babylonians have been more often praised for the former than for the latter. That false evaluation was due to two circumstances. First, the confusion between ancient Babylonian astronomy and the late

[42] Examples given in the Old Testament (1 Kings 7:23; 2 Chronicles 4:2) correspond to the same poor approximation ($\pi = 3$).

Since writing the above I have examined the paper by E. M. Bruins, "Quelques textes mathématiques de la mission de Suse," *Proc. Roy. Dutch Acad. Sci.* 53, 1025–1033 (1950) and his "Aperçu sur les mathématiques babyloniennes," *Revue d'histoire des sciences* 3, 301–314 (1950). He has investigated some very early Babylonian tablets found in Suse by R. de Mecquenem in 1936. They show that early Babylonian mathematicians investigated regular polygons of 5, 6, and 7 sides and that they found better approximations for what we call π than the Biblical $\pi = 3$; they found successive approximations such as the Heronian $3\frac{1}{8}$. As we have just seen, this is not the only relation between the early Babylonians and Hellenistic times. The stream of early Babylonian ideas emerging in Heron, in Diophantos (III–2), and later in Arabic algebra has been investigated by Solomon Gandz, "The origin and development of the quadratic equations in Babylonian, Greek and early Arabic algebra," *Osiris* 3, 405–557 (1937); "Indeterminate analysis in Babylonian mathematics," *Osiris* 8, 12–40 (1948).

Chaldean or Seleucid ones (the main discoveries were made by Chaldeans); second, the mathematical genius of the ancients was revealed to us only very recently by Neugebauer and Thureau-Dangin.[43]

However, the Babylonians built the mathematical foundations, without which there can be no scientific astronomy, and they began the long series of observations, without which later generalizations would have been impossible. They created the art of astronomical observations. A kind of transit instrument was used by the early Assyrian king Tukulti-Ninurta I (1260–1232) for the rebuilding of the Ashur palace; [44] by that time they were already familiar with a simple form of sundial (gnomon) and with a kind of clepsydra.[45]

Moreover, the Sumerians had invented the building of brick towers (ziggurat) for religious purposes (Fig. 18). The earliest tower was built in Nippur for the cult of the great god Enlil. As it was impossible then to build a narrow tower like a medieval belfry, the towers were in the shape of a succession of edifices of decreasing size placed on top of one another (somewhat like some of our latest skyscrapers), with a broad stairway or inclined plane turning around like a helix and enabling the priests and votaries to reach the very top. The general effect was

[43] The pioneer student of Babylonian astronomy was the Jesuit father, Franz Xaver Kugler, *Sternkunde und Sterndienst in Babel. Assyriologische, astronomische und astralmythologische Untersuchungen* (6 parts; Münster in Westfalen, 1907–1935) [*Isis* 25, 473–476 (1936)]. The best work on the subject is now done by Otto Neugebauer. For general orientation, see his article, "The history of ancient astronomy. Problems and methods," *J. Near Eastern Studies* 4, 1–38 (1945), with full bibliography. Note that Kugler and Neugebauer devote much of their efforts to the explanation of late Chaldean or Seleucid astronomy, which does not concern us in this volume.

[44] A. T. Olmstead, "Babylonian astronomy," *Am. J. Semitic Languages*, 55, 113–129 (1938), p. 117.

[45] For the clepsydra, see Neugebauer, *Mathematische Keilschrift-Texte*, vol. 1, p. 173.

Fig. 18. Ideal reconstruction of the ziggurat of Ur. [From Sir Leonard Woolley, *Ur excavations* (Oxford: Clarendon Press, 1939). With kind permission of the Museum of the University of Pennsylvania.]

pyramidal, but the building was very different in every respect from the Egyptian pyramids. That invention is immortalized by the existence of ziggurat ruins [46] and also by the tradition concerning the tower of Babel (Genesis 11:1–9). As the tower dominated the Mesopotamian plains, the priest sacrificing at its top was able to observe the whole sky without hindrance, if he were minded to do so. Some of them did and accumulated valuable observations, but the main astronomical work was not begun until much later.

The growth of astrology was as slow as that of astronomy proper, and the methods of divination favored by the early Babylonians were derived from the particularities of the liver and other terrestrial omina rather than from the observation of the stars. The sophisticated astrology that affected so deeply the Roman and medieval world was very largely a Chaldean (that is, a late) creation.

A civilization as complex as the Sumerian one implied the establishment of calendric rules. We have already spoken of the Babylonian year of 360 days and of the *nychthēmeron* divided into 360 equal parts; that was a very neat mathematical conception. Their calendar, however, was primarily based on the moon. They recognized months of 29 and 30 days, which succeeded one another with some regularity.[47] The average length of twelve lunar months (354 days) is too short for a solar year, while that of thirteen months (384 days) is too long. In order to harmonize the lunar and solar cycles the Babylonians used twelve months but intercalated a thirteenth one when necessary. This must have been done very early, for during the third dynasty of Ur (2294–2187) it had already been recognized that the insertions reoccurred in a cycle of eight years.[48] In one of the letters of Hammurabi to all his governors he orders the insertion of such a month. This Babylonian calendar was the model for the Jewish ones, as well as the Greek and Roman before the introduction of the Julian one (45 B.C.). Not only that, but it still influences the ecclesiastical calendar of our own days.[49]

On the other hand, an initiative often ascribed to the Babylonians is certainly of a much later date; I am referring to the invention of the week. Of course, a lunar month invites subdivision into shorter periods, punctuated by the phases of the moon. The Babylonians attached a special importance to the 7th, 14th, 21st, and 28th days of the month; for example, on those days certain things were forbidden to the king. Thus, they had subdivided the month into periods of seven days, but those Babylonian weeks were not continuous like ours and the first day of each month was the first day of a week. The invention of our seven-day continuous week (the weeks following each other independently of month and year) and of the astral names given to each day (curiously preserved by the Catholic Church in the Western European languages) was not completed until the last centuries preceding Christ's birth; it was due to a combination of the Jewish

[46] The best example of a ziggurat is the Sumerian one of Ur, the excavation of which, begun in 1854, was completed in 1933. For a full description, see Sir Leonard Woolley, *Ur excavations*. Vol. 5. *The ziggurat and its surroundings* (folio, 164 pp., 89 pls.; Oxford: Clarendon Press, 1939). The reconstruction of that ziggurat is taken with kind permission from that work.

[47] A strict alternation of 29- and 30-day months would have led to discrepancies between the a priori calendar and the observations of the first crescent; hence it was necessary sometimes to break the alternation.

[48] The *octaetēris*, the introduction of which in the Greek calendar was ascribed to Cleostratos (VI B.C.) and also to Eudoxus (IV–1 B.C.).

[49] The *epactai* (*hēmerai*) or intercalary days are the days denoting the excess of the solar year over 12 lunar months (365–354 = 11 days). The epact of a given year is the moon's age at the beginning of it; it increases by about 11 days year by year

sabbath and the story of the creation of the world (Exodus 20:11) with Egyptian hours and Chaldean astrology. That is a very complex and interesting bit of folklore, rather than science, which we shall tell in the next volume.[50]

It is typical of the Babylonian spirit that they did not think of equal continuous weeks, which are superfluous for astronomical purposes, but introduced the fundamental idea of equal hours, without which astronomic computations become hopeless. Our own hours are derived from the Babylonian *nychthēmeron* for their equality and from the Egyptian calendar for their number.

The most remarkable observations of the Babylonians concern Venus. Some Venus tables compiled during the rule of Ammiṣaduga (the tenth king of the Amurru dynasty, Hammurabi being the sixth king of the same dynasty) have come down to us and have exercised the ingenuity of many scholars.[51] The Babylonian astronomers of Ammiṣaduga's time (*c.* 1921–1901) noted the first and last appearance of Venus at sunset and sunrise and the length of its disappearance, adding predictions suitable to each case. For example (Fig. 19),

> If on the 21st of Ab Venus disappeared in the east, remaining absent in the sky for two months and 11 days, and in the month Arakhsamna on the 2nd day Venus was seen in the west, there will be rains in the land; desolation will be wrought. [7th year]
>
> If on the 25th of Tammuz Venus disappeared in the west, for 7 days remaining absent in the sky, and on the 2nd of Ab Venus was seen in the east, there will be rains in the land; desolation will be wrought. [8th year]
>
> If in the month Adar on the 25th day, Venus disappeared in the east, . . . [8th+9th year]

In those tables the months during which Venus is invisible are counted as of 30 days each. The Babylonian astronomers knew the synodic period of Venus (584 days) and were aware of the period of eight years during which Venus reappears five times in the same places (as seen from the earth).[52]

The early Babylonians made many other observations. They knew that the moon and planets do not move far away in latitude from the Sun's path (the ecliptic) and they observed the relative positions of planets and stars in that narrow zone (the zodiac); they had estimated the synodic period of Mercury with an error of only 5 days.[53] However, their main contribution was of a more general nature. They were really the founders of scientific astronomy; the admirable results obtained later by Chaldean and Greek astronomers were made possible because of

[50] However, I should justify at once my reference to "Egyptian hours." The fact that the order of the days is different from the natural order of the planets can be accounted for only on the basis that each hour of the day was dominated by a different planet. Each day was named after the planet dominating its first hour. The explanation implies a rotation of 168 hours per week, that is, the division of the day into 24 hours in the Egyptian manner, not into 12 hours in the Babylonian one. For further details, see Francis Henry Colson, *The week* (134 pp.; Cambridge, 1926).

[51] The latest and fullest translation and discussion of those tablets will be found in Stephen Langdon and J. K. Fotheringham, *The Venus tablets of Ammizaduga. A solution of Babylonian chronology by means of the Venus observations of the first dynasty. With tables for computation by Carl Schoch* (126 pp., folio; Oxford, 1928). The examples quoted below are taken from that book (p. 7).

[52] The synodic period of Venus is exactly 583.921 days. Hence from superior to inferior conjunction the mean interval is 292 days, so that in each year there are usually one superior and one inferior conjunction. Eight Julian years = 2922 days; 5 synodic periods of Venus = 2919.6 days, or 2.4 days less; 8 Babylonian lunisolar years including 3 intercalary months = 2923.5 days, that is, 4 days more than 5 synodic periods. See Langdon and Fotheringham, *The Venus tablets of Ammizaduga*, p. 105.

[53] Their determination being 111 days instead of 115.87, according to Ernst F. Weidner, *Alter und Bedeutung der babylonischen Astronomie* (Leipzig, 1914), p. 13.

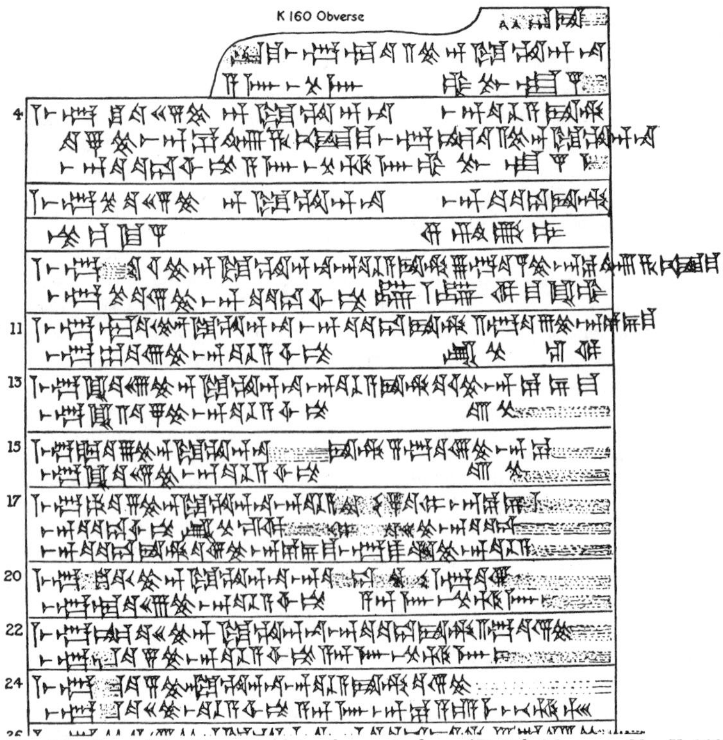

Fig. 19. One of the Venus tablets of Ammiṣaduga (British Museum, No. K 160; obverse, upper half). [From Stephen Langdon and John Knight Fotheringham, *The Venus tablets of Ammizaduga* (London: Oxford University Press, 1928).] For a translation, see the text.

the Babylonian foundation. It is probable that they influenced other oriental people — Iranian, Hindu, Chinese — but these are very moot questions, which are still too far from any convincing solution to be discussed here.[54]

TECHNOLOGY

The Sumerian culture was from its beginnings, as far as known to us, typical of the Copper Age. In the course of time pure copper was replaced by harder alloys with lead and antimony,[55] also with tin, that is, by various bronzes. In Hammurabi's time iron was still a rarity and it did not come into common use until

[54] Carl Bezold, *Sze-ma Ts'ien und die Babylonische Astrologie* (Hirth's Festschrift; Berlin, 1920, pp. 42–49). Apropos of Ssŭ-ma Ch'ien (II–2 B.C.), Bezold concludes that the Chinese became acquainted with Babylonian astrology probably before 523 B.C. Meissner, *Babylonien und Assyrien*, vol. 2, p. 398. Léopold de Saussure, *Les origines de l'astronomie chinoise* (594 p.; Paris, 1930) [*Isis* 17, 267–271 (1932), 27, 291–293 (1937)]. Ungnad, "China und Babylonien," *Reallexikon der Assyriologie*, vol. 2, pp. 91–93 (1938).

[55] Sumerian women were acquainted probably as early as their Egyptian sisters with stibnite ($Sb_2 S_3$), which they used as an eye cosmetic and collyrium. It is not difficult to obtain pure antimony from the trisulfide.

a thousand years later; the Assyrian king Sargon II (721–705) stored away in his place of Khorsabad lumps of wrought iron (a mass of *c.* 160,000 kg of excellent iron was excavated there!) — but we must not anticipate. The Sumerian goldsmiths handled gold, silver, lapis lazuli, ivory, etc. with astounding virtuosity.[56]

The plains of Mesopotamia are fertile only if properly irrigated. The greatest technical achievement of the Sumerians was the digging of a network of canals, which served to irrigate the land and were also used as means of communication and transportation between the different parts of the country. As political integration was gradually improved, the scope of these undertakings was increased. The burden of them was assumed by the state; the early rulers of Lagash were as proud of their irrigation projects as they were of their conquests. Traces of those early canals can be seen from the air, but it is not always easy or possible to distinguish them from those left by the capricious Euphrates when it changed its course. Therefore, archaeologists disagree on the details of the map, though they all recognize the vastness of those undertakings. Documentary evidence concerning them is found in many letters addressed by Hammurabi to provincial governors. Indeed, it was not enough to dig the canals, it was necessary to keep them in a proper state of repair and to clean them out at regular intervals. The silt dug out from the canal bed was piled up upon the banks, which rose higher every year; when the banks grew too high it became simpler to cut a new canal. Travelers in lower Mesopotamia often come across the remains of such embankments. In many cases water had to be raised from the canals to the higher level of the land; that was done by means of a shadoof such as is still used in Egypt today or some other contrivance. A discussion of this and of other agricultural tools, such as the plow, and of ships and chariots would take too much space, for the history of each tool would easily extend to a separate chapter.

The Sumerians and their Semitic collaborators and successors were great business men. It took business minds, if not to understand the need of irrigation, at least to organize it on a national scale. The main products of the country were agricultural – plants such as grain and dates, herds of domesticated animals producing meat, leather, and wool. The commercial methods are illustrated by a very large number of clay tablets — which are contracts duly sealed by both parties, pay rolls, inventories, accounts — and by a number of special regulations in the Code of Hammurabi, to which we shall return presently. In spite of their commercial sophistication, neither the Sumerians nor their successors invented the use of currency; the idea did not occur to them. They used pieces of precious metals for barter against other commodities; the earliest coins were made only in the seventh century in Assyria or in Lydia, and the Greek cities of Western Asia were quick to see the value of the invention and developed it splendidly.[57] It is not correct to say that the Greeks developed it because of their commercial needs or to imply that such needs did not exist before, for Babylonian trade was

[56] See examples reproduced by C. Leonard Woolley, *The development of Sumerian art* (New York: Scribner, 1935).

[57] In Babylonian times, if not before, some pieces of metal bore an official stamp indicating their weight; this obviated the need of repeated weighings for each transaction. Such stamped pieces of metal constitute a transition to proper coinage; see Meissner, *Babylonien und Assyrien*, vol. 1, p. 356. In the time of the Assyrian king Sennacherib (705–681) there is a reference to half-shekel pieces, called "Ishtar heads"; see A. T. Olmstead, *History of Assyria* (New York, 1923), p. 321. This brings us to the time of the Lydian invention.

extensive and complex enough to justify the innovation. The Sumerians and Babylonians did not think of it, that is all. It is rather amusing to think that there were so-called moneylenders among them, advancing "money" (or rather pieces of metal or other goods) at a high rate of interest, but there was no money *stricto sensu*. Needs are not always necessary and are never sufficient conditions for the creation of inventions.

On the other hand, the Sumerians' masterly solution of the problem of weights and measures has already been mentioned. In this field they surpassed all the other peoples of antiquity and in some respects they were not themselves surpassed until modern times. This is one of the most astounding anticipations in the whole history of the human mind.

Many actual weights have been discovered, though the earliest datable ones are by no means as early as one might expect from the cuneiform documents. Some weights were in the form of lions and ducks. The earliest duck weights are inscribed to Nabū-shum-lībur (1047–1039) and Erība-Marduk (802–763); the earliest lion weights are Assyrian ones of the eleventh century. Though the use of weights implies that of scales, no Mesopotamian scales, or representations of scales, have yet come down to us.[58]

We would expect the early people of Mesopotamia to engage in a variety of industries which more sophisticated ages would call "chemical industries," and which were that indeed, except for the lack of chemical consciousness. The most important of those industries would be the manufacture of pottery, glazes, and glass; then one might add the tincture of metals and the making of paints or dyes, drugs and remedies, soaps and cosmetics, perfumes and incense, beer and other fermented beverages. Such industries, or at least some of them, would naturally develop in any country, as soon as enough stability made them possible; the development would be natural and inarticulate. Artisans engaged in them have little time to become literate, and none to write; there is no reason why they should reveal their successful tricks and publish their secrets, even if they were able to do so and had time to spare.

However, there has come down to us an extraordinary chemical text dating from the rule of Gulkishar (1690–1636), sixth king of the first dynasty of the Sea-Land. That document, originating in Lower Mesopotamia in the seventeenth century, is in the form of a small cuneiform tablet preserved in the British Museum (Fig. 20).[59] Not only is it the earliest known record of actual recipes for the making of glazes but the next ones appeared only a thousand years later.

[58] The Accadian verb "to weigh" is shaqālu, which seems to go back to proto-Semitic, for it is found in all Semitic languages (Arabic, thaqala; Hebrěw, sheqel); from it comes "shekel," unless the verb comes from the noun. Since payments were made in gold, silver, or bronze, which had to be weighed, in Assyrian and Aramaic the verb also comes to mean "to pay." There are words in Assyrian and Sumerian for scales; these words generally occur in a dual form, as they also do in Hebrew, referring to the two scale pans. (Information kindly given by my Harvard colleague, Robert H. Pfeiffer, 26 September 1944). The Egyptian idea of scales of judgment occurs in Job 31:6.

[59] This tablet is of baked clay, $3\frac{1}{4} \times 2\frac{1}{16}$ inches, written on both sides; B.M. No. 120960. It was edited and translated by C. J. Gadd and R. Campbell Thompson, "A middle-Babylonian chemical text," *Iraq* 3, 87–96 (1936), 1 pl. [*Isis* 26, 538 (1936)]. For Babylonian chemistry see also R. Campbell Thompson, *A dictionary of Assyrian chemistry and geology* (Oxford: Clarendon Press, 1936), pp. xxiii, 197 [*Isis* 26, 477–480 (1936)]; "Survey of the chemistry of Assyria in the VIIth century B.C.," *Ambix* 2, 3–16 (1938). Ernst Darmstaedter, "Chemie," *Reallexikon der Assyriologie*, vol. 2 (1938), pp. 88–91. Thompson and Darmstaedter deal mainly with Assyrian (VIIth century) chemistry, very little with the earlier Babylonian efforts.

It describes the making of a glaze with copper and lead for pots, and the making of a green body with clay mixed with verdigris. Apparently, the author was torn between the desire of publishing his inventions and of protecting his own interests, between pride and jealousy. He solved the dilemma by describing his results in cryptic language. In this he was very different from his Assyrian successors of a thousand years later, but he was a forerunner of the medieval (and later) alchemists, who camouflaged their ideas or their lack of ideas with the most obscure jargon they could think of. On account of the singularity of that text we reproduce Gadd and Thompson's translation of it. We quote it completely though without the long annotations which are indispensable for a full appreciation of it but would not interest our readers.

To a mina of *zukû*-glass (thou shalt add) 10 shekels of lead, 15 shekels of copper, half (a shekel) of saltpetre, half (a shekel) of lime: thou shalt put (it) down into the kiln, (and) shalt take out "copper of lead."

To a mina of *zukû*-glass (thou shalt add) ⅙th (mina = 10 shekels) of lead, 14 (shekels) of copper, 2 shekels of lime, a shekel of saltpetre: thou shalt put (it) down into the kiln, (and) shalt take out "Akkadian copper."

(Thou shalt) green the clay (??) and (?) in vinegar and copper thou shalt keep it. At the third (day) of thy keeping it will deposit a "bloom," and thou shalt take (it) out. Thou shalt continually pour (it) off and it will dry and thou shalt make it. If it is (like) marble, be not troubled. "Akkadian" (copper) and (that of) lead thou shalt take in equal parts, and triturate it together. After thou hast melted it together, into a mina of the melt a shekel and and a half of *zukû*-glass, 7½ grains of saltpetre, 7½ grains of copper, 7½ grains of lead thou shalt triturate together and thou shalt melt and keep it (so) (for) one (day?) and shalt take it out and cool (it) . . .

[Not translated.]

Thou shalt pour, and shalt lay him in a stone sarcophagus (?).

[Not translated.]

Thou shalt dip and lift it up, and bake (?) it (?), (and) cool (it). Thou shalt look (at it) and if the glaze (is like) marble be

Fig. 20. A Babylonian text of the seventeenth century, explaining the making of glazes (British Museum tablet No. 120960, obverse and reverse). [Reproduced with kind permission of the Trustees of the British Museum and of *Iraq* 3, pl. 4, 1936.]

not troubled: thou shalt again put (?) it back into the kiln (and) take out . . . (?)

If thou takest out . . . (?) thou shalt again put (?) it back into the kiln: the "copper-clay" will become "copper-gum." Into a mina and 2 shekels of *zukû*-glass (put) 15 grains of copper, 15 grain of lead, 15 grains of saltpetre; lime thou shalt not bring near. First examine it (and then) in a winepourer of old skin thou shalt put (it) and shalt keep it.

Property of Liballiṭ(?)-Marduk son of Uššur-an-Marduk, priest of Marduk, a man of Babylon. Month of Ṭebet, 24th day, year after Gulkishar the king.

GEOGRAPHY

A good many geographic documents have come down to us, most of which concern what we call today "historical geography." They may be enumerations of countries, as in the list of Sargon's conquests, or geographical glossaries (Sumerian, Accadian) for the use of scribes, or itineraries, or documents for administrative purposes, like the list of places with which the temple of Lagash had dealings. As soon as a ruler governs a country of sufficient size, he needs various geographical tools to direct the work of his officers.

Another kind of geographic knowledge springs out of cosmography. The Babylonians (or let us say, some of them, very few) were anxious to know where their land was located with reference to other lands, or to the whole earth, or even to the universe: heaven and earth. A few tablets answer such intellectual needs. The Babylonians conceived the earth as an overturned gufa [60] floating on the ocean. The earth has seven floors and the whole of it is divided into four sectors, which in an early document are named after the four nearest countries, Elam in the south, Accad in the north, Subartu (later Assyria) in the east, and Amurru (Syria) in the west. In the course of time the business of war and peace acquainted the Babylonians with more distant countries, in particular Arabia and Egypt. The earth to their minds is the counterpart or image of heaven. Their gods dwell on the top of a mountain, the departed spirits in a kind of underworld (like the Egyptian Ṭuat, the Hebrew Sheol, and the Greek Hadēs).

To come back from dreams to reality, the best proof of their geographic ability is given by various maps. We reproduce two of them. The first (Fig. 21) is a map of the Sumerian city of Nippur, which is so faithful that it helped the archaeologists to conduct their excavations. The other (Fig. 22) is a map of the world with a descriptive commentary. Babylonia, Assyria, and neighboring districts are represented as a circular plain surrounded by the Persian Gulf. Near the center of this plain is marked the city of Babylon (every nation conceived its capital city as the hub or navel of the world) and to one side of it is the land of Assyria. The positions of other cities are indicated by little circles. The triangles resting on the circular zone and outside of it refer to foreign countries. This is very vague, to be sure, but not much more so than some Arabic maps or Christian *mappae mundi*.

NATURAL HISTORY

Familiarity of the Babylonians with a relatively large number of plants and animals is proved by various documents. Father Scheil, investigating tablets of the time of Samsuiluna, last ruler (1912–1901) of the dynasty of Larsa, was able to

[60] A gufa is a round boat made of wickerwork used in Mesopotamia from very early times unto our own days. The word occurs in the Arabic vernacular under the form quffa.

Fig. 21. Fragment of a Sumerian tablet containing a plan of Nippur. Babylonian Expedition of the University of Pennsylvania. [From H. V. Hilprecht, *Explorations in Bible lands during the nineteenth century* (Philadelphia, 1903), p. 518.]

write a paper enumerating the fish sold at the market of Larsa. Some thirty kinds were sold, twelve kinds by the piece, the others by the basketful. The prices quoted for the first are difficult to compare; those relative to the latter can be divided into six groups, the cheapest costing one-tenth as much as the dearest. The people who flourished in Larsa at the end of the twentieth century knew their fish! [61] The main source of names which may interest the naturalist is provided by the glossaries. For example, some tablets enumerate hundreds of animals; the names are written in cuneiform script in two columns, the first giving the Sumerian term and the other its Accadian equivalent.[62] Similar tablets enumerate various plants and medical tablets refer to many others. Some 250 plants have been distinguished, but relatively few identified with any certainty. That is, Assyriologists know that a certain name in Sumerian and its Accadian counterpart represent a definite plant, but they cannot be sure which plant is meant. Some of the names that we use today are derived from Sumerian names, but even in such cases it does not follow that the same plant was meant by the Sumerians, the Assyrians, and ourselves. Here are a few such names (the cuneiform word being quoted in parentheses): cassia (kasū), chicory (kukru), cumin (kamūnu), crocus (kurkānu), hyssop (zūpu), myrrh (murru), nard (lardu).[63]

Some of the lists evidence a kind of crude classification. For example, the animals are divided into fish (and other animals living in the water), articulata, serpents, birds, and quadrupeds. The larger groups are sometimes subdivided into smaller ones, the dogs, hyenas (?), and lions on one hand, the asses, horses, and camels on the other. The plants are divided into trees, potherbs, spices and drugs, cereals, etc. The plants bearing fruits that look somewhat alike, such as the fig, the apple, and the pomegranate, were put together.

It is probable that the early Babylonians had already recognized the sexuality of date palms. Such knowledge seems confirmed by Assyrian monuments of the

[61] V. Scheil, "Sur le marché aux poissons de Larsa," *Rev. d'Assyriologie 15*, 183–194 (1918).

[62] Benno Landsberger and Ingo Krumbiegel, *Die Fauna des alten Mesopotamien nach der 14. Tafel der Serie Ḫar-ra = ḫubullu* (158 pp.; Leipzig: Hirzel, 1934).

[63] As quoted by E. A. Speiser, *Some sources of intellectual and social progress in the ancient Near East* (Studies in the history of culture; Menasha, Wisconsin: American Council of Learned Societies, 1942), pp. 51–62, 55. Reginald Campbell Thompson, *The Assyrian herbal* (322 p.; London, 1924) [*Isis 8*, 506–508 (1926)]. Thompson rejects some of the identifications quoted above.

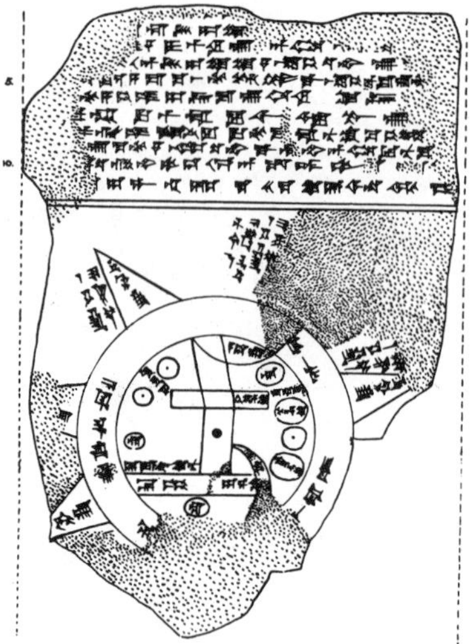

Fig. 22. Babylonian map of the world, explained in the text. [From *Cuneiform texts from Babylonian tablets,* part XXII (London, 1906), pl. 48. Reproduced with permission of the Trustees of the British Museum.]

ninth century B.C. but is presumably much older.[64] The events leading to that discovery may be reconstructed as follows. Date palms drink a great deal; as the Arabs put it, in order to thrive the palms must have their heads in the fire and their feet in water. When the supply of water was limited it was necessary to restrict the number of trees. Some farmer may have had the bright idea of pulling out all the "sterile" date palms (that is, the male ones), and if he did that thoroughly enough he had a painful awakening; for now no dates whatever were available. The "sterile" trees were necessary after all; without them the other trees would not bear any fruit. Later it was discovered that in order to insure fructification it was advisable to climb up the "sterile" tree, take its flowers and carry them up close to the flowers of the "fertile" trees — in fact, attach them to the latter. That very laborious procedure has been observed not only in Mesopotamia but in every country where dates are cultivated. Its discovery is immemorial, and in a country sophisticated as early as Mesopotamia we may assume that it goes back to the earliest times. The chain of experiments that we have outlined may have taken many centuries, or many millennia, but it was already completed in Babylonia, if not in Sumer. Of course, this does not mean that the sexuality of date palms was understood as such, though there is no reason why intelligent men should not have made comparisons between the bringing together of (what we would call) male and female flowers on one hand, and the congress of animals or of men on the other. This assumption is encouraged (though not proved) by the ascription of sexual terms to various plants; the Assyrians applied the term male to cypresses and mandrakes and the terms male and female to liquid amber.[65] It is highly probable that the Babylonians did not speak of the sexuality

[64] G. Sarton, "Artificial fertilization of date-palms in the time of Ashur-nasir-pal, B.C. 885–60," *Isis 21,* 8-13, 4 pl. (1934); *23,* 245-50, 251-52 (1935); *26,* 95-98 (1936).

[65] Thompson, *Assyrian herbal.* Of course, sexual terms may be given to plants because of analogies, as Greek *orchis* and English orchis (testicle), or Arabic khiṣyun or khiṣyatun.

of date trees except perhaps as a poetic metaphor, but they fully realized the necessity of bringing together the flowers of the sterile and the fertile trees in order to secure the fertilization of the latter. This is the outstanding example of application preceding theory; in this case the application was already completed by 2000 B.C., if not much earlier, the theory formulated only in A.D. 1694!

Frequent references have been made to two kings, Hammurabi and Ammisaduga, who were respectively the 6th and the 10th of the first dynasty of Babylon (or the Amorite dynasty). We often think of that age as the golden age of Babylonia, but though it lasted three centuries, that was only a beginning. It was followed by the first dynasty of the Sea-Land, which lasted 368 years, and then by the Kashshū (or Kassite) dynasty, which lasted nearly six centuries (1746–1171) and reëstablished the capital in Babylon. That dynasty came probably from the north and was connected with the Mitanni kings of Upper Mesopotamia. The ruling caste of Mitanni was apparently of Indo-Iranian origin, and it used horses.

It is true that individual horses were already known to Hammurabi but the "asses of the mountain," as they were known to the early Babylonians, were still a great rarity in his time. Under the Kassite dynasty they were introduced in larger numbers and were even exported to Egypt. Indeed, we learn from some of the 'Amārna letters that a Kassite king gave to the Pharaoh a present of lapis lazuli, five yokes of horses, and five wooden chariots. The craftsmen of Babylonia needed gold, and her more valuable exports, exchanged for Nubian gold, were lapis lazuli and horses.

The most astonishing of the Hittite documents discovered in the royal archives of Boghāzköy is a treatise on the training of horses written *c.* 1360 by a man called Kikkulish (or Kikkuli). It is written in cuneiform script but in the Hittite language, and its philological interest is increased by the presence in it of many Indian terms.[66] This text is so curious that a brief analysis of it may not be out of place.

The training, extending to six months, is described day by day and almost hour by hour. The best horses are selected after a trial run. Then they are made to fast and sweat under blankets in order to be rid of excess weight. They are trained to walk and to run for longer and longer distances, galloping or ambling. Special precautions are taken for the feeding and watering at regular times in prescribed quantities. Chopped straw should be mixed with the feed, probably to induce better chewing. Just imagine a treatise of that practical kind composed in the fourteenth century and remember that the earliest Greek hippiatrica did not appear until seventeen centuries later![67] The Hittite book could not have been composed in Anatolia much earlier than it was, for it almost coincides with the beginning of horse culture in Western Asia; yet we may be sure that it embodies a very old Indo-European tradition. It was destined to fall into oblivion together with the Hittite language and the kingdom of Mitanni; the latter vanished in the

[66] Bedřich Hrozný, "L'entraînement des chevaux chez les anciens Indo-Européens d'après un texte mitannien-hittite provenant du 14e siècle av. J.C.," *Archiv Orientální* 3, 431–461 (Prague, 1931) [*Isis* 25, 256 (1936)]. This includes a French translation of the first tablet out of five; a summary of the training is given on pp. 437–438. The date 1360 is Hrozný's tentative dating; see p. 433.

[67] Apsyrtos (IV-1), Hieroclēs (IV-2).

first half of the thirteenth century B.C. However, the Hittite methods of training horses were probably imitated by the Assyrians, later by the Medes and the Persians, and thus transmitted to the Western world.

THE CODE OF HAMMURABI

In 1901–02 the French archaeological mission sent to Persia under the direction of Jacques de Morgan discovered on the acropolis of Susa one of the most impressive monuments of antiquity. It is a block of black diorite, somewhat regularized

Fig. 23. The code of Hammurabi. The code was written on both sides of a diorite stela 245 cm high. We reproduce only the top of it, a bas-relief, which shows Hammurabi being charged by the god of justice, the Sun-God Shamash, to write his code, or else offering his code to the Sun-God. [Louvre.]

and very well polished, 2.45 m high, now preserved in the Louvre Museum.[68] At the top of the front part is a low relief representing the Sun God (Shamash) offering a code of laws to King Hammurabi (Fig. 23). The code itself is engraved below the bas-relief and also on the reverse. The monument had been originally erected in Sippar (Babylonia) but taken away as war booty by an Elamite conqueror, perhaps Shutruk-Nakhunte (*c.* 1200–1100 B.C.) and reërected by him in his own capital. Parts of the code had been erased, probably in order to make room for an inscription glorifying the conqueror, but it has been possible to reconstruct most of the lacunae, because there existed copies of the code on clay tablets and perhaps on other stones.[69]

This is the earliest code of laws that has come down to us in sufficiently complete shape, but it is far from primitive. It implies an already long evolution of legal thought.[70] It illustrates splendidly the legal aspect of the human genius, an aspect of it of such indispensability for the building up of any civilization, that the historian of science, no matter how much he tries to restrict his own field, is obliged to devote some attention to it.

Assyriologists are not yet agreed as to the date of Hammurabi, the crucial date of Babylonian history. It was first believed to be before 2000 and even as early as 2225; Meissner placed it at 1955 (the rule extending from 1955 to 1913). The present tendency is to reduce that number, but whether Hammurabi ruled in the twentieth century or at the end of the eighteenth century,[71] his code remains a monument of prodigious antiquity.

The code proper includes 282 articles. It is preceded by an invocation in which the king explains his greatness and his excellent purpose. He codified the existing laws in order "to cause justice to prevail in the land, to destroy the wicked and the evil, that the strong might not oppress the weak, to rise like the Sun over the black headed (people) and to light up the land." After having set forth all of his virtues and glory and enumerated his military and peaceful achievements, he concludes "When Marduk commissioned me to guide the people aright, to direct the land, I established law and justice in the language of the land, thereby promoting the welfare of the people." At the end of the code an epilogue restates similar ideas – "I, Hammurabi, the perfect king, was not careless or neglectful of the black-headed people . . ." – and calls down a rich variety of curses upon the people who would be reckless enough to alter or disobey his laws. It is clear that the great king did not believe in hiding his light under a bushel; it is equally

[68] On account of its tremendous importance many casts of it have been made which can be seen in the leading museums of archaeology. One of those casts is available in the Semitic Museum of Harvard University.

[69] The text was first published by Father Scheil in the *Mémoires de la Délégation en Perse* (Paris, 1902), vol. 4. An abundant literature has been devoted to it. The best English translation, by Theophile J. Meek, is included in Pritchard, *Ancient Near Eastern texts*, pp. 163–180. Quotations in this chapter are made from that translation with kind permission of the Princeton University Press. See also Edouard Cuq, *Etudes sur le droit babylonien, les lois assyriennes et les lois hittites* (530 pp.; Paris, 1929) [*Isis 15*, 268 (1931)]. Every history of Babylonia as well as every history of ancient law gives of necessity considerable space to this code.

[70] The Lipit-Ishtar code in Sumerian was certainly older than Hammurabi's in Accadian, perhaps two centuries older. Francis R. Steele, *The code of Lipit-Ishtar* (28 p., 6 fig.; Philadelphia: University of Pennsylvania Press, 1948) [*Isis 41*, 374 (1950)]. The most convenient survey of the early codes is in Pritchard, *Ancient Near Eastern texts*, pp. 159–223.

[71] According to the most recent calculations, Hammurabi ruled for 43 years, from 1728 to 1686; see Pritchard, *Ancient Near Eastern texts*, p. 163.

clear that he did not consider himself an innovator but rather the guardian and fulfiller of old traditions.

The laws might be roughly divided into six groups, concerning movable property, landed property, business, family, injuries, labor. The Babylonians were capitalists and businessmen; their society might be theocratic and their minds chockfull of magical fancies, but when material interests were at stake they looked at things in a very concrete and hard way. The code is on the whole very rational. We are not competent to discuss the details of it; it must suffice to sketch out rapidly some its contents. It deals with larceny, which is punished differently according to the place where it occurred — temple, palace, or private house; with abduction of minors or slaves, robbery with armed forces, arson; leased estates, estates without owners, damages to fields and gardens; torts and business conflicts; debts and deposits; regulations concerning taverns; marriage, adultery, abandonment, divorce, rights of widows, relations with concubines and slave girls; rights of children, adoption. The last part of the code deals with professional duties and crimes.

Though written in Accadian the code was partly derived from Sumerian usage which it sometimes abrogated and sometimes continued. The differences can be appreciated, because Sumerian laws have come down to us in the form of tablets, preserved in Philadelphia. On the other hand the Babylonian code was imitated and partly followed by the Hittites (fourteenth or thirteenth centuries), by the Assyrians (before the ninth century), and by the Hebrews. Comparison of these Oriental codes is very instructive because of the light it throws on the psychology of the nations involved, but it would require considerable space and is not our present task.

It is obvious that the qualities which we ascribe to the Romans because of their juridical achievements were already shared by the Babylonians some two millennia before them. In particular, the Babylonians had already imagined a series of fictions without which the law cannot be properly formulated. On the other hand, it must be admitted that much in the Babylonian code (and in other codes of the Ancient East) was hard and cruel, especially the *lex talionis* ("eye for eye, tooth for tooth, hand for hand, foot for foot," Exodus 21:24), which was the general guide in the reparation for injuries. Some contradictions are due to the fact that Hammurabi was legislating for a nation which, unified though it was on the surface, was extremely complex; he was obliged to combine and harmonize divergent traditions. Taking everything into account — even the primitive wish for exact retribution and the idea that injuries vary in gravity with the social rank of the victims — the king (or his legal adviser) did the work remarkably well. The code of Hammurabi is one of the outstanding landmarks in the history of mankind.

MEDICINE [72]

The study of Babylonian medicine is far more difficult than that of Egyptian medicine and the results are less certain. In the case of Egypt we have a series of large papyri datable within a few centuries, and an analysis of the two longest ones, the Smith and Ebers papyri, would suffice to teach the essentials. In the

[72] Georges Contenau, *La médecine en Assyrie et Babylonie* (228 pp., ill.; Paris: Maloine, 1938) [*Isis 31*, 99–101 (1939–40)], with bibliography, pp. 51–52, 207–227.

case of Babylonia we have to depend very largely upon documents of later date, chiefly those found in the library of Ashshurbānipal (now in the British Museum). That Assyrian king ruled in the seventh century (688–626), yet there can be no doubt that the knowledge compiled by his Accadian scribes is very largely of Babylonian and even Sumerian origin, and that the essence of it may be traced back to the third millennium.[73] That does not make their knowledge older than the Egyptian, which can also be traced back to much earlier times than those of the papyri that have come down to us.

We may assume that in both cases, Babylonia and Egypt, the bulk of the medical knowledge transmitted to us dates back to the third millennium, but there is the great difference that the actual texts were written in Egypt about the seventeenth and sixteenth centuries and in Assyria a thousand years later.

The Sumerian origin of most of the Assyrian documents is obvious enough. They are actually written in Sumerian, even in old Sumerian, with a maximum of pictographic signs.[74] The Assyrian physicians of the seventh century B.C. used Sumerian formulas, even as the French physicians of the seventeenth century used Latin ones, and for the same reason — tradition. Sumerian (or Latin) was a nobler language and it had the advantage of being restricted to an elite; the canaille could not understand it and admired the physicians more because of that very fact (*omne ignotum pro magnifico*); the physicians were aware of the prestige that accrued to them on account of the obscurity of their jargon and hence continued to use it. (Some people are still playing that game today.) Not only are the medical tablets written in Sumerian but they are generally very brief, bare statements without explanations. This suggests that the medical teaching was largely oral; the physician's knowledge was transmitted from master to disciple, perhaps from father to son. The tablets were used not so much for study as for recapitulation and remembrance, like cribs or ponies.

Moreover, while the main papyri give us large collections of facts, comparable to our textbooks, the tablets are *disjecta membra*. There are a few exceptions to this rule, the most notable being the so-called "tablet of Constantinople," which comes a little closer than most tablets to being a complete medical text, though it is very short.[75] It deals with the troubles caused by scorpion stings and the means — purely external ones — of curing them; the remedies combine drugs with incantations.

The most impressive document concerning Babylonian medicine is the code of Hammurabi described in the previous section. The code does not speak of physicians proper, but only of surgeons. It is possible that the physician, using

[73] An Accadian treatise on medical diagnosis and prognosis has been prepared by René Labat, *Traité akkadien de diagnostics et pronostics médicaux* (297 pp., album of 68 pl.; Collection de travaux de l'Académie internationale d'histoire des sciences, no. 7, Paris, 1951). It was my duty and privilege to examine the manuscript of it (July 1951). It has been incompletely preserved on 40 tablets which date back to various times, the earliest that of King Marduk-apal-iddin (722–711), the latest the eleventh year of Artaxerxes (453), yet all represent older Babylonian traditions. The treatise is divided into five parts: 1. When the exorcist goes to the patient's house. 2. When you come near the patient. 3. If, being ill during a day. . . 4. When you take the patient's hand. 5. When the woman is pregnant, the top of her forehead is yellow.

[74] That is, with more ideographic than phonetic signs. Examples given by Contenau, *La médecine en Assyrie*, p. 178.

[75] French translation in Contenau, *La médecine en Assyrie*, pp. 190–193. There are various texts of the same family.

Sumerian formulas, was a sacred personage not amenable to ordinary law; the surgeon, however, was a kind of artisan who was well rewarded if he did well, but punished if he failed. Various articles of the code explain that. It is worth while to quote them, not only because they are the earliest medical laws in existence, but also because they throw a revealing light on Babylonian culture in general.

215: If a physician performed a major operation on a seignior with a bronze lancet and has saved the seignior's life, or he opened up the eye-socket of a seignior with a bronze lancet and has saved the seignior's eye, he shall receive ten shekels of silver.

216: If it was a member of the commonalty, he shall receive five shekels.

217: If it was a seignior's slave, the owner of the slave shall give two shekels of silver to the physician.

218: If a physician performed a major operation on a seignior with a bronze lancet and has caused the seignior's death, or he opened up the eye-socket of a seignior and has destroyed the seignior's eye, they shall cut off his hand.

219: If a physician performed a major operation on a commoner's slave with a bronze lancet and has caused (his) death, he shall make good slave for slave.

220: If he opened up his eye-socket with a bronze lancet and has destroyed his eye, he shall pay one-half his value in silver.

221: If a physician has set a seignior's broken bone, or has healed a sprained tendon, the patient shall give five shekels of silver to the physician.

222: If it was a member of the commonalty, he shall give three shekels of silver.

223: If it was a seignior's slave, the owner of the slave shall give two shekels of silver to the physician.

The two following articles concern veterinary medicine:

224: If a veterinary surgeon performed a major operation on either an ox or an ass and has saved (its) life, the owner of the ox or ass shall give to the surgeon one-sixth (shekel) of silver as his fee.

225: If he performed a major operation on an ox or an ass and has caused (its) death, he shall give to the owner of the ox or ass one-fourth its value.

Babylonian medicine is full of incantations and imprecations. The code itself ends with extravagant praise of the righteous king, adjurations to his subjects to obey the perfect code he has given them, and terrible maledictions upon the man wicked and foolish enough to disobey it. Some of those imprecations are medical.

May Ninkarrak, the daughter of Anum, who commands favors for me in Ekur, cause to come upon his members until it overcomes his life, a grievous malady, an evil disease, a dangerous sore, which cannot be cured, which the physician cannot diagnose, which he cannot allay with bandages, and which, like the bite of death, cannot be removed! May he lament the loss of his vigor!

One would not be far from the truth if one called Babylonian medicine theocratic. The gods are the creators of everything good and of everything evil. Diseases are the marks of their inscrutable displeasure. Remedies may serve as palliatives, but the only sure way of curing a sickness is to appease the god who caused it. Hence the physician is a kind of priest. He seems to have been separate from the priest proper but they probably worked together, the priest physician and the physician priest, in order to obtain the patient's return to health. Certain gods were especially concerned with healing and were appealed to more frequently than others. Disease, impurity, sin, were confused in the patient's and the physician's minds. Babylonian medicine was in some respects comparable to Christian Science. Disease was caused by gods, but it might be caused also by demons, or by "the evil eye" [76] or the "animal magnetism" of other men. The power ascribed to demons or witches may seem incompatible with divine power, yet religious beliefs so close to superstition are necessarily inconsistent — and it is not our business here to unravel them.

Granted the divine or demoniac origin of diseases, we cannot expect diagnosis and prognosis to be based on physiological grounds. It is more logical to base them on divination, and in this the Babylonians were extremely consistent — not only they, but also their most distant Sumerian ancestors. One of the antediluvian kings, Emmeduranki, was credited with the discovery of the principles of divination (that is, the means of deducing the gods' intentions from various observations). By the twenty-eighth century Urukagina, king of Lagash, was obliged to punish diviners who were taking unjust fees; this shows that divination was already well established in that early period.[77]

The methods of divination were many, for every aspect of nature, every happening, is susceptible of mantic interpretation. The diviners whom we have just mentioned were practicing divination by oil. When oil was poured upon water, the shapes it assumed in spreading itself indicated the shapes of things to come. One might also observe the flight of birds, or interpret dreams. The circumstances of births were keenly noted, especially those that were exceptional or monstrous. Popular curiosity in dreams and in monsters (calves with six legs and two heads, etc.) is a witness of that immemorial interest, and modern chapbooks on dreams perpetuate methods of hoary antiquity.[78] Babylonian soothsayers also observed the stars, but the astrology transmitted to us by the Romans was a relatively late creation, as is suggested by the common appellation "Chaldean." The favorite method of Babylonian divination and the most interesting to historians of science was the examination of the liver, hepatoscopy; we shall come back to it presently.

The methods of divination dominated Babylonian life; we may assume that they were Babylonian (or rather Sumerian) inventions, but the general divinatory point of view was not exclusive to them. We find it all over the ancient world. As far as the Greco-Roman world is concerned, the reader may consult the monumental work of Auguste Bouché-Leclercq (1842–1923): *Histoire de la divination dans l'antiquité* (4 vols., Paris, 1879–1882), or more simply read Cicero's *De divinatione*.[79] That spirit is still very much alive in the underworld [80] of our time. If the premises of divination are accepted, the methods could not differ essentially from one nation to another. Hence comparisons that have been made between, let us say, Babylonian and Chinese omina do not necessarily prove, even if sundry details agree, that the latter borrowed from the former.[81]

Before considering the *extispicium* (inspection of entrails), or more particularly

[76] That superstition is universal and immemorial; Greek *bascania* = Latin *fascinum* (hence fascination); *maldocchio, iettatura*, etc.; Hebrew *qinah*, meaning envy. F. T. Elworthy, *Encyclopedia of religion and ethics*, vol. 5 (1912), pp. 608–615.

[77] Leonard W. King, *History of Sumer and Akkad* (London, 1910), p. 183.

[78] An Egyptian dreambook of the Twelfth Dynasty was edited by Alan H. Gardiner, *The library of A. Chester Beatty. Description of a hieratic papyrus with a mythological story, love-songs and other miscellaneous texts* (folio, 45 pp., 61 pls.; London, 1931) [*Isis* 25, 476–478 (1936)]. For persistence of interest in monsters see Sebastian Brant's broadside (Basel, 1496 [*Osiris* 5, 119, 171 (1938)], or the side shows of our circuses.

[79] There is an elaborate discussion in the edition of Arthur Stanley Pease (656 pp., Urbana, 1920–1923).

[80] Reference is made to the intellectual underworld, which cuts across all classes and conditions of people.

[81] Meissner suggested that much; *Babylonien und Assyrien*, vol. 2, p. 244.

hepatoscopy, let us ask ourselves how much anatomy the Babylonians knew. Our impression is that their knowledge was rudimentary, even more so than that of the Egyptians. It was derived from the cutting up of the animals, used either to appease the gods or to feed the people, and as far as human anatomy was concerned, from the accidents of war and peace. Lists of names in the glossaries are the only clear indications of detailed knowledge, and those lists are not very long.[82] The most fatidic organs (*exta*) from the Roman point of view were six, to wit, the spleen, the stomach, the kidneys, the heart, the lungs, and above all, the liver. The liver may have been given that supreme importance because of nonanatomic traditions, but that is doubtful; a purely anatomic explanation is far more plausible. The Romans were as deeply impressed by the liver as the Babylonians, and for the same obvious reasons. When a man loses blood he faints, and if the outflow is not stopped he soon dies. Blood is thus easily recognizable as the fluid of life. When a carcass is opened, the liver is by far the most conspicuous organ; it is also the bloody organ par excellence; one sixth of the blood in the human body is to be found in it. Hence it was natural enough to consider it as the essential instrument of life. The Babylonians also recognized the importance of the heart, and gradually they reached the point of considering the heart as the seat of intellect, and the liver as the seat of the emotions and of life itself. Moreover, the shape of the liver, its division by fissures into five lobes, gave ample opportunities for divinatory distinctions. The livers that they examined, or rather interrogated for soothsaying, were generally those of sheep or goats. They gave special names to various parts of the liver, but, assuming that the Assyriologists are sure of the exact meaning of each name, there is no point in discussing the details of those hepatoscopic fancies. Their haruspices or extispices may have become very familiar with the peculiarities of livers, but even that did not make anatomists out of them.

[82] Contenau, *La médecine en Assyrie*, pp. 65–67.

Fig. 24. Babylonian liver model in clay in the British Museum (Bu. 89-4-26.238). From the plate in *Cuneiform texts from Babylonian tablets*, part VI (London, 1898), pl. 1. For the deciphering of the inscriptions, by Theophilus Goldridge Pinches, see that publication, pl. 2–3.

Fig. 25. Hittite liver model in clay in Berlin Museum (VAT 8320). [After Alfred Boissier, *Mantique babylonienne et mantique hittite* (82 pp., 5 pls.; Paris: Geuthner, 1935).]

Babylonian hepatoscopy is represented by a large number of texts (some 640 were already published in 1938) and, what is more remarkable, by many clay models. Two such models are in the British Museum, one of which is particularly clear and is covered with inscriptions (Fig. 24). Other models,[83] found in Boghāzköy, include inscriptions in Hittite as well as Accadian (Fig. 25). Finally, a bronze model (126 mm long) was discovered in the Etruscan site of Piacenza (Fig. 26). It is possible that the mysterious Etruscans brought Babylonian hepatoscopy with them from Western Asia, and later transmitted it to the Romans. These three liver models are good symbols of the transmission of science to very distant places; it is a pity that the science which they symbolized was on such a low intellectual level, but that fact undoubtedly facilitated their transmission. Superstitions deemed useful, and not only useful but supremely useful, are more agile than pure knowledge, which few people are able to appreciate at any time.

The Babylonians did not restrict their attention to the liver, but examined the entrails surrounding that organ, chiefly the intestines.[84]

The main purpose of the Babylonian physician was to appease or circumvent the gods and to drive the devils out of a suffering body; this was done by means of prayers — supplications, imprecations, deprecations — sacrifices, magical rites, and so on. When divination had revealed the nature of the ailment, some magical or antidemoniac drugs might be applied, or else trouble might be averted by the wearing of amulets and talismans. However, when all the documents of that nature have been rejected, there remain not a few that may be said to represent more rational tendencies. Assyriologists (chiefly the late R. Campbell Thompson, 1876–1941) have been able to recognize a number of ailments — of the head (including mental troubles and baldness!), of the eyes, of the ears, of the respiratory and digestive organs, of the muscles, of the anus, for example, a description of hemorrhoids. They have deciphered tablets dealing with pregnancy and child-

[83] I saw in the Louvre (May 1948) some fifteen objects of this kind excavated in Māri (Tell-Hariri) in 1936. They date back to the beginning of the second millennium. See G. Conteneau, *Manuel d'archeologie orientale* (Paris: Picard, 1947) [*Isis 40*, 153 (1949)], pp. 1906–1911.

[84] For hepatoscopy in addition to Bouché-Leclerq and the books referred to in the illustrations apropos of the liver models, see Alfred Boissier, *Mantique babylonienne et mantique hittite* (82 pp., 5 pls.; Paris: Geuthner, 1935). Some 57 hepatoscopic tablets have recently been edited by Albrecht Goetze, *Old Babylonian omen texts* (Yale Oriental Series, Babylonian texts, 10; New Haven: Yale University Press, 1947). These tablets, preserved at Yale since 1913, are undated, but undoubtedly very ancient – some of them pre-Hammurabi. Goetze adds a list of other monuments of the same kind previously published.

Fig. 26. Etruscan liver model in bronze. It represents a sheep's liver; greatest length, 126 cm. It was found in 1877 in a field near Settina and is kept in the Civico Museo of Piacenza. [Reproduced from G. Körte, "Die Bronzeleber von Piacenza," *Mitt. kgl. deut. arch. Inst., Rom 20*, 348 (1906), pl. xii.]

birth, and with genital troubles, or describing treatments — the remedy being laid upon the suffering part, or introduced into the mouth or the anus. Some definite herbs and other drugs have been tentatively identified. The "scientific" prescription is generally followed by an incantation, but this may have been done, at least by the more advanced physician, simply to honor the tradition and satisfy the patient. It did no harm and increased the drug's efficacy. As most of the texts are seventh-century recensions, it is difficult to say how much of the more scientific aspect of the prescriptions is old and how much is new. A novelty might be cloaked in a Sumerian garb to make it appear less novel, less disturbing, more acceptable.

The Babylonians suffered not only from personal diseases but also, no doubt, from general ones, affecting at the same time many people. Fevers then as now were common in the low lands of Iraq, and some of those fevers passed from one person to another, as a forest fire passes from one tree to the neighboring ones. Some texts, speaking of the "devouring activity of the God," refer probably to epidemics.[85] Did the Babylonians realize the existence of contagious diseases? Their dark minds were probably familiar with the magical transference of a disease from the patient to an animal (a primitive notion of vast currency), but were they aware of the possibility of a physical contagion? I am afraid I cannot be as positive as I was a few years ago[86] when I alluded to their understanding or intuition of the transmissibility of leprosy. Was the contagious disease with which they were acquainted really leprosy?[87] Was it the same disease as the one referred to in the Old Testament? And was that Hebrew disease leprosy?

In addition to prophylaxis with talismans, did they originate the prophylaxis by segregation of the patients and their objects that we find described in the Bible? It is tempting to answer those questions in the affirmative, yet it is not possible to substantiate one's answers with unambiguous texts.

HUMANITIES

It is impossible to say whether culture began earlier in the country of the Two Rivers than in that of the Nile, for we would have to know what is meant by "the

[85] Contenau, *La médecine en Assyrie*, p. 40.
[86] In a review in *Isis 15*, 356 (1931).
[87] Ebeling, "Aussatz," *Reallexikon der Assyriologie*, vol. 1 (1932), p. 321.

beginning of culture." When does culture begin? Where does the rainbow begin? What is certain is that the Sumerian culture was the dominant one in the Near East from, say, 3500 to 2000; the "Egyptian empire" culminated only at the end of the sixteenth century.

It is equally certain that Mesopotamian "literature" anticipated the Egyptian, was indeed the earliest whose records have come down to us. According to Kramer,

> We are amply justified in stating that although practically all our available Sumerian literary tablets actually date from approximately 2000 B.C., a large part of the written literature of the Sumerians was created and developed in the latter half of the third millennium B.C. The fact that so little literary material from these earlier periods has been excavated to date is in large part a matter of archaeological accident. Had it not been, for example, for the Nippur expedition, we would have very little Sumerian literary material from the early post-Sumerian period.
>
> Now let us compare this date with that of the various ancient literatures known to us at present. In Egypt, for example, one might have expected an ancient written literature commensurate with its high cultural development. And, indeed, to judge from the pyramid inscriptions, the Egyptians in all probability did have a well-developed written literature in the third millennium B.C. Unfortunately, it must have been written largely on papyrus, a readily perishable material, and there is little hope that enough of it will ever be recovered to give a reasonably adequate cross-section of the Egyptian literature of that ancient period. Then, too, there is the hitherto unknown ancient Canaanite literature which has been found inscribed on tablets excavated in the past decade by the French at Rash-esh-Shamra in northern Syria. These tablets, relatively few in number, indicate that the Canaanites, too, had a highly developed literature at one time. They are dated approximately 1400 B.C., that is, they were inscribed over half a millennium later than our Sumerian literary tablets. As for the Semitic Babylonian literature as exemplified by such works as the "Epic of Creation," the "Epic of Gilgamesh," etc., it is not only considerably later than our Sumerian literature, but also includes much that is borrowed directly from it.
>
> We turn now to the ancient literatures which have exercised the most profound influence on the more spiritual aspects of our civilization. These are the Bible, which contains the literary creations of the Hebrews; the Iliad and Odyssey, which are filled with the epic and mythic lore of the Greeks; the Rig-veda, which contains the literary products of ancient India; and the Avesta, which contains those of ancient Iran. None of these literary collections were written down in their present form before the first half of the first millennium B.C. Our Sumerian literature, inscribed on tablets dating from approximately 2000 B.C., therefore antedates these literatures by more than a millennium. Moreover, there is another vital difference. The texts of the Bible, of the Iliad and Odyssey, and of the Rig-veda and Avesta, as we have them, have been modified, edited, and redacted by compilers and redactors with varied motives and diverse points of view. Not so our Sumerian literature; it has come down to us as actually inscribed by the ancient scribes of four thousand years ago, unmodified by later compilers and commentators.[88]

The Nippur expedition alluded to above is the one conducted by the University of Pennsylvania from 1889 to 1900. The American archaeologists succeeded in excavating an exceedingly large number of tablets, some 50,000 being now preserved in the Museum of that university in Philadelphia.[89] Some 3,000 tablets (more than two-thirds of them in Philadelphia), written in Sumerian, date from *c.* 2000 but represent earlier times. Those tablets are not yet completely deciphered, because

[88] Samuel N. Kramer, *Sumerian mythology. A study of spiritual and literary achievement in the third millennium* B.C. (Philadelphia: American Philosophical Society, 1944), p. 19 [*Isis* 35, 248 (1944)].

[89] In addition to the tablets given to the museum of Constantinople. For a brief account, see Sir E. A. Wallis Budge, *Rise and progress of Assyriology* (London, 1925), pp. 247–250.

the Sumerian language, being unrelated to any other known to us, resisted the efforts of philologists much longer than, say, Accadian or Egyptian. However, a sufficient number have been read or explained to justify Kramer's proud conclusion. They contain chiefly mythological texts, hymns to the gods, lamentations, proverbs and "wisdom," and cosmology.

The early Sumerians did not think of themselves as upstarts, but rather as late recipients of a glorious tradition. They originated the tale of man's golden age.

In those days there was no snake, there was no scorpion, there was no hyena,
There was no lion, there was no wild dog, no wolf,
There was no fear, no terror,
Man had no rival.
In those days the land Shubur (East), the place of plenty, of righteous decrees,
Harmony-tongued Sumer (South), the great land of the "decrees of princeship,"
Uri (North), the land having all that is needful,
The land Martu (West), resting in security,
The whole universe, the people in unison,
To Enlil in one tongue gave praise.[90]

In those days of long, long ago evoked in that tablet, there was universal peace and no "confusion of tongues," and all the people were happy and praised God. This strange idea that human society was perfect in the beginning and then degenerated (the very opposite of the idea of "progress") was immensely popular. Not only did the writers of antiquity, with few exceptions, share it, but it continued to obtain some currency until the seventeenth century and later.[91] The idea of progress did not have much chance of asserting itself until modern times and did not triumph until the nineteenth century.[92] Even today there are people who cannot accept it, because the evils of the world are so cruel and blatant that the goodness thereof is hidden from their eyes.

Though the Sumerian corpus that has come down to us is not much anterior in its actual state to 2000, internal evidence enables us to carry it back many centuries. A literary revival began under the first king of the Accadian dynasty, Sargon (2637–2582; or 2450–2350?). By the time of Hammurabi the period of creation was already over, but its prestige was so great that Sumerian was accepted as a classical language, the language of religion and the humanities. The Babylonian scribes and their followers did their best to preserve the Sumerian masterpieces and to interpret them. We have noticed an analogous situation in Egypt but with a conspicuous difference — the Egyptian script changed, but however much the language evolved it remained essentially the same language, while the Babylonians used a language radically different from Sumerian.

Two of the Nippur tablets, one in Paris, the other in Philadelphia,[93] testify strongly in favor of the Sumerians' humanism and of their literary consciousness. These two tablets bear lists of writings, or perhaps catalogues of libraries, being the oldest documents of their kind. The Philadelphia tablet lists 62 titles, the

[90] From a tablet (No. 29.16.422) in the Nippur collection in Philadelphia. See Kramer, *Sumerian mythology*, frontispiece, p. 107.

[91] A striking example is that of Simon Stevin of Bruges, 1605; see *Isis 21*, 259 (1934).

[92] John Bagnell Bury, *The idea of progress* (London, 1920) [*Isis 4*, 373–375 (1921–22)].

[93] They are so much alike that they may have been written by the same scribe. Samuel N. Kramer, *The oldest literary catalogue. A Sumerian list of literary compositions compiled about 2000* B.C. (Bull. American Schools of Oriental Research, No. 88, 1942), pp. 10–19; also *Sumerian mythology*, p. 14, pl. 2.

Louvre 68; 43 titles are common to both lists; hence the two tablets give us 87 titles of literary compositions, 28 of which are thus far identified.

The early Sumerian tablets, it must be admitted, are of greater interest to the historian of literature and religion than to the historian of science. Nevertheless, we find in them many short texts that are equivalent to the (later) Egyptian ones discussed in the previous chapter under the heading "The dawn of conscience." Not only did the human conscience wake up in Mesopotamia as brilliantly as in Egypt, but it made itself heard.

As the Sumerians did not imagine that their gods were perfect, they could eschew the problem of evil, but they tried to find out what was man's place in the universe — somewhere below the gods and above the animals. How did civilization begin? Many of their myths are meant to explain the development of culture, the shape of the things that they observed among themselves or the shape of the things to come, their own dreams and wishes. There is nothing at all that is very deep, but here and there a phrase already reveals the anxiety and the piety of human hearts, and this lifting of the veil is extremely moving.

Attempts have been made to decipher musical notations on early tablets, and it has been claimed that one of those tablets represented the harp accompaniment to the Sumerian hymn of the creation of man.[94] This may be farfetched, but it is certain that the Sumerians and their successors enjoyed music and were acquainted with a good variety of musical instruments, such as drums, rattles and bells, flutes, horns and trumpets, harps and lutes.

Cuneiform script being very difficult, only very few people (priests and scribes) were able to write. The great majority could neither write nor read, and yet many written communications passed between them. Public and private scriveners did whatever writing (and reading) had to be done. Even as a man dictates a letter to his secretary and then signs it, so did a Sumerian officer, landowner, or merchant dictate to his private scribe or to a public one, or in many cases he let the scribe redact the documents in the proper form, and then he impressed upon the fresh clay the seal (a cylinder seal) that he always carried with him. As every man of substance needed a personal seal, there was a great demand for such seals and large numbers of them have come down to us. Thanks to those thousands of cylinders — each of which when rolled upon the clay leaves a picture of some complexity — one is able to study the evolution of Sumerian, Babylonian and Assyrian art from, say, 3000 to within a few centuries before the birth of Christ. The engraving of these seals on stones (the best of them on hard stones such as lapis lazuli, serpentine, jasper, agate) called for great dexterity, and the technical difficulties put the artisans on their mettle. Some of these seals, especially the early ones, say of Sargon's time, are works of art of considerable merit. They may be studied from the purely artistic point of view or they may be considered as documents illustrating many aspects of Babylonian life. For example, some of them are clearly the seals of physicians, whose names can be read on them. One such seal, in

[94] Francis W. Galpin, *Music of the Sumerians* (quarto, 126 pp., 12 pls.; Cambridge: Cambridge University Press, 1937) [*Isis* 29, 241 (1938)].

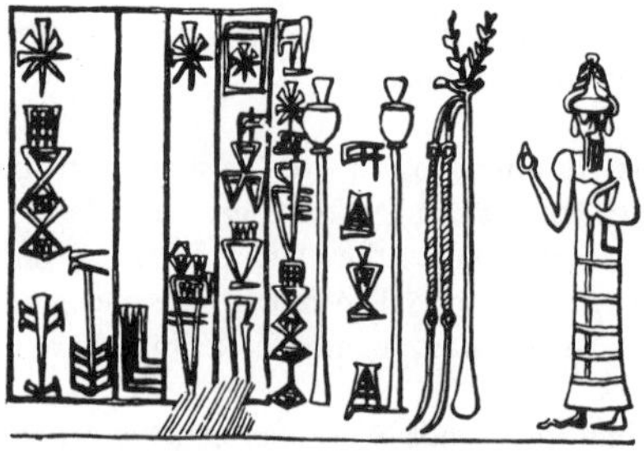

Fig. 27. Seal of the physician Ur-lugal-edina (Louvre). [After the drawing in W. H. Ward, *Seal cylinders of Western Asia* (Washington, 1910), Fig. 772, p. 255; courtesy of the Carnegie Institution.]

the Louvre, belonged to a physician called Ur-lugal-edina; it is of unusually large size (60 mm high, 33 mm in diameter) and bears an inscription in archaic style,[95] it may date from the middle of the third millennium (Fig. 27).

Most of the buildings have vanished, but much sculpture remains which can be admired in the great museums of the world. To speak only of the most ancient monuments, think of the fragments of the Stela of the Vultures dedicated to King Enannatum of Lagash (in the Louvre), the Stela of Naram-Sin (Louvre), a great grandson of Sargon, the many statues of Gudea.[96] The creations of Sumerian craftsmen are equally interesting and many of them are astounding. Consider the silver vase of Entemena of Lagash (Louvre), bearing inlaid on its surface a spread eagle which is the prototype of all the heraldic eagles down to the one adorning the blazon of the United States of America; the "Ram caught in a thicket"; the bull's head of gold and lapis lazuli (Philadelphia); the gold helmet of Mes-kalam-dug (Baghdād); the golden vessels from the royal cemetery of the first dynasty of Ur. I do not know what to admire most, the mathematical abstractions of the early Sumerians, their sexagesimal order, or the restrained shape of those vessels. If they happened to be Greek, one would fall into raptures with their purity of style and their refined discretion, but they were created by Sumerian goldsmiths who flourished almost three millennia before the Periclean age.

The Mesopotamian culture of which we have tried to outline the main aspects continued so long under different regimes — Sumerian, Babylonian, Assyrian, Chaldaean — that it is difficult to explain its influence upon other peoples with precision. At any rate, the accounts written by non-Assyriologists are full of ambiguities. One must think of that culture as a center of spiritual energy moving ahead for three or four millennia and starting radiations around itself all the time. Those radiations reached Syria, Egypt, the islands and perhaps the mainlands of the Eastern Mediterranean, Anatolia, Armenia, Persia, perhaps India and China. It is very important to know when each wave started.

In my own account I have tried hard to speak only of ancient achievements prior to 1000, most of them prior to 2000, some of them prior to 3000, all of them, even the latest, very much prior to Homer.

[95] William Hayes Ward (1835–1916), *Seal cylinders of Western Asia* (quarto, 460 pp., 1315 figs.; Washington, 1910) [*Isis* 3, 356 (1920–21)], p. 255. Contenau, *La médecine en Assyrie*, p. 41, illustrates two medical seals.

[96] Photographs of these monuments and many others may be seen in every good history of ancient art. See C. Leonard Woolley, *The development of Sumerian art*; Simon Harcourt-Smith, *Babylonian art* (76 pls.; London, 1928).

What kind of phenomena or reactions did those Babylonian waves excite in other lands? There are many traces of them in the Old Testament — the tower of Babel, the flood, much history and wisdom, perhaps also some poetry. Many other traces can be detected in other cultures, even that of our own day — sexagesimal fractions, sexagesimal division of the hours, degrees, and minutes, division of the whole day into equal hours, idea of a complete system of numbers with an infinity of multiples and submultiples, metrical system, position concept in the writing of numbers, astronomic tables. We owe them the beginnings of algebra, of cartography, and of chemistry. The training and use of horses came to us from India (?) and Cappadocia across Mesopotamia. The concepts of purity and prophylaxis set forth in Leviticus are perhaps of Babylonian origin. This hasty enumeration is more than sufficient to illustrate the magnitude of our debt to our Sumerian and Babylonian ancestors.

IV

DARK INTERLUDE

As our purpose is not to write a manual of archaeology but simply to outline the development of scientific knowledge in antiquity, there is no point in discussing on the same scale other early cultures than the Egyptian and the Mesopotamian, and the more so that we know practically nothing of the scientific achievements to be credited to other nations (Hindu, Iranian, Scythian, Chinese, etc.) in ancient, pre-Hellenic, times. It is possible that our ignorance will be alleviated later, but that is doubtful, at least as far as the Near East is concerned. The centuries preceding and following the year 1000 B.C. witnessed in that part of the world a tremendous upheaval caused by the introduction of iron, complicated migrations, and widespread turbulences. However, we must try to set forth the conditions obtaining in the Aegean area, which was the cradle of Greek culture.

THE AEGEAN AREA [1]

The Aegean culture flourished in the Archipelago with its southern and eastern outposts, Crete and Cyprus; in the Greek peninsula with the Ionian isles close to it; and in a small part of northwestern Anatolia, Troas. From that territory it radiated, as was unavoidable, to other Mediterranean shores, but we need consider only its original area as defined above. The geographic basis of that culture is the very one that is outlined in any introducton to the study of Greek culture. The Aegean sea might be compared to a great lake studded with islands. The Greek mainland itself is maritime in that no place on it is very distant from the sea, at least as the crow flies. The climate is that of the Eastern Mediterranean, characterized by hot dry summers and mild rainy winters, or let us say, whatever rain there is falls in the winter and early spring.[2] Human beings living in such an environment tend to become amphibian.[3]

[1] In addition to the works of the pioneers, Heinrich Schliemann (1822–90) and Sir Arthur Evans (1851–1941), consult their biographies: Emil Ludwig, *Schliemann of Troy. The story of a goldseeker* (336 pp., ill.; London: Putnam, 1931) and Joan Evans, *Time and chance. The story of Arthur Evans and his forebears* (422 pp., 16 ills.; London: Longmans, 1943) [*Isis 35*, 239 (1944)]. See also Harry Reginald Hall (1873–1930), *Aegean archaeology: An introduction to the archaeology of prehistoric Greece* (xxii+270 pp., 33 pls., 112 figs., 1 map; London, 1915); Gustav Glotz, *The Aegean civilization* (xvi+422 pp., 87 ills., 3 maps, 4 pls.; London, 1925); Pierre Waltz, *Le monde égéen avant les Grecs* (Collection Armand Colin no. 172; 206 p.; Paris, 1934), a popular but competent introduction to the subject.

[2] For a more detailed definition of Mediterranean geography and climate, see G. Sarton, "The unity and diversity of the Mediterranean world," *Osiris* 2, 406–463 (1936).

[3] Strabon (I–2 B.C.) used that very word in the astounding prolegomena to his *Geography* (I, 1, 16): "And to this knowledge of the nature of the land, and of the species of animals and plants, we must add a knowledge of all that pertains to the

The main products are wheat, barley, grapes, figs, and olives; the crops are never very generous and they may fail altogether if the rain is inadequate. Food shortage may then drive the men away to other places. The seaways are often easier to them than the land ones, for the fertile plains are few and small and the shores are hemmed in by mountains. In good weather the sky is perfectly blue and the luminosity and visibility are almost unconceivable to northern barbarians.

The Aegean people enjoyed all the geographic features that are adduced to account for the Greek miracle; this proves that the physical environment does not suffice to explain genius. Or may it be that the Aegean stage was necessary to bring the Greek genius to its splendid maturity?

What kind of people were the early inhabitants of the Aegean area? Anthropologists disagree. Whoever they were, and no matter how many migrations occurred, they cannot have all disappeared. The invaders never wish to obliterate, but rather to assimilate, the conquered people. There must have remained a fair proportion of Aegean chromosomes in the Greek cells.

The Aegean territory was (and is) a bridge between Asia and Europe, as well as between Europe and Africa – not a single bridge, but a hundred. Aristotle's remark,[4] that the Hellenic race, being situated between the races of Europe and of Asia, was intermediate in character, applied equally well to their Aegean predecessors. Whether the latter were the ancestors of the Greeks or not, they were their forerunners and their heralds.

sea; for in a sense we are amphibious and belong no more to the land than to the sea" (*amphibioi gar tropon tina esmen cai u mallon chersaioi ē thalattioi*) (Loeb Classical Library, vol. 1, p. 28).

[4] *Politica* 1327*b*.

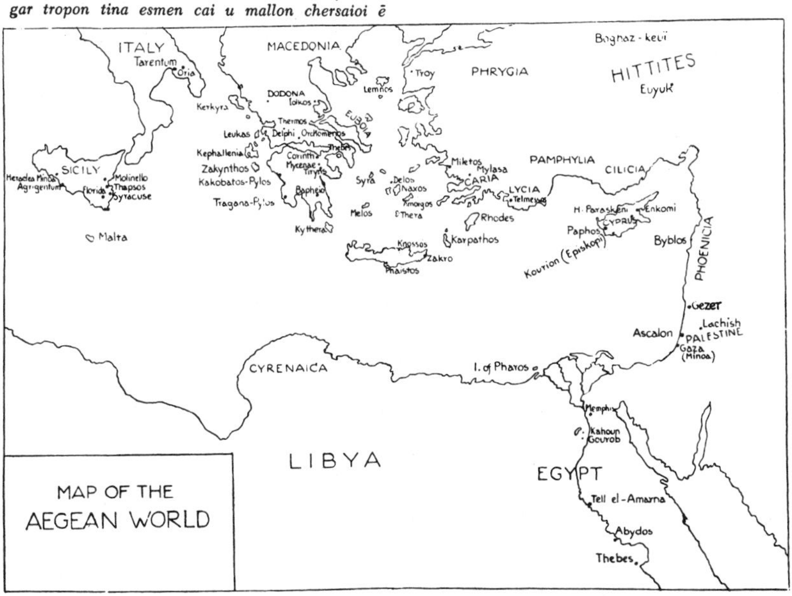

Fig. 28. Map of the Aegean world. [Reproduced with permission from Gustave Glotz, *The Aegean civilization* (London: Kegan Paul, 1925), map 3.]

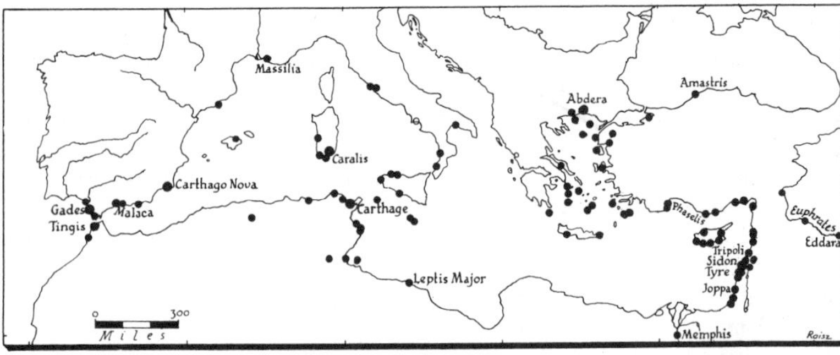

Fig. 29. Map of the Phoenician settlements or factories around the Mediterranean Sea. Outside of Phoenicia proper (at the easternmost shores of the Mediterranean), those factories or trading stations were separated from one another with very little, if any, hinterland behind each of them. The situation was the same for the Greek colonies and for most of the European factories in Asia in the early colonial days. There has been considerable discussion about the importance and even the reality of some of those Phoenician factories, but the purpose of the map is rather to illustrate their wide dissemination.

THE AEGEAN CULTURE

It was remarked in the previous chapter that the study of Mesopotamian archaeology was first (and is still generally) called "Assyriology" because of the accident that Assyrian antiquities were studied before Babylonian and Sumerian ones. A similar accident happened in the investigation of the Aegean culture; our first knowledge of it being due to Heinrich Schliemann's excavation of Mycenae (1876),[5] it was first called Mycenaean in spite of the fact (which could not then be realized) that Mycenae was a late rather than an early center of that culture. The same Schliemann had already done some digging at Hissarlik in Troas, and the exploration of that site was continued by himself in 1878 and by his assistant Wilhelm Dörpfeld in 1892. In the following year Arthur Evans began his own research in Crete and started it on a larger scale in 1899, the results being finally published in his monumental work *The Palace of Minos.*[6] It is now understood that Crete was the very cradle of the Aegean culture and that that culture developed there for a longer time and with more continuity than in any other part of the Aegean area. Thanks to half a century of studies by Evans and many other archaeologists, and in particular to the analysis of pottery and other relics all over that area, we have finally been provided with a rough chronology sufficiently well articulated with Egyptian chronology to inspire confidence (Fig. 30).[7]

The Aegean culture which flourished first in Crete and spread gradually over

[5] Aegean monuments had been found before 1876 in various places (for example, Thera and Rhodes, even Thebes) but not recognized as such. The Cyclopean walls of Tiryns and Mycenae, and in the latter place the "Treasury of Atreus" and the "Lion Gate" had been known even to the ancients and described by Pausanias (II-2), but Schliemann's excavation of Mycenaean graves excited world-wide interest. Antiquities that had been taken for granted were seen in a new light.

[6] Arthur Evans, *The palace of Minos* (4 vols.; London: Macmillan, 1921–1935; index, 1936). Schliemann died in 1890, Dörpfeld half a century later in 1940, Evans in 1941. The great gap in these dates is due to the fact that Schliemann died at 68 while his younger contemporaries lived to be 87 and 90.

[7] Table first published in *Isis 34*, 164 (1942–43).

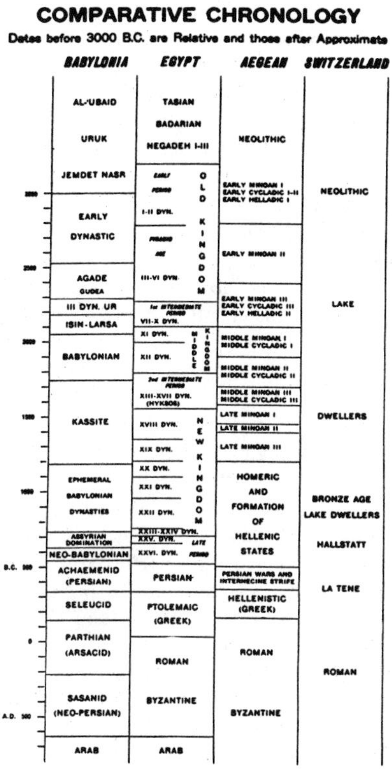

COMPARATIVE CHRONOLOGY
Dates before 3000 B.C. are Relative and those after Approximate
BABYLONIA
EGYPT
AEGEAN
SWITZERLAND
AL-'UBAID
URUK
JEMDET NASR
EARLY DYNASTIC
AGADE
GUDEA
III DYN. UR
ISIN-LARSA
BABYLONIAN
KASSITE
EPHEMERAL BABYLONIAN DYNASTIES
ASSYRIAN DOMINATION
NEO-BABYLONIAN
ACHAEMENID (PERSIAN)
SELEUCID
PARTHIAN (ARSACID)
SASANID (NEO-PERSIAN)
ARAB
TASIAN
BADARIAN
NEGADEH I-III
I-II DYN.
III-VI DYN.
VII-X DYN.
XI DYN.
XII DYN.
XIII-XVII DYN. (HYKSOS)
XVIII DYN.
XIX DYN.
XX DYN.
XXI DYN.
XXII DYN.
XXIII-XXIV DYN.
XXV. DYN.
XXVI. DYN.
PERSIAN
PTOLEMAIC (GREEK)
ROMAN
BYZANTINE
ARAB
OLD KINGDOM
MIDDLE KINGDOM
NEW KINGDOM
NEOLITHIC
EARLY MINOAN I
EARLY CYCLADIC I-II
EARLY HELLADIC I
EARLY MINOAN II
EARLY MINOAN III
EARLY CYCLADIC III
EARLY HELLADIC II
MIDDLE MINOAN I
MIDDLE CYCLADIC I
MIDDLE MINOAN II
MIDDLE CYCLADIC II
MIDDLE MINOAN III
MIDDLE CYCLADIC III
LATE MINOAN I
LATE MINOAN II
LATE MINOAN III
HOMERIC AND FORMATION OF HELLENIC STATES
PERSIAN WARS AND INTERNECINE STRIFE
HELLENISTIC (GREEK)
ROMAN
BYZANTINE
NEOLITHIC
LAKE DWELLERS
BRONZE AGE LAKE DWELLERS
HALLSTATT
LA TENE
ROMAN
B.C.
A.D.

Fig. 30. Comparative chronology as compiled by Richard A. Martin, Curator of Near Eastern Archaeology in the Chicago Museum of Natural History [*Isis 34*, 164–165 (1942).]

the whole area (mainland and islands) was a culture *sui generis*, very different from the culture of Egypt (to which it was occasionally indebted) and from that of Mesopotamia. Its existence, that is, its unity, may be surprising at first, considering the natural dispersion of that island world, but is explained by the circumstance that the Cretans had achieved sea power,[8] being the first to do so in the Mediterranean basin. Said Thucydides,

> Minos is the earliest of all those known to us by tradition who acquired a navy. He made himself master of a very great part of what is now called the Hellenic Sea, and became lord of the Cyclades islands and first colonizer of most of them, driving out the Carians and establishing his own sons in them as governors. Piracy, too, he naturally tried to clear from the sea, as far as he could, desiring that his revenues should come to him more readily.[9]

[8] It may be added that no culture is territorially continuous. It exists only in centers of sufficient human density, whence it percolates more or less slowly into the surrounding districts. Those centers are seldom close to one another; they are generally distant from one another, sometimes very distant. Any two centers may be separated by fertile land or desert, or by a stretch of river or sea; those differences are significant but not essential.

[9] Thucydides, I, 4 (Loeb Classical Library, vol. 1, p. 9). The Carians were a strange people, addicted to piracy, speaking a language unrelated to Greek, and having customs of their own such as matriarchy and a special mode of burial. Says Thucydides (I, 8), "When Delos [one of the Cyclades] was purified by the Athenians in this war

This Minos is a half-mythical figure, yet a good symbol of the Cretan hegemony for the period of, say, 1700 to 1400. Cretan thalassocracy had begun many centuries earlier (say by 2100) but "Minos" brought it to a climax. Sea power meant not only political but also cultural unity.

That unity was relative; Aegean culture was far from being uniform in its spatial or temporal distribution. For one thing, Cretan manners were always appreciably different from those of the Greek mainland, and each island had its cherished peculiarities, yet they all traded with one another.[10] In the course of time the cultural features never ceased to grow and change. In lieu of the dynastic stages recognized in Egypt and Mesopotamia, the diagnosis of ceramic and other cultural elements enables archaeologists to distinguish three great ages, Early, Middle, and Late Minoan and to subdivide each of those ages into three periods of unequal lengths. For example, what they call Late Minoan II was the golden age of Crete, corresponding to a part of the Eighteenth Dynasty of Egypt (1580–1350).

That culture had a script of its own, or various scripts, which have thus far resisted every attempt at decipherment[11] and will probably remain unreadable unless a bilingual text is discovered. It created artistic monuments that an expert eye recognizes immediately. The rulers built themselves palaces unlike in general structure and in many details the palaces of Egypt and Babylonia; large halls were provided for assemblies. Ingenious methods were devised for bringing fresh water into the living quarters and draining away the soiled water and excreta;[12] there were bathrooms in the palace of Cnossos even as in the earlier ones of Karnak. The beehive tombs and the Cretan coffins in terra cotta were typical. The Aegean people bequeathed no sculpture of large size but many little objects of rare and intriguing appearance — like the Snake Goddess in polychrome faience of the Ashmolean Museum, Oxford, or the one in gold and ivory of the Boston Museum (Fig. 31), or the gold and ivory statuette in the Royal Ontario Museum in Toronto (Fig. 32).[13] Such objects once seen are unforgettable; they constitute perhaps the best symbols of the culture that they immortalize. Similar remarks might be made with regard to the frescoes with which the walls were decorated and the scenes painted on pottery. Those paintings reproduce octopuses, flying fish, cockerels, wild ducks, and other animals, as well as various plants, with a very refreshing and astonishing realism. The palace of Minos, if we had been able to visit it when it was new, would have seemed very gay to us (at least the living rooms) and very modern.

After the golden age of Crete, about the sixteenth century, the Aegean culture was inherited by ungrateful Mycenaens who continued it somehow for a few more centuries (say from 1500 to 1200); then that splendid culture was submerged by northern barbarians (Dorian invasion). The Bronze Age, which had

[426 B.C.] and the graves of all who had ever died in the island were removed, over half were discovered to be Carians, being recognized by the fashion of the armor found buried with them, and by the mode of burial, which is that still in use among them."

[10] Obsidian objects are found all over the area, though the only source of it is the island of Melos, the most westerly of the Cyclades. Pottery objects of definite provenience are also widely distributed.

[11] This is the more tantalizing because some of the Cretan symbols are very much like hieroglyphics; examples are given in *Isis 24*, 377 (1935–36).

[12] The drain pipes of the palace of Cnossos were not the first of their kind; some 1300 feet of copper pipes were found in the pyramid temple of Abuṣīr (Fifth Dynasty = 2750–2625) built a thousand years before Cnossos!

[13] C. R. Wason, "Cretan statuette in gold and ivory," *Bull. Roy. Ontario Museum* (March 1932), pp. 1–12; 14 figs.

lasted some two millennia, was now brutally replaced by a new age — the Age of Iron.[14] The revolutionary period marking the transition between those two ages is the "dark interlude" mentioned in the title of this chapter. It is neither possible nor necessary to locate it exactly on the chronological scale, for its occurrence and length varied from place to place, but we may say that obscurity, confusion, and chaos obtained in various degrees according to localities during the centuries immediately preceding and following the year 1000. The iron industries had been invented about the middle of the second millennium by the Hittites, and from the Hittite territory in Anatolia they had reached Syria and Egypt in the South and Macedonia in the West. The rude Dorian invaders were probably able to establish their domination over the Aegean peoples because of their iron weapons and iron tools.[15]

[14] The earliest iron sword of the Aegean area was found in a tomb of Mouliana (northeastern Crete) dating from the very end of Late Minoan III, corresponding to the Nineteenth Dynasty (1350–1205). Glotz, *The Aegean civilization*, p. 389.

[15] The iron age reached central and western Europe somewhat later. The Hallstatt period, as it

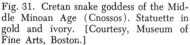

Fig. 31. Cretan snake goddess of the Middle Minoan Age (Cnossos). Statuette in gold and ivory. [Courtesy, Museum of Fine Arts, Boston.]

Fig. 32. Gold and ivory statuette from Crete; same age as that shown in Fig. 31, say the sixteenth century. The original height was *c.* 26 cm. [Reproduced with kind permission of the Royal Ontario Museum, Toronto; for a fuller description see the *Bulletin* of that Museum (March, 1932).]

Similar statuettes exhibiting the same kind of sophistication and modernity exist in the Fitzwilliam Museum in Cambridge and in the Museum of Cnossos, the last one in polychrome faience (reproduction in the Ashmolean Museum, Oxford).

The Dorian invasion and the other migrations that it released introduced endless confusion, reaching at some times and in some districts the limit of hopeless chaos, yet we should not take an exaggerated view of those phenomena. Thucydides warns us at the beginning of his history that migrations were of frequent occurrence but on a small scale. We may assume that those migrations were always incomplete and fragmentary; they concerned chiefly the more restless part of the population, those men who were not yet definitely settled, or were at odds with their neighbors, and who were always ready to move. Of course, those men displaced others, who might have preferred to stay where they were, but they never displaced all the people of the invaded territories. Hence the culture discontinuity caused by migrations, whether these were voluntary and quiet, or unexpected and violent, was never accompanied by a complete anthropological discontinuity.

The intimate knowledge of Aegean culture that we owe primarily to a great variety of monuments is confirmed by allusions in Egyptian, Hittite, and Babylonian records, by folkloric remains in the Aegean area, by reminiscences in the Homeric poems, and by vague references in later authors, such as Thucydides and Herodotos (V B.C.), Virgil and Strabon (I–2 B.C.), Plutarch (I–2), Pausanias (II–2). Both the vagueness and the scarcity of those references prove the depth of the discontinuity between the Aegean and Greek civilizations, yet the latter was to a large extent an unconscious inheritor of the former. The past, even the distant past, can never be completely obliterated.

THE EARLY GREEK AND PHOENICIAN COLONIES. INVENTION OF THE ALPHABET

The Aegean dispersion was accompanied toward its end by a Greek one, and when it ceased altogether it was continued by Greek colonization. In most cases the same populations were partly involved but Aegean patterns of culture were gradually replaced by Greek patterns. The mixture of the two types can be appreciated at its best in Cyprus, where the Minoan culture survived longer than anywhere else. As far as those dark events can be reconstructed at all, archaeologists agree that there were three main waves of early southward migrations. First, tribes coming from the western coast invaded Thessaly and displaced other tribes, which moved to Boeotia. Second, northern people, the "Dorians," overran a large part of the Peloponnesos and many of the islands as far south as Crete and as far east as Rhodes. Third, northwestern tribes moved from Epiros across the Ionian sea to Apulia, while others conquered the territories just above the gulf of Corinth and Elis in the northwestern part of the Peloponnesos. According to Thucydides,[16] the first two waves occurred some sixty and eighty years after the Fall of Troy. Those waves excited other waves, the outstanding ones being the Dorian migration (continuing the Dorian movements already mentioned), the Aeolian migration leading to the occupation of Tenedos, Lesbos, and Mysia (on the mainland opposite Lesbos), and the Ionian migration carrying displaced inhabitants from the northern Peloponnesos and Attica to the Cyclades, Chios, Samos, and parts of the mainland opposite (Halicarnassos, Cnidos).

It is well-nigh impossible to follow the details of those migrations in time and space; it suffices for our purpose to refer to them in their totality. During the

is called in European archaeology, lasted from *c.* 1000 to *c.* 500; it is so called after the main site in Hallstatt, Salzkammergut, Austria. It is characterized by the use of bronze and iron, agriculture, domesticated animals, and typical artifacts.

[16] Thucydides, I, 12.

Dark Age, many populations displace one another from one part of the old Aegean area to another, and maybe some of them transgress the early limits of that area. Greek colonization continues in a different way the old Aegean one.

In most cases the migrants or colonists were not opening new trails but following more assiduously and in greater numbers trails that were traditionally known to them. They were not plunging into the darkness but aiming at places of which vague but enticing reports had come to them. For example, we hear of Dorian colonies in Bithynia (southwestern angle of the Black Sea) and in Crimea; Ionian colonies were scattered all around that sea. This linkage of Russia with the Mediterranean was not by any means a novelty. Communications had existed between Russia and the Caucasus on one hand and between Russia and Egypt on the other.[17] Such communications were probably continued under Minoan patronage, and when the Minoan world crumbled to pieces, echoes of its fall must have reached Russia. The Greek unrest destroying Aegean culture was accompanied by a similar unrest destroying the Stone Age Tripolye [18] culture of southern Russia and replacing it by a new one. Nor was this the end. Human waves, like mechanical ones, never stop completely; that is, if new energy is added to them from time to time they go on for ever, the vibration passing from one system to many others. The violent waves of the Iron Age were transmitted across Scythia and beyond, all the way to China.[19]

Before leaving the shores of the Black Sea it is well to bear in mind that the use of iron originated probably with the Hittites, by whom or from whom it was transmitted in the middle of the second millennium to Mesopotamia and Egypt. When it reached the Aegean area and caused the revolution of the Iron Age, and when ripples of that upheaval disturbed the shores of the Black Sea, a remarkable cycle had been completed. The Hittites flourished mostly within the crescent of the Red River;[20] iron products were probably carried down that river to the Black Sea, and hence across the straits to the Aegean Sea. We have already remarked above that the Hittites spoke a language not very distant from ancient Greek, related to it through a common parentage. In short, an Asiatic Indo-European people discovered the value of iron metallurgy, and kindred European tribes developed that discovery to its first climax.

If the Greek upheaval of the Dark Age was caused by the use of iron (it was concomitant with the beginning of the Iron Age), we must give much credit for that to the Hittite predecessors.

To return to the Mediterranean Sea, it so happens that when the Minoan sea power came to an end, the Greek descendants were not, as one might have expected, the only heirs. Their inheritance was immediately disputed by a people

[17] Margaret Alice Murray, "Connexions between Egypt and Russia," *Antiquity 15*, 384–386 (Gloucester, 1941), 2 pls.

[18] Tripolye is the name of its principal site, some 50 miles from Kiev on the middle Dnieper River.

[19] Gregory Borovka, *Scythian art* (112 pp., 74 pls.; London, 1927), a fine collection of examples with an excellent introduction and references to the main Scythian publications.

[20] The largest river of Asia Minor, some 600 miles long, *Encyclopedia of Islam* (5 vols.; Leiden: Brill, 1908–1938), vol. 2, p. 1054. The name we give to it is the translation of its Turkish name, Qizil-Irmāq; the Greeks called it Halys.

of an entirely different ancestry, the Phoenicians, a Semitic nation established along the Syrian coast, north of Palestine.[21]

Those Phoenicians spoke a language closer to Hebrew than to any other language of the Semitic family. The mysterious Hycsos who invaded Egypt in the seventeenth or sixteenth century may have been identical with the Phoenicians (or Arabs?) or related to them.[22] At any rate, the Phoenicians themselves are revealed without ambiguity when the Pharaoh Aḥmose I (first king of the Eighteenth Dynasty; 1580–1557) invaded their country. From then on, but not for very long, they were subjected to the Egyptian rule. They are frequently mentioned in the 'Amārna cuneiform tablets; some of them were trying to throw off the Egyptian yoke and were intriguing with the Hittites, whose growing strength and apparent friendship increased their own hopes of liberation. After the rule of our old friend Amenhotep IV, or Ikhnaton (1375–1350), Egyptian power collapsed. Ramses II (fourth king of the following Nineteenth Dynasty; 1292–1225) reconquered Phoenicia as far as Beirūt and began the unforgettable series of inscriptions engraved on the rocks of the Nahr al-Kalb, just north of that city.[23] Under Ramses III (Twentieth Dynasty; 1198–1167), the Phoenicians took advantage of new foreign invasions to emancipate themselves from Egyptian control and they remained independent until the Assyrian conquest (*c.* 876).

Placed as they were along the shores of the eastern end of the Mediterranean Sea, it is not surprising that the Phoenicians developed very early a deep interest in navigation. Look at the map! It is as if they were standing on a balcony overlooking the whole of Mediterranean life. On a clear day they could see the hills of Cyprus; Egypt, which was still the outstanding center of culture as well as the largest market, was close to their left. However, as long as Minoan thalassocracy continued, Phoenician sailors were restrained or, if they ventured too far, were treated as pirates. About the twelfth century, when the Cretans lost the strength to rule the sea, Phoenician sailors were ready to succeed them and they did. Their readiness to take over and their efficiency are sufficient proof of a long preparation. As their liberation from Egyptian bondage coincided with the fall of Crete, they could exploit the situation fully. They soon became the masters of Mediterranean trade, with no rivals except the Greeks. It is probable that the trade of the Greek colonies was handled by Greek sailors, and hence Phoenicians were obliged to establish colonies or factories (that is, trading stations) of their own. The main center of Phoenician trade was the harbor of Tyre, whose glory is still reflected in the lines of Ezekiel (27:13–25). The Tyrians had factories[24] in Cyprus, Rhodes, Thasos, Cythera, Corfu, Sicily, Gozo (near Malta), Libya, Pantelleria, Tunisia, Sardinia, and other islands. Almost everywhere they were competing with Greeks, and their rivalry was not only commercial but naval. The Greeks hated them

[21] Georges Contenau, *La civilisation phénicienne* (396 pp., 137 ills.; Paris, 1926) [*Isis* 9, 179 (1927)]. Raymond Weill, *Phoenicia and Western Asia to the Macedonian conquest* (208 pp., London: Harrap, 1940).

[22] That tradition was related by Manethon (III–I B.C.), fragment 42 (Loeb Classical Library), p. 85.

[23] Franz Heinrich Weissbach, *Die Denkmaeler und Inschriften an der Mündung des Nahr el-Kelb* (Wiss. Veröff. des deutsch-türkischen Denkmalschutz-Kommandos, Heft 6, 16 figs., 14 pls.; Berlin, 1922). René Mouterde, S.J., *Le Nahr el Kelb* (Beyrouth: Imprimerie Catholique, 1932), small popular guide.

[24] It is better to say factory than colony, because the Phoenician settlements were essentially different from the Greek in that the latter were independent offshoots from the mother country (like swarms out of a beehive); while the former were more like branch offices controlled by the main administration in Tyre.

and accused them of greed and unfairness; these accusations and the hatred prompting them were probably reciprocated. The most famous of these Phoenician outposts was Carthage, their first settlement on the African coast, established in a strategic location, half-way across the sea, in the ninth century if not before. The rivalry between Greeks and Phoenicians which began in the twelfth century remained under one form or another one of the main themes of ancient history; the war between the Greeks and the Persians (499–478) was to a large extent a war between the Greek and the Phoenician navies; the Punic wars (264–146) between the Romans and the Carthaginians were the final tests, which ended with the victory of the Western power.[25]

To return to Phoenician colonization, it extended to Spain and even to the western coast of that country beyond the Pillars of Hercules.[26] According to Strabon[27] this was done soon after the Trojan war. The Tyrian merchants exported and distributed around the Mediterranean Sea an abundant selection of goods — glass and earthenware, metal objects made of Cyprus copper, textile fabrics, which they embroidered. Their main specialty, and, in fact, monopoly, seems to have been the dyeing of textiles with the purple obtained from murex.[28] Most of the wares sold by them were obtained from Egypt, Arabia, Mesopotamia, or the islands, but credit was often given to them for inventions (such as that of glassware) which they had not made but had simply helped in diffusing. The Phoenician arts were derived to a large extent from Egyptian models.

Indeed, the Phoenicians were not creators as the Greeks proved to be at a later time; they were primarily merchants, international brokers.[29] They were very active and intelligent and the development of the arts in the Mediterranean sea (the cradle of our civilization) was largely due to their ministrations.

The outstanding service that they rendered mankind — and its importance cannot be overestimated — was the invention of the alphabet; we may call it the masterpiece of brokerage. As we have explained in previous chapters, alphabetic or syllabic symbols were invented and used separately by Egyptians and by Sumerians, but there is an immense difference between the use of such symbols and the exclusive use of them. That invention was probably made independently by the Cretans and by Phoenicians or by some of the latter's neighbors (in Rās Shamrā or the Sinai). The Cretan syllabary cannot yet be interpreted and it left no descendant except the Cypriot syllabary of a much later time. The Asiatic in-

[25] The destruction of Carthage in 146 did not obliterate Phoenician culture in Tunisia and a Phoenician dialect continued to be used. St. Augustine (V-1) quoted Punic words in his sermons.

[26] The Pillars of Hercules, Heracles, or Melqart (in Phoenician, king of the city, a name of God), that is, the Straits of Gibraltar. There were early Phoenician settlements, for example, in Carthagena (New Carthage) and Onoba (Huelva), on the coast east and west of the Straits. Later (450–201) a large part of the Spanish Peninsula south of the Douro and Ebro rivers was under Carthaginian power.

[27] Strabon, I, 3, 2.

[28] *Murex trunculus* and *M. brandaris*, marine gastropods abundant along the Syrian coast.

[29] A charming letter written by Renan to Berthelot makes me realize that I am perhaps unfair to the Phoenicians. They were not only merchants, but manufacturers, creators of many goods. The letter is dated Sour (= Tyre), 12 March 1861 "Une chose bien curieuse, c'est que les restes de la civilisation phénicienne sont presque tous des restes de monuments industriels. Le monument industriel, chez nous si fragile, était, chez les Phéniciens, colossal et grandiose. Toute la campagne est parsemée des restes de cette industrie gigantesque, taillés dans le roc. Les pressoirs, sortes de portes composées de trois blocs superposés, ressemblent à des arcs de triomphe; les vieilles usines, avec leurs cuves, leurs meules, sont là, dans le désert, parfaitement intactes. Les puits dits de Salomon, près de Tyr, sont quelque chose de merveilleux et d'une profonde impression." *E. Renan et M. Berthelot, Correspondance, 1847–1892* (Paris, 1898), p. 254.

vention was completed certainly before 1000 and perhaps as early as 1500; the Phoenician alphabet, if it was not the very first, was the one that triumphed, the only one that emerged by the end of the eleventh century; after having suffered countless vicissitudes it survives in most of the alphabets used today. Let us consider it more carefully.

The Phoenician alphabet was consonantal; every one of its symbols stood for a consonant or for a long vowel (which might have a consonantal value, like *w* and *y*). There were no signs for short vowels; hence the symbol for *b* might be used for a final *b* or for a syllable like *ba, bi, bu, be, bo*. The same kind of alphabet is still used in Hebrew and Arabic and causes no great difficulty for people who know the words and their inflections sufficiently well. In the course of time, the Greeks imitated the Phoenician alphabet [30] and improved it by adding to it new symbols to designate short vowels.

The essence of the invention is the idea of representing every sound of the language with as few signs as possible and without ambiguity. The Phoenician scribe who invented the alphabet knew his own language perfectly well and tried to reduce the number of symbols to its minimum; as there was no ambiguity in his own mind concerning the vocalization, he thought that it was superfluous to indicate it; his error was corrected later by the Greeks. The Phoenicians were too economical, but we should pause before blaming them. The alphabetic economy so clear to them was not understood by other nations, and is not completely understood to this day by nations whose writing is alphabetic. The early Western printers did not realize at first that they could print every Latin book with a series of twenty-odd letters; as they tried to imitate the ligatures and abbreviations of the copyists they used over one hundred fifty different characters! The Arabic printers of today still need a far greater variety of type than the Arabic alphabet (28 letters) would require, for many letters must be written differently at the beginning, in the middle, or at the end of a word, or when they are used in conjunction with certain other letters.

This example illustrates the immense trouble involved in persuading men to accept a great invention, which would simplify their work and economize their strength. To summarize the story, we witnessed the tentative efforts of Egyptians and Sumerians; ineffectual inventions by the Cretans and other peoples; the Phoenician oversimplification imitated in other Semitic alphabets; the Greek perfect solution, followed by imperfect adaptations to other languages and by the futile overelaborations that exist to this day. Those inclined to underrate the Phoenician invention because it was not perfect should think of our own alphabets and especially of the English one which is a real monstrosity, and be a little humbler. The Phoenician alphabet did not indicate the vowels, but the English as often as not indicates the wrong ones; is that much better? Alphabetic economy consists in reducing the writing of a language to the combination of a minimum number of signs. The English alphabet is very small; in fact, it is too small, just as the Phoeni-

[30] According to Herodotos, v, 58, the alphabet was brought to Greece by the Phoenicians who came with Cadmos. Cadmos of Tyre, son of a Phoenician king, is one of the mythological personalities symbolizing Phoenician origins. A sufficient proof of the Semitic origin of the Greek alphabet is the fact that the first three letters of that alphabet have Semitic names (alpha, bēta, gamma; aleph, bēth, gīmel). The order of the letters in all the ancient alphabets (with one exception) is the same as the Semitic order. The exception is the Sanskrit (Devanāgarī) alphabet, the order of which is dominated by phonetic considerations.

cian was; and its use involves a large number of ambiguities, perhaps a greater number than in any other language. There is nothing in that to be proud of.[31]

Before abandoning this subject a final remark may be inserted. It should be possible to devise a single alphabet that would be suitable for transcribing phonetically every language. An international alphabet of that kind was proposed at the Copenhagen conference of 1925, and after a few modifications was accepted by the International Phonetic Association (last revision, 1951).[32] Unfortunately, it has not yet obtained any popularity and probably never will. The difficulties involved are very great and perhaps insuperable. A humbler purpose, but easier to attain, would be the invention of an unambiguous alphabet for each language. When the English-speaking peoples complete such a reform for their own language, English will have a far better chance of being adopted as a second language by the other peoples.

This digression may serve to illustrate the pregnancy of the Phoenician invention; it was so simple, yet so deep, that some of the most civilized nations of our days have not understood its implications.[33]

My account of that stupendous invention is of necessity oversimplified. Claude Schaeffer has discovered in Rās Shamrā an Ugaritic ABC that may be older than the Phoenician; at any rate, these two alphabets are closely connected, and the order of letters is the same. This order has remained the same throughout 3,000 years as in our own alphabet, except that *z* was moved to the end in Cicero's time.

When studying the art of alphabetic writing (or the art of writing in general), we must bear in mind that a high degree of illiteracy [34] continued for long periods of time in spite of the fact that the art had been discovered and was actually practiced by rare individuals. This was because mnemonic traditions were so satisfying that many people, including highly educated ones, did not feel the need of writing. For example, such traditions must have been very strong in the golden age of Hellenism; otherwise, Socrates' diatribe against the art of writing in *Phaidros* [35] would hardly be intelligible. Another curious fact, underlined by Max Müller,[36] is that in no Greek writer do we meet any expression of wonder at the most wonderful invention of antiquity, the alphabet. Of course, all the fundamental discoveries of early times were taken for granted, even as our own children take for granted the marvels of our own time.

The intense rivalry that obtained between the Greeks and the Phoenicians did not separate them so much that they could not influence one another. We have

[31] For more remarks on English "orthography" see G. Sarton, "The feminine monarchie of Charles Butler 1609," *Isis 34*, 469–472 (1943), 6 figs.

[32] Leonard Bloomfield, *Language* (New York: Holt, 1933), pp. 86–89; Louis Herbert Gray, *Foundations of language* (New York: Macmillan, 1939), p. 58. Thanks to my colleague Joshua Whatmough.

[33] Innumerable memoirs have been devoted to the alphabet and new ones appear every year. There are also many syntheses, of which it will suffice to mention two of the most recent: Hans Jensen, *Die Schrift in Vergangenheit und Gegenwart* (Hannover, 1925; much improved ed., Glückstadt, 1935) [*Isis 30*, 132–137 (1939)] and David Diringer, *The alphabet* (607 pp., ill.; London: Hutchinson, 1948) [*Isis 40*, 87 (1949)]; this is an abbreviation of the original Italian edition (867 pp.; Florence, 1937).

[34] This word is taken here *stricto sensu*, the inability to read and write, but illiteracy might be combined, and often was, with a high degree of education, even literary and poetic education. Many great poets were "illiterate."

[35] Plato, *Phaidros*, 274 C.

[36] "Literature before letters" (1899), reprinted in his *Last Essays* (1901), vol. 1, pp. 110–138, a very interesting essay.

just given the main proof of the influence of the latter upon the former; there is no doubt that the Greek alphabet is derived from the Phoenician. Moreover, a number of Phoenician (or at least Semitic) words are imbedded in the Greek language, not by any means rare words, like *chrysos* (gold), *cypros* (copper), *chitōn* (man's garment), *othonē* (fine linen), *baitylos* (meteoric stone), *byssos* (flax, linen), *gaylos* (a kind of ship), *mna* (mina, weight or sum of money), *myrra* (myrrh), *nabla* (musical instrument, with 10 or 12 strings), and — most important of all – *byblos* or *biblos* (papyrus, book; hence our word Bible).[37]

THE CONTINUITY OF ORIENTAL INFLUENCES

Before proceeding any further it is well to warn our readers once more that the Oriental influences should not be conceived as anticipating the Greek achievements, and then stopping short. Much of the Egyptian, Mesopotamian, and Phoenician work was obviously prior to Homer, but we should always remember that those ancient cultures continued in one form or another until the Roman conquests and that they even survived the latter. In addition to the pre-Hellenic influences, there were thus many others throughout the course of Greek history, or rather there was an endless give-and-take between East and West.

To understand the situation, ask yourself how you would answer the queries, "Have the French influenced the Italians?" and "Have the Italians influenced the English?" Obviously, the answers are not simple and easy. When two cultured nations flourish together there is tug of war between them; at one time one is dominant and the other imitative, at another time the situation is reversed, and so on.

In a way, every stream of thought, once started, continues to flow, and even when the flow has almost completely stopped, it deposits silt which recalls the past. In every language there are words that are like the fossil remains of earlier life. For example, in English, *Isidore*, *Susannah*, *megrim*, *ebony*, *gum*, *adobe* are witnesses of ancient Egypt.[38]

Egyptian ideas, arts, and customs were transmitted throughout the Dark Age not only by the Egyptians themselves, but by the Aegeans, the Phoenicians, and the Greeks trading or having any kind of intercourse with them. To be sure, the wars and revolutions destroyed many of those traditional links, but they could not destroy them all and enough remained to constitute in the hearts of men a kind of Egyptian model or mirage. The Egyptian traditions were kept alive by craftsmen, travelers, storytellers, and gossips, and from time to time they were given a new currency by great writers such as Herodotos in the fifth century, Plato, Aristotle, Theophrastos, Nearchos in the fourth, Agatharchides of Cnidos in the second, Caesar, Posidonios, Diodoros, Strabon and Vitruvius in the first, and even by various men of our own era, such as the author of the Periplos of the Red Sea, Dioscorides, Josephos, Columella, Tacitus, Lucanus, and above all Pliny in the first century, and Athenaeos and Zosimos in the third.

In the land of Egypt, the relations between Greeks and natives become more frequent and intimate during the Twenty-sixth (or Saitic) Dynasty (663–525)

[37] Some of those words are included by Glotz, *The Aegean civilization*, p. 386, in his list of Greek words retained in the Cretan dialects of historical times.

[38] *Chronique d'Egypte*, vol. 11 (1936), p. 406.

and during the Persian regime (525–331);[39] they became even deeper after Alexander's conquest. The consequences of that conquest, the Orientalization of the West as well as the Westernization of the East, are so ample and numerous that we need not insist upon them;[40] moreover, they concern a period later than that covered in this volume. We mention them here only to illustrate the continuity of Eastern-Western interchanges in every age. The interchanges never stopped, and continue to this day, but their intensity and their rhythm in either direction vary from time to time.

MATHEMATICAL TRADITIONS

Examples of the survival of pre-Homeric scientific ideas have already been quoted in the preceding chapters whenever it was found convenient to adduce them. In this and the following sections we will now try to bring all such examples together, whether already quoted or not, after having classified them broadly by subject. Some of those examples are relatively late in date but that does not matter, because if old Egyptian ideas survived let us say in Hellenistic times, they must have existed in a latent way during the whole of the intervening period, however long. This is especially true of written ideas, which may be forgotten, that is, the papyrus or tablet on which they were written may be lost and buried for centuries, then be rediscovered and take a new lease of life. The most ancient traditions, however, were very largely oral and oral traditions cannot be interrupted at all without dying.

Whether an ancient idea never ceases to live and circulate, or on the contrary disappears for a while, or seems to disappear, and pops up only at long intervals, in any case credit must be given to the early inventors. Many such ideas vanished in silence and obscurity, and managed to weather the vicissitudes of the Dark Age in the manner of tough-coated spores surviving unfavorable seasons, then reappeared in Homer and Hesiod or in the reported sayings of the Ionian philosophers, or later still.

When an old Egyptian idea is expressed in a Greek book, we must assume that it was reinvented by the Greeks or else that it was transmitted to them, regularly or irregularly, in the open or underground. If it is not expressed, we cannot conclude that it did not exist or that it was not transmitted. Arguments *a silentio* are always weak and often worthless. The kind of argument to be avoided is the one used by no less a man than Zeuthen [41] when he remarked that no pentagon or decagon is to be found in Egyptian monuments, and hence that Egyptian geometry could not have been very highly developed. It is highly probable that the Egyptians did not know the geometrical construction of a pentagon, for that would imply a relatively high degree of geometrical sophistication.[42] However, the fact that

[39] Dominique Mallet, *Les rapports des Grecs avec l'Egypte de la conquête de Cambyse 525 à celle d'Alexandre 331* (Mémoires de l'Institut français d'archéologie orientale, vol. 48, folio, xv+209 pp.; Cairo, 1922).

[40] Pierre Jouguet, *L'impérialisme macédonien et l'hellénisation de l'Orient* (Paris, 1926). Jouguet had admirably told one side of the story, but there is another side, the Orientalization of the West which is not as well documented perhaps as the first, yet can be read in Roman history. Sarton, "Unity and diversity of the Mediterranean world," *Osiris 2*, 424–432 (1936).

[41] H. G. Zeuthen, *Histoire des mathematiques dans l'antiquité et le moyen âge* (Paris, 1902), p. 5.

[42] It would imply knowledge of the so-called golden section, the division of a line in extreme and mean ratio (Euclid, II, 2) [*Isis 42*, 47 (1951)].

they did not use the pentagon in the arts does not prove such ignorance nor would their use of it prove their knowledge of its geometrical construction. Indeed, it is easy enough to divide a circle into five equal parts without geometrical consciousness of any kind. It may be added that pentagonal ornaments occur in Mycenaean art, that a regular dodecahedron of Etruscan origin was found on Monte Loffa, near Padua, and that no fewer than twenty-six objects of dodecahedral form and Celtic origin have been recovered.[43] In brief, elaborate geometrical ornaments can be drawn without explicit geometry; the lack of such designs proves only the lack of interest in them. Incipient geometers may have played with pieces of wood shaped like regular triangles and squares and built solid angles with them. The combination of those solid angles would lead them to the construction of regular polyhedra (except the dodecahedron). The base of a solid angle made of five regular triangles would naturally be a regular pentagon. Four solid pentagonal angles brought together would make a regular dodecahedron.

There are Babylonian prisms of pentagonal and even of heptagonal base, but we do not for that reason think of ascribing a knowledge of the geometrical construction of those bases to the Babylonian geometers.[44] The earliest treatise on the construction of the regular heptagon was probably the lost one of Archimedes (III-2 B.C.) preserved in the Arabic version of Thābit ibn Qurra (IX-2).[45]

Egyptian arithmetic. We have explained that the Egyptians had a preference for fractions the numerator of which was the unit and tended to express other fractions in terms of the former. The fractions of that simpler and preferred type, like 1/72, were called "part 72." The Greek representation of those fractions was equally simple; 1/72 was written *οβ′* or *οβ″* (as if we wrote 72′). The Egyptians had separate symbols for ½ and ⅔, and so did the Greek. These can hardly be coincidences. Moreover, we can detect Egyptian traces in Greek mathematics down to the beginning of medieval times.

According to Psellos (XI-2) — a late witness, I admit — Anatolios and Diophantos, who both flourished in Alexandria at the same time (III-2), wrote treatises on the Egyptian method of reckoning. Two late mathematical papyri, the Michigan papyrus No. 621 of the fourth century and the Akhmīm one of the sixth or seventh century, as well as Coptic ostraca from Wādī Sarga (near Asyūt) dating from the same period, contain unmistakable examples of Egyptian computation.[46] Moreover, Ptolemy (II-1),[47] and even Proclos the Successor (V-2), the most illustrious philosopher and teacher of his time, and one of the last heads of the Academy,[48] were still writing fractions in the Egyptian manner. For example, Proclos wrote ½ ⅓ 1/15 1/50 for 23/25.

[43] Sir Thomas Heath, *History of Greek mathematics* (Oxford, 1921), vol. 1, p. 160 [*Isis* 4, 532-535 (1922)].

[44] Professor Ferris J. Stephens, curator of the Babylonian collections of Yale University, kindly sent me (letter of 7 February 1945) drawings of such bases (four heptagons, one pentagon). They are sufficiently irregular to prove their empirical construction.

[45] Carl Schoy, "Graeco-Arabische Studien," *Isis* 8, 35-40 (1926).

[46] Louis C. Karpinski, "Michigan mathematical papyrus No. 621," *Isis* 5, 20-25 (1923), 1 pl. *Introduction*, vol. 1, p. 354. J. Baillet, *Le papyrus mathematique d'Akhmīm* (Mémoires de la Mission archéologique française au Caire, vol. 9, 91 pp., 8 pls.; Paris, 1892); *Introduction*, vol. 1, p. 449. W. E. Crum and H. I. Bell, *Wadi Sarga* (Coptica, vol. 3; Copenhagen, 1922), pp. 53-57.

[47] *Almagest*, I, 9.

[48] Proclos died in 485. The Academy was closed in 529 by Justinian's order.

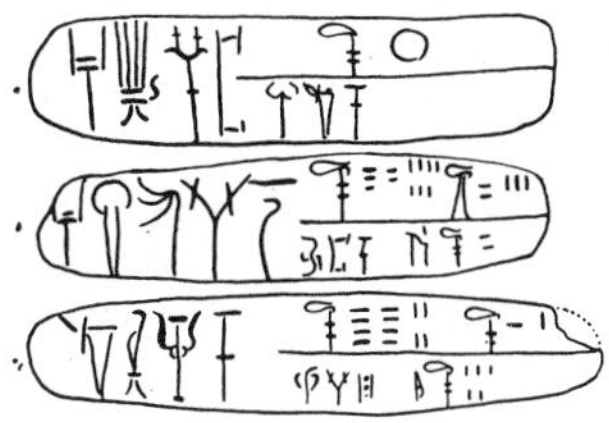

Fig. 33. Minoan arithmetic: percentage tablets. [From Sir Arthur Evans, *The Palace of Minos* (London: Macmillan, 1921–1935); see *Isis 24*, 375–381 (1936).]

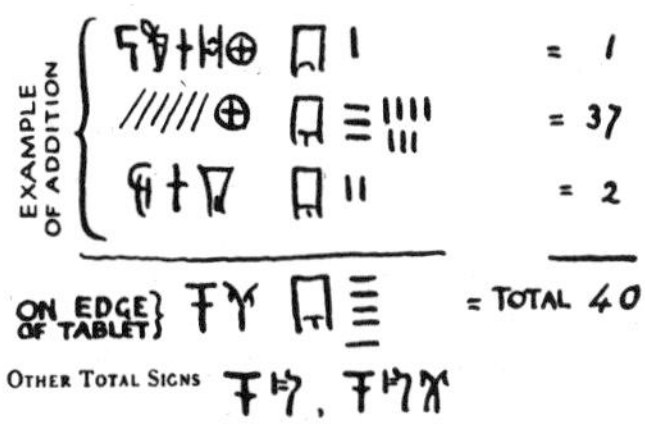

Fig. 34. Minoan arithmetic: example of signs of addition. [From Evans, *The Palace of Minos*.]

Minoan arithmetic.[49] Our knowledge of Minoan mathematics is very restricted because the Minoan scripts have not yet been deciphered. Yet it is clear that many tablets contain numerals, which it has been found possible to interpret.[50] Their numerals were different from the Egyptian ones, yet their methods of reckoning were definitely Egyptian. Both systems were decimal, but the Minoan symbols stopped at the thousands or ten thousands while the Egyptians reached the million level. The most interesting feature of Minoan computations was a system of percentages, amounts mentioned in many tablets being arranged so as to make up 100. For example, on one tablet, the two sums of the upper register 57 + 23 total 80; in the lower register we have 20 with the "throne" sign. Does that mean that the royal share was 20 percent? The Cretans seem to have developed an elaborate system of registration and accounting; they were as business-minded and as fussy about such matters as we are (Figs. 33 and 34).

The eventual decipherment of Minoan writings may give us more information about their mathematical or scientific ideas, whether original or of Egyptian derivation, but in any case Egyptian ideas might and did reach Greece also through other channels.

Egyptian geometry. The invention of geometry and its transmission to Greece was explained by Herodotos in terms often quoted:

This king[51] moreover (so they said) divided the country among all the Egyptians by giving each an equal square parcel of land, and made this his source of revenue, appointing the payment of a yearly tax. And any man who was robbed by the river of a part of his land would come to Sesostris and declare what had befallen him; then the king would send men to look into it and measure the space by which the land was diminished, so that thereafter it should pay the appointed tax in proportion to the loss. From this, to my thinking, the Greeks learnt the art of measuring land; the sunclock and the sundial and the twelve divisions of the day, came to Hellas not from Egypt but from Babylonia.[52]

[49] G. Sarton, "Minoan mathematics," *Isis 24*, 371–381 (1935–36), 6 figs., derived from Sir Arthur Evans, *The palace of Minos*.

[50] The situation is curiously the same as for Mayan archaeology. We cannot read Mayan inscriptions except in so far as they include numerals. The Mayas had developed early (say about the time of Christ) a vigesimal system of numbers.

[51] This king is named Sesostris and there were three kings bearing that name in the XIIth Dynasty (2000–1788). The Sesostris of Greek tradition, however, is a mythical personality which cannot be identified with any one of the known kings of Egypt. The text is quoted from A. D. Godley's translation (Loeb Classical Library).

[52] Herodotos, II, 109.

Of course, geometry was invented not only in Egypt but in other places, for the need of it was soon obvious to any civilized nation. The Egyptian account is plausible *grosso modo*; it was repeated by Strabon (I–2 B.C.) and by Proclos (V–2). Socrates in *Phaidros* made a wider claim.

> I heard, then, that at Naucratis, in Egypt, was one of the ancient gods of that country, the one whose sacred bird is called the ibis, and the name of the god himself was Theuth.[53] He it was who invented numbers and arithmetic and geometry and astronomy, also draughts and dice, and, most important of all, letters.[54]

Socrates then proceeds to explain that the most important of those inventions was that of letters *grammata*, that is, writing. Said Thoth to the king of Egypt, "This invention, O king, will make the Egyptians wiser and will improve their memories; for it is an elixir of memory and wisdom that I have discovered," but the king was not convinced and feared that the invention of writing would impair the memory instead of improving it and that the people would read without understanding.[55] This is one of the earliest criticisms of learning and technique as opposed to wisdom — a criticism that has been repeated time after time apropos of every great innovation.

The Egyptian invention of mathematical and physical sciences is implied in many of the Greek fragments referring to the Ionian philosophers. We shall come back to that when we speak of each of them. Egypt was generally considered by early Greek writers as the cradle of science, and the Greeks who had intellectual ambitions would try to visit that country and to spend as long a time as possible interrogating the men of learning and the priests. They were probably disappointed, because their hopes were a bit wild and because the priests could not or would not communicate much knowledge to infidels and barbarians. Nevertheless, the Greek visitors learned something and their ambitions were sharpened and focused. What does one get from teachers anyhow? Mainly inspiration and hints; real knowledge must be conquered by each man for himself; as to wisdom, if it be not in him, whence will it come?

The most curious reference to Egyptian mathematics is that of Democritos of Abdera (V B.C.), as reported unfortunately by a very late witness, one of the fathers of the church, Clement of Alexandria (155–220).[56] According to Clement, Democritos declared,

> I have roamed over the most ground of any man of my time, investigating the most remote parts. I have seen the most skies and lands and I have heard of learned men in very great numbers. And in composition no one has surpassed me, in demonstration not even those among the Egyptians who are called *harpedonaptai*, with all of whom I lived in exile up to eighty years.

[53] Theuth's name is now generally spelled Thoth.

[54] Plato, *Phaidros*, 274 C. English translation by Harold North Fowler (Loeb Classical Library).

[55] Thoth said, *mnēmēs te gar cai sophias pharmacon hēyrethē*. The conservative king answered, *ucun mnēmēs all' hypomnēseōs pharmacon hēyres.*

[56] *Strōmata* (Book I, chap. 15); Wilhelm Dindorf, ed., *Clementis Alexandrini Opera* (Oxford, 1869), vol. 2, p. 57. That whole chapter 15 deals with the barbarian origins of Greek philosophy, many ancient writers chiefly Plato being adduced in evidence. In the following chapter, Clement shows that the barbarians were inventors not only of philosophy but of almost every art. See also Book V, chap. 7, and Book VI, chap. 4. English version by William Wilson (2 vols.; Edinburgh, 1867–1869).

Who were those *harpedonaptai* or rope stretchers? Were they land measurers, or architects? It has been suggested [57] that they knew the art of drawing perpendiculars on the ground by means of a rope divided by four knots in the proportion 3, 4, 5. That is possible but there is nothing to prove it.[58] It is more probable that they were land measurers, and charged with the proper orientation of buildings, to which the ancient Egyptians attached a deep religious importance. The ceremony of "stretching the cord" (Egyptian term) was the astronomical determination of the axis of a temple along the meridian.[59] A priest or clerk took a sighting of the pole star through a cleft stick; another one stood in front of him with a plumb line and moved until the plumb line and the star were seen in the same direction.[60] Then each drove a stake in the ground, and a cord stretched between the stakes determined the meridian. It is possible that the perpendicular east-west direction was determined afterward by means of a 3, 4, 5 rope, as suggested above, or otherwise.[61] The coöperation of the rope stretchers might be asked for frequently during the construction of a large building, or any other architectural project. The same rope stretchers may or may not have been used also for the redetermination of land boundaries after the flood. It is remarkable that we do not hear of them any more in Greek literature.

Babylonian mathematics. It is relatively easy to discuss the survival of ancient Egyptian mathematics because there was no other. Later documents known to us only repeat less well the ancient ones. The Babylonian situation is very different; there was a great mathematical and astronomical revival in the last two or three pre-Christian centuries. The Chaldean (*Chaldaioi*) mathematicians of that later period did not disregard the ancient ideas but developed them so much that they created definitely new departures. The mathematics that influenced the Greek authors like Hypsicles (II-1 B.C.) and Geminos (I-1 B.C.) were definitely Chaldean. It is true that Heron of Alexandria (I-2) may have inherited older geometrical ideas, but his example is isolated.

As to algebra, some of it may have reached Hipparchos (II-2 B.C.) [62] and some of it did reach Heron of Alexandria (I-2) and Diophantos (III-2), but the

[57] Heath, *History of Greek mathematics*, vol. 1, p. 122.

[58] It is possible, however, that they had some knowledge of the proposition $3^2 + 4^2 = 5^2$ and similar ones. See the Kahun Papyrus 6619 of the Berlin Museum, in M. Cantor, *Vorlesungen zur Geschichte der Mathematik* (Leipzig, 1907), vol. 1, p. 95.

[59] T. Eric Peet, *The Rhind mathematical papyrus*, p. 32.

[60] The actual instruments used exist in very early exemplars. See Ludwig Borchardt, *Altägyptische Zeitmessung* (Berlin, 1920) [*Isis 4*, 612 (1921-22)], pp. 16-17.

[61] For example, let it be required to draw a perpendicular line to the meridian at O (Fig. 35). Let us measure on the meridian $OA = OB$, then take a rope much longer than AB and divide it into two equal parts by a knot at C. The rope is fastened at A and B and then the knot C is taken as far east as possible; the line OC is the perpendicular. This would be obvious to the Egyptians because of their intuitive understanding of symmetry. For verification, the operation would be repeated westward, and OC and OD should be collinear. The collinearity would be easily tested with three stakes or plumb lines.

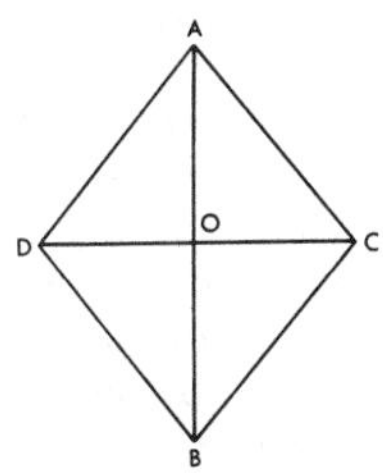

Fig. 35.

[62] *Isis 26*, 81 (1936).

inventions of Archimedes (III–2 B.C.) were probably his own.[63] When one tries to explain how Babylonian ideas could have reached Heron and Diophantos and yet have remained unnoticed by other Greek mathematicians, one keenly realizes how little the ancient mathematical traditions are clear to us; we get only a few glimpses of them here and there. Perhaps the wonder must be looked for in the opposite direction; is it not almost miraculous that so much of the highest ancient mathematics, which could never have interested more than a very few people, has been preserved for us?

The sexagesimal ideas go back to high antiquity, and though the Greeks may have obtained them from the Chaldeans, we may consider the Greek usage as a distant continuation of the Sumerian one. For example, Ptolemy divided the circle into 360 degrees[64] and divided the hour into 60 parts.[65] Now the division of the equator into 360°, comparable to that of the day into 360 *gesh*, is very ancient. The division of the ecliptic into 360°, on the contrary, dates only from Achaemenidian times.

The Greeks inherited the sexagesimal system from the Sumerians but mixed it up with the decimal system, using the former only for submultiples of the unit and the latter for multiples, and thus they spoiled both systems and started a disgraceful confusion of which we are still the victims. They abandoned the principle of position, which had to be reintroduced from India a thousand years later. In short, their understanding of Babylonian arithmetic must have been very poor, since they managed to keep the worst features of it and to overlook the best. This must have been due to deficient tradition rather than to lack of intelligence, or else to the fact that, as we should remember, intelligence is always relative. The Greeks used their intelligence in a different way and did not see simple things that were as clear as daylight to their distant Sumerian and Babylonian predecessors.

ASTRONOMIC TRADITIONS

The Greeks inherited immemorial Egyptian ideas; the stimulation that they received from Babylonia was at once much greater and much later. As far as we can judge, pre-Homeric astronomy was very largely of Egyptian origin. Yet consider the theory of the five ages of the world outlined by Hesiod (VIII B.C.) at the beginning of his *Works and Days*. The first age, according to him, was a divine golden age; evil increased in every following age until it reached a maximum in his own days, and the old poet lamented, "Would that I were not among the men of the fifth generation, but either had died before or been born afterwards. For now truly is a race of iron, and men never rest from labour and sorrow by day, and from perishing by night; and the gods shall lay sore trouble upon them."[66] This suggests two remarks. Why did Hesiod call the people of his own age "a race of iron"?[67] The Iron Age had begun many centuries before, but the introduction of iron was remembered by him as a new and ominous departure; he spoke of the Iron Age as we do of our own age when we call it the machine age, or the age of

[63] Yet see p. 74.
[64] Ptolemy, *Almagest*, I, 9.
[65] *Ibid.*, tables in II, 12.
[66] Hesiod, *Works and days* (ll. 174–178; Hugh G. Evelyn-White's version in the Loeb Classical Library.
[67] *Nyn gar dē genos esti sidēreon.*

steam and electricity. In the second place, does not his description of the first age remind us of the Sumerian tale of man's golden age quoted in the previous chapter?[68] It is true the same conceit might have developed independently in different places. The idea that everything goes from bad to worse is not unnatural to old men witnessing their own decay and feeling less and less able to cope with the changing world.

The art of astronomic observations was well developed both in Egypt and in Mesopotamia and some knowledge of it or sufficient hints may have reached the Aegean peoples from either side. Yet, the problems involved are so natural and the solutions so well determined that rediscovery of the same art is possible without need of imitation, or at least without consciousness of imitation. The Egyptian legacy was largely in the form of its decans and of the constellations and the stars pertaining to each; that legacy can be traced throughout the ages. The Egyptians, we may recall, divided the whole horizon into 36 decans; each division corresponds to one-third of a sign of the zodiac. The decanic division referred to the equator, while the later zodiacal one refers to the ecliptic, but as the extent in latitude of the decads and zodiacal signs was not clearly stated, the star groups and the star lore relative to them could easily flow from one system to the other.[69]

We must assume that some knowledge or awareness of the Babylonian tables also percolated westward. As to the calendar, it accompanied the Egyptian or the Babylonian merchants wherever they went. The ancient Greek calendar was lunar, but with some consideration of the seasons, that is, of the year. The only way of reconciling the lunar and solar cycles was to take into account common multiples of each. In this the Greeks had Babylonian examples or they could avail themselves of Babylonian experience.

We have seen that the Babylonians had also discovered the synodic periods of Venus and Mercury, and they originated the idea of the "great year," the cycle of 36,000 years which would reappear so curiously many centuries later in Plato's *Republic* (see p. 71). The conception of the *saros*, as a period of 3,600 years may also be of early origin, but when people use the word saros they think almost always of a much shorter period, of which neither the Babylonians nor the Greeks had any idea before the fifth or fourth century B.C.[70]

There are such deep and persistent misunderstandings on the subject that it is necessary to clear them up here and now. The early Babylonians, it was believed,

[68] For a technical discussion of that Hesiodian-Babylonian parallel see King, *History of Babylon*, pp. 302 ff.

[69] For the decan tradition see Wilhelm Gundel, *Dekane und Dekansternbilder. Ein Beitrag zur Geschichte der Sternbilder der Kulturvölker. Mit einer Untersuchung über die ägyptischen Sternbilder und Gottheiten der Dekane von Siegfried Schott* (*Warburg Studien 19*; 462 pp., 33 pls.; Glückstadt: Warburg Bibliothek, 1936; [*Isis 27*, 344–348 (1937)].

[70] The word saros is obviously not an original Greek word; its accentuation is uncertain, and it appears but very late in a Greek text, the *Assyriaca* of Abydenos, written about the beginning of our era; see Carolus Mullerus, *Fragmenta historicorum graecorum* (Paris, 1851), vol. 4, p. 280. Its meaning in that text is a cycle of sixty times sixty years or 3600 years. The word is derived from Sumerian *shar*=3600. Berossos (III–1 B.C.) was very probably the transmitter of that Babylonian idea. It is significant that the Babylonians distinguished three periods of years which were called (I quote the Greek transcriptions), *sōssos*=60 years, *nēros*=10 *sōssoi*, *saros*=60 *sōssoi*; we notice once more the typical mixture of decimal and sexagesimal factors. The misuse of the word saros to designate the 18-year period was introduced very late; perhaps as late as 1691, by Edmund Halley! See O. Neugebauer, "Untersuchungen zur antiken Astronomie. III. Die babylonische Theorie der Breitenbewegungen des Mondes; V. Der Halleysche 'Saros'," *Quellen und Studien zur Geschichte der Mathematik* (Berlin, 1938), Abt. B, Band 4, pp. 193–358, esp. p. 295; 407–411.

had discovered an 18-year period,[71] at the end of which the Sun and the Moon occupied the same relative positions as at the beginning. Each saros completed a cycle of possibilities of those relative positions, and hence the eclipses that occurred in one cycle would, or at least might, be repeated in every other. However, there is no mention of this saros in early Babylonian documents. Such a period would have been exceedingly difficult to discover, if only because it does not embrace a whole number of days but 8 hours more.[72] "To get eclipses at about the same time of day the period must be trebled; after 54 years [73] the visible eclipses return to a great extent in the same order. If we arrange the visible eclipses in series of 54 or of 18 years, it is then not difficult to establish the existence of the saros. But it is quite another matter to find or to discover this period. If someone who knew nothing of this period was given the task of finding from a complete list of lunar eclipses, e.g. from Oppolzer's *Canon*, a period after which they would return in the same manner, he would certainly find it a very difficult one." [74] For the early Babylonians, even if they had had complete lists of all the visible eclipses (which is very doubtful), the discovery of the saros was not simply difficult but impossible.

Scientific astronomy, by which we mean a system of rational explanations of the movements of celestial bodies, owed little to the early Babylonians and Egyptians, except experimental data and perhaps the means of obtaining more data. The desire for such explanations seems typically Greek and their elaboration occupied Greek minds for many centuries. The knowledge that some Greeks, like Hypsicles (II-1 B.C.), Geminos (I-1 B.C.), and Diodoros of Sicily (I-2 B.C.), obtained from Mesopotamia does not enter into account, for that was late knowledge, after the formulation of Hellenic astronomy. Scientific astronomy, we might say, was Greek or perhaps late Babylonian, Chaldean.

The "scientific astrology" that obtained so much popularity in the last centuries preceding our era was Chaldean and Egyptian; it was also Greek, being a perverse synthesis of all the rational and irrational knowledge which had accumulated up to that time. The success of astrology among intelligent and educated people was due to its scientific structure and appearance, while all the fables that it carried and its fantastic purpose appealed to the natural imbecility of men and their love of marvels. The purpose was as old as the hills. Man has always been anxious to know the future, and, with admirable inconsistency, if some misfortune was foretold, he hoped to avert it. Many fairy tales are based on that very theme:

[71] More exactly 223 synodic months, equalling 242 draconic months (6585⅓ days or 18 Julian years plus 11 days); after this period full and new moon return to the same position relative to the nodes.

[72] Theodor von Oppolzer, *Kanon der Finsternisse* (Vienna, 1887). O. Neugebauer has proved that the saros was not sufficient to predict eclipses of the Sun, though it may have sufficed to predict eclipses of the Moon. It is significant that the earliest Greek text on eclipses is the one by Philippos of Opus (*c.* 350 B.C.), restricted to lunar eclipses. See Neugebauer, "Untersuchungen zur antiken Astronomie." The matter has been very clearly explained by the Dutch astronomer Antonie Pannekoek, "The origin of the saros," *Dutch Academy, Proceedings of the section of sciences 20*, 943–955 (Amsterdam, 1918). In my own brief account, I have largely followed Pannekoek and even used his own wording, because it could not be improved upon.

[73] More exactly, 54 years 34 days. That is the cycle called later *exeligmos* by Geminos of Rhodes (I-1 B.C.) and by Ptolemy, *Almagest*, IV, 2. It is the shortest period containing whole numbers of synodic months and days and exact returns of the Moon to former positions. The word *exeligmos* was first used for military evolutions bringing soldiers back to their original places, then for the revolutions of celestial bodies.

[74] Pannekoek, "The origin of the saros," p. 944.

at the time of the hero's birth, soothsayers predict that he will die in a definite kind of accident; pains are taken to avoid the possibility of such an accident; yet it occurs and the hero dies as had been foretold.

The words Chaldean and Egyptian have retained an occult flavor, because of the astrology and other superstitions connected with them. We have already explained that Chaldean refers to a late epoch; the word Egyptian is more ambiguous, yet, in its occult meaning, it refers rather to Ptolemaic than to ancient Egypt. It was during the Ptolemaic period (which coincided roughly with the Chaldean) that the astrologic ideas which have come to us in Greek, Latin, Arabic, and almost every vernacular were first elaborated and clearly expressed.[75] The "Egyptian days" often quoted in medieval writings, for example, in Anianus (XIII-2?), are simply the dismal days (*dies mali*) of that age.[76]

Ptolemaic astrology was largely of Chaldean origin, but it embodied ancient Babylonian and ancient Egyptian conceits mixed up with Greek astronomy. The creation of that astrological *Weltanschauung* which dominated late ancient and medieval thought and is not yet extinct today proves the survival through the Dark Interlude of some astronomical ideas of immemorial antiquity.

BIOLOGIC AND MEDICAL TRADITIONS

Ideas concerning life and death, health and disease, and the means of lengthening life or of restoring health when it is lost, must be among the first to exercise human minds anywhere. We would expect such ideas, or at least some of them, the most agreeable and felicitous, to be transmitted from generation to generation for thousands of years. Unfortunately, they are not as tangible or specific as, say, astronomical ideas, and the existence of definite traditions is more difficult, if not impossible, to prove. Many of them are so simple and natural that they might (and did) occur independently in many places.

D'Arcy W. Thompson, the learned translator of Aristotle's *Historia animalium*,[77] has pointed out that many of the "vulgar errors" which escaped the master's critical mind must have been very ancient ones, errors so deeply rooted in his unconsciousness that he did not think of challenging them. The stories concerning "the goats that breathe through their ears, the vulture impregnated by the wind, the eagle that dies of hunger, the stag caught by music, the salamander which walks through fire, the unicorn, the mantichore" would not surprise us in a medieval bestiary, but we are shocked to find them in Aristotle. "Some of them," said Sir D'Arcy, "come through Persia from the farther East: and others (we meet them once more in Horapollo,[78] the Egyptian priest) are but the exoteric or allegorical expression of the arcana of ancient Egyptian religion." The mantichore can be easily recognized as Persian, for Aristotle had its story from Ctesias (V B.C.) and

[75] Carl Bezold and Franz Boll, "Reflexe astrologischer Keilschriften bei griechischen Schriftstellern," *Sitzber. Heidelberger Akad., Phil. Kl.*, No. 7, 54 pp. (1911). Franz Cumont, *L'Egypte des astrologues* (254 pp.; Brussels: Fondation Égyptologique Reine Elisabeth, 1937) [*Isis* 29, 511 (1938)].

[76] Of course, there were "unlucky days" in every age, like "Friday the 13th" in our own.

[77] In Aristotle, *Works. The Oxford Translation* (vol. 4, 1910). The remark that I quote was made by him in *The legacy of Greece*, p. 160, reprinted in his *Science and the classics* (Oxford: Oxford University Press, 1940) [*Isis* 33, 269 (1941-42)], p. 74.

[78] Horapollon of Nilopolis (IV-1), Egyptian archaeologist who wrote in Coptic a treatise on hieroglyphics, known to us in a poor Greek version.

its name is Avestan;[79] some of the other stories might be traced to Egyptian or other oriental sources, or they might not. The tradition of such fables might be purely oral; it would not be weaker on that account but would leave no traces. In any case, we can hardly imagine Aristotle inventing such stories; it was bad enough for him to give them a new currency and a kind of scientific prestige.

Another story told by him [80] has been traced to an Egyptian source in an unexpected way. Aristotle spoke of edible sea urchins whose ova grow in bulk at the time of the full Moon. That was then as now a part of the fishermen's folklore [81] and Aristotle tried to rationalize it. In 1924 an English zoölogist, H. Munro Fox, investigated the facts and established that the Mediterranean sea urchins do not "increase and decrease" with the Moon, but that their Red Sea cousins spawn regularly at each full Moon during the breeding season. In other words, the story is true in the Red Sea but false in the Mediterranean; it had passed from Egyptian to Aegean folklore, probably in very early times, and has remained there unchecked until our own day.[82]

Let us pass to medicine. The Egyptian Imhotep, who might possibly be identified with a vizier of King Zoser (Third Dynasty, beginning of thirtieth century), was glorified by his people and finally accepted by them as a medical god. The apotheosis of Imhotep prefigured that of Asclepios.[83] Medical usages being of immediate interest to any intelligent visitor or to any one whose health was jeopardized, we may assume that they had many chances of being transmitted to the Aegean people and to their Greek successors. The relations between Greece and Egypt increased considerably during the Twenty-sixth Dynasty (663–525), the so-called Saitic Renaissance. The capital was then in Sais in the western Delta (on the Rosetta arm of the Nile). One of the kings of that dynasty, Ahmose II (called Amasis in Greek; ruled 569–525), allowed the Greeks to build themselves a city in Naucratis (on the Canopic arm of the Nile), and they soon made of that city the most important commercial center in Egypt. That Greek center, not very far from the capital, was the means of abundant interrelations between Greece and Egypt.[84] The two cities, Sais and Naucratis, constituted an anticipation of Alexandria. Of course, all this occurred late in the sixth century, yet before the time of Herodotos and of Hippocrates.

Herodotos [85] noticed that "the practice of medicine is so divided among the Egyptians, that each physician is a healer of one disease and no more. All the country is full of physicians, some of the eye, some of the teeth, some of what pertains to the belly, and some of the hidden diseases." This piece of information is confirmed from Egyptian documents of the Old Kingdom (*c.* 3400–2475), wherein we find the hieroglyphic names of the medical branches mentioned in the Greek text.[86]

[79] "If we are to believe Ctesias," wrote Aristotle cautiously (*Historia Animalium*, 501A 25), but he did not hesitate to repeat the description of that fantastic animal. The name *mantichōras* or *mantichoras* means in its Old Persian (Avestan) form, manslayer.

[80] Aristotle, *De partibus animalium*, 680A, 32.

[81] The lore is extended to every kind of shellfish, which are supposed to increase and decrease with the Moon.

[82] G. Sarton, "Lunar influences on living things," *Isis 30*, 495–507 (1939); see p. 505.

[83] Jamieson B. Hurry, *Imhotep* (ed. 2, 228 pp., 26 ills.; Oxford, 1928) [*Isis 13*, 373–375 (1929–30)].

[84] Breasted, *History of Egypt*, pp. 590–591.

[85] Herodotos, II, 84.

[86] Hermann Junker, "Das Spezialistentum in der ägyptischen Medizin," *Z. Ägyptische Sprache 63*, 68–70 (1927).

Some of the temples of Egypt were very early applied to medical purposes. The sick and the afflicted, sterile women desiring children, and patients of every kind would spend the night in the temple, sometimes many nights and days, and try to obtain healing or comfort from the gods. Priests would look after them, pray with them, use incantations, and sometimes alleviate their troubles with "proved" remedies or gentle treatment. The very fact of resting a long while in the temple, of enjoying significant dreams, and of being bathed, as it were, in the divine effluvia did often suffice to quiet the patient's mind, to improve his condition, and even to cure him completely. Holy books and medical ones might be kept in such temples for the guidance of the intercessing and nursing priests. Indeed, two Berlin medical papyri (one dating from the Nineteenth or Twentieth Dynasty, 1350–1090, the other from Ramses II, 1292–1225) probably belonged to the temple of Ptah at Memphis. Greek travelers would visit such temples, and even if they were unable (as was likely) to understand the books or the formulas intoned by the priests, they could not help seeing the sick people lying in the temple yards or witnessing the priestly ministrations. Linguistic barriers could not impede the transmission of such knowledge. A good account of the healings due to Isis is given by Diodoros of Sicily (I–2 B.C.).[87]

The practice of incubation (encatheudein and many other Greek expressions) was popular in the Greek temples, chiefly those devoted to the Greek Imhotep, Asclepios. It was continued throughout the Middle Ages in Western and Eastern churches, and can still be witnessed today in the Aegean islands or in provincial churches of the Greek mainland.

The accumulation of knowledge was nowhere as slow as it was in the empirical study of the plants growing around us, the rejection of the dangerous ones and the recognition and adoption of those that were useful as food or medicines. That process of selection took place throughout the prehistoric ages, and the Egyptians and Sumerians of the first dynasties were already enjoying much knowledge of that kind bequeathed to them by their distant ancestors. In their times they must have bequeathed at least a part of their experience to all the peoples they were dealing with — Aegeans, Phoenicians, Greeks, and others.

In order to measure the debt of the Greeks, say of Homer's age, to their Oriental predecessors, we still lack an instrument of prime importance, to wit, a good etymological dictionary including lists of foreign Greek words grouped according to their several origins.[88] It is probable that such lists would reveal the Oriental provenience of many plant or animal names. From it one might conclude that the Greeks became acquainted with this or that herb or this or that animal from their contacts with Egyptians, Babylonians, Persians, and so on. However, one should be careful not to employ such a method too rigidly. The herbs that the Greeks would have appreciated first and most might have received new Greek names. Thus herbs could have been transmitted without their original names, and con-

[87] Hurry, *Imhotep*, pp. 49–56, 105–11. Mary Hamilton, *Incubation or the cure of disease in pagan temples and Christian churches* (234 pp.; London, 1906); Diodoros' account is given by her in English on p. 98.

[88] I am acquainted with the following partial lists; there may be others. Heinrich Lewy of Breslau, *Die semitischen Fremdwörter im Griechischen* (268 pp.; Berlin, 1895). At the end of Georg Curtius, *Principles of Greek etymology* (London, ed. 5, 1886), vol. 2, pp. 461–473, there are Sanskrit and Iranian indices.

versely, names might be transmitted without the herbs or attached by mistake to other herbs.[89]

TECHNICAL TRADITIONS

The Egyptians and the Babylonians were great builders and ingenious artisans and as such had been obliged to solve a good number of technical problems. The monuments created by them would be obvious to any visitor, and the objects exported to and by Aegean or Phoenician middlemen would diffuse technical ideas wherever they were carried. The Aegean builders may have taken lessons from their Egyptian forerunners, and they may even have borrowed Egyptian workmen.

Consider mining, in which the ancient people of the Near East had acquired considerable experience; the tradition of it was transmitted to the rest of the Mediterranean world by Phoenicians. At any rate, some local stories seem to confirm that hypothesis. The half-mythical Cadmos, son of a Phoenician king, is said to have brought the art of mining to the Greeks. He was the first to work the gold and silver mines of the Pangaion mountains in Macedonia. Another Phoenician prince, Thasos, worked gold mines in an island of the northern part of the Aegean Sea which was named after him, the island of Thasos.[90]

After the fall of Crete, Cyprus became the metallurgical center of the Aegean world, and on account of its proximity to the Syrian coast it received some of the earliest Phoenician settlements. The great Samian builders and engineers, the most famous of whom was Eupalinos (VI B.C.), may have derived their knowledge from very early sources, for Eupalinos himself came from Megara.[91]

Each separate invention would require a special study, which would confirm Greek dependence on Oriental examples or demonstrate Greek originality. Let us examine two cases. The invention of a way of soldering iron is traditionally ascribed to Glaucos of Chios (VI B.C.). It is difficult to believe that the early Hittite metallurgists would have overlooked that problem, the solution of which was often urged upon them by circumstances. The soldering of gold was perfectly done by Egyptians at the beginning of the first dynasty.[92] The people of Chios had the advantage of being able to use mastic[93] to keep the air away from the surfaces to be fused together. This may have enabled Glaucos to perfect the invention, if he did not make it.

The other case is that of the level. The invention of that tool and other tools used by masons and stonecutters was ascribed to Theodoros of Samos (VI B.C.). Now the Greek level (*diabētēs, libella*) is identical with the one used by the ancient Egyptians.[94]

Many recipes described in the book of Zosimos of Panopolis[95] (III-2) and in

[89] It would be worth while to make a fresh study of Dioscorides (I-2) from that point of view. See Max Wellmann, "Die Pflanzennamen des Dioskurides," *Hermes* 33, 360–422 (1898) and the indices at the end of his edition of Dioscorides (Berlin, 1914), vol. 3, pp. 327–358. The index of plant names taken out of the dictionary of Pamphilos (I-2) begins with a long list of Aegyptiaca.

[90] Herodotos, VI, 47.

[91] *Ibid.*, III, 60.

[92] Petrie, *Wisdom of the Egyptians*, p. 119.

[93] Resin exuded by the mastic tree (*Pistacia lentiscus*) abundant in Chios, one of the main sources of its prosperity throughout the ages.

[94] Clarke and Engelbach, *Ancient Egyptian masonry*, p. 224, Fig. 264; other Egyptian tools are illustrated.

[95] Panopolis or Chemmis on the Nile in Upper Egypt, modern Akhmīm.

the chemical papyri of Leyden and Stockholm (III–2) are of Egyptian origin, though it is not yet possible to determine their antiquity. (Some may be Ptolemaic, and Greek rather than Egyptian.) The excellence of the ancient Egyptian artisans and of their competitors in Western Asia suggests that they had made many experiments in the manipulation and combination of substances. Technical experience of that sort could easily be transmitted for thousands of years, from father to son, from master to apprentice, from place to place, without the formality of writing. We may safely assume that the Greeks inherited much of it in various ways.

Finally, we hear of an Achaean prince visiting the Hittite court about the fourteenth century to study the training of horses and the handling of chariots.[96] Other contacts between Hittites and Achaeans suggest that the latter may have drunk straight from Hittite sources instead of depending always on Phoenician brokers.

MYTHOLOGY

Though mythology is out of our field, it cannot be omitted in any study of the influences of their Oriental forerunners to which the ancient Greeks may have been submitted. In all times and places foreign cults have exerted a peculiar fascination upon certain people. It would seem that the Greeks, or let us say, some Greeks, were very early captivated by the gods of Egypt and Syria. As compared with scientific ideas which are often esoteric, and with technical ideas which are implied in objects but require a kind of rediscovery, the religious ceremonies and rituals were celebrated with a great wealth of public and private illustrations. No visitor could escape them, and if he had in him occult tendencies he was likely to be charmed and allured. Were not those Egyptian gods, worshiped in such spectacular manner, especially powerful? Would they not help his own salvation, or at least fulfill some of his desires? The visitor might return home half-converted, carrying in his heart new aspirations and new hopes.

In a previous section we spoke of incubation from the medical point of view, but incubation was primarily a religious ceremony. To the Egyptians, the sleeper was a temporary guest of the Other World, a temporary companion of the dead. While he was slumbering in the temple, he could commune with the gods and with the spirits. That idea can be traced in the Greek religion as well as in the Egyptian. It gave a peculiar value to dreams, especially to temple dreams. We may assume that the Greeks inherited it from the Egyptians.[97]

It is probably that the influence exerted by the Oriental religions was at first general and vague. Yet Isis began her foreign conquests in the seventh century, if not before. Herodotos[98] says that the women of Cyrene worshiped her. The diffusion of the Egyptian religion was much intensified when the Naucratis settlement was established in the Delta in the sixth century, and from that time on it increased steadily. Temples and inscriptions to Isis and other Egyptian gods can

[96] Georges Contenau, *La civilisation des Hittites et des Mitanniens* (Paris: Payot, 1934), p. 142.

[97] Adrian De Buck, *De godsdienstige opvatting van den slaap inzonderheid in het oude Egypte* (Leiden, 1939) [*Chronique d'Egypte 15*, 215 (1940)]. For mysteries, Greek and Oriental, see Franz Cumont, *Lux perpetua* (Paris: Geuthner, 1949) [*Isis 41*, 371 (1950)], pp. 235–274.

[98] Herodotos, IV, 186.

be found in many of the Islands, even in the sacred Delos. Gradually the Egyptian and Greek gods were brought together and sometimes assimilated. Herodotos identified Amon with Zeus, Isis with Demeter, Osiris with Dionysos, the cat-headed Pasht with Artemis, Thoth with Hermes, Ptah with Hēphaistos. He seemed anxious to trace Greek rituals and theology to Egyptian models. We have already explained that Asclepios was the Greek counterpart of Imhotep.[99]

One cannot appreciate Greek culture in its full complexity without attaching great importance to the sacred mysteries, the celebration of which satisfied the emotional needs of the people. Those mysteries which constituted the inner life of religion were to a large extent of foreign origin. They permeated not only the folklore at every level of society, but the arts, poetry, the drama, even philosophy. The Eleusinian mysteries originated probably in Egypt.[100] The main gods of Eleusis were Demeter, the glorification of motherly love (*cf.* Isis) and Triptolemos, god of cornsowing, inventor of the plow (*cf.* Osiris). The comparisons between Egyptian and Greek myths should not be carried too far; the transmission of inventions (whether religious or technical) is often restricted to a mere hint, but that hint is like a spark which may start a conflagration. The Eleusinian mysteries might be largely independent of the Egyptian religion, but they came back very close to it. Indeed, some of the feelings expressed in the Homeric hymn to Demeter, or in the writings of Pindar, Sophocles, Plato, or Plutarch might have been expressed by Egyptian priests. Let us recall only Sophocles' words, "Thrice blessed are the mortals who will go to Hades after having contemplated those mysteries; only they will know life immortal, for the others there will be nothing but sufferings." [101]

Orphism was of Thracian and Phrygian origin and the Dionysian mysteries were probably derived from Crete and Egypt. The "sacred heart" of Dionysos Zagreus symbolized immortality and the migration of souls. From the fifth century on, Orphism and the Dionysian mysteries tended to combine with the Eleusinian mysteries.

The religious influence of Egypt was greater upon the Old Testament than upon Greek literature. There is tangible evidence of it in the Books of Wisdom and in the Psalms. In the third century, thanks to the Septuagint, those Egyptian ideas rejoined in the Greek minds the Egyptian seeds that had been sown in them more directly centuries or even millennia earlier.

In the days of Hammurabi the old Sumerian god Enlil had been replaced by the god Marduk (or had been renamed Marduk), and associated with the latter was Ishtar,[102] goddess of beauty, of love and fruitfulness. Ishtar was a Moon

[99] The main Greek source on Isis and Osiris after Herodotos is the essay of Plutarch (I-2) *Peri Isidos cai Osiridos* which he wrote for one Clea, priestess at Delphi. This is a very late source, of course, but it embodies old traditions. See text in Plutarch's *Moralia* (Loeb Classical Library, vol. 5). Plutarch had visited Egypt but his knowledge of Egyptian matters had remained very superficial.

[100] Paul Foucart, *Les mystères d'Eleusis* (508 pp.; Paris, 1914). Martin P. Nilsson, *The Minoan-Mycenaean religion and its survival in Greek religion* (604 pp., 4 pls.; Lund, 1928). Georges Méautis, *Les mystères d'Eleusis* (92 pp., ill.; Neuchâtel: La Baconnière, 1934) [*Isis 26*, 268 (1936)]. Foucart had exaggerated the Egyptian influence; Nilsson ascribes the mysteries rather to Aegean influences; Méautis's little book is a popular but sound outline, very readable.

[101] Augustus Nauck, *Tragicorum graecorum fragmenta* (Leipzig, 1856), Sophocles, 753.

[102] West Semitic, Astarte; Greek, Aphroditē; Latin, Venus.

goddess who was able to influence the seas (tides) and women (menstruation). The Phoenicians introduced her worship in the islands, chiefly in Cyprus and in Cythera (southeast of the Peloponnesos). Later the Greeks believed that she had risen from the foam of the sea near Cythera (hence her name Aphroditē Cythēreia). Astarte's connection with the Moon had soon been transferred to another nature goddess of Asiatic origin, Artemis (Diana, to whom the famous temple of Ephesos was dedicated). The cult of Aphrόditē and Artemis was well established in Greece before the Homeric age.

It is unnecessary to extend this mythological digression. We may conclude in a general way that Greek religion was shot through and through with foreign — Egyptian and Asiatic — elements. The foreign gods brought with them foreign notions of many kinds which the Greek people accepted without repugnance and almost unconsciously. Does one doubt the gods?

THE DARKEST HOUR BEFORE THE DAWN

This chapter is suggestive and tantalizing, rather than instructive; it cannot throw much light upon the Dark Age. Even if that age was not intrinsically dark, it is very dark to us, and never more, it seems, than for the period that immediately precedes the Homeric dawn. We know little if anything for certain; we can only guess, but we must guess and there is no harm in that as long as we do not confuse our guesses with certainties. The reader will have noticed that many of our guesses are based on facts of relatively late date. As we have no texts dating from the Dark Age itself, we must depend on later ones, trusting that those late testimonies represent to some extent earlier conditions.

I think that out of all those guesses, which strengthen one another, one can build a fairly strong case for the reality of Oriental (and especially Egyptian) influences upon the creators of the new Greek civilization. We should be careful not to exaggerate those influences, in quality or quantity, but also not to minimize them overmuch, and we should always bear in mind the warning given previously, to wit, we should never conceive those influences as completely preceding Greek culture. Some of them did certainly precede that culture, but the Egyptian, Babylonian, and Greek cultures coexisted for centuries, and hence exchanges of influences could continue and did continue throughout the golden age of Greece and even beyond that, during Hellenistic and Roman days. In fact, they reached a climax during those later days, but those are outside the scope of this volume.

Scholars who would pooh-pooh Egyptian influences remark that the ancient Greek travelers never learned hieroglyphics [103] and had to depend upon the gossip of interpreters. That is probably true; it is also true that interpreters are undependable. Yet interpreters do sometimes tell the truth, or enough of it to put intelligent men on the right track. The stories written late enough by Herodotos, or those written six centuries later still, by Plutarch, contain many errors, but I am more tempted to marvel at the amount of truth that they managed to convey. In judging the past we should never forget the difficulties and uncertainties of any kind of tradition, even the safest. As to the hieroglyphics, the Greeks shared their ignorance of them with all but a very few of the Egyptian people; [104] yet

[103] The earliest text revealing some knowledge of them, rudimentary, is that of Horapollon (IV-1).

[104] It is hardly likely that every priest could read them. Bear in mind the ignorance exhibited by many of our own priests in the Middle Ages, and yet a knowledge of Latin was incomparably

for one Egyptian who could decipher the *Book of the Dead* there were thousands who knew the essential meaning of that book. They knew it by oral tradition, and they could transmit it in the same way. When the Greek-Egyptian osmosis began in good earnest in the sixth century, the decantation of knowledge from Egyptian into Greek vessels increased rapidly. We may be sure that one of the reasons of that swift increase was the slow incubation which had prepared it for a thousand years or more.

Uncritical friends of Hellas like to insist on the profound difference obtaining between the applied, empirical, adulterated knowledge of the Egyptians and Babylonians on one hand and the rational science of the Greeks on the other. I trust that those who have read my account, however brief, of the earliest Egyptian and Sumerian science will already be able to answer such criticism; much in that early science was genuine and admirable, some of it was on a higher level than early Greek science. It is unfair to exaggerate the irrational aspects of the early Oriental science and to compare them with the most rational aspects of Greek science, leaving Greek mysteries and other irrationalities in the darkness.

If the Greeks owed so much to their Oriental predecessors, asked the late John Burnet, how is it that the Greek progress was not more rapid? [105] That is a clever but two-edged query. As far as it can be answered at all, one might say that the Greeks did not receive immediately the best, nor the most complete tradition (how could they?), but only hints, and one might say also that the Greeks were not ready at once to receive such tradition, let alone to improve it. Teaching is always a two-way business; much depends on the teacher and quite as much on the pupil. The tradition of Oriental knowledge was, we may be sure, incomplete, corrupt, and capricious. Every tradition is like it, and therefore, however much we respect it, we must never respect it blindly. We must always be prepared to accept the best and reject the worst. The early Greeks were too unsophisticated to do that; teachers and pupils were equally crude. And then it was the usual vicious circle. One can learn well only what one already knows.

If pre-Homeric knowledge derived from foreign nations was still very vague and uncertain, if it amounted to little more even among the intellectual elite than an awareness of the existence of old and rich civilizations to the south and east of them, together with an itching curiosity, that was already far from negligible. Whenever the desire for knowledge has been awakened in good minds by a few teasing hints, the road to that knowledge is open, and progress toward it, however slow at first, is bound to accelerate.

It would seem that the burden of proof would now rest at least as much upon the shoulders of those who deny or belittle Oriental influences as upon those of their contradictors. Great civilizations, like the Egyptian and Babylonian, radiate outward, and it is hardly conceivable that, intelligent and eager as the early Greeks were, those radiations would have been completely lost on them. Those who deny such possibilities lack sufficient appreciation of the old Oriental cultures, and above all they lack anthropological experience. Both shortcomings were excusable a century ago; they are no longer now.

---

easier to attain than the ability to read hieroglyphic or hieratic texts. The clerical ignorance of Latin has been repeatedly illustrated by George Gordon Coulton, *Europe's apprenticeship* (London: Nelson, 1940).

[105] John Burnet (1863–1928), *Greek philosophy. Part 1 Thales to Plato* (London, 1924), p. 4.

During the Dark Age that preceded the Homeric dawn, the Greek people were not inactive. They were slowly imbibing the ideas that Aegean wanderers and Phoenician traders had distributed among them. In this respect that Dark Age resembles the Christian Middle Ages; both were periods of unconscious assimilation and preparation. Homer and Hesiod did not come out of nothing.

V

THE DAWN OF GREEK CULTURE. HOMER AND HESIOD

THE GREEK MIRACLE. The *Iliad*

It is necessary to divide our account into chapters for the reader's convenience, but it is well to bear in mind that such a division is somewhat artificial. The chapters are not mutually exclusive; the fields that they cover interlap and interlock. Thus, the period dealt with in Chapter IV brought us to the Mycenaean or Late Minoan age. The Homeric age followed immediately but its roots were Mycenaean and even pre-Mycenaean; we are thus obliged to think as much as possible in Mycenaean and Minoan terms if we would appreciate the Homeric flowering.

One often speaks of the Greek miracle, this being the simplest way of expressing one's wonder at the Greek achievements and one's inability to account for them. The wonder begins right at the very end of the Mycenaean age, at a time when the new Greek culture was not yet completely emancipated from its origins. The first and greatest gift of that age was a long epic in the Greek language, the *Iliad*.

MINSTRELS AND RHAPSODISTS

We trust that an analysis and description of that poem is superfluous. Should some of our readers need it they could easily find it in many places, or they could read the poem itself in their own language. According to ancient tradition the *Iliad* was composed by Homer, but to the query "Who was Homer?" one can hardly answer anything except "the author of the *Iliad*," and there seems to be no means of escaping from that vicious circle. At any rate, Homer's fame spread rapidly as the Greek civilization matured and nobody doubted his existence. He was conceived as a blind old man,[1] singing or reciting his own compositions; and seven Greek cities[2] claimed to be his birthplace. Such incompatible claims are the best proof of ignorance, even if it is disguised in the cloak of knowledge. They

[1] It is curious that *homēros* in the Cumaean dialect had the same meaning as *typhlos*, blind. On the other hand, in the Ionian dialect *homēreuō* means *podēgeō*, to lead or guide. The name might thus be a description, physical or mental, of the author, as if one said "the blind man," "the guide," "the poet."

[2] Smyrna, Rhodos, Colophon, Salamis, Chios, Argos, Athenai. These names are interesting; note that the majority are Ionian. The Homeric dialect is largely Ionian.

show that even at a relatively early time Homer had ceased to be known as an ordinary mortal. How could that happen? How could so great a poem exist and its author vanish?

The study of comparative literature [3] makes it much easier now to explain that mystery. The *Iliad* is unique because of its earlyness and of its beauty, but similar poems have been created from time to time by many peoples spread all over the earth. The same factors were apparently at work everywhere; the desire to explain their origins and to commemorate great events of the past inspired anonymous poets of many nationalities. Their compositions were almost always in metrical form, because of the innate love of rhythm that exists in every man; on the other hand, that form helped memorization and thus national archives could be transmitted indefinitely without being written down. Indeed, these poems were generally composed before writing had been invented by the nation concerned, or at least before it had become popular. Minstrels traveling from one court to another helped to create these poems and recited them for the entertainment and the edification of their hosts. After a while, certain poems that had been received with particular favor were standardized not only in their general form but also in their anecdotic and stylistic particularities. Not unlike the children of today, the ancient people loved the old stories best. There was an element of surprise, to be sure, in a new tale, and that was pleasant enough to the listeners, but their joy was greater when they recognized an old tale, when the minstrel evoked familiar heroes and described them in familiar terms. Striking epithets or metaphors or even whole verses, having once appealed to their imagination and tickled their fancy, were gradually expected and received with smiles or other marks of approval; [4] the skilful minstrel soon learned that it would not do to omit them; other characteristics of the poetic narrative, concerning its form or its matter, were gradually crystallized in the same way.

The majority of the minstrels, we may assume, were much like our musicians who move from place to place playing their repertory and adding little if anything to it. Their art consisted in remembering much and interpreting well. A few minstrels were more ambitious and were anxious to invent new ballads or else to reshape completely the older ones; they were like the virtuosi of today, who are not satisfied to interpret the compositions of the great musicians but are eager to play their own. There was scope for infinite variations — the whole gamut between a creative power that must find expression at one end and retentive passivity at the other — but the minstrels and troubadours of each people were alike in this, that they were exploiting the same stock of national memories; their creative and interpretive abilities were moderated as well as guided by the need of satisfying the same public, whose tendencies were, on the whole, conservative.

[3] Especially the splendid work done by the two Chadwicks, man and wife, Hector Munro Chadwick and Norah Kershaw Chadwick, *The growth of literature* (3 vols.; Cambridge: University Press, 1932–1940) [*Isis* 29, 196 (1938)]; vol. 1 (1932) deals with the ancient literatures of Europe; vol. 2 (1936), with the Russian, Yugoslav, Indian, and Hebrew literatures; vol. 3 (1940), with the Tatars, Polynesia, the Sea Dyaks, African peoples, and general survey. See also Solomon Gandz, "The dawn of literature," *Osiris* 7, 261–515 (1939).

[4] The number of phrases and lines repeated is very large, and no wonder, for the repetition being partly instinctive and partly systematic, everything conspired to bring back favorite sayings. See the concordance to the parallel passages in the *Iliad*, *Odyssey*, and hymns in Henry Dunbar, *Complete concordance to the Odyssey and hymns of Homer* (Oxford, 1880), pp. 391–419.

There was no better way of pleasing their patrons and obtaining their favor than by reciting the lays that were already loved. No matter how great their originality, the minstrels ended by doing as the virtuosi do who include old favorities in their program or in the encores. The poet[5] whom we call Homer was the most successful of these early minstrels. It is impossible to know how much he invented, but it is safe to assume that however much he created he inherited much more from his predecessors and helped to perpetuate their best compositions. It is possible that he was chiefly an "editor" of genius, who put together all the best lays available to him and harmonized them masterfully in a single whole. This hypothesis helps to explain the unity of the *Iliad*, as well as its occasional lapses, such as unnecessary repetitions or imperfect transitions.

The methods of these minstrels and of the later rhapsodists[6] are readily understood from the comparative study of various early literatures, and more vividly from the performances of their living representatives. This was done admirably by the late Milman Parry (d. 1935), a Harvard philologist who traveled in Yugoslavia, armed with recording apparatus, and collected two popular epics of great length from the very lips of the rhapsodists. Unfortunately, his life being cut short by an accident, he was not able to complete his task.[7] It is probable that the Homeric rhapsodist was not essentially different, as to outlook, temperament, methods, from the blind Yugoslavian poet Huso, whose recitations were immortalized by Parry's efforts.

The appreciation of oral traditions is a little difficult for us, for they implied an ability to memorize long poems that modern man has almost completely lost. Some men had that ability to a degree that would be incredible if we did not have abundant proof of it.[8]

[5] *Aoidos* = *vates*, poet, seer. The word occurs in the *Iliad*, XXIV, 721, and frequently in the *Odyssey* and in Hesiod.

[6] *Rhapsōdoi* = stitchers of songs or song sewers. The word as applied to reciters of Homeric poems occurs for the first time in Herodotos, (v, 67), but is probably of earlier coinage for it expresses the business of the early minstrels rather than that of the later reciters, whose initiative decreased as the epics were gradually canonized.

[7] As Parry died at the age of 35, before having been able to exploit his materials, his achievement has not received the attention and praise that it deserved, and therefore the following details may be welcome. He took more than 2,550 double-sided disks from the lips of 90 different singers. His recordings include two long epics of 13,000 and 12,000 odd lines (2,200 disks) and 300 other songs, so-called women's songs (350 disks). In many cases he recorded the same lays or songs from different singers, or twice from the same one after an interval of some days or weeks between the recordings. This enables one to measure individual variations and to understand better the regularities and irregularities of oral transmission. Parry's work was done at the eleventh hour, when the heroic recitations that he recorded were fast disappearing; but for him immemorial traditions might have been completely lost. These details were obtained from an article by the Hungarian composer Béla Bartók (*New York Times*, June 28, 1942), who investigated the Parry records, being especially interested in their musical aspects. See also Harry Levin, "Portrait of a Homeric scholar," *Classical J.* 32, 259–266 (1937), with bibliography of Parry's writings.

[8] Various examples of that ability, which seems almost uncanny to us, are quoted by Solomon Gandz, "The dawn of literature," *Osiris* 7, 304–308, 353, 384–385, 407 (1939). A few recent French examples are given by Sainte Beuve in his review of Grote's *Histoire de la Grèce* (Nouveaux lundis 10, 61, original date 1865). An almost perfect Vaidika (i.e., one who knows the Vedas) is described in a letter to Max Müller dated Bombay 1883; *Life and letters of Friedrich Max Müller* (London, 1902), vol. 2, p. 134.

By way of contrast, the following story illustrates the new point of view due to the diffusion of printing. An old "cantastórie" of Naples was found out by his audience to be blind; he pretended to read the *Orlando furioso* of Ariosto but instead recited the poem. That discovery finished him in their esteem. See Marc Monnier, *Les contes populaires en Italie* (Paris, 1880), p. 78. The event occurred in the seventies of the last century.

HOMER?

The question, "Who was Homer?" is futile, if it is taken to mean What kind of man was he? How different was he from the other minstrels? When and where did he live? and so on. But the question, "Was there a Homer?" is a very pertinent one and I think we can answer it in the affirmative. The remarkable unity of the *Iliad*, imperfect as it is, could not be explained otherwise. No matter how and when its several parts were composed, it took one supreme minstrel to put them together in a sequence which was probably not very different from that which has come down to us.

We shall come back to the method of its tradition later on. Let us first consider a more fundamental question. When was the *Iliad* completed? The Trojan War, episodes of which form its historical core, was variously dated by Greek chronologists from about 1280 to 1180. That ambiguity of a century does not matter to us, for a period many times as long must have elapsed between the events and the completion of the poem.[9] Some elements of it, such as the catalogue of ships or of the Greek expeditionary forces,[10] are of greater antiquity, or at any rate reflect conditions of an earlier, pre-Trojan, time; but the artistic integration of those elements cannot have occurred much before the tenth or ninth century.[11] If we have to name a single century, we shall not be far wrong in naming the ninth; that is, that dating tallies pretty well with previous and later events.

Further discussion would be out of place here, and the more so because it would never be convincing, however elaborate it was. Let me insist only on a single point. There is no reference to writing in the *Iliad* (nor for that matter in the *Odyssey*) except one that is incidental: "But Proetos sent Bellerophon to Lycia giving him baneful signs, on a folded tablet he had traced many signs poisoning the king's mind [against Bellerophon]." [12] I do not doubt that the words "baneful signs" represent a kind of writing, such as the Minoan writing that Sir Arthur Evans discovered in Crete. Lycia, incidentally, was a Cretan colony. That Homeric line might then be used to prove that some kind of writing was known in those days, but that proof is superfluous, for we have many actual examples of that writing, albeit undeciphered. Writing was known in the Aegean world; it was probably a Cretan invention. Its use was restricted to inscriptions, legal or magical records, inventories, accounts, and other very short technical texts. No minstrel ever thought of using it for literary purposes. This is not simply a local, Greek, fact but a general fact, which has been well established by anthropologists and by comparative philologists. An interval of time, which may extend to centuries, elapses between the invention of writing and its common use. On account of deeprooted traditions,

[9] For the sake of comparison: the *Chanson de Roland* (XI-2) was completed some three centuries after the events that inspired its creation.

[10] *Iliad*, II, 494–779.

[11] In terms of Egyptian chronology, the events described date from the Twentieth Dynasty (1200–1090) or the Twenty-first (1090–945); the poem dates from the Twenty second or Libyan Dynasty (945–745).

[12] *Iliad* VI, 168–169; *pempe de min Lyciēnde, poren d' ho ge sēmata lygra grapsas en pinaci ptyctōi thymophthora polla.* The word *grapsas* should not deceive us. The early meaning of *graphō* was to scratch; later, much later, it was taken to mean delineate, draw (Herodotos, II, 41), or write (Herodotos I, 125). The word, *anagignōscō*, meaning to know well, to recognize, was first used to mean reading by Pindar (*c.* 522–442), and *epilegomai* was first used with that same meaning by Herodotos (I, 124, 125, etc.). Before Pindar there was no word for reading. The Syrian word *biblion* was first used by Herodotos to mean paper, a letter, and by Aristotle to mean a book.

and perhaps also of the vested interests of minstrelsy, heroic poetry would not be among the first things to be committed to writing but among the last.

We may be sure that Homer was not conscious of writing except as a rare and somewhat mysterious method of communication which might be used in exceptional cases but did not concern men of letters. We may be sure that he never thought of writing his compositions. Moreover, how could he have done so? The invention of writing is worthless for literary purposes if it is not completed by the invention of writing materials. In Homer's time no such materials were suitable for long compositions. Papyrus did not become available in Greece until the beginning of the Twenty-sixth (Saitic) Dynasty, during the rule of Psametik I (663–609).

MORE ABOUT THE ILIAD

The *Iliad* is not only the earliest monument of European literature, the earliest monument of any size or quality, but it is also – and it is this that is really unexplainable, "miraculous" – of supreme excellence and of very large size.[13] There is no virtue in size, of course, but there is more virtue in a long poem than in a part of it. Moreover, it is astonishing to meet at the very threshold of European literature, not simply a few little pieces by means of which the oldest poets would have tried their strength, but an immense literary monument, representing the accumulated efforts of many men and of centuries. It is as if the earliest architectural monument to have come to us were already as large and as elaborate as one of the outstanding medieval cathedrals. The *Iliad* is in its mode and style so close to perfection that it has remained a model of excellence to this very day. We admire it not merely because of its antiquity, but irrespective of it. As a matter of fact, most critics would agree that it is the best of all the Western epics, with the possible exception of the *Odyssey*. And that epic – allow me to repeat it – does not appear toward the end of Greek culture or when that culture reached its climax, but at the beginning of it, or we might almost say, before its beginning.[14] Homer was truly a herald of Greek culture, of European culture, of

[13] The earliest Western epic is also the largest. It contains 15,693 verses. Here are a few figures concerning other epics for the sake of comparison. The *Odyssey* contains 12,110 lines, the *Aeneid* 9,895, the *Divina Commedia* 14,233, *Paradise Lost* 10,565. "The man put on trial for love" or "tormented by love" (*Erōtocritos*), composed probably in the first half of the sixteenth century and ascribed to *Bitzentzos ho Cornaros* (Vincenzo Cornaro) of Sitia in Crete, extends to 11,400 political verses (verses of eight plus seven syllables). The two Yugoslavian epics mentioned above contain 13,000 and 12,000 lines. It is remarkable that all those poems are of the same order of magnitude, the longest being about 50 percent longer than the shortest. It is true, the *Chanson de Roland* (XI-2) and the Byzantine epic *Digenēs Acritas*, prior to the fourteenth century are both somewhat shorter – less than 5,000 lines each. See Karl Krumbacher, *Geschichte der byzantinischen Literatur* (Munich, ed. 2, 1897), pp. 827–832, 870–871; Henri Grégoire, *Digenis Akritas* (New York, 1942) [*Isis 34*, 263 (1942–43)].

On the other hand, Eastern epics are considerably longer. The Mahābhārata measures *c.* 220,000 lines, the Rāmāyana *c.* 48,000, the Shāhnāma of Firdawsī (XI-1) 60,000, the Mathnawī of Jalāl al-dīn-i-Rūmī (XIII-2), 26,660 couplets. That is typical of Eastern extravagance. The size of the Western epics is more appropriate to the human size and to the length of human life.

[14] The contrast between Greek and Latin literature is great in that respect. Homer appears at the beginning or before the beginning of the Greek age; Vergil on the contrary lived from 683 to 734 U.C. (70–19 B.C.). The Romans had reached political maturity and obtained considerable international power before they were able to boast a literature worthy of a great nation. By the end of the second Punic war (201 B.C.) their literary achievements were still of an inferior nature; it was only after the conquest of Greece half a century later that their literary ambition was fully awakened.

Western culture – a herald of such gigantic stature that he is still overshadowing us. Is not that a miracle, or can one think of anything less easily explainable, more miraculous?

THE *Odyssey*. HOMER II

What is more, the miracle was not an isolated one, or if it was for a time it did not remain so very long. A second epic, the *Odyssey*, came gradually into being. We may be certain that it was completed later than the *Iliad*, perhaps as much as a century later or more. We must postulate the existence of an author or editor for that poem, even as we had to postulate it for the *Iliad*. In fact, both poems have been traditionally ascribed to the same author, Homer. In order to reconcile that tradition with the probabilities derived from internal evidence, I would suggest that the author of the *Iliad* be called Homer I and the author of the *Odyssey*, Homer II. This does not press the difference between them too far; it does not even completely exclude the (slight) possibility that Homer II may have been the same person as Homer I, but at a much older age.[15]

When one assigns different dates to the two epics it is well to remember that such dating is always ambiguous. For each poem contains stories, ideas, definite phrases or lines that represent different chronological strata; then for each poem there were different stages in the long process of amalgamation and standardization. Neither poem was completed at a definite date. Whether one studies the vocabulary, grammatical, rhetorical, or prosodic characteristics, one finds many elements that are common to the *Iliad* and the *Odyssey*.[16] Indeed, the outstanding qualities are common to both, that is, the simplicity of thought and diction, and the rapidity of development (in great contrast with the slowness, fantastic exuberance, and turgidity of the Oriental epics).

The differences in subject and mood between the *Iliad* and the *Odyssey* are considerable. The *Iliad* is a story of war; the *Odyssey* is a story of peace, of domestic life, of merchants, travelers, and colonists; it is full of romance but also of magic; it is at once more superstitious and more moral. The artistic unity of the *Odyssey* is deeper and the mood is gentler. It is a kind of novel, the first in world literature,[17] and moreover it has a moral purpose. Says Jaeger, "It is impossible to read the *Odyssey* without feeling its deliberately educational outlook as a whole, although many parts of the poem show no trace of it. That impression derives from the universal aspect of the spiritual conflict and development which moves parallel with the external events in the tale of Telemachus – which is in fact their real plot and leads to their real climax." [18] There lies between the two poems an unmistakable interval of peaceful culture and urbanization, though how long that interval may have lasted, nobody can say for certain. It might be a matter of a century or two; on the other hand, the natural difference between two successive generations, the older one more warlike, the younger one more peaceful, or perhaps

[15] The idea that the *Iliad* and the *Odyssey* are of separate authorship is not by any means new. It goes back to early Hellenistic times, say to the third century B.C., when the scholars entertaining it were called *hoi chōrizontes* (the separators), yet their opinion was generally rejected.

[16] For a detailed comparison see Carl Rothe, *Die Odyssee als Dichtung und ihr Verhältnis zur Ilias* (370 pp.; Paderborn, 1914).

[17] The Egyptians have left us short stories but no full-sized novel.

[18] Werner Jaeger, *Paideia, the ideals of Greek culture* (Oxford: Blackwell, 1939), vol. 1, p. 28 [*Isis 32*, 375–376 (1949)].

even the difference between the mellowness of an old man and the impetuosity of his own youth, might suffice to explain the contrast.

The best argument to my mind for a longer interval is the following. The *Iliad* mentions bronze fourteen times as often as iron, and the *Odyssey* only four times. That fact is significant, because the differences would not be deliberate; the poets would hardly think of that but would simply react each to his own environment. Both poems have their roots in the Bronze Age, but Homer II was more familiar with iron, and less familiar with bronze, than Homer I.

If we assumed that the *Iliad* was completed about the middle of the ninth century, we may perhaps assume also that the *Odyssey* was completed a century later, but that is at best a plausible guess. After having made that reservation, it will be simpler in what follows to abide by the ancient tradition and speak of "Homer" as the author of the Homeric poems in general. Those poems, and especially the *Iliad* and the *Odyssey*, are concrete realities. When we speak of Homer we simply mean those two epics.

EARLY HOMERIC TRADITIONS

The early story of the *Iliad* and the *Odyssey* is necessarily obscure. Both poems were kept alive by minstrels and rhapsodists who recited them at banquets or religious festivals. Homer's fame was already such by the middle of the sixth century (*c.* 540) that Xenophanes of Colophon could say, "From the beginning all men have learned from Homer." [19] In the time of Pindar, half a century later, some of the Rhapsodists were called Homēridai,[20] but we need not conclude with the scholiast that they were descendants of Homer, except in a spiritual way. The Homēridai were simply the followers of the old minstrels and especially of the most illustrious of them, Homer; they were in the fullest sense the keepers of the Homeric tradition. For practical purposes the Homeric canon was vulgarized [21] and Homer's national fame securely established during the fifth century. One of the guests of Xenophon declared, "My father being anxious that I should become a

[19] *Ex archēs cath' Homēron epei memathēcasi pantes.* Hermann Diels, *Die Fragmente der Vorsokratiker* (Berlin: Weidmann, ed. 5, 1934), vol. 1, p. 131, frag. 10.

[20] *Nemean* II, 1–2. *Homēridai rhaptōn epeōn aoidoi.*

[21] The first canonic text of Homer was established when Peisistratos was dictator of Athens. After his death in 527, that text was lost or neglected. Yet the Homeric poems were kept alive, by public and private recitations, e.g., in the national festivals, Panathenaia, held every year, but chiefly in the musical contests of the Greater Panathenaia held every fifth year (these Homeric recitations had been introduced by Peisistratos). The existence of that early text is proved by the many quotations made by Herodotos, Plato, Xenophon, quotations that can be readily (if not always literally) identified in our editions. Two other Hellenic editions (*diorthōseis*) are mentioned, the one prepared by the poet Antimachos of Claros (near Colophon, Ionia), who flourished toward the end of the Peloponnesian War, and the other prepared by Aristotle for Alexander the Great, who carried it with him in all his campaigns.

The scientific study of the text began only in the Hellenistic age, however. Zenodotos of Ephesos (III–1 B.C.), first librarian of the Museum of Alexandria, has been described as the "first" editor (*diorthōtēs*); he is said to have produced before 274 the "first" edition of the *Iliad* and the *Odyssey*. Zenodotos was certainly not the first editor, but he was a better philologist than his predecessors. It is probable that the division of each epic into 24 books was due to him. The third and fourth librarians of the Museum, Aristophanes of Byzantium (II–1 B.C.) and Aristarchos of Samothrace (II–1 B.C.), improved considerably Zenodotos' methods; the text familiar to us was established by them. Yet Didymos of Alexandria (I–2 B.C.) corrected Aristarchos' edition. And so on. The history of Homeric learning is a good cross section of the history of Greek scholarship.

good man made me learn all the poems of Homer,"[22] and the final consecration was given, albeit grudgingly, by Plato. The latter, referring[23] to the eulogists who called Homer the educator of Hellas, conceded that he was the greatest of poets and the first of tragedy writers, yet would banish him from the city. In spite of Plato's unreasonable and illiberal verdict, Homer has remained in the city and kept his position in the heart of every Greek. The validity of his title, "educator of Hellas" has been confirmed by the whole history of Greek-speaking peoples down to our own days; it has never been doubted except by Plato, and the very Christians have seldom allowed their antipagan prejudices to cool their admiration for him. In fact, Homer deserves a broader title; he was not simply the educator of Greece, but one of the educators of mankind. We shall come back to this presently.

WHAT DID HOMER TEACH?

What did Homer teach? In the first place, he taught the Greek language. His immortal works helped to standardize that language, or rather to lift it up to that level of excellence and dignity which can be reached only by means of literary masterpieces. His writings became for the Greek people a kind of Bible, to which they were always ready to listen and which gave them and their children patterns of honor, of good breeding, and of good language. In spite of its mythological contents, that Bible was a lay Bible; that is, there was nothing sacerdotal in it, and it was remarkably free of magic and superstitions. The Ionian poet was truly an ancestor of the Ionian scientists whose achievements will be explained later.

In the second place, the *Iliad* and the *Odyssey* taught history, the history of the Minoan and Mycenaean origins, which were in some respects dim and distant, yet in other respects near enough in the form of tools, usages, words, and folklore to be easily recognized and understood by the listeners. It is the very function of epic poetry to record the past for posterity and prevent its oblivion. It is impossible to itemize the historical contents of the Homeric poems without giving a course on Mycenaean culture. The reader will find a very brief characterization of that culture in the previous chapter, and sufficient bibliographic references to continue his study of it as far as he may wish. Note that every textbook of Minoan or Aegean archaeology is necessarily full of references to Homer. Homeric lines help to explain the monuments and these in their turn help to interpret Homer. The latest editors of Homer refer continually to Aegean antiquities. The pioneer exponent of the archaeological interpretation of Homer was Wolfgang Helbig (1884), who was followed by many others.[24]

Homeric poetry offers us a mirror of the Mycenaean age, which was then vanishing, yet was vividly and joyfully remembered by the old people and the minstrels. Like every epic, it was turned toward the past; it is thus a little paradoxical to call it the harbinger of a new age. It is a climax or an ending rather than a beginning yet it provided the new people, the Greeks, with a solid foundation upon which

[22] *Ho patēr epimelumenos hopōs anēr agathos genoimēn, ēnancase me panta ta Homēru epē mathein.* Xenophon, *Symposium*, III, 5.

[23] *Republic*, 606E.

[24] W. Helbig, *Das homerische Epos aus dem Denkmälern erläutert* (362 pp., ill.; Leipzig, 1884; 2nd ed., 480 pp., Leipzig, 1887). Martin P. Nilsson, *Homer and Mycenae* (296 pp., 52 ills., 4 maps; London: Methuen, 1933). Helbig's book was very imperfect especially because he mixed up Mycenaean antiquities with Greek and even Etruscan ones; Nilsson's book contains many debatable points, but the main thesis is beyond question. H. L. Lorimer, *Homer and the monuments* (575 pp., ill.; New York: Macmillan, 1950).

to establish a new culture; it provided them with a standard of propriety and a guide of conduct; it gave them pride and dignity.

To put it otherwise, I am more and more convinced that the Greek culture of Homeric days was not something radically new but rather a second growth of the Aegean culture which had been temporarily suppressed by violent upheavals and almost destroyed. Life is never completely destroyed, however; consider, for example, the rich growth of plants in a region devastated by volcanic eruptions or desiccated by a long drought. One might think that everything is dead, but it is not. Life is dormant and may remain so for a long time, yet let the blessed rain fall and the mercy of heaven permit it, and it will soon reappear as vigorous as ever. Much is lost in the process, of course, and new elements are mixed with the old ones. The new Greek culture was a revival of the old one; that revival was deliberate, at least from the point of view of the minstrels and their patrons. It was different in many respects from the Aegean one, for the conditions of life had been deeply modified. For one thing this was the age of iron; the days of bronze could never come back.

GEOGRAPHY

It would be tempting to analyze the Homeric poems from the point of view of each of the scientific categories of our own days, but that would be long and not very rewarding. Moreover, it would be very difficult, if not impossible, to date exactly the elements of that knowledge. How much is prehistoric, how much old Minoan, how much Mycenaean, how much neo-Greek? For example, when the *Iliad* was composed much geographic knowledge had already been accumulated by Phoenician and Aegean sailors and colonists; the Mediterranean and the Euxine world had been pretty well explored. Bold navigators had reached the Atlantic and introduced the conception of the great river Oceanos encircling the earth's disk and returning into itself.[25] That conception was mixed with the mythologic one of Oceanos, son of Uranos and Gaia, wedded to Thetys, the father of primeval water and of all the rivers.[26] Another story, that of the Argonauts (*Argonautai*) sailing on the ship Argo under Jason's leadership to capture the Golden Fleece in Colchis (on the southeastern shore of the Black Sea), perpetuates the remembrance of some of the earlier sea adventures. The minstrels told many other stories, equally marvelous, but they did not care about geographic accuracy or even about geographic consistency. Geography and mythology, facts and fancies, were inextricably mixed up in their tales. It is as futile to try to account exactly for the wanderings of Odysseus as for those of Sindbad the Sailor in much later times. The storytellers remembered the adventures and the wonders and forgot geographic realities. One reality, however, had impressed their minds – the four winds, Boreas, Euros, Notos, and Zephyros – which represented roughly the four cardinal points, north, east, south, and west; two of those directions were immemorially known because of the rising and setting of Sun and stars; the two others were suggested by the climateric regularities of the Aegean Sea. We may be sure that the early Greek sailors knew their Mediterranean localities pretty well, but they did not communicate much of their knowledge to Homer or the latter was not interested in it.

[25] *Ōceanos apsorroos. Iliad*, XVIII, 399; *Odyssey*, XX, 65.

[26] *Iliad*, XXI, 195–197.

MEDICINE. OTHER ARTS AND CRAFTS

The medical knowledge implied in the Homeric poems is such as we would expect among people intelligent and quarrelsome, having much experience of war wounds and of their healing. They had learned to anoint their limbs with oil (*aleiphō lipa* or *lip' elaiō*). The best observers among them had opportunities of realizing the various effects of special wounds, the peculiarities of fainting spells, the convulsive motions of dying men. Many good descriptions of such facts occur in the epics. There were professional physicians and these were appreciated – "a physician is worth many other men" [27] – but were not always available and the fighting men had to help one another in case of need. Much of the medical work was surgical. Yet the physicians were concerned with internal medicine as well as with surgery and used drugs of many kinds (*introi polypharmacoi* [28]). Some women also practiced the medical arts; they nursed patients, collected herbs and prepared drugs, such as the anesthetic and soothing potion (*pharmacon nēpenthes*) the secret of which Helen had received from an *Egyptian* woman.[29] The so-called "anatomical" vocabulary of Homer includes some 150 words. One bit of Homeric physiology is still embedded in our own language. The life spirit (*thymos*, *psychē*; cf. *anima*, *spiritus*) was placed in the midriff (*phrenes*); hence our words phrenetic, phrenology! But that localization should not be accepted literally. In Homer the words *phrēn*, *phrenes* refer to other organs, and especially to the heart or the parts about the heart and to the seat of the mind.[30] The early Greeks used the word *phrēn* as carelessly as we still use the word heart (as when we say "he has a good heart" to mean "he is kind").[31] Homeric anatomy should not be taken more seriously than Homeric geography.

The best technicians, then as now, were not learned people, masters of words, but craftsmen – smiths, potters, carpenters, leatherworkers, whose experience and folklore might be considerable. The woman spun and wove. The husbandmen knew the lore of beasts and plants; they had learned to use dung (*copros*) to fertilize their fields.[32] The craftsman (*dēmiurgos*) was often moving from place to place, so were the seer, the healer (*iētēr cacōn*), the builder, the minstrel.[33] Homeric science is simply Mycenaean folklore with a few novelties and variations.

The exercises of the body – gymnastics and choral dance – which the Greeks would later develop to such a high pitch in their Olympiads [34] and other festivals, were clearly of Cretan origin. Homer refers to the dancing floor (*choros*) "which once in broad Cnossos, Daidalos built for Ariadne with the beautiful locks." [35] Such dances are often pictured in Cretan frescoes. The early musical instruments were of the same origin.

[27] *Iētros gar anēr pollōn antaxios allōn. Iliad*, XI, 514.

[28] *Iliad* XVI, 28.

[29] *Odyssey*, IV, 220–221.

[30] The Latin word *praecordia* offers the same ambiguities.

[31] Such mistakes are easily explainable. We are tempted to localize our emotions not in the brain where they originate but in the heart where we actually feel them. Indeed, emotions modify the heart beat and may even cause distressing palpitations.

[32] *Odyssey*, XVII, 297.

[33] *Odyssey*, XVII, 383–386.

[34] The Olympiads were periods of four years separating the successive athletic festivals held in Olympia, Elis. The first Olympiad (776–773) was reckoned from the victory of Coroibos of Elis in the footrace of 776. The dating by Olympiads was not systematized until much later, by Timaios of Tauromenium in Sicily (III-1 B.C.).

[35] *Iliad*, XVIII, 590.

HOMER, THE FIRST EDUCATOR OF THE WESTERN WORLD. FÉNELON

Homer was the educator of Hellas. This must be understood in a broad way, humanistic rather than technical. One might say that he taught everything essential, but one might also say that he taught nothing. He certainly did not teach history, except vaguely. He gave the Greek-speaking people ideals of nobility, virtue, courtesy, poetry; thanks to him they were provided from the very beginning with a viaticum of humanities. He awakened or strengthened their literary and artistic sensibility, and whatever he did, he did with remarkable clearness and soberness, without unnecessary mysticism or too much hocus-pocus. The educational influence of the *Iliad* and the *Odyssey* has continued to our own day with hardly any interruption; there is no older and more persistent tradition in the Western world.[36]

From ancient times almost to our own the rhapsodists or reciters plied their trade. We find references to them in the papyri [37] and later in Byzantine and neo-Greek literature, as well as in the unwritten folklore of Greek lands. The Homeric tradition was at first restricted to the people who understood Greek, and thus it hardly touched the people of Western Europe before the fourteenth century. Indeed, that fundamental and essential part of Hellenism was not, like Hellenic science and philosophy, transmitted to us indirectly via the Syriac-Arabic channel,[38] and when the Catholic Church allowed the knowledge of Greek almost to die out in Western Europe, Homer was known only very imperfectly through the Latin literature of the Roman age and through various adaptations in medieval Latin and in vernacular poems or narratives.[39] The Greek revival of the fourteenth and fifteenth centuries brought back the original text to the attention of scholars, and the Greek *princeps* edited by Demetrios Chalcondyles (Florence, 1488) re-established it forever (Fig. 36). From that time on, Homer has been one of the educators of Western Europe with almost unbroken continuity.

It is not possible here to tell the history of that tradition, for even the swiftest outline of it would take too much place. Moreover, such an outline would be repetitious and tiring. It is more interesting to select one episode, which is familiar enough to French readers but less so to English ones. The abbé Fénelon (1651–1715), having been appointed by Louis XIV as tutor to his grandson the Duke of Burgundy, wrote for the latter the didactic novel entitled *Les aventures de Télémaque* (Fig. 37). That book, first printed without author's name in 1699,[40] immediately obtained considerable success (many editions appeared in France and the Low Countries during the first year). It evoked much criticism in the royal

[36] Unless some prophets of the Old Testament – Amos, Hosea, Micah, and Isaiah – anticipate Homer, but that is doubtful even in the case of Amos.

[37] There are not simply references to the rhapsodists in the papyri but actual Homeric texts, many of them. For example, see Paul Collart, "Les papyrus de l'Iliade" in Pierre Chantraine, P. Collart and René Langumier, *Introduction à l'Iliade* (304 pp.; Paris: Les Belles Lettres, 1942). Three hundred seventy-two papyri containing fragments of the *Iliad* are known, plus 35 containing commentaries, scholia, paraphrases; these 407 papyri date from the third century B.C. to the seventh after Christ. Their number increases to the third century after Christ, then decreases together with Egyptian Hellenism. See *Chronique d'Egypte*, No. 36 (1943), p. 315.

[38] The *Iliad* was translated into Arabic only in very recent times by Sulaymān al-Bustānī and first printed in Cairo in 1904. That is a curiosity of Arabic literature of no interest for the study of Homeric tradition.

[39] To be sure, the Vergilian tradition continued the Homeric one; our statement refers to Homer independently of Vergil.

[40] *Télémaque* was composed probably in 1693–94; its first publication in 1699 was due to the

μιχθεῖσ' ἐν φιλότητι κελαινεφέϊ κρονίωνι
σωτῆρας τέκε παῖδας ἐπιχθονίων ἀνθρώπων,
ὠκυπόρων τε νεῶν ὅτε τε σπέρχωσιν ἀέλλαι
χειμέριαι κατὰ πόντον ἀμείλιχον· οἱ δ' ἀπὸ νηῶν
εὐχόμενοι καλέουσι Διὸς κούρους μεγάλοιο,
ἄρνεσσιν λευκοῖσιν ἐπ' ἀκρωτήρια βάντες
πρύμνης· τὴν δ' ἄνεμός τε μέγας καὶ κῦμα θαλάσσης
θῆκαν ὑποβρυχίην· οἱ δ' ἐξαπίνης ἐφάνησαν
ξουθῇσι πτερύγεσσι δι' αἰθέρος ἀΐξαντες·
αὐτίκα δ' ἀργαλέων ἀνέμων κατέπαυσαν ἀέλλας·
κύματα δ' ἐστόρεσαν λευκῆς ἁλὸς ἐν πελάγεσσι
ναύταις σήματα καλὰ πόνου σφίσιν· οἱ δὲ ἰδόντες
γήθησαν· παύσαντο δ' ὀϊζυροῖο πόνοιο·
χαίρετε τυνδαρίδαι ταχέων ἐπιβήτορες ἵππων·
αὐτὰρ ἐγὼν ὑμέων καὶ ἄλης μνήσομ' ἀοιδῆς·

Εἰς ξένους

αἰδεῖσθε ξενίων κεχρημένον ἠδὲ δόμοιο·
οἳ πόλιν αἰπεινὴν νύμφης ἐρατώπιδος ἥρης
ναίετε ἄδην· ἧς πόδα νείατον ὑψικόμοιο
ἀμβρόσιον πίνοντες ὕδωρ ξανθοῦ ποταμοῖο
ἵβρον καλὰ ῥέοντος· ὃν ἀθάνατος τέκετο Ζεύς·

ΤΕΛΟΣ ΤΩΝ ΤΟΥ
ΟΜΗΡΟΥ ΥΜΝ
ΩΝ·

Ἡ τοῦ ὁμήρου ποίησις ἅπασα ἐντυπωθεῖσα πέρας εἴληφεν ἤδη σὺν θεῷ ἐν φλωρεντίᾳ, ἀναλώμασι μὲν, τῶν εὐγενῶν καὶ ἀγαθῶν ἀνδρῶν, καὶ περὶ λόγους ἑλληνικοὺς σπουδαίων βερνάρδου καὶ νηρίου ταναΐδος τοῦ νερϊλίου φλωρεντίνοιν· πόνῳ δὲ καὶ δεξιότητι δημητρίου μεδιολανίου κρητὸς, τῶν λογίων ἀνδρῶν χάριν καὶ λόγων ἑλληνικῶν ἐφιεμένων, ἔτει τῷ ἀπὸ τῆς χριστοῦ γεννήσεως χιλιοστῷ τετρακοσιοστῷ ὀγδοηκοστῷ ὀγδόῳ μηνὸς δεκεμβρίου ἐνάτῃ·

Fig. 36. *Editio princeps* of Homer (Florence, 1488); colophon, p. 439 b. [From the copy in the Boston Public Library.]

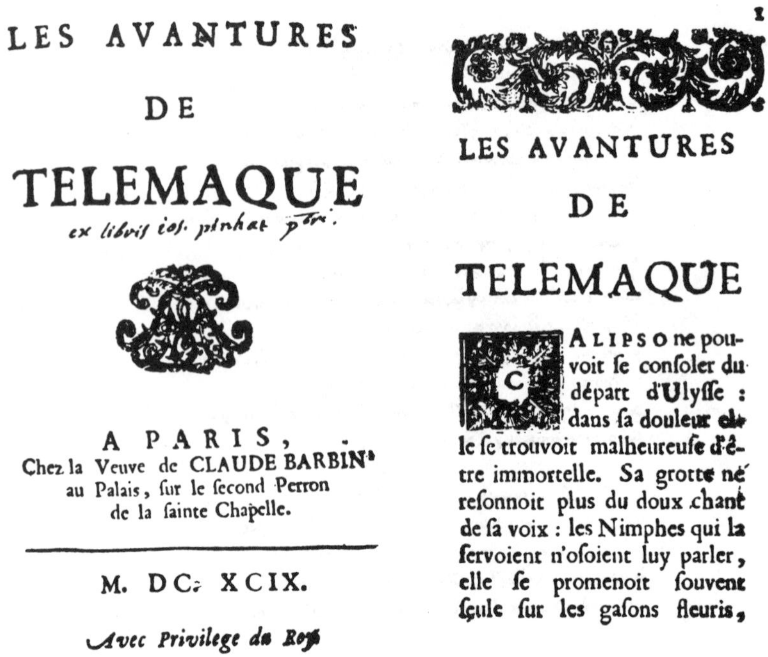
LES AVANTURES

DE

TELEMAQUE

ex libris ios. pinhat pbri.

A PARIS,
Chez la Veuve de CLAUDE BARBIN,
au Palais, ſur le ſecond Perron
de la ſainte Chapelle.

M. DC. XCIX.

Avec Privilege du Roy.

1

LES AVANTURES

DE

TELEMAQUE

CALIPSO ne pouvoit ſe conſoler du départ d'Ulyſſe : dans ſa douleur elle ſe trouvoit malheureuſe d'être immortelle. Sa grotte ne réſonnoit plus du doux chant de ſa voix : les Nimphes qui la ſervoient n'oſoient luy parler, elle ſe promenoit ſouvent ſeule ſur les gaſons fleuris,

Fig. 37. Title page and first page of the first edition of *Télémaque* (2 vols.; 145 mm high). Volume 1 contains on its last page (p. 216) the royal privilege, dated Versailles, April 6, 1699. [From the copy in the Harvard College Library.]

circle because of its satirical, utopian, and "liberal" tendencies and completed the author's disgrace. Its early diffusion was largely due to the foreign editions. It exerted a deep influence upon thought and letters during the eighteenth and a good part of the nineteenth century.[41]

LEGENDS

The story of Homer was beclouded with legends almost from the beginning. The early Greeks did not deny his existence but seven cities claimed him as their

indiscretion of a copyist. The first approved edition, not essentially different from the many previous ones, was published only in 1717, two years after the death of the Archbishop of Cambrai, by the care of a collateral descendant, the marquis de Fénelon.

[41] In the nineteenth century *Télémaque* had long ceased to be considered a liberal book and had become on the contrary very conservative and gradually even out of date. May I be permitted to tell the following anecdote? My paternal grandmother, who had been educated in a French conventual school, often told me that *Télémaque* was one of her main textbooks. The nuns had led her to believe that *Télémaque* contained every (proper) word in the French language! The main point is this, that while the *Abrégé de l'histoire sainte* (or some other such book) taught her the Hebrew and Christian traditions, *Télémaque* inculcated in her mind the Homeric and Greek ones.

Télémaque was translated into Japanese from the English in 1879, under the title *Heneromu monogatari*. It was written in the style of the ancient Japanese romances, a metrical prose with Chinese flavor. G. B. Sansom, *The Western world and Japan* (New York: Knopf 1950), pp. 400, 403 [*Isis* 42, 163 (1951)]. Thus did Greek thought interpreted by a Frenchman of the seventeenth century reach the Far East two centuries later.

son; seven different birthplaces are too many for a mortal, yet too few for a mythical hero. In the course of time, as the Homeric poems became the basis of Greek education wherever the language was spoken, the legends concerning their author increased and more birthplaces were invented for him. For example, Heliodoros of Emesa wrote in his youth (*c.* 220–240) [42] a famous novel, wherein he claimed incidentally that Homer was born in Thebes, the son of the god Hermes (= Thoth) by the wife of an Egyptian priest.[43] We gather from the papyri that Homer was very well known in the Greek circles of Egypt; it is possible that the Syrian Heliodoros obtained his Homeric lore from Egyptian sources. The very fact that a Greek author who became eventually a bishop in Thessaly could give credence to such a fable speaks volumes for the reality of Egyptian influence on Greek thought. If the Greeks of the third century were prepared to believe that their own Homer, the educator of Hellas, was an Egyptian, they must be ready to consider Egypt as the cradle of their culture.[44]

Such extravagances were not restricted to ancient and medieval ages; they pop up from time to time as late as the last century. The following illustration will probably amuse the reader as much as it does me. The Flemish magistrate Charles Joseph De Grave (1736–1805) devoted the leisure hours of a very active life to archaeological studies, the extraordinary fruits of which appeared soon after his death in a book entitled *République des Champs Elysées ou Monde ancien* (Fig. 38).[45] Inspired by *Télémaque* and by the *Atlantica* of Olof Rudbeck the Elder (1630–1702),[46] the austere scholar tried to reinterpret from top to bottom the story of our classical origins. Even as the Swede Rudbeck was anxious to place them in Sweden, the Fleming De Grave writing a century later was placing them in Belgium. That kind of illusion is common enough, but very few people will work as hard to establish it on so heavy a substructure. According to De Grave, Homer was a Belgian poet who had been celebrating the Belgian country. That seemed pretty obvious to him, but was not so obvious to other scholars, especially to those who had not been brought up in the bosom of sweet Flanders.

WOLF AND SCHLIEMANN

After this little intermezzo, we may return for a moment to the discussion of textual difficulties which was carried on throughout the seventeenth and eighteenth centuries by scholars of many countries; as those scholars were trained with more rigor, their discussions became gradually more critical and more exacting. That long travail culminated in the *Prolegomena ad Homerum of* Friedrich August Wolf

[42] My dating of the *Aethiopica* is based on the discussion of R. M. Rattenbury in the edition of it published by the Association Guillaume Budé (2 vols.; Paris, 1935–1938); it is conjectural. The identification of the author with the bishop is not certain.

[43] *Aethiopica,* III, 14.

[44] More critical writers like Pausanias (II-2) in his *Description of Greece,* x, 24, 3, and Philostratos of Lemnos (III-1) in his *Hērōicos,* XVIII, 1-3, dealing with the Trojan war, preferred to admit their ignorance concerning Homer's origin.

[45] Published in 3 vols. (Ghent, 1806). We reproduce the programmatic title page from the copy kindly lent by the Library of Congress. The same title page appears in each of the three volumes. The last line of the subtitle reads: "Que les poètes Homère et Hésiode sont originaires de la Belgique, &." For information on the author see the notice in vol. 1, pp. (9)-(16) and the article by Edm. De Busscher in *Biographie nationale de Belgique* (Brussels, 1876), vol. 5, pp. 114-127.

[46] Olaus Rudbeck, *Atlantica* (1679-1689). New edition by Axel Nelson published by the Swedish History of Science Society (Uppsala, 1937, 1938, 1941) [*Isis 30,* 114-119 (1939); *31,* 175 (1939-40); *33,* 71 (1941-42)].

RÉPUBLIQUE

DES CHAMPS ÉLYSÉES,

OU *MONDE ANCIEN*,

Ouvrage dans lequel on démontre principalement :

Que les Champs élysées et l'Enfer des Anciens sont le nom d'une ancienne République d'hommes justes et religieux, située à l'extrémité septentrionale de la Gaule, et surtout dans les îles du Bas-Rhin ;

Que cet Enfer a été le premier sanctuaire de l'initiation aux mystères, et qu'Ulysse y a été initié ;

Que la déesse Circé est l'emblême de l'Eglise élysienne ;

Que l'Elysée est le berceau des Arts, des Sciences et de la Mythologie ;

Que les Elysiens, nommés aussi, sous d'autres rapports, Atlantes, Hyperboréens, Cimmériens, &c., ont civilisé les anciens peuples, y compris les Egyptiens et les Grecs ;

Que les Dieux de la Fable ne sont que les emblêmes des institutions sociales de l'Elysée ;

Que la Voûte céleste est le tableau de ces institutions et de la philosophie des Législateurs Atlantes ;

Que l'Aigle céleste est l'emblême des Fondateurs de la Nation gauloise ;

Que les poëtes Homère et Hésiode sont originaires de la Belgique, &c.

OUVRAGE POSTHUME

De M. CHARLES-JOSEPH DE GRAVE, *ancien Conseiller du Conseil en Flandres, Membre du Conseil des Anciens, &c.*

Veterum volvens monumenta Deorum,
ô Patria! ô divum Genus!

TOME  PREMIER.

A GAND,
De l'Imprimerie de P.-F. DE GOESIN-VERHAEGHE,
rue Hauteporte, N°. 229.

1806.

Fig. 38. Title page of volume 1 of De Grave's *République des Champs Elysées* (3 vols.; Ghent, 1806). [From the copy in the Library of Congress.]

(1795) (Fig. 39),[47] which opened the modern phasis of the "Homeric question," that is, the series of misgivings concerning the existence of Homer and the integrity of the *Iliad* and the *Odyssey*. We have already referred to them and stated our own diffident conclusions.

Among the innumerable publications devoted to these topics, I would like to single out one which the average philologist would prefer to pooh-pooh. One of the great English writers of the last century, Samuel Butler (1835–1902), the author of *Erewhon* and *The Way of All Flesh*, published toward the end of his life (in 1897) a book, *The Authoress of the Odyssey* (Fig. 40), wherein he tried to prove that the *Odyssey* was written by a woman (a woman of Trapani in Sicily!). To use our own terminology, Homer I was a man, but Homer II – definitely – a woman. His arguments are not convincing, except the more general part of them which do but confirm the impression of every sensitive reader that

[47] We are remarkably well documented about the life and works of Friedrich August Wolf (1759–1824). See Wilhelm Körte, *Leben und Schriften Friedr. Aug. Wolf's, des Philologen* (2 vols.; Essen, 1833); J. F. J. Arnoldt, *Fr. Aug. Wolf in seinem Verhältnisse zum Schulwesen und zur Paedagogik* (2 vols.; Brunswick, 1861–62); Victor Bérard, *Un mensonge de la science allemande* (300 p.; Paris, 1917); Siegfried Reiter, *F. A. Wolf. Ein Leben in Briefen* (3 vols.; Stuttgart: Metzler, 1935), including autobiographical fragment (vol. 2, pp. 337–345).

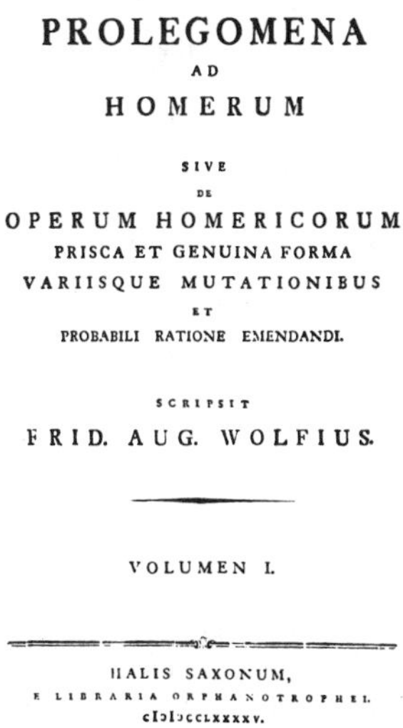
PROLEGOMENA
AD
HOMERUM
SIVE
DE
OPERUM HOMERICORUM
PRISCA ET GENUINA FORMA
VARIISQUE MUTATIONIBUS
ET
PROBABILI RATIONE EMENDANDI.
SCRIPSIT
FRID. AUG. WOLFIUS.
VOLUMEN I.
HALIS SAXONUM,
E LIBRARIA ORPHANOTROPHEI.
CIↃIↃCCLXXXXV.

Fig. 39. Title page of volume 1 (the only one published) of Wolf's *Prolegomena* (Halle a.d. Saale, 1795). [From the copy in the Harvard College Library.] This copy bears the following inscription: "To the celebrated Harvard University in Cambridge, new England. From the author, Fr. A. Wolf, Berlin, d. 21, April, 1817." Note that Wolf gave the book to Harvard 22 years after its publication, toward the end of his life; he died in 1824.

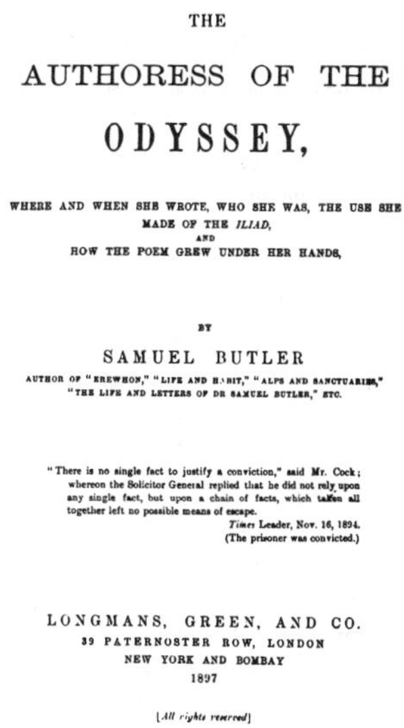
THE
AUTHORESS OF THE
ODYSSEY,
WHERE AND WHEN SHE WROTE, WHO SHE WAS, THE USE SHE MADE OF THE *ILIAD*,
AND
HOW THE POEM GREW UNDER HER HANDS,
BY
SAMUEL BUTLER
AUTHOR OF "EREWHON," "LIFE AND HABIT," "ALPS AND SANCTUARIES," "THE LIFE AND LETTERS OF DR SAMUEL BUTLER," ETC.

"There is no single fact to justify a conviction," said Mr. Cock; whereon the Solicitor General replied that he did not rely upon any single fact, but upon a chain of facts, which taken all together left no possible means of escape.
Times Leader, Nov. 16, 1894.
(The prisoner was convicted.)

LONGMANS, GREEN, AND CO.
39 PATERNOSTER ROW, LONDON
NEW YORK AND BOMBAY
1897
[*All rights reserved*]

Fig. 40. Title page of Samuel Butler, *The authoress of the Odyssey* (1897). [From the copy in the Harvard College Library.]

the literary atmosphere of the *Odyssey* is gentler, more domestic, and let us even admit, feminine than that of the *Iliad*. More than that Butler could not prove, but that much was easy enough.

Samuel Butler was an amateur of whimsical genius studying Homer for the mere love of it, as so many Englishmen have done and are still doing; he was simply diverting himself and refreshing his soul, but meanwhile professional philologists of many countries were displaying immense erudition and endless ingenuity in the search of the texts, line by line, word by word, analyzing, stratifying, classifying, disarticulating them in every possible way. While they were thus engaged, competing with one another, often quarreling about this word or that, it took an ex-business man, that is, a Philistine, to have the simple idea of checking Homeric words against monuments. The philologists were working day and night in their libraries, surrounded with dictionaries, editions, commentaries, and the dusty memoirs of their predecessors. Their task was endless and they often worked in a kind of fever. Time was precious. They could find none for adventure and had no inclination to wander in the places that the Homeric poems were supposed to describe or to refer to. Moreover, was not Homer simply a weaver of fairy tales? Was there any hope of finding traces of the ancient gods and heroes? Owing to his ignorance [48] and simplicity as much as to his enthusiasm and faith, Heinrich Schliemann (1822–1890) thought there was. Nay, he was sure of it, so sure that he was ready to stake his fortune and his life on that belief. Homeric poems were not spun out of thin air; there must be a material basis to them and he would go and unveil it. He visited Greece and Troy for the first time in 1868 and in that same year began his excavations in Ithaca. The following twenty years were largely devoted to excavations in Troy, Mycenae, Orchomenos, Tiryns, and he was truly the pioneer of Greek prehistoric archaeology. He was the first to excavate methodically and, though his methods have been refined in many ways, he is the founder of that line of research; [49] the first to improve his methods was his young assistant and successor, Wilhelm Dörpfeld (1853–1940).

Even as Wolf had opened a new era of philological discussion, Schliemann began the era of archaeological exegesis and made possible a new interpretation of Homeric poetry as a mirror of the Mycenaean age.

This could not affect one of the Homeric questions, the one that teases the average person most – the identification of Homer – but in a deeper sense it resurrected him (*Homēros anestē*) as the author or editor of poems celebrating the dawn of Greek culture. We shall never know the truth concerning the author (or the two authors, or the plural one), but that does not matter very much for we have the two poems, the *Iliad* and the *Odyssey*, presumably in their integrity, and these are indestructible treasures, the value of which can but increase in the future.

[48] "Ignorance" is written from the philologists' point of view. Schliemann was not a well-trained scholar but a self-taught amateur. Yet he knew Homer by heart; he knew the words and the things that they signified to Greek imagination. He had taken the trouble of mastering modern Greek and could discuss native lore unceasingly with his Greek wife (since 1869) and friends, with Greek schoolmasters, sailors, and shepherds, with the most learned men of Hellas as well as with the humblest. In these respects his equipment was immeasurably superior to that of the average philologist.

[49] He had to stand much criticism not only from the armchair philologists but also from archaeologists finding fault with his methods from the point of view of later improvements. For a fair appreciation by a professional archaeologist, see Stanley Casson (1889–1944), *The discovery of man* (London: Harper, pp. 226–227 [*Isis* 33, 302–303 (1941–42)].

HESIOD

In their splendid work on the *Growth of Literature* the Chadwicks have shown that the early literature of many nations is not concerned only with narratives or sagas, but deals also with other subjects. The *Iliad* and the *Odyssey* are the outstanding examples of epic poetry in world literature, but the early Greek minstrels were reciting occasionally poems of different content, the purpose of which was mainly didactic, gnomic (aphoristic "wise sayings," riddles), or mantic (divinatory, prophetic). This is not surprising, for why were there minstrels and why do we find them all over the world? Simply because people always craved for information, for knowledge of one kind or another. Personal, familial, or tribal gossip could not satisfy the more intelligent among them very long; they wanted to extend their horizon. They could not help asking themselves many troublesome questions. "Why were they doing what they did?" "Where did they come from and whither were they bound?" "Why were they living at all?" "Why was the world as it was?" Such questions introduce mythology and cosmology; they also introduce science, and the history of science is largely the history of the successive answers that have been proposed.

The peoples' historical curiosity was satisfied with the sagas that gave them consciousness of their own traditions, pride of race, humanities, nobility. That was good but left many important questions unanswered, not only the very deep ones that have just been mentioned, but simpler ones, more practical and more urgent. The husbandman's need of special knowledge is considerable and very diversified; the same can be said of sailors and craftsmen. In addition, all people need moral and social guidance such as is transmitted to them in the form of proverbs. Every proverb [50] is like a parcel of folk wisdom, standardized, hallmarked, and ready for transmission. For example, a saying like "If a man sow evil he will reap more evil" [51] is easy to remember, especially if it is put in rhythmic form or expressed with rhymes and alliterations; it is also easy to repeat and he who quotes it sententiously in the family circle or the market place obtains a measure of personal credit for the wisdom of his whole tribe (he deserves that credit for he helps to preserve that wisdom and to teach it).

The best didactic poetry of the Greeks is associated with the name of Hesiod, who flourished somewhat later than Homer; perhaps for that reason, his personality is more tangible. He was the first Greek poet who spoke in his own name and expressed the intention of delivering a personal message, "to tell of true things." [52] Like Homer, Hesiod originated from the Asiatic coast, but Homer was probably an Ionian, while Hesiod's father was established in Cyme, a harbor of Aeolis (just north of Ionia). Poverty obliged the father to leave Cyme and seek better fortune elsewhere; he crossed the Aegean and settled on the Greek mainland, in Ascra, in Boeotia. His sons Hesiod and Perses were perhaps born and certainly educated in the new abode. They were farmers, like their father, but their destinies were very different. Perses was an idler and good-for-nothing, while Hesiod, not content to do his duty as a husbandman, answered an impelling call from the Muses to sing

[50] *Paroimia, Cata tēn paroimian* = as the saying goes (Plato). A list of Greek proverbs will be found in Hermann Bonitz, *Index aristotelicus* (Berlin, 1870), p. 570.

[51] *Ei caca tis speirai, caca cerdea c'amēseien.* Hesiod, fragment in Loeb Classical Library ed., p. 74.

[52] *Ego de ce . . . etētyma mythēsaimēn. Works and days*, line 10.

and preach. Toward the end of his life he moved to Oenoë in Locris, where he was murdered.[53]

There is no reason to doubt the existence of the poet Hesiod and we may assume that he flourished somewhat after Homer II, say toward the end of the eighth century. He was a Boeotian and this may account for the occasional crudity of his verse, as compared with Homer's.[54] The two main poems ascribed to him that have survived, the *Works and Days* and the *Theogony*, are excellent specimens of their kind. Note that both are relatively short, 828 and 1022 lines, but that is not surprising, for didactic or aphoristic poetry does not lend itself to the long developments and digressions that the narrative style of the *Iliad* or the *Odyssey* encouraged. The storytellers are keenly aware that their audience is eager for detailed accounts (for example, of battles and banquets) and thrilling enumerations, and that it loves the tantalizing spinning out of dramatic episodes. On the contrary, husbandmen want brief advice, and the proverbs wherein the folklore is crystallized are naturally terse.

Works and Days

The *Works and Days* (*Erga cai hēmerai*) (Fig. 41) may be divided into four parts: (1) an exhortation to his younger brother Perses, (2) a collection of rules for husbandry and for navigation, (3) ethical and religious precepts, (4) a calendar of lucky and unlucky days. The first part contains allegories or fables explaining the conditions of man and the value of goodness. In the first of these allegories, useful emulation is contrasted with quarrelsomeness. The myth of Pandora which follows explains the origin of evil and the inevitability of labor (compare with the story in Genesis having the same purpose); the Fable (*ainos*, tale) of the Hawk and the Nightingale shows the wrongness of violence and injustice. The most interesting of these stories to us is that of the Five Ages of the World: [55] the age of gold which was the age of peace and perfection; the age of silver, less pure and less noble; the age of bronze; the fourth age, which seems to refer to the Minoan revival the glorious remembrance of which had inspired Homer; finally the age of iron, the present age of sorrow, hatred, and strife. Hesiod was living in an age like ours, when thoughtful men were contemplating the ruin, misery, and chaos that are the sequel of war and of moral decline, and when in their disillusionment they were tempted to say, "The world is getting worse every day, it must needs come to an end." This kind of social pessimism may strike us as modern, because some of our contemporaries are in a similar mood, but it suggests comparison also with more ancient times, for example, with the Sumerian hymn quoted above

[53] According to Thucydides, III, 96, the murder took place near the temple of Zeus in Nemea, Argolis, but this may be due to a misunderstanding. The memory of Hesiod's death is preserved in the following graceful lines written by Alcaios of Messena *c.* 200 B.C. "When in the shady Locrian grove Hesiod lay dead, the Nymphs washed his body with water from their own springs, and heaped high his grave: and thereon the goat-herds sprinkled offerings of milk mingled with yellow-honey: such was the utterance of the nine Muses that he breathed forth, that old man who had tasted of their pure springs." The first line of the Greek text (*Anthologia graeca*, VII, 55) is

Locridos en nemeï scierō necyn Hēsiodoio.

The word *to nemos* (*nemus*) means a wooded pasture, a glade; the proper name *Nemea* is derived from it. Thucydides may have confused a common name with a proper one.

[54] The Boeotians were supposed to be dull and slowwitted and the Athenians liked to make fun of them. That bad reputation, whether deserved or not, is perpetuated by the English words Boeotia and Boeotian.

[55] *Works and days*, lines 109–201.

(p. 96). In a sense, the idea that everything is slipping from bad to worse and that "the world is going to the devil" is an idea of all times, or rather it is bound to reoccur each time the social balance is violently disturbed by wars, revolutions or other calamities. Even when no calamities intervene, it may impress itself upon the mind of a man whose own body and mind are gradually deteriorating, or who lacks patience with the gradual emancipation and the waywardness (apparent or real) of the new generation.

It is clear that Hesiod had been moved to write his poem by the indiscipline of

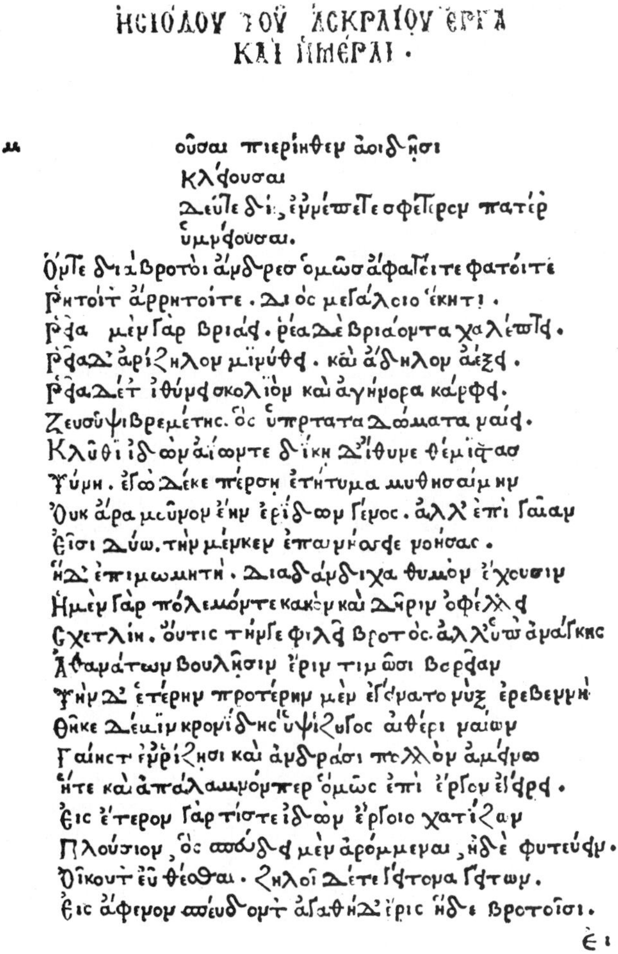

ΗΣΙΟΔΟΥ ΤΟΥ ΑΣΚΡΑΙΟΥ ΕΡΓΑ
ΚΑΙ ΗΜΕΡΑΙ.

μ ουσαι πιερίηθεν ἀοιδῆσι
κλείουσαι
δεῦτε δὴ ἐννέπετε σφέτερον πατέρ'
ὑμνείουσαι.
ὅν τε διὰ βροτοὶ ἄνδρες ὁμῶς ἄφατοί τε φατοί τε
ῥητοί τ' ἄρρητοί τε. διὸς μεγάλοιο ἕκητι.
ῥέα μὲν γὰρ βριάει. ῥέα δὲ βριάοντα χαλέπτει.
ῥεῖα δ' ἀρίζηλον μινύθει. καὶ ἄδηλον ἀέξει.
ῥεῖα δέ τ' ἰθύνει σκολιὸν καὶ ἀγήνορα κάρφει.
ζεὺς ὑψιβρεμέτης. ὃς ὑπέρτατα δώματα ναίει.
κλῦθι ἰδὼν ἀΐων τε δίκῃ δ' ἴθυνε θέμιστας
τύνη. ἐγὼ δέ κε πέρσῃ ἐτήτυμα μυθησαίμην
οὐκ ἄρα μοῦνον ἔην ἐρίδων γένος. ἀλλ' ἐπὶ γαῖαν
εἰσὶ δύω. τὴν μέν κεν ἐπαινήσειε νοήσας.
ἡ δ' ἐπιμωμητή. διὰ δ' ἄνδιχα θυμὸν ἔχουσιν
ἡ μὲν γὰρ πόλεμόν τε κακὸν καὶ δῆριν ὀφέλλει
σχετλίη. οὔ τίς τήν γε φιλεῖ βροτός. ἀλλ' ὑπ' ἀνάγκης
ἀθανάτων βουλῇσιν ἔριν τιμῶσι βαρεῖαν
τὴν δ' ἑτέρην προτέρην μὲν ἐγείνατο νὺξ ἐρεβεννή
θῆκε δέ μιν κρονίδης ὑψίζυγος αἰθέρι ναίων
γαίης τ' ἐν ῥίζῃσι καὶ ἀνδράσι πολλὸν ἀμείνω
ἥτε καὶ ἀπάλαμόν περ ὁμῶς ἐπὶ ἔργον ἐγείρει.
εἰς ἕτερον γάρ τίς τε ἰδὼν ἔργοιο χατίζων
πλούσιον, ὃς σπεύδει μὲν ἀρόμμεναι, ἠδὲ φυτεύειν.
οἶκόν τ' εὖ θέσθαι. ζηλοῖ δέ τε γείτονα γείτων.
εἰς ἄφενον σπεύδοντ'. ἀγαθὴ δ' ἔρις ἥδε βροτοῖσι.
ει

Fig. 41. *Editio princeps* of the *Works and days*, together with the *Idyls* of Theocritos (Milan, *c.* 1480); title of the *Works and days* (folio 33a). [From the copy in the Huntington Library.]

his foolish brother. He was trying to educate him, to shame him into a more decent conduct, and to buck him up (probably all in vain). The first part of his poem was a kind of mythological introduction which was meant to awaken in Perses' soul the love of tradition, the desire to be just and to work like a man.

The other parts require less explanation. The rules of husbandry and navigation;[56] (more than a third of the whole) are easier to read than to analyze. Let us quote a few lines. First the opening ones:

When the Pleiades, daughters of Atlas, are rising, begin your harvest, and your ploughing when they are going to set. Forty nights and days they are hidden and appear again as the year moves round, when first you sharpen your sickle. This is the law of the plains, and of those who live near the sea, and who inhabit rich country, the glens and dingles far from the tossing sea, – strip to sow and strip to plough and strip to reap, if you wish to get in all Demeter's fruits in due season, and that each kind may grow in its season. Else, afterwards, you may chance to be in want, and go begging to other men's houses, but without avail; as you have already come to me. But I will give you no more nor give you further measure. Foolish Perses! Work the work which the gods ordained for men, lest in bitter anguish of spirit you with your wife and children seek your livelihood amongst your neighbors, and they do not heed you. Two or three times, may be, you will succeed, but if you trouble them further, it will not avail you, and all your talk will be in vain, and your word-play unprofitable. Nay, I bid you find a way to pay your debts and avoid hunger.

Then these:

But when the artichoke flowers, and the chirping grass-hopper sits in a tree and pours down his shrill song continually from under his wings in the season of wearisome heat, then goats are plumpest and wine sweetest; women are most wanton, but men are feeblest, because Sirius parches head and knees and the skin is dry through heat. But at that time let me have a shady rock and wine of Biblis, a clot of curds and milk of drained goats with the flesh of an heifer fed in the woods, that has never calved, and of firstling kids; then also let me drink bright wine, sitting in the shade, when my heart is satisfied with food, and so, turning my head to face the fresh Zephyr, from the everflowing spring which pours down unfouled thrice pour an offering of water, but make a fourth libation of wine.[57]

That is not so Boeotian after all! Hesiod's immediate purpose was to explain to his brother how to work profitably and to escape want, but the poetry inherent in his subject was too much for him, or, to put it otherwise, the practical and censorious man was defeated by the poet in him. At his best he was so deeply moved by the gracious spectacles which surrounded him that he was lifted for a while to a higher level; he was then really a forerunner of the bucolic poets of a later time.[58]

Until 1951 it would have been correct to say that Hesiod's *Works and Days* were the first example of a "Farmer's Almanac." This is no longer true, because Samuel Noah Kramer of the University Museum of the University of Pennsylvania has discovered in Nippur and deciphered a cuneiform tablet of *c.* 1700, beginning with the line "In days of yore a farmer gave these instructions to his son"; there are 108 lines in all, explaining a farmer's duties throughout the year. Kramer has

[56] *Ibid.*, lines 383–694.

[57] *Ibid.*, lines 383–404, 582–596. The quotations are from the translation by Hugh G. Evelyn-White in the Loeb Classical Library, pp. 31, 47 (1914).

[58] The first editors of the *Works and days* realized this. Indeed the early editions included not only the *Works and days* but also the *Idyls* of Theocritos of Syracuse (fl. 285–270).

published a tentative translation entitled "Sumerian Farmer's Almanac." [59] Please note that the unknown Sumerian farmer who wrote or inspired that text flourished about a thousand years before Hesiod.

To return to the latter, the two final sections of his poem are very short (70 and 64 lines). The third offers homely advice for marriage and for good behavior in various circumstances, even some which are apparently very trivial (how to make water, *omichein* [60]); this includes superstitions that would interest the folklorist but upon which I have no time to insist. The precepts in the fourth part, dealing with lucky and unlucky days, are of course entirely superstitious, but we should bear in mind that similar fancies guided the farmer's activities until yesterday, that they still guide him today in many countries, and that there are among us so-called rational people who are afraid of "Friday the thirteenth." The poem ends with these lines:

> These days are a great blessing to men on earth; but the rest are changeable, luckless, and bring nothing. Everyone praises a different day but few know their nature. Sometimes a day is a stepmother, sometimes a mother. That man is happy and lucky in them who knows all these things and does his work without offending the deathless gods, who discerns the omens of birds and avoids transgression.[61]

The farmer was aware of many mysteries surrounding and threatening him; he was every day at the mercy of the elements and of luck. It was not enough for him to do his best in a practical way, he must be humble and full of awe.

Among the lost works of Hesiod was an astronomical poem of which only a few fragments remain. It describes the principal constellations and explained their names, that is, the myths connected with them. The fragments that have come down to us deal with the Pleiades, Hyades, the Great Bear, and Orion. These are the earliest texts of their kind in Greek literature.

DESCENT OF THE GODS. HESIOD II

The other extant poem, *Descent of the Gods* (Theogonia) is a summary of mythology, the history and genealogy of the gods, which should not detain us. It was originally followed by another poem, a catalogue of women and *eoiai*, that is, a list of heroines each of whom was introduced by the minstrel with the words *ē hoiē* (or like her). These women constituted the natural link between the world of gods and that of men, for the heroes, to whom a divine origin was generally ascribed, were brought to life by earthly mothers. After having set forth the intricate genealogy of the gods it was thus necessary to speak of the mortal women whom they had loved and by whom the heroes, the leaders of men, had been given to the world. This way of thinking helps to account for primitive matriarchy, but I must abandon that subject to anthropologists.

To a mythologically minded person (and that description applies to every Greek) the genealogy of gods and cosmology were related fields, for the origin of gods, the origin of the world, the procedure and details of creation were inextricably mixed. How had the poet obtained knowledge of the dark secrets that he revealed? He warns us in the prelude [62] that the daughters of great Zeus "plucked

[59] S. N. Kramer, *Scientific American* (New York, November 1951), pp. 54-55.

[60] *Works and days*, lines 727-732.

[61] *Ibid.*; Loeb, p. 65.

[62] *Descent of the Gods*, lines 29-34.

and gave me a rod, a shoot of sturdy olive, a marvellous thing, and breathed into me a divine voice to celebrate things that shall be and things that were aforetime."[63] This placing of the unknown past on the same plan as the future was natural enough. The genuine seer, like Calchas' son Thestor,[64] knows "the present, the future, and the past," and the timeless gods are not conscious of time. Remember the inscription in the Iseum of Sais; Isis says of herself, "I am everything which existed, which is now and will ever be, no mortal has ever disclosed my robe."[65]

Philologists agree that the two main Hesiodic poems are post-Homeric, though each contains elements that are or may be as old as anything embedded in the *Odyssey* or even in the *Iliad.* They would place the *Theogony* later, perhaps as much as a century later, than the *Works and Days.* The *Theogony* is thus ascribed to another writer, whom we might call Hesiod II.[66]

HESIODIC STYLE AND TRADITION

Though the *Works and Days* contains some graceful passages, Hesiod's style is generally inferior to Homer's. This may be because the subject lends itself less easily to poetic graces, or it may be due to the sterilizing fascination that Homer's greatness and popular success entailed. It is conceivable that the fame of the *Iliad* and the *Odyssey*, when these epics were finally perfected, discouraged other poets, including Hesiod, in the same manner that Michelangelo and Raphael created a kind of artistic desert around them.

The main reproach that one can make to Hesiod is that he lacks the rapidity and the fluidity of Homer and that many of his verses follow one another haltingly in a staccato rhythm, but this was often unavoidable, and I feel greater respect for an author who jumps readily from one idea to another when there is no real connection between them, than for one who forges artificial transitions. Hesiod's style is homely and naïve but that is not unpleasant, and his mood is rather stern and unromantic, but what would you have? He was in a more literal way than Homer a teacher, an instructor. People did not take as willingly to him as they did to the storyteller who by now had already assumed heroic grandeur.

It is not surprising then that the Hesiodic tradition was less glamorous and less universal than the Homeric one. Even today a hundred people know Homer for one who knows Hesiod, and so, I imagine, it always was. It would seem that the later poem, the *Theogony*, was the first to attract attention; it was commented upon by the founder of the Stoics, Zeno of Citium (IV–2 B.C.) and edited by Zenodotos of Ephesos (III–1 B.C.) and Aristophanes of Byzantium (II–1 B.C.). The first philologist to interest himself in the *Works and Days* was Dionysios Thrax (II–2 B.C.); curiously enough, the Greek text of that work was printed almost a decade before that of Homer.

Hesiod is not forgotten, however, and his words are still moving. He was close to the earth and to life. He explained the fundamental law of mankind, the necessity

[63] *Ibid.*, Loeb, p. 81.

[64] *Iliad,* I, 70.

[65] *Egō eimi pan to gegonos cai on cai esomenon cai ton emon peplon udeis pō thnētos apecalypsen.* Plutarch, *Isis and Osiris,* 354 c.

[66] Hesiod is mentioned by name in line 22 of the *Theogony.* This may be understood as a reference to the Hesiod who wrote the *Works and days* by the (different, later) author of the *Theogony.* Could it not be understood as well as a reference by the poet to himself?

of justice and honest labor; that law has not been abrogated and never will be. His stern advice is still applicable and a few idyllic traits of his are still warming our hearts.

BIBLIOGRAPHIC NOTES

Homer. The first edition of the Greek text of both the *Iliad* and the *Odyssey* we owe to Demetrios Chalcondyles; the colophon is dated Florence 9 December 1488, though the printing was not completed before 13 January 1489. See our facsimile of a page taken from the copy in the Boston Public Library. Descriptions in the British Museum *Catalogue of incunabula* (vol. 6, p. 678) and in Emile Legrand, *Bibliothèque hellénique* (Paris, 1885), vol. 1, pp. 9–15.

Edition of the *Iliad* by Walter Leaf (2 vols.; London, 1886–1888, 1900–1902), and by Jan Van Leeuwen (2 vols.; Leiden, 1912–13). Greek-English edition by Augustus Taber Murray in the Loeb Classical Library (2 vols.; London, 1924–25). Greek-French edition by Paul Mazon in the Collection des Universités de France (4 vols.; Paris, 1937–38). George Melville Bolling, *Ilias Atheniensium. The Athenian Iliad of the sixth century* B.C. (524 p.; New York: American Philological Association, 1951); an attempt to reëstablish the Peisistratian text; about 1000 of the 15,693 lines accepted by Wolf are here set at the foot of the page; see footnote 21.

Editions of the *Odyssey*, Books I–XII, by W. Walter Merry and James Riddell (Oxford, 1875, 1886), Books XIII–XXIV by David Binning Monro (Oxford, 1901); Books I–XXIV by Jan Van Leeuwen (Leiden, 1917). The *Odyssey* printed in Robert Proctor's type on Morris paper by the Oxford University Press in 1909 is a very beautiful book. Greek-English edition in the Loeb Classical Library by A. T. Murray (2 vols.; London, 1919). Greek-French edition by Victor Bérard in the *Collection des Universités de France* (3 vols.; Paris, 1924).

Hesiod. Editio princeps of the *Works and Days* together with the *Eidyllia* of Theocritos, printed by Bonus Accursius in Milano without date (*c.* 1478–1481, *c.* 1480). The title page of the *Works* (folio 33*a*) which we reproduce was obtained from the copy in the Huntington Library. Editio princeps of both works of Hesiod, with Theocritos and other works, printed by Aldus Manutius (Venice, February 1495/96). These two first editions are described in the British Museum *Catalogue of incunabula* (vol. 6, p. 757; vol. 5, p. 551).

Greek-English edition of Hesiod, together with the Homeric hymns and Homerica, by Hugh G. Evelyn-White (Loeb Classical Library, London, 1914).

Bibliophiles will enjoy the edition of the *Works and Days* in Greek and French by Paul Mazon printed in the Garamond type by Edouard Pelletan (Paris, 1912), together with woodcuts by Emile Colin and a long essay by Anatole France. This was the last book published by Pelletan. The Garamond type is so called after Claude Garamond (d. 1561); it was the type used by Robert Estienne (1503–1559) in his Greek editions after 1544. It is very pleasant to look at, but difficult to read because of numerous ligatures. Three sizes of it are still available at the Imprimerie Nationale, Paris.

VI

ASSYRIAN INTERMEZZO

We have already denounced the utter confusion that is caused by historians who deal with Mesopotamian science as if it were one single entity prior to Greek science. Matters are far more complex, and one should recognize at least three very different "groups" (not units): first, the "Babylonian" science which we have briefly described in Chapter III; second, "Assyrian" science to which this chapter is devoted; third, "Chaldean" science which was developed in Hellenistic, Seleucid, times.

"Babylonian" science is prior to the first millennium, that is, it is completely prior to Greek "historic" times, prior to Homer and Hesiod, not to mention the Ionian philosophers.

"Assyrian" science is chiefly a matter of the seventh century. It is contemporary with the beginning of Hellenic science, just a little ahead of it. Hellenic science was and remained independent of it.

"Chaldean" science is definitely post-Hellenic. It influenced late Hellenistic (or Roman) science and medieval science.

These three groups are separated by two intervals which lasted many centuries each; each group influenced its successor, yet the three groups are as different from one another as their chronological distance would suggest. To mix them up is just as silly as if one spoke of Bede, the two Bacons, Newton, and Rutherford, as if they all belonged to the same lot.

In our account of Babylonian science (Chapter III) we spoke of three kings; these were Sharrukīn (or Sargon), founder of the Accad dynasty (ruled 2637–2582), and two kings of the Amurru (or Amorite) dynasty, the sixth, the great lawgiver Hammurabi (1955–1913)[1] and the tenth, Ammiṣaduga (1921–1901). These names are mentioned simply to refresh the reader's memory, and emphasize once more the enormous temporal distance between Babylonian and Assyrian science.

Assyrian culture is Mesopotamian, but while the Sumerian and Babylonian cultures were centered upon the lower Euphrates, the Assyrian originated in the region of the upper Tigris. It was indebted not only to Sumerian and Babylonian examples but also to Hittite and Hurrian influences. It was often inferior to its

[1] According to the most recent calculations the dates of Hammurabi are now 1728 to 1686 and the other dates would have to be changed accordingly. The main point is that all those Babylonian kings were much anterior to Greek "historic" times.

models; for example, the Assyrian codes that have come to us are decidedly on a lower level than the code of Hammurabi.[2] We need not bother here about the beginnings of Assyrian history; the city of Ashshur[3] was already flourishing *c.* 2600. The first ruler of the Assyrian empire was Ashur-nasir-pal II (884–859), who extended his dominion to the Mediterranean Sea and obliged the Phoenician cities of the coast to pay tribute to him. His capital was Nimrūd (Kalakh, Biblical Calah), south of Mawsul.

Let me name a few other rulers with whom the reader is already familiar because of Greek or Biblical reminiscences.

Shammu-ramat (810–806), widow of one king and mother of another, is famous under the Greek name of Semiramis. Indeed, to the Greeks Semiramis was a kind of goddess; she and Ninos were taken to be the mythical founders of the Assyrian empire (the empire of Ninos or Nineveh). Many wonderful achievements were credited to her.[4]

Sharrukīn II (722–705), or Sargon II,[5] took Samaria, Carchemish, raided Urartu, reconquered Babylonia, and built a new capital near Nineveh, Dūr-Sharrukīn (Khorsābād).

Sin-ahē-erba (705–681), son and successor of the preceding, is the Biblical Sennacherib; he invaded Palestine but failed to take Jerusalem; in 689 he destroyed Babylon.

Ashur-bani-pal (668–625), called in Greek Sardanapalos, was a great ruler of a large part of the Near East, but not of Egypt. According to his enemies, he was a degenerate and a monster of cruelty, but it must be said to his credit that he was a patron of arts and letters and the preservation of what we call "Assyrian science" is largely due to his efforts. His capital was Nineveh (Quyunjiq, opposite Mawsul). He was the last ruler of the Assyrian empire, but he helped more than any other man to immortalize its memory. His crimes have left no trace, but his Babylonian library will endure for ever. It is because of him that we think of Assyrian learning, somewhat unjustly, as a creation of the end of the seventh century.

It has been worth while, I think, to weave our memories of Greek and Biblical traditions with those that the historian of science is bound to evoke, however briefly.

Assyrian art was made known to the world about the middle of the last century. In 1807, Claudius James Rich, British consul at Baghdād, was the first to refer to an Assyrian bas-relief and to indicate the archaeological possibilities of Quyunjiq, but excavations were not begun until 1843, by Paul Emile Botta (in Khorsābād),

[2] For a comparison of the Assyrian codes with earlier codes and bibliography on the subject, see James B. Pritchard, *Ancient Near Eastern texts* (Princeton: Princeton University Press, 1950), pp. 159-223 [*Isis 42*, 75 (1951)].

[3] Ashshur (or Ashur) on the upper Tigris, below Mawsul. The word Ashur appears in the names of many Assyrian kings and our word Assyrian is itself derived from it. The term "Assyriologist" is used to designate students not only of Assyrian antiquities, but also of Mesopotamian antiquities in general. This is due to the accident that Assyrian monuments and documents were the first to be discovered and investigated.

[4] There is, of course, a confusion between the real woman and the legendary one and, as always happens, all kinds of legends clustered around the mythical person. The name became proverbial. Margrete of Denmark (1353–1412), who ruled the three Scandinavian kingdoms, was called the Semiramis of the North (*Introduction*, vol. 3, p. 1021), and the same epithet was bestowed upon Catherine II of Russia (1729–1796).

[5] This is the Biblical Sargon. He is named Sargon II with reference to an earlier king of Assyria, Sharrukīn I (2000–1982), not to the king of Accad, Sharrukīn (2637–2582).

and soon afterward by Austen Henry Layard, Hormuzd Rassam, and others. The fruits of the French excavations are in the Louvre, the treasures dug up by the English archaeologists in the British Museum. Together they reveal a new art comparable to the best of Egyptian art, to Greek art, and to the early Persian art which continued to some extent the Assyrian tradition. In a history of art one would have to describe and discuss those masterpieces at considerable length, but in a history of science we cannot give much space to them. They help us to evoke the prodigious artistic background of Assyrian culture. Most of the Assyrian bas-reliefs were cut in a kind of soft limestone, and were originally colored in black, white, blue, red, and green. They concern the archaeologist as well as the artist, for they provide a large amount of information on the manners and customs, arts and crafts, religious and scientific ideas of the Assyrian people.[6]

The most fascinating of all those monuments for the historian of science are some mythological scenes from the time of Ashur-nasir-pal (884–859) which have been interpreted as representing the artificial fertilization of date palms. There are many such bas-reliefs in the British Museum, the Louvre, and other museums. It is probable that artificial fertilization had been practiced long before that time, perhaps in prehistoric days; it was old enough in the time of Ashur-nasir-pal to be an intrinsic part, not of science, but of mythology. If our interpretation is correct, it does not prove, of course, that the Assyrians knew the sexuality of plants; I would say that they did not know it; yet, they acted as if they knew it. This is an excellent example of an important application preceding by more than twenty-five centuries the scientific knowledge from which it might have been derived.[7]

The region of Mawṣul on the upper Tigris where the Assyrian capital was located was too far north for the cultivation of date palms, but the Assyrian empire extended almost to the Persian gulf and the Assyrians had inherited the whole of Sumerian lore.

The excavations made at Nimrūd revealed many other monuments of the age of Ashur-naṣir-pal, such as colossal lions, winged and human-headed, bas-reliefs of apes, and two statues of himself (one now in the Louvre, the other in the British Museum).

The development of Assyrian art can be followed from the ninth century down to the end of the seventh, a period of almost three centuries, because of the monuments due to other kings: Shalmaneser III (859–824), black obelisk and bronze bands decorating the gates of his palace; Tiglath-pileser III (745–727); Sargon II (722–705), the colossal winged and human-headed bulls found in his palace of Khorsābād; Sennacherib (705–681); and finally Ashur-bani-pal (668–625), of whom we must say a little more.

To begin with the art, the most famous Assyrian bas-reliefs were created during his rule and were excavated from the ruins of Nineveh (Quyunjiq). Those bas-

[6] For general guidance see the histories of ancient art. Cyril John Gadd, *The Assyrian sculptures* (78 pp., 18 pls.; London: British Museum, 1934). Georges Contenau, *Les antiquités orientales au Musée du Louvre* (Paris, 1928), pls. 5–20.

[7] The sexuality of phanerogams was clearly explained for the first time by Rudolf Jacob Camerarius in 1694. For the interpretation of the Assyrian bas-reliefs see G. Sarton, "The artificial fertilization of date palms in the time of Ashur-nasir-pal," *Isis 21*, 8–13, (1934), 2 pls. Also S. Gandz, *Isis 23*, 245–250 (1935); G. Sarton, *Isis 26*, 95–98 (1936). Nell Perrot, *Les représentations de l'arbre sacré sur les monuments de Mésopotamie et d'Elam* (144 pp., 32 pls.; Paris: Geuthner, 1937) [*Isis 30*, 365 (1939)].

reliefs, which are one of the glories of the British Museum, represent hunting scenes and other animal scenes which suggest that the king's palace included a kind of zoölogical garden. The sculptures prove a familiarity with the anatomy of animals, such as lions, that could hardly have been obtained during the brief excitement of the hunt. It is probable that wild animals were kept in cages, then released to provide the king and his familiars with easy sport. These amazing bas-reliefs show that the artists had observed lions and other animals in their full strength, also when wounded, bleeding, vomiting blood, and dying. One of them gives us an unforgettable vision of a lioness who has been wounded in the middle of her spine; she drags the lower part of her body which is paralyzed. Such representations remained almost unique in the history of art until the Renaissance and modern times.

These hunting scenes would be sufficient to immortalize the name of Ashur-bani-pal and the memory of the unknown artists whom he employed, but he has other and greater titles to the gratitude of scholars. In addition to the bas-reliefs, the ruins of Quyunjiq hid a large number of clay tablets, which constituted the king's library. It is a very fortunate circumstance that the library was found *in situ* at the very beginning of Assyriological investigations.[8]

It is probable that there were other royal libraries in Assyria prior to this one,[9] but the library of Ashur-bani-pal is the only one available to us, and hence all the knowledge that it has brought to us must be credited to his age.

This does not mean that that knowledgc was new knowledge, obtained by his own contemporaries. Far from it; it was new knowledge only in the sense that philological knowledge can be new. When one of our contemporaries discovers in a papyrus or in ancient codex an unknown text of Aristotle or Archimedes, that is a novelty, a great novelty, in spite of the fact that the text itself is very old. Let us put it this way: the discovery is new and startling, but the object discovered is old, and so is the knowledge that has suddenly been revealed.

That is exactly the situation for the tablets excavated in Quyunjiq. They show that the Assyrians of the seventh century, if not before, had realized the scientific value of the texts preserved in Sumerian and had made immense efforts to collect Sumerian tablets, to investigate and teach the Sumerian language, to edit Sumerian texts, and to translate them into Assyrian with necessary commentaries. The Assyrians did for the Sumerian texts what Chinese Buddhists did for Sanskrit and Tibetan texts, Japanese for Chinese texts, or our own Hellenists for Greek classics. It would be more correct to say the early Renaissance Hellenists were revealing those classics; very few Hellenists of our own day are privileged to do that; most of them must be satisfied to reëdit well-known texts for the hundredth time.

The Ashur-bani-pal library contained books on grammar, dictionaries, historical archives, Sumerian texts with interlinear Assyrian translations; many of those texts

[8] The majority of the clay tablets preserved in the world's museums were dug out by natives for sale to dealers in antiquities; in many cases their exact provenience is unknown. This considerably decreases their value, unless their provenience and date can be determined from the text that they carry.

[9] There are many tablets bearing the library mark of Ashur-bani-pal's great grandfather Sargon, but Sargon's library is lost. All the tablets of the royal library bore a label (just like books in our libraries). The simplest label read: "Palace of Ashur-bani-pal, King of the world, King of Assyria." As the number of tablets grew and the text of the labels was lengthened, stamps were prepared by means of which the whole label could be printed at once. Among the historical documents is an autobiography of Ashur-bani-pal, the greater part of which is devoted to his education; that is the only autobiography of an Assyrian king. See A. T. Olmstead, *History of Assyria* (New York, 1923), pp. 489–503.

were scientific – astronomic, astrologic, chemical, medical, and so on. The king was anxious to enrich his collection; we read in a letter, probably written by himself:

Word of the king to Shadunu: It is well with me; mayest thou be happy. When thou receivest this letter, take with thee these three men [names given] and the learned men of the city of Borsippa, and seek out all the tablets, all those that are in their houses, and all those that are deposited in the temple of Ezida . . .

The king continues with a list of the important works that he especially wants and then concludes:

Hunt for the valuable tablets which are in your archives and which do not exist in Assyria and send them to me. I have written to the officials and overseers . . . and no one shall withhold a tablet from thee, and when thou seest any tablet or ritual about which I have not written to thee, but which thou perceivest may be profitable for my palace, seek it out, pick it up, and send it to me.[10]

The tablets are so abundant that a good many scholars and scribes must have been enlisted to compose and write them. During the last half-century of its existence, Nineveh was the seat of a school of translators and philologists, what might be called a Sumerian Academy. The many bilingual texts that have come down to us have made it possible for our own Assyriologists to study and to master the Sumerian language. The Sumerian scholars of today are the pupils of the Assyrian philologists of the seventh century.

A great many of the scientific tablets have been edited by modern scholars, and some of them translated into European languages. The following list is exemplary rather than complete.

Magic. Leonard W. King, *Babylonian magic and sorcery, being the prayers of the lifting of the hand* (230 pp., 76 pls.; London, 1896). This is as far remote from science as possible but is mentioned to illustrate the superstitious background.

Medicine. Reginald Campbell Thompson, *Assyrian medical texts from the originals in the British Museum* (114 pp., folio; Oxford, 1923) [*Isis 7*, 256 (1925)]; Assyrian prescriptions for diseases of the feet (*J. Roy. Asiatic Soc.* (1937), 265–286 [*Isis 28*, 226 (1938)].

Botany. R. C. Thompson, *The Assyrian herbal, a monograph on the Assyrian vegetable drugs* (322 pp.; London, 1924) [*Isis* 8, 506–508 (1926)]; some 250 plants are dealt with and Assyrian ideas on plant sexuality discussed; *Dictionary of Assyrian botany* (420 pp.; London: British Academy, 1949) [*Isis 43*].

Chemistry and geology. R. C. Thompson, *Dictionary of Assyrian chemistry and geology* (314 pp.; Oxford: Clarendon Press, 1936) [*Isis 26*, 477–480 (1936)].

This short list may serve as a starter. The details of Sumerian-Assyrian knowledge cannot be discussed here, because they would take us away from the main stream of ancient science. Assyrian science does not really belong to that stream; it is exotic.

The works of Thompson are all in the analytical stage, of great value to Assyriologists, of little value to historians of science. It is not yet possible to determine whether Assyrian science was exclusively Sumerian, or whether the Assyrian scholars added new knowledge to the old that they were preserving and interpreting.

This chapter is called an "intermezzo" because that knowledge, whether purely

[10] As quoted by Edward Chiera (1885–1933) in his excellent book, *They wrote on clay* (Chicago: University of Chicago Press, 1938).

Sumerian or Assyrianized, did not influence Hellenic science. The Oriental influences to which Hellenic culture was submitted were real enough, but general, religious, philosophical, nontechnical. Astronomical information might be transmitted, but not much else. There is no evidence that any Greek author [11] was ever able to read cuneiform.

Though "Chaldean science" is outside the field of this chapter, a few words may be added concerning it, in order to guide the reader.

The Chaldean dynasty was the last Babylonian dynasty; its six kings ruled for 87 years, from 625 to 538. The founder, Nabopolassar (625–605) and his ally Cyaxares, king of Media, destroyed Nineveh in 612 and divided the Assyrian empire between them. From then on, Assyrian traditions were continued partly by the Chaldeans, partly by the Medes and Persians; for example, Achaimenid art reveals strong Assyrian influences. The second king Nebuchadrezzar [12] (605–561) conquered Judea and destroyed Jerusalem in 586; the Babylon admired by Greek historians is the new one built by him. In 538, Babylon was taken by Gobryas, a general of Cyros the Great and for two centuries (536–332) Babylonia was under Persian rule. The earliest Babylonian mathematicians and astronomers known by name to the Greeks belong to this Persian period, to wit, Nabu-rimanni (son of Balatu) who flourished in Babylon in 491, and Kidinnu who flourished a century later, c. 379.[13] Persian Babylonia was conquered by Alexander the Great in 332 and remained in his power until his death in the city of Babylon in 323. Babylonia was then governed by one group of his successors, the Seleucid dynasty (312–171).[14]

The term Chaldean science might refer to events occurring during the Chaldean dynasty, for example, to astronomic observations of the time of Nebuchadrezzar.[15] In general, the term Chaldean or Babylonian (neo-Babylonian) is used rather vaguely and confusedly for events of later, Seleucid, times, which are entirely outside the scope of this volume.[16] Many astounding results of "Babylonian" astronomy and mathematics are really Seleucid, Hellenistic. When Babylonian discoveries are described by historians of science (many of whom are entirely unfamiliar with the complexities of ancient chronologies), it is thus essential to ascertain their approximate date before any further discussion of their merit and of the influences that have led to them or proceeded from them. The significance of a discovery made *c.* 2000 B.C. is obviously very different from one made *c.* 200 B.C.

[11] Except perhaps such a man as Seleucos the Babylonian (II-1 B.C.)?

[12] Or Nebuchadnezzar. He was the second king of that name; the first, who belonged to the Second Isin dynasty, ruled from 1146 to 1123.

[13] Strabon called them Naburianos and Cidenas (*Geography* XVI, 1, 6).

[14] To complete this history, Babylonia was governed by the Parthians (Arsacid dynasty) from 171 B.C. to A.D. 226, then by the Sasanian dynasty, 226–641, which was dispossessed by the Muslims.

[15] *Introduction*, vol. 1, 71.

[16] At least one astronomer of the Seleucid period is known by name, Seleucos the Babylonian (II-1 B.C.), who is a good illustration of the chronological confusion caused by uncritical scholars. For this Babylonian was a follower of Aristarchos of Samos (III-1 B.C.). That is, far from being capable of influencing Greek science, he was himself influenced by an Hellenistic astronomer!

VII

IONIAN SCIENCE IN THE SIXTH CENTURY

THE ASIATIC CRADLE OF GREEK SCIENCE

Historians of science may complain that the three preceding chapters contain very little science as they understand it; they may remark also that earlier chapters contained much more – and wonder. These two observations are correct. The Homeric age was one of the greatest literary ages in the whole past, but it was not a scientific age; there was in it a strong interest in the decorative arts, which help to make life more beautiful, and in the practical arts, which help to make it prosperous, but we can hardly detect an interest in knowledge for its own sake. The comparison of Homeric culture with the Oriental cultures that preceded it is not quite fair, however. The Homeric age lasted only a few centuries, while the development of (pre-Homeric) Egyptian or Babylonian culture lasted ten times as long. As a matter of fact, that age was only the literary preface to the age of Greek science.

We have used the word "miracle" when we spoke of the sudden emergence of masterpieces like the *Iliad* and the *Odyssey*, as perfect and complete as Athene herself springing out of Zeus's head, fully armed, with a great shout.[1] The emergence and development of Greek science in the space of three centuries is not easier to explain, and hence we might use the word miracle[2] again to express our admiration and our puzzlement. Indeed, during that short period (sixth to fourth century) so many scientific deeds were accomplished, deeds so varied and unexpected and of such pregnancy, that we must devote to them the remainder of this volume.

This and the following chapter will deal with the birth of Greek science in the sixth century in Ionia (Fig. 42). The reader will remember that the *Iliad* was written in a dialect close to Ionian and that it reflected the manners and customs characteristic of the decline of the Minoan period. The connection between Ionia and the land of Minos was not accidental. The early Ionians were to a large extent

[1] Pindar, *Olympian ode*, VII, 36.

[2] The word is correct if one considers only its original meaning: *miraculum*, a wonderful or marvelous thing; it has become objectionable because of its use in the English Bible to designate a divine or prophetic sign (*oth*, *sēmeion*) or an act of divine power (*dynamis*).

settlers from Crete.[3] We described the Homeric age as a revival of the Mycenaean one; in the same way we might say that the Ionian philosophy, of which we shall speak in a moment, was the flowering of a long series of efforts, not only Greek but Minoan.

In other words, Ionian philosophy, as well as Homeric poetry, should be, or at least might be, considered as a climax rather than a beginning, but we need not quarrel about that, for in the first place every climax is a beginning, and in the

[3] John Burnet, "Who was Javan?" a paper read before the Classical Association of Scotland in 1912; *Essays and addresses* (London, 1929), pp. 84–101.

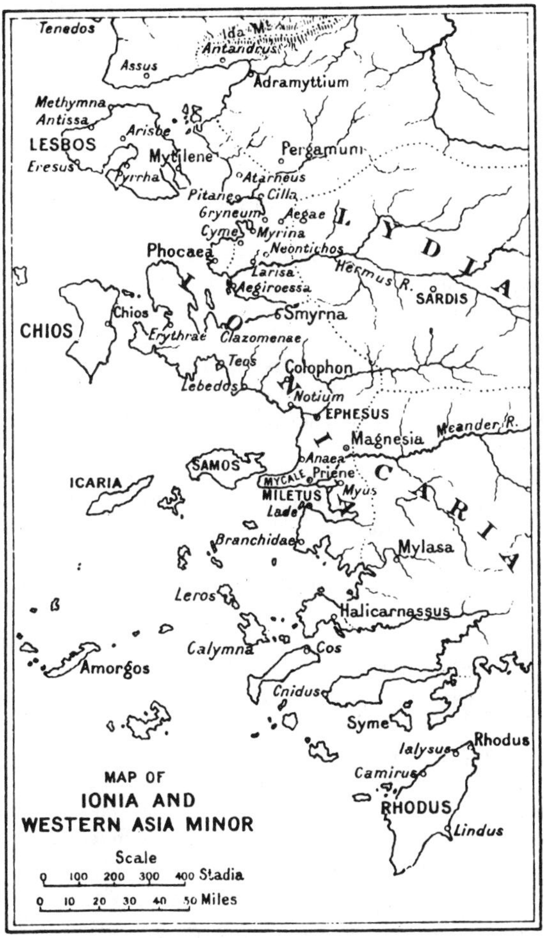

Fig. 42. Map of Ionia. [Borrowed with permission from the Loeb Classical Library edition of Herodotos, with English translation by Alfred Denis Godley (vol. 1, 1931).]

second place, take it as you please, the fundamental query remains the same. How did it come to pass that Greek science was born in Ionia? Geographic explanations are not sufficient, for the environment is very much the same on both sides of the Aegean sea. Racial explanations are not more satisfactory, for the same people or the same mixture of people might be found in various parts of that area. I shall venture to offer two social explanations. The first is that the Ionian colonists were a selected group of people living in new political surroundings which were largely of their own making, that is, of their own liking; they were likely to be brave, resourceful, spontaneous, and relatively free from restrictions. Their success is comparable to that of other colonists of a much later time, the Pilgrim Fathers who settled in New England in 1620, and can be explained partly in the same way. The Ionian pilgrims founded a New Crete on the western coast of Asia; that New Crete was to be the cradle of the New Greece. The second is that the western coast of Anatolia was an excellent place for the mixture of ideas and cultures and the resulting stimulation. As long as people stagnate in their ancestral villages they do not ask themselves many questions, for every query has been asked and answered a number of times and it is no use worrying about it any longer. On the contrary, when people of different races and with different traditions come together it must sooner or later occur to the most intelligent of them that there is more than one way of looking at things and of solving problems. If they are intelligent enough, they must even wonder whether their own traditional solutions are the right ones, or they may realize that things which they had never thought of questioning are questionable. The Ionian harbors were the terminals not only of Greek, Phoenician, and Egyptian sea lines but also of Anatolian caravan roads, connecting them step by step with the whole of Asia. Thus the conditions were exceedingly favorable for the development of science; all that was needed was people with enough native genius to improve them; the Ionians were such people; they had already proved their genius in poetry, the time had now come, at the end of the seventh century, to prove it again in a new field, natural philosophy, or, as they called it, "physiology,"[4] and they did so.

Their success, material and intellectual, was so great that for a long time the "Barbarians" (meaning the non-Greek-speaking peoples) used the word Ionian to designate all the Greeks, even as the Muslims later called the Latin Christians "Franks," and the South Americans called their northern neighbors "Yankees."

ASIA, THE HOME OF PROPHETS

Before examining the Ionian achievements, it is well to glance over the world as it was in this period, say the seventh and sixth centuries. We have already introduced to our readers the little Ionian world, as well as the Aegean, the Egyptian, the Babylonian, and others. All these worlds were different in many ways, yet none was absolutely different from the others. The phrase "one world" is not Wendell Willkie's invention. The whole world has already been one world to the

[4] The term *physiologia* has the same meaning as our phrase natural philosophy, or physics (in a broad sense). The names of our sciences have been derived from Greek in the most capricious way, and in many cases it is impossible to deduce their intended meaning from the etymology. Thus, geography is a science of the earth and geology is another, but astrology is a superstition. The meaning of physiology is now restricted to the study of the functions of living creatures, or even more so, to the study of the functions of the human body.

extent that communications existed between the parts;[5] in those days the communications were already fairly well organized in many directions (and had been for centuries or millennia), yet there were many differences. To use a physical comparison, the one world was not isotropic to social relations (it is not now and never will be). The speed and ease of communications were not by any means the same in every direction; thus some parts held closer together than others, and all kinds of groups and subgroups or multiple groups were naturally constituted.

It is thus well to ask ourselves what was happening in other parts of the world during the incubation of Greek science in Ionia. Let us first remark that the Mediterranean world is but a very small part of the world (look at a terrestrial globe) and that Ionia is but a very small fraction of that small part (on the globe it is reduced almost to nothing). We shall come back often enough to Ionia and to the Mediterranean later on; at present let us look around. The Egyptian and Babylonian moods have already been sketched, but there was a country, closer to Ionia than either Egypt or Mesopotamia, that was as foreign to the Greeks as either of these, if not more so — the land of Canaan, or Palestine. By the end of the seventh century many of the prophetic books of our Bible had already been composed: Amos, Hosea, Micah, Isaiah, Hezekiah, Zephaniah, Jeremiah, Nahum, Habakhuk; the Pentateuch (or Torah), and the books of Samuel were already completed. We shall come back to Samuel later; let us consider now only the Prophets and the Torah, and compare them with the Homeric writings. The difference between the respective languages, Hebrew and Greek, is small as compared with that between ways of thought. The Hebrew prophet was a seer;[6] the rhapsodist, a poet and storyteller. The latter referred sometimes to the gods and the heroes just as he referred to ordinary mortals, but the former spoke in God's name, in the name of the one God and of eternal justice. The contrast is so great that communication between the Hebrews and the Ionians was probably reduced to a minimum.

Among the caravans reaching Miletos or the boat parties floating down the lower reaches of the Maiandros river there must have been merchants who came from farther East or who had come across other merchants of the Halys region or the regions of the upper Euphrates and the upper Tigris and beyond. Some information may have thus percolated to them from Iran. There was (or had been) a great prophet in Iran, Zarathushtra (him whom the Greeks called later Zoroaster). Zarathushtra preached a monotheism different from the Hebrew one, yet as deeply impregnated with morality. The god of the Iranians, as well as the god of the Jews, was the personification, or rather the hypostasis, of goodness, justice, and purity. It is probable that the Ionians, even if they did receive his

[5] In extreme cases where no communications existed the oneness did not apply to the isolated parts, yet it existed potentially, for all men are built in the same way and have the same brains, the same passions, the same desires. For example, before 1492, the Americas were essentially cut out from the rest of the world and the Americans were then natural "isolationists." It is of great interest to compare their solutions of many problems with the solutions attained in the rest of the world. Those solutions were different, but not essentially different, for the American mind was a human mind and the American problems were human problems. New solutions occurred when the data of the problems were new, for example, when the native Americans domesticated or used plants and animals that did not exist elsewhere.

[6] The usual name for a prophet in the Old Testament is *nabi*, but the earlier name was *roeh*, or seer, as is explicitly stated in 1 Samuel 9:9, also *hozeh*, with the same meaning. The word always used in the New Testament is the same as ours, *prophētēs*.

message (which is doubtful), did not pay more attention to it than to the Jewish one. They were not interested in it at that time. This does not mean that whatsoever things are true, honest, just, pure, lovely, or of good report did not touch them, but they looked at those things from another angle.

Communication with India could be accomplished in various ways, the simplest one being along the Persian Gulf and the Euphrates. Two great prophets appeared in India in the sixth century, the Buddha and Mahāvīra, both of whom explained profound doctrines concerning the life of good men. Within the same period two more prophets appeared farther east, in China, Lao Tzŭ [7] and Confucius. It must suffice here to indicate these astounding simultaneities, for it would be impossible to do justice to Buddhism, Jainism, Taoism, or Confucianism in a few paragraphs. It is better to invite the reader to explore those subjects elsewhere as much as they deserve and as he desires.[8] The main point is that in the very period when "physiology" was developed in Ionia, prophets and seers, moral educators, were at work in Palestine, Iran, India, and China. Their territory was infinitely greater than that of the early physiologists, but their success was equal. All of them, the prophets and the early scientists, were working together (though they knew it not) to lift mankind up to a higher level, nearer to the gods, further away from the beasts.

The extent of the communication that may have obtained between the prophets of Asia and the Greeks was very small, and reduced at best to hints. Those hints are revealed to us in the form of words or phrases that have passed from one literature into another (for example, Egyptian images in the Psalms) or decorative motifs in the fine arts (for example, Egyptian motifs in the ivories of Samaria or in the Achaemenian monuments of Pasargada).[9] Some of the hints included in the Greek fragments will be mentioned below, when there is a special reason for mentioning them. They are not really needed for our argument. The only point to bear in mind is that Ionia was a great center of communications between East and West and that the Cretan colonizers of that Asiatic shore found there excellent conditions not only for their material prosperity but also for thought stimulation. Catalysts need not be bulky and their action is out of all proportion with their own mass; in Ionia the Greek genius was catalyzed by Egyptian and Asiatic ferments. Progress is always the result of a compromise between tradition and adventure. In Ionia the Aegean traditions were revitalized with *outre-mer* novelties, new freedoms, and new restraints.

MILETOS

Let us now focus our attention on the main harbor and the richest market of Ionia, Miletos.[10] It had been colonized by Cretans, for it was named after an earlier

[7] Doubts have been expressed concerning the reality of Lao Tzŭ, or his date, and many scholars consider the "classic" ascribed to him, *Tao tê ching*, as a much later creation. The kernel of Taoism, however, dates back at least to the sixth century. See Homer H. Dubs [1941; *Isis 34*, 238, 423 (1942–43)] and Arthur Waley, *The way and its power* (London: Allen and Unwin, 1934).

[8] First help in *Introduction*, vol. 1, pp. 66–70.

[9] Details in *Isis 21*, 314 (1934).

[10] The twelve main cities of Ionia, forming at times a confederation, were Miletos, Myos, Priene, Samos, Ephesos, Colophon, Lebedos, Teos, Erythrai, Chios, Clazomenai, and Phocaia. The first three were on the coast of Caria, the others on the coast of Lydia (north of Caria). Smyrna (of Aeolian, not Ionian, origin) was conquered by Colophon *c.* 688 and remained an Ionian city afterward.

Miletos, situated on the northeastern coast of Crete.[11] The "new" Miletos stood on a triangular limestone promontory between two gulfs not far from the mouth of the Maiandros river. In the course of time that river has deposited a prodigious amount of silt around its variable lower reaches and changed those gulfs into swamps; the present bed of the river almost surrounds the site of the ancient city. It is the old site with which we are concerned, however, and that site was excellent for navigation and trade; the city, jutting out into the sea like an enormous ship, was well protected by various islets and rocky ridges; it had four harbors, one or another of which could be conveniently entered by the boats sailing in from Rhodes or farther south, Phoenicia and Egypt, or from the west, across the Cyclades and the Sporades, or from Chios, Lesbos, and the Hellespont. Land communications, it is true, were less easy, but the sea trade boosted the Miletos market so much that caravans would find a way to it, at whatever risk or cost. Moreover, the agricultural resources of the neighboring fields and orchards were sufficient to feed the city and to permit, if not exportation of much food, at least the victualing of outgoing parties. The oil [12] and fig trades were probably important. Flax and wool were available at no great distance and the wool trade developed so well that it became famous. A Milesian type of pottery was already established in the seventh century.

The main caravan road did not terminate in Miletos, for it passed through Sardis, the outstanding market in the immediate hinterland, and from Sardis it was easier to proceed to other harbors – such as Cyme, Phocaia, Smyrna, or Ephesos, Miletos being a little too much to the south. Sardis, the capital of Lydia, had prospered so much that the wealth of one of her kings, the last one, Croesus, had become legendary and is still now.[13] Some of the wares reaching Sardis from Babylonia and Persia were deflected to Miletos.

It was the sea-borne trade, however, that was the spring of Miletos' wealth and greatness, and that trade was activated by the existence of many Milesian colonies along the shores of the Propontis and the Euxine (that is, the Sea of Marmara and the Black Sea) (Fig. 43). Some of those colonies dated from the eighth and seventh centuries. The city of Naucratis in the Nile Delta was also originally a Milesian colony; it may date back to the seventh century, but it did not assume much importance until its reorganization during the rule of the fifth king of the Twenty-sixth Dynasty, Ahmose II (Amasis of the Greeks) who ruled from 569 to 525. The Milesian merchants who had factories in Naucratis gathered all kinds of Egyptian and African wares, many of which were shipped to Miletos for further distribution. We shall come back to that presently.

Let us first complete our brief account of Milesian history. After Cyrus' defeat of Croesus (546) and conquest of Lydia, Ionia fell under Persian domination. Miletos received a more favorable treatment than the other cities and was allowed

[11] It is one of the few towns mentioned by Homer (*Iliad* II, 647) in "Crete of a hundred cities" (*Crētē hecatompolis*).

[12] The importance of olive oil in the Mediterranean economy of that age can hardly be exaggerated. Oil took the place of butter with us, to some extent that of soap, and it was used for lighting.

[13] Croesus (*Croisos*), son of Alyattes, was the last independent king of Lydia; he ruled from 560 to 546, when he was vanquished by Cyros. We still use his name to designate a very rich man, and his life, to illustrate an old adage, quoted to him by Solon, that no man should be deemed happy until his life is happily ended. Croesus was allowed to live by his conqueror, actually survived him, and accompanied the latter's son, Cambyses, in the expedition to Egypt (525).

to keep a modicum of independence. We understand these matters very well in the light of events in recent European history. The Persians expected better results from the "free" collaboration of Miletos than could be obtained from its submission; it was more advantageous to milk the old city than to kill it. As a matter of fact, Milesian prosperity continued for a while during the Persian regime, but it is easy enough to imagine the growing impatience of the Greek merchants with their Persian masters. An Ionian revolt led by Miletos was crushed in 494 and the city was then destroyed. It was liberated at Mycale (North of the Maiandros river) in 479 when the Persian fleet was defeated by the Greeks, but it never recovered its former glory.[14]

Let us return to the middle of the sixth century, to the time before the Persian conquest, when Miletos was the richest emporium of the Eastern Aegean and the main distributing center of goods between Ionia and the Greek islands, Phoenicia, Egypt, the Black Sea, and to a lesser extent Mesopotamia and the countries farther east. Milesian pottery of the seventh and sixth centuries has been found in Egypt, in the islands, in Anatolia, in South Russia.

The Milesian sailors and merchants must have obtained considerable knowledge of the parts of the world to which their business extended. They must have become familiar with a great variety of lands, peoples, religions, languages, and customs. A perfect stage was set and the actors were ready. Being what they were, Cretans or Greeks (call them as you please), keen, imaginative, and curious, as we know from Homer, it is not surprising that they asked themselves many questions; but

[14] This explains why Miletos, so very important in the history of science of the sixth century, ceases to attract our attention later.

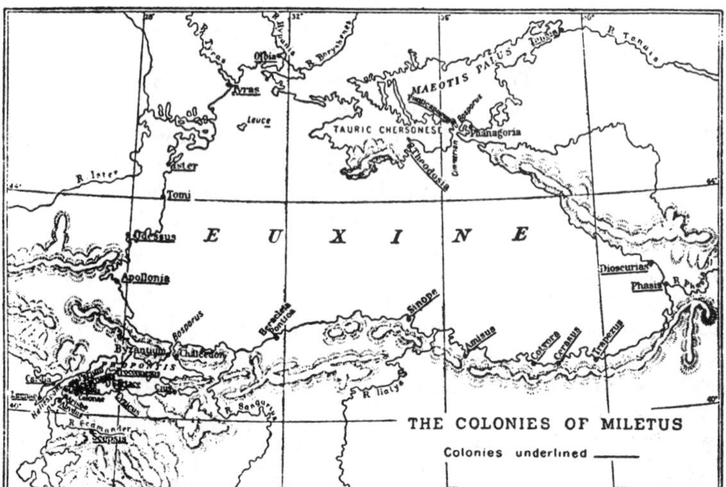

Fig. 43. Map showing the Milesian colonies in the Propontis and the Euxine. [Borrowed with kind permission from Adelaide Glynn Dunham, *The history of Miletos* (London: University of London Press, 1915), map 4.]

ΤΩΝ ΕΠΤΑ ΣΟΦΩΝ ΚΑΙ
ΤΩΝ ΣΥΝ ΑΥΤΟΙΣ ΚΑΤΑΡΙ
ΘΜΟΥΜΕΝΩΝ ΑΠΟΦΘΕΓΜΑΤΑ
συμβουλαὶ καὶ ὑποθῆκαι.

Σωσιάδου τῶν ἑπτὰ σοφῶν ὑποθῆκαι.

SEPTEM SAPIENTVM ET EOrum qui cum ijs adnumerantur, apophthegmata, consilia & præcepta.

Δεύτεραι φροντίδες σοφώτεραι.

PARISIIS, M. D. LIIII.

Apud Guil. Morelium.

Fig. 44. Title page of the first (?) Greek edition of the *Sayings of the Seven Wise Men* (Paris, 1554); see note 19. [From the copy in the Harvard College Library.] That edition is not mentioned, nor is any other book covering the same ground mentioned in the *Bibliographie hellénique . . . aux XVe et XVIe siècles* (4 vols.; Paris, 1885–1906) and the *Bibliographie ionienne* (2 vols.; Paris, 1910) of Emile Legrand (1841–1903).

in addition, they began to collect and classify the answers and were thus led to undertake new research in many fields – astronomy, physics, mathematics, geography, cartography, anthropology, biology, and medicine.

Our knowledge of those beginnings is naturally vague and uncertain. No treatises of the early physiologists have come down to us, only traditions, sometimes late and obscure ones. In that regard the contrast with Egypt and Babylonia is great, for our knowledge of science in those countries is derived from authentic, contemporary documents, papyri or clay tablets, that are immediately available to us. There is nothing we can do about that, except make the most of whatever information has filtered down. All the ancient sayings concerning Ionian thought, as well as all the direct or indirect quotations from their lost writings, have been assembled and criticized. In what follows, the doxography will be used and quoted from as far as necessary, and sometimes the nature and date of the traditions will be indicated (when it can be done briefly), but a criticism of the sources could not be offered to our readers without lengthening our survey far beyond the limitations of our space and of their patience.[15]

Excavations have been conducted at Miletos since 1899 by the Berlin Academy, under the leadership of Theodor Wiegand (1864–1936), and many reports have been published since 1906.[16]

[15] The criticism has been carried out almost to the limit by scholars such as Tannery, Burnet, Diels, etc. See bibliography at the end of this chapter.

[16] For a brief account of ancient Miletos, see Adelaide Glynn Dunham, *The history of Miletus down to the anabasis of Alexander* (164 pp. 4 maps; London, 1915).

THE SEVEN WISE MEN

Many of the traditions relative to early Ionian science were legendary at the very start. A good illustration is the legend of the Seven Wise Men, which appealed to the popular imagination and, like every successful legend, occurs in many forms (Fig. 44). Let us quote one form of it. There were flourishing about the beginning of the sixth century seven men renowned for their wisdom in philosophy or politics (*hoi hepta sophoi* – Thales of Miletos, Cleobulos of Rhodes, Bias of Priene, Pittacos of Mytilene, Solon of Athens, Periandros, tyrant of Corinth, and Chilon of Lacedaimon (Fig. 45). Note that this list contains four men of the Asiatic coast or islands (the first four in my enumeration) against three of the Greek mainland. The list varies from author to author;[17] it is always restricted to seven names, but only four seem to be constantly included, Thales,

[17] The earliest list is that given by Plato (*Protagoras*, 343); it is the same as the most popular one, which we have quoted, except that the tyrant Periandros is replaced by Myson of Chenae, a hardly known person of an unknown place. It was said that Plato rejected Periandros because of his being a tyrant.

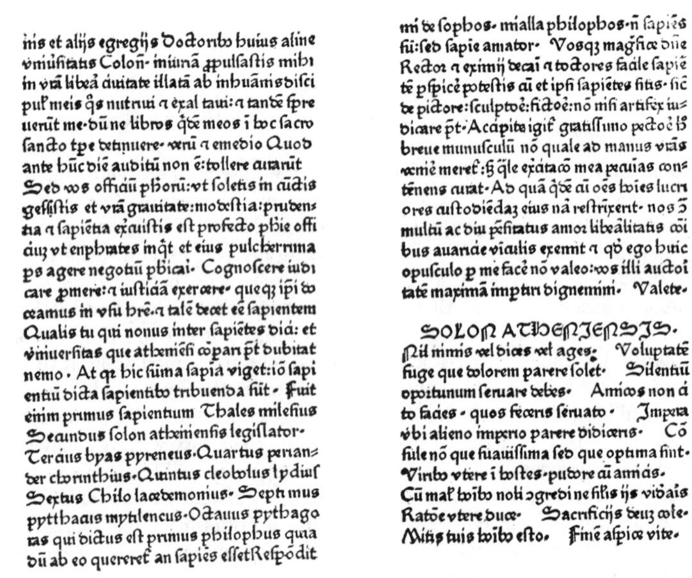

Fig. 45. Pages from the first Latin edition of the *Dicta septem sapientum Graeciae* (Cologne: Johann Guldenschaff, *c.* 1477–1487); see note 19. [Courtesy of the Pierpont Morgan Library, New York.] The pages selected are the two final ones containing the sayings of Thales and those of Chilon of Lacedaimon (fl. 560–556), who died of joy when his son gained a prize at the Olympic games. Plato was the first to include Chilon among the Seven Wise Men.

Bias, Pittacos, and Solon, that is, three Easterners against one Western.[18] Among the names included in other lists we notice the Scythian prince Anacharsis and the Cretan Epimenides, the Rip van Winkle of that age. Both are chronologically plausible, but other lists include men who lived at another time, like Epicharmos of Cos (540–450) or Anaxagoras (500–428), or mythical beings like Orpheus. As the Seven Wise Men, whoever they were, were supposed to represent ancient wisdom, and as the popular sayings (*gnōmai*, *apophthegmata*, *sententiae*) represented that wisdom in a different way, several of those sayings were early credited to them. Thus Thales was supposed to have invented the maxim "Know thyself" (*gnōthi sauton*); Solon, "Nothing too much" (*mēden agan*); Pittacos, "Seize the opportunity (*cairon gnōthi*); and so on.[19] Other traditions, reported by Herodotos,[20] connect some of the sages with Croesus, which does not fit with the chronology (Croesus belonged to the second third of the century) but is characteristic of the popular fancy; it was natural to bring the wisest men into the presence of the greatest king.[21]

One member of the group – we might call him a charter member, for he is never omitted and he generally heads the list – Thales of Miletos, is of very great interest to us, for he is the first of the Greek "physiologists,"[22] and we might perhaps say the first in the history of the world.

THALES OF MILETOS

Two of the wise men, Thales and Bias, sensing the danger to which the growing power of Persia was exposing their country, advised the Ionian cities to stand together and establish a general council in Teos. That story and others suggest that Thales was a practical man, a kind of early Franklin. He was said to be of Phoenician origin and that is not implausible, but we have only Herodotos' word for it.[23] He was born *c.* 624 and lived until 548 or 545, that is, he may have lived long enough to be obliged to witness the Persian conquest that he had tried to avert.

[18] Barkowski, "Sieben Weise," Pauly-Wissowa, ser. 2, vol. 4 (1923), pp. 2242–2264. Bruno Snell, *Leben und Meinungen der Sieben Weisen* (Tusculum Bücher; 182 pp.; München: Heimeran, 1938). Convenient collection of the traditions in Greek (or Latin) and German.

[19] In an early edition in the Harvard Library, *Septem sapientium et eorum qui cum iis adnumerantur apophtegmata, consilia et praecepta* (19 pp. in Greek only; Paris, 1554), I find a large number of sayings ascribed to the Seven Wise Men (list as quoted at the beginning of this section) and three others – Anacharsis, Myson, and Pherecydes of Syros (one of the Cyclades). For example, the sayings credited to Thales cover two pages. Is that edition the Greek princeps? The princeps of a similar collection in Latin, *Dicta septem sapientum Graeciae* (8 leaves), was printed at Cologne by Johann Guldenschaff *c.* 1477–1487; see the *Catalogue of books printed in the XVth century, now in the British Museum* (London, 1908), vol. 1, p. 256, and Arnold C. Klebs, "Incunabula scientifica et medica," *Osiris 4*, 1–359 (1938), No. 905.

[20] Herodotos, I.

[21] The stories of the Seven Wise Men (of Greece) should not be confused with those concerning the "seven sages" of Rome. The two cycles have points of contact but are not only independent but very different. The second is definitely of Oriental origin; its popularity East and West was considerable; witness the existence of versions in many languages. The literature on the subject is very large; the following items may suffice for general guidance. Killis Campbell, *A study of the romance of the seven sages with special reference to the Middle English versions* (108 pp.; Baltimore, 1898); *The seven sages of Rome* (332 pp.; Boston, 1907), edition of Middle English text with notes. Joseph Jacobs, *Jewish Encyclopedia*, vol. 11, p. 383 (1905). Carra de Vaux, "Sindibād-nāme, Syntipas," *Encyclopedia of Islam*, vol. 4, p. 435 (1927). Jean Misrahi, *Le roman des sept sages* (170 pp.; Paris: Droz, 1933), an early French text.

[22] Aristotle, *Metaphysics*, 983B.

[23] Herodotos, I, 170.

He might have obtained some of his knowledge and genius from his Phoenician ancestry; he could have obtained it as well from the Ionians, who were by this time a wealthy and sophisticated nation, familiar with many crafts, but probably lacking in unity. What could these prosperous and disunited people do against their totalitarian and warlike neighbors? There was already much to learn in Miletos, but that was not enough for the eager youth, who traveled to Egypt, where his attention was drawn to new astronomical and mathematical ideas.

His popularity must have been great, for he was considered one of the Seven Wise Men; his name was included in every list of them, and it was generally the first to be mentioned. Strangely enough, his fame rested mainly on an achievement that we are now obliged to discredit, though its genuineness was accepted as a cast-iron belief almost until our own day.

A legend that is almost indestructible (it is bound to reappear from time to time in uncritical books) deserves to be told. Indeed, it must be told, because we cannot scotch it before telling it first. It has a very old tradition and the first account of it occurs in Herodotos.[24] The Lydians and Persians had been at war for a long time, with ups and downs on both sides but no decisive victory on either. The two armies stood challenging each other in 585 when a solar eclipse, which Thales had predicted, came to pass (28 May) and the two kings were so impressed by it that they ceased fighting. Thanks to the efforts of two peacemakers, Syennesis the Cilician and Labynetos the Babylonian, the two kings were persuaded to conclude a sworn agreement and an exchange of wedlock between them. It is said that Thales was proclaimed a wise man by the oracle of Delphi in 582; that honor may have been due to the eclipse prediction credited to him.

That is a pretty legend, but it has become impossible to give credence to it. The theory was that the old Babylonians had discovered the saros, a period that enabled them to predict eclipses. Thales had heard of that in Egypt and might even have witnessed the Egyptian eclipse of 603 or been told of it. A new eclipse would then occur, or at least might occur, 223 synodic months or 18 years 11 days later, that is, in 585. As we explained above (p. 119), it is now agreed by historians of ancient astronomy that the Babylonians could not have discovered that period before the fifth or the fourth century. Hence, Thales could not have learned it from them. We must remember, however, that the Babylonian observations and perhaps the Egyptian ones also had been repeated for a very long period of time. Thales may have made a lucky guess? Even that can hardly be admitted. Herodotos' account is very sober: "Thales of Miletos had foretold this loss of daylight to the Ionians, fixing it within the year in which the change did indeed happen." Does this mean that Thales could determine only the year of the eclipse and not the day? But then the psychologic effect of his prediction would have been lost.

We must conclude that Thales did not predict the solar eclipse of 28 May 585, because he lacked the necessary knowledge, but he himself may have alleged that he had predicted it, or his companions may have been led to believe that he had. It is foolish for us now to claim that he predicted it; it is even more foolish to say that he understood the phenomenon. The explanation with which we are familiar

[24] *Ibid*, I, 74.

would have been incomprehensible to him, for he conceived the earth as a disk floating upon the ocean.

I return to the initial comparison of Thales with Franklin. Both were living in a stimulating environment, and both responded to it with open mind and natural genius. Both were inquisitive, quick to learn, ready to apply their knowledge to practical aims. Thales' journey to Egypt is like Franklin's to England; both observed eagerly what was done in the Old World and brought back the notions that they deemed useful. Franklin brought back a knowledge of electricity and Thales that of astronomy. That is not a slight achievement after all.

Thales was the first Greek mathematician as well as the first Greek astronomer. He learned in Egypt not only the periodic recurrence of eclipses but also a number of geometric facts. Practical as he was, he got hold of the facts and forgot the hocus-pocus, and then he tried to use the facts, that is, to solve such problems as how to measure the height of a building or the distance of a ship from the shore. We do not know exactly how he solved those problems, for various solutions are possible all of which involve the comparison of similar triangles. What is more noteworthy, Thales did not stop there but, being rationally minded as well as practical, he wanted to explain his solutions and this led him to the discovery of geometric principles, and of the science of geometry.

A number of geometric propositions are ascribed to him: (1) a circle is bisected by its diameter; (2) the angles at the base of an isosceles triangle are equal; (3) if two straight lines cut each other, the opposite angles are equal; (4) the angle inscribed in a semicircle is a right angle; (5) the sides of similar triangles are proportional; (6) two triangles are congruent if they have two angles and a side respectively equal. Did Thales know all and each of these propositions, or propositions equivalent to them? Was he able to prove them, and if not, how did he know them? There is no certainty on these points but we may perhaps say that Thales was the first man in any country to conceive the need of geometric propositions. This involves a kind of paradox, for we have insisted that Thales was, like Franklin, a practical man, and yet his chief intellectual merit was his recognition that it is not enough to solve problems, one must rationalize the solutions. The paradox is easy to remove. Thales was intelligent enough to realize that methods are more valuable than individual solutions, and methods imply principles or, as we say in geometry, theorems.

Another inexhaustible subject of discussion is this: Was Thales really the first geometer (in the scientific sense) or did the Egyptians anticipate him? The discussion involves too many uncertainties to be profitable; we do not know really how the Egyptians or how the Ionians thought out their geometric problems. The one thing clear is this. Greek tradition ascribed the earliest geometric propositions to Thales. By his time the Egyptian achievements had long been completed. His work, derived from theirs, opened new possibilities of development which led gradually to the *Elements* of Euclid and to all the marvelous geometric fruits of our own day.

According to Aristotle,[25] Thales said "that the magnet has a soul in it because it moves the iron." If that tradition is right, Thales knew one of the properties of the

[25] *De anima*, 405A.

loadstone. He may thus be called the founder of magnetism. The tradition that would make of him a founder of electricity is weaker and we prefer to omit it.

Thales' practical successes in the fields of astronomy, geometry, and magnetism may have increased his intellectual ambition. The first man of science in the Western world, he already anticipated the exaggerated optimism of the Victorian physicists. It was not encugh for him to rationalize geometric practice; he wanted to explain the world itself, not as his childish predecessors had done by means of myths, but in concrete verifiable terms. Would it not be possible, he thought, to determine the nature (*physis*) or substance of the world? What is the material world made of?

His conclusion that water is the original substance may seem fantastic on the surface, but it becomes far more plausible if one examines it more closely. Water is the only substance that is known to man without any difficulty in the three states, solid, liquid, and gaseous. It is easy to recognize that the steam boiling out of a kettle is the same substance as the water that gradually disappears from the kettle; that the ice or snow found in the mountains changes into water if brought into a warmer place; it is not difficult to connect clouds, fogs, dew, rain, hail, with the water of sea and rivers. Water seems to occur everywhere in one state or another; would it be overbold to imagine that it may occur also in hidden forms? Moreover, without water no life is possible, but as soon as water appears there may be life, an abundance of it. People living in moist climates may remain unconscious of the biological necessity of water, but along the Mediterranean shores, where everything dries up in the summer and where desert or semidesert conditions are familiar enough, the first merciful rains [26] create something like a resurrection of nature, the spectacle of which is awful and unforgettable. Finally, many old traditions led to the same conclusion. Thales, like Homer, thought of the earth as surrounded by the ocean; his physical views did not conflict with the oceanic myth or with Egyptian cosmology. He might conceive himself as rationalizing and explaining those ancient myths. There is also a possibility that he was influenced by the Babylonians, who regarded water as the uncreated first principle; the word chosen by them to represent water meant originally voice, loud cry (this suggests a comparison with the Greek *logos*, but we must not anticipate).[27]

While the Jews were postulating the moral unity of the cosmos, Ionian "physiologists" — of whom Thales was the first — were postulating its material unity. Thales' induction that its original substance was water was premature, but it was neither wild nor irresponsible. After having considered all the facts, Thales had concluded that if there was an original substance, ubiquitous and life-giving water was the best guess.

Philosophically minded historians will note with interest that a similar conclusion was reached more than twelve centuries later by the Muslim Prophet. Indeed God revealed to him, "We have made of water everything living." [28] It is not impossible that the Thalesian conceit had percolated into Muhammad's head, but it is not at all necessary to assume such a transmission. The Prophet had had at

[26] *Osiris* 2, 415–416 (1936).

[27] Stephen Langdon, "The Babylonian conception of the logos," *J. Roy. Asiatic Soc.* (1918), pp. 433–449 [*Isis* 4, 423 (1921–22)].

[28] Qur'ān 21:30.

least as many opportunities as Thales of witnessing desert sterility on one day, and life abundant after a rain on the morrow. Both men concluded in a similar way, but they expressed their conclusions differently according to their temperaments; Muhammad was a seer and a prophet (like his Jewish predecessors), Thales, on the contrary, was a man of science. It is typical of the Greek genius that though the second man was twelve centuries older than the first, he is much closer to us.

A final tradition is best quoted in Aristotle's own words:

> Thales knew by his skill in the stars while it was yet winter that there would be a great harvest of olives in the coming year; so, having a little money, he gave deposits for the use of all the olive-presses in Chios and Miletos, which he hired at a low price because no one bid against him. When the harvest-time came, and many were wanted all at once and of a sudden, he let them out at any rate which he pleased, and made a quantity of money. Thus he showed the world that philosophers can easily be rich if they like, but that their ambition is of another sort.[29]

Aristotle told the story as well as he could to exonerate his predecessor, but I do not like the idea of a philosopher who gets rich just to show that he can do it. That seems a little foolish and disingenuous. Is it not simpler to suppose that Thales took so much trouble because he wanted money, and became rich because that was his heart's desire? That was very Ionian, by the way, and very Greek. Judging from the other specimens, as well as from Thales, the "wise men" of the ancient Greeks were not otherworldly saints but rather practical and clever people. The Greeks have generally loved money and many of them have gathered large amounts of it and been very generous with it.[30] Aristotle's story describes Thales' greed but makes no mention of his generosity; that is why it fails to convince us. We might have loved him better if he had been more disinterested, but we should try to see him as he was.

ANAXIMANDROS OF MILETOS

Anaximandros (*c.* 610–545), son of Praxiades, was a fellow citizen and companion (*politēs cai etairos*) of Thales. He has been called the latter's disciple, but that can be understood only in a loose sense. We are not aware that Thales did any formal teaching, but Anaximandros, being some fifteen years younger, received from him some guidance and stimulation. As we shall see presently, their particular views were different, yet they had in common, as against the other citizens of Miletos, a deep curiosity and strong desire to explain the nature of things. In that sense, but in that sense only, it is true that Anaximandros continued Thales' work. Toward the end of his life he wrote a treatise *peri physeōs* (*de natura rerum*), the first treatise on natural philosophy in the history of mankind; it was still available in the time of Apollodoros of Athens (II–2 B.C.), but only very few lines of it have come down to us. Before discussing his philosophy or general physiology, it is pertinent to explain the more concrete achievements to which his life was devoted.

[29] Aristotle, *Politics* I, 1259A.

[30] It has always been the ambition of every noble minded son of Hellas that he might obtain enough money to help his people and be hailed and remembered as the benefactor (*evergetēs*) of his country or his village.

His best scientific work was done in the field of astronomy, by means of a single instrument, the gnomon. That instrument had been invented in Babylonia and in Egypt, but it is so simple that Thales or Anaximandros, or even earlier Greeks, may have reinvented it. It is simply a stick or a pole planted vertically in the ground, or one might use a column built for that purpose or for any other; the Egyptian obelisks would have been perfect gnomons if sufficiently isolated from other buildings. Any intelligent person, having driven his spear into the sand, might have noticed that its shadow turned around during the day and that it varied in length as it turned. The gnomon in its simplest form was the systematization of that casual experiment. Instead of a spear, a measured stick was established solidly in a vertical position in the middle of a horizontal plane, well smoothed out and unobstructed all around in order that the shadow could be seen clearly from sunup to sundown.[31] The astronomer (the systematic user of the gnomon deserves that name) observing the shadow throughout the year would see that it reached a minimum every day (real noon), and that that minimum varied from day to day, being shortest at one time of the year (summer solstice) and longest six months later (winter solstice). Moreover, the direction of the shadow turned around fron West to East during each day, describing a fan the amplitude of which varied throughout the year.

Anaximandros, or any other astronomer, Babylonian, Egyptian, Chinese, or Greek, making such observations day after day, must have asked himself many questions. Why does the noon shadow grow in six months' time from its shortest length to its longest and then reverse the process, and so on, year after year? And how do the azimuths of the shadows compare with their length? He observed that the extreme directions at sunrise (or sunset) correspond to the shortest and longest noon shadows (that is, to solstitial times). The extreme westward positions of the sunrise shadow at the two solstices could be marked, the middle position between those two extremes (due west) corresponding to the equinoxes. Similar observations could be made at sunset and would lead to similar conclusions, confirming the previous one. The direction of the sunset shadow at the time of the equinoxes would be collinear with, but opposite to, that of the sunrise shadow at the same times.

In short, the gnomon enabled the astronomer to determine the lengths of the year and of the day, the cardinal points, the meridian, real noon, the solstices, and later, the equinoxes and the length of the seasons (Fig. 46). A relatively large amount of precise information could thus be obtained with the simplest kind of tool. It requires some imagination to appreciate what could be done, and what could not be done, with a gnomon in Anaximandros' day. Indeed, our minds are so conditioned from childhood that we see ourselves standing on a sphere, our erect body pointing to the zenith and making a definite angle with the equatorial plane. Thus we see readily[32] that the gnomon would enable us to determine that angle (the latitude), but Anaximandros could not possibly have thought of that. He conceived the earth as a flat disk or tambourine (thickness about one-third of the diameter), suspended in space, surrounded by the ocean and by great anchor rings (solar, lunar, stellar).

[31] Similar observations were made by the Chinese at Yang-ch'êng (modern Kao-ch'êng Chên, Honan) during the Chou dynasty (c. 1027–256), a tower being used as a gnomon; *Isis 34*, 68 (1942–43).

[32] The diagram of Fig. 46 will refresh the

The idea of (terrestrial) latitude could not have occurred to him, but he was able to adumbrate our concept of the obliquity of the ecliptic. Indeed, he could observe that the Sun moved each day in a plane, and described a semicircle from east to west, culminating in the meridian at noon; the inclination of that plane to the horizon varied from day to day, being smallest at the winter solstice (when the noon shadow of the gnomon was longest) and largest at the summer solstice (when the noon shadow was shortest); the plane reached its half-way position at the times of the equinoxes (when the Sun rose due east and set due west). The angle between the two extreme positions of the solar plane (ecliptic) is twice as large as the one that we call the obliquity of the ecliptic. Anaximandros could possibly have measured that angle, but it would be highly misleading to say that he discovered the obliquity of the ecliptic (that is, the angle between the ecliptic and the equator), because he could not conceive the equator any more than the latitude.

Anaximandros apparently did not travel like Thales; at any rate, the traditions do not mention any travel of his. Yet he is said to have created the first *mappa mundi* (*pinax*). The Greek world was in the center of the map, other parts of Europe and Asia around it, and the Ocean formed the outside boundary.[33] It was probably to that map that Suidas (X–2) referred with the words "outline of geometry" (geōmetrias hypotypōsis) which were wrongly interpreted as a treatise on geometry (in the usual sense). We must beware of the Greek terms adopted in our language; for example, the words geography and geometry are etymologically close but represent two very different fields. Anaximandros' map might have been called perhaps a first attempt in geodesy, but it was unavoidably a very poor one.

We now come to the feature that occupies most place in the histories of Greek philosophy — his conception of the world. We have kept it for the end in order to emphasize the concreteness of his thought. We should imagine him as an astronomer trying his best to solve definite problems, succeeding sometimes and failing at other times, as is the fate of every honest scientist. Yet he wanted to go beyond that, to extrapolate his experience and knowledge and give his views on the universe. He explained those views in the treatise that he wrote at the age of sixty-four. He was probably stimulated in this by the example of his older contemporary, Thales. Thales' idea that water was the primary substance had much to commend itself (as we have shown above), yet it had obvious shortcomings. How could we

---

reader's knowledge. A gnomon is placed at O; its maximum and minimum shadows OS_1 and OS_2 are cast at noon of the two solstices. The corresponding angles a_1 and a_2 are the zenith distances of the sun at those times. As the sun travels equal distances north and south of the equator, the average of the two zenith distances is the angle between the equator and the zenith; this is also the declination of the zenith at O, or the latitude at O. Thus $\phi = \frac{1}{2}(a_1 + a_2)$. The obliquity of the ecliptic ω is given by the equation $\omega = \frac{1}{2}(a_1 - a_2)$.

[33] The geographic aspect of Anaximandros' work has been over-emphasized by William Arthur Heidel, "Anaximandros' book, the earliest known geographical treatise," *Proc. Am. Acad. Arts Sci. 56*, 237–288 (1921).

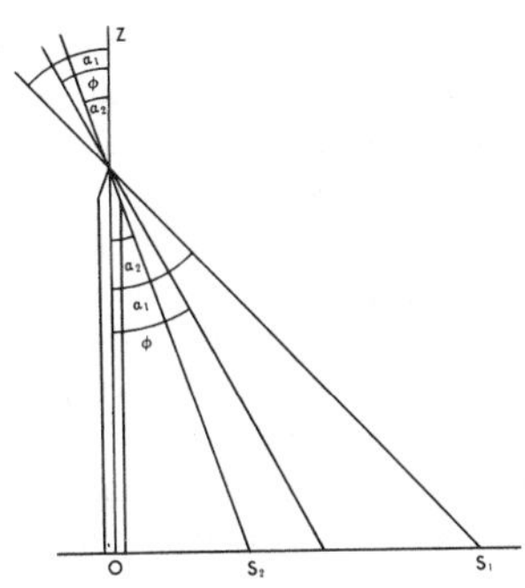

Fig. 46. The gnomon.

understand the transformation of water into earth, or wood, or iron? What other principle could one suggest? It is clear that if one had to choose among the kinds of matter familiar to one's senses, ubiquitous and protean water was incomparably the best. Water was the best, yet it would not do.

Anaximandros saved himself from that corner by taking refuge in an abstraction, in a word. Philosophers, and even a few scientists, have repeated that performance over and over again, to their satisfaction and apparently to the satisfaction of their readers. Anaximandros did not abandon the Thalesian idea of the substantial unity of nature, but since no tangible substance could serve as prime substance (*archē*),[34] he imagined one that was intangible and called it *apeiron*. There has been much discussion ever since as to the nature of *apeiron*; the word means infinite or indefinite, undetermined; it may also mean unexperienced.

Before giving our own guess, let us explain the main traits of Anaximandros' cosmology. We need not be very precise in our terminology; so little of his book is available and that little is so unclear and ambiguous that to explain his views with exact terms would be like weighing dirt with a balance of precision. Anaximandros conceived the world as a system in rotation, wherein the heaviest objects, rocks and earth, would fall to the lowest place, lighter ones like water would remain a little higher, fumes and vapors still higher. That circular motion is eternal and the source of universal power, creation and destruction. The primary substance *apeiron* is undetermined because it is potentially everything. The universe is of infinite duration in a boundless space. Anaximandros seems to have made a distinction between determination (as that of a definite substance) and indetermination, which is like nothing we know, or cannot be distinguished from anything else. For example, we know the difference between cold and hot, dry and wet, but where is the limit? When does a thing cease to be cold or dry, and become warm or damp? He seems to have been able also to distinguish between infinity and endlessness; one may never be able to reach the end of a thing because it has no end or because it comes back into itself, like a closed curve. He seems to consider time infinite but space boundless (in the second sense, like the surface of a sphere). It is futile to discuss his thoughts more elaborately, for we cannot discuss them without giving the few fragments that have come to us meanings more definite and more precise than they can possibly have without the lost context.

We must still say a few words of Anaximandros' theory of life. He thought that the first animals were created in water and were then surrounded by a kind of husk (*phloios*); later these animals found a new abode upon dry land, dropped their shells, and adapted themselves to the new circumstances (he had perhaps in mind the insects that issue from marine larvae). Man must derive from other animals, because his own period of immaturity is too long and too frail. In short, he had conceived not only a general cosmology but a theory of organic evolution. He was a distant forerunner (very distant indeed) of Darwin as well as of Laplace!

That such thoughts found utterance as early as the sixth century is almost incredible, yet the general meaning of the texts that have come to us is unmistakable. Scientists may object that gratuitous assertions, or assertions based on such flimsy

[34] According to Simplicios (VI-1), Anaximandros was the first to use the word *archē* with that meaning (preserved in English, as in archetype).

evidence, cannot be considered scientific achievements, and should be abandoned to metaphysicians or to poets. To be sure, such assertions would not be permissible today, but we must remember that Anaximandros made them before the purpose and methods of science had been formulated. His thoughts helped to prepare that formulation. He was neither a scientist nor a metaphysician in the modern sense of those terms; he was a philosopher or a physiologist in the Greek sense. He was the first to state some of the fundamental problems of science; his answers were too bald and premature, but not in their own background irrational.

ANAXIMENES OF MILETOS

The Milesian tradition — the search for a first principle or a primary substance — had been somewhat deviated by Anaximandros, but it was redressed by the latter's fellow citizen and successor, Anaximenes. This Anaximenes, son of Eurystratos, flourished about the end of Anaximandros' life and he died in the sixty-third Olympiad (528–525). No writings of his have reached us except three short fragments, and the doxography concerning him extends to only a few pages, yet Theophrastos attached so much importance to his thought that he had devoted a special treatise to him.

Anaximandros' expedient, his metaphysical conception of the prime substance, his escape from reality, was not to the taste of Anaximenes, who tried to reintroduce a physical principle. Water would not do because it was too tangible, too determinate. But what of wind or air, permeating everything?[35] Air (*pneuma*) is tangible enough (does one not feel a blast of wind?), and yet it can easily become almost intangible. It has biological properties, for men and animals cannot live without breathing, and what is breath but air? Moreover, air can be compressed or it can indefinitely expand. Air is material enough, yet it tends to become immaterial, and even spiritual. According to dictionaries, the spiritual meaning of *pneuma* is not older than the Septuagint,[36] yet it must have occurred to any thinking man long before that time, for the semantic passage from air to breath, hence to life and spirit, was a very natural one.

Air is the primary substance, but it may take all kinds of appearances by condensation or thickening (*pycnōsis*) or by rarefaction or thinning (*manōsis*). Anaximenes associated these qualitative changes with changes in temperature. He had persuaded himself with a crude experiment that rarefaction increased the temperature while compression decreased it: when we blow with open mouth, the air that we breathe out is warm; when we blow with the lips almost closed, the air is cool.[37] His assimilation of air with the breath of life was the occasion of his comparing the whole world with a single living organism, say a man. Breath is to the latter what wind is to the former. This introduced the concept of the little world versus the big world,[38] which was to play so large a role in medieval philosophy and is still misleading uncritical minds today.

[35] *Holon ton cosmon pneuma cai aēr periechei.* Anaximenes, frag. 2.

[36] The spirit of God, *pneuma theu* (Genesis 1:2). The Septuagint dates from the first half of the third century B.C. The word *pneuma* occurs frequently in the New Testament with the meanings of breath, spirit, ghost, life.

[37] The experience is curious but deceiving, and Anaximenes' conclusion was the opposite of the truth. As we know, adiabatic compression increases the temperature, while adiabatic dilatation decreases it.

[38] In a fragment (No. 34) of Democritos (V B.C.) we have *anthrōpos micros cosmos* and he is

Anaximenes still conceived the earth and other planets (including Sun and Moon) as disks supported by the air, but he was first among the Greeks to think of the stars as situated on a rotating sphere. This preserved the eternal rotation of Anaximandros. The planets are freely suspended, but the stars are attached to the sphere like nails. He rejected the (Egyptian) idea that the stars and planets pass under the earth, but claimed that they turned around like a cap around the head; they disappear from our view when they pass behind mountains at the edge of the world.

The essentials of Anaximenes' thought were a restatement of the material unity of nature, his choice of air as the primary substance, and his explanation of all the accidents of nature by rarefactions and condensations of that substance. The great rhythm of the cosmos was somewhat like the respiratory rhythm of our own life.

It is typical of the Milesian spirit of "natural philosophy" that Anaximenes' hypotheses were preferred to those of Anaximandros and that his views were considered as the climax of previous speculations. These fell gradually into oblivion and Milesian philosophy came to mean the philosophy of Anaximenes. We shall come back to that when we deal with a later Ionian, the last one, Anaxagoras of Clazomenae (V B.C.).

CLEOSTRATOS OF TENEDOS

We may now leave the physiologists of Miletos, and even leave Miletos itself, but we must remain close to the Asiatic coast. Thales, Anaximandros, Anaximenes were all interested in astronomy. That interest might have been spontaneous, for the phenomena to be observed every night in the sky are too obvious and too impressive not to challenge the curiosity of thoughtful men. It is very probable, however, that their curiosity received further stimulation from Oriental sources. The sailors and merchants who reached Miletos brought with them Babylonian and Egyptian ideas. A few examples of such transmission have already been mentioned; here are two additional ones.

Cleostratos flourished not in Miletos but in Tenedos, a little island off Troas, near the mouth of the Hellespont. According to one tradition, Thales died in Tenedos, and hence Cleostratos might have obtained the Milesian doctrines in his own island home, either from the first teacher or from one of his disciples; of course, it would not have been very difficult for him to obtain them even if that tradition is false, for Tenedos was not far away from Ionia and must have been familiar to Milesian travelers en route to the Propontis. We have seen that Anaximandros had some understanding of what we call the obliquity of the ecliptic. Pliny tells us [39] that Anaximandros discovered it in the fifty-eighth Olympiad (548–545), that is, toward the end of his life. Thales lived until about the same time and the discovery of the obliquity of the ecliptic might be considered as the climax of the early Ionian astronomy. Some time afterward (say *c.* 520), Cleostratos, who was making astronomical observations in Tenedos and tried to determine the exact

said to have written treatises entitled *megas cosmos* and *micros cosmos*. The idea of microcosm and macrocosm was probably not uncommon after that, yet the terms were used by Latin writers rather than Greek ones. *Microcosmos* is listed in H. Stephanus, *Thesaurus graecae linguae* (Paris: Didot, no date), vol. 5, p. 1052 (orig. pub. Paris: Stephanus, 1572).

[39] *Natural history*, II, 6, 31.

time of a solstice, recognized the signs of the zodiac, especially those of the Ram and the Archer. Now the zodiac, an imaginary belt in the heavens on both sides of the ecliptic,[40] had been distinguished by Babylonian astronomers a thousand years earlier. Indeed, it was impossible to observe the trajectories of moon and planets for any length of time without realizing that those celestial bodies travel in a relatively narrow belt and are never very distant in latitude from the Sun (or, as we would have put it, from the ecliptic). What Cleostratos did then was probably to recognize the constellations through which the Sun, Moon, and planets pass during the year, and perhaps to divide those constellations into twelve equal lengths of the ecliptic, the twelve "signs" of the zodiac.[41] These constellations, and perhaps others, may have been described, their risings and settings indicated, in his lost poem on the stars (*astrologia*).

Another invention is ascribed to him, namely, that of an eight-year cycle (*octaetēris*) of intercalations, a period containing an exact number of days, lunar months, and solar years:

$$365\tfrac{1}{4} \times 8 = 2922 \text{ days} = 99 \text{ months.}$$

Now such a cycle was also known to the Babylonians, and Cleostratos may have borrowed it from them, or their determinations of months and years may have enabled him to rediscover it for himself. This was but the first of various other cycles introduced from to time by Greek astronomers to serve calendrical purposes.

We can never be certain in such matters, but the balance of probabilities favors the hypothesis that the astronomy of the Ionians in general, and that of Cleostratos in particular, was stimulated by the reception of Babylonian knowledge. This does not decrease the value of Cleostratos' achievements; he was one of the founders of Greek astronomy.

J. K. Fotheringham, "Cleostratus," *J. Hellenic Studies 39*, 164–184 (1919); *40*, 208–209 (1920) [*Isis* 5, 203 (1923)]. E. J. Webb, "Cleostratus redivivus," *J. Hellenic Studies 41*, 70–85 (1921) [*Isis* 5, 490 (1923)].

XENOPHANES OF COLOPHON

The city of Colophon, where Xenophanes was brought up, was one of the twelve Ionian cities; it was an opulent market town but had often been sacked by foreign invaders or by pirates. When Cyrus conquered it, Xenophanes preferred to leave and he spent the rest of his life wandering; he is said to have traveled for sixty-seven years. He may have visited Egypt and this would help to account for some of his ideas, but the traditions refer only to his journey westward to Sicily. He visited Zancle (= Messina) and Catania, and settled for a time in Elea, on the

[40] The zodiac is generally conceived as a belt of about 16° of latitude, divided in two by the ecliptic. The exact width does not matter.

[41] Our word signs or the Latin *signa* is a translation of the Greek *sēmeia*, meaning signs of the gods, *omina*. It is possible that Cleostratos was first to use that word in its technical zodiacal meaning, especially with reference to Aries and Sagittarius. The word zodiac, *zōdiacos* (*cyclos*) refers to the living signs; the usual Latin translation was signifer, "signifero in orbe qui Graece *zōdiacos* dicitur" (Cicero, *De divinatione* II, 42, 89). The term "signs of the zodiac" is ambiguous, for it may refer to twelve divisions of the zodiacal belt, extending to 30° of longitude each, or it may refer to the constellations characteristic of each division. In the absence of texts we cannot say which of these two ideas was foremost in Cleostratos' mind. We cannot say whether he recognized twelve signs, or only two, or some.

western coast of Lucania.[42] Note that he thus helps us to cross two barriers — we pass with him from the sixth to the fifth century (he lived with the period 570–470) and from the Aegean sea to the Tyrrhenian, or from the Eastern to the Western Mediterranean.

The most curious of his ideas was a kind of monotheism or pantheism that may well have been of Egyptian origin. At any rate, such sayings of his as, "A single God and the greatest among gods and men," "God is one and is all," "God is the cause of motion," suggest a new theosophy essentially different from Milesian physiology and its relative positivism. Xenophanes had been duly influenced by his Milesian neighbors, however, and this appears in the most remarkable of the fragments ascribed to him. It is worth while to quote it in full:

> And Xenophanes is of opinion that there had been a mixture of the earth with the sea, and that in the process of time it was disengaged from the moisture, alleging that he could produce such proofs as the following: that in the midst of earth and in mountains shells are discovered; and also in Syracuse he affirms was found in the quarries the print of a fish and of seals, and in Paros the print of an anchovy in the bottom of a stone, and in Malta parts of all sorts of marine animals. And he says that these were generated when all things were originally embedded in mud, and that an impression of them was dried in the mud, but that all men had perished when the earth, precipitated into the sea, was converted into mud; then again that it originated generation, and that this overthrow occurred to all worlds.[43]

This is astounding. On the basis of it we may call Xenophanes the earliest geologist and the earliest paleontologist. Should one object that the extract is known only by a very late quotation and that its genuineness is thus far from certain, there is little that one could say in defense of it. Yet why should Hippolytos have invented it? There was nothing to be gained in that. Moreover, the statement would be even more astounding in the third century after Christ than it was at the turn of the sixth century before Christ, for that century had been, in Ionia at least, an age of unusual freedom and adventure, a golden age. In Xenophanes' mouth such words were astonishing, to be sure, but not more so than many others ascribed to Thales, Anaximandros, and Anaximenes. Along the coast of Asia Minor Greek science had begun in a wonderful manner. The Ionian physiologists were the worthy descendants of the Homeridae.

EGYPTIAN INTERLUDE. NECHO, KING OF EGYPT (609–593)

An effort has been made in the previous sections of this chapter to explain the birth of Greek science in Ionia. The speed of my narrative should not deceive the reader. The development of what might be called the school of Miletos (or of Ionia) took well over a century; Thales and Anaximandros were born in the last quarter

[42] Elea is south of Paestum; its modern name is Castellammare di Veglia (or della Bruca). The tradition of Xenophanes' stay in Elea, not to mention his foundation of the Eleatic school, is very weak. He had a good reason for going to Elea, however, for a colony of Phocaeans had been established there (*c.* 543, 536?) soon after the Persian conquest of Ionia. It would have been very tempting for him to go and see his countrymen, who were political refugees like himself.

[43] Quoted after Arthur Stanley Pease, "Fossil fishes again," *Isis* 33, 689–690 (1942). The reader should be warned that this extract is of relatively late tradition, being taken from that rich source of ancient knowledge, the Philosophical Subjects (*ta philosophumena*) of St. Hippolytos (III–1). The idea of a general flood belongs to the folklore of many nations; for the Greek people it was represented by the myths of Deucalion and Pyrrha, who were saved from destruction and became the original ancestors of the Hellenic race.

of the seventh century, Xenophanes died in the first third of the fifth century. The men dealt with were concerned with physiology, that is, physics, biology, astronomy, "philosophy of nature." Before explaining another remarkable feature of Milesian science — the development of geographic thought — we should return for a moment to Egypt and move backward a century or so to the beginning of the period of this chapter.

The Twenty-fifth (or Aethiopian) Dynasty of Egypt, having lasted barely half a century, came to an end in 663.[44] The last Aethiopian king was defeated by Ashur-bani-pal (king of Assyria, 668–626), and the whole of Egypt was then for a few months an Assyrian province. One of the native governors of that province, Psametik, son of Necho (Nekaw) of Sais, succeeded in reorganizing a kind of national unity and with the help of Greek and Carian mercenaries "in brazen armor"[45] he delivered his country from Assyrian bondage and founded the Twenty-sixth (or Saitic) Dynasty. He was a strong and able ruler, and his dynasty, the last national one, was truly a period of renaissance. He looked back for his models (in religion, art, epigraphy) to the classic periods of the Old and Middle Kingdoms, when the glory of Egypt had reached its climax. That renaissance did not last very long (only 138 years, hardly more than four or five generations) because it was artificial. Psametik could evoke prosperity, but such as it was, it was dependent on the protection of foreign mercenaries and the trading abilities of foreign merchants. Then (even as today) intense nationalism was strangely combined with military impotence. In spite of its glorious aspect, the Saitic kingdom was essentially unstable and it fell like a house of cards as soon as Cambyses appeared at the gates of Pelusium[46] in 525.

It had been Psametik's mistake — a generous one – to place culture above strength and to devote all his efforts to the development of the arts of peace under the noses of aggressive and greedy neighbors. He repaired the irrigation works of the Delta, favored the establishment of Greek colonies, revived trade not only with Greek people of many origins, but also with Carians, Syrians, Phoenicians, Israelites. There were Greek and Carian quarters in Memphis. Psametik had established his capital in his own birthplace, Sais, on the (western) Rosetta branch of the Nile, and the Delta became the dominant part of Egypt.

Thanks to Psametik's archaeological and patriotic enthusiasm, the arts were revived. Our museums contain many gracious objects of the Saitic period, especially in bronze and fayence, but no large monuments have survived.[47] The lords of the Delta built probably with mud, not with stone, and their dwellings vanished. Psametik and his successors encouraged the scribes to copy the old books of their nation and many such copies have come down to us, written in a new and faster kind of hand, a "streamlined" form of the hieratic cursive, known to us as demotic (popular). It was not possible to resuscitate all the old gods, but Osiris and Isis became the favorites, and Imhotep was deified. The influence of the Greeks upon Egypt was commercial and material; on the contrary, that of Egypt upon Greece

[44] From that time on Aethiopia (or Abyssinia) was definitely separated from Egypt.

[45] *Hoplisthentas chalcō*. Herodotos, II, 152.

[46] The fortified city of Pelusium, east of the easternmost mouth of the Nile, was the key to Egypt on the northeast side.

[47] The artistic masterpiece of the age is probably the head of a man with broken nose, in green basalt, now in the Berlin Museum and often reproduced. It makes one think of a monument of the Old Kingdom.

was spiritual. The Greek interest in the gods of Egypt, and especially in those that we have just named, dates largely from this time when the opportunities for Greek-Egyptian anastomosis were frequent. A curious little example of Egyptian prestige is given us by the tyrant of Corinth, Periandros (ruled (625–585), who gave his nephew and successor the name Psammetichos (or Psammis, Greek forms of the Egyptian name Psametik). Periandros, it may be remembered, was one of the Seven Wise Men. This homage to Egypt is particularly significant in his mouth.

To return to the original Psametik, his son Necho succeeded him in 609 and was probably so much impressed by the grandeur and beauty of the kingdom which he had fallen heir to that he did not imagine its essential weakness and precariousness. The Assyrians were then involved in a deadly struggle against the Babylonians and the Medes. Taking advantage of their great peril and of the strength of his Greek mercenaries, Necho invaded Palestine in 609, defeated and slew Josiah (king of Judah, 638–609) at the battle of Megiddo. Four years later he was himself defeated at Carchemish on the Euphrates by Nebuchadrezzar (king of Babylon 604–562) and then lost all the territories that he had conquered in Asia.[48] After Megiddo, Necho sent to the Branchidai[49] near Miletos and there dedicated to Apollo the garments in which he had won that victory. Thus we remain close to that city after all. The Egyptians paid homage to Greek gods, while the Greeks worshiped Isis and Osiris.

What we have said of Necho is already sufficient to arrest the attention of the historian of ideas, for did he not establish bonds of union between Egypt, Greece, Israel, Chaldea? Yet we have more direct reasons to be interested in him — two great achievements of which the historian of geography must take notice.

In the first place, he continued the digging of a canal connecting the Nile with the Red Sea. A canal had been built during the Middle Kingdom (2160–1788) between Bubastis on the Tanitic arm of the Nile and Lake Timsāh. Necho reëxcavated it and continued it to the Bitter Lakes and the Gulf of Suez (baḥr al-qulzum). It was dug wide enough for two triremes to pass each other and its length (from Bubastis, I take it) was four days' sailing. Herodotos, to whom we owe most of our information on this subject, tells us that 120,000 Egyptians perished in the undertaking, which had to be abandoned before completion.[50] Why was it abandoned? According to Herodotos, because of an oracle foreboding evil from the barbarians (that is, the foreigners; that evil was realized in the following century); according to Diodoros of Sicily (I–2 B.C.), because Necho's engineers discovered that the Red Sea was higher than the Delta and feared to flood Egypt with salt water; the decisive reason may have been the increasing difficulty of raising laborers and supplies. The canal was completed a century later by Darius (king of Persia and Egypt, 521–486). Necho deserves praise for having understood the need of a communication between the Red Sea and the Mediterranean, which, if he had been fortunate enough to complete it, would have enhanced the prosperity of his kingdom. Alas! that would not have saved it but would have increased the greed of its neighbors and its own mortal danger.

[48] We hear echoes of this in the Old Testament: Jeremiah 46:1–12; 2 Kings 24:7.

[49] The Branchidai were the descendents of Branchos, himself the son of Apollo and of a Milesian woman. They were hereditary priests administering the oracle of Apollo Didymaios at Didyma, near Miletos. Xerxes (king of Persia, 485–65) exiled them to Bactria or to Sogdiana, across the Oxos river.

[50] Herodotos, II, 158.

In the second place, being anxious to promote foreign trade, Necho ordered Phoenician ships to sail around Libya (Africa). The idea was a natural one, at any rate to Greek minds, because of their belief in an ocean surrounding the earth, yet it required unusual imagination and courage to implement it as Necho did. Herodotos' account of this achievement is so clear and withal so short that we cannot do better than to reproduce it:

> For Libya shows clearly that it is encompassed by the sea, save only where it borders on Asia; and this was proved first (as far as we know) by Necos king of Egypt. He, when he had made an end of digging the canal which leads from the Nile to the Arabian Gulf, sent Phoenicians in ships, charging them to sail on their return voyage past the Pillars of Heracles till they should come into the northern sea and so to Egypt. So the Phoenicians set out from the Red Sea and sailed the southern sea; whenever autumn came they would put in and sow the land, to whatever part of Libya they might come, and there await the harvest; then, having gathered in the crop, they sailed on, so that after two years had passed, it was in the third that they rounded the Pillars of Heracles and came to Egypt. There they said (what some may believe, though I do not) that in sailing round Libya they had the sun on their right hand.[51]

It is a pity that Herodotos does not go into more detail but such as it is his account inspires confidence. The very fact which he could not believe confirms the story, for when the Phoenicians sailed westward around the Cape of Good Hope the sun was always in the north, that is, on their right.[52]

Necho was in many respects a great king. We have seen that Periandros admired his father; he himself was admired by another of the wise men, and more famous one, Solon of Athens (VI B.C.). When the latter visited Egypt he studied Necho's laws, and after his return introduced some of them into the new Athenian code. The essential weakness of the Saitic kingdom increased, but Necho could stave off the coming storm. The last king but one of his dynasty, Aḥmose II, has already been mentioned. During his rule (569–525), the Greek merchants had obtained so much power that they were allowed to build, or to reorganize, the city of Naucratis,[53] on the Canopic branch of the Nile, not very far west of the capital, Sais. That city became the main center of Greek commerce in Egypt (somewhat like Alexandria in later Ptolemaic days). Its main sanctuary, appropriately called Hellenion,[54] was decorated with the offerings of many Ionian, Dorian, and Aeolian cities; in addition, some cities, like Miletos, had temples of their own. Aḥmose sent liberal gifts to the Greek temples of Europe and Asia, and he had formed an alliance with the powerful tyrant, Polycrates of Samos. This is the Polycrates whose good luck was legendary, yet who was crucified in 522. In the mean-

[51] *Ibid.*, IV, 42.

[52] H. F. Tozer, *History of ancient geography*, ed. 2 by M. Cary (Cambridge: University Press, 1935), pp. 98–101. Tozer is unconvinced, however, and thinks that an ingenious storyteller might have deliberately invented that fact to give credence to his story. I can not believe that Herodotos and his informants were such sophisticated liars. For medieval stories of African circumnavigation see *Introduction*, vol. 2, p. 1062; vol. 3, pp. 803, 1892); those stories are less convincing than Herodotos'. Note that the medieval circumnavigations took place, if they took place at all, in the opposite direction. The same is true of the first turning around the Cape of Good Hope eastward by Bartholomeu Dias in 1488 and the first (almost complete) circumnavigation by Vasco da Gama in 1498.

[53] There are no ruins to be seen today in Naucratis (nor in Sais), but Naucratis was excavated by Flinders Petrie and many small objects were brought to light. See his report, *Naukratis* (2 vols.; London, 1886–1888).

[54] *To Hellēnion*. It was perhaps more than a sanctuary, but the whole or a part of the Greek factory including temples to the Greek divinities, *theoi Hellēnioi*.

time the peril increased considerably, for a new power had appeared in the East, Cyros, founder of the Persian empire. Cyros had defeated Croesus in 546 and the Babylonians in 539; he died in 529. Aḥmose lived until 525, but his son Psametik III was vanquished in the same year by Cyros' son, Cambyses. This was the end of an independent Egypt, but in a sense it had already lost its independence, for the Saitic kingdom was already Greek in many ways and the dynasty of Psametik (663–525) was a kind of anticipation of the Ptolemaic one a few centuries later (332–30).

During that period (seventh to sixth centuries) the Near East was submitted to a deep, unceasing turmoil. Its various elements – Greek, Asiatic, African – were repeatedly mixed. The main ferment was Ionian, but that ferment was excited by Egyptian and Babylonian examples. Physical contacts do not suffice without sympathy and understanding. There was enough sympathy between the Egyptians and the Greeks to bring results in both nations; unfortunately, the Egyptian influence, extensive as it was (the necessary contacts were abundant enough), could not reach very deep, for the demotic writings were even less legible and more forbidding than the hieroglyphics. Greeks and Jews must have met in Palestine and elsewhere but there was not enough sympathy between them to cause mutual admiration and emulation. We can detect plenty of Egyptian traces in Greek art,[55] letters, and knowledge, but hardly any Jewish. The best of the Jews and the best of the Greeks were pursuing their own purposes independently; they could not yet come together in Miletos or in Naucratis, as they would two or three centuries later in Alexandria.

HECATAIOS OF MILETOS, THE FATHER OF GEOGRAPHY

If Necho's circumnavigation of Africa was accomplished, the news of that almost incredible event must have spread among the Phoenicians and have passed from them to the Milesians, either directly, or indirectly through Egyptian officers of the Saitic court. Even if that event did not actually happen, we may be sure that other stories were told by the Phoenician and Greek sailors. Milesian ships were often sailing around the Mediterranean and the Euxine, and collecting wares and news of every kind. No information was more likely to be treasured than what we would call the geographic in the broadest sense (*géographie humaine*). A place like Miletos in the sixth century was of necessity a geographic clearing house, even as a Portuguese harbor would be twenty centuries later. To be sure, information is not safely preserved, classified, standardized, unless a man of outstanding ability makes himself responsible for doing so. The success of Sagres was due to the genius and the devotion of Henry the Navigator; in the same way, the opportunities for geographic and anthropologic knowledge that were available in Miletos were capitalized and exploited by Hecataios.

Hecataios, son of Hegesandros, belonged to an ancient family of Miletos and was born about the middle of the century, that is, about the time of the Persian conquest. Thus he was brought up as a Persian subject; his family was probably

[55] Let me just indicate (I cannot do more here, but must do that much), the evident Egyptian influences in the so-called Archaic Greek sculpture. The early *curoi* stand erect like Egyptians, with the characteristic thrusting forward of the left foot. Compare an album of Egyptian sculpture with Gisela M. A. Richter, *Kouroi. A study of the development of the Greek kouros from the late seventh to the early fifth century* (New York: Oxford University Press, 1942).

ready to "collaborate" with the Persians and share their prosperity, but the common people were less easy to deal with and by the end of the century the air was rife with thoughts of insurrection. Hecataios tried in vain to avoid the rebellion and when war had become inevitable he realized that only a very bold strategy would save his countrymen. His advice was rejected in both cases, being considered too timid in the first, and too adventurous in the second. It all ended with the sack of Miletos in 494; Hecataios lived long enough to witness the battle of Mycale (479) and the liberation of his country;[56] he died *c.* 475.

He is said to have traveled extensively and his travels probably occurred toward the end of the century when his presence at home was obnoxious to the people. According to Herodotos, he not only visited Egypt but traveled as far south as Thebes. Such a visit would have been facilitated by the fact that after 525 Egypt was a Persian province. Hecataios was a Persian subject traveling from one province of the empire to another.

Two works have been ascribed to him, a historical one called *Geneaologies* and a geographical one entitled *Periodos gēs*, "description of the earth" or "descriptive geography." Both works are lost, and are known to us only by some 380 fragments, most of which are very short. The first of these books is less known and less important to us than the other, but we may pause a moment to consider its *incipit*, preserved by the famous Demetrios Phalereys:[57] "Hecataios of Miletos speaks thus. I have written these things because they seem true to me. The narrations of the Greeks are many and, in my opinion, foolish."[58] These words, we should remember, replaced the title, and perhaps they were meant also to replace the jacket blurb of an enterprising publisher and to challenge the reader's attention at the very outset; we should not judge them too severely.

The great majority of the 331 extracts from Hecataios' geography are culled from Hermolaos' epitome of the geographic dictionary of Stephanos of Byzantium (VI–1); hence they are very short indeed, as dictionary quotations are (often less than five words), yet they are sufficient to illustrate the general scope of the work. When Hecataeos was growing up in Miletos, he must have heard discussions concerning the views of the great physiologists, Thales, Anaximandros, and Anaximenes. What was really the main substance of the universe? Knowing the Greek temperament, we can easily imagine those discussions which were of their nature endless and sterile. They may have discouraged a young man whose ambition was humbler and more tangible. Hecataios may have said to himself (as any true scientist would), "Before settling the enigma of the universe, let us take careful stock of the things around us." One of the most obvious and alluring tasks was to

[56] The destruction of Miletos in 494 shocked the Greek people profoundly; it united and strengthened them. They defeated a Persian army in Marathon in 490, delayed another one at the pass of Thermopylae in 480, and won the naval victory of Salamis in the same year. The Persians were finally beaten on land at Plataea and their fleet at Mycale in 479. The naval victory of Mycale, so close to Miletos, was the best revenge of the sack of that city fifteen years previously.

[57] This Demetrios of Phaleron (one of the harbors of Athens) was an orator who obtained such popularity that the Athenians erected 360 statues to him. Later they tired of him and condemned him to death. He fled to Egypt, where he helped the first Ptolemy to create the library of Alexandria; the second Ptolemy, Philadelphos (ruled 285–247) exiled him to Upper Egypt where he died of a snake-bite. His treatise on explanation (*peri hermēneias*), whence our quotation is taken, may be the work of another Demetrios of Alexandria.

[58] Müller, fragment 332 (1841). *Hecataios Milēsios hōde mytheitai; tade graphō, hōs moi alēthea doceei einai; hoi gar Hellēnōn logoi polloi te cai geloioi, hōs emoi phainontai, eisin.*

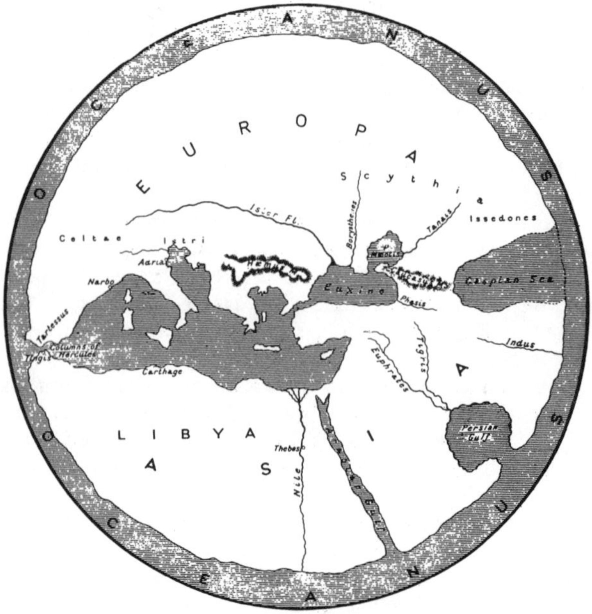

Fig. 47. Schematic map illustrating Hecataios' general view of the flat world. [Reproduced with kind permission from H. F. Tozer, *History of ancient geography* (Cambridge: University Press, ed. 2, 1935), map II. A more elaborate map, including many more Hecataian names, is appended to R. H. Klausen, *Hecataei Milesii fragmenta* (Berlin, 1831). Klausen's book contains a geographic index to Hecatios; in Müller's edition that index is mixed up with many others.]

collect the scraps of geographic and anthropologic information that were brought home continually by sailors and merchants and to put them in good order, together with the observations and reminiscences of his own travels. This was the first attempt of its kind and because of it its author deserves to be called "the father of geography." His *Periodos* was divided into two main parts, Europe and Asia (the latter including Libya). Look at the schematic map which shows that division and justifies it (Fig. 47). The flat earth was conceived as round, surrounded by the Ocean, and roughly divided by the Mediterranean, the Euxine, and the Caspian into two halves — the upper or northern one being Europe, and the low or southern one Asia and Africa.[59] The map makes it unnecessary to describe other features. Note that the Mediterranean, the Red Sea, the Persian Gulf, the Caspian Sea, and the Nile all connect with the circumambient Ocean; that was natural enough for the first three, but wrong as regards the fourth. We shall come back to the Nile presently. Hecataios' survey was very largely restricted to the coasts, which is not

[59] This summary is derived from the fragments and from Herodotos IV, 36, assuming that the geographic views he makes fun of are Hecataian.

surprising for his information was received from merchants and sailors, and the Milesian and other Greek colonies were generally restricted to harbors with little if any hinterland. He was interested not only in cities, but in peoples and even in animals; according to Porphyry (III–2), Herodotos' descriptions of the phoenix, the hippopotamus, and the hunting of crocodiles were taken from Hecataios.[60]

Did Hecataios actually draw a map? That is quite possible and it is even said that he improved Anaximandros' map. A statement of Herodotos may be taken to mean that many maps had been made;[61] another one unmistakably refers to a map.[62] At the time when Miletos was in great jeopardy, its tyrant Aristagoras went to Sparta to beseech the help of King Cleomenes;[63] "he brought with him a bronze tablet on which the map of all the earth was engraved, and all the sea and all the rivers." This happened in Hecataios' time and he may have seen that bronze map – in fact, he might well have been the author of it.

A word about the Nile. The Greeks visiting Egypt could not help asking themselves questions concerning the greatest wonder of that country, the river Nile. The Ionians would certainly notice one important feature, the formation of the immense delta, because of their own experience on a much smaller scale, for example, the alluvia of the Maiandros. Other features were harder to understand. How came it that the Nile flooded the country in the summer, when the Greek rivers were drying up? Herodotos, who is our guide in this as in many other matters, explains various Greek opinions on the subject.[64] The first opinion, probably Thales', was that the river was caused to rise by the Etesian winds[65] which prevented its flowing out to the sea; according to the second opinion, probably Hecataios', the increase of the river was due to its connection with the Ocean;[66] according to the third opinion, Anaxagoras', the river rose because snow melted in the mountains of Libya. This last-named opinion came much nearer to the truth, yet Herodotos rejected it as well as the others to give his own worthless one.[67] Hecataios' explanation of the Nile floods is interesting in spite of its gross erroneousness; it shows how his mind was dominated by the Homeric Ocean.

This general view, let us remark, was on the whole a correct one. The continents, we now know, are large islands surrounded by seas. Geographers give separate names to those seas according to their location, but all those seas are but parts of one ocean. If one restricts oneself to the Old World, the Homeric concept is even truer. For Europe-Asia-Africa constitute as it were a single continent surrounded by a single ocean. The Homeric-Hecataian view was essentially true, but the ancient

[60] Müller, fragments 292–294.

[61] *Gelō de horeōn gēs periodus grapsantas pollus hēdē* (Herodotos, IV, 36), "I laugh to see how many have drawn maps of the earth." In this context *periodos gēs* means map rather than verbal description, and *graphō*, to draw rather than to write.

[62] Herodotos, V, 49.

[63] This Cleomenes was king of Sparta from 520 to 491; Aristagoras visited him before 499 (the Spartans refused to help him, but the Athenians did). Aristagoras obtained some temporary success and captured Sardis in 499, but after that the Persians had the upper hand. He fled to Thrace, where he was slain in 497, before the destruction of Miletos.

[64] Herodotos, II, 19–25.

[65] *Etēsiai anemoi*, periodic winds blowing from the northwest during the summer; or, in the Aegean Sea, for 40 days from the rising of the Dog Star (Sirius). The word *etēsiai* in this context is equivalent to monsoon (Arabic *mawsim*, *mawāsim*, season).

[66] See the map, or Müller, fragment 278.

[67] The true explanation was given by Aristotle (IV-2 B.C.). The inundation of Egypt is caused by tropical rains in the highlands of the Blue and White Nile, occurring in the spring and early summer. On this subject see *Introduction*, vol. 1, p. 136; vol. 3, p. 1844.

Greeks could not possibly have realized the extent of that continent north, east, and south.

Hecataios was a poor theorist (there is no trace of mathematical geography in his work, or none has come to us), but his effort to put together and to organize the available knowledge of the tangible world was a good step in the right direction. He is one of the founders of geography.

The best edition of what remains of Hecataios' work is included in the *Fragmenta historicorum Graecorum*, edited by Charles and Theodore Müller of Paris (Paris, 1841), vol. 1, pp. ix–xvi, 1–31, with Latin translation. Our knowledge of ancient geography has so much increased since 1841 that a new edition is very desirable.

GREEK TECHNICIANS OF THE SIXTH CENTURY

Much of our knowledge of Greek technology in the sixth century is of a legendary nature, but the core of those legends is generally substantiated by indirect information and sometimes by monuments. The collateral information is chiefly Egyptian; technical processes practiced in Egypt would draw the attention of the Greek colonists established in Naucratis or roaming across the country and would be imported into the Greek islands almost as easily as the objects that they had helped to create. In most cases, however, it is hardly possible to say whether a Greek method is an invention or an importation from Egypt or elsewhere. The line between imitation and invention is hardly visible; from servile imitation to pure invention there are numberless intermediate steps.

The legendary history of inventions in Greece introduces a very curious personality, the Scythian prince Anacharsis, who came to Athens *c.* 594. His intelligence, gentleness, and good humor, as well as the simplicity of his manners, won him the approval and sympathy of his neighbors. He became a disciple and friend of Solon, and was counted as one of the "seven wise men" (not in the most common lists, however). Various wise sayings were ascribed to him as they were to the other "sages." For example, he compared the laws to spiderwebs which catch small insects but let the larger ones pass through. When he returned to his native country he brought back with him Greek customs and religion,[68] and because of that impiety was killed by his brother Saulios, king of the Scythians. This Anacharsis is doubly interesting to us, first because of his origin, second because he established himself in Athens. This suggests that the legend is of relatively late formation, for it would have been more natural in 594 for a Scythian "inventor" to go to Miletos than to Athens. For one thing, Milesian ships would have brought him to Ionia rather than to Attica. Be that as it may, he has the distinction of being the first Athenian in the history of science as well as the first Scythian. To put it otherwise, it is remarkable that the first Athenian in our quest, next to Solon, should be a Scythian, or – in modern language, with a bit of stretching — a Russian!

Many inventions were credited to him: a two-armed anchor, bellows, potter's wheel.[69] These particular inventions are certainly earlier than the sixth century,

[68] He is said to have introduced the religion of the Cretan goddess Rhea, wife of Cronos, mother of Zeus and other gods, later identified with the Phrygian "Great Mother." One can easily imagine that that bold innovation scandalized and frightened the Scythians. Indeed, Anacharsis was smuggling in with Rhea – *in potentia* – the whole of Greek mythology.

[69] Bellows were used in Egypt at least as early as the Eighteenth Dynasty, the potter's wheel as early as the First Dynasty. See Alfred Lucas, *Ancient Egyptian materials and industries* (Lon-

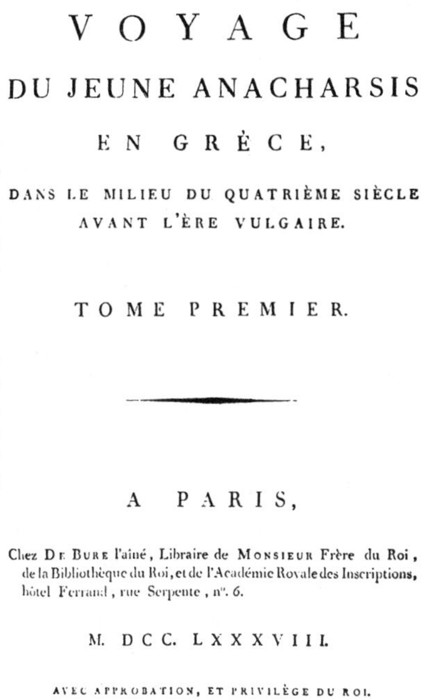
VOYAGE

DU JEUNE ANACHARSIS

EN GRÈCE,

DANS LE MILIEU DU QUATRIÈME SIÈCLE AVANT L'ÈRE VULGAIRE.

TOME PREMIER.

A PARIS,

Chez De Bure l'aîné, Libraire de Monsieur Frère du Roi, de la Bibliothèque du Roi, et de l'Académie Royale des Inscriptions, hôtel Ferrand, rue Serpente, nº. 6.

M. DCC. LXXXVIII.

AVEC APPROBATION, ET PRIVILÈGE DU ROI.

Fig. 48. Title page of volume 1 of the first quarto edition of the *Voyage du jeune Anacharsis*. [From the copy in the Harvard College Library.] When the work was first published in 1788 it was offered in two editions, one in four volumes quarto, the other in six volumes octavo. To each edition was added a "Recueil de cartes géographiques, plans, vues et médailles de l'ancienne Grèce, relatifs au voyage du jeune Anarcharsis," edited by Jean Denis Barbié du Bocage (1760–1825). The second and third editions (Paris 1789, 1790) appeared also in two sizes, octavo and duodecimo, and the sixth (Paris, 1799) in two sizes, quarto and octavo, like the first. Barthélemy revised every edition down to that sixth one, called the fourth, which was edited posthumously by his nephew, Barthélemy de Courcay, and included many additions and corrections. Later editions generally reproduced that of 1799.

much earlier, and were probably made in more than one place. Anacharsis may have imported them from Egypt or elsewhere, or he may have reinvented them independently, or he may have improved them in various ways.

The reader may allow here a little digression which is not irrelevant to our general purpose. The most important work for the diffusion of Hellenism in France at the very end of the seventeenth century was Fénelon's *Télémaque*; in the same manner, the best vehicle of Hellenism a century later was *Le voyage du jeune Anacharsis*, by the abbé Jean-Jacques Barthélemy (Fig. 48).[70] The title of the book was certainly inspired by the wise Anacharsis of whom we have just spoken, for its hero is a Scythian, but Barthélemy placed the story in the middle of the fourth century, because he wanted to give a survey of Greece during that golden

---

don: Edward Arnold, ed. 3, 1948), p. 246 (*Isis 43*); Flinders Petrie, *Wisdom of the Egyptians* (London: British School of Archaeology in Egypt, 1940), p. 133 [*Isis 34*, 261 (1942–43)]. For the anchor, see F. M. Feldhaus, *Die Technik* (Leipzig, 1914), p. 930; Albert Neuburger, *The technical arts and sciences of the ancients* (London, 1930), p. 493.

[70] Elaborate biography by Maurice Badolle, *L'abbé Jean-Jacques Barthélemy (1716–95) et l'hellénisme en France dans la seconde moitié du XVIIIe siècle* (414 pp.; Paris, 1927). Barthélemy was born in Cassis, Provence, but spent most of his life in Paris; he never visited Greece! He was not only a very distinguished Hellenist, but an Orientalist as well. He was one of the founders of numismatics (1750), deciphered a Palmyrene inscription (1754), and was the first interpreter of Phoenician (1758). He was a professional numismatist, being the director of the Cabinet Royal des Médailles, whose holdings were more than doubled during his administration. His popular fame is based exclusively upon the *Voyage*, to which half of his life was devoted; his scientific fame is based upon many memoirs published by the Académie des Inscriptions and upon the royal collection of coins and medals.

age.[71] He spent more than thirty years preparing it, and when the book finally appeared (Paris, 1788) its success was prodigious.[72] The first edition was followed by a good many others, complete or abridged. Before the end of the century it had been translated into German, Italian, English, and Danish; within the first twenty years of the nineteenth century it was translated into Dutch, Spanish, and Greek; in 1847 it was translated into Armenian; the last French reprint appeared as late as 1893; abridgments were still published after that date. In every large library Anacharsis needs many shelves for his accommodation.

The popularity of *Télémaque* may be difficult to understand to our contemporaries, whose taste is hopelessly corrupted by the radio and the movies, but that of *Anacharsis* is truly incomprehensible. This is indeed a heavy manual of Greek archaeology, complete with an atlas of maps and plates; a feeble narrative is but the pretext for an endless series of dissertations on the landscapes and monuments of Greece, its public and private antiquities, arts, literature, philosophy, religion.[73] The French readers who had absorbed the *Encyclopédie* and Buffon's *Histoire naturelle* (a great many had actually read those works volume by volume as they appeared) had truly a great appetite for learning, and their interest in Greece had steadily increased during the second half of the eighteenth century, reaching a climax in 1770 and a new one in the revolutionary period.[74] The success of Barthélemy's book was largely due to its timeliness.

To return to Ionia in the sixth century. The invention of the art of soldering iron (*sidēru collēsis*) was ascribed to Glaucos of Chios, and that of various tools needed in the art of building — level, square, lathe, key – to Theodoros of Samos. This Theodoros, son of Telecles, is rather a mysterious personality; he was a technician, architect, brassfounder, goldsmith, gem engraver,[75] and flourished *c.* 550–530. He invented ways of polishing precious stones and brought the art of bronze casting from Egypt to Greece (that art was much practiced during the Saitic dynasty). All these inventions would call for remarks similar to those offered above concerning the bellows and the potter's wheel, and the history of each would take us too far

[71] "Young Anacharsis" leaves Scythia in 363, and travels to Byzantium, Lesbos, and Thebes (in Boeotia), arriving in Athens a year later. He visits Athens and various parts of Greece, attends Olympic games, and so on. From 354 to 343 he travels in Egypt and Persia, then comes back to Mytilene, where he meets Aristotle. He then returns to Athens but after a while makes a new journey to Asia Minor and the Greek islands, attending the Delos festival. After the battle of Chaironea (338) he goes back to his native country.

[72] Anacharsis' fame at the end of the eighteenth century is illustrated by another amusing fact. The eccentric Baron de Clootz, born in the duchy of Cleves in 1755, a defender of Islam, a French revolutionary, "l'orateur du genre humain," assumed Anacharsis' name! This modern Anacharsis was guillotined in 1794. I do not know exactly when he assumed that name, whether it was before the publication of Barthélemy's book or was a consequence of that publication.

[73] Numismatic investigations are an excellent training for accuracy, and Barthélemy's erudition was of sterling quality; that is, it was as good as it could be at that time, but his work was badly composed. It was too learned and oratorical for a novel, too irregular and chaotic in structure for a manual. It was neither fish nor fowl, yet the public liked it. The imposing display of erudition, gently placed within its reach, flattered its self-esteem.

[74] The Greek fashion in France was largely due to a single author, Plutarch (I-2), who was read in French translations, the favorite being that of Jacques Amyot (1513–1593). The love of classical antiquity was partly due to a revulsion from the Middle Ages, and, at the time of the Revolution, to a revulsion from the "Ancien régime" and a return to nature, or to antiquity assumed to be closer to nature.

[75] According to Herodotos, III, 40–42, he had made the emerald ring which Polycrates of Samos threw into the sea in order to appease the gods who might be jealous of his good fortune; a few days later the ring was found in the belly of a fish and brought back to Polycrates. The data concerning Theodoros of Samos are collected in Pauly-Wissowa, ser. 2, vol. 10, pp. 1917–1920 (1934).

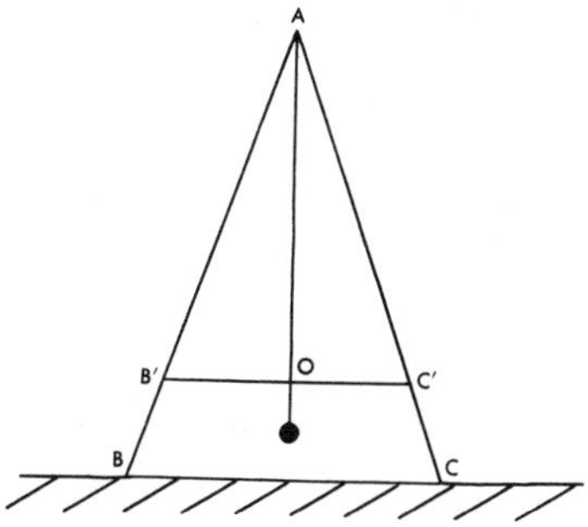

Fig. 49. The *diabētēs.*

out of our path. Let us say a few words of the level. The instrument "invented" by Theodoros was probably the *diabētēs*[76] mentioned in inscriptions (Lesbos). The principle of it is as simple as it is ingenious (Fig. 49). In the triangle *ABC*, probably made of wood, the distances *AB* and *AB′* are respectively equal to *AC* and *AC′*. The middle point *O of B′C′* is marked, and a plummet (*staphylē*) is hung from *A*. If one stands the *diabetes* vertically on a stone and the plumb line is opposite *O*, then the lines *B′C′*, *BC*, and the stone are horizontal. Such an instrument and others involving the same idea (the determination of the horizontal with a plumb line) were used by the Egyptians for astronomic purposes. Not only do we know that, but an exemplar of it has been found in a Theban tomb of the Twentieth Dynasty, and is preserved in the Cairo museum.[77]

The inventiveness of the Greeks or their readiness to make the best use of foreign inventions must have been greatly stimulated in the sixth century by rising architectural and engineering needs, which had to be satisfied. Necessity is the mother of invention. One of the most ambitious creations of that age was the building, or rebuilding, of the Artemision of Ephesos. Ephesos, one of the outstanding Ionian cities, was the center of a cult to the Asiatic nature goddess whom the Greeks called Artemis; in the sixth century that cult had become exceedingly popular and a gigantic temple was planned for the celebration of its rites.[78] Its building called for the solution of many difficulties. Theodoros of Samos is sometimes quoted as the chief architect and it is said that he found means of establishing solid foundations on marshy ground. It is a fact that that fundamental problem had to be solved in the marshes of Ephesos; it is equally certain that it was solved, for otherwise the temple would have fallen; yet it stood for centuries. About the same time, say the middle of the century, a Cretan, Chersiphron of Cnossos, came to assist Theodoros in the realization of that immense project. Chersiphron invented a method of mov-

[76] *Ho diabētēs.* Curiously enough the same word was used by Aretaios (II-2) to name the disease diabetes which he was the first to describe.

[77] Photographs of that Egyptian level and other tools may be examined in Somers Clarke and R. Engelbach, *Ancient Egyptian masonry* (Oxford, 1930), Figs. 263-267.

[78] Artemis = Diana of the Romans. "Great is Diana of the Ephesians" (Acts 19:34). The Artemision was burnt by Herostratos of Ephesos, who hoped thereby to immortalize himself, on the very night that Alexander the Great was born (356); it was rebuilt on a magnificent scale. The foundations of the old Artemision were discovered by John Turtle Wood in 1869; *Isis 28*, 376-384 (1938).

ing huge columns; his son Metagenes continued his father's activities and improved his methods.[79]

The island of Samos, not very far northwest of Miletos, was one of the most important Ionian colonies, and many of its sons or adopted citizens enjoyed fame as architects and engineers. We have already named Theodoros of Samos, but the greatest of all was Eupalinos. Says Herodotos:

> I have written thus at length of the Samians, because they are the makers of the three greatest works to be seen in any Greek land. First of these is the double-mouthed channel pierced for an hundred and fifty fathoms through the base of a high hill; the whole channel is seven furlongs long, eight feet high and eight feet wide; and throughout the whole of its length there runs another channel twenty cubits deep and three feet wide, wherethrough the water coming from an abundant spring is carried by its pipes to the city of Samos. The designer of this work was Eupalinos son of Naustrophos, a Megarian. This is one of the three works; the second is a mole in the sea enclosing the harbor, sunk full twenty fathoms, and more than two furlongs in length. The third Samian work is the temple, which is the greatest that I have seen; its first builder was Rhoicos son of Philes, a Samian. It is for this cause that I have written at length more than ordinary of Samos.[80]

Eupalinos was of Megara but is immortalized by the water conduits that he built in Samos, probably during the rule of Polycrates (*c.* 530–522). The remains of the tunnel described by Herodotos were found in 1882; it is about 1000 m long and 1.75 m high and wide; at the bottom of the tunnel there is a trench, about 60 cm wide and reaching at the south end a depth of 8.3 m, wherein the clay pipes were embedded.

This was indeed a great achievement but not the earliest of its kind. Not to speak of Egyptian and Cretan water supplies, a very remarkable one had been established in Jerusalem under Hezekiah (VIII B.C.), king of Judah from 719 to 690. Its main feature was the tunnel of Siloam, in the village of that name, outside Jerusalem near the southeast corner of that city. It is an underground aqueduct more than 500 m long and of semicircular shape.[81] The Siloam and the Samos tunnels were started at both ends; we can be sure of this for the place of junction can be observed in both cases. Indeed, the junction was imperfect, more so even in the Samos tunnel than in the one built in Jerusalem almost two centuries earlier. How did Hezekiah's engineer and Eupalinos solve the mathematical problems involved? We can only guess; did they have instruments to measure azimuths and differences of levels? The problem involved was solved theoretically for the first time in the treatise on *dioptra* by Heron of Alexandria (I–2).[82]

[79] Their methods are described by Vitruvius (I-2 B.C.), *De architectura*, x, 11–12.

[80] Herodotos, III, 60.

[81] Remains of that tunnel can be examined today. The undertaking was recorded in an inscription which is now in the museum of Constantinople. The Siloam inscription is the oldest Hebrew inscription of any length. See also 2 Chronicles 32:30. Other tunnels were dug to tap subterranean supplies of water in Transjordan in a place like Shobek, and in Palestine, in places such as Megiddo, Lachish, Gezer. Some of these early tunnels are of great size and represent remarkable engineering feats. Nelson Glueck, *The other side of the Jordan* (New Haven: American Schools of Oriental Research, 1940), p. 17 [*Isis 33*, 279–281 (1941–42)]. Glueck does not attempt to date these monuments, which are prehistoric.

[82] In chapter 15 of that treatise. See edition of the *Peri dioptras* by Hermann Schöne in *Heronis opera* (vol. 3, Leipzig, 1903), pp. 239–241. Curt Merckel, *Die Ingenieurtechnik im Altertum* (Berlin, 1899), pp. 499–503, 619. Wilhelm Schmidt, "Nivellier-instrument und Tunnelbau im Altertume," *Bibliotheca Mathematica 4*, 7–12 (1903). Neuburger, *The technical arts and sciences of the ancients*, pp. 416–417, 420–421.

The noun for tunnel is *hyponomos* and the verb *diorussein*.

As Hezekiah's engineer is unknown, Eupalinos may be called the first known civil engineer in history.

Now let us say something of the first known bridgemaker, another son of Samos, Mandrocles, who flourished about 514, that is, a generation later than Eupalinos. Here again Herodotos is our source,[83] but his text is too long to be quoted verbatim. When Darius I (King of Persia, 521–485) made his expedition against the Scythians (*c.* 514 or before) he ordered Mandrocles to build a bridge across the Bosporos to enable his immense army to pass into Europe. Mandrocles was able to satisfy him, for, as Herodotos put it, "Darius, being well content with his bridge of boats, made to Mandrocles the Samian a gift of ten of every kind.[84]

The number of men mentioned in this section is remarkable, especially when one considers that the majority of engineers and other technicians worked anonymously, or at any rate that their personalities were lost in their achievements. Those whom we have been able to name represent a host of forgotten ones. It is equally remarkable that they originate in many places – Scythia, Chios, Crete, Samos, Megara. Scythia was outlandish, but the other places were natural enough, for they were centers of Aegean and Ionian culture. The main localities where those men were employed, Ephesos and Samos, were both in Ionia.

CADMOS OF MILETOS

Cadmos, son of Pandion, is often called the earliest Greek historian. His fellow citizen, Hecataios, of whom we have already spoken in our account of Milesian geography, was also a historian but a little younger than he. Indeed, Cadmos was already active about the middle of the century (or *c.* 540), when Hecataios was born. His Phoenician name is typical of the mixed Milesian culture.

By the middle of the century the achievements of the Ionians, and particularly of the Milesians, were already sufficiently considerable to suggest the value of records. Local pride may have felt the need of such records more keenly after the Persian conquest (546). The Milesians were naturally anxious to show to their conquerors the greatness of their own nation. Cadmos accomplished their purpose, writing in prose an account of the foundation of Miletos (*ctisis Milētu*), and of the history of Ionia. His work must have been of some size, for it was divided into four books, yet hardly anything remains of it.

A similar task was done a little later (*c.* 510) by Eugeon of Samos, who wrote the annals of his native island (*hōroi Samiōn*).[85]

Thus we may say that Greek historiography was born in Ionia, as well as natural philosophy, or otherwise that Ionia was (for the Greeks) the cradle of human history, as well as that of natural history. The Ionians laid the foundations of Greek science in the fullest sense.

We should bear in mind that the Greeks were not alone in composing annals of their past. Not to go farther east, it will suffice to remember that their relatively close neighbors, the Jews, were engaged in similar activities. The Book of Judges

[83] Herodotos, IV, 87–89.

[84] Herodotos, IV, 88. The word for bridge is *schedia*, the meaning of which is not quite clear – raft, float, pontoon, bridge of boats; it must have been some kind of floating bridge. *Edōrēsato pasi deca* means a great gift, abundant gifts.

[85] Cadmus Milesius in Charles Müller, *Fragmenta historicorum graecorum* (Paris, 1848), (vol. 2, pp. 2–4). Eugeon Samius, *ibid.*, p. 16.

and the Books of Kings may have been composed about the sixth century, and the Books of Samuel were somewhat earlier.

THE RELIGIOUS BACKGROUND AND THE SUPERSTITIOUS UNDERGROUND

At the end of this, our first, chapter devoted to Greek science, the reader should be reminded that then, as at every other time, the number of scientists and scholars was very small as compared with the total number of citizens or inhabitants, whose main business was agriculture, trade, this or that craft or profession. There were farmers, merchants, sailors, public officers of many kinds, priests and servants of the temples, poets, artists, men of science. The last named constituted the smallest group. The reader should be warned also of the great importance of religious beliefs. Then as now, those beliefs were the substance of life, and then as now they ran the whole gamut from the highest and purest kind of faith and symbolism to the crassest kind of superstition.

The second warning is especially necessary because the Greeks are often praised for their rationalism, which is just as foolish as if the Christians were praised for their sanctity. Indeed, there have always been among the latter a few, very few, saints; even so, a few, very few, Greeks were the founders of rationalism and of science. The people as a whole are as good as circumstances permit, and their behavior is very largely irrational. Rationalism and religion, it should be recalled, are not mutually exclusive; rationalism and superstition are, but it is sometimes difficult to draw the line between superstition and religion.

The main difference between Greece and, say, Palestine is that the Greeks had no sacred writings comparable to the Old Testament, no definite dogmas to which subservience or at least acquiescence was required. The nearest things to Scriptures were the Homeric poems, and these were definitely lay scriptures, not sacred ones. To be sure, Homer often referred to the gods, but that was incidental and he did it with a poet's license. Nevertheless, the *Iliad* and the *Odyssey* exerted a deep influence on Greek religion, for they helped to standardize the myths and to popularize them. Moreover, they humanized the gods and the heroes, sometimes to a degree that may scandalize the modern reader, but did not disturb the Greek listener. The latter knew that the gods were extremely powerful but he did not expect them to be perfect. Such as they were, Homer and Hesiod did not invent them but made them more familiar, consecrated their existence and their peculiarities. The Homeric epithets were easy to remember and were soon engraved on every heart.

The historian of Greek thought is constantly aware of two contradictory tendencies – the poetic or mythopoeic one and the rationalistic one. The intensity and popularity of the former can be judged from the bewildering richness of Greek mythology. The second was far less popular, though not by any means restricted to men of science; the Greek merchants, we may be sure, were sufficiently matter-of-fact and did not mythologize their business. The two tendencies occurred together and not necessarily in different groups. Men of science might accept the myths as poetical descriptions of things that were not susceptible of scientific explanation.

The religious life of the Greeks was not rigid, but it was exceedingly complex and varied. It is perhaps that very complexity that saved them from dogmatism and from religious tyranny. At the beginning there were local gods in every city and every state, gods for every phenomenon and for every occasion. In the course of time some of those gods acquired more importance.[86] The diocese of each god might increase or decrease as the political domains of his servants waxed or waned, or for many other reasons. Sanctuaries might grow in popularity and eventually acquire national, even international, ascendency. It is almost impossible to disentangle the mixed motives that led to the abandonment of certain gods, or the success of others. The caprice of little men might have as much power in the end as the political schemes of the great. Moreover, as the gods obtained national stature, there was an opposite propensity to reindividualize them and to attach different degrees of prestige to each of their avatars or of their sanctuaries.[87] There was thus a kind of rhythmic growth and decrease of the gods, an ebb and flow in their power and their scope.

The Greeks had an abundance of gods, yet such was their ardor for worship and their love of mystery that they were irresistibly attracted by foreign gods – Isis and Osiris of Egypt, the Magna Mater of Phrygia, the Phoenician Astarte, and many others. Greek mythology is permeated with Egyptian and Asiatic elements. We can easily imagine that the Greek colonists in Asia and Africa contributed not a little to that religious diffusion. Everything conspired to perfect the syncretism – their own fears and hopes, their love of the unknown and the occult, their desire to conciliate foreign associates, the frank proselytism of their neighbors. Being unrestricted and unprotected by any clear orthodoxy (as were the Jews) they saw no reason why they should not honor the exotic gods and sacrifice in their temples.

The love of magic was apparently strong in their hearts, or at any rate not weaker than it is in the hearts of men, even thinking men, all over the world. They knew well enough the awful powers of nature in all their manifestations (sun and moon, winds, rain, thunder, earthquakes) and were anxious to propitiate them with suitable rites and exorcisms. Special ceremonies were invented to promote fertility, health, longevity, communion with the immortal gods, salvation. The monotony of their lives was periodically relaxed by holidays in their own temples, athletic or musical contests, quiet festivals or orgiastic ones.

Their hospitable religion did not amalgamate only foreign cults; here as elsewhere it appropriated all the local folklore, the beliefs in sacred stones, caves, springs, trees, even animals. The cult of animals was never as popular or as intense as in Egypt or India, but it existed nevertheless; witness the owl of Athene, the eagle of Zeus, the snakes of Asclepios, the bear dances of Athenian girls, and chiefly the Black Demeter of Phigalia (in Arcadia), she who was represented with a

[86] In this case also Homer helped a lot. A line like

Zeu te pater cai Athēnaiē cai Apollon

(Iliad, II, 371) put Zeus, Athene, and Apollo in the front line; a kind of superior trinity was constituted.

[87] Comparisons with the Catholic religion help us to understand the vicissitudes of the Greek gods. Why was Santiago de Compostela gradually superseded by Loretto, and Loretto by Lourdes? As the cult of the Virgin Mary became more popular, there was a gradual tendency to single out particular shrines and to treat different apparitions of Our Lady almost as if they were different persons. The faithful would pray not to Our Lady, but to a madonna who was closer and presumably more accessible – such as Notre Dame de Hal or Notre Dame de Chartres, Nuestra Señora del Pilar or Nuestra Señora de Guadelupe. Or they would abstract a quality and pray to Notre Dame des Sept Douleurs, to the Madonna della Misericordia, the Madonna dell' Umiltà, or the Immaculate Conception, even as the Greeks would have prayed for victory to Athene Salpinx, for health to Athene Hygieia, or for wisdom to Pallas Athene.

horse's head. Greek mythology is a fantastic hodgepodge of every kind of irrationality, but the wiser men did not take it except with plenty of salt. While the physiologists of Miletos were earnestly trying to explain natural phenomena in rational terms, their neighbors, that is to say, the overwhelming mass of the people, were content to mythologize them and to invent new sacrifices of propitiation or deprecation, rites for the conservation of good things and for the destruction of bad ones, blessings and curses.

We have already come across two great religious centers, Didyma and Ephesos, both in Ionia, but there were many others, the most famous being Delos in the Cyclades, and Delphi, so close to the center of Greece that it was believed to be the navel (*omphalos*) of the world.[88]

The existence of those sanctuaries was due to the innate desire of holiness and salvation, and, on the other hand, it helped to strengthen that desire and to spread it. The Greeks loved sanctity as they loved beauty, and they soon developed a rich casuistry relative to the causes of losing it, the means of restoring it, the rites of purification, the ways of interrogating the gods and interpreting their answers. Their love of beauty, of pageantry, of the drama inspired the organization of festivals and games, some of which had already acquired national fame in the sixth century. The Panathēnaia[89] were celebrated in Athens from very early times, the Olympia in Olympia from 776 on, the Pythia near Delphi from 586, the Isthmia in Corinth since 582, the Nemeia in Argos since 573. The dates given, which are traditional, are probably too early, because people like to deepen the antiquity of their institutions, or to count the age of those institutions from too meager a beginning; but after all, is not every birth modest and obscure and is not every baby very small? These festivals included not only athletic contests, but musical and dancing ones as well. There were competitions in playing the lyre and the flute, in singing to the accompaniment of those instruments, in composing music in a definite mode (for example, the Pythian mode, *pythicos nomos*) and in reciting Homeric poems (*rhapsōdeō*). Finally, there were dramatic festivals, especially those dedicated to Dionysos, which are of great literary significance, because they were the cradles of Greek tragedy. In many sacred places, oracles were rendered in various ways, for example, the oracle of Zeus in Dodona (near the lake and town of Ioannina, Epiros) by the wind rustling through the leaves of oaks or beech trees; the oracle of Apollo in Delphi, by the frenzy of a female medium, the Pythian prophetess.[90] These oracles were regulated by the servants of the temples; their administration might imply a certain amount of conscious or unconscious fraud, especially when political questions were involved, but perhaps less than most people imagine. It is foolish to think that everybody believed in divination, except the priests whose business it was to prophesy or to interpret. There were probably a few cynical and skeptical priests, greedy and corruptible; the majority were faithful and honest, otherwise the institution that they served could not have functioned

[88] That belief was already established in the time of the poet Pindar (*c.* 518–442) but was probably earlier.

[89] In English it is more common to speak of the Panathenaic games or festival; but in Greek usage Panathenaia often refers to everything – the festival, the games, the musical contest, the sacrifices. The same remarks apply to the Olympia (instead of Olympic games, etc.), the Pythia (instead of Pythian games, etc.), the Isthmia, the Nemeia.

[90] Herbert William Parke, *History of the Delphic oracle* (465 pp., ill.; Oxford: Blackwell 1939) [*Isis* 35, 250 (1944)].

as long or as well as it did.[91] The oracles helped to standardize the rites and customs; they often were a kind of moral arbitration, as by an impartial and superior conscience, and hence they might strengthen private and public morality.

The most impressive rites were the mysteries (*mystēria*), secret ceremonies of initiation and progressive edification. The purpose of those elaborate ceremonies, taking place in a hidden part of the temple (for example, the *telestērion* at Eleusis), was to establish in the mind of the initiated a state of awe, religious fervor, and enthusiasm.[92] The national festivals generally included mysteries; or rather they were public forms of rejoicing of which the local mysteries were the occasion (just as Christian pilgrimages are focused upon special Masses). For example, in Delphi Apollo overcame the dragon Python and that victory was celebrated periodically in the Pythia.[93] It was a kind of sacred drama the celebration of which in a magnificent and awful landscape must have stirred religious emotion to the highest pitch.

Among other mysteries it will suffice to mention the Orphica, dedicated to the Thracian hero, poet, and musician, Orpheus, and solemnized in many places; those dedicated to the Pelasgic Cabiri (*cabeiria*)[94] in the island of Samothrace; those connected with Demeter and celebrated in Attica, the Thesmophoria by women only and the Eleusinia by men and women alike in Eleusis at the seashore not far from Athens. The Eleusinian mysteries are perhaps the best known to educated readers who are not students of mythology. Those complicated mysteries relative to Demeter, Persephone, and Triptolemos are really nature myths concerned with fertility and immortality; they had been introduced from Crete by the "wise man" Epimenides in 596. The Eleusinian and other mysteries contain an abundance of Pelasgic, Thracian, Asiatic, and Egyptian ideas. It is as if all the beliefs and religions that had grown up in the countries surrounding the Eastern Mediterranean had been put in a crucible for centuries and millennia; the most sacred rites of Hellas were like the residue and the quintessence of that alchemy.

At their best the mysteries emphasized the sanctity of life, and they enhanced a man's deepest religion, his feeling of partnership with his fellow men in the secret purpose of nature. They were a combination of poetry and drama with pantheism and with the cult of individual gods and heroes. They did no harm to the wise men and women, and sanctified them in the same way as the Mass exalts and strengthens the faithful members of the Catholic and Orthodox churches. Nor was participation in the mysteries necessarily incompatible with the search for the truth and the love of science. On the other hand, their influence on simple people was a mixture of good and evil; they helped them to be virtuous, yet increased their superstitious tendencies. The Greek mysteries, like all the religious mysteries, helped good

[91] My own confidence in the intrinsic honesty of the average priest and diviner is largely due to the reading of Plutarch (I-2).

[92] The word enthusiasm (*enthusiasmos*) is here used in its original sense. It is derived from *entheos*, full of the god, inspired, possessed, and thus means divine inspiration.

[93] It is after that very dragon that a group of snakes are named, the Pythonidae, including the largest species. The original Python had been produced from the mud left over after the flood; he lived in a cave of Mount Parnassos. Apollo's slaying of the Python may symbolize the victory of good over evil, or of sunshine and spring over darkness and winter. One recognizes familiar patterns, occurring under different forms in the mythology of many nations.

[94] The Pelasgi (*Pelasgoi*) were the earliest inhabitants of Greece, but there is no agreement as to their original location: northern Greece, Asia Minor, Crete? The Cabiri were their divinities. The adjective Pelasgic might be replaced by "prehistoric."

people to be better by exalting their good nature, but made the bad people worse, by adding self-righteousness and hypocrisy to their other vices.

In short, the Greeks were more inclined to poetic myths than to theology; they had no sacred writings and no dogmas, yet were intensely religious; most of them attended the festivals whenever they could, and many celebrated the mysteries with genuine fervor. A few men managed to combine rationalism with "enthusiasm" (and why not?); the great mass was abandoned to divination and superstitions of every kind.

The final paradox is this: the ancient Greeks did not have any kind of systematic theology, yet they created the logical instruments that were needed for the development of the three dogmatic religions of the West – Judaism, Christianity, and Islam. In each of these religions there is a woof of scripture and tradition, but the warp is Greek. The ancient Greeks had no theology of their own, but they were the founders of theology.

BIBLIOGRAPHY

Paul Tannery (1843–1904), *Pour l'histoire de la science Hellène* (Paris, 1887); rev. ed. by A. Dies (Paris, 1930). The revision was very insufficient, but much of the old text retains its importance.

—— *Recherches sur l'histoire de l'astronomie ancienne* (Paris, 1893).

John Burnet (1863–1928), *Early Greek philosophy* (London, 1892; ed. 2, 1908; ed. 3, 1920).

Theodor Gomperz (1832–1912), *Griechische Denker* (3 vols.; Leipzig, 1896–1909); *Greek thinkers* (4 vols.; London, 1901–1912).

Hermann Diels (1848–1922), *Die Fragmente der Vorsokratiker* (Berlin, 1903; ed. 3, 3 vols., 1912–1922; ed. 4, anastatic reprint, 1922; ed. 5, Berlin, 1934–35).

Kathleen Freeman, *The pre-Socratic philosophers* (500 pp.; Cambridge: Harvard University Press, 1946). This is derived from Diels, the chapters being numbered as in Diels' fifth edition. All in English!

VIII

PYTHAGORAS

WHO WAS PYTHAGORAS?

Our previous chapter ended with a tantalizingly brief account of Greek religion. That account was too brief for any purpose except to make the reader realize the importance of religion in Greece, the cradle of science. The historian of science, even of Greek science, should never disregard religion. It would not be right to say that the luxuriant development of religion in all its forms, a development that reached a kind of climax during the sixth century, helped science, nor yet that it harmed it. Then as now the two developments, scientific and religious, were parallel, contiguous, interrelated in many ways; they were not necessarily antagonistic; they often took place in the same minds.

One curious aspect of that sixth-century efflorescence of religion is that it occurred in the western part of the Greek domain, rather than in its eastern part as we would expect, but that may be accidental. The Ionian physiologists represented, it is true, a rationalistic wing, but how many were there? or rather, how few? The Oriental Greeks or the Greek Orientals were on the whole religious minded, fond of rites and miracles. When the Persian menace, and later the Persian terror, drove them westward, some of them did not stop in Greece, or at least did not stay there, but continued farther west and found an asylum in the Ionian colonies of Sicily and Magna Graecia.[1] We have already spoken of one of those eastern refugees, Xenophanes of Colophon; another and more illustrious one was Pythagoras.

What kind of man was he? It is difficult to say because the biographies that have come down to us were written late and are full of impurities. They were composed by Diogenes Laërtios (III–1), Porphyry (III–2), and Iamblichos (IV–1), the work of the latest being the most popular as well as the most spurious. What is more disquieting, the older traditions, such as those of Herodotos, Aristotle, and his pupils, were already fabulous to a degree. For example, Herodotos, the nearest witness in point of time, was already combining Pythagorean ideas with Egyptian, Orphic, and Bacchic ones,[2] and he mixed up the story of Pythagoras with that of

[1] The term Magna Graecia is used because it is more accurate than South Italy, but it was unknown in the sixth century. Magna Graecia or Graecia Major (*hē megalē Hellas*) referred to the Greek colonies of South Italy, never to the whole of that country. Polybios (II–1 B.C.) was the first to use the Greek term, and Livy (I–2 B.C.), the Latin one; Strabo (I–2 B.C.) extended it to the Greek colonies of Sicily. T. J. Dunbabin, *The Western Greeks. The history of Sicily and South Italy from the foundation of the Greek colonies to 480* B.C. (518 pp.; Oxford: Clarendon Press, 1948) [*Isis 40*, 154 (1949)].

[2] Speaking of the Egyptians, Herodotos remarks (II, 81), "Nothing of wool is brought into temples, or buried with them; that is forbidden.

Zalmoxis, thus explaining *obscurum per obscurius*.[3] According to the story, which he tells with some diffidence (and we should not be more credulous than he was), Zalmoxis was a Thracian who had been a slave of Pythagoras, son of Mnesarchos. Having obtained his freedom, wealth, and some familiarity with the Ionian way of life, he returned to his native country, where he built a great hall and entertained his neighbors. He expounded to them the ideas of immortality and paradise and, in order to convince them, disappeared for three years in an underground chamber. While they were still mourning him he reappeared to them alive in the fourth year, and they could not disbelieve him any longer. This story shows that by the fifth century Pythagoras was almost as mythical as Zalmoxis himself.

Yet there is a small substratum of fact that we may perhaps accept as true. Pythagoras, son of Mnesarchos, was born in Samos and flourished there during the rule of Polycrates (executed in 522). According to Aristoxenos of Tarentum (IV–2 B.C.) – who is not too late a witness as ancient traditions go – he left Samos to escape Polycrates' tyranny; that is plausible, or he may have been driven away, like so many others, by the fear of the Persians. It would have been natural enough for him to seek refuge in Egypt, where the Samians had many representatives (they had a temple of their own in Naucratis). If we may believe Iamblichos, he first went to Miletos, where Thales recognized his genius and taught him all he knew, then visited Phoenicia, where he remained long enough to be initiated into the Syrian rites. This increased his desire to go to Egypt, which was then considered the fountainhead of esoteric knowledge; he spent there no less than twenty-two years, studying astronomy and geometry as well as the mysteries. When Cambyses conquered Egypt in 525, Pythagoras followed him back to Babylon, where he spent twelve more years studying arithmetic, music, and other disciplines of the Magi.[4] He then returned to Samos, being in the fifty-sixth year of his age, but soon resumed his wanderings through Delos, Crete, and Greece proper, and finally reached Croton,[5] where he established his famous school. After he had obtained considerable popularity and power, which he may have abused, political enmities or local jealousies drove him out and he spent the last years of his life in Metapontion.[6]

We have told the story at some length, though we attach little credence to Iamblichos. Whether the details are correct or not, the substance is plausible.[7] Was

---

In this they follow the same rule as the ritual called Orphic and Bacchic, but which is in truth Egyptian and Pythagorean; for neither may those initiated into these rites be buried in woollen wrappings." There was some truth in Herodotos' confusion, for Orphism and Pythagoreanism had been mixed up long before his time. The "golden lamellae" found in tombs in Italy and Crete and believed to be Orphic are Pythagorean. F. Cumont, *Lux perpetua* (Paris: Geuthner, 1949), pp. 248, 406.

[3] Herodotos, IV, 95. He writes *Salmoxis* but the spelling *Zalmoxis* is more usual; *Zalmos* is a Thracian word for skin.

[4] The word Magi is the one used by Iamblichos. *Magos* (derived from Old Persian *magush*) designated at first the Iranian, Zoroastrian, priests and interpreters; later, Chaldean priests, and magicians. The word magic, by the way, is derived from the same root, *hē mageia, hē magicē technē*, the science or art of the Magi. Joseph Bidez and Franz Cumont, *Les mages hellénisés* (2 vols.; Paris: Les Belles Lettres, 1938) [*Isis 31*, 458–462 (1939–40)].

[5] Croton (*Crotōn*) or Crotona was then a Greek colony of old standing, having been founded by Achaians and Spartans *c.* 710. Metapontion was another Achaian colony in the same neighborhood. It is located near Tarentum at the bottom of the gulf, while Croton is at the southwest entrance to it.

[6] He died in Metapontion *c.* 497. When Cicero visited that city *c.* 78 B.C. he was shown the house where Pythagoras died; *De finibus*, v, 2, 4.

[7] The chronology is not unacceptable. If Pythagoras was 56 in 510, he was born in 566 and may well have known Thales, who lived until *c.* 548. This would leave very little time for his activities in Croton, however, for he is said to have died in 497. According to the Sicilian his-

Pythagoras actually Thales' disciple or not? Did he devote thirty-four years to graduate studies in Egypt and Babylon? We cannot even be sure that he traveled much on his way from Samos to Croton. The story accounts for the Egyptian and Babylonian roots of Pythagoras' thought, but a man as intelligent and inquisitive as he was might have gathered a considerable amount of Oriental lore even without visiting those countries, or at any rate without spending there as many years as Iamblichos says he did. Surely Pythagoras did not need thirty-four years to learn what was then to be learned there and was assimilable to his fertile and eager brain. The intention of Iamblichos, or of his informant, was to show that Pythagoras had not simply visited Egypt and Babylonia as many Greeks did, for business or pleasure, but that he had remained in those countries long enough to study with their doctors, to drink deep from their wisdom, and even to be initiated into their mysteries.

THE PYTHAGOREAN BROTHERHOOD AND THE EARLY PYTHAGOREAN DOCTRINES

One of the aspects of the religious revival that occurred in many places during the sixth century was the growth of communities of people sharing a new revelation and occult doctrines of various kinds. Such communities would naturally take the form of brotherhoods, for the men and women sharing eschatologic secrets would be like members of a family, like brothers and sisters defending their common heritage against outsiders. Pythagoras and his immediate disciples imitated that practice in Croton. Some of their teachings were scientific doctrines that will be explained in the following sections, but others were of a more general nature, and it was probably to these that the order owed its popularity. Pythagoreanism was primarily a way of life.

The Pythagoreans conceived a new kind of holiness, the attainment of which implied ascetic exercises and the observation of taboos, for example, the abstention from certain articles of food, such as meat, fish, beans, and wine, and the avoidance of woolen clothes.[8] Women were admissible as well as men and seemed to have played an important part in the early community. Members of the order wore distinctive garments, went barefooted, and lived simply and poorly.

They fancied that the soul can leave the body either temporarily or permanently and that it can inhabit the body of another man or of an animal, but whether Pythagoras derived that belief from Hindu or other Oriental sources is of course impossible to say. Granted the intuitive feelings that a soul leaves the body with the last breath and that there is some kinship between men and animals,[9] feelings

---

torian Timaios of Metapontion (III-1 B.C.), Pythagoras spent twenty years in Croton, the revolt against him and his school occurred in 510 or soon after, and it was then that he moved to Metapontion. He spent perhaps less time in Egypt and Babylonia than Iamblichos tells us.

[8] Wool (as distinguished from linen) was taboo as being an animal product. That particular taboo was already referred to in footnote 2. It is interesting to note that while the Pythagorean mystics were forbidden to wear wool, the Muslim ones of a later age were invited to do so; the Arabic term ṣūfī applied to them means woolly!

Some of the Pythagorean taboos are quoted by Plutarch in his life of Numa (ch. 14).

[9] Those feelings still exist among us. As to the second, we recognize various animals in ourselves and in our neighbors. When we call one of them a lion or a lamb, a monkey or a fox, a bull or a pig, our meaning is clear and can be transmitted to others without ambiguity. To be sure, we do not press the comparison as far as our ancestors did.

Pythagoræ Philiſophi Aurea uerba

Qua tranſiui?ꝗd egi?ꝗd quod agendũ fuerit,prætermiſi?à primo incipi
ens,diſcurras ad reliqua· Cũ turpe ꝗd feceris,te ipſum crucia· Cnm uero
bona pfeceris,tibi cõgratulare. Hæc exercere,hæc meditari, hæc te amare
oportet·hæc te ĩ diuinæ uirtutis ueſtigiis collocabũt p eũ. ꝗ animo nrõ
quadruplicẽ fontẽ ppetuo fluẽtis naturæ tradidit·Exi ad opus,cũ diis uo
ueris. Nã iſta ſi tenebis,cognoſces ĩmortaliũ deoꝝ,mortaliũue hominũ
cõditionẽ,qua procedunt.& cõtinentur oĩa ·Cognoſces quãtũ fas ẽ,na-
turã circa oĩa ſimilẽ·ne te ſperare cõtingat ,quæ ſperãda non ſũt·neq; te
ꝗcꝗ̃ lateat. Cognoſces hoĩes,cũ ſuoꝝ ſint maloꝝ cã miſeros eſſe·Qui bo
na,ꝗ̇ prope ſunt,nec uidẽt,nec audiũt. Solutionẽ uero maloꝝ pauci ad-
modũ ĩtelligunt· Tale fatũ lædit mẽtes hoĩũ,ꝗ reuolutiõibus ꝗbuſdam
ex aliis ad alia ferũtur,ĩfinitis malis obnoxii lætifera diſcordia ĩſita latẽ-
ter obeſt ,eam tu cedẽdo deuita,& poſtꝗ̃ uenerit,ne exaugeas· O Iupiter
pater,uel a malis hoĩes libera,uel oſtẽde illis,quo dæmone utantur·At tu
cõfide,quoniã diuinũ genus hoĩbus ĩeſt·his·n·ſacra natura proferẽs uni-
uerſa demõſtrat. Quoꝝ ſiꝗd tibi fuerit reuelatũ,abſtinebis,ab iis,à ꝗbus
abſtinẽdũ iubeo· Quod ſi medicinã adhibueris,aĩam ab his laborib⁹ li-
berabis· Veꝝ abſtine à mortalibus,ꝗ̇ ſupra diximus ĩ purgatiõe ſolutio-
nẽq; animæ. Recto iudicio conſydera ſingula. Optimã deinde ſententi-
am tibi uelut aurigam præpone·

¶ Corpore depoſito cum liber ad æthera perges,
¶ Euades hominem, factus deus ætheris almi·

¶ Symbola,pythagore phyloſophi.

¶ Cum ueneris in templum adora,neq; aliquid interim,quod ad uictũ pertineat,aut dicas,aut agas, Ex itinere præter propoſitũ nõ ẽ ĩgrediendũ ĩ templũ,neq; orãdũ,neq; etiã ſi prope ueſtibulũ ipſum trãſiueris. ¶ Nudis pedibus ſacrifica,& adora· ¶ Populares uias fuge,p diuerticula uade. ¶ Ab eo,quod nigram caudam habet abſtĩe,terreſtrium enim deorum eſt· ¶ Linguã in primis coherce deum imitans. ¶ Flantibus uentis echon adora· ¶ Ignem gladio ne ſcalpas ¶ Omne acutum abs te dimoue· ¶ Viro,qui põdus eleuat auxiliare·nõ tamẽ cũ eo deponas,ꝗ deponit. ¶ In calceos dextrũ præmitte pedẽ,ĩ lauacrũ uero ſiniſtrũ· ¶ De reb⁹ diuinis abſq; lumine ne loquaris. ¶ Iugum ne tranſilias. ¶ Stateram ne tranſilias. ¶ Cum domo diſceſſeris,ne reuertaris, furiæ enim congredientur. ¶ Ad ſolem uerſus ne mingas· ¶ Ad ſolem uerſus ne loquaris. ¶ Oleo ſedem ne abſtergas· ¶ Gallum nutrias quidem,ne tamẽ ſa-

X iii

Fig. 50. The "Golden words" and the "symbols" of Pythagoras. In September 1497 the great Venetian publisher Aldo Manucci il Vecchio (1449–1515) published a small folio (30 cm high) containing Iamblichos' *De mysteriis Aegyptiorum, Chaldaeorum et Assyriorum,* and a dozen other texts translated by the Florentine Platonist Marsilio Ficino (1433–1499). Fewer than three pages of that book concern Pythagoras, but those pages constitute the first Pythagorean printed publication. They contain his "Golden words" (sayings ascribed to him) and his "symbols." The page that we reproduce shows the end of the "aurea verba" and the beginning of the "symbola"; these are chiefly in the nature of taboos. [From the copy in the Harvard College Library.]

shared by many nations, whether primitive or sophisticated, the concept of transmigration of souls might occur (and did occur) independently in many places.[10]

The religion of the Pythagoreans was otherworldly to the extent that they regarded this life, the life this side of death, as a kind of exile (*apodēmia*). Like every other religion, it was pure enough at its highest level and the opposite at its lowest. For example, many of their rules (as has already been remarked) were simply taboos,[11] that is, irrational interdictions due to the fact that certain classes of things were considered sacred and were forbidden because of their purity or impurity; it was unlucky to meddle with them. Those rules were called *acusmata* and the humblest members of the order, the *acusmaticoi*, were the poor bigots for whom the taboos took the place of doctrines, for they were unable to understand much else (Fig. 50).[12] On the contrary, the fully initiated attached more importance to eschatology and theology, or to the scientific ideas that constituted the very core of their thinking. It is impossible to know much about those doctrines, or to know them with precision, for the members were pledged to silence (*echemythia, echerrhēmosynē*) and even to secrecy.

Political ideas were added gradually to the others, for the order was a little society immersed in a larger one yet kept jealousy separated from it. Conflicts were bound to appear between the groups, and should the little Pythagorean group try to obtain power in order to escape those difficulties, its troubles would increase. It is certain that the Pythagoreans were thwarted and annoyed and that Pythagoras was forced "to leave town" and go to Metapontion. The followers who remained in Croton, Metapontion, and other places suffered greater persecutions after his death and some of them were even massacred (some of those persecutions occurred perhaps as late as 450).

The martyrdom of his disciples increased Pythagoras' prestige. He was soon regarded as a saint or even (in the Greek way) as a hero, intermediary between gods and men, and the late accounts of his life and deeds were composed in the spirit of hagiography. Under those circumstances, is it at all surprising that the early doctrines are obscure and the founder himself largely unknown? To know the facts about him is as hopeless as to know those concerning St. Gregory the Wonder worker or St. George the Martyr.

ARITHMETIC

In his lost book on the Pythagoreans, Aristotle wrote that "Pythagoras son of Mnesarchos first worked at mathematics and arithmetic and afterwards, at one

[10] That concept was called *palingenesia* or *metensōmatōsis* rather than *metempsychōsis*, much used in English writing. It was not by any means rare; many people shared it more or less – primitive people, Hindus and Buddhists, Egyptians, Greeks and Romans, Jews, Celts, and Teutons; *Encyclopedia of Religion and Ethics*, vol. 12 (1922), pp. 425–440. For a fuller discussion of Pythagoreanism in general than is possible here, see John Burnet, *Ibid.*, vol. 10 (1919), pp. 520–530.

[11] The very use of the word taboo implies an anthropological explanation which was not available until last century. The word taboo or tabu was introduced into English by Captain Cook (1728–1779), who had met with it, and the thing signified, at Tonga (South Pacific); the explanation of its meaning developed slowly during the nineteenth century. See article by R. R. Marett, *Ibid.*, vol. 12 (1922), pp. 181–185.

[12] Here are a few of the Pythagorean taboos: not to pick up what has fallen, not to touch a white cock, not to break bread, not to eat from a whole loaf, not to stir the fire with an iron poker, not to allow swallows to nest under one's roof. We need not smile or feel superior about that, for other taboos, neither better nor worse, inhibit the lives of our contemporaries, if not our own lives.

time, condescended to the wonderworking practiced by Pherecydes."[13] Aristotle's hypothesis is plausible, though it does not tally with the traditions concerning Pythagoras' oriental education. It is possible that Pythagoras' first independent thinking was centered on mathematics and that the mystical tendencies of his youth reasserted themselves later in life. (He was not by any means the last mathematician to become a mystic in his old age!) At any rate, in order to develop a mystical theory of numbers it was necessary first of all to obtain a sufficient knowledge of them. Pythagoras was in all probability the founder of the great mathematical school bearing his name.

Here are a few examples of the speculations that are of sufficient antiquity to be ascribed to him. The first is the distinction between even numbers (*artios*) and odd ones (*perissos*), the former being divisible into two equal parts, the latter, not. That was of immediate value, for one often wishes to divide a group into two smaller ones as fairly and symmetrically as possible. If one builds a temple, the number of columns at the entrance should be even, otherwise a column would face the middle of the door, spoil the inward or outward view, and obstruct the traffic; the number of columns on the sides might be odd or even.[14]

Pythagoras' arithmetic was based on the use of dots drawn in sand, or of pebbles, which could be grouped easily in different ways. He was then able to make many arithmetic experiments concerning the number of pebbles that would fill a given pattern. If pebbles are arranged in such a manner that they form triangles (Fig. 51), the numbers of pebbles in the triangles (1, 3, 6, 10, . . .) are the triangular numbers. Pythagoras probably saw that those numbers were the sums of one or more natural numbers beginning with one. Did he generalize the result?

$$\sum_{1}^{n} i = \tfrac{1}{2}n(n+1).$$

Probably not, but he experimented far enough to see how these numbers derived each from the preceding one:

$$\begin{array}{l} 1 = 1 \\ \quad + 2 = 3 \\ \qquad\quad + 3 = 6 \\ \qquad\qquad\quad + 4 = 10 \\ \qquad\qquad\qquad\quad + \ 5 = 15 \\ \qquad\qquad\qquad\qquad\quad + \ 6 = 21 \\ \qquad\qquad\qquad\qquad\qquad\quad \cdots \end{array}$$

the successive additions being made, not with numerals as we have just done, but with pebbles. The fourth triangular number, a triangle with four pebbles on each side, interested Pythagoras specially. This was the so-called quaternion or *tetractys*

[13] As quoted by Sir Thomas Heath, *History of Greek Mathematics* (Oxford, 1921), vol. 1, p. 66, Pherecydes of Syros, son of Babys. "Wise man," cosmologist or physiologist of the sixth century; sometimes quoted as Pythagoras' teacher. Kurt von Fritz, Pauly-Wissowa, vol. 38, pp. 2025–2033 (1938).

[14] In the Parthenon there are 8 columns at each end and 17 along each side, that is, 46 in all.

Fig. 51. Triangular numbers.

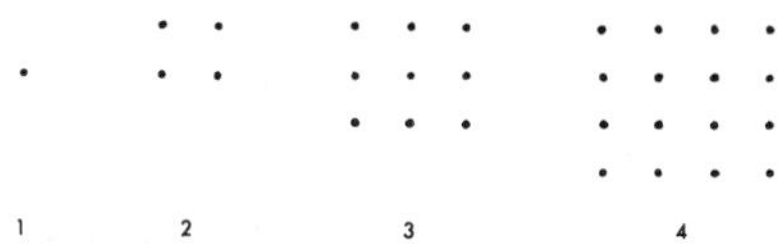

Fig. 52. Square numbers.

$(1 + 2 + 3 + 4 = 10)$, to which the school attached marvelous properties.[15] The Pythagoreans swore by it!

Square numbers were investigated in the same manner. How does one pass from one to the next? To pass from No. 3 to No. 4, for instance (Fig. 52), one adds a number of pebbles enveloping No. 3 at two sides of one corner. That two-sided series of pebbles, called a gnomon,[16] was necessarily an odd number. Hence the obvious rule: a square plus an odd number makes another square;

$$n^2 + (2n + 1) = (n + 1)^2.$$

More concretely, consider the series of odd numbers 1, 3, 5, 7, 9, . . . The first is also the first square; by adding one by one each of the following odd numbers to it one will obtain all the square numbers:

$$\begin{aligned} 1 &= 1 \\ +3 &= 2^2 \\ +5 &= 3^2 \\ +7 &= 4^2 \\ +9 &= 5^2 \\ &\dots \end{aligned}$$

Hence each square number is the sum of all the odd numbers smaller than twice its root:

$$1 + 3 + \dots + (2n - 1) = n^2.$$

This is as beautiful as it is simple. One can imagine Pythagoras' delight when he discovered these particles of the universal truth; if he had mystical tendencies, such as he would have easily acquired in Egypt and Asia, his growing exaltation was natural.

We have spoken of pebbles, because Pythagoras had no numerals as we have. It is probable that the literal numerals were not yet in use in Pythagoras' time.[17] If

[15] Pythagoras knew that the fourth triangular number is ten. It was very tempting to develop the mystical consequences of that fact; how much of that elaboration was already due to him, how much to later Pythagoreans, it is impossible to say. The elaboration of Pythagorean arithmetic can be followed for a thousand years, glimpses of its maturity occurring in Nicomachos of Gerasa (I-2) and in Iamblichos (IV-1). In the latter's arithmetical theology, *Theologumena tēs arithmēticēs* (note the title!), the sacredness of the tetractys is emphasized. The decad represents the universe; are there not 10 fingers, 10 toes, etc.? Martin Luther D'Ooge, *Nicomachos* (New York, 1926), notes on pp. 219, 267 [*Isis* 9, 120–123 (1927)]. The reference to the decimal basis of numeration was implicit; it is remarkable that no Pythagorean thought of making it explicit.

[16] The same word *gnōmōn* was used previously to designate an astronomical instrument, the vertical index of a sundial. The new mathematical meaning was derived from the fact that the word was used for a carpenter's square (Latin, *norma*).

[17] The earliest occurrence of them is in a Halicarnassian inscription of 450 B.C.; Heath, *History of Greek mathematics*, vol. 1, p. 32. They may have been used for humbler purposes before that, though the Greeks probably made all their

Pythagoras wrote the numbers at all, he probably used decimal symbols similar to the Egyptian ones, but that was simply an adaptation of abacus methods to writ-

---

computations with some kind of abacus or with pebbles. Whichever the method of reckoning, the Greek number words prove that the number base and the abacus were decimal. The Greek word for pebble was *psēphos*. Herodotos uses the expression *psēphois logizesthai* to mean "calculate" in the sentence, "The Greeks write and calculate by moving the hand from left to right" (II, 36). The verb *psēphizo* expresses the same idea. Compare our words "calculus," "calculate," derived from *calculus*, a pebble. For the abacus, see note 20 below. The use of pebbles is of course much older than that of the abacus; the abacus is a machine devised for a better utilization of the pebbles (or their equivalent).

I	II	III	IV	V	VI	VII	VIII	IX	X
II	IV	VI	VIII	X	XII	XIV	XVI	XVIII	XX
III	VI	IX	XII	XV	XVIII	XXI	XXIV	XXVII	XXX
IV	VIII	XII	XVI	XX	XXIV	XXVIII	XXXII	XXXVI	XL
V	X	XV	XX	XXV	XXX	XXXV	XL	XLV	L
VI	XII	XVIII	XXIV	XXX	XXXVI	XLII	XLVIII	LIV	LX
VII	XIV	XXI	XXVIII	XXXV	XLII	XLIX	LVI	LXIII	LXX
VIII	XVI	XXIV	XXXII	XL	XLVIII	LVI	LXIV	LXXII	LXXX
IX	XVIII	XXVII	XXXVI	XLV	LIV	LXIII	LXXII	LXXXI	XC
X	XX	XXX	XL	L	LX	LXX	LXXX	XC	C

Fig. 53*a*

Fig. 53. Tables of Pythagoras. (*a*) Roman; the Egyptian (Roman) system requires only five different symbols. (*b*) Greek; the Greek system requires 27 different symbols; the accents following each numeral have been left out. (*c*) Hindu-Arabic; the Hindu system requires 10 different symbols; its practical value lies in the fact that it adapts abacus methods to script in a deeper manner than the Egyptian.

The three tables are decimal, because no other basis was thought of, except the (Babylonian) sexagesimal for fractions and that only much later (Ptolemy, II–1), the duodecimal in exceptional cases (division of the day, of the pound), and other oddities in metrology and coinage (such as exist to this day in the English systems); see *Isis* 23, 206–209 (1935).

α	β	γ	δ	ε	ϛ	ζ	η	ϑ	ι
β	δ	ϛ	η	ι	ιβ	ιδ	ιϛ	ιη	κ
γ	ϛ	ϑ	ιβ	ιε	ιη	κα	κδ	κξ	λ
δ	η	ιβ	ιϛ	κ	κδ	κη	λβ	λϛ	μ
ε	ι	ιε	κ	κε	λ	λε	μ	με	ν
ϛ	ιβ	ιη	κδ	λ	λϛ	μβ	μη	νδ	ξ
ζ	ιδ	κα	κη	λε	μβ	μϑ	νϛ	ξγ	ο
η	ιϛ	κδ	λβ	μ	μη	νϛ	ξδ	οβ	π
ϑ	ιη	κζ	λϛ	με	νδ	ξγ	οβ	πα	ϙ
ι	κ	λ	μ	ν	ξ	ο	π	ϙ	ρ

Fig. 53*b*

1	2	3	4	5	6	7	8	9	10
2	4	6	8	10	12	14	16	18	20
3	6	9	12	15	18	21	24	27	30
4	8	12	16	20	24	28	32	36	40
5	10	15	20	25	30	35	40	45	50
6	12	18	24	30	36	42	48	54	60
7	14	21	28	35	42	49	56	63	70
8	16	24	32	40	48	56	64	72	80
9	18	27	36	45	54	63	72	81	90
10	20	30	40	50	60	70	80	90	100

Fig. 53*c*

ing. Let us assume, however, that the literal symbols were already available, for this will give us an opportunity of discussing them.

The Greek numerals are 27 in number, divided into three groups of nine each, the first nine designating the units from 1 to 9, the second group the decades from 10 to 90, the third group the centuries from 100 to 900. The symbols used are simply the Greek letters (with an accent to the right of each) in their alphabetic order; but since there are only 24 letters in the Greek alphabet, three obsolete letters were added, one in each group, to wit, digamma or stigma for 6, koppa for 90, and swampi for 900. Moreover, the first ten letters (including stigma) were used also to designate the thousands from 1000 to 10000 (in that case the accent is put to the left of the letter, below the line). Not only were the Greeks obliged to remember three times as many symbols as we are, but many simple relations were hidden by that multiplicity. Consider the fundamental distinction between odd and even numbers. It is easy for us to remember that even numbers end in 0, 2, 4, 6, 8. How was it for the Greeks? An odd number might end with any one of the 27 symbols! (Fig. 53).

The table of multiplication that is called in many languages the Pythagorean table (*mensula Pythagorae*) was certainly not devised by Pythagoras. The earliest example of it known to me occurs in the *Arithmetica* of Boethius (VI–1), printed in Augsburg in 1488.[18] There may be earlier ones in manuscripts written possibly with Roman numerals, for the Hindu-Arabic numerals were hardly introduced into the West before the twelfth or thirteenth century, and there was so much resistance to their use that they did not obtain any popularity until much later.

The table of Pythagoras in Hindu numerals is very clear. One sees at once that the lines (or columns) 2, 4, 6, 8, 10 contain only even numbers, that in line (or column) 5 every number ends with 5 or 0 (it is true that in the Greek script half of the numbers end with ε). Neither Pythagoras nor even the latest Pythagorean of antiquity knew of Hindu numerals (or of equivalent ones); hence it is probable that the *mensula Pythagorae* is but a late medieval creation, perhaps not much older than the printed Boethius.[19]

The early Pythagorean ideas on numbers were almost certainly restricted to such as could be illustrated by means, not of numerals, but of counters or pebbles. That simple method brought to light facts of transcendent meaning. Pythagorean arithmetic is the root not at all of our arithmetic or of the art of reckoning, but rather of the present theory of numbers.

The reader, especially the one interested in the sociology of science or in the materialistic interpretation of history, may object that our conclusion does not tally with all that we know of the early and strong proclivities of Greek people to trade. After all, trading and every form of business transaction necessitate plain arithmetic in our sense; from the point of view of sellers and buyers (that is, the whole population), the theory of numbers is a luxury. One might answer that religion, philosophy, and the humanities are also luxuries from the mercantile point of view. Moreover, arithmetic (reckoning) was developed and intensely cultivated by the Greeks, but in an empirical manner. We may be sure that the

[18] Facsimile reproduction in *Osiris* 5, 138 (1938).

[19] Johannes Tropfke, *Geschichte der Elementar-Mathematik* (Berlin, ed. 3, 1930), vol. 1, p. 144. David Eugene Smith, *History of Mathematics* (Boston, 1925), vol. 2, p. 124 [*Isis* 8, 221–225 (1926)].

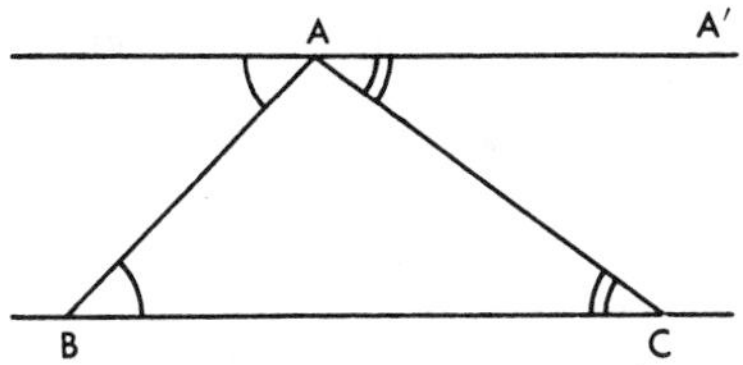

Fig. 54. Angles between parallels.

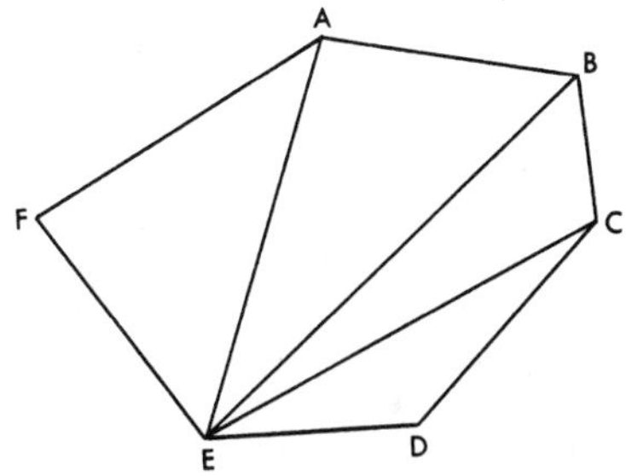

Fig. 55. Internal angles of a polygon.

average Greek dealer knew how to count rapidly and exactly, either in his head or with the help of some kind of abacus.[20] However efficient he might be in that art, he never imagined that he was doing mathematical work; on the other hand, the ancient mathematicians never thought of reckoning as a part of their own field. Even today it is only ignorant people who confuse mathematics with reckoning or accounting, or who mistake tellers for mathematicians.[21]

GEOMETRY

Among the geometric achievements of the Pythagorean school that seem early enough to be creditable to Pythagoras himself I would choose the following.

The interior angles of a triangle are equal to two right angles; this could be proved almost immediately if one but knew that when a line cuts two parallels the alternate angles are equal (Fig. 54). If AA' is parallel to BC, the three angles of the triangle are equal to the two right angles in A. Pythagoras may have extended that proof to polygons of more sides (Fig. 55). In the hexagon $ABCDEF$, join EA, EB, EC. The sum of the internal angles of the hexagon is equal to that of the internal angles of the four triangles, or eight right angles. More generally, for a polygon of n sides the sum of the internal angles is equal to $(2n - 4)$ right angles. The sum of the external angles (each being the supplement of an internal one) is equal to $2n - (2n - 4) = 4$ right angles; it is thus independent of the number of sides.

[20] The best history of the abacus is in Smith, *History of Mathematics*, vol. 2, pp. 156–195. He distinguishes three different types of abacus – the dust board, the table with loose counters, and the table with counters fastened to lines. The word abacus derives from the Greek *abax*, which is clearly a foreign word, probably Semitic (the Hebrew *abaq* means dust). The first use of *abax* occurs in Aristotle (*Atheniensium respublica*, last chapter), where it refers to a reckoning board for counting votes. Sextos Empiricos (II–2) mentions in his treatise against mathematicians (IX, 282) an *abax* which is a board sprinkled with dust for drawing geometric diagrams. It is probable that some kind of abacus was already used by Babylonians and by early Chinese. No Greek monument has come down to us except the white marble abacus (1.49 × 0.75 m) found in the island of Salamis and kept in the Epigraphical Museum at Athens (Smith 2, 162–164). It is not datable and its large size suggests public, ceremonial (?) use. Heath has argued (1, 51–64, 1921) that the Greeks had little need of the abacus for calculations and shows how these could be made with Greek numerals. See also Carl B. Boyer, "Fundamental steps in the development of numeration," *Isis* 35, 153–168 (1944). Heath's and Boyer's arguments fail to convince me.

[21] The most blatant form of that confusion occurs with reference to the "lightning calculators" who give exhibitions of their prodigious ability. Newspaper reporters and other people will often speak of the "mathematical genius" of those calculators. It is mathematical, if you please, but of a relatively low order.

Common experience in flagging or tile flooring helped to show that the only regular polygons by which the space can be filled without gaps are the equilateral triangle, the square, and the regular hexagon. The proof was easy, for each angle of these regular polygons measures, respectively, two, three, or four thirds of a right angle. The space around a point in one plane, equaling four right angles, can be filled with six triangles, four squares, or three hexagons (Fig. 56).

Did Pythagoras know the "Theorem of Pythagoras," that is, that the square built upon the hypotenuse of a right-angled triangle equals the sum of the squares built upon the two other sides of that triangle?[22] Why not? This can be seen almost intuitively in various ways.

For example, suppose we have two unequal squares (Fig. 57) such that the smaller one EF^2 is inscribed in the larger one AB^2 (that is, until its four apices touch the four sides of the larger square). It is clear that the four triangles *EAF*, . . . outside the smaller square are equal. Draw EE' parallel to *AB* and FF' parallel to *BC*, intersecting in *O*; we thus divide the square AB^2 into four parts – two equal rectangles and two squares EO^2 and FB^2. Then the area of the largest square AB^2 can be expressed in two ways:

$$\begin{aligned} AB^2 &= EF^2 + 4 \text{ triangles} \\ &= EO^2 + FB^2 + 2 \text{ rectangles.} \end{aligned}$$

But each of the rectangles equals two of the triangles, hence

$$EF^2 = EO^2 + FB^2 = AF^2 + AE^2.$$

Q.E.D.

The demonstration is so easy that it may have been made previously and independently by the Egyptians, the Babylonians, the Chinese, and the Hindus. The possibility of Egyptian priority has been considered in Chapter II; we need not consider the other possibilities, for they never come close to certainty. Pythagoras may have been the first to prove the proposition (not simply to see that it was true), or his proof may have been more conscious, or more rigorous, using a method equivalent to Euclid's. It was said that Pythagoras had sacrificed an ox to celebrate that discovery, or was it perhaps to celebrate the discovery of particular triangles (sides $3n$, $4n$, $5n$) wherein the geometric proof was easily completed by a numerical verification?

[22] Euclid, I, 47.

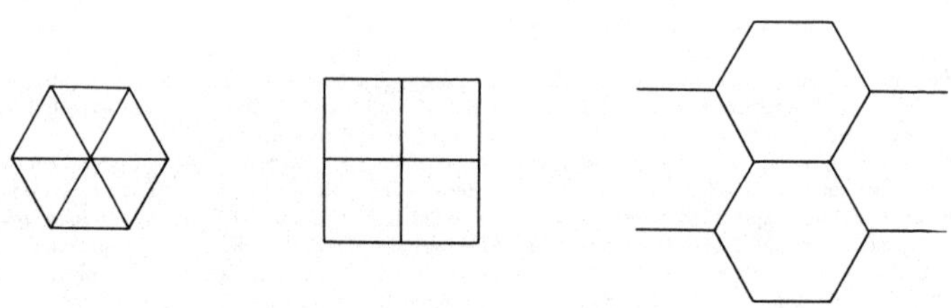

Fig. 56. Fitting of regular plane polygons.

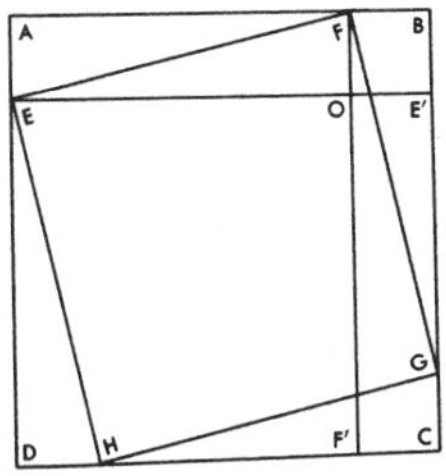

Fig. 57. The Pythagorean theorem.

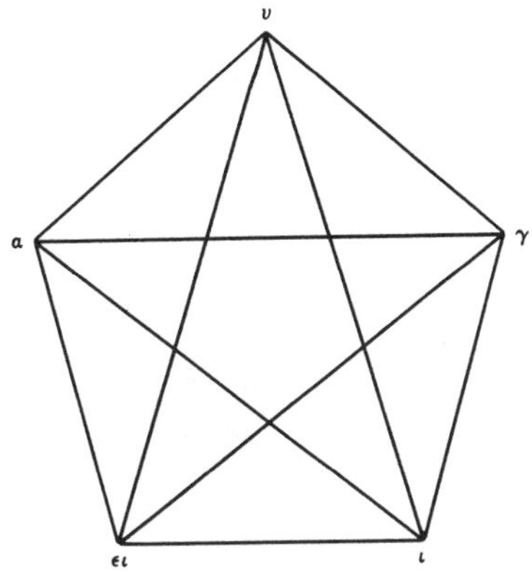

Fig. 58. The Pythagorean pentagram.

He may have initiated the geometric problems concerned with the finding of an area equal to another (say a square equal to a parallelogram) or with the application of areas (*parabolē tōn chōriōn*), the one exceeding the other (*hyperbolē*) or falling short of it (*elleipsis*) by a given quantity. In the course of time those problems led to the geometric solution of quadratic equations, and curiously enough the Greek terms just quoted, probably later than Pythagoras, were afterward applied to the three different species of conic sections.

The geometric ideas and theorems that we are tempted to ascribe to Pythagoras could not have been proved easily, in spite of their simplicity, without the use of letters to designate the lines involved. We have used letters in our own explanations, without thinking of it, because it is very difficult to do otherwise. It does not follow that Pythagoras used letters. For example, he might have proved the theorem bearing his name by drawing lines on the sand and pointing to the lines and areas with his fingers. It is only when the proof is written that letters (or other symbols) become indispensable.

According to a tradition of which we hear a late echo in Lucian (120–180), the Pythagoreans used as a symbol of mutual recognition the pentagram,[23] to which they gave the name "health." [24] The five letters of that name (ὑγίεια, *hygieia*) were the five apices of that symbol (Fig. 58).[25] This is perhaps the oldest example of the application of letters to various points (or other parts) of a geometric figure. It may be older than the use of letters for ease in geometric demonstrations, or it may have been suggested by that very use.

Pythagoras or his immediate disciples were already acquainted with some regular solids; the cube and pyramid (tetrahedron) were easy enough to conceive

[23] The pentagram is a concave polygon of five sides, a five-pointed star. The regular pentagram is easily derived from the regular pentagon by drawing the diagonals of the latter. In medieval and later times the pentagram was often called pentaculum (pentacle) and pentalpha.

[24] Lucian: A slip of the tongue in salutation (*Hyper tu en tē prosagoreusei ptaismatos*). See Lucianus, ed. Carl Jacobitz (Leipzig, 1836), vol. 1, p. 448, or the English translation by H. W. Fowler and F. G. Fowler (Oxford, 1905), vol. 2, p. 36. The figure is called *to pentagrammon*. The same chapter contains a reference to the Pythagorean quaternion (*hē tetractys*) used as a sacred oath.

[25] The diphthong *ei* counting for one letter.

or to build, the octahedron not difficult. Their knowledge of the pentagram does not prove that they were able to construct a regular pentagon; but even if they did not know the geometric construction, they could always divide a circumference empirically into five equal parts. Moreover, if, after having built a regular pyramid and a regular octahedron, they continued to play with equilateral triangles and put as many as five together (one apex being common to all five) they thus built one of the solid angles of an icosahedron. Even if they did not complete the icosahedron they must have recognized that the base of that solid angle was a regular pentagon. Playing with regular pentagons, they may have succeeded in building a dodecahedron. There is much guesswork in all this, however, and we shall postpone further discussion of the regular solids, the "Platonic figures," until later.

ASTRONOMY

We must be at least as discreet when we discuss Pythagorean astronomy as we were for geometry. It cannot be our purpose to catch the new ideas in their embryonic stage, as it were, for that is naturally impossible. It is safer to wait until they have become sufficiently clear and definite. Thus in this section we shall indicate only a few general ideas probably prior to Philolaos (V B.C.), to whom the earliest Pythagorean astronomic writings are ascribed.

The idea that the earth is a sphere is probably as old as Pythagoras. How did he reach such a bold conclusion, one may wonder? He may have observed that the surface of the sea is not flat but curved, for as a distant ship approaches one first sees the top of its mast and sail and the rest appears gradually. The circular edge of the shadow cast in an eclipse of the Moon would also suggest the spherical shape of the Earth, but that is a more sophisticated kind of observation, implying an understanding of eclipses that had not yet been attained in the sixth century. It is more probable that as soon as the hypothesis of a flat earth had been rejected, the sphericity of the earth was postulated rather wildly, on insufficient experimental grounds. The earth cannot be flat, therefore it ought to be spherical. Was not the starry heaven visibly part of a sphere? Were not the disks of Sun and Moon circular? And was any volume or surface comparable in symmetry and beauty to those of the sphere? This fundamental Pythagorean idea was an act of faith rather than a scientific conclusion. Does not every scientific hypothesis start that way? This hypothesis made the theory of eclipses possible, and, conversely, the development of that theory, the observations that it suggested, repeatedly confirmed the initial assumption.

The dogma of spherical perfection and its cosmologic consequences may be considered the kernel of early Pythagorean science. It was postulated that the celestial bodies are of spherical shape and that they move along circular paths, or as if they were attached to spheres. The earth, naturally enough, was supposed to be immobile in the middle of all, its center being the center of the universe. The movement of all the spheres is uniform, like that of the heavens. How could it be other than uniform?

The Babylonians had been satisfied to describe as accurately as possible the movements of the planets, and to account for them by numerical tables. Pythagoras, acquainted with Milesian physiology, was no longer satisfied with descriptions.

He wanted to explain the phenomena, to justify them. The planets cannot be "errant"[26] bodies; they must have circular and uniform movements of their own. That opinion is ascribed to Alcmaion as well as to Pythagoras; whoever held it first, it represents a great step forward in the Pythagorean doctrine. The stars as seen from a position north of the equator move clockwise and with clocklike regularity; the planets (meaning the Sun, Moon, and our planets) do not wander erratically but they have counterclockwise motions of their own. If one could but analyze those complicated motions they would be reduced to uniform circular ones. The whole of Greek astronomy grew out of that arbitrary conviction.[27]

Another conviction established itself gradually in the same obscure, mystical manner. Out of Milesian monism emerges a new kind of dualism. There is a substantial difference between the celestial world on the one hand, eternal and divine, perfect, unchangeable, the elements of which move in circles without angular acceleration, and the sublunar one (*ta hypo selēnēn*) on the other, subject to endless changes, decomposition, decay, and death, and wherein the motions are capricious and irregular. The superlunar world is the world of the immortal gods and perhaps of the souls. The sublunar one is the abode of things either lifeless or mortal.[28]

This Pythagorean dualism influenced scientific thought until the time of Galileo and even later. Its influence upon religion was hardly less important; we shall discuss some aspects of it apropos of the *Epinomis* later on. It will suffice now to remark that the sidereal religion, which was to be the core of astrology, derived straight from those Pythagorean fancies, added to the Chaldean ones.

MUSIC AND ARITHMETIC

The stories told about Pythagoras' musical experiments are hard to believe, except one. Bearing in mind that in his time the Greeks and other ancient peoples had already acquired considerable familiarity with stringed instruments, his experiments with strings are quite plausible.[29] Of course, every citharist would know that he could obtain different sounds and pleasing combinations of sounds by pinching the strings at certain places or changing the length of their vibrating parts. Pythagoras may well have repeated such experiments more methodically and with the detachment of a scientist rather than with the intuitive subjectivity of an artist, and he may have discovered that the uniform strings the lengths of

[26] As their Greek names implied; *planaō* means to cause to wander, to mislead; *planētēs* is a wandering, erratic, misleading, body.

[27] The conviction was arbitrary as far as the nature of the planetary motions was concerned. The Babylonian ephemerides, however, had proved that those motions were not erratic but predictable.

[28] The distinction between celestial mechanics and sublunar mechanics was criticized by a few medieval thinkers, such as Buridan (XIV-1) and Oresme (XIV-2), but it was not completely invalidated except by Newton. Then it took another form, the distinction between theoretical and practical mechanics. One of the founders of thermodynamics, Rankine, found it necessary as late as 1855 to show the artificiality of that distinction (*Introduction*, vol. 3, p. 1843).

[29] Two stringed instruments are mentioned in Homer, the *phorminx* and the *citharis* (the form *cithara* is later). A third word, *lyra*, is post-Homeric. It is probable that these three words represented essentially the same kind of instrument. Terpandros of Lesbos, the "father of Greek music" (fl. *c.* 700–650), is said to have increased the number of strings up to seven, or to have canonized the heptachord and the musical system based upon its use. The great antiquity of those stringed instruments in Greek lands (not to speak of Babylonia and Egypt) is proved by the ascription of their invention to gods, the lyre to Apollo and the cithara to Hermes. Empty tortoise shells were primitively used to subtend the strings, or, being covered with skin, to function as a resonance box.

which were in the relation 1:¾:⅔:½ (or 12:9:8:6) produced harmonious sounds. The ratios of the vibration numbers 12:6, 12:8, and 8:6 are the intervals that we call octave, fifth, and fourth (in Greek, *diapasōn*, *diapente*, and *diatessarōn*).[30]

That discovery directed Pythagoras' thought to the ratios themselves, that is, to the theory of means and proportions. Or was it the other way round, and did his familiarity with proportions draw his attention to the musical intervals? Pythagoras was certainly not the first to think of arithmetic means; and geometric means ($a{:}b = b{:}c$) were natural enough to be conceived very early. He was perhaps the introducer of a new kind of mean, called "harmonic" (*harmonicē analogia*) wherein the three terms are such that "by whatever part of itself the first exceeds the second, the second exceeds the third by the same part of the third." [31] More clearly, if b is the harmonic mean of a and c, we can write $a = b + a/p$, $b = c + c/p$; hence $a/c = (a - b)/(b - c)$, or $1/c - 1/b = 1/b - 1/a$. (If b were the arithmetic mean of a and c, we would have $a - b = b - c$. One sees why the harmonic proportion was also called subcontrary, *hypenantia*.)

The numbers 12, 8, 6 quoted above form a harmonic proportion. The cube was called a "geometric harmony" (*geōmetricē harmonia*) because it has 12 sides, 8 angles, and 6 faces.[32] The theory of means was susceptible of many extensions, which were fully exploited by the Pythagorean arithmeticians in later times.

The idea of harmonic proportion was soon extended to astronomy. The heavenly spheres were supposed to be separated by musical intervals and the planets emitted different notes in harmony. According to Hippolytos (III–1), "Pythagoras maintained that the universe sings and is constructed in accordance with harmony; and he was the first to reduce the motions of the seven heavenly bodies to rhythm and song." [33] St. Hippolytos is a very late witness and not a reliable one. Those mathematical fantasies were potentially in Pythagoras' mind; it is improbable that he formulated them as neatly as Hippolytos put it, but the formulation took place in the fifth or fourth century, in or before Plato's time.[34]

MEDICINE. ALCMAION AND DEMOCEDES

The earliest medical center of Greece that might be called a school, a theoretical school, was perhaps the one that developed in Croton. Its origin may be prior to Pythagoras, but it more probably synchronized with the Pythagorean school. The writings of the first teacher, Alcmaion of Croton, son of Peirithoos, are lost, but as far as we can judge from the fragments and the doxography, he was a disciple of Pythagoras. Some medical ideas are ascribed to Pythagoras himself, yet it is simpler to consider Alcmaion as the medical teacher of the sect.

The title of Alcmaion's treatise *peri physeōs* suggests a Milesian influence, and he may have been a Milesian (or Ionian) refugee, like so many of his contempo-

[30] *Hē diapasōn* (*hē dia pasōn chordōn symphōnia*), *hē dia pente*, *hē dia tessarōn*.

[31] Thus defined by Porphyry in his commentary on Ptolemy's Harmonics. Diels: Vorsokratiker (1[3], p. 334). Compare Plato's definition of the harmonic and arithmetical means in Timaeos (36 A).

[32] Statement ascribed to Philolaos by Nicomachos (I–2) in his *Introduction to arithmetic*, II, 26, 2, Martin Luther D'Ooge's edition (New York 1926), p. 277.

[33] The fact that there are seven planets and seven tones in the heptachord must have deeply impressed the early Pythagoreans and increased their faith in hebdomads. See next section.

[34] Hippolytos, *Philosophumena*, I, 2, 2. Plato, *Republic*, 617B (Myth of Er); *Timaios*, 325B Aristotle, *Metaphysics*, A 5, 986 A 1; *De caelo*, 290 B 12. Aristotle refutes the theory.

raries whom the fear of the Persians or local tyrannies had driven out. He investigated sense organs, especially those of vision, and, if we may believe Chalcidius (IV–1), was the first to attempt a surgical operation on the eye.[35] He claimed that the brain was the central sensorium and that there were some pathways or passages (*poroi*) between it and the sense organs; if those passages were broken or stopped, say by a wound, communication was interrupted. These pregnant views – the first seeds of experimental psychology – were amplified in the following century by Empedocles and the atomists.

Alcmaion may have been the initiator of another psychologic doctrine to which later Pythagoreans attached more and more importance. The souls are comparable to the celestial bodies and thus move eternally in circles. Circularity and immortality are equated. On the other hand, men die because they cannot return to their beginning,[36] the cycle of life is not a circle but an unclosed curve. Life, we would interpret, is a running-down process; the stars and souls do not run down but turn eternally around.

Alcmaion's main medical theory was that health is an equilibrium of forces (*isonomia dynameōn*) in the body; when one of the forces dominates the equilibrium is upset and we have a state of monarchy (*monarchia*) and disease.

Another physician of Croton, Democedes son of Calliphon, obtained considerable fame. He was for a time in the service of Polycrates, tyrant of Samos (d. 522), and later flourished in Susa at the court of Darios (king of Persia, 521–485). The great king had dislocated his foot in alighting from a horse; Democedes succeeded in healing it after Egyptian physicians had failed to do so and he used his prestige to beg the lives of his unfortunate colleagues who were about to be impaled. Then he cured Darios' wife, Cyros' daughter, Atossa,[37] who was frightened by a tumor growing on her breast. He took advantage of a political mission forced on him by the king to sail away from Sidon (in Phoenicia) and return to his native country. Persian emissaries tried to persuade the magistrates of Croton to surrender the fugitive in order that they might bring him back to their lord. Democedes was finally allowed to remain because of his wedding with the daughter of the athlete Milon, who was the most illustrious son of Croton.[38] It is typical of Greek life to come across this reference to athletics mixed up with the beginnings of nonanonymous medicine.

The first eleven chapters of the Hippocratic treatise on hebdomads, *De hebdomadibus* (*Peri hebdomadōn*) set forth a number of cosmologic, embryologic, physiologic, medical remarks on the importance of the number seven: the embryo takes a human form on the seventh day, some diseases are dominated by hebdomadal

[35] Chalcidius' commentary on Timaios, ch. 244, *primus exsectionem aggredi est ausus*, F. G. A. Mullach, *Fragmenta philosophorum graecorum* (Paris, 1867), vol. 2, p. 233. Of course *exsectio* might refer to an anatomic dissection, but then why *ausus*? There was nothing venturesome in the dissection of a dead eye.

[36] Aristotle, *Problemata*, 916 A 33. *Tus anthrōpus phēsin Alcmaiōn dia tuto apollysthai oti u dynantai tēn archēn tō telei prosapsai.*

[37] Atossa is the queen immortalized as the main character in Aischylos' play *The Persians*, the action of which takes place in Susa, residence of the Persian kings.

[38] Herodotos, III, 125, 129–138. Milon of Croton was one of the most famous athletes of ancient Greece, whose exploits became legendary. He was six times the champion wrestler in the Olympic games, and six times also in the Pythian ones. His countrymen admired him so much that they put him in command of the army that defeated the Sybarites in 511 and utterly destroyed their city. Sybaris was a Greek colony, in the Gulf of Tarentum, north of Croton. The Sybarites' love of pleasure and luxury is immortalized in the English words sybarite and sybaritic.

cycles, there are seven planets, etc. That text is of early date, not later than the sixth century,[39] yet it is not Pythagorean but definitely Ionian (Cnidian?). This would show that mystical extensions of the idea of number were not restricted to Magna Graecia. But why should they be? Mesopotamia might well have been the cradle of such fancies. We should not forget that Pythagoras himself was a Samian.

For bibliography on the *Peri hebdomadōn* see *Introduction,* vol. 1, p. 97. That text is not available in Greek, except a fragment, but has come down to us in an Arabic translation by Ḥunain ibn Isḥāq (IX–2)[40] and in a barbaric Latin translation. The Latin text may be found in Littré, *Oeuvres complètes d'Hippocrate* (10 vols.; Paris, 1839–1861), vol. 8, pp. 634–673; vol. 9, pp. 433–466. The Arabic text was translated into German by Christian Harder, "Zur pseudohippokratischen Schrift *Peri hebdomadōn sive To prōton peri nusōn to microteron,*" *Rheinisches Museum 48*, 433–447 (1893), and from German into Italian by Aldo Mieli, *Le scuole ionica, pythagorica ed eleata* (Florence, 1916), pp. 93–115 [*Isis* 4, 347–348 (1921–22)]. See also Joseph Bidez, *Eos* (Brussels: Hayez, 1945), pp. 126–133 [*Isis* 37, 185 (1947)]; the idea of the microcosmos adumbrated in that treatise is probably of Iranian origin.

NUMBERS AND WISDOM

If one puts together the discoveries that may be ascribed to Pythagoras, or at least to the early brotherhood, in the fields of arithmetic, geometry, astronomy, and music, one is startled by the predominance of numerical concepts. Would one not expect that predominance to be even more startling for those early cogitators than it is for us? And mystically minded as they undoubtedly were, is it very surprising that they finally jumped to a bold and great conclusion? Numbers are immanent in things. To the Ionians, who had postulated a single material basis of nature, and to Anaximandros, who postulated a metaphysical basis, the indefinite, Pythagoras could now triumphantly retort: Numbers are the essence of things. We need not try to investigate that idea more deeply, because it is not likely that Pythagoras had carried it very far, and chiefly because it does not bear analysis. It is valid only as long as it remains in the nebulous form that Pythagoras gave to it. Later Pythagoreans established all kinds of relations between definite numbers and indefinite ideas, but those elaborations were of their nature arbitrary and illusive, while the general concept was (and remains) very impressive.

That numerical philosophy had far-reaching consequences, which are still felt today, in two directions, good and evil. It initiated the quantitative study of nature on the one side, and number mysticism, numerology, on the other. One might claim that the physicists of all ages, the natural philosophers, have been constantly allured by the hope of finding new numerical relations. It is as if they had heard old Pythagoras whisper into their ears: The number is the thing. We would rather say that mathematical relations reflect, if they do not reveal, the essence of reality. As to number mysticism, it is the caricature of the same concept, its reduction to absurdity by the extravagance of ignorant and foolish men.

[39] According to W. H. Roscher; but Franz Boll would place it later, not before 450 (*Introduction,* vol. 1, p. 97). W. H. S. Jones, *Philosophy and medicine in ancient Greece* (Baltimore: Johns Hopkins University Press, 1946), pp. 6–10 [*Isis* 37, 233 (1947)].

[40] It is not mentioned in Ḥunain's bibliography edited by Gotthelf Bergsträsser, *Ḥunain ibn Ishāq über die syrischen und arabischen Galen-übersetzungen* (Leipzig, 1925) [*Isis 8*, 685–724 (1926)] but may have been translated by a member of Ḥunain's school.

THE PURSUIT OF KNOWLEDGE IS THE GREATEST PURIFICATION

If numbers are the essence of things, the better we understand them the better shall we be able to understand nature. The theory of numbers is the basis of natural philosophy. It would seem that the Pythagorean brotherhood drew that conclusion early. Vulgar people deal with numbers only because of the need for measuring and counting salable objects and of computing profits, but Pythagoras taught that there was a far deeper reason for being interested in numbers. We should try to penetrate the secrets of nature. Such disinterested efforts raise human life to a higher level, closer to the gods.

The desire for purification and the desire for salvation are innate in the best of men.[41] They had been cultivated before Pythagoras' days in the Orphic mysteries and other religious ceremonies, but Pythagoras was probably the first to associate them or even to confuse them with the desire for knowledge, especially for mathematical knowledge, symmetry, and music. According to the greatest musicologist of antiquity, Aristoxenos of Tarentum (IV–2 B.C.), the Pythagoreans used music to purge the souls even as they used herbs to purge the bodies. We may safely assume that that remark applies to Pythagoras himself or to his earliest (and most scientific) disciples. He went even further when he proclaimed that the pursuit of disinterested knowledge is the greatest purification. The highest kind of life is the theoretical or contemplative.[42] These views are the seeds of others fully set forth in the *Phaidōn* and in the *Nicomachean Ethics*; they are also the seeds of pure science. It was the strange destiny of Pythagoras to be at one and the same time the founder of science and the founder of a religion. He was the first to assert that science is valuable, irrespective of its usefulness, because it is the best means of contemplation and understanding. He was the first to connect the love of science with sanctity. He might well be feted as the patron saint of men of science of all ages, the pure theorists, the contemplators.

[41] The words designating them are all ancient words: *sōtēria, lysis, apallagē* for salvation; *catharsis, catharmos, lysis* for purification.

[42] We have to use two terms for one in Greek, *theōrein*, used for the contemplation of a spectacle, such as the Olympic games, or the contemplation of truth; *theōrēma* may mean a spectacle but also a speculation; *theōria* is a viewing or a theory. Our words theorem, theory, theoretical have lost the early concrete senses and preserved only the abstract ones.

PART TWO

THE FIFTH CENTURY

IX

GREECE AGAINST PERSIA. THE GLORY OF ATHENS

THE PERSIAN WARS

The eight preceding chapters represent a long series of centuries – indeed, many millennia – and many countries – the whole of the ancient world. The rest of this volume, some two-thirds of it, deals with only two centuries, and the narrative is centered upon a single country, a very small one, Attica, or rather upon its main city, Athens.

That city existed long before the sixth century, and we have already referred to it; yet it was one of the last city-states to appear in Greek history, and it might be considered a kind of upstart, for example, by the men of Sparta, where the Dorian type and tradition had preserved their greatest purity.[1] At any rate, Athens developed rapidly, and within little more than a century it had become eminent and strong enough to be the protagonist of the Hellenic world in its life-and-death struggle against Persia; after the victory it was for half a century the leading nation in that world, but what is far more important, it has remained ever since the best symbol of Hellenic culture. When we think of that culture we are thinking most of the time of Athens, and the words Athens and Greece are almost interchangeable in our grateful reminiscences.

These things require a bit of explanation. By the end of the sixth century the Achaimenian empire[2] dominated the best part of the known world. It included the whole of Western Asia (except Arabia) and even Egypt.[3] Persian trade was well organized and ramified in many directions; the competition with Greek settlements was especially intense in the Black Sea, the straits leading to it, and the

[1] One might say roughly that the Greek people with whom we are dealing in this volume were a mixture of Mediterranean men (Cretans, Achaians, etc.) with various invaders, chiefly the Dorian invaders who had come from the north. That question is very complicated and perhaps insolvable. Good summary by A. J. B. Wace in *Companion to Greek studies* (Cambridge, ed. 3, 1916), pp. 23–34.

[2] Lack of space and the need of dramatic unity preclude an account of Achaimenian culture in this book. It will suffice to recall that the first of the Achaimenidai was Cyros (ruled 559–529) and the last, Darios III, defeated by Alexander the Great in 331. The dynasty lasted 228 years. One would have to speak of its achievements in a history of ancient art, and even in a history of ancient education (though Persian education as explained in Xenophon's *Cyrupaideia* is largely fictional and utopian), but the historian of science may overlook it without loss, at any rate on the scale of this book. See the history written by the late Albert Ten Eyck Olmstead (1880–1945), *History of the Persian empire* (596 pp., ill.; Chicago: University of Chicago Press, 1948).

[3] Egypt was under Persian rule from 525 to 332.

Eastern Mediterranean. The Persians were able to combine the extensive caravan trade of Asia and North Africa with the sea trade of the Phoenicians. In their rivalry with Greece and their growing hatred of her, the Phoenicians were the natural allies of Persia. Now Phoenician colonies extended from one end of the Mediterranean to the other. Thanks to them, Persian trade covered the whole sea, as is witnessed by the discovery of golden darics (Persian coins) in many places around it. The Greek colonies were numerous and still flourishing, but were everywhere outflanked or encircled by Persian or Phoenician outposts. That situation was ominous, though perhaps less so for the contemporary Greeks, who could not measure its gravity as easily as we can when we contemplate the excellent maps that we owe to the accumulated efforts of many investigators.[4]

The pressure was especially severe in the Ionian colonies, whose hinterland was under Persian control and where border incidents were bound to occur repeatedly and to cause mutinies and repressions. The Ionian rebellion began in 499. In the following year Sardis (capital of the satrapy of Lydia) was taken by surprise and destroyed by the Greeks, who were duly punished on their return march, near Ephesos. The revolt spread to other colonies in Cyprus and Asia. Its main center was in the illustrious city of Miletos, which was captured by the Persians "in the sixth year of the revolt" (494) and utterly ruined. In 493 Chios, Lesbos, Tenedos were overrun. The situation was becoming dangerous, and Themistocles (*c.* 514–*c.* 460), one of the first Athenian statesmen to realize its gravity, persuaded his fellow citizens to prepare for defense by the building of a permanent fleet and the establishment of a naval arsenal at the Peiraieus (Athens' harbor). We need not tell the rest of this story, which is so complex that an intelligible summary of it would take considerable space. It must suffice to recall the heroic deeds of Marathon where the Persian army of Darios was defeated in 490,[5] the glorious rearguard defense of the pass of the Thermopylae in 480 (Leonidas and his 300 Spartans were annihilated), and the naval victory of Salamis in the same year, where the Persian fleet was completely defeated by the Athenian one; the Persian king, Xerxes, witnessed the tragedy from the throne erected for him on a hill of the Attic shore. In the following spring, the Persians revenged themselves by invading Attica; they sacked Athens and set fire to the Acropolis including the old Parthenon. In the summer, they suffered a new defeat at Plataiai (in Boeotia, near the Attic border) and at about the same time (August 479) another Persian fleet was defeated by the allied Greek fleet off Mycale (on the Ionian coast, opposite Samos). The independence of Greece was now secured.

The importance of that conflict between Asia and Europe can hardly be exaggerated; it is one of the greatest conflicts in the history of the whole world and one of the most pregnant; the final victory of the Greeks determined the future. (If the Persians had won, the future would have been very different; it is not possible, nor would it be profitable, to imagine what might have happened.) To call it a conflict between Asia and Europe, however, or between East and West, however true on

[4] G. Sarton, "The unity and diversity of the Mediterranean world," *Osiris 2*, 406–463 (1936), chiefly pp. 422–423.

[5] A Greek soldier ran from Marathon to Athens, to bring the happy tidings. In remembrance of those heroic deeds (including the runner's), long distance "marathon races" take place in many countries, for example, in Boston, every year. The distance is 26 miles 385 yards (Webster); this is supposed to be the distance of Marathon from Athens; I do not know how it was computed.

the surface, is misleading. Many of the Greeks had lived for generations in Asia or Egypt, and on the other hand the Phoenicians, the naval allies of Persia, were scattered all over the Mediterranean and could threaten the Greeks from the west. Neither was it a conflict between Aryans and Semites, for the Persians were as Aryan as the most Aryan Greeks, while their allies, the Phoenicians, were Semites. The Achaemenian empire was a conglomeration of all the races and nations of Western Asia, which had been blended repeatedly for millennia. The main language of the empire was Aramaic, a Semitic language. It is more correct to consider the conflict one between Asiatic despotism and Greek democracy; democracy was vindicated, and though that first attempt did not last very long, it remained an example that the world never forgot.

The freedom of Greece had not been defended by all the Greek nations but only by a few of them, primarily the Ionian colonies, Athens and Sparta (remember that the martyrs of Thermopylae were Spartans). Athens emerged as the leader. How shall we account for that? Were the Athenians a special and superior race of Greeks? At the beginning, they were mainly or seemed to be autochthonous and they wore a golden cicada in their hair to proclaim that fact,[6] yet the location of Attica on the easternmost part of the Greek peninsula was extremely favorable to every kind of commerce, especially with the colonies of Ionia and the Aegaean islands. Ionians streamed into Athens, and Athenian culture was very strongly influenced by Ionian models. To my mind, that is the main explanation of the Athenian supremacy — Ionian intelligence and versatility grafted upon the old Attic stock (history gives many examples of such graftings and of their fruitfulness). Moreover, Attica beckoned other foreigners, and they came to her from many places and races and were gradually amalgamated. The very language of the Athenians betrayed their cosmopolitanism,[7] and that language in its turn was another means of cultural unity. The national prestige of Athens was already recognized before the end of the sixth century, in spite of the fact that other cities were more powerful. After Salamis this prestige increased considerably; Athens became the leading city, and its goddess, Pallas Athene, the best symbol of Hellenism. Athens was the main political, commercial, and cultural center, but not by any means the only one. Others flourished in Thebes, Corinth, Sicyon, Megara, even in Macedonia, Ionia, Cyrenaica, Italy, and Sicily. The Greek world was very wide and diversified, and in the course of time every corner of it produced great men of its own. Yet more and more of these men, if they were not born in Athens, were impelled to come to her for their education, or to accomplish their purpose, exert their influence, and obtain the final consecration of their merit.

FIFTY YEARS OF RELATIVE PEACE

During the fifty years between Salamis and the beginning of the Peloponnesian War, the supremacy of Athens increased considerably and seemed to be established

[6] *Introduction*, vol. 3, p. 1188.

[7] Curious remarks on the Athenian dialect are made in the *Constitution of Athens*, II, 8, an extremely interesting book formerly ascribed to Xenophon, but somewhat earlier (*c.* 431–24). Says the unknown author, "As they have opportunities of listening to many dialects, the Athenians have borrowed from each. Whereas the other Greek peoples have adopted each of them its own language, mode of living and dress, the Athenians use a mixed language the elements of which have been borrowed from other Greeks and barbarians." See Greek–English edition of that text with commentary by Hartvig Frisch, *Constitution of Athens* (Copenhagen: Gyldendal, 1942).

for ever. Athens was the head of the Ionian league, which was gradually transformed into her own maritime empire. The Athenian and Attic festivals were the most famous and the most popular of Greece. In spite of its national eminence and its cosmopolitanism, the Athenian culture remained original and spontaneous. It was animated by pride in the present and faith in the future, naïve patriotism, and a good deal of self-conceit, mitigated by the love of discussion, such as is possible in times of peace and prosperity. Those fifty years were the golden age of Athens; we might compare them with the Elizabethan age of England, a period of about equal length (45 years, 1558–1603) and of equal enthusiasm. The last thirty years of that period were dominated by the personality of a great statesman, Pericles (499–429), and therefore it is sometimes called the Periclean age. It is better not to do so, however, for the Periclean age was not the whole of the golden age; it was the most fastuous part and perhaps the most creative, yet the original gold was already beginning to tarnish; spontaneity was being replaced by sophistication, naïve conceit by skepticism, and dark clouds were gathering in the offing.

The outstanding political fact is the creation of the Ionian (maritime) league and the Athenian hegemony. For a time Athens ruled the world and Athenian culture dominated all the other Greek cultures. Maritime power was the only kind of power that could unite the amphibian Hellenic states; the use of it was a tremendous stimulus to international commerce, whether that commerce dealt with material goods or with ideas. At the beginning, the center and the treasury of the Ionian league were in Delos (the smallest of the Cyclades in the Aegaean sea), the most holy place for the worship of Apollo. The island was so well protected by its sanctity that the Persian sailors on their way to Salamis did not venture to loot it. As the Athenian domination increased, the treasury of the league had been transferred from Delos to Athens, but on the other hand every precaution was taken to increase the sanctity of that holy place. For example, all human and animal remains were taken out, and efforts were made to prevent its pollution by the occurrence of births and deaths. It is sad to have to record that in later times the sanctity of Delos was polluted in a deeper way. The festivals in Apollo's honor and the Delian games attracted crowds of people, and in between the games and festivals there came the sacred embassy (*theōria*) sent by Athens every year, and there came also a good many pilgrims from every part of the Greek world. Like every other sanctuary, Delos was a great market place — no harm in that, but it also became a slave market, the greatest of its kind in that age. Fancy combining religious festivals with the slave trade! Delos was severely punished for that incredible degradation during the Mithridatic war against Rome; one of the generals of Mithridates[8] took Delos in 84 B.C. and butchered the men, but permitted the women and children to live in slavery.

Let us glance for a moment at another part of the Greek world which was also helping to accomplish Hellenic unity, Delphi in Phocis, a sanctuary established in an admirable and awful site, on the slope of Mount Parnassos. It was supposed to be the navel (*omphalos*) of the earth. Zeus had determined the position of that "navel"

[8] Many satraps or kings of Pontos (northeast Asia Minor, south of the east end of the Black Sea) were called Mithridates, a name derived from that of the Iranian sungod Mithras. This particular one was Mithridates VII Eupator, or the Great, king of Pontos from his early youth *c.* 120 to 63, next to Hannibal the most dangerous enemy of the Romans, a cruel brute, yet interested in arts and letters.

by releasing two eagles, one at the western and the other at the eastern end of the world. They flew with equal speed and met at Delphi. A pretty story, yet a bit primitive! A marble stone (navel stone) was set up in the middle of the temple.[9] That sanctuary was very ancient; the first temple, having been burnt as early as 548, was rebuilt with greater splendor by means of contributions obtained from every part of Greece and even from the Greek colonies in Egypt. The Pythian games were celebrated at Delphi, but the overwhelming attraction of the place was the chasm (*chasma*) through which intoxicating vapors rose from the underworld. A prophetess, the Pythia,[10] sat on a tripod over the chasm, fell into a trance, and gave oracles to which superstitious reverence was paid by almost every person, whether educated or not. The Delphic oracle was one of the formative elements in the development of Greek culture.[11] At the religious festivals orations were delivered, which sometimes took the nature of political speeches and eulogies of the leaders.[12] The power of Athens was based largely upon the financial contributions of her allies, but also, to an extent that we cannot measure but that must have been considerable, upon the skillful use of all the resources that such places as Delos and Delphi offered for general persuasion and the strengthening of national unity.

The supremacy of Athens might have lasted a long time but for the festering jealousies of her rivals, especially of Sparta. It was clearer every year that the unity of Greece was artificial; it had lasted as long as the Persian danger existed; in spite of the festivals and games it could not last much longer. All the Greeks were united against the barbarians, but when those barbarians had been discouraged and the danger removed, unity yielded to suspicion and antagonism. The growing tension led to the civil wars (431–404), to which we shall come back presently.

Our main task in this chapter is to illustrate the beauty and nobility of the Athenian golden age (480–431); the following chapters will be devoted to the philo-

[9] Two marble omphaloi have actually been found in Delphi by the French archaeological mission. See article "Omphalos" by W. J. Woodhouse, Encyclopedia of Religion and Ethics, vol. 9 (1917), p. 492. The idea that the navel of the earth (*omphalos tēs gēs*) is in one's own city or territory is a form of naive egocentrism and parochialism not by any means exclusive to the Greeks. The people of Boston used to believe that Boston was the "hub of the universe." The idea is the same though the metaphor is different. I prefer the "navel" image, which is organic, to the "hub" one, which is mechanical.

[10] *Pythia* (*hiera*), priestess of Pythian Apollo. The Pythiae were probably women with exceptional mediumistic powers.

[11] All this may seem very irrational, but we should bear in mind that the events of ancient history (for example, the political and military events) were largely dominated by faith in omens and oracles. Plutarch's *Parallel lives* are full of references to divination; those references increased the popularity of his work in earlier times (down to the eighteenth century), and they are now probably one of the main causes of its unpopularity. No matter how foolish divination was, as long as people believed in its validity, they were influenced by it. The belief was wrong but the influence was real. As far as Delphi is concerned and the directive power of its Pythiae, see Auguste Bouché-Leclerc, *Histoire de la divination dans l'antiquité* (4 vols.; Paris, 1879–1882), chiefly vol. 2, pp. 39–207), and Herbert William Parke, *History of the Delphic oracle* (465 pp., ill.; Oxford: Blackwell, 1939) [*Isis* 35, 250 (1944)]. The Delphic revelations were generally cryptic, negative (Thou shalt not . . .) or restraining, conservative. Modern statesmen might sometimes wish to be able to justify their decisions or indecisions by reference to a divine oracle! This would provide them with unbeatable alibis.

[12] This is witnessed by one of the fossils of our language: the word panegyric, meaning a laudatory speech or writing, is derived from *panēgyris*, meaning a national assembly, generally in the nature of a religious festival, such as those that met at Delphi and Delos. The festive orations were called *panēgyricoi*; as those orations became gradually more and more laudatory of the leaders, any oration in praise of a person was called *panegyricus*, as for example the *Panegyricus* of Pliny the Younger (61–114), a bombastic eulogy of the emperor Trajan (emperor, 98–117).

sophic and scientific achievements; in this one we must speak, however briefly, of the literary and artistic creations, which are more obvious and help us better than any others to appreciate the glory of Athens.

LYRIC POETRY

The earliest aspect of that glory is given to us by the lyric poets, who appeared even before the Persian Wars and were the first after the Homeric and Hesiodic ages to voice the highest aspirations of Hellas. The best of those poets were really the mouthpieces, or one may call them the interpreters and "commentators," of the whole public; the national games and panegyrics afforded them excellent opportunities of singing the joys and the pride of the Greek-speaking people, of uttering the unformulated conclusions of the public conscience, of expressing the purest thoughts in words so well chosen, so harmonious, that they would easily fly from mouth to mouth, be treasured in the people's hearts, and be endlessly repeated. Such winged words were more effective than the vulgar headlines of our newspapers.

Poetry was not yet dissociated from music; the poet was also the composer; poetic and musical composition occurred together in his mind and excited one another. Prosody and melody were combined; the poet's recitation or psalmody was accented with the accompaniment played by himself on a lyre or by somebody else on a flute.

The lyric poems were of many kinds: religious hymns, songs to accompany processions or ritual dances, odes celebrating the winners of national games, poems recited at the end of a banquet to thank the host, eulogies (*encōmion*) of notable men, elegies or dirges (*thrēnos*), epigrams and epitaphs, not to speak of the more personal pieces expressing the poet's own passions. The poet was not explaining facts, though he might refer to them; his purpose was rather to express the emotions of his brothers. He did it well; sometimes he did supremely well.

The outstanding examples are Simonides (556–467) of Ceos (one of the Cyclades), his nephew Bacchylides of Ceos, and above all the Theban, Pindar (*c.* 518–438). Note that the three of them, though born in the sixth century, cover a good part of the century that is at present engaging our attention.

Our readers may have been shocked by our references to divination and to oracles. What? Those Greek people, who we are told were so wise, allowed themselves to be bamboozled by soothsayers and hysterical women! The Greeks were guided also by poets who were oracles of another kind. In the darkness that surrounded them, emotional words had the power of swaying their minds, words that seemed divine either because of the extraordinary conditions of their utterance (for example, in the Pythian chasm) or because of their unusual cadence and beauty. Great poets are the best and not the least mysterious of the soothsayers.

Simonides was brought up in Athens, but he traveled in Thessaly and other parts of Greece, and even as far as Magna Graecia, and his fame was such that King Hieron [13] invited him to come to Sicily and treated him with munificence. Let me just quote a short fragment of his, to give an idea (necessarily very incomplete) of his poetry; it is an extract from his ode on Thermopylae.

[13] Hieron was tyrant of Syracuse from 478 to the time of his death in 467. He was an enlightened patron of letters, and welcomed at his court Aischylos, Simonides, Pindar, Bacchylides, and others.

Of those who at Thermopylae were slain,
Glorious the doom, and beautiful the lot;
Their tomb an altar: men from tears refrain
To honor them, and praise, but mourn them not.
Such sepulchre, nor drear decay
Nor all-destroying time shall waste; this right have they.
Within their grave the home-bred glory
Of Greece was laid: this witness gives
Leonidas the Spartan, in whose story
A wreath of famous virtue ever lives.[14]

According to a fragment preserved by Plutarch, Simonides considered a hundred or even a thousand years as only a point or a prick (*stigmē*) between two infinite lines, the past and the future.

Simonides' nephew Bacchylides, who was about forty years younger, followed his example, traveling all over Greece and writing odes and other lyrics for the people who welcomed him. He spent some time in the Peloponnesos and also at Hieron's court. Until the end of the last century we knew very little of his poetry, but then nineteen poems of his were discovered in a papyrus. Instead of a hundred lines we had now some 1400 lines, and it was possible to appreciate his genius. This is a good example of the progress of knowledge due to modern scholars. One would think that our history of ancient Greek literature would be complete, yet until 1897 one of the greatest poets was very imperfectly known.[15]

Pindar (518–438), who came halfway between the two poets of Ceos,[16] surpassed them both and all others. According to Quintilian (I–2), "Of the nine lyric poets Pindar is by far the greatest,"[17] and he has remained to this day the symbol of lyric poetry in the golden age. He did not invent any new form of poetry, but he did better what others had done before him, and he did it on a larger scale; his genius was of a higher potential and more fruitful. He came from the vicinity of Thebes and was educated in Athens (this confirms that Athens was already a literary center at the beginning of the century). At the time of Marathon he was almost thirty and hence the years of his maturity coincided with the national exaltation which he was able to express in the most adequate language; his words are at once splendid and solemn, swift and sound. He had traveled even more than his rivals, for we find him not only in his native Thebes, in Athens, and in other cities of Greece proper, but also in Macedonia, in Cyrene, and in Sicily.

These lyric poets represent a kind of pan-Hellenic preface to the Athenian culture. Their restlessness took them all over the Greek countries, and though they

[14] Translation by John Sterling. For the Greek text see F. G. Schneidewin, *Simonidis Cei carminum reliquiae* (Brunswick, 1835), p. 10.

[15] Frederick G. Kenyon, *The poems of Bacchylides from a papyrus in the British Museum* (300 pp.; London, 1897). The British Museum published in the same year a complete facsimile of that papyrus. Various editions and translations of Bacchylides have appeared since in many countries. Thus 1897 is the date of Bacchylides' rebirth.

[16] His activity covers almost exactly the first half of the fifth century; his earliest extant poem dates from 502, the latest from 452.

[17] Quintilian, *Institutio oratoria*, (x, 1, 61); Loeb Classical Library, vol. 4, p. 35. The "nine lyric poets" were, in chronologic order, Archilochos of Paros (720–676), Alcman of Sparta, born in Sardis (seventh century), Sappho of Lesbos (fl. 600), Ibycos of Rhegium (fl. Samos 540), Anacreon of Teos (563–478), Pindar, Bacchylides, Philetas of Cos (d. *c.* 280), Callimachos of Cyrene (fl. 260–240). Note their dispersion in time, eighth to third century, and in space. One only, the very greatest, came from the mainland – Pindar; four from the Aegaean islands – Archilochos, Sappho, Bacchylides, Philetas; two from Asia – Alcman and Anacreon; one each from Graecia Magna – Ibycos – and Cyrene – Callimachos.

owed much to Athens, they considered themselves not Athenians but Hellenes. They wrote and sang poems for the courts or communities that received them. It has been said of Simonides that he was the first to accept payment for his work. Such a statement is difficult to understand, for we know that the ancient rhapsodists, who also wandered from one end of Greece to the other, were rewarded for their pains and feasted by their hosts. It may be that the reference is to payment in money as contrasted with payment in kind, but if so, it only expresses a change in economic conditions. Simonides was perhaps one of the first poets to be paid in money, because there was more money in circulation in his day and people were more ready to use it rather than to barter their talent for other goods.

Simonides and Bacchylides came from Ceos, Pindar from Thebes; all rambled across the Greek-speaking lands; Simonides died in Syracuse, Pindar in Argos (in the Peloponnesos). The most famous odes of Pindar dealt with Pythic victories, and hence his glory began in Delphi and echoed all over Greece together with other Delphic memories. His own utterances were Delphic in their somber greatness.

At the end of his ode in honor of a young athlete, Aristomenes of Aegina, who won the wrestling match in 446, he exclaimed:

> Short is the space of time in which the happiness of mortal men groweth up, and even so, doth it fall on the ground, when stricken down by adverse doom. Creatures of a day, what is any one? what is he not? Man is but a dream of a shadow; but, when a gleam of sunshine cometh as a gift of heaven, a radiant light resteth on men, aye and a gentle life.[18]

Thanks to his genius, and partly also to his association with the "navel" of the earth, Pindar's fame was already great within his lifetime, and he became a classic very soon after his death.

The pan-Hellenism of all these poets is increased by the fact that they did not write in their own dialect, but in a kind of artificial language, a literary Dorian dialect, hardly used except by themselves.[19] They symbolize the natural unity of Hellenes, created by their Homeric traditions, by their mysteries and national games, the panegyrics, theories, and pilgrimages — a unity older than the political unity of the Ionian league or of the Athenian empire, and superior to it.

THE ARTS

The development of lyric poetry was to a large extent independent of empire and prosperity, because it did not cause any large expenditures. The poets took part in the public and private festivals and the only additional cost that their presence entailed was the cost of their own entertainment and of the royal gifts that they deserved (but did not necessarily receive). It is true that their genius was partly induced by public enthusiasm; we express the same thing when we say that they were the spokesmen of the people, and hence they were bound to sing louder and more beautifully in days of triumph and expansion. The building of temples and of other public monuments was, on the contrary, very expensive. In the case of

[18] *Pythian ode,* VIII; translation by Sir John Sandys (1844–1922) in the Loeb edition of Pindaros' *Odes* (1919), p. 269.

[19] This is less extraordinary than may first appear. Poetry is essentially different from the everyday language; hence it is not strange that poets are led gradually to the use of a vocabulary and grammar of their own. Compare the use of Galician, which is closer to Portuguese than to Castilian, by the king of Castile, Alfonso X el Sabio (*Introduction,* vol. 3, pp. 343–344).

sanctuaries, like Delos, Delphi, Eleusis, the necessary funds were brought by the pilgrims or solicited by their faithful congregations from everywhere. When Athens became the center of the Ionian league, she received contributions from her allies and her financial resources increased with her trade. Moreover, silver mines of the Laurion (in Southern Attica) were state property, farmed out to capitalists and worked by slaves. The silver that was extracted from those mines was used at first (upon Themistocles' advice) to strengthen the navy; later a substantial part was appropriated for the rebuilding of Athens and its adornment with glorious monuments.

The outstanding artistic creations were due to the initiative of Pericles and of his assistant Pheidias (born in the year of Marathon, 490; died in prison, 432). The latter was not simply the greatest sculptor of his age (and one of the greatest of all ages) but he had been entrusted by Pericles with the general direction of all his artistic undertakings. His main works as a sculptor, the gigantic chryselephantine statues of Pallas Athene in Athens and of Zeus in Olympia, are lost, but much remains of the decoration of the main buildings of the Acropolis, a part of the Propylaia and the Parthenon. To most people the glory of Greece is the glory of Athens during a couple of centuries, and the glory of Athens is symbolized by the new Parthenon, which was completed during the years 447–434. That monument associates three great men in its pure splendor, Pericles the master mind, Ictinos the architect, and Pheidias the sculptor. The people made no mistake about it; it is really the best symbol of Greek culture, and like other works of art (as against literary and scientific achievements) it can be appreciated by any person worthy of it in a single intuition. The best literary expression of the Parthenon's greatness was given by Ernest Renan in his "Prière sur l'Acropole quand je fus arrive a en comprendre la parfaite beauté," itself one of the masterpieces of French prose.[20]

Greek sculpture was already well developed in the sixth century, and some of the most admired statues date from that time. In the first half of the fifth century, Ageladas of Argos, whose own work is lost, instructed three famous pupils, Pheidias, Myron, and Polycleitos. These three men represent the maturity of Greek sculpture; many people today prefer the less mature, more naïve, production of the previous century, but we may accept the Greek verdict which was united in its praise of Pheidias and Pindar.

At about the same time as Ageladas the painter Polygnotos flourished. He was born in Thasos (an island just south of the Thracian coast) but came early to Athens. The three great sculptors were also living in Athens, except when commissions obliged them to establish themselves temporarily in other places. The most famous of Polygnotos' wall paintings were to be seen in the lesche[21] of Delphi; they represented the Sack of Troy and Ulysses in the Underworld, and as far as we can judge from early descriptions they were colored very simply, without any play of light and shade, and without landscapes in the background; yet they were impressive in their austerity and dignity. These paintings are lost, but we are

[20] Renan conceived it when he visited Athens in 1865; he rewrote it later and published it only in May 1876 (*Revue des Deux Mondes*); later, he included it in his *Souvenirs d'enfance et de jeunesse* (1883).

Parthenōn means the virgin's chamber. It is the temple of *Athēnē Parthenos*, the virgin goddess of wisdom.

[21] A *leschē* is a place where people gather (*legō*) for conversation, generally a kind of arcade (*stoa*).

given some idea of the graphic ability of Polygnotos' contemporaries by the fairly abundant drawings preserved on the Greek vases (Attic vases of the fifth century are characterized by the so-called red-figure style).

TRAGEDY

We have not yet spoken of the most significant feature of Athenian life in the fifth century – throughout that century but with growing emphasis – the drama. This was a novelty, yet the continuation and amplification of an old tradition. The people love to dance and sing, they love to listen to the recitation of poems. That love went back to Homeric days and the lyric poets of the sixth to the fifth century had given a new form to it; on the other hand, religious mysteries and other ceremonies had introduced dramatic performances. According to popular legend, the inventor of the tragedy proper was Thespis [22] (fl. 560–535) of Icaria (near Marathon), who came to Athens and planted his seeds in the most fertile ground. The great victories over the Persians and the national exaltation that followed increased the need not only for lyric poetry but for dramatic poetry – the solemn utterance of people's emotions, the pooling of their ebullient feelings. The tragedy was a sort of public rite, the highest form of rite that any nation ever celebrated.

Tragic poetry developed in such an astounding manner because of the social mood, which favored it, and the miraculous occurrence of three men of genius. It gradually replaced lyric poetry, because it served the same needs more completely. To lyric poetry and music it added choral recitation and dramatic exchanges of views. It was lyric poetry dramatized and multiplied, combined with religious mysteries and transformed into a public performance that was self-contained. The early tragedies were extremely simple, even naïve in their grandeur; toward the end of the century they became more sophisticated even as the people witnessing them (pure lyricism was gradually subordinated to the drama), yet they fulfilled the same purpose. The theater was a school of decorum, earnestness, and piety; it helped ordinary men to share the common triumphs and humiliations with dignity, and to think nobly. That is of course what lyric poets like Pindar were doing, but the playwrights could do it more effectively, and they could reach a larger audience.

Our readers are familiar wth these masterpieces, but it is well to evoke briefly the three creators, Aischylos, Sophocles, and Euripides. All three were connected with Salamis (480), where the new Greece had awakened to freedom and glory. The oldest, Aischylos, was then forty-five and he actually took part in the battle. Sophocles, a very handsome boy of fifteen, was selected as the coryphaios of the youthful chorus celebrating the triumph; he walked naked ahead of it holding a lyre and singing the paean. Euripides' part was more passive, but auspicious; he was born on the very day of the victory.

Aischylos was born in Eleusis, the most sacred place of Attica, about 525. He took part in the two immortal battles, Marathon and Salamis. His epitaph records his part in the first battle, and his first tragedy, *The Persians* (472), was a celebration of the second. Only seven plays of his (out of some 80) remain, and they are

[22] We know little about Thespis, but his name is preserved in our language: "Thespian art", or "a Thespian" to mean, jocosely, an actor. It is said that he introduced an actor (*hypocritēs*, hence our word hypocrite, one who plays a part) to answer (*hypocrinomai*) the chorus. The invention of tragedy would then consist in adding individual action to the lyric chorus.

all very austere and solemn; the drama is still on the Thespian level of simplicity, and lyricism dominates. He reminds one of Pindar. The fundamental idea of his plays is fatality lurking in the darkness, then revealing itself slowly; human greatness causes divine jealousy, the pride of men (*hybris*) is soon followed by their delusion (*atē*) – the gods reduce to madness and blindness those who are too proud.[23] The display of pride and its punishment is the main event; but it is so awful that it takes a religious aspect. Lyricism is as natural here as it would be in a sacred hymn. The play is, as it were, a vision which unfolds itself gradually before our eyes like a ritual action or mystery. The vision is unfolded by the chorus and sometimes interrupted by the dialogue; the dialogue helps to explain what is happening, and at the same time breaks the rhythm and suspense which might become unbearable. Though Aischylos necessarily spent most of his life in Athens, he went thrice to Sicily and was at one time the guest of the tyrant Hieron; he died at Gela, on the southern coast of Sicily, in 456.

The second playwright, Sophocles, was born near Athens in 495, a full generation later than his exemplar. He was even more industrious than the latter, for it is said that he composed no fewer than 130 plays. We should not think of him, however, as an infant prodigy; Greek moderation, mixed with irony, was not as easily fooled as we are by precocious deeds; it realized that promises of youth may be as beguiling as the double blossoms of some trees, which bear no fruit. Sophocles began to write early, but his success was relatively late; some 81 of his plays were written after his fifty-third year. Only seven of his plays remain, all of which belong to that late period of his life; the earliest of the extant plays, *Antigone*, dates from 442.

It is often claimed that Sophocles improved the tragedy; it is more prudent to say that he increased its complexity. The most obvious changes were the introduction of a third actor, the increase of the chorus from twelve men to fifteen, and the use of painted scenery in the back of the stage (*scēnographia*). More profound were the changes in the play itself: the sufferers are no longer the victims of inexorable destiny, their fate is partly determined by their own moderation (*sōphrosynē*) or lack of it. The play, therefore, becomes more nearly human; it is closer to our sensibility. The dramatic psychology is more complex than in Aischylos. The part of lyric poetry is reduced, for more room is needed for the dialogue.

Sophocles seems to have spent the whole of his life in Athens, sharing with his fellow citizens the joys of the golden age and the anxieties and miseries of the iron one; he drank these to the bitter end, for he lived until 406; yet he left the memory of a happy man.

In point of time Euripides is twice as near to Sophocles as the latter is to Aischylos, but the moral distance is much greater. Euripides was a child of Salamis (480), hence fifteen years younger than Sophocles, yet they died in the same year, 406. One essential difference between them was reported by Sophocles himself, "who said that he drew men as they ought to be, and Euripides, as they were." [24] The plays of Sophocles were more human than those of Aischylos, but those of Euripides were more human still; human passions have become his main interest;

[23] That idea was a commonplace of Greek poetry. It can be traced back to Homer, and is expressed by all the old tragedians, for example, by Sophocles in *Antigone* (l. 620). Most people remember it in its Latin form (late translation of a line ascribed to Euripides): *Quem* (or *quos*) *vult perdere Iupiter dementat prius.*

[24] Aristotle, *Poetica*, 25.

his view of men is more realistic than that of his predecessors, yet equally grim. As the tragic events become more intense and more complex, the chorus is no longer subordinated to the dialogue; it has ceased to have any dramatic importance and is included simply as a lyric accessory and in obedience to tradition. The gods are still there, however, not in the center of the stage as in Aischylos, yet around it; indeed, one of Euripides' weaknesses was the excessive use of divine intervention (*theos apo mēchanēs, deus ex machina*) to untie difficult knots and end the play.

Euripides was more sophisticated than either Aischylos or Sophocles; it is significant that he was one of the first Athenians to boast a library of his own; he did not take part in public affairs but was simply a student, a man of letters and somewhat of a philosopher; he had been influenced by Heracleitos and Anaxagoras and was a friend of Herodotos and of Socrates. His knowledge of things and of men was vaster than that of Sophocles but he paid dearly for that advantage; his life was unhappy, he was disillusioned and restless, less loyal to Athens, less religious in the old sense. He had more versatility and imagination than Sophocles, and was more lively, more brilliant, sometimes even more gracious; on the other hand, he was less discreet and reverent and occasionally he scandalized his audience with strange philosophic ideas. He wrote fewer plays than Sophocles and even than Aischylos; yet we know his work much better than theirs, for a quarter of it (eighteen plays out of seventy-five) has come down to us; we have more plays from him alone than from the two others together. Toward the end of his life, he left Athens and went to Magnesia in Thessaly, and later to Macedonia, where he was warmly received by Archelaos,[25] king of that country; he died there in 406.

It is very instructive to compare these three men. In spite of their differences, which were tangible but partly due to their differences of age, they had many qualities in common: grandeur, soundness, and moderation. How did it happen that these three men were contemporaries and formed a constellation unique in the history of letters? One is almost tempted to conclude with Goethe [26] that their genius was, in part at least, the genius of their time and place. It is futile to try to classify them and to say, this one is the greatest, and so on. Let us leave that idle game to schoolmasters and pedants. Each was great in his own way, and his own environment. The oldest, Aischylos, is more solemn, he makes one think of the Hebrew prophets; Sophocles, the middle one in time, represents also the middle point in human and dramatic qualities; Euripides is more concerned with individual psychology, he is more pathetic and more modern. Sophocles is certainly the best symbol of Athenian moderation of the golden age; we would place him close to Pindar and to Pheidias; of the three tragedians he was the most completely loyal to Athens. Aischylos fought in Marathon and Salamis and was fortunate enough to die in the middle of the golden age; Sophocles and Euripides witnessed the glory of that age but also the political decadence and the downfall that followed. Sophocles managed to preserve his serenity, while Euripides became a sadder if not a wiser man. Sophocles remained in his native land and held public

[25] Archelaos, king of Macedonia from 413 to 399, was a patron of arts and letters. His palace was decorated by Zeuxis, one of the most famous painters of ancient Greece. The history of Macedonia is very complicated; Alexander the Great was the twelfth king (including four usurpers) after Archelaos.

[26] Conversation with Eckermann, 3 May 1827.

offices even during the gloomy days of confusion and defeat. The two others abandoned their mother Athens and ended their lives in exile, Aischylos in Sicily and Euripides in Macedonia.

COMEDY

The story of the Athenian drama, which we have told in three sections – Aischylos, Sophocles, Euripides – must be completed with a fourth section dealing not with the tragedy but with the comedy. This is not a new story, however, but a continuation of the preceding one. The comedy is as old as the tragedy, for they stem out of the same cycle of popular entertainments; Dionysiac rites gave birth to both. The comedy sprang out of the rustic festivals of harvest and vintage, thanksgiving holidays, merry processions in honor of the gods of fertility, to whom men owe the good things of life. Though tragedy and comedy grew in the same cradle, the second developed much later.[27] This was probably because tragic festivals needed some direction to be as solemn and stately as they should be, while the merry entertainments could more easily take care of themselves. At any rate, the only representative of the "ancient comedy" whose works have come to us did not make his appearance until the last quarter of the century: Aristophanes the Athenian (448–386). With him we are already moving into the fourth century, yet it is proper to speak of him now. Of his forty-four plays (eleven of which are extant), most were written in the fifth century.

Aischylos, Sophocles, and Euripides were contemporaries, and so were Sophocles, Euripides, and Aristophanes, but the lapse of time between the last two (of the four) was as great as that between the first two.[28] Each of these men influenced his followers, but we must bear in mind that that process was sometimes reversed, for younger men challenge their elders. Thus, Euripides exerted some influence upon Sophocles, and Aristophanes upon Euripides. There were irreducible dissimilarities, however, between the two last named. It has been claimed that Euripides was the father of comedy, because his subtle analyses of character sometimes verged on satire, but what an immense difference in purpose there is between them. Both were Atticists and primarily men of letters; yet in spite of his greater sophistication Euripides is still a follower of Sophocles. Aristophanes, on the contrary, started something radically new. He is an aggressive critic of men and manners, sparing nobody, not even the most powerful and the most respectable men of the city. He attacks the warmongers, the statesmen, the politicians, the sophists, the communists, above all, the flatterers of the people, and the stupid people itself (*dēmos*) which allows itself to be flattered and tricked by the demagogues. He attacks not only public men like Cimon and Pericles, but poets like Euripides and philosophers like Socrates. Beyond the men he even attacks the institutions – the senate and assembly, the tribunals and magistracies. His satires are bold and exaggerated like those of a cartoonist, because he realized that the only way to get them across was to simplify and amplify as the cartoonist does; his style is blunt and strong to the point of vulgarity and obscenity; yet it

[27] With the exception of the satyr play, not a farce but a "playing tragedy," *paizusa tragōdia*. Poets competing for the Dionysia had to submit a group of four plays (*tetralogia*), to wit, three tragedies (*trilogia*) plus a satyrical drama (*satyricon drama*). The *Cyclops* of Euripides (derived from the *Odyssey*, IX) is a satyrical drama, the only one of his that has survived.

[28] The differences between their birthyears are 30, 15, 32 years.

was not offensive (except perhaps to the victims), because its roughness was redeemed by good humor, buffoonery, and ready wit. Political instinct was natural to him as it was to every educated Athenian, but he had no prepossessions; he was guided not by any *parti pris* but by his robust common sense and his sense of fun; he wanted the people to laugh with him and be on its guard against its would-be deceivers and its own foolishness. Like every good satirist he was abreast of the times, sensitive to everything that happened around him, somewhat cynical and skeptical. Sometimes he would praise the good old days in order to bring out the miseries of his own; thus, strangely enough, he defended Aischylos against Sophocles. He was neither religious nor antireligious but was less concerned about religion than about justice and peace. His plays combine incredible fantasies with realism and truth (*Dichtung und Wahrheit*); however grotesque his characters, there is always enough verisimilitude in them to attract and retain the attention, and to prove his point. He had a strong feeling of nature and of humanity in the raw. Some of his verses were derived from popular ditties; his language was familiar [29] and racy, intensely alive; it was the most telling kind of language for his own audience, but the modern reader should know Greek exceedingly well (in a living way) if he would appreciate its niceties.

Aristophanes was the first specimen in world literature of the comic satirist, the distant ancestor of such men as Erasmus, Molière, Voltaire, and Anatole France. He criticized democracy, because it was his privilege to live in the midst of the first that ever existed and because it was his misfortune to witness a period of infinite tragedy and anarchy, when democratic ideals were tried beyond endurance. He saw the evil and corruption of his time and attacked boldly the political and spiritual leaders who must bear the responsibilities as well as the honors. Such criticism as his was healthy in spite of its violence, and it afforded the best proof of the validity and genuineness of the Athenian democracy. Democracy cannot exist without self-criticism; excessive criticism is better than none.

We will understand better the value of Aristophanes' work in his time, if we ask ourselves a few questions. Could one conceive the possibility of such criticism as his in contemporary Sparta or in Persia? Or to come closer to us, would it have been possible to produce a play in Berlin, say in 1941 (and have it crowned!) making fun of Hitler's messianism and showing up the divinely inspired leader who guides his people to the abyss? And what about a play in Washington in the same year, accusing the president and his secretaries of being warmongers and clamoring for peace? Would it have been possible to produce in Moscow, in 1951, a play "debunking" Stalin?

These very things were possible in Athens during the anxieties of the Peloponnesian Wars. Glory to Athens and to Aristophanes! On account of his poetic sincerity and his courage, he deserves the epitaph composed in his honor (it is said by Plato): "The Charites,[30] trying to find an imperishable temple, have chosen Aristophanes' soul."

[29] Sometimes too familiar to our taste. For example, he indulged in silly puns which are not as funny to us (even when the footnotes have made them clear) as they were to his contemporaries.

[30] The *Charites* or *Gratiae* (the Graces) were three daughters of Zeus, named Euphrosyne (cheerfulness), Aglaia (brightness), and Thalia (bloom), whose mission it was to enhance the amenities of life. Would that the Charites were still with us today, for we need their aid more than ever.

THE FIFTH CENTURY ITSELF A TRAGEDY

In this brief account of the artistic and literary achievements of the golden age – achievements that have never been repeated or equaled anywhere in any other century – the reader may have noticed references to the terrible events that replaced enthusiasm and hope with misfortune and disillusionment and all but destroyed the majesty and the glory of Athens. We must say a few more words about that without entering into details which are not in themselves very interesting. For a while – from a long distance, it now seems such a short while – Greece has been magnificently united under Athenian hegemony. Unfortunately, the Greeks are jealous people; that was then, it has always been, and it is now their main weakness. Older cities than Athens found it hard to be subordinated to her; for one city in particular – proud Sparta – this was well-nigh unbearable. Spartan jealousy was increased by profound differences in outlook, differences that could not be compensated or bridged in any way. Athens was a democracy, Sparta's outlook was aristocratic and totalitarian; the difference between the two cities in the fifth century was as great as the difference between London and Berlin in 1940. In both cases no solution was possible except war, and war came with all its horrors. It is not necessary to describe the Peloponnesian War, or rather the two wars, which devastated the Greek world between 431 and 421 and again after a short armistice between 414 and 404 and ended with the complete victory of Sparta. These civil wars had become world wars comparable in their relative size and intensity and the pregnancy of their consequences to the Persian Wars out of which united Greece had emerged so full of hope at the beginning of the century, comparable to the two world wars that have darkened our own days.

To the miseries of the war were added for five long years (430–425) the unspeakable agonies and fears of the plague. Well might the Athenians feel that the end of the world was near; truly enough their own gay world had come to an end, never to return. And yet, throughout those frightful years, the cultural life was never completely stopped, and in particular the tragedies of Sophocles and Euripides and the grim comedies of Aristophanes continued to be played – new plays each year, which were submitted to competition as usual, the best being crowned.

The year 404 was the year of final humiliation. Athens was obliged to surrender. The walls of the Peiraieus (her harbor and naval arsenal) and the long walls between the city and the harbor were demolished; the democratic government was overthrown and its power yielded to the Thirty Tyrants. We need not describe these atrocious deeds, which seemed to put an end to the noble city for ever. Yet, Athens revived, as we shall see, and it assumed a new kind of glory and spiritual hegemony. It remained a great city, one of the great cities of the ancient world; Greece revived also, but it never recovered its unity, nor its peace, nor the innocent buoyancy of its first golden age.

In the course of time, a new Atticism conquered the ancient world – the Atticism of Plato and Aristotle, which is living to this day. That new Atticism was more international than that of the fifth century and more self-conscious, but it was less pure. The immense difference between the first golden age and the second is revealed at once by the contrast between the work of Pheidias on the one hand and that of Scopas and Praxiteles on the other – but we must not anticipate.

To return to the fifth century, when we consider it from the height of our own time, across twenty-five centuries, we realize that it was itself like an Aischylean tragedy, beginning with so great a pride that the gods were jealous and angry; it ended with their vengeance and the Athenians' folly and ruin.[31]

THE DANGER OF COMPARING THE PAST WITH THE PRESENT

This chapter must end with a warning. We speak of the glory of Athens, but we should not forget that this was but one side, the brilliant and happy side, of the medal; the obverse was not so nice. Our general impressions of the past are necessarily one-sided: we remember only the greatness and the beauty, the things that deserve to be remembered or rather that need no remembrance at all, because they have never ceased to exist; we forget the things that were evil, ugly, mean, transitory, perishable, for why on earth should we burden our memory with them?

It cannot have been very pleasant to live in Athens during the Peloponnesian Wars, and even before their outbreak periods of unadulterated peace were brief and few. We should always bear that in mind when we compare the past with the present (as we may and should). We are sometimes *laudatores temporis acti* and unjust to our own contemporaries, because the horrors and mediocrities of today are very obvious to us – we suffer from them – while the horrors of the past are forgotten, or else they have lost their sting.

Should we try to recall the sad and seamy side of the fifth century? Certainly not in detail, for what would be the good of that? why should we allow ourselves to be distracted by evils long past? The evils of today are sufficient. It is well to know, however, that men and women have experienced all kinds of misery, everywhere and always, with but brief intermezzi of peace and happiness. The consciousness that a certain amount of evil and pain has always obtained, even in the most glorious periods of the past, should help us to bear the evils of today with more equanimity.

Our duty is to discern as clearly as possible the evils of our own day in order that we may be able to cure or remove them; there is no need of seeing as well the evils of the past, for they are no longer curable and Father Time has already removed them. Yet we must remember them in a general way, and our praise of the past should always be tempered with that remembrance, for the sake of fairness.

It must always be clear to us that what we admire in the past (and we could not admire it too much) is not by any means the whole of the past, but only a small part, the best, of it. We should not idealize the past as Renan did in his "Prière sur l'Acropole," but see it whole, and admire only that which was so good that it never died. We do not love the past except that part of it which is not past and never will be.

[31] The comparison with a tragedy is the more apposite because Sparta would not have won the war without the financial help she received from Persia. Thanks to Sparta's treason, Persia, which had been completely defeated in 479, was able to dictate the peace in 404. Could a more tragic reversal of fortune be imagined? A more detailed history of the political background would reveal many minor tragedies which helped to create the supreme tragedy of the Athenian defeat. Two of the saviors of Greece, the Athenian Themistocles and the Spartan Pausanias, ended their lives as traitors and outcasts.

Obviously, not all of the Athenians were on the spiritual level of the Parthenon and only the best of them could appreciate Sophocles and Pheidias. Yet those few were the leaven, and it is because of the encouragement of those few as well as their own genius that great men like Pheidias and Sophocles have been able to create their masterpieces. Those great men have survived while the rest died away; they alone remain to symbolize the eternal values of a golden age.

X

PHILOSOPHY AND SCIENCE TO THE DEATH OF SOCRATES

While the lyric poets, the tragedians, and the artists shared the people's feelings, and tried at one and the same time to express and to guide them. a few other men, whom the Greeks called physiologists (students of nature) or philosophers (lovers of wisdom), tended to withdraw from the crowd, to commune with themselves, and to make their own souls. The former group could enjoy more completely the Hellenic games and festivals and could share with relative freedom the people's interest in myths and omens. The philosophers could not do that to the same extent, if at all, for meditations were engrossing their thoughts; they were doing their best to understand the nature of things, of men, and of gods; not only could they not share popular superstitions and fancies, the very freedom of their thoughts was unavoidably challenging these ideas. It was their role then, and it is their role today.

The poetic and artistic creations were publicized and acclaimed, and the poets and artists who had distinguished themselves became popular heroes; the activities of the philosophers were more esoteric and easily gave rise to suspicions and jealousies. Instead of being praised and worshiped, the philosophers might become public enemies and scapegoats.

On the other hand, as knowledge of things was becoming more abundant and more precise, the philosophers were driven to limit the sphere of their own meditations and to think more deeply. That process was very gradual. We might say that it was hardly perceptible before 450. The philosophers of the first half of the fifth century are still very much like those of the preceding century, though they are already very different from the "prophets."[1] After the middle of the century, some of them are closer to what we still call "natural philosophers." The great men of science, like the two Hippocrates, and the great historians, like Herodotos and Thucydides, belong definitely to the second half of the century.

Athens was the center of intellectual life; yet philosophers did not need to be as

[1] We are thinking chiefly of the Hebrew prophets whose utterances are collected in the Old Testament; they lived probably in the period extending from the ninth to the sixth century. There were many other "prophets" in Asia, first Zoroaster (VII B.C.?), whose thoughts percolated across Asia Minor and reached the Greek world through the Magi [J. Bidez and F. Cumont, *Les mages hellénisés* (Paris: Les Belles Lettres, 1938)]; then in India, the Buddha and Mahâvīra, and in China, Confucius and Lao tzŭ (all of them, strangely enough, in a single century, the sixth).

near to it as did the artists. As always, they were moved by contrary impulses; the wish to find a proper audience and worthy disciples would draw them to the leading city, while the desire for quietness and solitude would entice them away from it. Moreover, Athens was not by any means the only center of attraction; the glory of Hellenism was much increased by emulation between many cities scattered far and wide. Most of the philosophers shared the wanderlust of the poets and traveled considerably across the Greek world; of course, all of them visited Athens at one time or another and probably more than once, but relatively few settled there permanently. For one thing, the political vicissitudes were too many and peace too precarious for permanence.

The thoughts of the early philosophers are very imperfectly known to us, for their works are lost, and we have only fragments and the sayings of doxographers,[2] indirectly and poorly transmitted. Often we have only a series of obscure sayings to depend upon, and much ingenuity has been spent in their interpretation. In a book like this one, it would be a waste of time to do so. Suppose we found a new interpretation, how could we be sure that it represents the author's original meaning? However plausible, it would remain uncertain. We might as well discuss the Pythian oracles. Our task is more modest: we shall evoke these early philosophers, without trying to explain their views with more precision than our very scanty information warrants.

In this chapter we shall focus the reader's attention on a dozen men: four Ionians – Heracleitos, Anaxagoras, Melissos, and Leucippos – and the other eight coming two by two from four other parts of Hellas – Parmenides and Zenon from Magna Graecia (southern Italy), Empedocles and Gorgias from Sicily, Democritos and Protagoras from Thracia, Antiphon and Socrates from Attica (note that only one in six came from the country around Athens). Of these twelve, only three (Heracleitos, Parmenides, and Zenon) may be said to belong to the first half of the century, and three to the second half (Melissos, Democritos, and Socrates); the six others flourished chiefly in the middle part.

HERACLEITOS OF EPHESOS

The chief of the twelve Ionian cities (*dōdecapolis*) on the western coast of Asia Minor, Ephesos, was famous all over the ancient world because of its great temple to Artemis.[3] It is there that Heracleitos was born and, as far as we know, spent most of his life. He had traveled extensively in his youth but came back to his native city, and we are told (by Diogenes Laërtios) that when he had com-

[2] The doxographers were scholars who wrote histories of the philosophers and compiled extracts from their writings. The main ones were Aristotle and Theophrastos; doxographic books of the latter are known only indirectly through later extracts. Collections of philosophical opinions, *Placita philosophorum*, were ascribed to Plutarch (1-2), Stobaios (V-2), and others, but the outstanding collector was probably one Aëtios, of whom nothing is known, but who flourished probably at the end of the first century after Christ. Most opinions are known indirectly, and often through quotations made by adversaries such as the skeptics or Christian polemists who tried to discredit paganism. This very difficult subject has been cleared up as much as was possible by Hermann Diels (1848–1922), *Doxographi graeci* (Berlin, 1879; editio iterata, 864 pp., 1929). For a briefer statement of doxographic difficulties see P. Tannery, *Pour l'histoire de la science hellène* (1887; new ed., 1930), pp. 19–29 [*Isis 15*, 179–180 (1931)].

[3] For the temple of Artemis (Diana) see G. Sarton and St. John Ervine, "John Turtle Wood, discoverer of the Artemision 1869," *Isis 28*, 376–384 (1938), 4 figs. Ephesos was one of the sacred places of classical antiquity; it was later one of the earliest sanctuaries of Christendom. Remember St. Paul's visit to it and his Epistle to the Ephesians.

pleted his great book *On the whole* (*Peri tu pantos*) he deposited it in the temple of Artemis; it is also said that he made it as obscure as possible and therefore he was called Heracleitos the Dark One (*ho scoteinos*). His book was said to have been divided into three parts, dealing respectively with the universe, politics and ethics, and theology. That is possible and the 130 fragments that remain of it can be (and have been) arranged in three groups corresponding to that division.[4] Even when the whole of it was available, the book was so difficult to understand that the king of Persia, Darios, son of Hystaspes, invited Heracleitos to come to his court and give him the necessary explanations; Heracleitos declined the invitation, saying that he had a "horror of display and could not come to Persia, being content with little, when that little is to my mind." The two letters are quoted *in extenso* by Diogenes Laërtios, and I mention them because they help us to locate Heracleitos in the time sequence. Darios I ruled from 521 to 485; Heracleitos' book was thus written before 484, and we may say that he flourished in the beginning of the century.

The two letters are plausible. We know that Heracleitos was contemptuous of men, including kings and even philosophers. For he remarked that "much learning does not teach understanding, or it would have taught Hesiod and Pythagoras, as well as Xenophanes and Hecataios."[5] Like the other Ionian philosophers, he assumed that in spite of appearances there must be some unity of substance in the universe, and he postulated that the primordial substance was fire. Why fire? Probably because of what we might call his second principle: the eternal flux of things (*panta rhei*);[6] that was perhaps his dominating idea: everything is always changing, up or down. Now fire, which flares up and goes down, and changes its appearance at every moment, is a good symbol of the ceaseless universal change; moreover, look at the Sun, the great source of everlasting and everchanging fire. His third principle was that the apparent disharmony of the world hides a profound harmony, for every change happens in accordance with a universal law.[7] Each quality implies its opposite; the existence of each thing implies its nonexistence somewhere else. These opposites are reconciled in the general scheme of nature. "God is day and night, winter and summer, war and peace, surfeit and hunger."[8] This tallied with Heracleitos' other view that it is the invisible harmony that matters, not the visible discordance and ugliness. Most men are too stupid to see the beauty that is hidden below the surface. Heracleitos was a sad man, for he saw the relativity and vanity of all things; we cannot hold fast to anything, because everything runs away. Popular tradition considered him the typical pessi-

[4] Edition and translation of Heracleitos, *On the universe*, by W. H. S. Jones at the end of vol. 4 (1931) of Jones' edition of Hippocrates in the Loeb Classical Library. That volume also includes a translation of Heracleitos' life by Diogenes Laërtios (III-1).

[5] Fragment 16.

[6] "You could not step twice into the same river; for other waters are ever flowing on to you" (fragment 41; see also fragment 81).

[7] The invisible harmony is superior to the visible (fragment 47). The Greek original of that principle was very appropriately engraved upon the medal dedicated by the French Academy of Sciences to the memory of the great mathematician, Henri Poincaré (1854–1912). The medal was reproduced and described in *Isis* 9, 420–421 (1927). Compare also fragment 45: "They understand not how that which is at variance with itself agrees with itself. Harmony of the tension as in bow and harp," and fragments 56, 59.

[8] This is the beginning of fragment 36, but I should quote the rest of it to illustrate the enigmatic nature of his sayings: "But he [God] undergoes transformations, just as fire, when it is mixed with spices, is named after the savor of each." What remains of his book *On the whole* is a collection of riddles.

mist, as opposed to the typical optimist, Democritos; the first was always crying and the second laughing.

Heracleitos, we may conclude, was a philosopher and poet in the old Ionian style, not much of a man of science, even less so than Xenophanes. Yet, in his book *On the whole*, he begins with physics or with nature, then considers political questions, and finally discusses theological ones – a good order. We may end our account with one of his political maxims: "People must fight for their laws as for the city walls." [9] That is worthy of the Parthenon.

ANAXAGORAS OF CLAZOMENAE

With Anaxagoras, the last of the Ionians, we enter more definitely into the scientific field. The contrast with Heracleitos is astonishing, for the latter spoke like a poet and seer, and Anaxagoras like a cool-headed physicist. His main work was a treatise on nature (*Peri physeōs*) of which seventeen fragments remain; we have no reason to doubt the genuineness of those fragments, which cover about three pages of printing.

Anaxagoras was born at the beginning of the century at Clazomenae, one of the twelve Ionian cities, situated about the middle of the western coast of Asia Minor, somewhat north of Ephesos. As Ephesos was a great center of pilgrimage, it is highly probable that young Anaxagoras went there and met Heracleitos. At any rate, he moved to Athens soon after the Persian Wars, being the first Ionian philosopher to do so. This shows once more that Athens had become a focus of attraction. Anaxagoras was fortunate in that he gained the friendship of Pericles, who was then the most powerful man in the city. Pericles' admiration is so well described by Plutarch that it is worth while to repeat the latter's description verbatim.

> But the man who most consorted with Pericles, and did most to clothe him with a majestic demeanour that had more weight than any demagogue's appeals, yes, and who lifted on high and exalted the dignity of his character, was Anaxagoras the Clazomenian, whom men of that day used to call "Nus," either because they admired that comprehension of his, which proved of such surpassing greatness in the investigation of nature; or because he was the first to enthrone in the universe, not Chance, nor yet Necessity, as the source of its orderly arrangement, but Mind (Nus) pure and simple, which distinguishes and sets apart, in the midst of an otherwise chaotic mass, the substances which have like elements.
>
> This man Pericles extravagantly admired, and being gradually filled full of the so-called higher philosophy and elevated speculation, he not only had, as it seems, a spirit that was solemn and a discourse that was lofty and free from plebeian and reckless effrontery, but also a composure of countenance that never relaxed into laughter, a gentleness of carriage and cast of attire that suffered no emotion to disturb it while he was speaking, a modulation of voice that was far from boisterous, and many similar characteristics which struck all his hearers with wondering amazement.

A little further in the same *Life*, Plutarch remarks:

> Moreover, by way of providing himself with a style of discourse which was adapted, like a musical instrument, to his mode of life and the grandeur of his sentiments, Pericles often made an auxiliary string of Anaxagoras, subtly mingling, as it were, with his rhetoric the dye of natural science.[10]

[9] Fragment 100.

[10] Plutarch, *Life of Pericles*, IV, V, VIII; translations by Bernadotte Perrin, Loeb edition of the *Lives*, vol. 3, pp. 11, 21.

We shall come back presently to the discussion of Anaxagoras' ideas, but what astonishes one is the impression conveyed by Plutarch that it was Anaxagoras who enhanced the prestige of Pericles and not vice versa. This is a great tribute to the importance that the Ionian philosopher had obtained in Athens; it is also a great tribute to the Athenian public of that time. Would our own people have more respect for a philosopher than for a leading statesman? It is said also that Euripides was Anaxagoras' pupil. We must think of Anaxagoras as the first teacher of natural philosophy in Athens, the forerunner of Plato and Aristotle.

According to him, there is neither coming into being nor ceasing to be, but there are only commixtures (*symmisgesthai*) and decompositions (*diacrinesthai*). The universe was originally a chaos of innumerable seeds (*spermata*) to which Mind (*nus*) gave order and form by a movement of rotation (*perichōrēsis*). Note that the "seeds" are not elements, for each is as complex as the whole, nor atoms, for there is no limit to the subdivision of matter – and that their number is undetermined. The two main points are, first, the introduction of mind as contrasted with matter, mind being power gradually transforming chaos into cosmos; second, the idea of an initial and eternal vortex by means of which the organization of matter takes place. The introduction of *nus* originated the contrast between mind and matter, but it would be an exaggeration to call Anaxagoras the founder of philosophic dualism. His *nus* is not well defined and may be interpreted as a physical force as well as a spiritual one.[11] The initial vortex and its use for the gradual organization of the universe suggest the cosmologic theories of Kant and Laplace, but is only a vague adumbration of them. Nevertheless, the fact that such a comparison occurs to our minds is much to the honor of the first Athenian philosopher.

Anaxagoras' compromise between the naïve Ionian monism and Pythagorean pluralism is remarkable. The whole of the universe and its parts, however small, are homogeneous; their differences are only differences in size, not in composition.[12]

Let us quote the first fragment[13] to illustrate the tone of his prose (so different from the poetic one of Heracleitos):

> At the beginning all things were confused, infinite in number as well as in smallness, for the infinitesimally small existed. But all things being together, none was apparent because of its smallness [none was large enough to be perceived]; everything was occupied by air and aether[14] both of which are infinite, because of all things, it is these [two] which are greatest in number and size.

[11] The best definition occurs in Hermeias, Christian critic of the pagan philosophers, who flourished in the fifth century or later. Hermeias, VI; Diels, *Doxographi graeci*, 1879, p. 652). "Intelligence (*nus*) is the principle, cause and ruler of all things, it gives order to the things which are out of order, motion to the things which are immobile, it separates the things which are mixed, and makes a cosmos of the chaos." Should we accept that definition, magnificent in the mouth of an adversary, we would be obliged to say that Anaxagoras was the father of philosophical dualism, but we are not as confident as was Hermeias. To go to another extreme, *nus* might be translated by energy, but it is better to keep the Greek term and to admit our ignorance of its exact meaning.

[12] See fragments 15, 16 in Tannery, numbered 3, 6 in Diels. The seeds or *spermata*, we should repeat, are not simpler than the rest nor essentially different from it in composition. To use a modern image (which I admit is a dangerous procedure), the seeds are like points of initial (fortuitous?) "organization" which serve eventually as ferments for the general organization. Lucretius called those seeds *homoiomeria* (*De rerum natura*, I, 830 ff.).

[13] First in Diels and Tannery; it was probably the beginning of Anaxagoras treatise.

[14] The distinction between air (*aēr*) and aether

Such depth and subtlety of thought as appear in the fragments of Anaxagoras with so little basic knowledge to support it is as amazing as the Parthenon, which was being built in the same period. How could Anaxagoras do it?

Our astonishment increases when we realize that his scientific knowledge was not only meager but mostly wrong. His cosmologic views were forward, yet his astronomic knowledge was decidedly backward as compared with that of the Pythagoreans. One cannot give him much credit for his explanation of eclipses of Sun and Moon by the interposition of Moon, Earth, or other bodies, because the explanation was not a novelty and because it was combined with crude ideas such as that the Earth and other planets are flat, that the Sun is larger than the Peloponnesos, and so on. He suggested that the Moon was a body like the Earth, with plains and ravines, and was inhabited. The great meteoric stone that fell in 467 at Aegos Potamoi (the "goats' river" in the Thracian Chersonese, or Gallipoli Peninsula, the northern shore of the Dardanelles) was said by him to have fallen from the Sun; this is the first dated meteorite in the world's history.[15]

Anaxagoras was deeply interested in anatomy and medicine. He is said to have studied the anatomy of animals and to have made experiments on them. He dissected the brain and recognized the lateral ventricles. He ascribed the occurrence of acute diseases to the penetration of bile (black or yellow) into the blood and the organs.

He attempted to square the circle and wrote a book on scenography, the application of perspective to the designing of stage backdrops and properties. He would thus be one of the founders of the mathematical science of perspective. That story is plausible because the contemporary importance of the drama created a need for good (if very simple) scenery, and it was natural enough for dramatists to apply for that to a man of science; it would have been especially natural for Euripides to consult his teacher Anaxagoras.[16]

The learned men of Greece were relatively well acquainted with Egypt and its great river, so utterly different from the miserable rivers or torrents of their own country, and they speculated on the cause of the annual innundation, thanks to which the land of Egypt might be called a gift of the river (*dōron tu potamu*). Anaxagoras claimed that the flood is due to the melting of snow during the summer on the mountains in the interior of Lybia. Herodotos reported that explanation and rejected it. The correct explanation was first given by Aristotle and Eratosthenes: the flood is due not to the melting of snow but rather to the tropical

(*aithēr*) is not quite clear. Anaxagoras was already aware of the corporality of air, which is somewhat like steam; aether is much subtler, somewhat like the substance of the shining blue heaven (the empyrean, *empyros*). The word *aithēr* derives from the verb *aithō*, to light up, to burn or blaze. His idea seems to be that the universe is largely made up of two substances, one of which is rare (tenuous), the other much more so. Other forms of matter must be due to extraordinary condensations.

[15] According to Pliny the Elder (I-2) in his *Natural history* (II, 149), Anaxagoras was enabled by his astronomical knowledge to prophecy that in a certain number of days a rock would fall from the sun, and that the fall occurred in the daytime . . . That, of course, is nonsense, but Pliny adds that "the stone is still shown, it is of the size of a wagon-load and brown in color" (qui lapis etiamnunc ostenditur magnitudine vehis, colore adusto). Thus, the stone could still be seen in Pliny's time (23–79).

[16] The story is plausible, but we have it from a very late witness, Vitruvius (I-2 B.C.), in the preface to book VII, "Interior decoration," of his *Architecture*. Vitruvius ascribes mathematical writings on perspective to Democritos as well as to Anaxagoras, and what he says applies equally to both. The first idea of scenography he ascribes to Agatharchos of Samos (V B.C.), one of their contemporaries.

rains that fall during the spring and early summer about the upper waters of the Blue and White Niles. Anaxagoras' explanation was not quite correct, but it was rational, and he was the first to assert that the flood originated in the mountains where the Nile begins its course.[17] It took thousands of years before people generally accepted the true explanation, for the solution was found and lost many times. The story of ideas concerning the Nile floods is a good example of the difficulties of establishing and preserving truth before modern times.

We do not discuss Anaxagoras' astronomic ideas, because the discussion of each item would require considerable space, and it is not worth while. He was an astounding cosmologist, but he was not an astronomer. He was somewhat of a mathematician and might perhaps be called a theoretical physicist. He was a genuine man of science, asking himself scientific questions and trying to find rational answers. Though the Athenians had begun by admiring him, they were repeatedly shocked, not so much by definite assertions, as by his general attitude of mind, the attitude of a rationalist who brushes superstitions aside; such an attitude seems blasphemous to the bigot.[18] This would be a sufficient explanation of the charge of impiety that was leveled against him, but it is possible that his prosecution was partly caused by the wish to harm his patron, Pericles, who had become highly unpopular at the beginning of the Peloponnesian War. Many friends of Pericles were indicted; the most illustrious of them, Pheidias, was condemned to imprisonment and actually died in prison. Euripides had shown more foresight and had abandoned Athens *c.* 440, before the situation had become as grave as it was to be ten years later. Pericles managed to save Anaxagoras from prison but not from exile.

Whatever were the real causes of Anaxagoras' accusation – his friendship with Pericles or perhaps Persian leanings[19] – the pretext was religious. Anaxagoras was indicted for rationalism (*c.* 432). He was certainly not the first victim in the incessant war between bigotry and science, but he is the first known one. We may not call him a martyr of science, because his sentence was simply one of banishment, but he was the first man in history who was punished for thinking freely, for following the dictates of his reason and his conscience rather than the opinions of the community. We do not know the details of his life in exile, but he finally established himself in Lampsacos, a city of Mysia, on the southern shore of the Dardanelles. Why did he select that place of retirement? He simply joined other refugees. When the glorious city of Miletos (the cradle of Ionian philosophy and the leader of the Ionian revolt) was destroyed by the Persians in 494, many of the Milesians took refuge in Lampsacos. Later another refugee (or call him a traitor), Themistocles, had settled there. This was less attractive, but we may assume that the Milesians had created in Lampsacos a tradition of Hellenism and philosophy. This would appeal to Anaxagoras, who there spent the last years of his life and died in 428. It is not probable that he had time to establish a school of philosophy, but his presence must have strengthened the Hellenic tradition of

[17] H. F. Tozer, *History of ancient geography* (Cambridge: The University Press, 1935), p. 63, app. xi.

[18] The people called him in derision *ho nus* (the mind), as is recalled by Plutarch in the text quoted above. This was typical enough. Anaxagoras' references to "intelligence" rather than to the gods of the city was to them the proof of his impiety.

[19] This is suggested by A. T. Olmstead in his *History of Persia* (Chicago: University of Chicago Press, 1948), p. 328.

that locality, which was to be in the following century the birthplace of Anaximenes, one of the companions and historians of Alexander the Great.

THE ELEATIC SCHOOL. PARMENIDES AND ZENON OF ELEA. MELISSOS OF SAMOS

When Phocaea, the northernmost of the Ionian cities, was taken by the Persians, many of its inhabitants established new homes at Elea (or Velia), on the western coast of South Italy. It is possible that another Ionian, Xenophanes of Colophon, settled for a time in that city and awakened the philosophic spirit of some of its children. At any rate, a great philosopher, Parmenides, one of the founders of metaphysics, was born there, and may have been the pupil of Xenophanes when the latter was an old man.

Parmenides is the typical metaphysician: he is passionately concerned not with appearances, but with the means of reaching the truth beyond them; the necessary means he believes to be not observational, experimental, as a man of science would, but purely logical. One should be able to reach the absolute truth, he seems to think, by pulling on one's own logical bootstraps. We should not blame a man of the fifth century for entertaining such illusions, inasmuch as almost every metaphysician has shared it down to our own day.

Parmenides tried to develop Ionian monism as rigorously as possible against pluralism or Pythagorean dualism. This he does like a mathematician more interested in rigor than in common-sense reality. His "what is" (*to eon*) or Being fills the totality of space; the non-Being is pure space, emptiness (absolute vacuum). This (non-Being) cannot exist, yet it can be thought and expressed (as we have just done). Starting from that premise, Parmenides concludes that the universe must be one, and limited, yet must fill the whole of space; for reasons of symmetry it must be spherical; vacuum is unthinkable, for the universe is equally full in all its parts; the universe of Being is eternal, changeless, motionless. Change and motion are unreal. Note that the conclusions are exactly the opposite of those reached by his Ionian contemporary, Heracleitos. Parmenides' premise was wrong; hence, he could not possibly reach correct conclusions; it does not follow that Heracleitos' conclusions were right.

The formulation of Parmenides' metaphysics (for this is definitely metaphysics, not science) was continued by one of his disciples, Zenon of Elea, and completed by another, Melissos of Samos.[20] It would seem that Eleatic philosophy was already constituted before Parmenides' departure to Athens at the age of 56. According to Plato, Parmenides conversed with Socrates, who was then very young. This would place his arrival in Athens about the middle, and his birth about the beginning, of the century. We shall not discuss the transcendental monism of the Eleatic school, but it was necessary to indicate its occurrence and to introduce Parmenides and Zenon, whose astronomical and mathematical views will be dealt with in the next chapter.

Parmenides' thought is fairly well known, because many lines of the poem in which he summarized it have been preserved. The poem begins with a preamble

[20] There is a curious reference to Melissos in the Hippocratic treatise on *Nature of man*, I. "In my opinion such men [philosophers] by their lack of understanding overthrow themselves in the words of their very discussions, and establish the theory of Melissos"; . . . *ton de Melissu logon orthun*.

and is divided into two parts dealing with truth (*ta pros alētheian*) and with opinion (*ta pros doxan*). The old Pythagorean dualism is replaced with a new logical dualism, truth versus opinion. His ideas were deep, or at any rate obscure; to do justice to them one would have to repeat them *in extenso* and examine them verbatim; even then one could not be sure of reaching a clear understanding of them.

Zenon completed Parmenides' "demonstration" by showing the logical absurdities to which one is led if one assumes that plurality and change are real. It was probably because of his systematic use of the *reductio ad absurdum* that Aristotle called him the discoverer of dialectics.

If we accept the statements that Zenon was born in 488 and that he was forty-four years old when he accompanied his master to Athens, the visit of both occurred in 444; the date is plausible, though I would prefer to say that they were in Athens about the middle of the century.

As to Melissos, he was the admiral of the Samian fleet and obtained a measure of success against Pericles, yet could not prevent the final defeat of his native island in 440. Did he go to Athens in or soon after that year and become Parmenides' disciple in that city? He it was who carried transcendental monism to the extreme. He declared that the changes of the phenomenal world are only illusions of our senses and that reason cannot recognize the reality of Being under any of its changing forms.[21] The real cannot be finite and spherical, as Parmenides taught; it must be infinite, because if it were not, there would be empty space outside of it. It is strange to think that Ionian monism, transplanted into the Pythagorean climate of South Italy, had blossomed out in such an intransigent and paradoxical fashion.

We shall come across Parmenides and Zenon later on, but for the present we must abandon them, since we are dealing with the history of science, not with that of metaphysics.

EMPEDOCLES OF AGRIGENTUM

The philosophers of whom we have spoken thus far — Heracleitos, Anaxagoras, Parmenides, Zenon — as far as we know them or can read between the lines of their writings, were strange personalities, but none was quite as strange as the Sicilian with whom we are going to deal now. Empedocles was born at Agrigentum (on the south coast of Sicily) *c.* 492. He was not only a philosopher but a poet, a seer, a physicist, a physician, a social reformer, a man of so much enthusiasm that he would easily be considered a charlatan by some people, or become a legendary hero in the eyes of others. His birthplace was one of the most beautiful cities of the ancient world, but the Carthaginians destroyed it *c.* 406 and it never recovered its lost splendor. During Empedocles' lifetime, it was still a very wealthy and sophisticated center of Greek culture, and Empedocles belonged to one of its prominent families. Its wealth and amenities had attracted many distinguished men, such as Pindar and Simonides, perhaps also Bacchylides, Xenophanes, Parmenides. When the Pythagoreans were driven out of Croton some

[21] Comparisons with the Hindu ideas represented by the Sanskrit words māyā and avidya suggest themselves, but we cannot do more here than indicate them. Māyā means illusion, unreality; avidya, spiritual ignorance, ignorance combined with nonexistence, illusion (personified as Māyā). The terms are used by Buddhists as well as by Hindus.

of them found refuge in Agrigentum. The sea view from the hills is magnificent and the lowlands around the city include sulfur and salt mines, hot springs, and other marvels which could not fail to stimulate inquisitive and ready minds. There is nothing to prove that Empedocles traveled in Egypt and in the Orient, as has been suggested, but he traveled in the Greek world, and the Greek world came to his native city. He could not help partaking in the ferment of ideas – philosophic, religious and scientific – that was then occurring everywhere the Greek language was spoken.

His writings included purification songs (*Catharmoi*), three books in verse on nature (*Peri physeōs*), and a poem on medicine (*Iatricos*). Some 450 verses (of all of his writings) have come down to us; these are but a fraction of the whole, yet they are sufficient to give us definite ideas about his style and his thoughts.

He postulated the existence of four elements or roots (*rhizōmata*) – fire, air, water, earth – and two moving forces, the one centripetal, love (*philotēs*), the other centrifugal, strife (*neicos*). Everything that exists is made up of these elements, which are themselves unchangeable and eternal and which may be united or reunited by love, or else separated and disintegrated by strife. The theory of the four elements was a strange compromise between Ionian monism on the one hand and complete pluralism on the other.[22]

"Why four elements?" one might have asked, but apparently nobody ever bothered about that, except that a fifth element was eventually added by Plato and by Aristotle. In spite of its arbitrariness, that hypothesis had a singular fortune, for it dominated Western thought in one form or another almost until the eighteenth century.[23]

These cosmologic views lasted so long, because it was equally impossible to prove or to disprove them before the birth of modern chemistry. On the other hand, astronomic ideas were more tangible. Those of Empedocles seem very crude; he conceived the heavens as an egg-shaped surface made of crystal, the fixed stars being attached to it but the planets free.

He was capable, however, of making physical observations and even experiments. There is one experiment to his credit that would suffice to give him an honorable and permanent place in the history of science. That experiment, with a clepsydra, enabled him to prove the corporeality of air. He was probably led to it by the arguments concerning the reality or the nonreality of empty space. The ordinary clepsydra was a closed vessel the base of which had one or many small holes; there was also a hole at the top. Now, if that upper hole was closed with one's finger and the clepsydra dipped into water, it would not fill itself, while as soon as the finger was removed the water rushed in. Various other experiments, equally simple, would have pointed to the same conclusion. For instance, if one tries to push an empty vessel with a broad opening under water, bubbles of air emerge from the surface; these bubbles, which can be seen and heard, must

22 There is no reason to believe that the atomic hypothesis had occurred to Empedocles or that he even heard of it. The first atomist known to us was Leucippos, whom we place about the middle of the century or later (see below).

23 The crystallization of Greek and Western fancies around the number four is the more curious when one compares it with Chinese physiological views based on five [*Isis* 22, 270 (1934–35)], and with the Hindu ones based on three [tridoṣa; *Isis* 34, 174–177 (1942–43)]. These classifications might be used to symbolize three outstanding cultural patterns – triangular (India), quadrilateral (Europe and Islamic Asia), pentagonal (Far East).

represent a material reality. Incidentally, the reference to Empedocles' use of a clepsydra is the earliest in Greek literature, but the Greeks must have used clepsydras of one kind or another before that time, for that instrument was already known to the Egyptians of the Eighteenth Dynasty and to the early Babylonians. The Greek theory of it is rather late, however, as we cannot trace it back further than Cleomedes (I-1 B.C.).[24]

Empedocles also made a series of observations concerning vision and light. How is it that we see an object? According to Aëtios, Empedocles seems to have reached a compromise: some emanations (*aporroai*) are emitted by the luminous bodies and are met by rays issuing from the eyes. The compromise suggests that other Greek thinkers had already tried to answer that riddle. Pythagoras and his followers claimed that vision was caused by particles emanating from the body; others, that it was the eye which emitted feeling rays. These fancies may seem vain to the modern reader, but he should consider that they imply a definite advance upon the men who took vision for granted and made no attempt to explain it; it did not even occur to them that an explanation was needed.[25]

Empedocles' speculations concerning the speed of light were equally hazardous but more fortunate, for they were confirmed by observations made by the Danish astronomer, Roemer, twenty-one centuries later (in 1676)[26] and by experiments that were completed only within the last century. Empedocles argued that light must have a finite velocity. This was not of course a result of observation but of pure reasoning. Aristotle is a good witness for that, because he made similar statements twice;[27] it is worth while to quote the first (and longest) of these statements:

> Empedocles says that the light from the Sun arrives first in the intervening space before it comes to the eye, or reaches the Earth. This might plausibly seem to be the case. For whatever is moved [in space], is moved from one place to another; hence, there must be a corresponding interval of time also in which it is moved from the one place to the other. But any given time is divisible into parts; so that we should assume a time when the sun's ray was not as yet seen, but was still travelling in the middle space.

Various anatomic and physiologic "discoveries" are ascribed to Empedocles. He recognized the labyrinth of the ear and said that respiration takes place not only through the movement of the heart but also through the whole skin. He showed the importance of blood vessels, blood being the carrier of innate heat. Blood issues from the heart and flows back to it; this is not an anticipation of circulation but rather of the tidal theory, which was developed by Galen (II-2) and was accepted with various qualifications until the time of Harvey (1628) and even somewhat later. It would seem that that tidal theory was already extended by Empedocles to the whole world; there are cosmic tides (or call it a cosmic

[24] See A. Pogo, "Egyptian water clocks," *Isis* 25, 403-425 (1936), ill. For Babylonian clepsydras, see p. 75. According to Diogenes Laërtios, IX, 46, one of Democritos' "mathematical" works was entitled *Conflict of the water-clock (and the heaven)*, but the book is lost and the title unclear.

[25] Similar discussions were indulged in by Hindus of the Nyāya school of philosophy. It is not necessary to assume that those ideas, preserved in Sanskrit texts, influenced the Greek thinkers, or vice versa; on account of the impossibility of dating those texts (say within a few centuries), neither assumption could be proved. D. N. Malik, *Optical theories* (Cambridge, 1917), pp. 1-2.

[26] I. Bernard Cohen, "Roemer and the first determination of the speed of light, 1676," *Isis* 31, 326-379 (1940), ill.

[27] *De sensu*, 446A26-B2; *De anima*, 418B 21-23.

breathing), even as there are tides (or breathing, blood pulsations) in the individual bodies. That view tallied with the idea of alternations between the two cosmic forces: love and hatred. It, too, enjoyed much popularity for centuries; it reappeared over and over again in the writings of many thinkers (for example, Leonardo da Vinci and even Goethe).

His medical views were equally prophetic: health is conditional on the equilibrium of the four elements in the body, disease is caused when their balance is upset. That theory of health and disease was often modified and amplified,[28] but it continued to be accepted just as long as the four elements themselves; it even survived them and is not completely eradicated to this day.

Other "anticipations" have been read in his obscure writings, such as views of the unity of nature, of organic evolution and adaptation, or "remembrances" such as those concerning the transmigration of souls.[29]

This portrait of Empedocles, in spite of its rich variety, is not yet complete, for there was in him also, and perhaps uppermost, a social reformer and missionary. The marshy lands around Agrigentum were very unhealthy; he drained some of them at his own expense. He used to travel from town to town preaching and singing his verses, purifying the souls of men and healing their bodies; he is even said to have brought back to life a woman of Agrigentum. He was a kind of salvationist and wonderworker. His fame was already considerable (if not of the best kind) during his lifetime, and he became a hero soon after his death. Legends gathered fast around his memory, as they did in the case of Pythagoras and of the early saints. They were exuberant enough to obliterate the truth, and we do not know the circumstances of his death. According to one set of legends, he threw himself into the crater of Aetna (Etna), or slipped into it while he was observing its activity; it is even added that the volcano vomited one of his sandals (this is the kind of circumstantial evidence that is attached to every legend in order to accredit it in the minds of uncritical listeners). According to another story, he had fallen into disgrace – not an unusual fate, for popular approval is as fickle as it is intense – and was obliged to leave Sicily. He first went to Italy; there is good reason to believe that he was in Thurii (Lucania) soon after its foundation (445), and afterward passed to the Peloponnesos and was in Olympia in 440; his *Purification songs* were sung by a rhapsodist at the Olympic games of that year (= Olympiad 85.1). After that his tracks are lost. Did he go to Athens? There is nothing to prove it, and it is not very probable. A provincial thaumaturge would hardly have been welcome in Athens and might have fared very badly. Anaxagoras, less enthusiastic and less eccentric than Empedocles, had been driven out of the city, and not many more years would pass before the condemnation of Socrates. It is more plausible that Empedocles remained in the Peloponnesos, wandering from place to place, with a young friend, Pausanias son of Anchitos. It was to this Pausanias that he dedicated his *Physics* (see the preamble of it), and therefore we may assume that that work was written during these years of exile. According to

[28] The four elements became the four qualities, the four humors, the four temperaments, but it was always the same Empedoclean conceit masquerading under those disguises; *Isis 34*, 205–208 (1943).

[29] The idea of metempsychosis was also ascribed to the Pythagoreans and the Orphics. It was probably of Oriental origin. The Hindu saṃsāra may have reached Greece via Persia, and it may have been confirmed by similar ideas coming from or through Egypt. Cumont, *Lux perpetua*, pp. 197–200, 408.

a lovely tradition, he died somewhere in the Peloponnesos *c.* 435–430. His friends, including Pausanias, were gathered around him at a feast. When the darkness had fallen, the guests of this last supper heard a loud voice calling Empedocles, the heavens were illuminated, and he disappeared.[30]

Short as it is, this sketch reveals that Empedocles the Sicilian was very different from the other Greek philosophers, except perhaps Pythagoras and the Orphic poets. There is something Oriental in him, oddly combined with genuine scientific velleities. The Oriental ingredients may have filtered through to his receptive mind from Persia, Babylonia, or Egypt, or even from India, or they may have been simply original aspects of his own mysterious nature. Empedocles was so great and rare a man that he left no school; none of his admirers or disciples, not even the faithful Pausanias, was able to continue the master's work.

THE ATOMISTS. LEUCIPPOS AND DEMOCRITOS[31]

After that Sicilian interlude, we may now return to the Greek mainland and to Greek rationalism, and witness the development of a new general explanation of the world: the atomic theory. Returning to Greece does not mean escaping the Orient, however, for Oriental influences had pervaded for centuries the Eastern Mediterranean world. In order to appreciate the significance of the new theory, let us try to forget all that we know and ask ourselves how the universe is constituted. There are two possible answers: it is made out of one stuff or of many. The first answer had been given by the Ionian physiologists, but even in the beginning it showed weaknesses that could not be cured except by adding qualifications implying a denial of the original monism. Thus, Anaximenes postulated that the *Urstoff* was air, and the plurality of aspects was explained as caused by thickening or thinning out. It is easy enough for us to accept that explanation, because we know that air is composed of innumerable particles which may be brought together or on the contrary removed farther and farther away, but without that image it is impossible. How could one understand the rarefaction or the condensation of a substance if it is made of one piece? One might thus say that Anaximenes was already a pluralist in disguise.

A similar statement might be made apropos of Pythagoras and his followers, who accepted the concept of empty space. True monism, as Parmenides and the Eleatics had clearly seen, implies a plenum.

The philosophies of Anaxagoras and of Empedocles were definitely an abandonment of the monist impasse. They walked out of it, and mankind with them, forever. Anaxagoras, postulating the existence of a governing intelligence, introduced dualism; Empedocles, with his four elements and his couple of forces, completed a kind of pluralism. The next step was taken by the atomists, who postulated the existence of an infinity of separate particles scattered in the infinity of empty space.

The ancients (for example, Aristotle, Theophrastos) were already agreed that the atomic theory had been invented by Leucippos, who flourished about the middle of the fifth century, and developed some thirty years later by Democritos. We must first make the acquaintance of these two extraordinary men.

[30] Joseph Bidez, *Biographie d'Empédocle* (176 pp.; Gand, 1894).

[31] The best general account is that of Cyril Bailey, *The Greek atomists and Epicuros* (630 pp.; Oxford: Clarendon Press, 1928) [*Isis 13*, 123–125 (1929–30)].

Very little is known about the first, not even his birthplace, which is variously given as Elea, Abdera, and Miletos. Miletos is the most plausible, and we shall call him Leucippos of Miletos. The two other places, Abdera and Elea, were probably suggested, the first because of a confusion with Democritos, the second because Leucippos began his life as a disciple of the Eleatic school and was actually a student of Zenon (at any rate, such a tradition was current). It is quite possible that he visited Elea, and highly probable that he stayed in Abdera. One can imagine the birth of atomism as a reaction against the fantastic ideas of Parmenides. It is said that Leucippos explained the atomic theory in a book curiously entitled *The great world system* (*Megas diacosmos*), but that book is also ascribed to Democritos, as well as a smaller one called *The small world system.* Leucippos' writings are lost, but one saying is definitely credited to him: "Nothing happens in vain [without reason], everything has a cause and is the result of necessity." [32]

Democritos is far better known.[33] To begin with, there can be no doubt about his birthplace, Abdera in Thrace, or about his time, for he tells us himself that he was a young man in Anaxagoras' old age, being forty years his junior. That tallies well with another tradition, according to which he was born in the 80th Olympiad (460–457); it also tallies with what we are told of his relation with Leucippos. We shall not be far wrong if we place their *floruits* respectively in 450 and in 420; or, to put it otherwise, the atomic theory was constituted in the third quarter of the century, in Abdera.

The mention of Abdera may surprise the reader, though he must be aware by this time of the ubiquity of genius in the Greek world. Abdera, at the northern end of the Aegean Sea, may seem far away; yet it was an ancient and flourishing city. Curiously enough, it acquired the reputation of being the abode of stupid people; [34] yet it gave birth not only to Democritos, but also to Protagoras and to Anaxarchos; [35] if it was (as we believe) the cradle of the atomic theory, few cities in the world deserve as much glory as Abdera. Athens might be the center of the Greek world, but it was not by any means the whole of it, nor the only fountain of merit; it was rather – by the middle of the century – the place where merit might hope to obtain its best reward. That reward was not always forthcoming. Democritos went to Athens and saw Socrates, but was too shy to introduce himself to him. He said, "I came to Athens and no one recognized me." If he came very late in the century, it is probable that the Athenians had not much use for him. He wrote a good many books of which the titles (but hardly more) have come down to us, arranged in groups of four.[36] As far as can be judged from those titles, they con-

[32] The Greek is clearer: *Uden chrēma matēn ginetai, alla panta ec logu te cai hyp' anagcēs.*

[33] There is an abundant literature on Democritos, because the endless discussions on "atomism" and "materialism," reoccuring every century in a new shape, always rebounded on him. For example, Karl Marx (1818–1883) wrote in his youth a thesis on the differences between Democritos and Epicuros (1841); hence a deep Russian interest in Democritos! See *Isis 26,* 456–457 (1936).

[34] One made jokes about the stupidity of the Abderites as one did about the Boeotians, just as the French do about the people of Pontoise and Charenton, and the Americans about the denizens of Brooklyn or Kalamazoo.

[35] Anaxarchos was said to be a member of the school of Democritos, which would suggest that that school continued for some time after its founders' deaths. He was a companion of Alexander in Asia, and after Alexander's death (323) was executed by the king of Salamis in Cypros. Anaxarchos was called "the optimist" (*ho eudaimonicos*), which confirms Democritan affiliations.

[36] The list of his writings has been transmitted to us by Diogenes Laërtios, IX, 46; he remarks that their grouping in tetralogies had been done by one Thrasylos, who had done the same for the works of Plato (grouping preserved in most editions of the latter's works). That habit was prob-

firm the stories concerning Democritos' education. On his father's death, he resolved to spend his inheritance (which was considerable) in research abroad. Such a decision was not a novelty in Greece; we have seen that the philosophers and the poets traveled as much as they could. Most of them, however, were satisfied to move about in the Greek-speaking territory; a few were attracted by the mysterious East, which they all confidently believed was the source of ancient wisdom. Democritos traveled on a larger scale and for a very long time. Wherever he went, he sought out the learned men and studied under their direction. He spent five years in Egypt, learning mathematics, and he went as far as Meroë (on the upper Nile). At that time (after 449) peace between Greece and Persia made it possible for a Greek to travel in Asia Minor.[37] Democritos took advantage of that to proceed to "Chaldea" (he actually went to Babylon, being the first Greek philosopher to do so), and from there to Persia, even, possibly, to India. The main point was that Democritos was not a mere sight-seer or tourist, nor a business man, but a philosopher in search of knowledge. How much knowledge was he able to gather? Could he read hieroglyphics or cuneiform? Probably not, but he was an intelligent man, alert, inquisitive, capable of checking the information that he received from one source with that obtained from another. He was certainly able to learn many things from his Egyptian, Chaldean, or Persian informants. How many? Must we conclude that he brought back the atomic theory from the East? We shall come back to that presently.

Before discussing that theory, we must complete our portrait of Democritos. He was not only one of the fathers of the atomic theory, but his was an encyclopedic mind, interested in all the branches of philosophy, and also in all the branches of science. His knowledge of mathematics, astronomy, and medicine will be examined in other chapters; here we must restrict ourselves to his views on psychology and ethics. He was the first to attempt a scientific explanation of enthusiasm, the state of a human soul possessed by God – it may be called divine inspiration, but it is also artistic creation, genius, and folly[38] – and this led him to the study of many kinds of psychologic and even metapsychic problems. His interest in ethics may be inferred from the collection of apothegms (*gnōmai*) ascribed to him. Are these sayings genuine? Who can tell? Some are like proverbs which, even if they come to us in the form coined by him, could not be called his; they represent not his own wisdom but the accumulated wisdom of his people. They are the earliest collection of their kind in European literature, and are remarkable on that ground alone. Here are a few examples:

Do not try to know everything if you do not wish not to know anything.

Courage is the beginning of action, but chance is master of the end.[39]

ably connected with the old traditions of the Athenian stage. A dramatist was supposed to offer four plays at a time, either four tragedies, or three tragedies plus a satyric play.

[37] A peace with Persia had been negotiated in 449 by Callias son of Hipponicos; see Olmstead, *History of Persia*, p. 332.

[38] Armand Delatte, *Les conceptions de l'enthousiasme chez les philosophes présocratiques* (79 pp.; Paris: Les Belles Lettres, 1934); Joseph Bidez, *Eos* (Brussels: Hayez, 1945), pp. 136 ff [*Isis* 37, 185 (1947)].

[39] This saying and the following are quoted as translated by Cyril Bailey, *The Greek atomists and Epicurus*, pp. 187–213 [*Isis* 13, 123 (1929–30)]. Bailey also gives the Greek text, which naturally rings better than the English, for the Greek is genuine, while the English is only a pale copy.

The great pleasures are derived from the contemplation of beautiful works.

Cheerfulness (*euthymia*) comes to man through moderation in enjoyment and harmony of life; excess and defect are apt to change and to produce great movements in the soul.

It is a great thing in misfortune to think aright.

He who does wrong is more unhappy than he who is wronged.

It is better to take council before acting than to repent.

[Yet] Repentance over shameful deeds is the salvation of life.

It shows a high soul to bear an offense meekly.

He who gets a good son-in-law finds a son, but he who gets a bad one loses a daughter.

The man who has not one good friend does not deserve to live.

Learn the statesman's art as the greatest of all and pursue those toils from which great and brilliant results accrue to men.

A man should consider state affairs more important than all else and see to it that they are well managed. He must not be contentious beyond what is fair or clothe himself in power beyond the common good. For a well-managed state is the greatest of all successes and everything is included in it. If it is preserved, so is all besides; if it is lost, all is lost.

Most of those ethical, economic, and political maxims were commonplaces among gentle people of Democritos' time; some are a little in advance of it and one can already hear in them a Socratic or Platonic, even a Christian, ring. Democritos insisted not only on moderation but on cheerfulness, and this was especially meritorious during the evil days that he must have witnessed. As he died in very old age, perhaps a centenarian, his life extended into the second quarter of the fourth century.[40]

Let us now consider the atomic theory, which Democritos received from Leucippos but which he developed into a consistent and fairly complete explanation of the world.

As against the universal flux of Heracleitos, Democritos postulated the relative stability of being, and as against the static unity of Parmenides, the reality of motion. The world is made of two parts, the full (*plēres, stereon*) and the empty, the vacuum (*cenon, manon*). The fullness is divided into small particles called atoms (*atomon*, that cannot be cut, indivisible). The atoms are infinite in number, eternal, absolutely simple; they are all alike in quality but differ in shape, order, and position.[41] Every substance, every single object, is made up of those atoms, the possible combinations of which are infinite in an infinity of ways. The objects exist as long as the atoms constituting them remain together; they cease to exist when their atoms move away from one another. The endless changes of reality are due to the continual aggregation and disaggregation of atoms. As the atoms themselves are indestructible, one may consider the theory as an adumbration of the principle of conservation of matter.

But how do the atoms move? How do they come together or separate? Why are they grouped in one way or another? An infinity of such questions can be asked which Democritos could not answer and in many cases could not even formulate;

[40] Hence, he was an older contemporary of Plato, who was influenced by him, yet never named him. J. Bidez, *Eos* (Brussels: Hayez, 1945), p. 134.

[41] Aristotle, *Metaphysica*, 985b14: "These differences, they say, are three – shape (*schēma*), order (*taxis*), and position (*thesis*). For they [the atomists] say the real is differentiated only by "rhythm" (*rhysmos* = *rhythmos*), intercontact (*diatigē*) and turning (*tropē*); and of these rhythm is shape, intercontact is order, and turning is position; for *A* differs from *N* in shape, *AN* from *NA* in order, Ｈ from *H* in position. The question of movement – whence or how it is to belong to things – these thinkers, like the others, lazily neglected."

the exact formulation of these questions has been very slowly and painfully established by the chemists of the nineteenth and twentieth centuries, and their work is not yet done and never will be. The atomic theory is determinist and mechanical. As far as man's will and freedom are concerned, determinism is limited by man's ignorance and by the infinite complexity of causes. Democritos did not conceive a spirit distinct from matter, but some groups of atoms he thought were subtler than others, and he conceived a whole gamut of such groups from the heaviest and most earthly to the lightest and most ethereal. The soul (or vital principle, *psychē*) is corporeal but made of the lightest atoms (like fire) and the most mobile (spherical in shape for greater mobility). There is a share of those lightest atoms (that is, of souls) in everything; this idea enabled the early atomists to explain sensations, thoughts, and psychologic phenomena of every kind. The word *psychē* occurs repeatedly in the Democritan fragments and may mean "mind" as well as "soul." There is some kind of *psychē* everywhere, or, to put it otherwise, the whole universe is animated (besouled), but there are no gods, there is no Anaxagorean *nus*, no Socratic providence. The superiority of the soul over the body, or of the less corporeal groups of atoms upon the more corporeal, was for Democritos a matter of such conviction that he did not discuss it but reaffirmed it many times; his "materialism" was thus dominated by a very genuine kind of idealism. Furthermore, he conceived some of the lightest groups of atoms, which he called *eidōla* (hence our word idols, but the meaning here is simulacra, images, phantoms, fancies), as scattered everywhere and capable of influencing our fate. This was an ingenious expedient to explain the facts implied in dreams, visions, divination, and other mysteries. The apparent hardness of his theory was compensated by its vagueness and its elasticity. The theory was extremely comprehensive; it could give some interpretation of most facts, from the most material to the most immaterial.

As Bailey remarks,

> Democritos was neither a skeptic, nor a rationalist, nor a phenomenalist, he does not fit into any of the modern categories; he neither denied nor affirmed the truth of *all* sensation nor of *all* thought; but built for himself a "theory of knowledge," subtle and almost paradoxical, but based directly on his atomic conception of the world. The final realities of the universe, the atoms and the void, are real and are capable of being known by the mind. Phenomena are built up of the final realities and retain the primary properties of size and shape: as such they are real and can be known by the senses. The mind may safely make its deduction from phenomena, both because the phenomenon as a unity of these primary properties is real and because sensation – the mere perception of the real phenomenon – is the same as thought. But once go beyond these primary properties, beyond the reality of the phenomenon, and you are attributing to the object what is really the subjective experience of your senses, and thought based on those "conventions" will lead you nowhere.[42]

There have been controversies as to the source of the atomic theory, by scholars to whom the Greek origins (Pythagorean, etc.), to which reference has been made before, seemed insufficient. Atomic theories were developed in India by philosophers of the Nyāya and Vaiśeshika schools, at a time that cannot be determined but that is almost certainly later than the time of Christ.[43] Assuming that those

[42] Bailey, *The Greek atomists and Epicurus*, p. 185.

[43] See the elaborate discussion of this by Arthur Berriedale Keith, *Indian logic and atomism. An exposition of the Nyāya and Vaiçeṣika systems* (291 pp.; Oxford, 1921) [*Isis 4*, 535–536 (1921–22)].

philosophies were preceded by earlier speculations, much earlier (Brahmanic, Buddhist, or Jaina), did the Greeks know of the earlier ideas? Could they have been influenced by them? That is not impossible, and Democritos himself may have heard of them while he was in Persia or in India(?). Such assumptions are unproved and gratuitous. The atomic hypothesis was one that wise men, trying to reconcile the unity and relative stability of nature with its ceaseless changes, were bound to make sooner or later. How could one harmonize monism with pluralism? It is not surprising that the hypothesis occurred to Hindu minds, and also independently to Greek ones. The Greeks were quite capable of reaching that solution by themselves, and so were the Hindus.[44]

One of the traditions concerning the Oriental origin of the atomic theory must be mentioned here, because it is so unexpected. The theory was ascribed by Posidonios (I-1 B.C.) to a Phoenician, Mochos of Sidon, and by Philon of Byblos [45] to another Phoenician, Sanchuniaton of Beirūt, whose works he translated into Greek; a part of that translation was preserved by Eusebios (IV-1). Both Mochos and Sanchuniaton were supposed to have flourished before the Trojan War; of the latter it was said more specifically that he flourished in the time of Semiramis.[46] As far as we can judge from the text of Eusebios, their doctrines were very remote from the atomism of Leucippos and Democritos. The Phoenicians, who were very clever dragomans and middlemen, may have transmitted some Hindu theory; they may even have invented a theory, but that would be for them a unique achievement.

Knowing the Greeks and the Phoenicians as we do, we are not at all surprised that the former invented the atomic theory; we would be exceedingly surprised if the latter had done so.[47] The Phoenician stories are not convincing: Oriental influences of many kinds impinged upon the eager mind of Democritos while he was living in the East, but the invention of the atomic theory was ascribed not to him but to his teacher Leucippos.

When judging the Greek atomic theory, we must beware of two exaggerations; the one consists in equating it to the modern theory invented by Dalton at the beginning of the nineteenth century, and the other in rejecting it altogether from the history of science, because of its vagueness. There is, of course, an immense difference between the Greek idea and the Daltonian, all the difference that exists between a philosophic conception that could not be tested and a scientific hypothesis, inviting a long series of experimental verifications. On the other hand, there is no doubt that the theory of Democritos, as revived by Epicuros and popularized

[44] Many Greek ideas in science and philosophy are duplicated in India. It is very interesting to compare those duplications, though it is seldom (if ever) possible to establish the precedence of one or the other or to prove the dependence of one upon the other. The duplications help to prove the essential identity of the human mind. Given definite problems that admit of only a few solutions, it is not astonishing that wise men of Greece, India, China, etc., hit independently upon the same solution.

[45] Also called Herennios Byblios, a Roman grammarian, who flourished in Byblos, Phoenicia, under Vespasian (emperor, 70-79). His writings are lost.

[46] Legendary queen of Assyria, probably identical with Sammuramat, wife of Shamshi Adad V (824-812).

[47] Bailey, *The Greek atomists and Epicurus*, pp. 64-65. Georges Contenau, *Manuel d'archéologie orientale* (Paris, 1927), vol. 1, pp. 316-319 [*Isis* 20, 474-478 (1933-34)]. Per Collinder, *Historical origins of atomism* (Lund: Observatory, 1938) [*Isis* 32, 448 (1947-49)].

by Lucretius, remained an intellectual stimulant throughout the centuries. It was driven underground by Jewish and Christian teachers, but it never died. The account of its vicissitudes is one of the most remarkable in the history of knowledge.

THE SOPHISTS. PROTAGORAS OF ABDERA, GORGIAS OF LEONTINI, AND ANTIPHON OF RHAMNOS

Let us return again to Athens and try to consider the intellectual landscape from the point of view of a well-educated man who lived in that city in the second half of the century and tried to understand the world around him. Not to speak of political conditions, which were getting worse every day, he must have been exceedingly perplexed by the conflicting doctrines that were discussed around him. Would he believe Heracleitos or Parmenides, Anaxagoras or Empedocles, or would he follow the atomists? Or would it not be simpler and safer to attend the mysteries and panegyrics, to do his duty as a citizen and share the popular superstitions? Where could the truth be found? In the midst of such perplexities (aggravated by economic and political unrest), a good man might be forgiven if he were driven into bigotry, skepticism, or any other form of despair. What is the good of it all? Is there any truth? And if there is any, can mortal man attain it? The most perplexing question of all was this one. If he had growing sons to look after, to whom would he intrust their education?

The need of teachers was felt acutely, and it was now satisfied by a new class of them (there have always been teachers of some kind or another, for no civilization could continue otherwise) who were called sophists. In the usage that established itself about the end of the fifth century, a "sophist" (*sophistēs*) was a professional teacher of grammar, rhetoric, dialectics, eloquence, a man who taught young men to behave themselves, to be wise and happy. Some of these sophists were good men, perhaps most of them were, yet others, more conspicuous, were moneymakers and hypocrites (that is unavoidable). It would seem that the bad teachers increased in numbers as time went by and the name sophist acquired gradually the bad meaning that it has preserved to this day.

There is nothing to be gained in the company of rascals, but it is worth while to get acquainted with three illustrious sophists of the golden age: Protagoras, Gorgias, and Antiphon. The first two are immortalized by Platonic dialogues, bearing their own names as titles and giving portraits of them that are convincing and not unattractive.[48]

Protagoras of Abdera. Protagoras was born in Abdera (Democritos' birthplace) about 485, and when he was thirty he set out on his travels all over Greece as well as in Sicily and Magna Graecia, lecturing and teaching. He was the first to be called a sophist, and he reaped the first harvest. He was enormously successful and during his forty years of teaching he accumulated ten times as much money as the sculptor Pheidias. He came many times to Athens, and some of his stays may have been long enough to make him well known in the city; he obtained Pericles' favor. The tale of his material success is unpleasant and ominous; other men must have been stimulated by it and have been led by it to embrace a career that could

[48] The *Gorgias* (against rhetoric) and the *Protagoras* (against the sophists), both dating from Plato's maturity.

be so lucrative; any profession that offers such rewards is terribly jeopardized. The new profession began too well; no wonder that it went from bad to worse and that dialectics and sophistry acquired a sinister reputation. Protagoras' success may have been facilitated by the fact that his philosophy was a kind of Heracleitian relativism, and such a philosophy was acceptable in an age of increasing disillusionment. In one of his books, dealing with truth, he said that "man is the measure of all things"; hence, there cannot be any absolute truth. Another saying of his was even more indiscreet: "As to the gods, I cannot say whether they exist or not. Many things prevent us from knowing, in the first place the obscurity of the matter, then the brevity of human life." This was too much for Athenian democracy, which was very sensitive on religious matters and whose nerves had been unhinged by various profanations.[49] In 411, Protagoras was accused of impiety. The people who had bought his books were told by the town crier to bring them to the agora, where they were to be burned; [50] he was exiled, or else he was condemned to death but managed to escape. He could cheat the Athenian judges but not nemesis: the ship that was carrying him to freedom was wrecked and he perished.

One more remark must be added. The sophists were teachers of good speech; this implied grammar, and Protagoras, the first sophist, was *ipso facto* the first grammarian; he called attention to the genders of words and distinguished various tenses and moods of the verbs; he was also, of course, the first teacher of practical logic. We shall come back to that later on, but it was worth while to witness the birth of Greek grammar.[51]

Gorgias of Leontini. While the first and most illustrious of the sophists was a Thracian, his greatest rival, Gorgias, came from Sicily. He was born at Leontini (not far from Syracuse) *c.* 485. The exact date of his birth is unknown, but he was already growing old (*gērascōn*) in 427, when he was sent as an ambassador of his native city to Athens, and he is said to have survived Socrates and to have died a centenarian. It is also said that he was a disciple of Empedocles. Like Protagoras, he traveled considerably but spent many years in Athens. He earned much money and spent it fastuously. He was essentially the same type of man as Protagoras, the first sophist, but a bit worse. As far as can be judged from the few remaining extracts, one feels that both of them were skeptically minded, but that Protagoras was more of a philosopher, while Gorgias was already the full-fledged sophist of evil memory, the man who claims that what is likely is more worth while than what is true, and that he can make small things look great and vice

[49] Pierre Maxime Schuhl, *Essai sur la formation de la pensée grecque* (Paris: Presses Universitaires, 1949), p. 368 [*Isis 41,* 227 (1950)].

[50] This is the first recorded example of a burning of the books, the date being 411 B.C. It suggests that there was already an established book trade in Athens at that time. There have been many other outrages of the same kind ever since in many countries. It will suffice to mention two infamous examples, the Burning of the Books ordered by the First Emperor, Shih Huang-ti (III-2 B.C.), and at the other end of the chronologic scale the one ordered by Hitler on 10 May 1933.

[51] We might have written the "birth of grammar," for Greek grammar was probably the first grammar to be born and constituted, the only possible rival being Sanskrit. We do not know the beginnings of grammatical consciousness in India, but the first Sanskrit grammarian was Pāṇini (IV-1 B.C.), who flourished before the existence of any full-fledged Greek grammarian. On the grammatical proclivities of Protagoras see Gilbert Murray, *Greek studies* (Oxford: Clarendon Press, 1946), pp. 176–178 [*Isis* 38, 3 (1947–48)].

versa – the dialectician, the orator who thinks more of the form of his speeches than of their content. He affected a good Attic dialect and loved to use archaic words and rare metaphors. Yet Plato was rather indulgent to him in the dialogue called after him. The *Gorgias* was written at about the same time as the *Republic* (*c.* 390–387) when Plato was preparing to open the Academy, but the action can be dated 405, when Socrates was sixty-four and Gorgias, at the top of his fame, an old man of eighty.

Gorgias wrote rhetorical essays, recited sportive poems, and delivered festive speeches in Olympia and Delphi, preaching peace and unity; but who would allow himself to be persuaded by a man whose main intention was to be elegant and persuasive, yet who could speak just as well (and everybody knew it) on the opposite side? In order to convince other people, one must be convinced oneself, and Gorgias was not. Granted his dialectical background, he was not a dishonest man, but his vision was dimmed by his success.

Antiphon of Rhamnos. The third sophist represents a type different from that of the two others and helps us to realize that there were various species in that genus. He was born in Rhamnos (not far from Marathon) at about the same time as the two others (*c.* 480), and became a professional orator.[52] He was the leader of a school of rhetoric,[53] his most illustrious pupil being Thucydides. Some fifteen of his speeches are preserved, all of which were written for other people or for the sake of exercise. Of his many speeches only one was delivered by himself, the one prepared for his own defense in 411, and that speech, which should have been the best and the most moving, is unfortunately lost. He was a politician and took part in the government of the Four Hundred (in 411); after the abolition of that oligarchy, he was put to death.

In addition to his speeches he composed a little book called *The art of avoiding grief* (*Technē alypias*), which was the earliest book of a very popular genre – the "consolations." Men are afflicted in many ways, and there is none but knows grief and sorrow. They all need comfort, and a good book of consolation is sure to be welcomed. Antiphon has had imitators in almost every country and every age; it may suffice to mention Boethius and Joshua Liebman.[54]

Protagoras, Gorgias, and Antiphon were the best kind of sophists and even they are not very attractive and would hardly deserve eternal remembrance, except that they and their kind help us to understand the intellectual climate in the second half of the century. The problems raised by the activities of the sophists are familiar to us, because they are the problems of education. When a society becomes more sophisticated, as was the case with Greek society by the middle of the

[52] He was the most ancient of the ten Attic orators listed in the Alexandrian canon. The ten orators are in chronological order (the dates following each name are sometimes approximate): Antiphon (480–411), Lysias of Athens (459–378), Andocides (440; after 390), Isocrates of Athens (436–338), Isaios (420–348), Hypereides (400–322), Lycurgos of Athens (396–323), Aischines (389–314), Demosthenes (385–322), Deinarchos of Corinth (361, d. very old). Their lives cover two centuries, the fifth and the fourth.

[53] He was not the first rhetorician. The first was the Sicilian, Corax, who became the leading man in Syracuse after the expulsion of the tyrant Thrasybulos in 467. He wrote the earliest treatise on rhetoric (entitled *Technē, the art*), a book mentioned by Aristotle, Cicero, and Quintilian.

[54] Joshua Loth Liebman (1907–1948), American rabbi, author of a best seller called *Peace of mind* (New York: Simon and Schuster, 1946).

century, there is an unavoidable tendency to change the old education into something new that will transmit the newly acquired refinement and disillusionment to the next generation. There is then a conflict between the old people and the young; that is of course the eternal conflict between succeeding generations, but greatly intensified by sudden cultural progress. Moreover, no kind of education, even if it is the very best, is good for everybody. One might claim that while it improves good boys it quickens the deterioration of the bad ones. Even so today some men gain nothing in college except snobbishness which accentuates their stupidity. It is clear that even the best sophists could not prevent the evil tendencies of an Alcibiades, but it is equally true (a fact of repeated experience) that any kind of education is good only for the boys who are attuned to it and may be harmful to others who are not. The contemporary Greek criticism of the sophists was vividly illustrated in some of the plays of Aristophanes, for example, in the *Banqueters* (*Daitaleis*; lost), acted in 427, or the Clouds (*Nephelai*), produced at the great Dionysia of 423. It would not be difficult, I imagine, to compile a long list of plays, from the days of Aristophanes down to our own, that would illustrate the perennial disgust of old men with new education and the real dangers of such education even at its best. In Athens the conflict was terribly embittered by the vicissitudes of a losing war, by demagogic excesses and economic anxieties. The conservatives had apparently good reasons for blaming the newfangled educators and the average good man was frightened by the increase of skepticism and impiety, the gradual abandonment of the old rites, and the discredit of the popular beliefs.

SOCRATES OF ATHENS

Among the sophists castigated by Aristophanes were Euripides and Socrates. We are already acquainted with the former, and we are now ready to be introduced to the latter, one of the noblest men in the whole history of mankind. Aristophanes' presentation of him as a "poor wretch" [55] was not only malicious but foolish; he confounded him unjustly with the mercenary sophists who made "the worse appear the better reason," or with the sophisticated ones who attached more importance to things in heaven (*ta meteōra*) or under the earth (*ta hypo tēs gēs*) than to the duties of men. Socrates was not at all a "meteorosophist," [56] but he was in the eyes of the Athenians a sophist, a teacher of youth, and as such he must share a measure of their resentment. This explains Aristophanes' conduct without excusing it, for he at least should have known better.

Socrates was born in Athens in 470; his father, Sophroniscos, was a sculptor and his mother, Phainarete, a midwife. They were simple people but not poor and could provide for him as good an education as was then available. He was trained in his father's profession but early showed an interest in philosophy. Such an interest could easily be stimulated and satisfied in Athens, where philosophic discussions were taking place all the time at the theater, at the agora, or in the streets. He had acquired some knowledge of arithmetic, geometry, and astronomy, and as to politics, it was in the Athenian air, far more pervasive than philosophy, in fact unescapable except for the dumb. He was drafted as a soldier, and was many times engaged in warfare; he took no part in public life, except on two occasions

[55] *Ho cacodaimōn Sōcratēs* (*Clouds*, 104). Socrates is one of the dramatis personae in that play.

[56] *Meteōrosophistēs; Clouds*, 360.

when he displayed civic courage of the highest order. His appearance was singular because of his picturesque ugliness; he had a snub nose and heavy lips, and reminds one (if the London statuette can be trusted) of an old-fashioned and kind muzhik.[57] He was robust and capable of enduring fatigue, hardship, and the inclemencies of weather in a measure that astonished his companions; he dressed in the plainest manner, walked always barefoot, and his diet was extremely frugal. This was not asceticism, for it implied no self-denial; Socrates lived very simply because he liked it best that way.

The peevishness of his wife, Xanthippe, has become proverbial, but one wonders whether it has not been exaggerated in order to bring out in stronger relief his own kindness and forbearance. She bore him three sons, the eldest of whom was already adult at the time of his father's death; [58] the two others were much younger (this would suggest that he was married relatively late).

He left no writings of his own, and we know him only through the books of two of his disciples, Plato and Xenophon. The portraits painted by them agree as to the essentials, but are colored the first by Plato's idealism and the second by Xenophon's more earthy common sense. In the Platonic dialogues wherein Socrates appears and speaks, it is impossible to determine just how much of his speeches must be ascribed to him and how much to Plato; [59] we cannot give to the former without taking something away from the latter. Yet Xenophon provides an excellent means of comparison and correction. When he and Plato agree, we are on safe ground, and leaving out some details without importance, the image of Socrates that has come down to us seems to be a very good likeness. There is no man of antiquity whom we know better, for thanks to Plato's art and to Xenophon's goodheartedness we can almost see him and hear him talk.

Though he spent his life teaching the youth, Socrates was different from the sophists in that he never was a regular schoolmaster, did not keep a school or a regular meeting place, did not deliver lectures, and did not expect any payment. The contrast between the wealth acquired by such men as Protagoras and Gorgias and the poverty of Socrates is very impressive; he was obviously another kind of man. Moreover, he despised the sophists and never failed to denounce their skepticism and their levity. It is that very fact which makes Aristophanes' accusation so odious; he was selecting as an example of the sophists their best adversary. How could a man as well informed as Aristophanes was commit such an outrage?

The following extract from Xenophon's *Memorabilia* gives one a good general idea of Socrates' personality, and incidentally a glimpse of Xenophon's own.

> Moreover, Socrates lived ever in the open; for early in the morning he went to the public promenades and training-grounds; in the forenoon he was seen in the market; and the rest of the day he passed just where most people were to be met; he was generally talking, and anyone might listen. Yet none ever knew him to offend against piety and religion in deed or word. He did not even discuss that topic so favored by other talkers, "the Nature of the Universe": and avoided speculation on the so-called "Cosmos" of the Professors, how it works, and on the laws that govern the phenomena of the heavens: indeed he would argue that to trouble one's mind with such problems is sheer folly. In the first place, he would inquire, did these thinkers

[57] George Sarton, *Portraits of ancient men of science* (Uppsala: Lychnos, 1945), p. 254.

[58] See extract from *Phaidōn*, quoted below.

[59] We can trust only the early Platonic dialogues; later on, Plato introduced Socrates only as his own mouthpiece. As explained in Chapter XVI, this was a real betrayal.

suppose that their knowledge of human affairs was so complete that they must seek these new fields for the exercise of their brains; or that it was their duty to neglect human affairs and consider only things divine? Moreover, he marveled at their blindness in not seeing that man cannot solve these riddles; since even the most conceited talkers on these problems did not agree in their theories, but behaved to one another like madmen. As some madmen have no fear of danger and others are afraid where there is nothing to be afraid of, as some will do or say anything in a crowd with no sense of shame, while others shrink even from going abroad among men, some respect neither temple nor altar nor any other sacred thing, others worship stocks and stones and beasts, so is it, he held, with those who worry with "Universal Nature." Some hold that *What is* is one, others that it is infinite in number: some that all things are in perpetual motion, others that nothing can ever be moved at any time: some that all life is birth and decay, others that nothing can ever be born or ever die. Nor were those the only questions he asked about such theorists. Students of human nature, he said, think that they will apply their knowledge in due course for the good of themselves and any others they choose. Do those who pry into heavenly phenomena imagine that, once they have discovered the laws by which these are produced, they will create at their will winds, waters, seasons and such things to their need? Or have they no such expectation, and are they satisfied with knowing the causes of these various phenomena?

Such, then, was his criticism of those who meddle with these matters. His own conversation was ever of human things. The problems he discussed were, What is godly, what is ungodly; what is beautiful, what is ugly; what is just, what is unjust; what is prudence, what is madness; what is courage, what is cowardice; what is a state, what is a statesman; what is government, and what is a governor; – these and others like them, of which the knowledge made a "gentleman," in his estimation, while ignorance should involve the reproach of "slavishness." [60]

This description in Xenophon's homely but pleasant style is very interesting, because it evokes the philosophic and scientific puzzles that the Athenians were expected to solve and that caused Socrates' rebellion. The word is not too strong. Socrates, who was depressed, as every good citizen was, by the vicissitudes of incessant wars, by the political intrigues, by economic difficulties, was irritated by the childish discussions and the empty discourses of the sophists, and also by the unwarranted hypotheses of the philosophers and the cosmologists. Before trying to account for the cosmos, would it not be better to put our own house in order? Instead of trying to understand inaccessible objects, should we not make clear the things that we can control? We are men; should we not try to know ourselves and other men before everything else? This reminds one of a story told by Aristoxenos of Tarentum (IV–2 B.C.). A Hindu sage met Socrates in Athens and asked him, "You call yourself a philosopher; what do you concern yourself with?" Socrates answered that he was studying human things, whereupon the Hindu began to laugh and remarked that it was impossible to understand anything about human things as long as one did not know the divine things. The anecdote is doubly interesting, in that it shows a very definite contrast between the Socratic and the Hindu manner of thinking and also because it is one of the very few definite examples of intercourse between Greek and Hindu philosophers. The presence of the latter in Egypt and Greece is plausible enough.[61]

The objection of the Hindu philosopher was partly met in the Platonic dialogue, *Alcibiades I*, a dialogue between Socrates and Alcibiades when the latter was

[60] Xenophon, *Memorabilia*, I, I, 10–17. Translation by Edgar Cardew Marchant, Loeb Classical Library (1923), p. 7.

[61] For other examples, see A. J. Festugière, "Trois rencontres entre la Grèce et l'Inde," *Revue de l'histoire des religions* *125*, 32–57 (1942).

eighteen years old; the time of it would thus be 432. In the third and last section they discuss the Delphic maxim, "Know thyself," and Socrates argues that to know oneself one must consider one's soul and especially the divine part of it. He concludes, "This part of the soul resembles God and whoever looks at this and comes to know all that is divine, will gain thereby the best knowledge of himself." [62] Is the *Alcibiades I* genuine? Some scholars consider it to be genuine but an early work written at the very beginning of the fourth century. Others claim that it is apocryphal and Bidez [63] adduces this very passage in order to prove it. Even if the dialogue is genuine, does the section to which reference has been made represent Socrates' thought or Plato's? The dialogue might be genuine and yet the words ascribed to Socrates apocryphal, as such.

Socrates' objection to astronomy as expressed by Xenophon is hardly above the level of the old Yankee gibe, "People are always talking of the weather, but they do nothing about it." It was thus foolish and unfair to call him, as Aristophanes did, a meteorosophist, for he was just the opposite. The end of Xenophon's quotation (above) is particularly good; it sums up the main tendency of Socrates' teaching. We might put it this way: "Let us be humbler than the physiologists, and more honest than the sophists. The knowledge that we must try to obtain is determined by our needs, personal and social; the main thing is to know how to live happily and honorably and to be good citizens."

This required the use of a special method, the application of which would be to all men what his conscience is to each individual. Morals and politics must be built carefully on a sound basis. Metaphysics must be subordinated to ethics. If we wish to discuss profitably, we must analyze our propositions, define the terms that we are using, and know exactly what we are talking about. We must classify the things we are dealing with, that is, we must try to know them in relation to other things; this again requires description and definition of each of them. One might then proceed by induction (*epagōgē*), that is, by enumeration of all the particulars to be considered, and reach a logical conclusion. The dialectical art used by Socrates he called obstetric (*maieuticē*), in remembrance of his mother's profession. By astute questions he brought out from the people he interviewed the admission of their errors, and the recognition of the truth. In a conversation with the courtesan Theodote he was even bolder, and explaining to her "how to make friends" he called himself a procurer.[64] This episode is a good illustration of Socratic irony and of his eagerness as a teacher. Indeed, he was ready to talk with everybody whom he might meet in the streets or in a friendly house, to engage them in a discussion, bring out his own favorite ideas, and oblige them to admit their validity.

Socrates was the first semanticist,[65] explaining to the people with whom he talked the danger of using "big words" or abstract words of which they did not grasp the meaning.

Virtue, he insisted, is largely a matter of knowledge, and hence it can be taught. The outstanding virtue is temperance (moderation). His notion of God was very different from the abstract intelligence (*nus*) of Anaxagoras; it came close to our

[62] Plato, *Alcibiades I*, 133c.

[63] Bidez, *Eos*, p. 122.

[64] *Memorabilia*, III, XI.

[65] The semantic point of view, in its relation to common speech, is beautifully illustrated by S. I. Hayakawa, *Language in action. A guide to accurate thinking, reading, and writing* (250 pp.; New York: Harcourt, Brace, 1941) [*Isis 34*, 84 (1942–43)]. Socrates would have loved that book.

ΠΛΑΤΩΝΟΣ, ἈΠΟλογία Σωκράτους,

Ἠθική.

ΟΤΙ μὲν ὑμεῖς, ὦ ἄνδρες Ἀ- A
θηναῖοι, πεπόνθατε ὑπὸ
τῶν ἐμῶν κατηγόρων, οὐκ οἶ-
δα· ἐγὼ δ' οὖν καὶ αὐτὸς ὑπ'
αὐτῶν ὀλίγου δεῖν ἐμαυτοῦ
ἐπελαθόμην· οὕτω πιθανῶς ἔλεγον. καίτοι ἀλη-
θές γε (ὡς ἔπος εἰπεῖν) οὐδὲν εἰρήκασι. μάλιστα δὲ αὐτῶν ἓν ἐθαύμασα τῶν πολλῶν ὧν ἐψεύ-
σαντο, τοῦτο, ἐν ᾧ ἔλεγον ὡς χρῆν ὑμᾶς εὐλα-
βεῖσθαι μὴ ὑπ' ἐμοῦ ἐξαπατηθῆτε, ὡς δεινοῦ
ὄντος λέγειν. τὸ γὰρ μὴ αἰσχυνθῆναι ὅτι αὐτίκα
ὑπ' ἐμοῦ ἐξελεγχθήσονται ἔργῳ, ἐπειδὰν μηδ' B
ὁπωστιοῦν φαίνωμαι δεινὸς λέγειν, τοῦτό μοι ἔδο-
ξεν αὐτῶν ἀναισχυντότατον εἶναι. εἰ μὴ ἄρα δεινὸν
καλοῦσιν οὗτοι λέγειν, τὸν τἀληθῆ λέγοντα. εἰ μὲν
γὰρ τοῦτο λέγουσιν, ὁμολογοίην ἂν ἔγωγε οὐ κατὰ τούτους εἶναι ῥήτωρ. οὗτοι μὲν οὖν, ὥσπερ ἐγὼ λέγω, οὐδὲν ἀληθὲς εἰρήκασιν· ὑμεῖς δέ μου ἀκούσεσθε
πᾶσαν τὴν ἀλήθειαν. οὐ μέντοι μὰ Δία, ὦ ἄνδρες
Ἀθηναῖοι, κεκαλλιεπημένους γε λόγους (ὥσπερ οἱ
τούτων) ῥήμασί τε καὶ ὀνόμασιν, οὐδὲ κεκοσμη-
μένους, ἀλλ' ἀκούσεσθε εἰκῇ λεγόμενα τοῖς ἐπι- C
τυχοῦσιν ὀνόμασι. πιστεύω γὰρ δίκαια εἶναι ἃ λέγω,
καὶ μηδεὶς ὑμῶν προσδοκησάτω ἄλλως. οὐδὲ γὰρ
ἂν δήπου πρέποι, ὦ ἄνδρες, τῇδε τῇ ἡλικίᾳ, ὥσπερ
μειρακίῳ πλάττοντι λόγους, εἰς ὑμᾶς εἰσιέναι. καὶ
μέντοι καὶ πάνυ, ὦ ἄνδρες Ἀθηναῖοι, τοῦτο ὑμῶν δέο-
μαι καὶ παρίεμαι ἐὰν διὰ τῶν αὐτῶν λόγων ἀ-
κούητέ μου ἀπολογουμένου δι' ὧνπερ εἴωθα λέγειν καὶ ἐν ἀγορᾷ καὶ ἐπὶ τῶν τραπεζῶν, (ἵνα ὑ-
μῶν οἱ πολλοὶ ἀκηκόασι, καὶ ἄλλοθι) μήτε θαυ-
μάζειν μήτε θορυβεῖν τούτου ἕνεκα. ἔχει γὰρ οὑτω- D
σί· νῦν ἐγὼ πρῶτον ἐπὶ δικαστήριον ἀναβέβηκα,
ἔτη γεγονὼς πλείω ἑβδομήκοντα. ἀτεχνῶς οὖν ξέ-
νως ἔχω τῆς ἐνθάδε λέξεως. ὥσπερ οὖν ἂν εἰ τῷ
ὄντι ξένος ἐτύγχανον ὤν, ξυνεγιγνώσκετε δήπου ἄν
μοι εἰ ἐν ἐκείνῃ τῇ φωνῇ τε καὶ τῷ τρόπῳ ἔλεγον ἐν

Me diuersum ab illis oratorẽ esse fatear. Nam illi &c.

PLATONIS, APOlogia Socratis,

Ad mores spectans.

QVONAM [a] modo, Athenienses, animi vestri accusatorum meorum oratione affecti fuerint atque comparati, nescio: equidẽ ipse illorum sermonibus ita sum concitatus, vt meipsius propemodum fuerim oblitus. ita accommodatè ad fidem & persuasionem dixerunt. quanquã nihil certè (vt vno verbo cõplectar) veri dixerunt. At inter multa quæ falsò protulerunt, illud vnum sum demiratus, quòd vos monuerint vt diligenter caueretis ne à me seduceremini, quasi dicendi facultate plurimùm valeam. Quòd autem minimè illos puduit eius rei, in cuius manifesto mendacio re ipsa à me continuò deprehendentur, vbi primùm à vobis ad dicẽdum planè ineptus comperiar: illud verò omnium mihi visum est impudentissimũ. Nisi fortè eum dicendi facultate valere dicant, qui verè dicit. Si enim huiusmodi intelligant, equidẽ (non ex illorum tamen sententia) me oratorem esse profitear. Atqui illi, quemadmodum prædixi, nihil veri dixerunt: [b] vos contrà à me omnem veritatem percipietis. Non tamen per Iouem, Athenieñses, elegantibus verborum sententiarúmq; ornamentis contextam orationem meam (cuiusmodi est illorum oratio) sed temerè & de medio petitos vulgaribúsque vocabulis enuntiatos sermones audietis. AEqua tamen & iusta esse credo ea quæ sum dicturus: neque aliter vestrûm quisquã expectet. Neque enim consentaneum esset me, hac ætate, tanquam puerũ fictis sermonibus ad vos prodire. [c] Verùm hoc vnum à vobis etiam atque etiam peto & efflagito, Athenienses, si eorundem illorum sermonũ via & ratione qua cõsueui dicere & in foro & in nummulariorum mensis (quibus in locis & aliis vestrûm plerique me conspexerunt) defendentem me audiatis: illud, nec vobis mirũ videatur, nec hac de causa tumultũ excitetis. Sic enim omnino se res habet. nũc ego primùm in iudicium prodeo, annos plus sexaginta natus, vt in hoc quidem forensi genere dicendi planè sim peregrinus. AEquè igitur acsi reuera peregrinus essem, ita mihi veniam dandam statuitote, qui nimirum & iisdem verbis & eadem loquendi forma vtar, in

a Genus causæ δικανικόν. Socrates enim in foro coram iudicibus capitis defensionem instituit: suo tamen modo, vt nimirũ philosophus, & quidẽ is, qui ab inflato & grandi dicendi genere de industria abhorreat. In proœmio (vt in defensionibus, rei ipsius vi & natura, vsitatum est) conflat inuidiam aduersariis: qui non solum dolosa eloquentia audientiũ animos percelluerint, sed & impudenti mẽdacio conati sint ipsũ ipsiúsque causã suspectam inuisamque reddere.

b Pollicetur se vera dicturum, vt beneuolentiam attentionémque sibi conciliet.

c Προκατάληψις, qua veniam petit pro ea dicendi ratione, qua sit vsurus, foro quidẽ illa inusitata, sibi tamen longo consuetudinis vsu inoleta & familiari. quam quidem vt necessariam excusationem prætexit, consuetudinis reuerentia, quæ quasi legis habet valorem.

Fig. 59. Beginning of the *Apology of Socrates* in the Greek-Latin edition (3 vols., folio; Geneva?: Henricus Stephanus, 1578). [From the copy in the Harvard College Library.] The Greek was edited by Henri Estienne = Henricus Stephanus (1528–1598); the Latin translation is by Jean de Serres = Johannes Serranus (1540–1598). The works are grouped in six syzygies (conjugations). The pagination of this edition is preserved in every scholarly edition. For example, this is page 17 of volume 1; in the Loeb Classical Library edition, at the beginning of the *Apology*, one will find in the margin St. 1, p. 17. For the title page, see Fig. 81.

idea of Providence (*pronoia*). We must care for our soul, and be grateful for it to the divine Providence; our consciousness of it is our true self. Piety is one of the essential virtues, and one of the first conditions of it is an impulse toward the divine. There was thus in Socrates a certain amount of mysticism,[66] a mysticism not of the Hindu kind but tempered with rationalism and common sense. He was also somewhat of a missionary, for he believed that he had received a definite mission, to care for the souls of his fellow citizens, and to teach them truth and goodness, and it was his duty to obey that divine command. As he put it in his proud *Apology* (Fig. 59).

> Know that the God commands me to do this, and I believe that no greater good ever came to pass in the city than my service to the God. For I go about doing nothing else than urging you, young and old, not to care for your persons or your property more than for the perfection of your souls, or even so much; and I tell you that virtue does not come from money, but from virtue come money and all other good things to man, both to the individual and to the state. If by saying these things I corrupt the youth, these things must be injurious; but if anyone asserts that I say other things than these, he says what is untrue. Therefore I say to you, men of Athens, either do as Anytos[67] tells you, or not, and either acquit me, or not, knowing that I shall not change my conduct even if I am to die many times over.[68]

In the *Gorgias*, Socrates explained that it is better to suffer than to do wrong, and that the unjust man is less unhappy if he is punished. The *Gorgias* was Plato's own apology, but there is no reason to doubt the authenticity of the ideas that he ascribes to Socrates. Our doubts must be restricted to the lights and shades, for no witness can give a testimony but he colors it with his own feelings.

One would like to go on and to quote more of Socrates' sayings, as one would quote from the Gospels, but it is better to send the reader back to the early dialogues of Plato and to Xenophon, for all those sayings are more luminous in their context, and we must try to evoke Socrates in his wholeness. We realize that he was utterly different not only from the sophists, but also from the philosophers who had preceded him, even from the wise Democritos. He introduced something radically new in human experience – lay wisdom combined with sanctity; he made of ethics and politics parts of religion.

His was an eccentric personality, often rough and cynical, very rational in spite of the strange mystical tendencies that have already been mentioned. He often referred to the divine voice that was guiding him, and his singular charm and personal magnetism were explained in mystical terms, for example, in Alcibiades' speech in the *Symposion*, and more elaborately in the Platonic dialogue *Theages*.[69] The best way to illustrate his singular greatness is to tell the story of his death.

[66] Apropos of this, it would be interesting to discuss Socrates' belief in his daemon (*to daimonion*) who guided him, or as we would put it, in his divine inspiration. He was a believer and an enthusiast. One should also consider his faith in divination (*manticē*), which he shared with all the ancients, but that would carry us much too far.

[67] Anytos was the most important of Socrates' three accusers, bitterly hostile to the sophists. The part he had just taken in expelling the Thirty Tyrants had increased his authority. Horace called Socrates "Anyti reus" (*Satirae*, II, IV, 3).

[68] Plato's *Apology of Socrates* (30A). Translation by Harold North Fowler (Loeb Classical Library).

[69] "Theages, or on wisdom: obstetric" was not written by Plato, however. It is a relatively late production (say II B.C.); yet it found its way to the library of Alexandria and was included in the earliest Platonic canon, compiled by the astrologer, Thrasyllos of Alexandria (d. A.D. 36), and therefore in many editions of Plato's works (Stephanus, pp. 121–131; Loeb, vol. 8).

His activities as described by his disciples were innocent enough; yet it is easy to imagine that his irony must have wounded the vanity of many people, and the simplicity of his life was an implied disavowal of the men whose main purpose was to obtain more money (honestly or not) and to indulge themselves. Socrates was in fact a living reproach to them all, and we can hardly expect them to have liked that. In spite of his benevolence, he made enemies who were bent on his destruction. The Athenian democracy was devout to the point of superstition, and Socrates' rationalism, though tempered as it was by mysticism, was shocking to them; his very mysticism was so different from their bigotry that it created an additional grievance. The enemies of Socrates and those who were jealous of him and could not bear his self-righteousness any longer welcomed Aristophanes' calumnies, embellished them and circulated them. Soon after the expulsion of the Thirty Tyrants in 399, he was indicted in the following terms: "Socrates is guilty of rejecting the gods acknowledged by the state and of bringing in strange deities; he is also guilty of corrupting the youth." He was condemned to put himself to death by drinking a concoction of hemlock. It must be added to the credit of the Athenian democracy that he was almost acquitted, for his guilt was decided by a majority of only 30 out of 501 judges. That majority would easily have been won in his favor, if he had taken any trouble to obtain the assembly's good will by cautious words, if he had defended himself in good earnest; he did the opposite and his apology was a masterpiece of irony, the kind of speech that was bound to turn narrow-minded people against him.[70]

It so happened that his condemnation was pronounced on the morrow of the day when the sacred ships had sailed for Delos, and it could not be carried out without profanation until their return a month later. He had thus a whole month to spend in prison, a month which Athenian generosity enabled him to pass in the communion of his friends and family.[71] The conversations of Socrates with them are preserved in Plato's dialogues,[72] especially in two immortal ones, *Criton* (on duty) and *Phaidon* (on the soul). Criton was a lifelong friend of Socrates, a man of substance, who visited him in his prison and tried to persuade him to escape. It is possible that the judges themselves would have welcomed that solution, but Socrates rejected it. The first duty of a citizen, he insisted, is to obey the laws of the city, even if their application is unjust. Injustice cannot be corrected with injustice. If the city has condemned him to die, any escape from the ordeal would be a kind of treason; he must die. This dialogue is the noblest defense of the law that has ever been written. It is Plato's, of course, but it represents Socrates' views, for the latter, in fact, did not try to escape.

[70] The condemnation of Socrates was very probably political. At the close of the Peloponnesian War he was accused of educating the people who had betrayed democracy and who had conspired with the enemy to bring about the downfall of Athens. It will suffice to recall the names of the treacherous oligarchs Alcibiades, Critias, and Charmides, all of whom were his disciples. According to Popper, Socrates had only *one* worthy successor, Antisthenes; see K. P. Popper, *The open society,* (London: Routledge, 1945), vol. 1, pp. 168, 171.

For a discussion of Socrates' trial from the legal point of view, see Sir John Macdonell, *Historical trials* (Oxford, 1927), pp. 1–18.

[71] Could one imagine any one of the modern dictators showing such magnanimity to his victims? He would do the very opposite and put them in solitary confinement, or perhaps cause them to be afflicted and "questioned" under torture. This shows how much we have progressed since 399 B.C.!

[72] Four Platonic dialogues deal with the trial and death of Socrates, to wit, *Euthyphron* (on holiness), the *Apology,* or the *Defense of Socrates at his Trial, Criton,* and *Phaidon.*

The *Phaidon* (Figs. 60, 61, 62) records the conversation of eight people in the prison during Socrates' last days, and mentions a great many more. The philosopher is glad to die; ideas are eternal, and his soul will continue to live afterward. The *Phaidon* ends with a description of Socrates' death, which must be quoted *in extenso*:

When Socrates had said this, he got up and went into another room to bathe; Criton followed him, but he told us to wait. So we waited, talking over with each other and discussing the discourse we had heard, and then speaking of the great misfortune that had befallen us, for we felt that he was like a father to us and that when bereft of him we should pass the rest of our lives as orphans. And when he had bathed and his children had been brought to him – for he had two little sons and one big one – and the women of the family had come, he talked with them in Criton's presence and gave them such directions as he wished; then he told the women to go away, and he came to us. And it was now nearly sunset; for he had spent a long time within. And he came and sat down fresh from the bath. After that not much was said, and the servant of the eleven came and stood beside him and said: "Socrates, I shall not find fault with you, as I do with others, for being angry and cursing me, when at the behest of the authorities, I tell them to drink the poison. No, I have found you in all this time in every way the noblest and gentlest and best man who has ever come here, and now I know your anger is directed against others, not against me, for you know who are to blame. Now, for you know the message I came to bring you, farewell and try to bear what you must as easily as you can." And he burst into tears and turned and went away. And Socrates looked up at him and said: "Fare you well, too; I will do as you say." And then he said to us: "How charming the man is! Ever since I have been here he has been coming to see me and talking with me from time to time, and has been the best of men, and now how nobly he weeps for me! But come, Criton, let us obey him, and let someone bring the poison, if it is ready; and if not, let the man prepare it." And Criton said: "But I think, Socrates, the sun is still upon the mountains and has not yet set; and I know that others have taken the poison very late, after the order has come to them, and in the meantime have eaten and drunk and some of them enjoyed the society of those whom they loved. Do not hurry; for there is still time."

And Socrates said: "Criton, those whom you mention are right in doing as they do, for they think they gain by it; and I shall be right in not doing as they do; for I think I should gain nothing by taking the poison a little later. I should only make myself ridiculous in my own eyes if I clung to life and spared it, when there is no more profit in it. Come," he said, "do as I ask and do not refuse."

Thereupon Criton nodded to the boy who was standing near. The boy went out and stayed a long time, then came back with the man who was to administer the poison, which he brought with him in a cup ready for use. And when Socrates saw him, he said: "Well, my good man, you know about these things; what must I do?" "Nothing," he replied, "except drink the poison and walk about till your legs feel heavy; then lie down, and the poison will take effect of itself."

At the same time he held out the cup to Socrates. He took it, and very gently, Echecrates, without trembling or changing color or expression, but looking up at the man with wide open eyes, as was his custom, said: "What do you say about pouring a libation to some deity from this cup? May I, or not?" "Socrates," said he, "we prepare only as much as we think is enough." "I understand," said Socrates; "but I may and must pray to the gods that my departure hence be a fortunate one; so I offer this prayer, and may it be granted." With these words he raised the cup to his lips and very cheerfully and quietly drained it. Up to that time most of us had been able to restrain our tears fairly well, but when we watched him drinking and saw that he had drunk the poison, we could do so no longer, but in spite of myself my tears rolled down in floods, so that I wrapped my face in my cloak and wept for myself; for it was not for him that I wept, but for my own misfortune in being deprived of such a friend. Criton had got up and gone away even before I did, because he could not restrain his tears. But Apollodoros, who had been

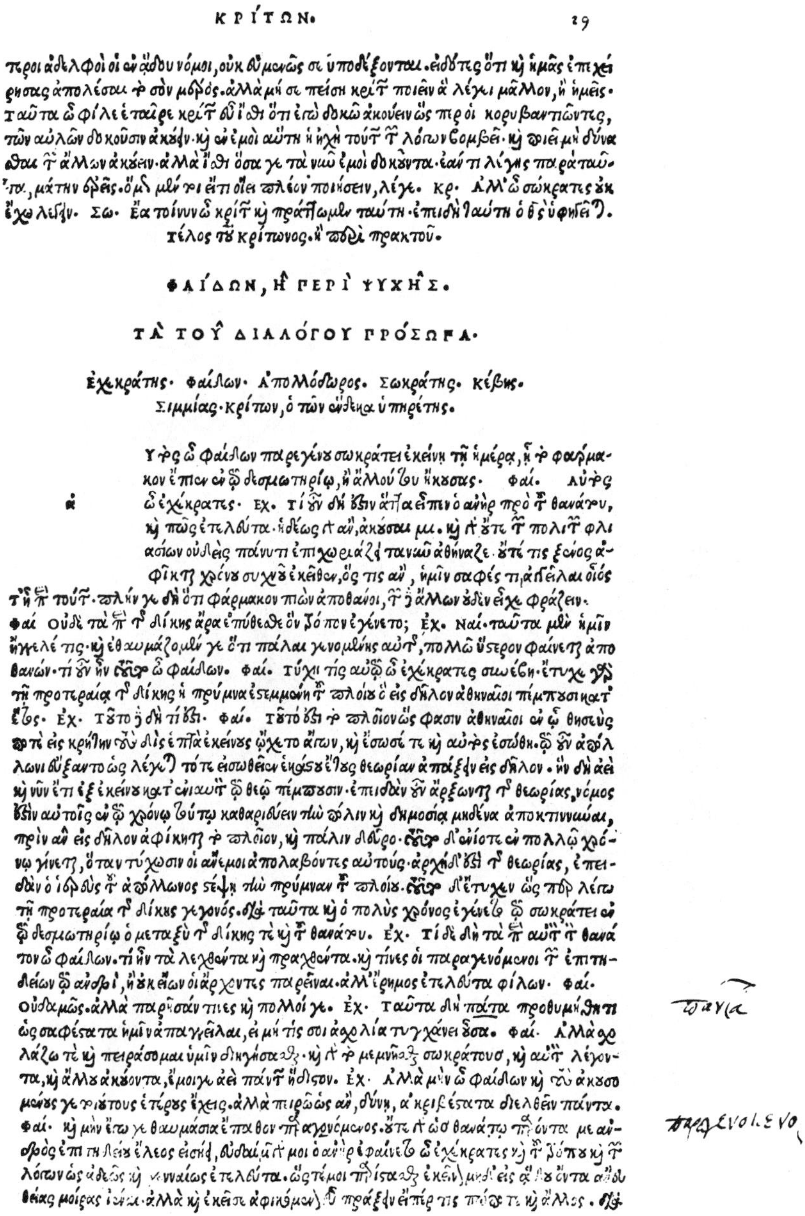

ΚΡΙΤΩΝ. 29

περὶ ἀδελφοὶ οἱ ἐν ἅδου νόμοι, οὐκ εὐμενῶς σε ὑποδέξονται. εἰδότες ὅτι καὶ ἡμᾶς ἐπιχείρησας ἀπολέσαι τὸ σὸν μέρος. ἀλλὰ μή σε πείσῃ κρίτων ποιεῖν ἃ λέγει μᾶλλον, ἢ ἡμεῖς. ταῦτα ὦ φίλε ἑταῖρε κρίτων εὖ ἴσθι ὅτι ἐγὼ δοκῶ ἀκούειν ὥσπερ οἱ κορυβαντιῶντες, τῶν αὐλῶν δοκοῦσιν ἀκούειν. καὶ ἐν ἐμοὶ αὕτη ἡ ἠχὴ τούτων τῶν λόγων βομβεῖ. καὶ ποιεῖ μὴ δύνασθαι τῶν ἄλλων ἀκούειν. ἀλλὰ ἴσθι ὅσα γε τὰ νῦν ἐμοὶ δοκοῦντα. ἐάν τι λέγῃς παρὰ ταῦτα, μάτην ἐρεῖς. ὅμως μέντοι εἴ τι οἴει πλέον ποιήσειν, λέγε. ΚΡ. ἀλλ' ὦ σώκρατες οὐκ ἔχω λέγειν. ΣΩ. ἔα τοίνυν ὦ κρίτων καὶ πράττωμεν ταύτῃ. ἐπειδὴ ταύτῃ ὁ θεὸς ὑφηγεῖται.

τέλος τοῦ κρίτωνος, ἢ περὶ πρακτοῦ.

ΦΑΙΔΩΝ, Η ΠΕΡΙ ΨΥΧΗΣ.

ΤΑ ΤΟΥ ΔΙΑΛΟΓΟΥ ΠΡΟΣΩΠΑ.

ἐχεκράτης. φαίδων. ἀπολλόδωρος. σωκράτης. κέβης. σιμμίας. κρίτων, ὁ τῶν ἕνδεκα ὑπηρέτης.

Αὐτὸς ὦ φαίδων παρεγένου σωκράτει ἐκείνῃ τῇ ἡμέρᾳ, ᾗ τὸ φάρμακον ἔπιεν ἐν τῷ δεσμωτηρίῳ, ἢ ἄλλου του ἤκουσας. ΦΑΙ. αὐτὸς ὦ ἐχέκρατες. ΕΧ. τί οὖν δή ἐστιν ἅττα εἶπεν ὁ ἀνὴρ πρὸ τοῦ θανάτου, καὶ πῶς ἐτελεύτα. ἡδέως γὰρ ἂν ἀκούσαιμι. καὶ γὰρ οὔτε τῶν πολιτῶν φλιασίων οὐδεὶς πάνυ τι ἐπιχωριάζει τανῦν ἀθήναζε. οὔτε τις ξένος ἀφῖκται χρόνου συχνοῦ ἐκεῖθεν, ὅστις ἂν ἡμῖν σαφές τι ἀγγεῖλαι οἷός τ' ἦν περὶ τούτων. πλήν γε δὴ ὅτι φάρμακον πιὼν ἀποθάνοι. τῶν δ' ἄλλων οὐδὲν εἶχε φράζειν. ΦΑΙ. οὐδὲ τὰ περὶ τῆς δίκης ἄρα ἐπύθεσθε ὃν τρόπον ἐγένετο; ΕΧ. ναί. ταῦτα μὲν ἡμῖν ἤγγειλέ τις. καὶ ἐθαυμάζομέν γε ὅτι πάλαι γενομένης αὐτῆς, πολλῷ ὕστερον φαίνεται ἀποθανών. τί οὖν ἦν τοῦτο ὦ φαίδων. ΦΑΙ. τύχη τις αὐτῷ ὦ ἐχέκρατες συνέβη. ἔτυχε γὰρ τῇ προτεραίᾳ τῆς δίκης ἡ πρύμνα ἐστεμμένη τοῦ πλοίου ὃ εἰς δῆλον ἀθηναῖοι πέμπουσι. ΕΧ. τοῦτο δὲ δὴ τί ἐστι. ΦΑΙ. τοῦτ' ἔστι τὸ πλοῖον ὥς φασιν ἀθηναῖοι ἐν ᾧ θησεύς ποτε εἰς κρήτην τοὺς δὶς ἑπτὰ ἐκείνους ᾤχετο ἄγων, καὶ ἔσωσέ τε καὶ αὐτὸς ἐσώθη. τῷ οὖν ἀπόλλωνι ηὔξαντο ὡς λέγεται τότε εἰ σωθεῖεν ἑκάστου ἔτους θεωρίαν ἀπάξειν εἰς δῆλον. ἣν δὴ ἀεὶ καὶ νῦν ἔτι ἐξ ἐκείνου κατ' ἐνιαυτὸν τῷ θεῷ πέμπουσιν. ἐπειδὰν οὖν ἄρξωνται τῆς θεωρίας, νόμος ἐστὶν αὐτοῖς ἐν τῷ χρόνῳ τούτῳ καθαρεύειν τὴν πόλιν καὶ δημοσίᾳ μηδένα ἀποκτιννύναι, πρὶν ἂν εἰς δῆλον ἀφίκηται τὸ πλοῖον, καὶ πάλιν δεῦρο. τοῦτο δ' ἐνίοτε ἐν πολλῷ χρόνῳ γίγνεται, ὅταν τύχωσιν οἱ ἄνεμοι ἀπολαβόντες αὐτούς. ἀρχὴ δ' ἐστὶ τῆς θεωρίας, ἐπειδὰν ὁ ἱερεὺς τοῦ ἀπόλλωνος στέψῃ τὴν πρύμναν τοῦ πλοίου. τοῦτο δ' ἔτυχεν ὥσπερ λέγω τῇ προτεραίᾳ τῆς δίκης γεγονός. διὰ ταῦτα καὶ ὁ πολὺς χρόνος ἐγένετο τῷ σωκράτει ἐν τῷ δεσμωτηρίῳ ὁ μεταξὺ τῆς δίκης τε καὶ τοῦ θανάτου. ΕΧ. τί δὲ δὴ τὰ περὶ αὐτὸν τὸν θάνατον ὦ φαίδων. τί ἦν τὰ λεχθέντα καὶ πραχθέντα. καὶ τίνες οἱ παραγενόμενοι τῶν ἐπιτηδείων τῷ ἀνδρί, ἢ οὐκ εἴων οἱ ἄρχοντες παρεῖναι. ἀλλ' ἔρημος ἐτελεύτα φίλων. ΦΑΙ. οὐδαμῶς. ἀλλὰ παρῆσάν τινες καὶ πολλοί γε. ΕΧ. ταῦτα δὴ πάντα προθυμήθητι ὡς σαφέστατα ἡμῖν ἀπαγγεῖλαι, εἰ μή τίς σοι ἀσχολία τυγχάνει οὖσα. ΦΑΙ. ἀλλὰ σχολάζω τε καὶ πειράσομαι ὑμῖν διηγήσασθαι. καὶ γὰρ τὸ μεμνῆσθαι σωκράτους, καὶ αὐτὸν λέγοντα, καὶ ἄλλου ἀκούοντα, ἔμοιγε ἀεὶ πάντων ἥδιστον. ΕΧ. ἀλλὰ μὴν ὦ φαίδων καὶ τοὺς ἀκουσομένους γε τοιούτους ἑτέρους ἔχεις. ἀλλὰ πειρῶ ὡς ἂν δύνῃ, ἀκριβέστατα διελθεῖν πάντα. ΦΑΙ. καὶ μὴν ἔγωγε θαυμάσια ἔπαθον παραγενόμενος. οὔτε γὰρ ὡς θανάτῳ παρόντα με ἀνδρὸς ἐπιτηδείου ἔλεος εἰσῄει. εὐδαίμων γάρ μοι ὁ ἀνὴρ ἐφαίνετο ὦ ἐχέκρατες καὶ τοῦ τρόπου καὶ τῶν λόγων ὡς ἀδεῶς καὶ γενναίως ἐτελεύτα. ὥστε μοι παρίστασθαι ἐκεῖνον μηδ' εἰς ἅδου ἰόντα ἄνευ θείας μοίρας ἰέναι. ἀλλὰ καὶ ἐκεῖσε ἀφικόμενον εὖ πράξειν εἴπερ τις πώποτε καὶ ἄλλος. διὰ

Fig. 60. Beginning of the *Phaidon*, p. 29 of the Greek *princeps* of Plato's works (2 parts folio; Venice: Aldus and Marcus Musurus, 1513). [From the copy in the Harvard College Library.] All the dialogues are arranged in tetralogies, and the four dialogues dealing with Socrates' trial and death form the first tetralogy. See Fig. 80.

Fig. 61. Frontispiece of the first English translation of the *Apology* and *Phaidon*. (London, 1675.) [From the copy in the Harvard College Library.]

weeping all the time before, then wailed aloud in his grief and made us all break down, except Socrates himself. But he said, "What conduct is this, you strange men! I sent the women away chiefly for this very reason, that they might not behave in this absurd way; for I have heard that it is best to die in silence. Keep quiet and be brave." Then we were ashamed and controlled our tears. He walked about and, when he said his legs were heavy, lay down on his back, for such was the advice of the attendant. The man who had administered the poison laid his hands on him and after a while examined his feet and legs, then pinched his foot hard and asked if he felt it. He said "No"; then after that, his thighs; and passing upwards in this way he showed us that

PLATO his

APOLOGY of *SOCRATES*;

AND

PHÆDO or Dialogue concerning the

Immortality of Mans Soul,

AND

Manner of SOCRATES his Death;

Carefully tranflated from the *Greek*,

AND

Illuftrated by Reflections upon both the *Athenian* Laws, and ancient Rites and Traditions concerning the Soul, therein mentioned.

Quintilianus inftitut. Orator. lib. 10 cap. 5.
Vertere Græca in Latinum *veteres noſtri Oratores optimum judicabant. Id ſe* L. Craſſus *in illis* Ciceronis *de Oratore libris dicit factitaſſe. Id* Cicero *ſua ipſe perſona frequentiſſime præcipit: quin etiam libros* Platonis [Timæum *nempe, quem inſcripſit* de Univerſitate] *atq;* Xenophontis *edidit hoc genere tranſlatos.*

LONDON,
Printed by *T. R. & N. T.* for *James Magnes* and *Richard Bentley* at the *Poſt-Office* in *Ruſſel-ſtreet* in *Covent-Garden*, 1675.

Fig. 62. Title page of the first English translation of the *Apology* and *Phaidon*; the translator is not named. [From the copy in the Harvard College Library.]

he was growing cold and rigid. And again he touched him and said that when it reached his heart, he would be gone. The chill had now reached the region about the groin, and uncovering his face, which had been covered, he said – and these were his last words – "Criton, we owe a cock to Aesculapius. Pay it and do not neglect it." "That," said Criton, "shall be done; but see if you have anything else to say." To this question he made no reply, but after a little while he moved; the attendant uncovered him; his eyes were fixed. And Criton when he saw it, closed his mouth and eyes.

Such was the end, Echecrates, of our friend, who was, as we may say, of all those of his time whom we have known, the best and wisest and most righteous man.[73]

For comparison wtih Plato's, this is Xenophon's final estimate of their Master:

All who knew what manner of man Socrates was and who seek after virtue continue to this day to miss him beyond all others, as the chief of helpers in the quest of virtue. For myself, I have described him as he was: so religious that he did nothing without counsel from the gods; so just that he did no injury, however small, to any man, but conferred the greatest benefits on all who dealt with him; so self-controlled that he never chose the pleasanter rather than the better course; so wise that he was unerring in his judgment of the better and the worse, and needed no counselor, but relied on himself for his knowledge of them; masterly in expounding and defining such things; no less masterly in putting others to the test, and convincing them of error and exhorting them to follow virtue and gentleness. To me then he seemed to be all that a truly good and happy man must be. But if there is any doubter, let him set the character of other men beside these things; then let him judge.[74]

As this book is written for men of science, it is well to add a few medical remarks:

Plato's description of the death of Socrates is a classic clinical account. It is startlingly in accord with what one would observe under similar circumstances now. The poison hemlock is conium, the dried fullgrown but unripe fruit of *Conium maculatum.* Conium is generally powdered after drying and it contains not less than 0.5 percent of coniine. This alkaloid was discovered in 1827 by Gieseke. It is the simple chemical, propylpyridine. Related alkaloids occur in conium, but they all have the same biological action. This is essentially a paralysis of the peripheral endings of motor nerves. The action comes on from the periphery and quickly comes to the diaphragm, which when it ceases to move, causes asphyxia and death. There is further evidence that there is paralysis of the sensory fibers but this is not as marked as the paralysis of motor nerves. Hayashi and Muto have shown that the phrenic nerve is more susceptible than other nerves. The phrenic controls the movements of the diaphragm (*Arch. Exp. Path. Pharmakol. 48*, 1901). You will find descriptions of the action of coniine in any good pharmacology.[75]

The condemnation of Socrates was inexcusable, but the manner of his execution was decent and compassionate. When we compare it with the sordid, clandestine, inhuman executions committed within our lifetime in many countries, not by individual murderers but by government orders, we are thoroughly ashamed.

His death was dignified in the extreme. There was no bitterness in his words, no anger, no denunciation. It was the death of a just and noble man. In its restraint and grace, it is a counterpart to some of the sepulchral monuments of that time.[76]

[73] Translation by Harold North Fowler taken from the Loeb edition of Plato, vol. 1, pp. 395–403.

[74] This is the final paragraph of the *Memorabilia* as translated by E. C. Marchant (Loeb Classical Library, 1923), p. 357.

[75] I owe them to the kindness of my friend, Dr. Chauncey D. Leake, pharmacologist and dean of the School of Medicine, University of Texas, Galveston (his letter of 22 October 1945).

[76] Percy Gardner, *Sculptured tombs of Hellas* (278 pp., 30 pls.; London, 1896). Maxime Collignon, *Les statues funéraires dans l'art grec* (412 pp., ill.; Paris, 1911). Alexander Conze, 1831–1914), *Die attischen Grabreliefs* (4 vols., atlas; Berlin, 1893–1922). The first two volumes are

It is certain that the circumstances of Socrates' death helped considerably in the establishment of his fame. In the first place, they inspired the veneration of his immediate disciples and sanctified him, and later they helped to fire the enthusiasm of Plato and Xenophon, who preserved his thoughts and transmitted them to posterity. Socrates' death is a magnificent climax to the efforts that Greek philosophers had made for more than a century in order to reach the truth. It consecrated the wisdom which he had attained partly because of their investigations and partly because of his own genius and sanctity.

Among the friends who attended his last moments were Echecrates of Phlius, one of the last of the Pythagoreans, Phaidon of Elis, Apollodoros of Phaleron, Cebes [77] and Simmias, both Thebans, Criton of Athens and his son Critobulos, Aischines "the Socratic," Antisthenes of Athens, and Euclides of Megara. It is remarkable that five of Socrates' immediate disciples (three of them present at his death) were themselves the founders of philosophic schools, to wit, Phaidon, who started a school in his native town Elis, Euclides, founder of the Megarian school, Antisthenes, founder of the Cynic school, and the two who were not present, Aristippos of Cyrene, chief of the Cyrenaic school, and Plato. The absence of the last named was due to illness, and we may well believe it, as it is stated in the *Phaidon*, by himself. One may say that the whole development of Greek philosophy after the fifth century was influenced by Socrates. We must not forget that during his long activity as an itinerant teacher and counselor Socrates must have made a deep impression upon the minds of people who were not philosophers or writers, yet were able to transmit his ideas, bad but powerful men like Critias and Alcibiades, and a great many others who were not so conspicuous for their qualities or their vices and whose names have not been recorded. Socrates was the first Greek philosopher to establish a moral system and to give priority to moral values. From then on, moral and political ideas were given more importance, and it would be no exaggeration to say that all the Western treatises on the subject derive directly or indirectly from his teaching. His life and death have helped to govern the ethics of the Western world. Their influence has not been cancelled or diminished by the progress of Christianity.

This is a history not of philosophy but of science, and it has sometimes been claimed that however good Socrates' influence may have been on philosophy, it was bad on science. Bearing in mind his rebellion against the astronomers, the meteorologists, and all the men who considered the heavens and the underground instead of the ordinary level of human life, some critics would call him a reactionary. Olmstead goes so far as to state that Socrates' influence on science was catastrophic.[78] We answer: on the surface, yes, but in reality, no. The men of science who have followed me so far and have read my account of the pre-Socratic philosophers have probably become as impatient and rebellious as Socrates himself. The

very readable and well-illustrated accounts of Greek tombs in general; Conze's work is a corpus of Attic sepulchral monuments. The reader's needs will be satisfied best by reading the relevant chapters in Gardner or Collignon.

[77] This Cebes is not the author of the *Tablet* (*Pinax*), a curious allegory of human life, as was formerly believed. The *Pinax* was written by a namesake who lived much later and was acquainted with Peripatetic and Stoic, as well as Pythagorean, doctrines. The first Greek writer who referred to the *Pinax* is Lucian of Samosata (*c.* 125–90), who believed it to be ancient, though it was probably not much prior to his own time.

[78] Olmstead, *History of Persia*, p. 446.

scientific method of those philosophers was bad, their speculations, based upon insufficient knowledge, were futile, their astronomic views were often silly; they were altogether on the wrong track. Even if one allows (as I do) that those adventures were unavoidable and necessary, they had lasted long enough. The philosophers of the fifth century seem to have exhausted all the fantasies of their age. There was something admirable in their boldness; but there had been enough of that. It was necessary to call a halt and Socrates did it. If he went too far, at any rate somebody had to do what he did, and perhaps no one else would have done it as well.

Moreover, some of his ideas were positive contributions which were necessary for the future development of science. First was his insistence upon clear definitions and classifications. There is no point in discussing if we do not know as correctly as possible what we are talking about. That is fundamental in science, even more than in philosophy. Second, he used a good method of logical discovery (his maieutics) and dialectics. Scientists must be trained to argue without logical flaws; otherwise, they will reach erroneous conclusions. Third, he had a deep sense of duty and respect for the law. The healthy growth of science requires moral purity, truthfulness, individual and social discipline; the bad citizen cannot be a good scientist. Fourth, his rational skepticism provides the basis of scientific research. The scientist must be ready to free the ground of prejudices and superstitions before he can begin to build. Of course, Socrates' doubting was not sufficiently thorough with regard to such matters as divination, but that was the fault of his environment. Our skepticism is always conditioned by the beliefs, however absurd, that are widely accepted by the majority of our neighbors.

Earlier philosophers were hardly aware of the fundamental importance of those four points; Socrates was fully aware of it, and he insisted upon them repeatedly, with great vigor; for that reason alone he would deserve a very high place in the history of science. His rebellion against sophistry and premature declarations of any kind is one in which every man of science would gladly join. In particular, the refusal to make unwarranted statements is the beginning of scientific wisdom.

Socrates' distinction between useful and nonuseful knowledge was not so happy; that was reactionary. When he decided that it was ridiculous to study the stars or the "so-called 'Cosmos' of the Professors,"[79] he was simply shutting a door that should have remained wide open. One may condemn bad scientific methods or futile controversies, but it is impossible to decide a priori which investigations are useful and which are not. The whole history of science is there to prove it; nothing might have seemed more silly to Socrates than to investigate the behavior of objects in close proximity to a piece of magnetic iron or of rubbed amber; and yet that was the way to the knowledge of magnetism and electricity and to all the electrical industries which have changed the face of the world. Socrates originated the endless quarrel between "pure and applied science," which may be settled by the remark that the latter would not grow or even exist without the former; he originated the other quarrel between "common sense" and the scientific paradoxes; as we now know, our common sense is often wrong, and the "paradoxes" disguise the real truth. He cannot be blamed too much, however, for making such mistakes at a time when the scientific experience of mankind was still rudimentary.

[79] *Ho calumenos hypo tōn sophistōn cosmos, Memorabilia,* quoted above.

THE BOOK OF JOB

This chapter, devoted to philosophy of the fifth century, is already very long, though it was restricted to the achievements of a relatively small nation, the Greek-speaking peoples. Within a century, they formulated some of the fundamental problems of philosophy; they did not solve them, but those problems are still teasing the intelligence of men. It would be rewarding to examine the philosophic ideas that were discussed in other nations of the same century, but that would carry us too far. For instance, it would be interesting to evoke K'ung Chi (V B.C.), a grandson of Confucius and the reputed author of two of the "Four Books,"[80] the *Doctrine of the mean* and possibly the *Great learning*, and Mo-ti (V B.C.), who combined utilitarian views with extreme altruism, and is sometimes called the founder of Chinese logic. Parallels with contemporary Hindu philosophy are made impossible, however tempting, by very doubtful chronology. There is one comparison, however, in which we may be permitted to indulge briefly, and this is one with the Book of Job.

That comparison is the more permissible, because it does not oblige us to go to countries as remote as India or China; it suffices to go to a country that was very close to the Greek world, though it remained curiously separated from it. The date of the composition of the Book of Job is uncertain, but the fifth century (or fourth) is the most probable.[81] The author was either a Jew or an Edomite,[82] in any case a Palestinian, and Palestine was closer to Attica than were many Greek outposts. He was familiar perhaps with Babylonian sources[83] and certainly with Egyptian ones; that is, he had drunk from the same sources as some of his Greek contemporaries, and yet the fruit of his meditation was very different from theirs. Consider that mystery for a moment: the Hebrew and the Greek imitated Egyptian models and produced respectively Hebrew and Greek masterpieces. What is imitation? Every man imitates his predecessors (the process of education is to a large extent a method of imitation of accepted models), but he imitates them according to his own genius; if he has any genius, he creates something new.

The Book of Job[84] is one of the masterpieces of the world's literature. Tennyson called it "the greatest poem of all times." Its subject is one that must always

[80] The Confucian tradition is based upon the Five Classics (wu ching) and the Four Books (ssŭ shu). (The numbers between parentheses in the following list indicate the pages of my *Introduction*, vol. 3, where the Chinese characters will be found, also references to the pages of the *Introduction* where more information concerning each is available.) The Five Classics are: 1. I-ching, Book of changes (2117); 2. Shu-ching, Book of history (2129); 3. Shih-ching, Book of poetry (2128); 4. Li-chi, Record of rites (2121); 5. Ch'un-ch'iu, Spring and autumn (2110). The Four Books are: 1. Ta hsüeh, Great learning (2131); 2. Chung yung, Doctrine of the mean (2110); 3. Lun-yü, Confucian analects (2123); 4. Mêng-tzŭ, Mencius (2123).

The Ta hsüeh and the Chung yung are parts of the Li chi and edited with it by Legge, *Sacred books of the East* (Oxford, 1885), vols. 27–28. Chinese–Latin–French edition of both by Seraphin Couvreur, *Les quatre livres* (Ho Kien Fou: Mission Catholique, 1910).

[81] The folktale upon which the Book of Job is based is, of course, of much greater antiquity; that is, Job may be a thousand years older than the Book of Job!

[82] The Edomites or Idumaeans were the descendants of Esau or Edom, brother of Jacob. They were a separate Hebrew tribe which had remained nomadic and was on a lower cultural level than the Israelites. The land of Edom is south of the Dead Sea.

[83] There is a "Babylonian Job," for which see Robert William Rogers (1864–1930), *Cuneiform parallels to the Old Testament* (New York, 1912), pp. 164–169.

[84] In my study of it, I was much helped by Robert H. Pfeiffer, *Introduction to the Old Testament* (New York: Harper, 1941), pp. 660–707 [*Isis 34*, 38 (1942–43)], a thorough analysis with full bibliography.

exercise the thoughts of man and vex his soul. How can one explain undeserved punishment, why do the wicked prosper and the good suffer? The theologic problem implied is called theodicy (so named by Leibniz), the vindication of the justice of God which permits natural or moral evil. How can the existence of evil be reconciled with the goodness and omnipotence of God? Job (meaning the author of the Book of Job) realizes that in view of God's inscrutable transcendance and of man's puny understanding the problem cannot be solved. A man's miseries engross his thoughts, but they are insignificant in the scheme of things. How dare we judge? Job's anxious questioning is deeply moving, because we know no more than he does.

The integrity of the Book of Job is questionable, and its composition heterogeneous.[85] We should not pay too much attention to its inconsistencies and ambiguities, for they are natural enough in the language of passionate men and the glory of poetic diction. The Book of Job is a poem, not a scientific treatise. The man who wrote it was a poet of genius, who described with vigorous terseness the marvels of creation and the wisdom of God. He combined knowledge and realism with a vivid imagination, his language is magnificent, and he uses images that have seldom been equaled.[86]

Out of the immemorial wisdom of the Ancient East, the Hebrew prophets had developed the idea of monotheism, endowing a national God with universal jurisdiction, and making of him the living symbol of moral perfection and of absolute justice; out of the same longing, the Greek philosophers had tried to explain the unity of the world on the basis of positive knowledge, and their idea of God was related less to morality than to physics and cosmology. Strangely enough, Job's God is in some respects closer to the Greek examples than to the Jewish one. He never refers to him by a personal name; his God is not a national but a cosmic one. This coincidence, however, is accident. There is no reason to assume that the author of Job was influenced by Greek models in any way (or vice versa). It is, therefore, highly significant that comparisons have been made between the Book of Job and Aischylos' *Prometheus bound.* This proves once more the unity of human genius, which is one of the forms of the unity of nature, and the image of the unity of God.

[85] This is discussed fully by Pfeiffer (pp. 667–675). The Book of Job includes inconsistencies which may be due to disarrangements, omissions, and interpolations. For example, the tendency of recent criticism is to regard the magnificent poem on divine wisdom (chapter 28, out of 42) as an interpolation. We cannot go into that and must take the Book as it is, assuming its integrity.

[86] I do not know Hebrew well enough to appreciate the original style and must base my judgment upon the English translations. Remember such phrases as these: I know that my Redeemer liveth (19:25); the eyelids of the dawn (3:9); when the morning stars sang together, and all the sons of God shouted for joy (38:7). The author used the largest vocabulary of any Hebrew writer, being in this sense (says Pfeiffer), the Shakespeare of the Old Testament. No poet of the Old Testament had a keener appreciation of nature.

XI

MATHEMATICS, ASTRONOMY, AND TECHNOLOGY IN THE FIFTH CENTURY

It is convenient to divide this chapter into three sections – Mathematics, Astronomy, Technology – in spite of the fact that this may oblige us to come back twice or thrice to the same personalities.

MATHEMATICS

ZENON OF ELEA

Students of early Greek mathematics are kept in a state of continual amazement by two complementary (or contradictory) facts: the neglect of plain arithmetic, and the singular depth of mathematical thought. The early Pythagoreans did not pay attention to ordinary methods of reckoning, yet their geometric ideas were largely based on numbers. A point for them is simply a unit having position; any geometric figure, beginning with the straight line, can be conceived and represented as a number of points. This raises the problems of continuity and infinite divisibility, or, more exactly, it raised those problems in the Greek minds, because those minds were ready for philosophical discussion. We have many evidences of Greek genius, but none is stronger and more startling than the mathematical thinking of this time, incited by logical difficulties that the average man of today (twenty-five centuries later) would hardly notice. As a first approximation, one is tempted to say that the more intelligent people are the quicker they are in their understanding; but one is very soon obliged to abandon that statement and almost to reverse it. Foolish people understand quickly, or believe they do, because they are not capable of imagining the difficulties and hence have no hurdles to leap over. The immense difference between Egyptian and Babylonian mathematics, on the one hand, and Greek mathematics, on the other, consists in that the former did not even conceive of some of the difficulties with which the Greeks were now beginning to struggle.

We may recall that Zenon visited Athens with his master Parmenides about the middle of the century. It was perhaps in Athens that he came across mathematicians like Hippocrates who were then trying to reduce geometric knowledge to a closely knit system. Being primarily a philosopher and a logician, Zenon perceived conceptual difficulties that would never have occurred to practical mathematicians (even Greek ones!). These would consider a straight line as made up of points.

How can we reconcile that notion with the line's continuity? The line is not a series of points or, to put it otherwise, a series of holes; it is a continuous whole. The practical mathematician would say: The points can be brought as near to one another, and the holes be as small, as you please; if the distance between two points is too large to satisfy you, well, divide it into a thousand or a million parts and imagine additional points in all those parts. The logician would demur and answer: The actual distance between any two points does not affect the argument; however small that distance may be, the two points remain separate and are different from the line or space joining them. Similar difficulties concerned the divisions of time (should we conceive it as continuous or discontinuous?) and motion (the passing of a body from one place to another in a given time). The paradoxical results of Zenon's meditations on these riddles are known to us through the Physics of Aristotle,[1] who called them fallacies, yet was unable to refute them, partly through the commentary on Aristotle by Simplicios (VI-1), and they cut so deep that they have exercised the minds of philosophers and mathematicians to our own day. Those questions are so subtle that to give a complete and accurate account of them would take considerable space. It must suffice here to indicate their general nature. Following Cajori's model, we shall call Zenon's four arguments against motion by the names "dichotomy," "Achilles," "arrow," and "stade," and summarize them as he does:

1. *Dichotomy.* You cannot traverse an infinite number of points in a finite time. You must traverse the half of any given distance before you traverse the whole, and the half of that again before you can traverse it. This goes on *ad infinitum,* so that (*if space is made up of points*) there are an infinite number in any given space, and it cannot be traversed in a finite time.

2. *Achilles,* the second argument, is the famous puzzle of Achilles and the tortoise. Achilles must first reach the place from which the tortoise started. By that time the tortoise will have got on a little way. Achilles must then traverse that, and still the tortoise will be ahead. He is always nearer but he never makes up to it.

3. *Arrow.* The third argument against the possibility of motion *through a space made up of points* is that, on this hypothesis, an arrow in any given moment of its flight must be at rest in some particular point.

4. *Stade.* Suppose three parallel rows of points in juxtaposition:

Fig. 1	Fig. 2
A	← *A* ....
B	*B* ....
C	*C* .... →

One of these (*B*) is immovable; while *A* and *C* move in opposite directions with equal velocity so as to come into position represented in Fig. 2. The movement of *C* relatively to *A* will be double its movement relatively to *B*, or, in other words, any given point in *C* has passed twice as many points in *A* it has in *B*. It cannot, therefore, be the case that an instant of time corresponds to the passage from one point to another.[2]

The four arguments seem to be directed against the belief held by most people at that time (including the Pythagoreans and Empedocles), and still held by the

[1] Zenon is one of the dramatis personae in Plato's *Parmenides.* Plato does not discuss Zenon's mathematical paradoxes, however, but only his arguments against plurality; he tends to belittle Zenon as compared with Parmenides. In *Phaidros,* 261D, Plato remarked that Zenon understood how to make one and the same thing appear like and unlike, one and many, at rest and in motion.

[2] Florian Cajori, "The purpose of Zeno's arguments on motion," *Isis* 3, 7–20 (1920). The article summarizes the controversy up to and including Tannery, whose conclusions Cajori shares. According to Tannery, Zenon opposed the idea that a point was unity in position. See also Philip E. B. Jourdain, "The flying arrow. An anachronism," *Mind* 25, 42–55 (Aberdeen, 1916) [*Isis* 3, 277–278 (1920)]. T. L. Heath, *History of Greek mathematics* (Oxford, 1921), vol. 1, pp. 271-283 [*Isis* 4, 532-535 (1921-22)], including many quotations from Bertrand Russell, who greatly admires Zenon.

majority of people of our own time, that space is the sum of points and time the sum of instants. Zenon argued that motion was not conceivable on a pluralistic basis.

DEMOCRITOS OF ABDERA

Democritos was born about thirty years later than Zenon. The dates of his birth and death are uncertain, but we shall not be far wrong if we write *c.* 460, *c.* 370. It does not follow that the mathematical speculations of Democritos were later than those of Zenon and that Democritos was acquainted with Zenon's perplexities. At any rate, those perplexities, or others of the same kind, were unavoidable as soon as one began to think rigorously on continuity and infinity, and the Greeks – not one of them, but many – were doing just that. In the catalogue of Democritos' works published by Diogenes Laërtios (III–1), five mathematical works are listed: (1) on the contact of a circle and a sphere, (2 and 3) on geometry, (4) on numbers, (5) on irrationals. We shall come back to the last one when we discuss that topic presently. The titles of items 2 to 4 are too vague to be useful. As to the first item, if we assume the title to mean the contact between a sphere and a tangent plane, we are drawn to the consideration of an infinitesimal angle. If we consider the simpler case (as Democritos probably did) of the angle between a circle and a tangent, the inherent difficulties obtrude rapidly. First, it was necessary to define the tangent; Democritos' mind was sharp enough to realize that the tangent and circle had only one point in common, though this could not be illustrated by any drawing. Then the angle must be considered. It had to be exceedingly small, because if the tangent were turned ever so little around its point of contact it would include a second point of the circle, and it would cease to be tangent.

Plato ignored Democritos, but Aristotle spoke very warmly of his ideas on change and growth. A century later Archimedes referred to Democritos' greatest mathematical discovery, that the volumes of a cone and a pyramid are one-third of the volumes of the cylinder and prism having the same base and height, adding that the proofs of Democritos' theorems were given not by him, but later by Eudoxos.[3] How did Democritos discover them? It is probable that he used a crude and intuitive method of integration, dividing the pyramid (or cone) into a large number of parallel slices. We shall come back to that when we discuss Eudoxos' discovery and use of the method of exhaustion.

The beginnings of perspective as applied to the designing of stage scenery were ascribed by Vitruvius to Democritos as well as to Agatharchos and Anaxagoras. These ascriptions are plausible but unproved. It is certain that problems of perspective had to be solved by the designers of scenery, but good solutions could be found in an empirical way.

HIPPOCRATES OF CHIOS

We now come to the greatest mathematician of the century, the first man to illustrate the name of Hippocrates. Almost every educated person is familiar with that name, but it evokes in his mind the memory of another man, the father of

[3] Aristotle's praise of Democritos occurs in *De generatione et corruptione*, 315A 34 ff.; Archimedes', in his *Method*. The relevant text is quoted by Heath, *Manual of Greek mathematics* (Oxford, 1931), p. 283.

medicine, Hippocrates of Cos. The name Hippocrates is not uncommon in Greece,[4] but it is remarkable that the two most illustrious bearers of it were contemporaries and came from the same group of islands, the Sporades off the coast of Asia Minor. The mathematician, who was the older man, was born in Chios and flourished in Athens during the third quarter of the fifth century. The physician belonged to the following generation; he was still a child when the mathematician was in his maturity, and he was active at the turn of the century. He came from the island of Cos (one of the Southern Sporades, the group also called Dodecanese islands).[5] We shall give him all the space that he so fully deserves in another chapter, but it was necessary to introduce him here and to place him for a moment near his older contemporary. I much hope that the readers of this book will remember that there were two Hippocrates', whose achievements were equally outstanding but so different that no comparison between them is possible. One could certainly not say that the second was a greater man than the first, yet he alone is remembered by most people while the older man is almost forgotten. Well, never mind.

The reason for Hippocrates' arrival in Athens about the middle of the century is traditionally believed to have been the loss of his possessions and his attempt to recover them. According to one story, he was a merchant whose ship had been captured by pirates; according to another story (told by Aristotle),[6] he was a geometer robbed of much money by the customs collector at Byzantium "owing to his silliness." Of course, mathematicians (from Thales down to Poincaré) have often been accused of being incompetent in ordinary life, but those stories are interesting in other respects; they help us to realize the existence of other sides of Greek life: the merchants, the pirates, and the wicked customhouse officers. Apparently, Hippocrates had started to be a merchant as well as a mathematician; the combination was not incongruous in the Greek community. Having lost his belongings, he applied himself to mathematics and was one of the first to teach for money; why should he not have been remunerated as well as the sophists; he might be called a sophist himself, though he specialized in the mathematical field.

Before explaining his work, we must recall another story which is very typical of the intellectual climate of that time. Three famous problems were then exercising the minds of the Athenian mathematicians: (1) the quadrature of the circle, (2) the trisection of the angle, (3) the duplication of the cube. How did these three problems emerge? The first is very ancient, and it was as yet impossible to know that an exact solution of it cannot be obtained. The two others are less natural. At least two legends, both reported by Eratosthenes, were circulated to account for the third problem. It will suffice to tell one of them. The people of Delos, suffering from a pestilence, were ordered by the oracle to double a certain altar which had

[4] The verb *hippocrateō* means to be superior in horse; Hippocratēs might thus be a suitable name for a cavalry officer!

[5] Chios is about 335 mi²; it gave birth not only to one of the greatest mathematicians of antiquity but also to another great one, Oinopides, to the historian Theopompos (378–after 305) and, so the people of the island claim, to Homer. Fustel de Coulanges in his "Mémoire sur l'ile de Chio," *Arch. Missions scientifiques* 5, 481 (1856), reprinted in his *Questions historiques* (Paris, 1893), pp. 213–339, provides a lot of information but makes the following blunder (p. 318): "Un certain Hippocrate de Chio est cité souvent par les anciens comme mathématicien, astronome et géomètre." This shows that "un certain Fustel de Coulanges," however distinguished in other ways, was not a mathematician, nor a historian of science.

Cos, south of Chios, is much smaller, very small indeed (111 mi²). It gave birth to only one illustrious man, the father of medicine.

[6] Aristotle, *Eudemian ethics*, VII, 14, 1247A.

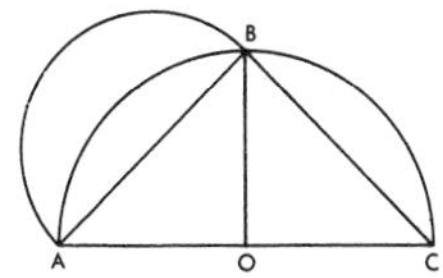

Fig. 63. The lunes of Hippocrates of Chios.

the form of a cube; therefore, that problem was called the Delian problem. The legend has all the earmarks of an invented *post factum*, and, as far as I know, there never was a cubic altar in Delos or anywhere else.[7] A simpler explanation is that some mathematicians may have wished to generalize a problem of plane geometry. To double a square it suffices to draw a new square on its diagonal. Could not a similar rule be found for the cube? It was not as easy as it looked. The emergence of these three problems, out of an infinity of others, is a new proof of the Greek genius, for they all combine apparent simplicity with inherent difficulties of a very high order.[8] They are insoluble except approximately; the second and third cannot be solved by simple geometric methods (that is, with ruler and compass); yet they were solved theoretically by the Greek mathematicians of the fifth century.

Hippocrates did not deal with the second problem, but we owe to him incomplete solutions of the two others. His attempts to square the circle led him to the discovery of lunes which could be squared; strangely enough, he discovered three out of the five species of lunes that can be squared in a simple way. This must have been very exciting, for it proved that at least some curvilinear figures were susceptible of quadrature.

Here is the simplest example of Hippocrates' lunes. Consider the half square *ABC* inscribed in the semicircle with center *O* (Fig. 63). Let us draw another semicircle on *AB* as diameter. The two semicircles are to each other as the squares of their diameters: $AC^2 = 2AB^2$. Hence, half of the larger semicircle is equal to the smaller one. Take out the segment common to both areas, and the remaining areas, to wit, the lune and the triangle *ABO*, are equal.

This is simple enough, yet it implies knowledge of the proposition that circles are to one another as the squares of their diameters.[9] If Hippocrates found the area of that lune, we must assume that he knew that proposition. His knowledge of it might have been intuitive; according to Eudemos, he was able to prove it, but if so, we do not know how he did it.

Hippocrates' work on the squaring of the lunes is very important in another way: it is the only fragment of Hellenic (pre-Alexandrian) mathematics that has been transmitted to us in its integrity, but the transmission was very indirect and slow.[10]

[7] Constantine G. Yavis, *Greek altars* (Saint Louis: Saint Louis University Press, 1949), pp. 169–170, 245), mentions altars that are almost cubical, not in Delos, however, but in Cypros. Two in the Vouni Palace may date from the fifth century; the dimensions of their bases are 1.95 × 1.70 m and 2.70 × 1.54 m – very far from square.

[8] In 1767, Johann Heinrich Lambert proved that π is irrational; in 1794, Legendre proved that π^2 is also irrational; in 1882, Ferdinand Lindemann proved that π is transcendental; see *Osiris 1*, 532 (1936). The three problems were investigated in the light of modern mathematics by Felix Klein (1849–1925), *Vorträge über ausgewählte Fragen der Elementarmathematik* (Leipzig, 1895; English trans., Boston, 1897; revised, New York, 1930) [*Isis 16*, 547 (1931)].

[9] Euclid, XII, 2.

[10] It was included in the history of geometry by Eudemos (IV–2 B.C.) preserved in the commentary on Aristotelian physics by Simplicios (VI–1). Almost a millennium elapsed between the latter and Hippocrates! The text is easily available in a Greek–French edition by Paul Tannery, *Mém. Soc. sci. Bordeaux 5*, 217–237 (1883); reprinted in his *Mémoires scientifiques* (Toulouse, 1912), vol. 1, pp. 339–370. Greek–German edi-

This illustrates once more how hard it is to know the facts of early Greek mathematics and how prudent the historian must be.

His solution of the third problem, the duplication of the cube, is equally interesting in its implication, for it shows that he had a clear understanding of compound ratios. That knowledge was derived from numbers and applied intuitively to lines.

If the side of the given cube is a, the problem is to determine x, such that $x^3 = 2a^3$. It is solved by finding two mean proportionals in continued proportion between a and $2a$: $a/x = x/y = y/2a$; for then $x^2 = ay$, $y^2 = 2ax$; hence $x^4 = 2a^3x$ or $x^3 = 2a^3$.

By the middle of the fifth century, so many geometric theorems had already been established and so many problems solved that it became increasingly necessary to put all these data in good logical order. This implied not only the classification of the results already obtained but, what is more important, the strengthening of the proofs. In many cases (as illustrated above about the proposition in Euclid) knowledge was intuitive, or the proof, if it had been found, had failed to be transmitted. If every item were put in its logical place, the gaps would be detected. The geometric edifice as far as it could be built would be stronger, and one would know more definitely what to do in order to bring it nearer to completion and logical perfection. It would seem that Hippocrates was one of the first to attempt that task, that is, he was the first forerunner of Euclid, not only as a discoverer of individual propositions, but as a builder of the geometric monument that was to be called later the *Elements*.

If the text of Hippocrates relative to the quadrature of lunes transmitted to us by Simplicios is really as he wrote it, then he is the first mathematician known to us who used letters in geometric figures and thus made possible the unambiguous description of such figures.[11] The manuscript tradition was thus greatly facilitated, for the figures, which may be difficult to draw neatly, can be omitted. They are not indispensable, for the reader can easily reconstruct them on the basis of the text. We are not surprised to find that Hippocrates' use of letters was not yet quite as simple and clear as Euclid's, but it was a very important beginning, almost necessary for the future progress of mathematics.

Hippocrates writes "the line on which is *AB*," or "the point upon which is *K*," while Euclid and we ourselves simply write "the line *AB*," "the point *K*." Such differences occur repeatedly in the history of mathematical notations, and, we might say more generally, in the history of science. The inventor is seldom able to express his invention in the simplest manner, and it takes another man, or many men, less intelligent but more practical than himself, to complete the invention. Hippocrates' invention might have been perfected, for example, by other teachers or even by students who would use the shorter phrase "the line *AB*" out of sheer laziness.

If Hippocrates actually wrote the first textbook of geometry, which is not only possible but plausible, he was obliged to tighten the proofs, and we may believe Proclos' statement that he invented the method of geometric reduction (*apagōgē*),

tion by Ferdinand Rudio (194 pp.; Leipzig, 1907).

[11] The Pythagorean pentagon bearing the letters *hygieia* (p. 211) is probably older than Hippocrates, but the use of lettered figures to facilitate geometric discourse is a very different matter from the use of letters for a symbolic purpose.

that is, the passage from one problem or theorem to another, the solution of which entails the solution of the former. We shall discuss that later on.

The achievements of Hippocrates of Chios were considerable, so great indeed that he might be called the father of geometry with as much justice as Hippocrates of Cos was called the father of medicine. It is better to avoid such metaphors, however, for there are no absolute fathers except Our Father in Heaven.

OINOPIDES OF CHIOS [12]

According to Proclos (V–2), Oinopides was a little younger than Anaxagoras; he places him ahead of Hippocrates and Theodoros. We may assume that Oinopides flourished in the third quarter of the century. It is interesting to note that he was not only the contemporary of Hippocrates, but also his fellow citizen. They must have known each other either in Chios or in Athens. Whether he was a little younger than Hippocrates or not hardly concerns us, for what matters is the chronologic order of discoveries, which is different from the chronologic order of birth dates; some men do their best work when they are young, others in old age.

Oinopides is more important as an astronomer than as a mathematician, and we shall devote more space to him in the second section of this chapter. His mathematical contributions are modest, yet significant. He was the first to solve the following problems: (1) to draw a perpendicular to a given straight line from a given point; (2) at a given point of a straight line to construct an angle equal to a given angle.

As everybody could solve these problems roughly, the ascription of their solution to Oinopides means that he was the first to show how to solve them rigorously with ruler and compass. Such problems had to be solved to make possible the writing of the *Elements*, yet Proclos says that Oinopides solved the first problem for astronomic reasons; he also says that Oinopides used the old name for perpendicular (*cata gnōmona* instead of *orthios*). All of which illustrates the transitional nature of this period: geometric knowledge is gradually ordered and crystallized, the *Elements* are in the making.

HIPPIAS OF ELIS

Hippias came from Elis,[13] a small country in the northwest corner of the Peloponnesos, renowned for horse breeding and almost sacred to the Greek people on account of the Olympic games which took place every fourth year in the plains of Olympia. He was born *c.* 460 and is much better known than his seniors Hippocrates and Oinopides, because he traveled considerably all over Greece giving public lectures and teaching; he was a kind of wandering sophist whose activities were dominated by the love of fame and of money. He was ready to discuss any subject but was especially interested in mathematics and science. When he reached Sparta he was disappointed, because the Spartans did not care enough about science to reward scientific lectures. He is immortalized by two Platonic dialogues, the *Hippias major* and the *Hippias minor*, wherein he appears as a sophist, vain and arrogant. This is not attractive, yet his mathematical fame is secure because of a single discovery, which is truly astonishing.

[12] Elaborate account by K. von Fritz in Pauly Wissowa (1937), vol. 34, pp. 2258–2272.

[13] Pyrrhon (IV–2 B.C.), founder of the Skeptical school, also hailed from Elis.

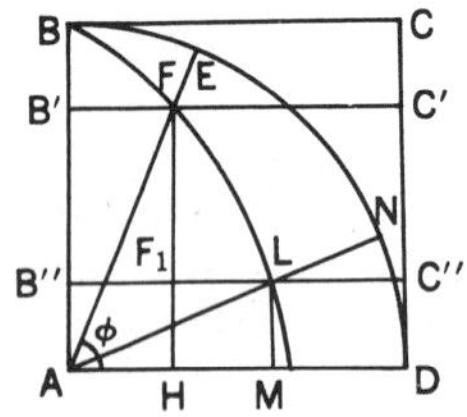

Fig. 64. The quadratrix of Hippias of Elis.

In order to solve the problem of the trisection of the angle, Hippias invented a new curve, the first example in history of a higher curve, a curve that could not be drawn by any instrument but only point by point. That is, at the very time when the best mathematicians are working hard to consolidate geometric knowledge into a well-ordered edifice, he is bold enough to jump out of it and begin exploring the mysterious unknown outside of it.

The curve discovered by Hippias was called the quadratrix (its name will be justified later on) and generated as follows (Fig. 64). Suppose we have the square *ABCD* (side a) and within it the quarter of a circle of radius a with center at *A*. Let us assume that the radius turns with constant speed from the position *AB* to *AD*, and that in the same time the side *BC* moves down to *AD* with constant speed, remaining parallel to itself. The locus of the points of intersection (such as *F* and *L*) of the two lines is the quadratrix. Now $< BAD$: $< EAD$ = arc *BD*: arc *ED* = *BA*:*FH*. Consider the vector *AF* connecting the center *A* with a point *F* of the curve; let its length be ρ and let it make an angle ϕ with *AD*; then $a/(\rho \sin \phi) = (\pi/2)/\phi$.

The curve can be used to trisect any angle such as ϕ. Let us divide the line *FH* into two parts in the ratio 2:1, so that $FF_1 = 2F_1H$. Then draw the line *B″C″* which cuts *FH* in F_1 and the curve in *L*, and join *AL*. The angle *NAD* will be one-third of ϕ.

The curve could be used equally to divide any angle in any ratio; it would suffice (in our example) to divide the line *FH* in that ratio and continue the construction as before.

The same curve was used a century later by Deinostratos (IV–2 B.C.) and others for the quadrature of the circle, and it is for that reason that it received the name quadratrix (*tetragōnizusa*).

THEODOROS OF CYRENE

The mathematician Theodoros of Cyrene [14] is best known to us because of the beginning of Plato's *Theaitetos*, in which he is introduced as a famous master. He

[14] We say the mathematician Theodoros, because the words Theodoros of Cyrene evoke in the minds of most men (mathematicians excepted) another man, more famous, sometimes called Theodoros the Atheist, disciple of Aristippos of Cyrene, who was himself a disciple of Socrates. Theodoros the Atheist, was banished from Cyrene and flourished for a time in Alexandria; toward the end of his life he was permitted to return to his native city where he died, probably at the end of the fourth century. In short, the two Theodoroses of Cyrene were not contemporaries; the mathematician belongs to the second half of the fifth century, the philosopher to the second half of the fourth century. Cyrene, main city of Cyrenaica, was an important cultural center. It gave birth not only to Aristippos and the two Theodoroses, but also to the poet Callimachos (d. *c.* 240) and the bishop Synesios (V–1).

was then (399)[15] an old man, hence we might say that he was born *c.* 470. It is said that Plato had visited him in Cyrene; at any rate, Theodoros was in Athens about the end of the century; he belonged to the Socratic group and was (or may have been) the mathematics teacher of Plato. A single mathematical discovery is ascribed to him, but it is a startling one. He is said to have proved the irrationality of the square roots of 3, 5, 7, . . ., 17.

It is significant that the discovery of the irrationality of $\sqrt{2}$ is not ascribed to him, which can only mean that it was known before him. Indeed, such knowledge is ascribed to the early Pythagoreans. The discovery of the irrationality of $\sqrt{2}$ was a shocking surprise, and the Pythagoreans seem to have thought for a time that it was an exception.

The square root of 2 appears very naturally and simply, because it is the diagonal of the unit square (side and area equal 1). How did the early Pythagoreans discover the irrationality of $\sqrt{2}$?

We must first introduce another man, Hippasos of Metapontum,[16] an early Pythagorean about whom some extraordinary stories circulated. They said that he had been expelled from the Pythagorean school for having revealed mathematical secrets. According to one tradition, he had revealed the construction of a dodecahedron in a sphere and claimed it as his own; according to another, he had revealed the discovery of irrational quantities – and this would very probably refer to $\sqrt{2}$ or to $\sqrt{5}$. Before we abandon Hippasos, one more mathematical thing may be said of him. The early Pythagoreans distinguished three kinds of mean – arithmetic, geometric, and subcontrary.[17] Hippasos suggested that the name harmonic be given to the third, a name well applied because of the importance of the harmonic means in musical theory, and he defined three other medieties (means). Let us now return to the discovery of the existence of irrational quantities, which was for the mathematicians of the sixth and fifth centuries a kind of logical scandal.

An irrational number (*alogos*) is one that cannot be expressed as a quotient of two integers; the discovery was made geometrically when it was realized that the diagonal of a unit square could not be measured in terms of the side or in terms of any of the equal parts, however small, into which that side can be subdivided.

How could one prove that irrationality? The traditional proof is referred to by Aristotle;[18] it is a *reductio ad absurdum.* The argument is so short and easy that we reproduce it.

Consider the square whose side is a and diagonal c. We must prove that a and c are incommensurable. Let us assume that they are not, and that their ratio c/a is expressed in the simplest manner by γ/α. Then $c^2/a^2 = \gamma^2/\alpha^2$; but $c^2 = 2a^2$; thus $\gamma^2 = 2\alpha^2$. Hence γ^2 is even and γ is even, and α must be odd. If γ is even, we may

[15] The dialogue is supposed to have taken place in the year of Socrates' death, 399, but it was written only some thirty years later, in 368–67.

[16] No article was devoted to him in my *Introduction,* because his floruit is too uncertain. He might belong to the sixth century as well as to the fifth. I call him Hippasos of Metapontum, but two other birthplaces are ascribed to him, Sybaris and Croton. Those three places are in the same region, however, around the Gulf of Taranto, the "shank" of Italy.

[17] To refresh the reader's memory: the number b is the arithmetic mean of a and c if $b = (a + c)/2$; it is their geometric mean if $a/b = b/c$; it is their harmonic mean if $a/c = (a - b)/(b - c)$ or $1/c - 1/b = 1/b - 1/a$. The three numbers a, b, c are said to be respectively in arithmetic, geometric, or harmonic progression.

[18] Aristotle, *Analytica priora,* 41A, 26–30.

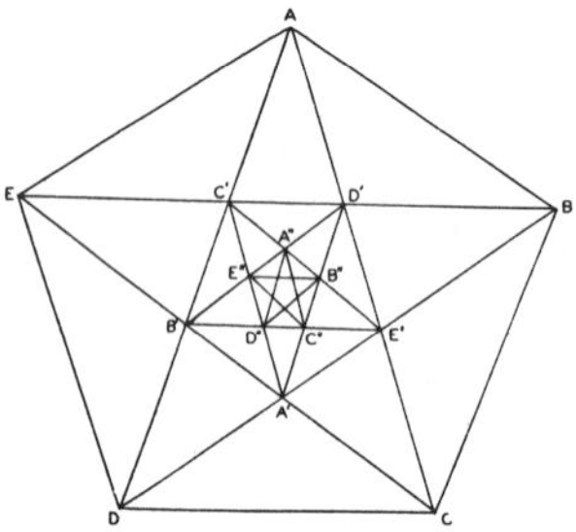

Fig. 65. Pentagons and pentagrams.

write $\gamma = 2\beta$; then $\gamma^2 = 4\beta^2 = 2\alpha^2$, so that, $\alpha^2 = 2\beta^2$. Hence α^2 is even, and α is even. We have found that α is both even and odd, which is impossible; hence a and c are incommensurable.

It is quite possible that the first irrational quantity was discovered by Hippasos (if not before), but one cannot prove that. One is tempted to assume it because of the tradition reported above, and because this leaves a little time for the theory of irrationality to develop. Yet the proof of the irrationality of $\sqrt{2}$ that has just been given, simple as it is, implies a degree of abstraction that can hardly have obtained as early as Hippasos' time, say the beginning of the fifth century. But another tradition ascribed to Hippasos some knowledge of the dodecahedron, a regular solid whose twelve faces are regular pentagons. An interest in the pentagon was natural enough to a Pythagorean, whose symbol was a pentagram (a regular pentagon whose sides have been extended to their points of intersection).

Now Kurt von Fritz has made the very interesting suggestion [19] that Hippasos' interest in pentagrams and pentagons and the numbers and ratios incorporated in them would have led him to the notion of incommensurability. How would a craftsman try to find the common measure of two lines a and b? He would try to measure the longer a in terms of the shorter b, and if that failed, he would try to measure it in terms of fractions of b. Now, such a method could not be applied in this case, because of the coarseness of physical measurements. Yet, if Hippasos had considered the pentagon with all its diagonals, he would have seen that the diagonals constitute a pentagram and enclose a smaller pentagon (Fig. 65). The same procedure might be continued, and this would be tempting enough. In practice, one could not continue it very long, but it is obvious that in theory it might be continued indefinitely, and this means that the diagonals and the side are irreducible to a common measure, are incommensurable.

The discovery of incommensurable quantities might have been made intuitively by Hippasos before their existence had been completely proved. It is possible even that Greek mathematicians had begun before the end of the century to consider more complicated cases. In *Hippias major* (303 B.C.) occurs the remark that just as an even number may be the sum of two even or two odd numbers, even so the sum of two irrationals may be either rational or irrational. A good example is that of the rational line cut in extreme and mean ratio; the three ratios of those segments and of the whole line are irrational.

[19] Kurt von Fritz, "The discovery of incommensurability by Hippasus of Metapontum," *Ann. Math.* *46*, 242–264 (1945). Our figure is borrowed with kind permission from his paper.

Fig. 66. Simple construction of various incommensurable quantities.

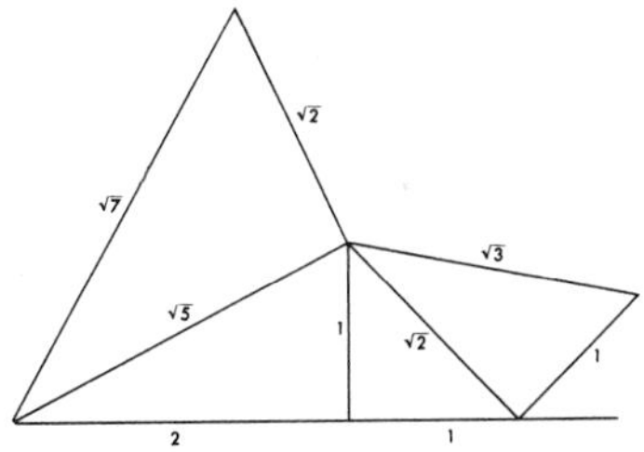

Assuming that Hippasos had discovered the irrationality of $\sqrt{2}$ and $\sqrt{5}$, how did Theodoros find the other surds up to $\sqrt{17}$? He might have constructed many of them in an easy way, as shown in Fig. 66. Once the possibility of irrational quantities was realized and admitted, it was not difficult to find new ones. The main difficulties were of another kind: if there were numbers that could not be represented by any ratio such as n/m, then the Pythagorean similitude between numbers and lines or between arithmetic and geometry could not be maintained any longer – or could it? We have no reason to suppose that these deep difficulties were solved before the fourth century, but a long period of ferment of ideas represented by Hippasos and Theodoros [20] prepared the age of Theaitetos and Eudoxos. We shall continue our discussions of this topic when we reach that age.

The Greek genius had intuitions of mathematical truth just as it had intuitions of beauty. It seems to have understood, if not at the very beginning, at any rate very early, that mathematics cannot be built up with logical rigor without solving many problems implying infinity. The uncanny depth of that genius may be better appreciated if we bear in mind that there are many educated people, well-educated even, say physicians or grammarians, who would hardly understand such things today, not to mention discover them. This chapter has already given many examples of Greek intuitions concerning infinity, to wit, the views explained by Zenon, Democritos, Hippasos, Theodoros, and now we add two more, Antiphon and Bryson.

ANTIPHON THE SOPHIST [21]

Antiphon was an Athenian, who flourished at the same time as Socrates, and was to some extent the latter's rival as an educator of youth. He was a sophist, interested in many sciences but also in divination and the interpretation of dreams. We should never forget that divination and particularly oneiromancy [22] were then legitimate parts of science, attracting the intelligent curiosity of some of the best minds, for the limitations of knowledge could not be as keenly realized then as

[20] Even by Democritos, for one of his treatises dealt with irrational quantities and solids (atoms?), *Peri alogōn grammōn cai nastōn*, but we must not forget that he lived until late in the fourth century. The title is enigmatic. Did he try to connect irrationals with atoms?

[21] Not to be confused, as he has often been, with his contemporary, Antiphon the Orator, who flourished also in Athens (*c.* 480–411), and is more important in literary and political history, but does not concern the historian of science at all.

[22] For a general introduction see the article on divination by Arthur Leslie Pease in the *Oxford classical dictionary* (Oxford: Clarendon Press, 1949), pp. 292–293, with long bibliography. The many articles in the *Encyclopedia of Religion and Ethics* permit a comparative study of divination in many countries; see vol. 4 (1912), pp. 775–830.

they are now. He deserves our attention, however, because he invented a new method for the solution of the old problem, the quadrature of the circle.

Antiphon suggested that a simple regular polygon, say a square, be inscribed in the given circle. Then an isosceles triangle could be built on each side, its vertex being on the circumference. A regular octagon would thus be constructed, and continuing in the same manner one would easily construct regular polygons of 16, 32, 64, . . . sides. Now it is obvious that the area of each of those successive polygons comes closer to the area of the circle than the preceding one, or, to put it otherwise, as more complex polygons are inscribed in the same circle, the area of that circle is gradually exhausted. Now the areas of these polygons can be exactly computed, or the polygons can be "squared"; they increase gradually though they cannot possibly increase beyond a given limit, the area of the circle itself.

This method was criticized by Aristotle, his commentators, and others on the ground that no matter how many times the number of sides of each polygon is doubled, the area of the circle can never be completely used up.

BRYSON OF HERACLEA [23]

Bryson was the son of the logographer or mythologist Herodoros of Heraclea Pontica.[24] He was a pupil of Socrates and also of Socrates' disciple, Euclid of Megara. Bryson thus belongs to a later generation than Antiphon and must have flourished in the first half of the fourth century, but we must speak of him here, because his work completes so well that of Antiphon.

While Antiphon's method consisted in drawing in the circle a series of polygons with 4, 8, 16, 32, . . . sides, Bryson proposed to construct a series of polygons circumscribed about the same circle. The areas of the circumscribed polygons diminish gradually. The area of the circle is the upper limit of the inscribed polygons and the lower limit of the circumscribed ones. Of course, Bryson was subject to the same criticisms as Antiphon and he was duly criticized by Aristotle, Simplicios, and many historians of mathematics.

It would seem to me that the modern historians (such as Rudio [25] and Heiberg) have been unduly severe to Antiphon and Bryson. The procedure of the latter lacked rigor but it derived from a sound intuition, and it led eventually to the method of exhaustion (formulated by Eudoxos) and to the integral calculus.

One cannot deny Bryson a definite discovery: that the area of the circle is a limit to the increasing areas of inscribed polygons and to the decreasing areas of the circumscribed polygons, and as the number of sides of those two series of polygons are increased, their areas approach closer to the area of the circle on both sides of it. The method was actually applied by Archimedes (III–2 B.C.), who measured the areas of two inscribed and circumscribed polygons of 96 sides each and reached the conclusion that $3\frac{10}{71} < \pi < 3\frac{1}{7}$ ($3.141 < \pi < 3.142$).

[23] Not to be confused with another Bryson of a much later date, Bryson the Neo-Pythagorean, who flourished probably in Alexandria or Rome in the first or second century after Christ. His *Economics* has been edited by Martin Plessner in 1928; see *Isis 13*, 529 (1929–30).

[24] There were many Greek cities in Europe and Asia named Heraclea, but this particular one, Heraclea Pontica, was situated on the southwestern shore of the Black Sea, in Bithynia. It was the native city of Heracleides of Pontos (IV–2 B.C.) and perhaps also of the painter Zeuxis (born *c.* 445).

[25] Ferdinand Rudio (1856–1929) *Das Bericht des Simplicius über die Quadraturen des Antiphon und des Hippokrates* (194 pp.; Leipzig, 1907; contains all the relevant texts in Greek and German.

Before concluding this section it is worth while to observe that the men whose mathematical ideas have been reviewed (excepting perhaps Hippocrates) were not mathematicians in the restricted sense of today; they were philosophers and sophists who realized the fundamental importance of mathematics and tried to understand it as well as possible. Note that they came from many parts of the Greek world. Zenon hailed from Magna Grecia, Hippocrates and Oinopides from Ionia, Democritos from Thrace, Hippias from the Peloponnesos, Theodoros from Cyrenaica, Bryson from the Black Sea; Antiphon was, as far as we know, an Athenian, the only one among them. If we had spoken of Archytas, who is astride of both centuries[26] and will be dealt with later, another country would have to be added to the list, Sicily. This shows that the mathematical genius was as widely distributed across Hellas as was the artistic or literary genius. That genius was not Athenian, or restricted to any locality; it was the genius of Greece.

ASTRONOMY

In our review of astronomic ideas in the fifth century we may leave out those of the philosophers like Heraclitos, Empedocles, Anaxagoras and restrict ourselves largely to the Pythagoreans. Indeed, the Pythagorean school was the leading school of astronomy in that century and the most progressive. Their mathematical mysticism had its useful side, for it helped to assume regularities in the celestial motions and to discover planetary laws. As Plato put it,[27] "as the eyes are designed to look up at the stars, so are the ears to hear harmonious motions, and these are sister sciences as the Pythagoreans say." This expresses beautifully the Pythagorean concept of the unity of mathematics, music, and astronomy, which influenced astronomic thinking down to the time of Kepler.

When we speak of Pythagorean astronomers, we do not mean only those who were fully initiated in all the Pythagorean mysteries, but also those who accepted, if only in part, the Pythagorean views on the system of the world. Thus, we shall begin our account with Parmenides (who was not a Pythagorean but the founder of the Eleatic school), and then deal with Philolaos, Hicetas, and a few others.

The Pythagoreans were the first to call the world cosmos (implying that it is a well-ordered and harmonious system) and to say that the earth is round. These designations are ascribed to Pythagoras himself and also to Parmenides; it is not easy to separate Parmenidean inventions from older Pythagorean doctrines, but we need not worry too much about that. The first section of our account may be understood as representing not only the views of Parmenides but also those of the Pythagoreans about the middle of the century. By that time, some points of Pythagorean cosmology were already determined: the universe is a well-ordered system; the most perfect shape is the sphere and the earth is round;[28] the planets are not "errant" bodies but have regular motions; those motions are uniform. It is possible that other ideas were already accepted, such as the divinity of the stars and planets, and the essential duality of the world – superlunar (perfect) and sub-

[26] Born *c.* 430, he was still alive in 360.

[27] Plato, *Republic*, VII, 530.

[28] I use the word round to represent the Greek term *strongylos*, opposed to *platys*, flat, and to *euthys*, straight. The word is less precise than the word spherical, but the general idea is the same.

lunar (imperfect).[29] Such ideas take us away from astronomy into mythology and religion. Their coexistence with the other, more scientific, ideas explains the paradox that the Pythagorean school was at one and the same time the cradle of mathematical astronomy and the cradle of astrology. These two aspects are apparently incompatible, yet they reappear repeatedly throughout the history of science (at least down to the seventeenth century). One cannot understand the development of ancient and medieval astronomy unless one bears that essential polarity constantly in mind.

PARMENIDES OF ELEA

Parmenides came to Athens about the middle of the century, but he was then in his fifties and it is possible that his astronomic views were already crystallized. He was the first to assume that the spherical earth was divided into five zones, but those zones were not well defined, and he conceived the central zone, torrid and uninhabited, to be twice as broad as it really is. We cannot attach much importance to those zones, which were too speculative. As to the spherical shape of the earth, we do not know how the early Pythagoreans, Parmenides among them, reached that conclusion. It is probable that it was at first an a priori conception but that it was promptly and repeatedly confirmed by observation of the stars. The world known to the Greeks extended from at least latitude 45° N (top of the Black Sea) to the Tropic of Cancer or even farther – a belt that was 20° to 25° of latitude wide. Now, this was more than enough to observe considerable changes in the starry heavens. As one traveled northward certain stars became circumpolar; on the other hand, a very bright star (Canopus), invisible in Greece proper, was just visible above the horizon in Crete and it rose higher as one proceeded to Egypt and sailed up the Nile. Moreover, travelers must have noticed the lengthening days as one went northward; this was enough to lead to the idea of zones. Parmenides was the first to conceive the universe as a continuous series of spheres or crowns (*stephanai*) concentric with the earth, which is at rest at the center. We need not recall his other astronomic views, some of which were not new (for example, that the Moon derives its light from the Sun) or were mere figments (for example, that the Moon and Sun are fragments from the Milky Way). It is remarkable, however, that a pure metaphysician, as he was, could guess so much of the truth, his vague anticipation of geographic zones is almost as astonishing as Democritos' anticipation of atoms.

PHILOLAOS OF CROTON[30]

Philolaos came from Croton or from Tarentum (both places are in the same region, the Gulf of Tarantum). It was in Croton that Pythagoras had established his school; it is not surprising then that Philolaos is quoted as a Pythagorean. He was a contemporary of Socrates, but so was Parmenides; hence, we cannot conclude that he was much younger than the latter. He was probably born after Par-

[29] These views were at least in part of Oriental origin: Mazdean, Babylonian, perhaps even Egyptian. Louis Rougier, *L'origine astronomique de la croyance pythagoricienne en l'immortalité céleste des âmes* (152 pp.; Cairo: Institut français d'archéologie orientale, 1933) [*Isis 26*, 491 (1936)].

[30] What we know of him is largely derived

menides and before Socrates, for he was the teacher of Simmias and Cebes in Thebes, and these two were among the last disciples of Socrates.[31]

His astronomy was Pythagorean, and he is often described as the first exponent of Pythagorean astronomy, a statement that must be qualified in two ways. First, Parmenides, who was not a full Pythagorean, however, was possibly older than he. Second, he represents a second (or third) and more sophisticated stage in the evolution of Pythagorean astronomy. His writings are unfortunately lost, except for a very few fragments.

How sophisticated his views were will appear in a moment. They illustrate once more the theoretical boldness of the early Greek men of science, unfettered as they were by religious prejudices or by common-sense restrictions. The whole question for them is to give a consistent explanation of reality, and no hypothesis is too daring if it provides such an explanation. Philolaos did not hesitate to reject the geocentric idea which earlier Pythagoreans had taken for granted. The universe is spherical and limited. In the very center is the central fire (Hearth of the Universe, Watchtower of Zeus, etc.) which is also the central force or the central motor. Around it rotate ten bodies: first, the counterearth (*antichthōn*) which always accompanies the Earth and shades the fire from it, second, the Earth itself, then the Moon, the Sun, and the five planets, finally the fixed stars. We do not see the counterearth because the Earth is always turning its back to it, that is, to the center of the universe. This implies that the Earth rotates around its own axis, while it turns around the center of the world.[32]

The boldness of that conception is staggering. Not only did Philolaos reject the geocentric hypothesis, but he did not hesitate to consider the Earth a planet like the others, and to assume that it rotates around the center of the world and also (perhaps) around its own axis. Moreover, he postulated the existence of another planet which remains always invisible! This seems exceedingly artificial. Why did Philolaos introduce the counterearth? According to Aristotle, he did so to account for eclipses and particularly for the greater frequency of lunar eclipses as compared with solar ones.[32a]

If the Earth turns around the center of the world, then the apparent movements of the stars might be accounted for by the rotation of the Earth in an opposite direction. In spite of that, Philolaos assumed that the sphere of fixed stars was rotating like the other spheres. This is a good example of great boldness combined with timidity (a phenomenon which is so common in the history of science that we might consider it the rule rather than the exception). It would have been much simpler, indeed, to assume that the outside sphere did not move. Philolaos could

from the *Peri arescontōn* (*De placitis*) of Aëtios, edited by Hermann Diels, *Doxographi graeci* (Berlin, 1879), pp. 45-69, 178-215, 267-444. Diels prints in parallel columns the *Placita* of Aëtios and the *Eclogae* of Stobaios (V-2). The date of Aëtios is very uncertain; his *Placita* were ascribed to Plutarch (I-2) and are probably a little later; we might place him tentatively at the end of the first century or in II(1). See note 2, p. 239.

[31] Simmias and Cebes, both Thebans, were warm friends of Socrates. They are the main speakers, besides Socrates himself, in the *Phaidon*; they are both mentioned in *Criton* and Simmias alone in *Phaidros*. Cebes is *not* the author of the *Pinax* bearing his name (*Cebētos Thebaiu pinax*).

[32] It is not certain, however, that Philolaos was aware of that implication. For example, the Moon always turns the same face to us and the ancients considered that it did not rotate around its axis; they did not realize the contradiction that this involved.

[32a] According to Burch, the counterearth could be interpreted as the antipodes. George Bosworth Burch, "The counter-earth" (to be published in *Osiris 11*, 1953).

not bring himself to do that – because all the spheres move . . . The gratuitous extra complication that he thus introduced did not necessarily conflict with reality. As the radii of the spheres increased, their angular speed decreased, and one might always determine the angular speeds of both the Earth and the fixed stars in such a manner that the apparent motion of the stars was exactly compensated. The very slow motion ascribed to the outside sphere might have been introduced in order to account for the precession of the equinoxes, but in spite of long centuries of Egyptian and Babylonian observations that phenomenon was then unknown; it remained unknown until the time of Hipparchos (II-2 B.C.).[33]

HICETAS OF SYRACUSE

The world system that has just been described was ascribed to Philolaos by Aëtios,[34] but Diogenes Laërtios ascribed it to Hicetas, and Aristotle, to the Pythagoreans in general.

Even if the system was invented by Philolaos, it is possible that Hicetas improved it. For example, Hicetas may have drawn out the implication that the Earth turns around its own axis, and may have abandoned the fantastic and gratuitous conception of central fire and counterearth. This is substantiated by Cicero (I–1 B.C.); he is a late witness, but he was using a text of Theophrastos (IV–2 B.C.), who was much closer to the event. The date of Hicetas is unknown; we may assume that he was a younger contemporary of Philolaos. "Hicetas the Syracusan, as Theophrastos says, believes that sky, Sun, Moon, stars, and in fine all heavenly bodies are at a standstill, and that no body in the universe except the Earth is in motion; and that as this turns and revolves round its axis with great velocity, all the phenomena come into view which would be produced if the Earth were at rest and the heavens in motion."[35]

Cicero's statement that nothing in the universe is moving except the Earth is of course wrong in any case, but the exaggeration is understandable in the mouth of a man who was not an astronomer and who overemphasized the thought expressed by Hicetas and Theophrastos: it is the Earth that turns around its axis every day, not the starry heavens.

On the strength of that tradition, it is permissible to ascribe to Philolaos the system making the Earth a planet like the others turning around the central fire at the same speed as the counterearth, and to Hicetas the system replacing the Earth in the center and explaining the apparent rotation of the stars by a real rotation of the Earth around its own axis.

ECPHANTOS OF SYRACUSE

In order to complete this story, we must say a few words of Ecphantos, though he belongs probably to the following century. As he was a Syracusan and a Pythagorean like Hicetas, we may assume that he was a direct or indirect disciple of the

[33] Otto Neugebauer, "The alleged Babylonian discovery of the precession of the equinoxes," *J. Am. Oriental Soc. 70*, 1-8 (1950). The Babylonian discovery was supposed to have been made by Kidinnu (Cidenas) *c.* 343 B.C., which is a century later than Philolaos anyhow. Paul Schnabel, "Kidenas, Hipparch und die Entdeckung der Präzession," *Z. Assyriologie 3*, 1-60 (1926) [*Isis 10*, 107 (1928)].

[34] Aëtios' text translated into English by T. L. Heath, *Greek astronomy* (London: Dent, 1932), p. 32–33 [*Isis 22*, 585 (1934–35)].

[35] *Academicorum priorum liber* II, 39, 123. Edition by James S. Reid (London, 1885), p. 322, translation by the same (London, 1885), p. 81.

latter. According to the *Placita* of Aëtios,[36] "Heraclides of Pontos and Ecphantos the Pythagorean move the earth, not however in the sense of translation but in the sense of rotation, like a wheel fixed on its axis, from west to east, about its own center." Thus, Ecphantos at least (if not Hicetas before him), affirmed unambiguously the daily rotation of the Earth. The fact that Aëtios associates him with Heraclides and even names the latter before him suggests that they were contemporaries (Heraclides of Pontos was born *c.* 388 and died *c.* 315–310).[37] Ecphantos is said to have combined Pythagorean with atomistic doctrines; this also would place him in the fourth century, even in Heracleides' time.

THE ASTRONOMIC VIEWS OF LEUCIPPOS AND DEMOCRITOS

The founders of the atomic theory were great cosmologists but poor astronomers. To consider Democritos alone,

> He said that the ordered worlds are boundless and differ in size, and that in some there is neither Sun nor Moon, but that in others both are greater than with us, and in yet others more in number. And that the intervals between the ordered worlds are unequal, here more and there less, and that some increase, others flourish and others decay, and here they come into being and there they are eclipsed. But that they are destroyed by colliding with one another. And that some ordered worlds are bare of animals and plants and of all water. And that in our cosmos the Earth came into being first of the stars and that the Moon is the lowest of the stars, and then comes the Sun and then the fixed stars; but that the planets are not all at the same height. And he laughed at everything, as if all things among men deserved laughter.[38]

These statements are made by St. Hippolytos (III–1); assuming that they represent Democritos' thoughts, they are remarkable because of their boldness and their gratuitousness. It is clear that Democritos could base them on nothing, and yet his intuitions have been confirmed by modern science. For example, we now know that the number of universes is, if not infinite, at least so large as to stagger the imagination; we also know that the stars are of many different kinds and are in different stages of evolution, some going up, others down. That is not science, of course, but poetic fancy. Some of his cosmologic views, however, were as prophetic as his atomic theory. How could he make such guesses? One cannot help wondering; and why in his abysmal ignorance did he venture to speculate on such matters?

On the other hand, Democritos did not believe that the Earth was round (the conception of a spherical Earth was apparently a Pythagorean monopoly, one that the people of other sects did not care to infringe). He had spent some time in the East and his astronomic ideas were definitely Babylonian. One of his tetralogies deals with uranography, geography, polography, and meteorology. The first part can be reconstructed from Vitruvius;[39] it was possibly accompanied with planispheres "on which were pictured in imitation of the Babylonian terms, the human and animal figures which have come to represent the constellations."[40] In spite of

[36] *Placita*, III, 13, 3.

[37] In my *Introduction* I placed Ecphantos in IV–1 B.C. and Heracleides in IV–2 B.C. That was somewhat arbitrary. Heracleides flourished in the middle of the century; Ecphantos probably flourished at the same time, perhaps a little earlier.

[38] St. Hippolytos (III–1) in his *Philosophumena* I, 11; translation by F. Legge, *Philosophumena* (London, 1921), vol. 1, p. 48. I left the last sentence in the quotation in spite of its irrelevancy to the rest, because it reechoes the old tradition describing Democritos as the laughing philosopher in contrast with the sad Heracleitos.

[39] *Vitruvius*, IX, 4.

[40] For details on the Babylonian origin of Democritos' astronomy see A. T. Olmstead, *History of Persia*, pp. 333–341.

his idea that the earth is flat, "disk-like laterally, but hollowed out in the middle,"[41] he accepted the possibility of "zones," but in the Babylonian manner. The Babylonians divided the celestial sphere into three concentric zones: The Way of Anu, above the pole, circumpolar stars; the Way of Enlil, the middle part, or zodiac; the Way of Ea, god of the deep, far down. Democritos abandoned that division into three and replaced it by a division into two hemispheres, northern and southern. The hypothesis that there existed southern constellations different from the northern ones was plausible enough, for when one traveled southward, across the Mediterranean and up the Nile, new constellations gradually appeared. But how could he reconcile these views with the flatness of the earth? The earth is flat but not perpendicular to the axis of the celestial sphere. This was not promising, though Democritos' descriptions prepared those of Eudoxos (IV–1 B.C.), and later of Aratos of Soli (III–1 B.C.), which enjoyed a long popularity.[42]

Democritos was also acquainted with the Greek astronomic views, especially those of Anaxagoras, whom he followed. There is a curious difference between them, however, concerning the order of the planets. Anaxagoras had put them in this order: Moon, Sun, five planets, stars; Democritos put Venus between the Moon and the Sun. That is, he recognized Venus as an "inferior" planet, though not Mercury, and to that extent he paved the way for Heracleides of Pontos.

OINOPIDES OF CHIOS

The mathematician Oinopides, who was a younger contemporary of Anaxagoras, is credited with two astronomic discoveries. The first is that of the obliquity of the ecliptic. That idea had been adumbrated by Anaximander of Miletos; indeed, it was possible not only to deduce the idea from the observations that he made with a gnomon (the simplest of astronomic instruments), but also to measure the obliquity. Yet, even if Anaximander measured the obliquity, one could hardly say that he understood it. On the other hand, if Oinopides was acquainted with Pythagorean astronomy (as he very probably was), it became possible for him really to understand the obliquity of the ecliptic, that is, to discover it.

The early measurement of the obliquity known to Euclid (24°; real value, 23° 27′) was not made by Oinopides but by other astronomers following him. It has been suggested that Euclid took an interest in certain mathematical problems because of their astronomic applications, and Proclos gives as an example Euclid's construction of the regular polygon of fifteen sides.[43] "For when we have inscribed the fifteen-angled figure in the circle through the poles, we have the distance from the poles both of the equator and the zodiac, since they are distant from one another by the side of the fifteen-angled figure."[44]

His second discovery is that of the "great year" (*megas eniautos*) of 59 years, or he introduced it from Babylonia. Assuming the lengths of the year and the month to be 365 and 29½ days, 59 is the smallest integral number of years containing an

[41] Heath, *Greek astronomy*, p. 38.

[42] The tradition of Aratos may be summarized as follows: Hipparchos (II–2 B.C.), Cicero (I–1 B.C.) Achilles Tatios (III–1), Theon of Alexandria (IV–2), Avienus (IV–2), Sahl ibn Bishr (IX–1). Apropos of the latter, see Ernest Honigmann, *Isis 41*, 30–31 (1950).

[43] Euclid, *Elements*, IV, 6.

[44] Proclos as quoted by Heath in *Euclid's elements* (Cambridge, 1926), vol. 2, p. 111; 24 × 15 = 360.

integral number (730) of months.[45] This is very puzzling, for if it is true that the Egyptians had known that length of the year (365 days) since the Third Dynasty (thirtieth century), the Babylonians had known a 19-year cycle since 747. That cycle included hollow and full months of 29 and 30 days alternately, plus 7 intercalary months; that was better than the Egyptian year.[46] The 8-year cycle (octaëtēris) of Cleostratos of Tenedos implied a year of 365¼ or 365 7/16 days. How did Oinopides come to insist on 365 days? According to Censorinus (III–I) Oinopides made the length of the year to be 365 22/59. Tannery's explanation of that contradiction is as follows: Having found the number of months in the great year, 730 [= 365 × 2], he had now to determine the number of days and did so on the basis of the Athenian calendar, recording the exact lengths of the synodic months (from full moon to full moon, or from first moonlight to first moonlight). This number was 21,557 days, which divided by 59 gives 365 22/59 days as the length of the year. It should be noted that Oinopides as well as Philolaos had a pretty accurate knowledge (correct within 1 percent) of the periods of revolution of Saturn, Jupiter, and Mars; such knowledge might have been obtained from Babylonia.[47]

Oinopides had traveled to Egypt soon after 459, and his calendar reform reestablishing the Pythagorean great year of 59 years was published on a large bronze tablet exhibited in Olympia in 456. Thus, all the visitors to the Olympic games could know of Oinopides' reform if they cared enough about it. Judging from the results, they did not care very much.

METON AND EUCTEMON

The first accurate solstitial observations were made by Meton and Euctemon in Athens in 432. These observations enabled them to determine the length of the seasons with greater precision. They introduced in that same year a new cycle, called the Metonic cycle, a period of 19 solar years, equivalent to 235 lunar months; this implied a year of *c.* 365 5/19 days, that is, 30 min 10 sec too long, a much better approximation than those of Cleostrotos and Oinopides, as the following table shows:

	Length of the year			
Cleostratos	365 days	10½ hr		
Oinopides	365	9		
Meton	365	6	18 min	56 sec
Mean tropic year	365	5	48	46

Our knowledge of the observations made by Meton and Euctemon is obtained from a papyrus (now in the Louvre) called "The art of Eudoxos" (or the papyrus of Eudoxos). It is probably the notebook of a student who flourished in Alexandria *c.* 193–190.

[45] 730, not 729 as Philolaos (and Plato) implied when they said that the cube of 9 equals the number of months of the Great Year ($9^3 = 729$). Such numerical coincidences were very pleasant to Pythagorean minds.

[46] We have a Babylonian calendar of 425 B.C. that is remarkably accurate, a "modern almanac." Richard A. Parker and Waldo H. Dubberstein, *Babylonian chronology 626* B.C. *to* A.D. *45* (Chicago: University of Chicago Press, 1942) [*Isis 34*, 442 (1942–43); 39, 174 (1948)]. Olmstead, *History of Persia*, p. 329.

[47] For the value of Philolaos' determinations as calculated by Giovanni Virginio Schiaparelli, *I precursori di Copernico nell'antichità* (Milan, 1873), p. 8, see Heath, *Aristarchus* (Oxford, 1913), p. 102; see also p. 132.

We must not continue this story because we cannot devote too much space to the calendar, which takes us away from the history of astronomy, to a mixed field wherein astronomic knowledge is dominated by religious and political needs.[48]

TECHNOLOGY AND ENGINEERING

The history of the arts and crafts and of various forms of engineering and architecture is almost endless, and we must restrict ourselves to a few significant examples.

ARTACHAIES THE PERSIAN

One of the outstanding engineering undertakings of the century was the digging of a canal across the Athos peninsula [49] by order of Xerxes (king of Persia, 485–465). Navigation is so dangerous around that mountainous peninsula that the great king ordered the digging of the canal in order to insure the safety of his fleet. Details are given by Herodotos.[50] Two Persians, Bubares son of Megabazos and Artachaies son of Artaios, were in charge (*epestasan tu ergu*). Artechaies was high in the king's favor and he stood very high in his own sandals, for he was the tallest man in Persia (8 ft high!). He died during the work or soon after; he was mourned by the king and the army and given a funeral and burial of great pomp. The isthmus is 2500 yd across, and traces of the digging can still be seen, or could be seen a century ago. "The canal forms a line of ponds from 2 to 8 ft deep and from 60 to 90 broad. It was cut through beds of tertiary sands and marls, being probably where it was deepest not more than 60 ft below the natural surface of the ground, which at its highest point rises only 51 ft above the sea level (Rawlinson)." [51]

AGATHARCHOS OF SAMOS[52]

It was said of Anaxagoras (p. 243) that he had written a book on scenography. Agatharchos, who was born *c.* 490 and flourished in Athens from 460 to 417, was a painter who actually practiced that new art and produced scenery or stage settings for Aischylos. He is the earliest painter known to us who used perspective on a large scale (that is, in wall painting or scenery, as opposed to vase painting). He may have done that before Anaxagoras wrote his book and rationalized the art, for Anaxagoras is associated with Euripides. Agatharchos did not simply practice the arts but he also wrote a technical memoir (*hypomnēmata*) concerning it. How his

[48] The literature concerning the calendar is very considerable. The standard work is still Friedrich Karl Ginzel (1850–1926), *Handbuch der mathematischen und technischen Chronologie* (3 vols.; Leipzig, 1906–1914); vol. 2 (1911) deals with the Greek calendar. Summary in Heath, *Aristarchus*, pp. 284–297. William Kendrick Pritchett and Otto Neugebauer, *The calendars of Athens* (127 pp.; Cambridge: Harvard University Press 1947) [*Isis* 39, 261 (1948)].

[49] The Athos Peninsula is the easternmost of the three peninsulas, the three-pronged fork, of the Chalcidice. Xerxes' canal was at the top of it; it runs north and south (not east and west). It was on that Peninsula lower down that the Mount Athos monasteries were established in Byzantine times; it then became the Holy Mountain.

[50] Herodotos, VII, 22 ff., 117.

[51] As quoted in W. W. How and J. Wells, *Commentary on Herodotus* (Oxford, 1912), vol. 2, p. 135. They add that the work was relatively easy; hence Stein's comparison with the Corinth canal, where there is a mile of rock and the land rises 255 ft above the sea, is misleading. The remains of the canal are now called Provlaka (derived from *proaulax*). Nearby is a tumulus that is probably the tomb of Artachaies built by order of Xerxes. H. F. Tozer, *Researches in the highlands of Turkey* (London, 1869), vol. 1, p. 128.

[52] J. Six, "Agatharchos," *J. Hellenic Studies* *40*, 180–189 (1920) [*Isis* 5, 204 (1923)].

writing would compare with that of Anaxagoras or Democritos we have no means of judging, as all have vanished. At any rate, it is significant that three men of this time, Agatharchos, Anaxagoras, and Democritos, are associated with scenography and hence we may safely assume that the art began in this time, which was natural enough, considering that it was the golden age of tragedy.

HIPPODAMOS OF MILETOS [53]

Another remarkable symbol of Greek maturity is given to us by the appearance of the first town planner. Hippodamos was an architect who planned the construction of Athens' harbor, Peiraieus (before 466), and of the Athenian colony of Thurii [54] in 443, but was not responsible for the building of Rhodes (in 408). We may thus say that he flourished shortly after the middle of the century.

He was concerned not only with the physical structure of cities (streets, squares, location of public buildings, and so on) but also with their moral structure, and was one of the forerunners of Plato in political thinking. He tried to establish an ideal constitution, which Aristotle criticized without sympathy. But Aristotle's introduction of him is interesting and picturesque.

> Hippodamos, the son of Euryphon, a native of Miletos, the same who invented the art of planning cities, and who also laid out the Peiraieus, — a strange man, whose fondness for distinction led him into a general eccentricity of life, which made some think him affected (for he would wear flowing hair and expensive ornaments; but these were worn on a cheap but warm garment both in winter and summer); he, besides aspiring to be an adept in the knowledge of nature, was the first person not a statesman who made inquiries about the best form of government.
>
> The city of Hippodamos was composed of 10,000 citizens divided into three parts — one of artisans, one of husbandmen, and a third of armed defenders of the state. He also divided the land into three parts, one sacred, one public, the third private: the first was set apart to maintain the customary worship of the gods, the second was to support the warriors, the third was the property of the husbandmen. He also divided laws into three classes, and no more, for he maintained that there are three subjects of lawsuits — insult, injury, and homicide.[55] (Then follows a long description and discussion.)

What shocked Aristotle most was that Hippodamos had no political experience as a statesman or administrator, but was simply a Pythagorean dreamer. Yet some of his dreams were more practical than Aristotle realized. For example, Hippodamos wanted his city to include farmers cultivating their own land for their private benefit. "What use are farmers to the city?" asked Aristotle. Well, Hippodamos must have believed that a "garden city" was a healthier place for every citizen to live in than a city of houses and shops, and was he not right? He was a dreamer indeed but a good dreamer, the distant ancestor of such men as Patrick Geddes

[53] Aristotle, *Politica*, II, 8; pp. 1267B–1269B). Pierre Bise, "Hippodamos," *Arch. Geschichte Philosophie* 35, 13–42 (1923) [*Isis* 7, 175 (1925)].

[54] His planning of the Peiraieus and of Thurii was done under the auspices of Pericles. Thurii was built near the site of the ancient Sybaris (on the Gulf of Tarentum, Lucania), which had been destroyed. I call it an Athenian colony because it was initiated by Pericles, but its purpose was Panhellenic. The early colonists included Herodotos, the orator Lysias, and his brothers. Thurii (or Thurium) grew rapidly and attained a high degree of prosperity; its location was marvelous. It was typical of the Greek spirit that the early colonists took a town planner with them. The Pilgrim Fathers who sailed on the Mayflower in 1620 (2063 years later!) to establish a colony in America did not think of town planning.

[55] Aristotle, *Politica* II, 8 p. 1267B.

(1854–1932) of our own time, who tried to harmonize the physical requirements of town planning with the moral and social aspects.[56]

THE SILVER MINES OF LAURION[57]

A little before reaching the Sunion promontory, which is the southern extremity of Attica, one passes through the Laurion region, rich in minerals. That region of about 80 km^2 has been mined from time immemorial (say from the early Iron Age on). The Greeks were working it mainly to obtain argentiferous galena, an ore that includes 65 percent of lead, and other metals were available, such as zinc and iron; there was even gold, but too little of it to be extracted by the old methods. Thanks to the Laurion mines, Attica was the sole producer of lead in the Greek world. Yet the main purpose of the Athenians was to obtain silver. About the beginning of the fifth century richer bodies of ore were discovered. The state took charge of the exploitation,[58] which was so profitable that about 483 each citizen received a bonus. Themistocles, who sensed the Persian danger ahead of others and realized the need for a strong navy, persuaded the Athenian government to devote the Laurion revenues to that urgent purpose. The victory of Salamis (480) was the fruit of that policy. Later the Laurion silver enabled Pericles to rebuild Athens in a magnificent way. When we admire the Parthenon we should always bear in mind the Laurion mines and the slave labor that made it possible; it is not pleasant to think that genius was not enough but that mining and slavery were needed to create that masterpiece, but we cannot dismiss those painful thoughts without hypocrisy.

The mines were overexploited in the fifth century. By the middle of the following century only old workings were open and there was no more prospecting. Xenophon, who makes that remark,[59] "proposes a socialist system of exploitation and that the state should hire out slaves as required, owing to the scarcity of private capital. But the orators show that there was much money available at Athens for investment in mercantile and other ventures, so either the mines no longer paid well, or the more important deposits had been discovered and there was greater risk of failure in a new prospect." Efforts were made to revive the mines in the third and second centuries, but those efforts were jeopardized by labor troubles and stopped by a slave revolt in 103. In the time of Strabo (I–2 B.C.), the Athenians were already obliged to work out the stones and slag that had been thrown aside; in the time of Pausanias (II–2) the mines were completely abandoned. Since 1860 better methods and new goals have made it profitable to reëxploit the mines and tailings, not for silver, but for lead, cadmium, and manganese. Remains of the ancient exploitation can still be observed *in situ*: narrow shafts, galleries, furnaces, cisterns, washing tables, and other equipment.

Of course, mining and metallurgy were not a novelty in the fifth century; they had been practiced for thousands of years by the Egyptians and other peoples. Neither was the state monopoly a novelty, nor the use of ore for military and monu-

[56] Philip Boardman, *Patrick Geddes, maker of the future* (Chapel Hill: University of North Carolina Press, 1944) [*Isis* 37, 91–92 (1947)].

[57] Elaborate account by Edouard Ardaillon, *Les mines du Laurion dans l'antiquité* (Bibliothèque des Ecoles françaises d'Athènes et de Rome, fasc. 77, 218 pp., ill., map; Paris, 1897). Oliver Davies, *Roman mines in Europe* (Oxford: Clarendon Press, 1935), pp. 246–252 [*Isis* 25, 251 (1936)].

[58] It was farmed out to contractors who used slave labor. The slaves were not state slaves.

[59] Xenophon, *On the means of improving the revenues of Athens*, IV, 3–4, a book written in his old age, *c.* 353. The quotation is taken from Davies, *Roman mines in Europe*, p. 249.

mental purposes. It was in the nature of things that rulers finding such riches would use and abuse it for their needs. The fifth-century exploitation of the Laurion mines is the earliest, however, to be known with some archaeologic, political, and economic details. It is very important to remember that the glory of Athens in the fifth century was based not only on the Greek genius but also on the exploitation of silver mines. The human spirit is never separated from bodies, or beauty from hard labor and suffering, nor other spiritual creations from villenage and innumerable agonies.

There were other mines in the Greek world than those of Attica. Herodotos refers to mines near Mount Pangaios (in Macedonia), in Thrace, and in the islands of Siphnos and Thasos.

As to mining in Palestine and Western Asia, we discover a faint echo of them in the Book of Job.

> Surely there is a vein for the silver, and a place for gold where they fine it. Iron is taken out of the earth, and brass is molten out of the stone. He setteth an end to darkness, and searcheth out all perfection: the stone of darkness, and the shadow of death.[60]

This implies a certain amount of mining, and even metallurgic, experience. Such experience would be available at this time in many countries, all over the world, but the miners and metalworkers were illiterate people who lacked the desire and the power to describe it. More than any other craft, mining has always been combined with an extraordinary amount of ignorance and superstition.[61]

[60] Job 28:1–3.

[61] Good introduction to mining folklore by A. E. Crawley, "Metals and minerals," *Encyclopedia of Religion and Ethics*, vol. 8 (1916), pp. 588–592.

XII

GEOGRAPHERS AND HISTORIANS OF THE FIFTH CENTURY

GEOGRAPHY[1]

The word geographers in our title is correct but misleading and requires an explanation. We shall deal mainly with four men,[2] the leaders of naval expeditions; these men were explorers, adventurers, not geographers in the narrow sense. The purpose of their expeditions was political and economic, but their main result from our point of view was an increase in the knowledge of the earth's surface. The reality of those four expeditions is plausible but not certain.

Of these four navigators two, Scylax and Sataspes, sailed under Persian auspices; the two others, Hannon and Himilcon, were Carthaginians, who were *de facto*, if not *de jure*, the allies of Persia against Greece, for there was an intense rivalry all around the Mediterranean Sea between the Greek settlements on the one hand and the Phoenician and Carthaginian settlements on the other. The explorations that we are going to describe represent the scientific front in the fifth century of the perennial conflict between the East and the West.

SCYLAX OF CARYANDA

Let us listen to Herodotos:

> Most of Asia was discovered by Darios. There is a river Indus, in which so many crocodiles are found that only one river in the world has more. Darios, desiring to know where this Indus issues into the sea, sent ships manned by Scylax, a man of Caryanda, and others in whose word he trusted; these set out from the city Caspatyros and the Pactyic country, and sailed down the river toward the east and the sunrise till they came to the sea; and voyaging over the sea westward, they came in the thirtieth month to that place whence the Egyptian king sent the Phoenicians afore-mentioned to sail round Libya. After this circumnavigation Darios subdued the Indians and made use of this sea. Thus was it discovered that Asia, saving the parts toward the rising sun, was in other respects like Libya.[3]

[1] Henry Fanshawe Tozer (1829–1916), *History of ancient geography* (1897); second edition with notes by M. Cary (Cambridge: University Press, 1935) [*Isis 26*, 537 (1936)]. E. H. Warmington, *Greek geography* (London: Dent, 1934) [*Isis 35*, 250 (1944)], anthology of Greek and Latin extracts translated into English. J. Oliver Thomson, *History of ancient geography* (Cambridge: University Press, 1948) [*Isis 41*, 244 (1950)].

[2] Plus of course two of the historians, Herodotos and Ctesias, whose works are full of geographic information.

[3] Herodotos IV, 44. We quote the whole chapter from A. D. Godley's translation (Loeb Classical Library, vol. 2, p. 243), because it is a good

This Scylax was thus a man of Caryanda,[4] who flourished in the time of Darios I (king of Persia, 521–485). One would like to know how this Carian found himself in Afghānistān (a long way off), but there is nothing improbable in that. The Persian governor established in the region of the upper Indus may have wished to know exactly where the river opened into the sea and how it was connected with the Western world. If Scylax was lucky (with regard to the monsoons),[5] the navigation from the Indus delta to the head of the Red Sea was difficult and tedious enough but not impossible, even for very small ships. Arab dhows [6] have repeated it innumerable times. The possibility of Scylax's voyage is confirmed by an inscription of Darios in Suez, wherein the king declares that he had dug a canal from the Nile to the Red Sea and had given orders for ships to sail from Suez to Persia.[7]

Scylax's voyage is quite plausible. An account of it may even have been written and transmitted to later writers, for example, to the author of the *Periplus of Scylax of Caryanda*. The *Periplus* describes the navigation all around the Mediterranean Sea, the Black Sea, etc. We might call the author Pseudo-Scylax, for the *Periplus* is certainly a later compilation; it can be dated *c.* 360–347. The existence of that apocryphal work confirms that of the original Scylax of Caryanda and of the latter's navigation across the Arabian Sea.

We may add that whatever doubts might exist in our minds would concern Scylax the man rather than the navigation, for we may take it as certain that many people had sailed down the Indus, across the Arabian Sea, and up the Red Sea before the fifth century. Scylax is the earliest of those navigators to be recorded.

SATASPES THE ACHAIMENIAN

According to Herodotos, Sataspes was a Persian belonging to the royal family, his mother being Darios's sister. Having ravished a noble girl he was condemned to be impaled, but his mother besought the new king, Xerxes (king from 485 to 465), to change the punishment into another, which according to her was heavier,

for he should be compelled to sail round Libya, till he completed his voyage and came to the Arabian Gulf. Xerxes agreeing to this, Sataspes went to Egypt, where he received a ship and a crew from the Egyptians, and sailed past the Pillars of Heracles. Having sailed out beyond them, and rounded the Libyan promontory called Soloeis,[8]

illustration of Herodotos as well as our only source concerning Scylax. Pactyica was west of the Indus, the country around Jalālābād in northeastern Afghānistān. Scylax could not sail down the Indus "toward the East," because the general direction of the river flow is southwest. Herodotos' geography was generally vague; would our own idea of distant countries be much clearer if we had no maps? The "Phoenicians afore-mentioned" is a reference to Necho, who was not a Phoenician and was later than Scylax, but Herodotos' chronology was often incorrect; he had no chronologic table on his desk.

[4] In Caria, the southwest corner of Asia Minor. Caryanda was on a little island not far from Halicarnassos, Herodotos' birthplace. Herodotos may have heard local traditions concerning Scylax.

[5] Knowledge of the monsoons was not really available until the time of Hippalos, who flourished in the first century before or after Christ; Thomson, *History of ancient geography*, pp. 176, 298.

[6] Dhow or dow; see Henry Yule and A. C. Burnell, *Hobson-Jobson. A glossary of colloquial Anglo-Indian words and phrases* (new ed. by William Crooke, London, 1903), p. 314. Dhow navigation as practiced today has been beautifully described by Alan Villiers, *Sons of Sinbad* (New York: Scribner, 1940). See also Richard LeBaron Bowen, Jr., Arab dhows of Eastern Arabia (64 pp., 37 ills.; Rehoboth, Massachusetts: Privately printed, 1949) [*Isis 42*; 357 (1951)].

[7] Claude Bourdon, *Anciens canaux, anciens sites et ports de Suez* (Cairo, 1925), pp. 12–30, pl. 1. Darios' inscription is on the stela of al-Kabrīt now in the gardens of the Suez Canal Co. in Ismailia.

[8] Probably Cape Cantin, 32°36′ N, Arabic Rās al-hudik(?), a promontory on the Moroccan coast about the latitude of the Madeira Islands (32°40′ N).

he sailed southward; but when he had been many months sailing far over the sea, and ever there was more before him, he turned back and made sail for Egypt. Thence coming to Xerxes, he told in his story how when he was farthest distant he sailed by a country of little men, who wore palm-leaf raiment; these, whenever he and his men put in to land with their ship, would ever leave their towns and flee to the hills; he and his men did no wrong when they landed, and took naught from the people but what they needed for eating. As to his not sailing wholly round Libya, the reason (he said) was that the ship could move no farther, but was stayed. But Xerxes did not believe that Sataspes spoke truth, and as the task appointed was unfulfilled he impaled him, punishing him on the charge first brought against him.[9]

Herodotos' story is full of tantalizing details. In the first place, Sataspes' mother definitely spoke of the circumnavigation of Africa, and she did not exaggerate when she spoke of it as a very hard undertaking. All the Mediterranean sailors feared the mysterious dangers of the ocean. Second, Sataspes is said to have hired an Egyptian ship and crew; it is probable that he hired in Egypt a Phoenician ship and crew; there had been commercial relations between the two nations from time immemorial and Phoenician galleys were already sailing up the Nile under Thutmose III (fifteenth century). Third, how far down the west coast of Africa did Sataspes actually sail? After leaving Soloeis he sailed southward "for many months," until his ship "could move no farther, but was stayed." Did he reach the equatorial doldrums on the latitude of Cape Verde? Or was he stopped by the tradewinds and northward current off the Guinea coast? One reason for believing that he reached the Guinea coast is the strange account of the "little men who wore palm-leaf raiment." At any rate, even if he went as far as that (say 10° N), he was still exceedingly far from the goal, but the ancients could not possibly imagine the immensity of the African continent.[10]

About the beginning of the fifth century, the Carthaginian government decided to explore the Ocean, or rather the Oceanic coasts, and to send two expeditions which were to sail out of the Straits of Gibraltar and proceed respectively to the left and right. The first expedition was intrusted to Hannon and the second to Himilcon.

HANNON OF CARTHAGE

Sataspes' navigation along the west coast of Africa took place during the rule of Xerxes (486–465). It is remarkable that a similar expedition was undertaken at about the same time, if not a little earlier, under Carthaginian auspices.[11]

The suffete [12] Hannon left Carthage with a fleet of sixty 50-oared ships carrying 30,000 men and women.[13] The Carthaginians' plan, therefore, was not simply exploration, but colonization; they probably meant to continue their old method of

[9] Herodotos, IV, 43, as translated by A. D. Godley in the Loeb edition (vol. 2, p. 241).

[10] The latitude of the Cape of Good Hope is 34°22′ S. Even Henry the Navigator (1394–1460) could not imagine the size of Africa, and he believed that the ancients had managed to circumnavigate it.

[11] The Carthaginians had sent an armed expedition commanded by Hamilcar to Sicily. The expedition was defeated and Hamilcar killed in 480. It was first assumed that our Hannon was the son of Hamilcar, and on that basis his own expedition was dated *c.* 470. That assumption was gratuitous; Annon (Annōn) was a common name in Carthage. It is better to stick to the synchronism of the two expeditions (Hannon's and Himilcon's); Himilcon's took place at the beginning of the century.

[12] "Suffete" is a Punic (Carthaginian) term designating a high magistrate; cf. the Hebrew word *shophet*. Punic is a dialect of Phoenician; Phoenician and Hebrew are sister languages.

[13] The two figures 60 and 30,000 do not tally, for the penteconters (50-oared galleys) could not carry 500 people each.

establishing in convenient harbors a string of trading stations (factories), which would insure their commercial needs and supremacy.[14] After his return to Carthage, Hannon wrote in the Punic language an account of his journey which was inscribed on a tablet placed in the temple of Melkarth. A Greek version of that Punic text has come down to us under the title of the *Periplus of Hannon.*

Their first important landing place was the island of Cerne, which was about as far distant from the Straits as Carthage was on the other side; this helps us to identify it with the island now called Herne, at the mouth of the Rio de Oro. Having established a base in Cerne, the Carthaginians made two other expeditions from there, the first of which took them to the Senegal River and the second to Cape Verde (Dakar) and the Gambia River, the Bay of Bissagos, and Sherbro Sound (in Sierra Leone; 7°30′ N). I quote modern names, not those used in the *Periplus*, for the identification of each term would require separate discussion, and we are not concerned with topographic details. The main point is that Hannon followed the West African coast, about 2,600 miles of it, perhaps as far as Cape Palmas, where the coast turns eastward. Did Hannon go farther south than Sataspes? Possibly, but it does not matter. We may credit both navigators (or at least one of them) with the reconnaissance of the northwest coast of Africa. To measure their achievement it suffices to remember that the exploration of the African coast was not continued farther south until almost 2000 years later, by the Portuguese navigators about the middle of the fifteenth century.

How can we believe Hannon's report? Simply because it includes facts that tally with modern observations and could not possibly have been invented. It is true that no identification of a place or river is completely certain, but all the identifications constitute a coherent pattern which we may trust to be substantially accurate. The anthropologic facts are equally convincing (as a whole), including the reference to bush burnings and to hairy people, called gorillas in the Greek text (Pygmies or Negritos, or real apes?); they captured three females and flayed them for the skins. The report was too short and was misunderstood by later writers, for example, Pliny (I–2), who told that Hannon had sailed all the way to Arabia. That misinformation was generally accepted even by prudent men such as Henry the Navigator and Richard Hakluyt.[15]

HIMILCON OF CARTHAGE

Himilcon is known to us only through a brief reference by Pliny (I–2),[16] who brackets him with Hannon, and a Latin poem of Avienus (IV–2), which is the translation of the Greek poem of Dionysios Periegetes (I–2). Pliny and Dionysios take us back to the first century and this leaves a large gap in the tradition; yet we have no reason to doubt the reality of Himilcon's voyage. One of the ultimate sources of Avienus and Dionysios was probably the account of a Massiliote captain who had visited Tartessos[17] toward the end of the sixth century and had some

[14] The same method was continued by the European nations at the beginning of the colonization period, the example being given by Portugal. The Portuguese empire in Asia in the sixteenth century was hardly more than a collection of trading stations, scattered along the coasts of India, Further Asia, China, and the islands.

[15] Richard Hakluyt (1552–1616), English historian of navigation; see *Isis 38*, 130 (1947–48).

[16] Pliny, *Natural history*, VII, 197.

[17] Tartessos, Phoenician colony near the mouth of the Guadalquivir in Andalusia; perhaps equivalent to Tarshīsh (Ezekiel 27:12, Jeremiah 10:9). It was a very prosperous colony until its de-

knowledge of the Spanish coast. Himilcon's voyage took place soon after the destruction of Tartessos, that is, about the beginning of the fifth century.

He was sent to explore the west coast of Europe, and he reached a group of islands called the Oistrymnides and a cape of the same name, this is, the Armorican peninsula (Bretagne) and some of the islands off it. He refers to the industry and skill of the islanders, who are excellent sailors in spite of the fact that they have no wooden ships (like the Phoenicians) but only "vessels with skins sewn together" (coracles); they sail to the islands of the Hibernians and of the Albions (Ireland and England). Phoenician sailors used to go to those islands for trade (the tin trade).[18] It may be that on his way to Brittany and beyond, or on his way back, Himilcon was blown away to a part of the ocean where there were no winds and where "amidst the swirl of waters seaweed rises up straight and often holds back a ship as brushwood might." [19] This has been understood by some historians as a reference to the Sargasso Sea, a large body of relatively still water in the Atlantic where weeds accumulate as they would under similar conditions in a river; it is a little difficult to accept that, because the Sargasso Sea is very distant from Europe.[20] It is possible that Phoenician navigators reached the Fortunate Islands,[21] but it is difficult to believe that they reached the Azores and the Sargasso beyond it.[22]

To sum up, these four accounts of navigation across the Arabian Sea and along the Atlantic coast of Europe and North Africa are more interesting than surprising. Such achievements as have been described above are far less remarkable than, say, Greek cogitations on infinity or arithmetic irrationality. What the Greeks did in the mathematical field was truly astounding, for they proved themselves superior not only to their contemporaries but also to a great many of our own. On the other hand, we fully expect the early navigators, especially the Phoenicians and their offspring the Carthaginians, to do many things comparable to those described or even more daring, and to have done them not only in the fifth century but long before. To consider only the sailing along the Moroccan coast and the planting of factories in Soloeis and other places, it is clear that this required more courage than knowledge. The Carthaginian art of navigation was quite sufficient for such purposes and it would have sufficed even to carry them step by step much farther south along the African coast, and to anticipate the Portuguese achievements of the fifteenth century. The development of Carthaginian colonization was stopped, however, by the life-and-death struggle between Carthage and Rome, which immobilized the Carthaginian navy in the Mediterranean or close to it, and ended with the death of Carthage in 146.

One final remark: what is most surprising in the four accounts is not so much the

struction *c.* 500, and was superseded by another Phoenician colony in the same region, Gades (Cadiz).

18 The details of the Phoenician tin trade are very obscure, largely because the Phoenicians wanted to hide their business. The location of the Tin Islands (*Cassiterides nēsoi*, *Cassiterides insulae*) is highly problematic: some of the British Isles? or some other islands of the Atlantic coast?

19 R. F. Avienus (IV-2), in his poem "Ora maritima," line 120.

20 The Sargasso Sea occupies the region between lat. 20° and 35° N, long. 40° and 70° W. It is surrounded by currents turning clockwise. The Bermuda Islands are located near its western end, the Azores at some distance from the northeastern corner.

21 *Fortunatorum insulae* (*ai tōn macarōn nēsoi*); the Islands of the Blessed. These were the Canary or the Madeira islands.

22 What makes me hesitate in my denial is a similar reference in the pseudo-Aristotelian *Mirabilia* (136, end of 844A). However, both Aristotle and Avienus refer to a place where the sea bed is scarcely covered over by the shallow water, and this cannot be the Sargasso Sea.

achievements that they report as the fact that they reached us. We must assume that many other attempts of the same kind, or superior to them, were made in ancient times. The tales did not reach us, because the adventurers died and did not return, or because they did not want any publicity or were not sufficiently articulate to tell their story. The psychology of sailors and adventurers is very different from that of writers; in fact, most of them could not write at all or compose a clear account. Scylax and Sataspes, Hannon and Himilcon must be regarded as the few representatives of a much larger body, the surviving symbols of ancient navigation.[23]

Two of the reports were saved for us, thanks to Herodotos, and his history contains a good many other facts of geographic interest; a few of these will be discussed when we speak of him presently. The main geographic event of the century occurred at the very end of it (in 401), when Xenophon led the Ten Thousand Greek mercenaries who had been left stranded on the upper Tigris across the mountains of Armenia and Cappadocia to Trapezus (Trebizond) on the Black Sea.[24] That retreat, so vividly described by Xenophon, is one of the outstanding events of its kind in the memory of mankind. Xenophon's *Anabasis*, written *c.* 379–371, is one of the masterpieces of historical and geographic literature. Though the purpose was not geographic, the *Anabasis* is the earliest description, sufficiently elaborate, of a large district and of the people living in it; it is not only one of the best books of its kind, but the first one.[25]

THE HISTORIANS: HERODOTOS, THUCYDIDES, AND CTESIAS

The second half of the century witnessed the birth of historiography, that is, the birth of a new branch of science, concerned with the accurate description of man's experience. Some people have argued that historiography cannot be called a science, because historical truth is too uncertain and elusive, and have blamed me for giving so large a place to it in my *Introduction*. Their objections are unfounded, because what characterizes scientific efforts is the purpose to find the truth, as much of it as is available, and to come as close to it as circumstances permit. The approximation that is attainable or is actually attained varies from field to field. The

[23] This geographic section might include a discussion of the early ideas concerning the floods of the Nile, but this topic has already been dealt with in our section on Anaxagoras.

[24] The Ten Thousand were Greek mercenaries who had been hired by the Younger Cyros, a Persian satrap who plotted against his brother and King Artaxerxes Mnemon (ruled 405–359). He set out from Sardis in the spring of 401 and was defeated and killed by Artaxerxes in the plain of Cunaxa, north of Babylon, between the two rivers. The Greek mercenaries, having obtained a safe-conduct from Artaxerxes, marched to the Tigris and followed its left bank until they reached its tributary, the Greater Zab. There their general and other officers were treacherously seized, and they found themselves without chief and guide. Xenophon was elected as their general and guided most of them to safety and home. The title of the book, *Anabasis*, is a bit misleading, for there was much going down as well as up, and the final march down to the Black Sea was very long.

The *Anabasis* has been discussed by travelers who tried to follow Xenophon's footsteps, the Englishman, H. F. Tozer (1881), and the German, Eduard von Hoffmeister (1911), and by the French "armchair traveler," Arthur Boucher (1913). Dates refer to their publications; see *Introduction*, vol. 1, p. 123.

[25] It has been objected that Xenophon's description is not accurate enough to enable one to draw his itinerary exactly on a map. That is hardly fair, because travel in a difficult country like the mountains of Armenia cannot be described with great precision in the absence of very definite (human) landmarks. Moreover, Xenophon describes well enough the region that he and his army traversed, if not the very paths. One cannot draw his itinerary on a very large map, but one can draw it on a small one, and this has been done repeatedly.

scientific nature of our efforts is determined by the quality of our purpose and methods, not by the degree of approximation of our results. Historical facts are uncertain, yet in the fifth century they were less obscure and less wobbly than the great majority of physical facts.

HERODOTOS OF HALICARNASSOS

Herodotos, son of Lyxes and Dryo, was born at Halicarnassos in Caria, *c.* 484.[26] Caria (in the southwest corner of Asia Minor) had been colonized by the Dorians but much influenced by the more advanced culture of the Ionian cities of the neighborhood; in the fifth century the Greek-speaking people of Caria were speaking the Ionian dialect. About the time of Herodotos' childhood the dynasty ruling Caria was enfeoffed to the Persian empire. Political troubles obliged the young Herodotos to leave his native country; he spent some time in Samos, then traveled considerably. He visited Athens and got to know Pericles and Sophocles. He spent the end of his life in Thurii (founded in 443) and died there about the beginning of the Peloponnesian War (431–404), say *c.* 426. In ancient times (down to the third century of our era) he was often called Herodotos of Thurii.

His travels were extensive: he visited Egypt and went up the Nile as far as Aswān and Elephantine; [27] he probably went to Cyrene; he was in Gaza and Tyre and traveled down the Euphrates to Babylon. He was acquainted with the northern Aegean as far as Thasos. What is most remarkable of all, he visited Scythia, the country north of the Black Sea, and must have spent some time at Olbia, near the mouth of the Hypanis (Bug), and higher up the river. Many of the facts mentioned by him were witnessed by himself; others were obtained by hearsay. In some places, like Athens and Delphi, he would come across people from every part of the Greek world.

Cicero called him the father of history; [28] that title of honor has stuck to him ever since and is well deserved. This does not mean that he was the first to write history. Not to mention Hebrew historians like the author of the Books of Samuel (VII B.C.), there had been many chroniclers in the Greek world. We have already spoken of another Persian subject, Hecataios of Miletos, whom Herodotos criticized as freely as he used him, and there were other "logographers," meaning annalists, chroniclers. Herodotos was the first, however, to write a book that was well composed and readable; in fact, he composed the first masterpiece of Greek prose (Fig. 67).[29]

[26] The name of Halicarnassos is familiar to most readers because of the mausoleum of Halicarnassos. This was the splendid monument built by Artemisia II to immortalize the memory of her brother and husband Mausolos, satrap of Caria from 377 to 353. The city was destroyed by Alexander in 334; remains of the monument, discovered by Sir Charles Newton in 1857, are preserved in the British Museum. Though the monument was destroyed, Artemisia succeeded in her purpose, as the word "mausoleum" has become a common word to designate a magnificent tomb. Each time that we use the word, we pay tribute to Mausolos and to herself.

Halicarnassos was the birthplace of two historians, Herodotos and Dionysios (I–2 B.C.).

[27] He does not mention Philae, the so-called pearl of Egypt, because its oldest monuments date only from *c.* 370.

[28] *De legibus* I, end of 1: "quamquam et apud Herodotum, patrem historiae, et apud Theopompum sunt innumerabiles fabulae," Theopompos of Chios (IV–2 B.C.) has sometimes been called the founder of psychologic history, the Greek predecessor of Tacitus (I–2).

[29] It is interesting to note the lateness of the first masterpiece in prose as compared with the masterpieces in verse; the date of the *Iliad* is uncertain, but parts of it existed three or four centuries before Herodotos' history.

ὀλίγον ἰδόντες ὑπὲρ Αἰγὸς ποταμοῦ, ἀλεξόμενοι χρόνον ἐπὶ συχνὸν, οἱ μὲν ἀπέθανον, οἱ δὲ ζῶντες ἐλάμφθησαν. καὶ συνδήσαντές σφεας οἱ Ἕλληνες ἦγον ἐς Σηστόν· μετ' αὐτῶν δὲ καὶ Ἀρταΰκτην δεδεμένον, αὐτόν τε καὶ τὸν παῖδα αὐτοῦ. καί τεῳ τῶν φυλασσόντων λέγεται ὑπὸ χερσονησιτέων ταρίχους ὀπτῶντι, τέρας γενέσθαι τοιόνδε· οἱ τάριχοι ἐπὶ τῷ πυρὶ κείμενοι, ἐπάλλοντό τε καὶ ἤσπαιρον, ὅκως περ ἰχθύες νεάλωτοι. καὶ οἱ μὲν, περιχυθέντες, ἐθώμαζον. ὁ δὲ Ἀρταΰκτης ὡς εἶδε τὸ τέρας, καλέσας τὸν ὀπτῶντα τοὺς ταρίχους, ἔφη· ξεῖνε ἀθηναῖε, μηδὲν φοβέο τὸ τέρας τοῦτο· οὐ γὰρ σοὶ πέφηνε, ἀλλ' ἐμοὶ σημαίνει ὁ ἐν Ἐλαιοῦντι Πρωτεσίλεως, ὅτι καὶ τεθνεὼς, καὶ τάριχος ἐὼν, δύναμιν πρὸς θεῶν ἔχει τὸν ἀδικέοντα τίνεσθαι. νῦν ὦν ἄποινά οἱ τάδε ἐθέλω ἐπιθεῖναι· ἀντὶ μὲν χρημάτων τῶν ἔλαβον ἐκ τοῦ ἱροῦ, ἑκατὸν τάλαντα καταθεῖναι τῷ θεῷ· ἀντὶ δ' ἐμεωυτοῦ καὶ τοῦ παιδὸς, ἀποδώσω τάλαντα διηκόσια ἀθηναίοισι περιγενόμενος. ταῦτα ὑποσχόμενος, τὸν στρατηγὸν Ξάνθιππον οὐκ ἔπειθε· οἱ γὰρ Ἐλαιούσιοι τῷ Πρωτεσιλάῳ τιμωρέοντες, ἐδέοντό μιν καταχρησθῆναι· καὶ αὐτοῦ τοῦ στρατηγοῦ ταύτῃ ὁ νόος ἔφερε. ἀπαγαγόντες δὲ αὐτὸν ἐς τὴν ἀκτὴν, ἐς τὴν Ξέρξης ἔζευξε τὸν πόρον, οἱ δὲ λέγουσι ἐπὶ τὸν κολωνὸν τὸν ὑπὲρ Μαδύτου πόλεος, σανίδι προσπασσαλεύσαντες, ἀνεκρέμασαν. τὸν δὲ παῖδα ἐν ὀφθαλμοῖσι τοῦ Ἀρταΰκτεω κατέλευσαν. ταῦτα δὲ ποιήσαντες, ἀπέπλεον ἐς τὴν Ἑλλάδα, τά τε ἄλλα χρήματα ἄγοντες, καὶ δὴ καὶ τὰ ὅπλα τῶν γεφυρέων, ὡς ἀναθήσοντες ἐς τὰ ἱρά. καὶ κατὰ τὸ ἔτος τοῦτο, οὐδὲν ἔτι πλέον τουτέων ἐγένετο. τούτου δὲ τοῦ Ἀρταΰκτεω τοῦ ἀνακρεμασθέντος προπάτωρ Ἀρτεμβάρης ἐστὶ, ὁ Πέρσῃσι ἐξηγησάμενος λόγον, τὸν ἐκεῖνοι ὑπολαβόντες, Κύρῳ προσήνεικαν, λέγοντα τάδε·

Ἐπεὶ Ζεὺς Πέρσῃσι ἡγεμονίην διδοῖ, ἀνδρῶν δὲ σοὶ Κῦρε, κατελὼν Ἀστυάγεα, φέρε· γῆν γὰρ ἐκτήμεθα ὀλίγην, καὶ ταύτην τρηχέην· μεταναστάντες ἐκ ταύτης, ἄλλην σχῶμεν ἀμείνω. εἰσὶ δὲ πολλαὶ μὲν ἀστυγείτονες, πολλαὶ δὲ καὶ ἑκαστέρω· τῶν, μίαν σχόντες, πλέοσι ἐσόμεθα θαυμαστότεροι. εἰκὸς δὲ ἄρχοντας ἄνδρας τοιαῦτα ποιέειν. κότε γὰρ δὴ καὶ παρέξει κάλλιον, ἢ ὅτε γε ἀνθρώπων τε πολλῶν ἄρχομεν, πάσης τε τῆς Ἀσίης;

Κῦρος δὲ ταῦτα ἀκούσας, καὶ οὐ θαυμάσας τὸν λόγον, ἐκέλευε ποιέειν ταῦτα· οὕτω δὲ αὐτοῖσι παραίνεε κελεύων παρασκευάζεσθαι, ὡς οὐκέτι ἄρξοντας, ἀλλ' ἀρξομένους.
» φιλέειν γὰρ ἐκ τῶν μαλακῶν χώρων, μαλακοὺς γίνεσθαι· οὐ γάρ τι τῆς αὐτῆς γῆς
» εἶναι, καρπόν τε θαυμαστὸν φύειν, καὶ ἄνδρας ἀγαθοὺς τὰ πολέμια. ὥστε συγγνόντες Πέρσαι, οἴχοντο ἀποστάντες, ἑσσωθέντες τῇ γνώμῃ πρὸς Κύρου. ἄρχειν τε εἵλοντο λυπρὴν οἰκέοντες μᾶλλον, ἢ πεδιάδα σπείροντες, ἄλλοισι δουλεύειν.

ΤΕΛΟΣ ΤΩΝ ΙΣΤΟΡΙΩΝ ΗΡΟΔΟΤΟΥ.

ΑΑ· ΒΒ. ΓΓ· ΔΔ· ΕΕ· ΖΖ· ΗΗ. ΘΘ· ΙΙ· ΚΚ· ΛΛ. ΜΜ· ΝΝ·
ΞΞ· ΟΟ· ΠΠ· ΡΡ· ΣΣ·

Ἅπαντα τετράδια, πλὴν τοῦ τελευταίου, ὅπερ ἐστὶ δυάδιον.

Ἐνετίῃσι παρ' Ἄλδῳ τῷ Ῥωμαίῳ ἔτει χιλιοστῷ πεντακοσιοστῷ δευτέρῳ, μαιμακτηριῶνος τετάρτῃ ἱσταμένου, κατὰ τὴν φθίνοντος, οὐ μὴν γε ἀπὸ προοιμίου.

AA. BB. CC. DD. EE. FF. GG. HH. II. KK. LL.
MM. NN. OO. PP. QQ. RR. SS.

Sunt Quaterniones omnes, præter ultimum duernionem.

Venetiis in domo Aldi mense Septembri. M.DII. et cum priuilegio ut in cæteris.

Fig. 67. Herodotos, *editio princeps* (folio; Venice: Aldo Manuzio, September 1502); page showing colophon and register of gatherings. That register explains that the book is made of 17 quaternions marked AA, BB, . . . , RR and one duernion marked SS, thus making a total of $(16 \times 17) + (8 \times 1) = 280$ pages; this includes the title page and the last page containing the printer's name, Aldus, and his mark. The page reproduced is the last page but one. [From the copy in the Harvard College Library.]

Let us now examine that great work.

It is an account of the past and present of Greece, Egypt, and Asia Minor. The main purpose is to explain the gigantic conflict between Asia and Greece, from the time of Croisos (king of Lydia, 560–546) to that of Xerxes and the end of the Persian wars, more exactly to the capture of Sestos (479–78).[30] The *History* was divided into nine books, each bearing the name of one of the muses; [31] that division was probably made by Alexandrian grammarians; it existed already in the time of Lucianos (120–200). When Herodotos referred to his own work, he never mentioned any book, but called his work *logos*.[32]

His general purpose is well explained in his own words, the first paragraph of the work:

> What Herodotos the Halicarnassian has learnt by inquiry is here set forth: in order that so the memory of the past may not be blotted out from among men by time, and that great and marvelous deeds done by Greeks and foreigners and especially the reason why they warred against each other may not lack renown.

That simple statement is as impressive as it is instructive. His purpose is to record for posterity the great deeds accomplished not only by the Greeks but also by the barbarians.[33] This is the more remarkable, because some of the foreigners dealt with were or had been recently the enemies of Greece in a terrible war which was ending at the very time when Herodotos was engaged in the composition of his work. What was the matter with him? Did he lack patriotism? He was a civilized man, trying honestly and gently to understand the men of other nations. It must be added that his cosmopolitan point of view was more natural to him than it would have been, say, to a Theban or an Athenian, because he belonged to a nation, Caria, founded by Dorians but submitted to strong Ionian and Persian influences; it was half Orientalized.[34] The ruling dynasty was non-Hellenic; Queen Artemisia I, of whom Herodotos speaks favorably,[35] was a vassal of Xerxes, and she accompanied him in his expedition with five ships which were reputed the best in his fleet, next to those of Sidon. Plutarch (I–2) wrote a book (*De malignitate Herodotis*) in which he accused the father of history of being *philobarbaros*, which would mean almost the same as "cosmopolitan" in the Soviet nations of today. He accused him of being unfair, because Herodotos was not sufficiently prejudiced. He reminds us of some fanatics of our own day who suspect every man whose patriotism is not as blatant as theirs. Add this new item to the list of Greek miracles. The

30 Sestos, the best harbor in the Dardanelles, on the northern (European) side. It was there that Xerxes had moved his army, from Asia to Europe, on a bridge of boats; it was the first town to be freed from Persia by the Athenian fleet in 479. Thucydides began his own retrospective account (*ta meta ta Mēdica*) at that time.

31 My references to Herodotos are always to book and chapter (say vii, 103), which will permit the reader to use almost any edition or translation.

32 The word *logos*, story or history, tallies with the word *logographos* applied to the earlier chroniclers.

33 The Greek word *barbaros* has not necessarily the evil connotation of our word "barbarian" derived from it. *Barbaroi* was the Greek equivalent of "foreigner" in English, or *gōyim* in Hebrew, *gentiles* in Latin. In the case of uneducated or narrow-minded people, all these words have bad meanings; foreigners are enemies and barbarians. Herodotos used the word *barbaros* as a civilized American uses the word "foreigner," without malice.

34 The names of Herodotos' parents, as given by Suidas (X–2), are very strange – Lyxes and Dryo. These names are unique in my experience. They might be Oriental names more or less Hellenized. If so, Herodotos would be himself a *barbaros*, at least in part. We should remember that "pure" Greeks must have been relatively rare in Asia.

35 Herodotos, vii, 99; viii, 103.

first Greek history was written by a man who had witnessed many episodes of the terrible conflict between Persia and Greece, yet was able to speak of it gently, fairly, without race prejudice.[86]

After having underlined, as was proper, this fundamental virtue of Herodotos' mind, let us consider more carefully his purpose and methods.

First, a few more words about his sources. The main source was, of course, the information he had collected during his travels in the three continents.[87] He was as critical as people could be in his time. We could not expect him, for instance, not to believe in divination, but he shows that such belief was not unconditional; oracles were not always accepted at their face value; one might consult many oracles and choose among them. Divination, then as now, was often a kind of thinking aloud and mutual suggestion. Herodotos often expressed his incredulity, or he covered himself with such remarks as "I told the story as it was told to me"; sometimes he quoted different traditions, leaving it to the reader to choose between them. He was a great storyteller; it has been suggested that he earned his living that way, but there is nothing to prove that. How he did earn it we do not know; perhaps he was a merchant (*emporos*); he was certainly interested in trade, as most Greeks were.[88] His book is full of anecdotes and short stories which could be detached from it, pleasant digressions that he loved to add in the manner of a good raconteur. It is possible that he handled documents and saw inscriptions, but he depended chiefly on oral tradition, and was skillful in cross-examining witnesses and checking their testimonies. He helps us to see those witnesses and to hear their very words, then spreads his own reflections, which are good-natured, often shrewd, and make us sometimes think of Montaigne.

His book is a treasury of Greek and Near Eastern folklore. It is comparable to the books of other great travelers, such as Marco Polo (XIII–2) and Ibn Baṭṭūṭa (XIV–2), and his fate was comparable to theirs. The stories they told were so extraordinary that many people refused to believe them; they would smile and say, "se non è vero . . ." Uncritical readers swallowed miracles and myths without demur, but the true stories seemed incredible to them. We shall give a few examples presently.

Herodotos was a master of simple Greek prose, the first author to make the Greek people realize that prose might be as beautiful and as moving as poetry. This was observed by another Halicarnassian, Dionysios (I–2 B.C.). His style is very easy without mannerisms of any kind; his tales are straightforward; he loves digressions and introduces them deliberately, as Homer did. He was influenced by the latter, as every Greek was, but also by the tragedians. Generous, candid, moderate, prudent, he could be as curious and naïve as a child. Picturesque details appealed to his fancy; he indulged in them in some of his digressions, chiefly in his catalogue

[86] Interesting commentary on Herodotos' toleration and generosity by Theodore Johannes Haarhoff, *The stranger at the gate* (Oxford: Blackwell, 1938, 1948), pp. 20, 22 [*Isis 41*, 75 (1950)]. Haarhoff knows full well the implications of race prejudice, because he is a professor of classics in the University of the Witwatersrand, Johannesburg.

[87] The three continents, or rather the three parts (*tria moria*) – Europe, Asia, Libya – were recognized at the beginning of the fifth century, if not before. Herodotos objected to this (II, 17), saying that one ought to add a fourth part, Egypt, otherwise, since the Nile divides Asia from Libya, Egypt would be half Asiatic and half Libyan. His ideas concerning the relative sizes of these parts were unavoidably wrong.

[88] W. W. How and J. Wells, *Commentary on Herodotos* (Oxford, 1912), vol. 1, p. 17, give good reasons for believing that Herodotos was a merchant.

of all the nations impressed into Xerxes' army and fleet, the men being armed and dressed differently, each according to his race and customs. That catalogue covers no less than thirty-eight chapters,[39] beginning with the Persians and ending with Queen Artemisia I and the Carian ships.

His philosophy of history was the same as that of the great poets and playwrights of his time. The fundamental idea is that of the vicissitudes of fortune; it is illustrated throughout his work, which begins fittingly with the history of Croisos and ends with that of Xerxes. In each case, we witness the implacable nemesis chastening insolent pride (*hybris*). The idea of providence also occurs in Herodotos,[40] as it did in Sophocles and Euripides.[41] Therefore, in spite of his simplicity and good nature, Herodotos was very much in earnest. I would like to complete my portrait of him with a comparison that surprised me: "Herodotos suffered the fate which befell Mozart. His charm, wit and effortless ease have diverted attention from the note of profound sadness and pity sounded not seldom in his history." [42] A comparison between two men as distant in time, space, and manners as Herodotos and Mozart is of course very hazardous; yet this one appealed to my imagination, for I love them both.

The history of the ancient Near East is extremely complex, and, even for us who have maps, synoptic tables, and dictionaries to guide us at every step, it is sometimes very hard to unravel the skeins of events and to understand what happened. We cannot expect an early historian to explain such complicated matters to us clearly and exactly. Herodotos' history contains a large number of important data, but it is not, and could not possibly be, a history comparable to those available today, which are the fruit of centuries. In particular, his history of Egypt is a mess; it begins to be valuable only when he deals with the Twenty-sixth (or Saitic) Dynasty (663–525), founded by Psametik I (ruled 663–609), and with the Persian conquest. Egypt remained a province of the Persian empire from 525 to Alexander the Great (332). It was natural enough for Herodotos, born a Persian subject, to visit Egypt, and the innumerable wonders of that country excited his curiosity. He was impressed by the enormous temples, covered with long inscriptions which he could not read, and was at the mercy of dragomans who could not read them either yet were ready to explain them. His account of Egypt is nevertheless exceedingly precious, because it is the only one we have that was composed by a Greek witness, an intelligent outsider, full of sympathy.

His account of Babylonia deserves similar criticism. His knowledge of Babylonian antiquity must have come close to that of an educated Babylonian of that time, who might have definite ideas about the traditions of his people, yet could not possibly know the history of the old dynasties as well as we do.

The story that Herodotos tells [43] about Psametik is typical of his credulity mixed with criticism. Some people claimed that the Phrygian culture [44] was older than the

[39] Herodotos, VII, 61–99.

[40] *Ibid*, III, 8.

[41] The "vicissitudes of fortune" was a commonplace of Greek literature. The metaphor of the wheel of fortune (*trochos theu, rota fortunae*) occurs already in a fragment of Sophocles (No. 871, A. C. Pearson's edition, vol. 3, p. 70). The idea of divine providence (*tu theu hē pronoia*) was illustrated best by the name under which Athena was worshiped in Delphi, *Pronoia Athēna*.

[42] John Dewar Denniston, *Oxford classical dictionary* (Oxford: Clarendon Press, 1949), p. 423.

[43] Herodotos, II, 2.

[44] Phrygia was the western part of the great central tableland of Asia Minor. Its ancient glory is symbolized by the legendary King Midas and by Midas II, who ruled from 738 to 696.

Egyptian. In order to find the truth, Psametik entrusted newborn children to a shepherd to be brought up among his flock. The children were to be well fed, but nobody was to speak to them. Finally, one of them uttered the word *becos* (meaning bread), which is Phrygian. Psametik concluded that the Phrygian culture was the older. Herodotos collected other traditions relative to the same story in Memphis, Thebes, and Heliopolis. He heard many other stories concerning the gods, but remarked,[45] "I do not wish to relate them, save only the names of the gods, for I believe that no man knows about the gods more than another."

The philosophic and religious pattern at the back of Herodotos' mind was a combination of Pythagorean with Oriental ideas. He ascribed the belief in metempsychosis [46] to the Egyptians, adding that some of the Greeks, whom he could name, shared it. This is quite plausible, but those Greeks were more likely to have obtained it directly or indirectly from India (samsāra) than from Egypt. He confused Demeter and Dionysos, rulers of the lower world, with Isis and Osiris, but that was natural.

He had no mathematical training and his astronomy was poor. He noticed the luxuriance of the Egyptian's astrology and divination,[47] appreciated their measurement and division of the year, $(30 \times 12) + 5 = 365$ days of 24 hours each.[48] Yet one of his own calculations [49] would give to the year an average length of 375 days! He describes [50] an eclipse that was seen before the battle of Salamis, yet no eclipse occurred in that year (480).

As his history is encyclopedic, there is no end to the remarks that could be made concerning his mention (or omission) of this or that item of the three kingdoms of nature.[51] We must restrict ourselves to a few samples.

He had noticed the method used in Babylonia for insuring the fructification of palm trees, and he had also observed the caprification of the fig trees. In his account [52] he mixes the two methods; this proves that he had heard of both and perhaps witnessed both without clear understanding, and later his memory had betrayed him.[53] The matter was better explained by Theophrastos (IV-2 B.C.). It is one of the most interesting stories in the whole history of science, combining folklore with religion, and evidencing the curious inertia of the human mind. It will suffice to remark that the sexual theory of the fertilization of higher plants was not clearly explained in scientific terms until 1694, and was not generally accepted without considerable resistence. The caprification of the fig tree was not explained until much later.[54]

In his description of the Scythian rivers Herodotos speaks [55] of the "great spineless fish, called sturgeons," [56] found at the mouth of the Hypanis, which are used for

45 Herodotos, II, 3.

46 *Ibid.*, II, 123.

47 *Ibid.*, II, 82, 83.

48 *Ibid.*, II, 4.

49 *Ibid.*, I, 32.

50 *Ibid.*, VII, 37.

51 A "chemical" analysis of Herodotos' work has been made by E. O. von Lippmann, "Technologisches und Kulturgeschichtliches aus Herodot," *Chem. Zeit.*, Nos. 1, 7, 8/9 (1924). It is divided into: elements, mineral substances, metals, organic substances.

52 Herodotos, I, 193.

53 See Chap. VI, note 6.

54 The fertilization of high plants was first explained in scientific terms by Rudolf Jacob Camerarius in 1694. The first sufficient explanation of the caprification of figs was given only in 1820 by Giorgio Gallesio. Ira J. Conduit, *The fig* (Waltham: Chronica Botanica, 1947) [*Isis 40*, 290 (1949)].

55 Herodotos, IV, 53.

56 *Cētea te megala anacantha ta antacaius caleusi* . . . D'Arcy W. Thompson, *Greek fishes* (London: Oxford University Press, 1947), p. 16 [*Isis* 38, 254 (1947-48)]. For salted fish see

salting, but he does not mention the caviar obtained from them, though it is difficult to believe that the Scythians or the Greek colonists had not yet discovered some form of it.

Herodotos observed the Nile and the land of Egypt and concluded with his often-repeated phrase, that Egypt is a gift of the river (*dōron tu potamu*), a statement which he justified. He could not explain properly the yearly flood but noticed the yearly deposit of silt. He observed petrified sea shells in the hills and concluded from them and from the coat of salt on the ground that those parts had once been covered by sea water.[57] Lower Egypt was once under water; the river brings more and more sediments and its delta projects into the sea.[58] He observed displacements of land and water also in Thessaly, and he ascribed the formation of the defile of Tempe (in northern Thessaly) to an earthquake.

> The Thessalians say that Poseidon made this passage whereby the Peneios flows; and this is reasonable; for whosoever believes that Poseidon is the shaker of the earth, and that rifts made by earthquakes are that god's handiwork, will judge from sight of that passage that it is of Poseidon's making; for it is an earthquake, as it seems to me, that has riven the mountains asunder.[59]

That is excellent, a curious combination of precocious geologic wisdom with mythology. He recognized that the landscape can be modified by earthquakes, but those earthquakes were caused by Poseidon. This is less astonishing, perhaps, when one recalls the many geologic oddities of the Greek region – hot and mineral springs, narrow gorges, underground rivers, earthquakes – yet most people take the wonders of nature as a matter of course and make no attempt at explanation. Herodotos combined scientific with mythologic interpretation; many men do that even today; their rationalism is always conditional and limited.

Herodotos was not a geographer in the scientific sense; for one thing, his mathematical knowledge was insufficient for true geographic understanding, and his mind was bent in another direction. Yet he had traveled considerably in the three continents, and his experience, completed by the experience of others, enabled him to have a pretty good idea of the inhabited world (the *oicumenē*). He did not wish to generalize and rationalize that knowledge, and remarked:

> It makes me laugh when I notice that many before now have drawn general maps of the earth, but nobody has set the matter forth intelligently; for they draw the Ocean flowing all round the earth, which they make as circular as if fashioned with compasses, and they draw Asia equal in size to Europe.[60]

This first book of history might be called also the first book of human geography, for it includes geographic descriptions of the known earth in general and of many of its parts. These descriptions always take the people into account, for Herodotos had more curiosity about them than about abstractions; he was more interested in

Koehler, "*Tarichos*," *Mém. Acad. St. Pétersbourg* (1832), pp. 347–488. Article "Salgama (*halmaia*) in Daremberg and Saglio, *Dictionnaire des antiquités grecques et romaines* (Paris, 1877–1919), vol. 4, p. 1014. The history of caviar remains to be written, though Koehler devotes a short chapter to it (pp. 410–417); according to him the only ancient author referring to caviar was Diphilos of Siphnos (fourth and third centuries), as related by Athenaios of Naucratis (III-1).

[57] Herodotos, II, 5; II, 12.

[58] *Ibid.*, II, 12.

[59] *Ibid.*, VII, 129.

[60] *Ibid.*, IV, 36, as translated in Eric Herbert Warmington, *Greek geography*, p. 229. This anthology contains long extracts from Herodotos, illustrating his views on the general outlines of the inhabited earth and the characteristics of various parts.

human than mathematical geography, and more in human than in natural history. As he had no maps, certainly no good ones, it is not surprising that his account is frequently erroneous; it is rather surprising that the errors are not deeper and more frequent. In many cases, he was aware of his lack of information and did not wish to commit himself. For example, he says:

> Concerning the farthest western parts of Europe I cannot speak with exactness; for I do not believe that there is a river called by foreigners Eridanos issuing into the northern sea, whence our amber is said to come, nor have I any knowledge of Tin Islands [Cassiterides] whence our tin is brought. The very name of the Eridanos bewrays itself as not a foreign but a Greek name, invented by some poet; nor for all my diligence have I been able to learn from one who has seen it that there is a sea beyond Europe. This only we know, that our tin and amber come from the most distant parts.[61]

His most curious and worst errors concern the general course of the Danube and of the Nile. As the Danube flows across Europe from west to east, he thought that the upper Nile also flowed in the same direction, and moreover he confused it with the Niger. His confusion is more excusable if we remember that it continued in various forms until the end of the eighteenth century.[62] Nowhere is the value of atlases and of the accumulation of knowledge that they represent more obvious. At present, any child looking at a good but simple map of Africa can follow the courses of the great rivers – Nile, Niger, Congo – from their sources to the sea, and appreciate immediately their mutual relation. There is for him neither difficulty nor ambiguity.[63]

The Persian empire was divided into twenty satrapies or provinces. The Royal Road of Persia from Sardis to Susa is described in detail.[64] The total length was 450 parasangs, or 13,500 stadia (1 parasang = 30 stadia), or 90 days (150 stadia to a day).[65] Resting stages were provided. The distance from Ephesos to Sardis is 540 stadia. Hence, the total distance from the "Hellenic sea" to the capital was 14,040 stadia or 93 days. Herodotos' description contains various errors, but such as it is, the existence of a royal road crossing the empire and divided into definite stages suggests that some kind of postal service was already organized; indeed, without such service, restricted to official use and combined with "spying and intelligence," the government of so large an empire would have been impossible. The road described by Herodotos was far longer and more circuitous than it might have been, partly because it followed the course of much older roads (Hittite).[66]

Herodotos' account of India, the remotest satrapy of the empire, was of indirect origin; it hardly extended beyond the Indus and was very incomplete; yet, such as

[61] Herodotos, III, 115. The identification of the Eridanos and the Tin Islands is a good illustration of the haziness of ancient geography. The Eridanos has been identified with the rivers Po, Rhône, and Rhine; the Tin Islands (*Cassiterides insulae*) with the Scilly Islands, Cornwall, and islands off the coast of Brittany or of Spain.

[62] For a summary of views concerning the great African rivers Nile, Niger, Senegal, even Congo, see *Introduction*, vol. 3, pp. 1158–1160, with bibliography.

[63] Modern man has another advantage. It is possible from an airplane to follow the course of a river, say the Nile, from the beginnings to the end and to see it in its reality, immediately.

[64] Herodotos, v, 52–53.

[65] It is Herodotos who says 150 stadia to a day (v, 53). The length of a stadion varied from time to time and place to place. If we assume the lengths of 7.5 stadia and 10 stadia per mile, 150 stadia per day equal respectively 20 and 15 miles per day. For the length of the stadion, see Aubrey Diller, "The ancient measurements of the earth," *Isis 40*, 6–10 (1949).

[66] Discussion of the road by H. F. Tozer, *History of ancient geography*, pp. 90–91, XIV. For ancient and Oriental postal services see *Introduction*, vol. 3, p. 1786.

it is, being the earliest in Greek literature, it is very interesting.[67] Perhaps the most interesting item is the first mention of cotton.[68] In India, says he, "there grows on wild trees wool more beautiful and excellent than the wool of sheep; these trees supply the Hindus with clothing." "The Hindus [in Xerxes' army] wore garments of tree-wool."

The main glory of Herodotos, however, lies in his description of men of different nations and of their manners and customs. He may not be the father of history, but he is certainly the father of ethnology.[69] The main value of his description is ethnologic, for considering his sources of information (direct observation, oral tradition) the chances of error were smaller in that field than in the recording of ancient historical events or of complex geographic relations (say the general lay of rivers and mountains). When he speaks of barbarians, he observes the type of food they eat, their marriage and other sexual customs,[70] the nature of their dwellings, their language and religion. The best example of ethnologic description is that of the Scythians, dwelling north of the Black Sea; the description is very detailed and is as fundamental a document for the history of Russia as the description that Tacitus (I-2) gave us five and a half centuries later is for the history of Germany. Herodotos begins with a general survey of the country and the climate; then he tells us about their gods, giving the names in the Scythian language (which we hardly know otherwise),[71] describes religious rites and sacrifices, military usages, methods of divination, the manners of medicine men, execution of criminals, methods of burial. Herodotos' descriptions have been checked by ethnologists and archaeologists and confirmed on every point. His account of the burial of Scythian kings and of the things buried with them has been proved correct by the modern excavation of such graves. The Scythians used hemp very much as other people use flax, and throwing hemp seeds on red-hot stones they enjoyed intoxicating steam baths.[72] This is the first reference to a plant (*Cannabis sativa, indica*) which has been used and abused extensively by the men of many nations (especially in the Near and Middle East) from the most ancient days down to our own. The history of cannabis is one of the longest chapters in the study of man's desire for intoxication.

Let us consider briefly a few other examples. A new branch of prehistoric archaeology was initiated in 1854 by the Swiss Ferdinand Keller – the study of lake dwellings (lacustrine archaeology).[73] Now, Herodotos described lake dwellings in Lake Prasiad, Macedonia, and the manner and customs of lake dwellers; a shorter note on lake dwellers in Colchis (eastern end of the Black Sea) was written by his contemporary, Hippocrates of Cos.[74]

[67] Herodotos, III, 95, 98, 100; IV, 44.

[68] *Ibid.*, III, 106; VII, 65.

[69] According to Isidic usage, I prefer to use the word ethnology for the study of the manners and customs of mankind, reserving the term anthropology for physical anthropology, the study of their anatomic and genetic differences.

[70] Marriage by capture or by purchase, communal marriage, *droit du seigneur*, exogamy, polyandry, religious prostitution, prenuptial unchastity, etc.

[71] Scythian was probably a form of Iranian, the northwestern branch of it. A. Meillet and Marcel Cohen, *Les langues du monde* (Paris, 1924), pp. 36, 42, 176, 285 [*Isis 10*, 298 (1928)].

[72] Herodotos, IV, 74, 75. The plant *Cannabis* is causing much trouble in our own country today under the Mexican name *marijuana*.

[73] Ferdinand Keller of Zurich (1800–1881); see *Isis 26*, 308–311 (1934), with portrait. The oldest lake dwellings date from the stone age, but they continued to be used in successive prehistoric ages and even in historic times.

[74] Herodotos, V, 16; Hippocrates, *Airs, waters, places*, 15. Both texts are given in English in my note on "The earliest representation of the remains of prehistoric pile dwellings apropos of Conrad

Herodotos mentions Pygmies in Lybia.[75] This was not a novelty, but Herodotos' account is more complete and convincing than previous ones. The existence of pygmy races (negrillos) has been fully and repeatedly proved by modern explorers (Du Chaillu, Schweinfurth, Stanley).[76]

He notices blood covenants: "These nations [the Lydians and the Medes] make sworn compacts as do the Greeks; moreover, they cut the skin of their arms and lick each other's blood." [77] This custom has been frequently observed by modern ethnologists.[78]

He speaks of sacred tattooing: "There was on the coast [near the Canopic mouth of the Nile] and still is, a temple of Heracles, where if a servant of any man take refuge and be branded with certain sacred marks in token that he delivers himself to the god, such a one may not be touched." [79] One might argue that branding should be distinguished from tattooing.

He describes the Egyptian worship of animals.[80] The stories related are not fables; they have been confirmed by archaeologic evidence and by studies on totemism, a branch of ethnology that dates only from the last quarter of the last century.[81]

There is no need of multiplying these observations. The ethnologic remarks constitute the most original part of Herodotos' work, so original that they were not properly appreciated until our own day. The very best commentators of the last century overlooked them, because ethnology did not yet exist, or was not sufficiently formulated, and whatever existed of it was not known to them. They were classical scholars, archaeologists, students of ancient politics and religion, and could not recognize the value of ethnologic facts when they came across them. The facts that ethnologists of today classify readily under such topics as animism, tabu, totem, lake dwellings, and so forth,[82] were discarded as oddities or inventions. Herodotos had laid the foundation of a science which fell into neglect soon after his death; it is not that the Greeks lacked interest in men; they were deeply interested in the riddles of life, but under the influence of Socrates and Plato, they were led to pay more attention to man's inner nature, to his ethical and political problems, and to neglect the study of his manners and customs. How do men live and solve their everyday problems? How do they feed themselves? What kind of garments do they make and wear and what kind of houses do they build? What are their sexual habits and family connections? Why do they behave as they do? How do they pass from childhood to adolescence, from celibacy to marriage, from manhood to old age?

Witz's painting of 1444," *Isis* 26, 449–451 (1936), 1 pl.; 32, 116 (1947–49). W. R. Halliday, "The first description of a lake-village," *Discovery* 1, 235–238 (1920) [*Isis* 4, 127 (1921–22)]. Robert Munro, *Encyclopedia of Religion and Ethics*, vol. 7 (1915), pp. 773–784.

75 Herodotos, II, 32.

76 Paul Monceaux, "La légende des pygmées et nains de l'Afrique équatoriale," *Revue historique* 47, 1–64 (1891); *Introduction*, vol. 3, pp. 1227, 1860.

77 Herodotos, I, 74.

78 P. J. Hamilton-Grierson, "Artificial brotherhood," *Encyclopedia of Religion and Ethics*, vol. 2 (1910), pp. 857–871.

79 Herodotos, II, 113.

80 *Ibid.*, II, 64–75.

81 The subject was cleared up by John Ferguson McLennan (1827–1881) and by James George Frazer (1854–1941), *Totemism* (Edinburgh, 1887), *Totemism and exogamy* (4 vols.; London, 1910). Note that Sir James died in 1941, and how close that is to us.

82 For introduction on these topics see *Encyclopedia of Religion and Ethics*: Goblet d'Alviella on animism, vol. 1 (1908), pp. 535–537; R. R. Marett on tabu, vol. 12 (1922), pp. 181–185; E. Sidney Hartland on totemism, vol. 12 (1922), pp. 393–407. These subjects, which were so controversial half a century ago, are now commonplaces in every textbook on ethnology.

How do they treat the sick and the insane? How do they dispose of the dead? . . . Herodotos tried to answer such questions, but few of his successors did. Some interest in "ethnology" was developed in the eighteenth century, but the science of ethnology was hardly established before the end of the last century and the beginning of our own. Many of the facts related by the father of history, which seemed irrelevant to our grandparents, have been verified by modern ethnologists and, being the first examples of their kind, have obtained considerable value. As one of the leading ethnologists of our time remarked: "Hérodote gagne de jour en jour." [83] The father of history has often been called the "father of lies," but many of the lies imputed to him were not inventions of his but gaps in our own knowledge. His stature increases in proportion as our ignorance of ethnology decreases.

THUCYDIDES OF ATHENS

We have hardly spoken of Sparta, because it is possible to write the history of Greek science without mentioning Sparta and without essential loss. It is well to speak of it briefly, however, not for its own sake, but for the better understanding of its great rival and enemy, Athens.

Sparta (or Lacedaimon), in Laconica, was the chief city of the Peloponnesos. It had been invaded by Dorians, who became the leading class, reducing the natives to secondary importance and even the mass of them to slavery. At the time of the Persian invasion they were the strongest nation in Greece, yet the victory was largely due to Athenian initiative, and it helped to increase the growth of Athens. During the half century of relative peace that followed Salamis (480), the Athenian empire developed and her moral supremacy was established. This became gradually unbearable to the Lacedaimonians, and was the main cause of the Peloponnesian War (431–404).

Or one might say that the main cause was deeper – absolute incompatibility of temperament and ideals between Sparta and Athens. It was a conflict between Ionians and Dorians, or between democracy and oligarchy, or between maritime and territorial power. The two rivals tried to bolster their strength by enlisting some of their neighbors as allies, and gradually two systems of alliances covered all of Greece and Ionia; the world was divided into two hostile groups and the difference of potential between them was steadily increasing; a discharge was bound to occur sooner or later, and the war would be on. It was a war to the death which crippled both and, in the end, destroyed Greek independence. We have no space to enter into many details, but the story of the war might be summarized as follows.

At the beginning, Athens seemed to hold all the winning cards; her empire was held together by overwhelming sea power. That initial advantage was lost, because of the occurrence of the plague (430–29) which decimated the Athenian population and left the survivors demoralized. The first ten years of continuous war (431–421) were concluded by the Peace of Nicias.[84] That peace was to last fifty years, but it proved to be hardly more than a suspicious and precarious armistice. The Sicilian expedition undertaken by the Athenians in 415 (134 triremes carrying

[83] Arnold Van Gennep, *Religions, moeurs et légendes* (Paris, 1909), vol. 2, p. 174.

[84] Nicias (*c.* 470–13), Athenian aristocrat and general who labored for peace and finally obtained in 421 the conclusion of the peace named after him. He disapproved of the Sicilian expedition, but it was decided against him, and he was appointed the leader of it. He was executed by the Syracusans in 413.

4000 hoplites) ended with complete disaster to the Athenian navy and army at Syracuse in 413. The last ten years of the war, 413–404, led to the surrender and humiliation of Athens.

Athens was prostrate and Sparta triumphant. In the light of eternity, however, Sparta did *not* win and Athens is immortal. The victory of Sparta did not stop the intellectual development of Athens (as we shall show in the following chapters), and Athens remained the school of Greece and of Europe. The glory of Greece is the glory of Athens, not of Sparta.

Moreover, the Spartans did not retain their material hegemony very long, for they were beaten by the Thebans at Leuctra in 371, and a generation later the disunited Greeks were obliged to yield to the Macedonians, when they were beaten by Philip II at Chaironea in 338.

We might thus say that while the Persian Wars saved Greece from barbarism, the Peloponnesian War prepared its decadence and fall.

The first wars had inspired Herodotos; the second introduced another great historian, one of the greatest of all time, Thucydides.

Thucydides, son of Oloros, was an Athenian. We know his character exceedingly well, but not the circumstances of his life. We cannot even say exactly when he was born and when and where he died. The most likely dates are *c.* 460, *c.* 400 (or a little later for both, 455–395). He suffered from the plague but recovered; this shows that he was in Athens in 430–29. We may assume that he was a man of substance, for he possessed the right of working gold mines in Thrace,[85] and he must have enjoyed sufficient economic independence to devote his life to the writing of his history. Part of his time must have been devoted to political and military affairs, for in 424 he was appointed *stratēgos* (general). He did not keep that office very long, for in the same year he failed to relieve Amphipolis and was banished for twenty years; [86] this gave him leisure for historical work; he may have spent part of those twenty years traveling for documentation; it is possible that he remained most of the time at Scapte Hyle, where he felt at home and was sufficiently remote from the war to consider it with some degree of aloofness and work in peace. If he wrote there the history of the civil war, as we believe he did, Scapte Hyle is a sacred place. We gather from his own words, however, that he had undertaken his work soon after the beginning of the war (431) and was still engaged in it after the Athenian disaster (404). Thus, even if he spent the years 424 to 404 (or the greatest part of them) in Scapte Hyle, his history was begun before and completed after his exile.

The history begins as follows (Fig. 68):

Thucydides, an Athenian, wrote the history of the war waged by the Peloponnesians and the Athenians against one another. He began the task at the very outset of the war, in the belief that it would be great and noteworthy above all the wars that had gone before, inferring this from the fact that both powers were then at their best in preparedness for war in every way, and seeing the rest of the Hellenic race taking sides

[85] Whether he owned property there or not, we cannot tell, but the exploitation of some of the mines had been farmed out to him. The mines were located in Scaptē Hylē, on the Thracian coast, opposite the island of Thasos, a little to the west of it (modern Eski Kavala, or old Cavalla). Cavalla, we may recall, was the first European landing of St. Paul, and in 1769 the birthplace of Muḥammad 'Alī pāshā, founder of modern Egypt; see *Isis 31*, 97 (1939–40).

[86] Thucydides, v, 26.

ΘΟΥΚΥΔΙΔΟΥ ΞΥΓΓΡΑΦΗΣ ΠΡΩΤΗΣ.

Ουκυδίδης Ἀθηναῖος ξυνέγραψε τὸν πόλεμον τῶν Πελοποννησίων καὶ Ἀθηναίων, ὡς ἐπολέμησαν πρὸς ἀλλήλους, ἀρξάμενος εὐθὺς καθισταμένου, καὶ ἐλπίσας μέγαν τε ἔσεσθαι καὶ ἀξιολογώτατον τῶν προγεγενημένων, τεκμαιρόμενος ὅτι ἀκμάζοντές τε ᾖσαν ἐς αὐτὸν ἀμφότεροι παρασκευῇ τῇ πάσῃ, καὶ τὸ ἄλλο Ἑλληνικὸν ὁρῶν ξυνιστάμενον πρὸς ἑκατέρους, τὸ μὲν εὐθύς, τὸ δὲ καὶ διανοούμενον. κίνησις γὰρ αὕτη μεγίστη δὴ τοῖς Ἕλλησιν ἐγένετο, καὶ μέρει τινὶ τῶν βαρβάρων, ὡς δὲ εἰπεῖν, καὶ ἐπὶ πλεῖστον ἀνθρώπων. τὰ γὰρ πρὸ αὐτῶν, καὶ τὰ ἔτι παλαιότερα, σαφῶς μὲν εὑρεῖν διὰ χρόνου πλῆθος ἀδύνατον ἦν. ἐκ δὲ τεκμηρίων, ὧν ἐπὶ μακρότατον σκοποῦντί μοι πιστεῦσαι ξυμβαίνει, οὐ μεγάλα νομίζω γενέσθαι, οὔτε κατὰ τοὺς πολέμους, οὔτε ἐς τὰ ἄλλα. φαίνεται γὰρ ἡ νῦν Ἑλλὰς καλουμένη, οὐ πάλαι βεβαίως οἰκουμένη, ἀλλὰ μεταναστάσεις τε οὖσαι τὰ πρότερα, καὶ ῥᾳδίως ἕκαστοι τὴν ἑαυτῶν ἀπολείποντες, βιαζόμενοι ὑπό τινων αἰεὶ πλειόνων. τῆς γὰρ ἐμπορίας οὐκ οὔσης, οὐδ' ἐπιμιγνύντες ἀδεῶς ἀλλήλοις, οὔτε κατὰ γῆν, οὔτε διὰ θαλάσσης, νεμόμενοί τε τὰ αὑτῶν ἕκαστοι ὅσον ἀποζῆν, καὶ περιουσίαν χρημάτων οὐκ ἔχοντες, οὐδὲ γῆν φυτεύοντες, ἄδηλον ὂν ὁπότε τις ἐπελθών, καὶ ἀτειχίστων ἅμα ὄντων, ἄλλος ἀφαιρήσεται, τῆς τε καθ' ἡμέραν ἀναγκαίου τροφῆς πανταχοῦ ἂν ἡγούμενοι ἐπικρατεῖν, οὐ χαλεπῶς ἀπανίσταντο. καὶ δι' αὐτὸ οὔτε μεγέθει πόλεων ἴσχυον, οὔτε τῇ ἄλλῃ παρασκευῇ. μάλιστα δὲ τῆς γῆς ἡ ἀρίστη αἰεὶ τὰς μεταβολὰς τῶν οἰκητόρων εἶχεν· ἥ τε νῦν Θεσσαλία καλουμένη, καὶ Βοιωτία, Πελοποννήσου τε τὰ πολλά, πλὴν Ἀρκαδίας. τῆς τε ἄλλης, ὅσα ἦν κράτιστα. διὰ γὰρ ἀρετὴν γῆς, αἵ τε δυνάμεις τισὶ μείζους ἐγγιγνόμεναι, στάσεις ἐνεποίουν, ἐξ ὧν ἐφθείροντο, καὶ ἅμα ὑπὸ ἀλλοφύλων μᾶλλον ἐπεβουλεύοντο. τὴν γοῦν Ἀττικὴν ἐκ τοῦ ἐπὶ πλεῖστον διὰ τὸ λεπτόγεων ἀστασίαστον οὖσαν, ἄνθρωποι ᾤκουν οἱ αὐτοὶ αἰεί. καὶ παράδειγμα τόδε τοῦ λόγου οὐκ ἐλάχιστόν ἐστι, διὰ τὰς μετοικίας ἐς τὰ ἄλλα μὴ ὁμοίως αὐξηθῆναι. ἐκ γὰρ τῆς ἄλλης Ἑλλάδος οἱ πολέμῳ ἢ στάσει ἐκπίπτοντες, παρ' Ἀθηναίους οἱ δυνατώτατοι, ὡς βέβαιον ὄν, ἀνεχώρουν· καὶ πολῖται γιγνόμενοι εὐθὺς ἀπὸ παλαιοῦ, μείζω ἔτι ἐποίησαν πλήθει ἀνθρώπων τὴν πόλιν· ὥστε καὶ ἐς Ἰωνίαν ὕστερον, ὡς οὐχ ἱκανῆς οὔσης τῆς Ἀττικῆς, ἀποικίας ἐξέπεμψαν. δηλοῖ δέ μοι καὶ τόδε τῶν παλαιῶν ἀσθένειαν οὐχ ἥκιστα. πρὸ γὰρ τῶν Τρωϊκῶν, οὐδὲν φαίνεται πρότερον κοινῇ ἐργασαμένη ἡ Ἑλλάς. δοκεῖ δέ μοι οὐδὲ τοὔνομα τοῦτο ξύμπασά πω εἶχεν, ἀλλὰ τὰ μὲν πρὸ Ἕλληνος τοῦ Δευκαλίωνος, καὶ πάνυ οὐδὲ εἶναι ἡ ἐπίκλησις αὕτη. κατὰ ἔθνη δὲ ἄλλα τε, καὶ τὸ Πελασγικὸν ἐπὶ πλεῖστον, ἀφ' ἑαυτῶν τὴν ἐπωνυμίαν παρέχεσθαι. Ἕλληνος δὲ καὶ τῶν παίδων αὐτοῦ ἐν τῇ Φθιώτιδι ἰσχυσάντων, καὶ ἐπαγομένων αὐτοὺς ἐπ' ὠφελείᾳ ἐς τὰς ἄλλας πόλεις, καθ' ἑκάστους μὲν ἤδη τῇ ὁμιλίᾳ μᾶλλον καλεῖσθαι Ἕλληνας· οὐ μέντοι πολλοῦ γε χρόνου ἐδύνατο καὶ ἅπασιν ἐκνικῆσαι. τεκμηριοῖ δὲ μάλιστα Ὅμηρος· πολλῷ γὰρ ὕστερον ἔτι καὶ τῶν Τρωϊκῶν γενόμενος, οὐδαμοῦ τοὺς ξύμπαντας ὠνόμασεν, οὐδ' ἄλλους, ἢ τοὺς μετ' Ἀχιλλέως ἐκ τῆς Φθιώτιδος, οἵπερ καὶ πρῶτοι Ἕλληνες ἦσαν. Δαναοὺς δὲ ἐν τοῖς ἔπεσι, καὶ Ἀργείους, καὶ Ἀχαιοὺς ἀνακαλεῖ. οὐ μὴν οὐδὲ βαρβάρους εἴρηκε, διὰ τὸ μηδὲ Ἕλληνάς πω, ὡς ἐμοὶ δοκεῖ, ἀντίπαλον ἐς ἓν ὄνομα ἀποκεκρίσθαι. οἱ δ' οὖν ὡς ἕκαστοι Ἕλληνες κατὰ πόλεις τε ὅσοι ἀλλήλων ξυνίεσαν, καὶ ξύμπαντες ὕστερον κληθέντες, οὐδὲν πρὸ τῶν Τρωϊκῶν δι' ἀσθένειαν καὶ ἀμιξίαν ἀλλήλων ἁθρόοι ἔπραξαν, ἀλλὰ καὶ ταύτην τὴν στρατείαν, θαλάσσῃ ἤδη πλείω χρώμενοι, ξυνῆλθον. Μίνως γὰρ παλαίτατος ὧν ἀκοῇ ἴσμεν, ναυτικὸν ἐκτήσατο, καὶ τῆς νῦν Ἑλληνικῆς θαλάσσης ἐπὶ πλεῖστον ἐκράτησε, καὶ τῶν Κυκλάδων νήσων ἦρξέ τε, καὶ οἰκιστὴς πρῶτος τῶν πλείστων ἐγένετο, Κᾶρας ἐξελάσας, καὶ τοὺς ἑαυτοῦ παῖδας ἡγεμόνας ἐγκαταστήσας, τό τε λῃστικὸν ὡς εἰκὸς καθῄρει ἐκ τῆς θαλάσσης ἐφ' ὅσον ἐδύνατο, τοῦ τὰς προσόδους μᾶλλον ἰέναι αὐτῷ. οἱ γὰρ Ἕλληνες τὸ πάλαι, καὶ τῶν βαρβάρων οἵ τε ἐν τῇ ἠπείρῳ παραθαλάσσιοι, καὶ ὅσοι νήσους εἶχον, ἐπειδὴ ἤρξαντο μᾶλλον περαι-

A A

Fig. 68. Thucydides, *editio princeps* (folio; Venice: Aldo Manuzio, May 1502). Note that the first Greek editions of Herodotos and Thucydides are both "Aldine" editions of the same year, 1502. We reproduce the first page of the text proper, beginning with the famous words translated in our own account: "Thucydides, an Athenian, wrote the history of the war. . ." The blank space, top left, was left to allow the limner to add a large ornamental initial, a small theta being printed for his guidance. [From the copy in the Harvard College Library.]

with one state or the other, some at once, others planning to do so. For this was the greatest movement that had ever stirred the Hellenes, extending also to some of the barbarians, one might say even to a very large part of mankind.

The author fully realized the great importance of his task; he could realize it at the very outset, because the war had been so long in preparing. It was more than the civil war of one nation, for it involved many other nations (the Spartans finally won with the help of Persia). To the philosopher, every war is a civil war; this was especially true of the Peloponnesian War, which cut mankind asunder. After 404, Thucydides revised his work and wrote a kind of new preface:

The history of these events has been written by the same Thucydides, an Athenian, in the chronologic order of events, by summers and winters, up to the time when the Lacedaemonians and their allies put an end to the dominion of the Athenians and took the Long Walls and Peiraieus. Up to that event the war lasted twenty-seven years in all; and if anyone shall not deem it proper to include the intervening truce in the war, he will not judge aright. For let him but look at the question in the light of the facts as they have been set forth and he will find that that cannot fitly be judged a state of peace in which neither party restored or received all that had been agreed upon . . . so that, including the first ten-years' war, the suspicious truce succeeding that, and the war which followed the truce, one will find that, reckoning according to natural seasons, there were just so many years as I have stated, and some few days over. He will also find, in the case of those who have made any assertion in reliance upon oracles, that this fact alone proved true; for always, as I remember, from the beginning of the war until its close, it was said by many that it was fated to last thrice nine years. I lived through the whole war, being of an age to form judgments, and followed it with close attention, so as to acquire accurate information.[87]

The history remained incomplete, for in spite of the statement just quoted, Thucydides did not carry his account beyond the year 411. The division of it into eight books was probably made by Alexandrian scholars. The genuineness of book VIII has been questioned; the text as we have it has been ascribed to Thucydides' daughter, to Xenophon, and to Theopompos of Chios. It is certain that the two last-named wrote *Hellenica*, in continuation of Thucydides; Theopompos' lost work told the story from 411 to 394; Xenophon's work, which is preserved, deals with a longer period, from 411 to the second battle of Mantinea, 362. Book VIII has all the marks of Thucydides' authorship, except that it does not include speeches.

The first twenty-three chapters of book I are an introduction dealing with archaeology, describing rapidly the events of the years 479 to 440, connecting his history with that of Herodotos, and explaining the genesis of the new war. The rest of the work is devoted to the war itself, the vicissitudes of which are told with moderation and objectivity in strict chronologic order. The first year of the war (431) is determined by naming the eponymic magistrates of Athens and Sparta, but this being done, the years are no longer identified except as the first year, the second year, etc., of the war, and the Athenian months are not named. The various calendars in use in Thucydides' time were a source of confusion, and he paid no attention to them. For each year, he distinguishes the good season (*theros*) and the bad one (*cheimōn*), and when further precision is needed he refers to agricultural events such as the awakening of spring, the wheat in the blade, the shooting of it into the ear, the vintage, the last beautiful days. His account is completely included within

[87] *Ibid.*

this rigid chronologic frame. He is often obliged to move suddenly from one part of Greece to another, and this annoys the reader, but we recognize that he was right; he subordinated topographic unity to chronology, which is the best guide and protection of the scientific historian. I use the word scientific deliberately, for Thucydides was in the full sense of the word a scientific historian, the first in the world's history. His book is the first masterpiece of Attic prose (Herodotos' was written in an Ionian dialect), but it is more than that: it is the first attempt to describe a war, its causes and vicissitudes, as a trained man of science would do, or say, as a physician would describe the ups and downs of a disease. He avoided the fables and ambiguities; as he proudly put it:

It may well be that the absence of the fabulous from my narrative will seem less pleasing to the ear; but whoever shall wish to have a clear view both of the events which have happened and of those which will some day, in all human probability, happen again in the same or a similar way – for these to adjudge my history profitable will be enough for me. And, indeed, it has been composed, not as a prize-essay to be heard for the moment, but as a possession for all time.[88]

The last words in the English translation, corresponding to the Greek *ctēma es aiei*, have often been quoted, or rather they have been misquoted as if the word used were *mnēma* (memorial) and as if Thucydides had exclaimed like Horace, "Exegi monumentum aere perennius." It is not that at all. Thucydides is not thinking of his own glory, but, like a good scientist, of the validity of his work; he has taken immense pains to obtain results of permanent value.

His sources were his own experience and the knowledge he had obtained from other witnesses; in some cases he made use of definite documents which he inserted into his narration; for example, the treaty of Nicias is quoted *in extenso*,[89] and also a treaty of alliance between the Athenians, Argives, Mantineans, and Eleans.[90] A part of that treaty was found in 1877 by the Archaeological Society of Athens upon a marble slab near the Acropolis; the text of that inscription tallies, as far as it goes, with the text given by Thucydides and is a magnificent justification of him. In spite of his devotion to Pericles, he was not a party man; or let us put it this way: his partisanship was moderate and he always remained capable of hearing and understanding the views of the other side and of explaining them with honesty and even with sympathy. The liberal education given by the sophists had trained Athenians to see both sides of a question, and the many sides of each personality. Of course, not every Athenian could benefit from such training, but the soul of Thucydides was exceedingly well adapted to it.

His essential purpose was to be as truthful as possible at all costs. He knew the feelings of a scientist who must describe unfortunate experiments; the failure is vexing, yet there is some pleasure in explaining it truthfully. He draws excellent portraits of the leading men; his portrait of Pericles is our best source for the study of the latter's character and policy, chiefly during the final years (433 to 429); it shows us a man who seemed to be able to achieve the impossible, for he could restrain the people without limiting their freedom;[91] that is, he could inspire them to accept the needed discipline as if they had chosen it themselves. It was pleasant for Thucydides to explain the political genius of Pericles, whom he admired so much,

[88] *Ibid.*, I, 22.
[89] *Ibid.*, V, 23.
[90] *Ibid.*, V, 47.
[91] *Ibid.*, II, 65.

but he could do justice to the men whom he did not like so well, describe the violence of Cleon, the timid honesty mixed with superstition, of Nicias, the brilliant recklessness of Alcibiades. His opinion of men was largely independent of their success or lack of it; a good man might be out of luck, yet his character would tell.

His nonpartisanship, objectivity, and honesty appear at their best in the discussion of the fundamental issue: the comparative values of Athenian democracy and Lacedaimonian totalitarianism. Democracy is magnificently defended in the funeral oration made by Pericles,[92] one of the noblest political discourses ever made. It is an immortal credit to the memory, not only of Pericles who pronounced it but also of the Athenians who listened to him, and to their mother, the city of Athens. How great were the men, fit to be given and to receive such a generous message! It is too long for full quotation but I cannot resist offering a few samples:

> For we are lovers of beauty yet with no extravagance and lovers of wisdom yet without weakness. Wealth we employ rather as an opportunity for action than as a subject for boasting; and with us it is not a shame for a man to acknowledge poverty, but the greater shame is for him not to do his best to avoid it. And you will find united in the same persons an interest at once in private and in public affairs, and in others of us who give attention chiefly to business, you will find no lack of insight into political matters. For we alone regard the man who takes no part in public affairs, not as one who minds his own business, but as good for nothing.[93]

And his final words:

> I have now spoken, in obedience to the law, such words as I had that were fitting, and those whom we are burying have already in part also received their tribute in our deeds; besides, the state will henceforth maintain their children at the public expense until they grow to manhood, thus offering both to the dead and to their survivors a crown of substantial worth as their prize in such contests. For where the prizes offered for virtue are greatest, there are found the best citizens. And now, when you have made due lament, each for his own dead, depart.[94]

Americans cannot read those sublime words without thinking of the Gettysburg address of Lincoln, and it is to the eternal honor of both statesmen – so distant in time and space – that their two funeral orations are so close in nobility and equanimity.

The other side of the argument was given by Thucydides in the words of

> Cleon son of Cleainetos who had been successful in carrying the earlier motion to put the Mytileneans to death. He was not only the most violent of the citizens, but at that time had by far the greatest influence with the people.[95]

Said Cleon:

> On many other occasions in the past I have realized that a democracy is incompetent to govern others,[96]

and he went on to explain that democracy and empire are incompatible. Thus, the Athenians of the end of the fifth century were facing the same dilemma as the British, the French, the Dutch, and the Americans of our own day. It is poignant to read Pericles and Cleon today when democracy is facing a new test, greater than

[92] *Ibid.*, II, 35–46.
[93] *Ibid.*, II, 40.
[94] *Ibid.*, II, 46.
[95] *Ibid.*, III, 36.
[96] *Ibid.*, III, 37.

any previous one. We ought to meditate the immortal words of Pericles but pay some attention to the conservative warnings of Cleon.

Thucydides helped his contemporaries, and he still helps us, to understand the fundamental differences obtaining between men, some of which are innate, others due to circumstances, yet deep seated. His own duty was to compare the two inveterate enemies, Athens and Sparta. The Athenians were characterized (for example, in the funeral oration) by their intellectual eagerness and curiosity, their expansiveness, hospitality, elegance and taste, generosity, and restlessness; the Lacedaimonians were comparatively poor, earnest, self-centered, slow and quiet, conservative, cautious, jealous, tenacious, and patient; it was terrible to have such men (who might be good men in their own way) as enemies. The two types still exist, and the fight between Athens and Sparta is not yet finished, and perhaps never will be. The scientific description which Thucydides has provided is more dramatic than one which might have tried to be more impressive but would have been like a lawyer's plea, less objective and less impartial. In the long run there is nothing more moving than the truth.

One may regret that Thucydides was so determined in his purpose that he left all the rest out, and gave no account of the society of his time, nor of the incomparable achievements of the Greek artists and thinkers. This was one of the golden ages, and how precious would have been the description of it by a contemporary as intelligent and sensitive as Thucydides. He was a man of science, however (I cannot help harping on this), who knew that a scientific investigation must have an object that is not too large and is clearly limited. Thucydides did not give us a picture of Athens' golden age, but he gave us, with as much precision and truthfulness as were in his power, a description of her life-and-death struggle with an implacable enemy. That was his job and nothing must distract him from it.

It has been argued that Thucydides' method or outlook changed during the thirty years of composition; philologists have tried to prove that by internal analysis. If one bears in mind that Thucydides kept revising his work, and that a part of book I may have been revised as recently as a part of book VII, it is clear that such analysis is unreliable. Yet we are willing to accept the general statement. Thucydides was clearly mature when he began his work but his experience increased, and the failure of Nicias' peace and of the Sicilian expedition must have disillusioned him; he cannot have been quite the same before and after those awful deeds. It must have been with him as it is with every scholar engaged in a long undertaking: he cannot help changing as life passes and his work grows.

We must come back for a moment to the first chapters of Thucydides' book, the archaeologic introduction. It is highly significant that he considered it necessary to write it. The point is that Thucydides (like Hippocrates of Cos, as we shall see later) was a modern; he was just as conscious of his modernity and up-to-dateness as we are of ours, and he was conscious of the long past that had gradually created the present situation. Therefore, a summary of past experience was necessary, but we are astonished to find that he made it to some extent (considering his means) as we would do ourselves. For example, he assumed that Homer's account of the Trojan War, however embellished it might be by his poetic imagination, must be based on realities. Speaking of the Aegean islands, he remarks:

> Still more addicted to piracy were the islanders. These included Carians as well as Phoenicians, for Carians inhabited most of the islands, as may be inferred from the fact that, when Delos was purified by the Athenians in this war and the graves of all who had ever died on the island were removed, over half were discovered to be Carians, being recognized by the fashion of the armor found buried with them, and by the mode of burial, which is that still in use among them.[97]

Thucydides is the only ancient writer who used archaeologic evidence to illustrate Greek origins. He might be called the father of archaeology, even as Herodotos was called the father of ethnology. The introduction throws some light also on his philosophy of history. For his account reveals a conception of progress opposite to the idea of regress expressed by Hesiod, which was the more usual point of view down to the seventeenth century. Yet his statement [98] quoted above suggests the possibility of repetitions in human affairs; Thucydides does not expand that suggestion and therefore one has no right to compare it with the Platonic idea of the recurrence of cycles (or periods) and of eternal return. He might mean simply what the scientist means: if similar circumstances recur we may expect similar results. Among the circumstances that the historian must necessarily consider are the passions of men, and these do not vary greatly from time to time or from place to place. The study of the past may thus help historians to foresee the consequences of human conflicts even as the study of clinical reports helps physicians to foresee the probable evolution of diseases.

The impartiality and objectivity of Thucydides extended to himself: he hardly spoke of his own condemnation and exile, made no apology. Was that disdain? Or the reaction of a good and proud conscience? Or scientific objectiveness? Perhaps in this case it was the three combined, but chiefly the third.

Where did Thucydides obtain his scientific outlook? The qualities of objectivity and impartiality that made that outlook possible were no doubt in himself, yet they might have been encouraged or discouraged by external circumstances. They were favored by his education. He sat at the feet of Antiphon of Rhamnos and of other sophists. Sophistry has become so obnoxious to us that we may find it difficult to realize its helpfulness in the fifth century. To begin with, we must remember that most Athenians had of necessity a forensic conception of truth. The members of a public assembly must determine the respective validities of different pleadings. How will they do it? How will they choose between two orators defending opposite points of view in a political controversy? It is rare that one party is pure white and the other pure black. Matters are not so simple as that. Party men of course would vote blindly for their own party. Now, the sophists, at least the better ones, were educating young people to avoid party prejudices and other preconceived opinions, to despise lies and superstitions. This was good preparation for rational and scientific thought. The men who taught the relativity of truth were not necessarily cynics or skeptics; thanks to their political experience, they were keenly aware of the special difficulties caused by prejudice and by the lack of open-mindedness. In purely scientific controversies it is relatively easy to follow the right path, but in political matters the first condition for discovering the truth is sufficient objectivity, tolerance, and sympathy for the adversary. Thucydides was eminently

[97] *Ibid.*, I, 8.

[98] *Ibid.*, I, 22.

well prepared by his own genius to understand such teaching, and he became as open-minded and objective as it was possible to be.

His love of truth enabled him to see the actualities, to record them candidly, and to classify them (as any man of science classifies his observations and reduces them to order); he was able to see things as they are, yet *sub specie aeternitatis*. In general, he did not consider the morality of events; it sufficed to describe them. Yet he reported the corruption that followed the plague, and the one caused by the other vicissitudes of the endless conflict – a theme familiar enough to the students of any war.

His style was as honest and austere as his mind; he wrote with earnestness, brevity, precision, clearness, and vigor. The details were as accurate as he could have them and the general account well balanced. Macaulay, himself one of the greatest English historians, did not hesitate to declare, "There is no prose composition, not even the *De corona*,[99] which I place so high as the Seventh Book of Thucydides. It is the *ne plus ultra* of human art" (that book VII deals with the ill-fated Sicilian expedition, the main cause of Athens' ultimate defeat). What more could one say? And who could say it with greater authority?

One aspect of Thucydides' method of composition has been discussed with painful iterativeness and prolixity by every critic, that is, his habit (which he shared with other historians of antiquity) of including actual speeches in his narrative. Let us listen to him:

> As to the speeches that were made by different men, either when they were about to begin the war or when they were already engaged therein, it has been difficult to recall with strict accuracy the words actually spoken, both for me as regards that which I myself heard, and for those who from various other sources have brought me reports. Therefore the speeches are given in the language in which, as it seemed to me, the several speakers would express, on the subjects under consideration, the sentiments most befitting the occasion, though at the same time I have adhered as closely as possible to the general sense of what was actually said.[100]

Is not that clear enough? Once it was understood that the speeches should not be taken literally, it made little difference whether they were written directly or indirectly, with or without quotation marks. The writing of the speeches as such was a convention that deceived nobody. It was a necessary convention, or at least a justifiable one, because the ancients could not know the *ipsissima verba* unless they had been present and had a retentive memory; the convention would be unjustifiable today, because literal reports of speeches can easily be obtained.[101]

A final question that the thoughtful reader might well ask is this: How was it possible for a patriotic Athenian to describe with such impassivity the tragic events that led to the abject defeat of his country? The answer has already been given, or at least a part of it. Thucydides was a patriot, to be sure, an enthusiastic lover of Athenian democracy, but he was also a man of science; his loyalty to truth was greater than any other. In the second place, so deep was his faith in democracy

[99] The *Peri stephánu* (On the crown) is the most famous oration of the greatest of Athenian orators, Demosthenes (385–322). He delivered it in 330 in justification of his 14-year struggles against Philip II of Macedon. Philip won the battle of Chaeronea (in 338), which was the end of Greek independence; he died in 336. Demosthenes continued the fight against Alexander, but lost.

[100] Thucydides, I, 22.

[101] It is even possible now to record the speech as it was spoken and to preserve it for posterity, almost as if it were a living thing.

that he did not accept the defeat of Athens as final. After that defeat Athens remained – or could remain – what she had been hitherto: the school of Greece (*tēs Hellados paideusis*[102]). As Pericles had explained it in the funeral oration, the main fruit of democracy was not efficiency but education. In spite of terrible vicissitudes Athens did continue to educate Greece and the whole Western world; the faith of Pericles and Thucydides was amply justified.

The Plague of Athens (*430–29*). A year after the beginning of the war, the invasion of Attica by the Spartans had driven its inhabitants into Athens. The city was overcrowded, sanitation was poor, and conditions were as favorable as could be for the spread of an epidemic. The epidemic occurred and was terrible. Let us quote Thucydides' account, the first elaborate description of a plague in world literature:

At the very beginning of summer [430] the Peloponnesians and their allies, with two-thirds of their forces as before, invaded Attica, under the command of Archidamos, son of Zeuxidamos, king of the Lacedaimonians, and establishing themselves proceeded to ravage the country. And before they had been many days in Attica the plague began for the first time to show itself among the Athenians. It is said, indeed, to have broken out before in many places, both in Lemnos and elsewhere, though no pestilence of such extent nor any scourge so destructive of human lives is on record anywhere. For neither were physicians able to cope with the disease, since they at first had to treat it without knowing its nature, the mortality among them being greatest because they were most exposed to it, nor did any other human art avail. And the supplications made at sanctuaries, or appeals to oracles and the like, were all futile, and at last men desisted from them, overcome by the calamity.

The disease began, it is said, in Ethiopia beyond Egypt, and then descended into Egypt and Libya and spread over the greater part of the King's territory. Then it suddenly fell upon the city of Athens, and attacked first the inhabitants of the Peiraieus, so that the people there even said that the Peloponnesians had put poison in their cisterns; for there were as yet no public fountains there. But afterwards it reached the upper city also, and from that time the mortality became much greater. Now any one, whether physician or layman, may, each according to his personal opinion, speak about its probable origin and state the causes which, in his view, were sufficient to have produced so great a departure from normal conditions; but I shall describe its actual course, explaining the symptoms, from the study of which a person should be best able, having knowledge of it beforehand, to recognize it if it should ever break out again. For I had the disease myself and saw others sick of it.

That year, as was agreed by all, happened to be unusually free from disease so far as regards the other maladies; but if anyone was already ill of any disease all terminated in this. In other cases from no obvious cause, but suddenly and while in good health, men were seized first with intense heat of the head, and redness and inflammation of the eyes, and the parts inside the mouth, both the throat and the tongue, immediately became blood-red and exhaled an unnatural and fetid breath. In the next stage sneezing and hoarseness came on, and in a short time the disorder descended to the chest, attended by severe coughing. And when it settled in the stomach, that was upset, and vomits of bile of every kind named by physicians ensued, these also attended by great distress; and in most cases ineffectual retching followed producing violent convulsions, which sometimes abated directly, sometimes not until long afterwards. Externally, the body was not so very warm to the touch; it was not pale, but reddish, livid, and breaking out in small blisters and ulcers. But internally it was consumed by such a heat that the patients could not bear to have on them the lightest coverings or linen sheets, but wanted to be quite uncovered and would have liked best to throw themselves into cold water – indeed many of those who were not looked after did throw them-

[102] Thucydides, II, 41.

selves into cisterns — so tormented were they by thirst which could not be quenched; and it was all the same whether they drank much or little. They were also beset by restlessness and sleeplessness which never abated. And the body was not wasted while the disease was at its height, but resisted surprisingly the ravages of the disease, so that when the patients died, as most of them did on the seventh or ninth day from the internal heat, they still had some strength left; or, if they passed the crisis, the disease went down into the bowels, producing there a violent ulceration, and at the same time an acute diarrhoea set in, so that in this later stage most of them perished through weakness caused by it. For the malady, starting from the head where it was first seated, passed down until it spread through the whole body, and if one got over the worst, it seized upon the extremities at least and left its marks there; for it attacked the privates and fingers and toes, and many escaped with the loss of these, though some lost their eyes also. In some cases the sufferer was attacked immediately after recovery by loss of memory, which extended to every object alike, so that they failed to recognize either themselves or their friends.

Indeed the character of the disease proved such that it baffles description, the violence of the attack being in each case too great for human nature to endure, while in one way in particular it showed plainly that it was different from any of the familiar diseases: the birds, namely, and the four-footed animals, which usually feed upon human bodies, either would not now come near them, though many lay unburied, or died if they tasted of them.[103]

This is not the end of the description, but we have given the essential, as far as the medical side is concerned. Note that the Athenians at first ascribed the plague to deliberate poisoning of cisterns by the enemy; that is a feature which recurs in many accounts of plagues down to the seventeenth century.[104] The medical description provided by Thucydides may seem clear to laymen, yet it is not sufficient for certain diagnosis. The epidemic was possibly new, that is, it was caused by the occurrence of a new microbe or virus against which the Athenian bodies were not yet prepared; this would explain its violence and deadliness (though overcrowding, semistarvation, and uncleanliness would explain much of it even if the microbe were not a newcomer). We know that diseases penetrating a virgin territory cause terrible havoc, as with the Black Death of the middle of the fourteenth century, syphilis at the end of the fifteenth,[105] the epidemic of smallpox among the Aztecs in 1520,[106] the European pandemic of cholera in 1831–32, the epidemic of measles in the Fiji Islands in 1875. Similar examples might be borrowed from the history of epidemics affecting plants and animals, such as the sudden and disastrous arrival of the gypsy moth in Massachusetts in 1889, the San José scale in the eastern states of the United States in 1893, the cotton boll weevil in Texas in 1894, and so on. It is quite possible that the plague of Athens was the first of its kind, and that it was never exactly duplicated; indeed, the reaction of a people virgin to a disease could never be duplicated by a people that has lost its virginity and has acquired some degree of inurement and immunity.

Many attempts have been made to identify the plague of Athens, and the multiplicity of the identifications illustrates their precariousness. None is convincing; it is a matter of a greater or lesser probability, not of certainty. Was it bubonic plague, smallpox, typhus fever, typhoid fever? The latest investigation, made by Shrewsbury [107] suggests measles and is very plausible. His paper contains a long

[103] *Ibid.*, II, 47–49.
[104] *Introduction*, vol. 3, p. 1656.
[105] *Isis* 29, 406 (1938).
[106] *Isis* 37, 124 (1947).
[107] J. F. D. Shrewsbury, "The plague of Athens," *Bull. History of medicine* 4, 1–25 (1950); commentary by William MacArthur, *ibid.* 5, 214–215 (1950).

bibliography but does not mention the *Thucydides* of Finley.[108] In this otherwise excellent book, the suggestion is made (or repeated) that the plague was not an infectious disease but ergotism.[109] Measles is perhaps the best guess, but how could one be sure?

It is typical of the nonscientific-mindedness of many historians (even our own contemporaries) that they have considered Thucydides' medical description of the plague as a kind of digression. To Thucydides' scientific mind, this was no digression at all but the very heart of his subject. The physical consequences of the plague were terrible, the moral consequences worse; one might say that the plague was the earnest of the ultimate defeat of Athens. Was it not worth while then to find out what the plague was? How did it come about? How did it stop? Here was a clear case of *prophasis, diagnōsis, therapeia* (the search for causes, diagnosis, and treatment). It was not Thucydides' fault that his analysis was not more useful; at any rate he did his duty, the duty of a scientific historian.

It is typical that the greatest philosophic poet of antiquity, Lucretius (I–1 B.C.), recognized the intrinsic importance of that description. The terrifying account of the plague of Athens with which he brings the *De rerum natura* to a close [110] follows the text of Thucydides.

The story of the plague has been told with some detail in the author's own words, because it is almost the only part of his history that interests the historian of science immediately. A reference to fire signals across mountains [111] may interest historians of technology, but such simple telegraphy must have been practiced long before that time.[112] Indeed, we know that many primitive people are accustomed to transmit messages by the use of fires or drums; in particular, the playing on drums permits the transmission of signals of great complexity.

Thucydides' history also contains references to three eclipses – the solar eclipse of 3 August 431,[113] the annular solar eclipse of 21 March 424,[114] and the lunar eclipse of 27 August 413; [115] these eclipses, which actually occurred, help to prove the author's trustworthiness.

HERODOTOS AND THUCYDIDES

Having become acquainted with the two earliest and greatest historians of Hellas, we may pause a moment and compare them. Each was the prototype of his kind. It is remarkable that the same nation should have given both to mankind within the same half century.

Their spans of life were almost equal (both were sexagenarians at the time of their deaths) and they followed each other at an interval of twenty years; they were contemporaries in the sense that fathers and sons are. Twenty years made a difference in that heroic age, but not a very great one. The main difference between them, as far as outside circumstances are concerned, is that Herodotos was a child of the Persian War, while Thucydides was a witness of the Pelopponesian

[108] J. H. Finley, Jr., *Thucydides* (Cambridge: Harvard University Press, 1942).

[109] For ergotism see *Introduction*, vol. 3, pp. 1650, 1668, 1860, 1868; George Barger, *Ergot and ergotism* (London: Gurney & Jackson, 1931).

[110] Lucretius, *De rerum natura*, VI, 1138–1286.

[111] Thucydides, II, 94.

[112] There are similar references in Herodotos, IX, 3; VI, 115, 121, 124, Xenophon, and other historians. Tozer, *History of ancient geography*, pp. 328–334. Wolfgang Riepl, *Das Nachrichtenwesen des Altertums* (492 pp.; Leipzig, 1913), deals chiefly with Roman times.

[113] Thucydides, II, 28.

[114] *Ibid.*, IV, 52.

[115] *Ibid.*, VII, 50.

one; also, Herodotos was a Carian, writing in Ionian, while Thucydides was an Athenian, the founder of Attic prose. The former came from the borderland of Hellenism, the latter from its very heart.

Herodotos' early training had been practical and perhaps commercial. Thucydides was a disciple of the Athenian sophists; as compared with his predecessor he was somewhat like a college man.

The contrast between their personalities, however, was much greater than that between their circumstances. Indeed, each of them had opportunities of experiencing the circumstances of the other. Thrace was as much a borderland as Caria. One war was as bad as another. Both had traveled and were acquainted with men of many types.

Of course, Herodotos had traveled considerably more and the frame of his travels set the frame of his work. He dealt with a much longer past and a wider world (the whole *oicumenē*, in fact), and painted on a much larger scale. Thucydides is to him like a painter of miniatures to a painter of immense frescoes; he dealt with the Greek world only and with a period of twenty-seven years; barring the introduction, his large book covers not more than twenty years – twenty years as against thousands of years, and Greece instead of the whole inhabited world.

Herodotos was a good-natured and remarkably well-informed storyteller, curious and childish, Pythagorean, half-Oriental, loving marvels and oddities; his style is easy, flowing, and delightful. Thucydides not only chose a small subject but restricted himself jealously to it. His mind and his style are equally austere; laughter is irrelevant. He is a political realist, a positivist, a man of science.

Their standards of accuracy were widely different. Herodotos did take some trouble to find the truth and he told it candidly, not without criticism; but how could anybody know the human geography of the whole world and the ancient history of the Near East? On the other hand, it was possible, if not easy, to relate exactly the military and political vicissitudes of the two leading nations of Greece during the brief interval of thirty years. Both were deeply interested in men – Herodotos in the way of an educated traveler, Thucydides more like a sophist and a politician.

The final result is curious. Herodotos' history contains many more items of interest to the historian of science, while Thucydides' history is of greater interest to the student of political history. The historian of science might be tempted to neglect it, but he would be wrong to do so. In its totality, the history of Thucydides is a monument of historical science, the application of scientific method to the study of the past, the first of its kind, and to this day one of the very noblest.

Leaving out of account the mathematical ideas and the medical research, the history of Thucydides is the greatest scientific achievement of that golden age.

CTESIAS OF CNIDOS

It is well to introduce a third historian, Ctesias of Cnidos, far less important than either Herodotos or Thucydides and far less known, for while the works of these two have come down to us in their integrity, we have only fragments of Ctesias'; but his is a very remarkable personality in many respects. To begin with, he helps us to realize that Persia and Greece, however different and even inimical, were not completely separated, nor was Persia isolated from India. People were passing from each of these countries to the others, as they pass today, in spite of restrictions, from Russia to the West and vice versa.

Moreover, Ctesias was a physician. He was born in Cnidos,[116] where a brilliant school of medicine was flourishing; not only was he a professional physician, but his father and grandfather had been so before him. He was taken prisoner by the Persians about 417, and became archiater at the Persian court. He was physician to Darios II (ruled 424–404) and to Artaxerxes II Mnemon (ruled 404–358). His main patron was Parysatis, queen and half sister of Darios, who remained very powerful as queen-mother. He assisted Artaxerxes at the battle of Cunaxa[117] in 401, and soon afterward was sent as an envoy to the Greek rulers of Cypros.[118] From Cypros he did not return to Persia but to his Cnidian home (*c.* 398), which was relatively near. It was in Cnidos that he wrote his works and probably spent the last years of his life. His works were thus written at the beginning of the fourth century, but we speak of him in this chapter, because they were the fruits of his Oriental experience, fruits gathered in the preceding century.

His main works are the *Persica*, a history of Assyria and Persia in twenty-three books, and the *Indica*, a single book dealing with India (Fig. 69). These books have been partly preserved by Diodoros of Sicily (I–2 B.C.), Nicholas of Damascus (I–2 B.C.), and others, but chiefly by Photios of Constantinople (IX–2). This may seem a very late witness, but the lateness does not matter so much, for it is obvious that Photios had the original writings in his hands. In his *Bibliotheca* or *Myriobiblon* (finished before 857) he gathered summaries or reviews of some 280 works, many of which are lost. For example, his article on the *Persica* begins thus: "Read a work of Ctesias of Cnidos, *Persica*, in twenty-three books. The first six, however, tell the history of Assyria and other events prior to the Persian ones." This review covers in the Greek text about 850 lines.

His account of the other work begins in the same way: "Read the *Indica* by the same author in a single book. In the writing of it he makes greater use of the Ionian dialect." This review is shorter; the Greek text of it covers about 442 lines.

A very convenient Greek-French edition of Photios' summaries has been given recently by Henry,[119] but we really need a new critical edition of all the Ctesias fragments and of the doxography relative to him.[120]

The first six books of the *Persica*, devoted to Assyrian history, are more or less preserved by Diodoros of Sicily; to Nicholas of Damascus we owe the account of the defeat of Astyages, king of Media, by Cyros in 549 and the beginning of Persian domination. All the rest of the history of Persia (to 398) is summarized by Photios, who compared the author with Herodotos. Ctesias' knowledge of the history of Persia was derived from Herodotos, whom he often criticized, but to that frame-

[116] Cnidos is on a narrow peninsula at the southwestern corner of Asia Minor. It is very close to Halicarnassos and to Cos.

[117] For the battle of Cunaxa in 401, see note 24. Xenophon and Ctesias were both present but in the opposite armies.

[118] Cypros had been ruled by the Persians and the Phoenicians, but in 411 a Hellenic rebirth was led by Evagoras (435–374) of Salamis (Salamis was the principal Greek city in Cypros, on the east coast within sight of Syria). Many Greek refugees joined Evagoras, the most notable being the admiral Conon (444–392) of Athens, who reorganized the Persian fleet and destroyed the Spartan fleet at Cnidos in 394.

[119] R. Henry, *Ctesias, la Perse, l'Inde, les sommaires de Photius* (Brussels: Office de Publicité, 1947) [*Isis* 39, 242 (1948)].

[120] The best edition of the *Persica* is by John Gilmore (London, 1888), in Greek only but well annotated and indexed. For the *Indica*, see the English translation of J. W. McCrindle (Calcutta, 1882), no Greek, but well annotated and indexed.

ΕΚ ΤΩΝ ΚΤΗΣΙΟΥ, ΑΓΑΘΑΡ-
ΧΙΔΟΥ, ΜΕΜΝΟΝΟΣ
ἱστορικῶν ἐκλογαί.
ἈΠΠΙΑΝΟΥ Ἰβηρικὴ καὶ Ἀννιβαϊκή.

Ex Ctesia, Agatharchide, Memnone excerptæ historiæ.
Appiani Iberica. Item, De gestis Annibalis.

Omnia nunc primùm edita. Cum Henrici Stephani castigationibus.

EX OFFICINA HENRICI
Stephani Parisiensis typographi.

AN. M. D. LVII.

Fig. 69. Ctesias, *editio princeps* (Paris: Henri Estienne, 1557); small size. This is the title page; as can be read on it, this is the first Greek edition not only of Ctesias but also of extracts from Agatharchides of Cnidos (II-1 B.C.), Memnon of Heracleia Pontica (first century?) and Appianos of Alexandria (II-2). Henri II Estienne (Paris 1531-Lyon 1598), editor as well as publisher of the book, was the scion of an illustrious French family of printers, humanists, and booksellers. [From the copy in the Harvard College Library.]

work he added much information that he had obtained during his long residence at the court of Persia. We may imagine that stories were told to him by the king or his assistants, or by the domineering Parysatis and her ladies. Much of that, however, is nothing but gossip. His work is so lacking in criticism that we might call him not the father of history (like his rival) but the father of historical novels, which is not so good. We must make the best of historical novels, however, when purer materials are not available. The data collected by Ctesias are often very interesting, and when he contradicts Herodotos, we should not hastily conclude that the latter is right, though he is in general far more dependable.

A good example of Ctesias' utter lack of criticism is his account of the Behistūn inscription.[121] It was engraved in 516 and related the victory of Darios I over his

[121] Diodoros of Sicily, II, 13. Behistūn, modern Bīsutūn [*Encyclopedia of Islam*, vol. 1 (1912), p. 734], is in western Iran, near Kirmānshāh. The name used by Ctesias was *to Bagistanon oros*, which is derived from the old Persian Bāgastāna, the place of the God (i.e., Mithras). The decipher-

rebellious vassals; it was written in cuneiform script, in three languages – Persian, Elamite, and Accadian. That monument is of immense importance to philologists, for the parallel inscriptions helped to decipher unknown languages; it has been called the Rosetta stone of cuneiform (or of Assyriology). Now Ctesias, who flourished hardly a century after the building of that monument, when traditions concerning it were still fresh, said that it was written in Syrian (Assyrian) letters and ascribed it to the Assyrian queen Semiramis! One would think that they would have known better at the Persian court, but the legendary Semiramis was the main heroine of his Assyrian romance.

Herodotos described the royal road of the Persian empire from Ephesos to Susa; Ctesias continued the account beyond Susa, to Bactria and India (his account is lost).

Another story told by Ctesias, which is trustworthy, is the one concerning the presence of asphalt and petroleum in Babylonia:

> Although the sights to be seen in Babylonia are many and singular, not the least wonderful is the enormous amount of bitumen (*asphaltos*) which the country produces; so great is the supply of this that it not only suffices for their buildings, which are numerous and large, but the common people also, gathering at the place, draw it out without any restriction, and drying it burn it in place of wood. And countless as is the multitude of men who draw it out, the amount remains undiminished, as if derived from some immense source. Moreover, near this source there is a vent hole of no great size but remarkable potency. For it emits a heavy sulphurous vapor which brings death to all living creatures that approach it, and they meet with an end swift and strange; for after being subjected for a time to a retention of the breath they are killed, as though the expulsion of the breath were being prevented by the force which has attacked the processes of respiration; and immediately the body swells and blows up, particularly in the region about the lungs. And there is also across the river a lake whose edge offers solid footing, and if any man, unacquainted with it, enters it he swims for a short time, but as he advances towards the center he is dragged down as though by a certain force; and when he begins to help himself and makes up his mind to turn back to shore again, though he struggles to extricate himself, it appears as if he were being hauled back by something else; and he becomes benumbed, first in his feet, then in his legs as far as the groin, and finally, overcome by numbness in his whole body, he is carried to the bottom, and a little later is cast up dead.[122]

This is confirmed by Herodotos' account [123] of the asphalt deposit of Is.[124]

The description of India is even more fabulous than that of Persia. At least Ctesias had lived many years in Persia, among Persians; he never visited India and his stories represent India as seen from a Persian window. India means chiefly the region of the Indus and the Hydaspes. It is curious that Ctesias did not refer to Taxila, which was then already the main city of that region (Punjab). The *Indica* are important, nevertheless, because they were for a long time the main source of Hindu lore in the West.

---

ment of the Babylonian cuneiform by Sir Henry Rawlinson (1847) was the foundation of Assyriology. Leonard William King and Reginald Campbell Thompson, *The sculptures and inscription of Darius the Great* (London, British Museum, 1907).

[122] Diodoros of Sicily, ii, 12; translation by Charles Henry Oldfather in Loeb Classical Library.

[123] Herodotos, I, 179.

[124] Is (modern Hit) was at eight days' distance from Babylon, near the Euphrates, west of it. It was the quarry for the bitumen used in the walls of Babylon.

To come back to the physician, there is a chapter concerning the hellebore [125] borrowed from Ctesias in the medical collection of Oribasios.[126] The substance of it is:

> My father and grandfather did not dare prescribe hellebore, because they had no idea of how to do it and of the correct dosage; if one gave somebody hellebore to drink, he advised the patient first to make his will. Among the people who tried it many were suffocated (*apepnigonto*) and few remained alive. At present its use has become quite safe.

This is highly instructive, for it reveals the growth of pharmocologic knowledge in Cnidos within the course of three generations; the physicians of Cnidos were making medical experiments and keeping track of the results.

Judging from the number of references to him in many Greek and Byzantine works, Ctesias seems to have been a favorite author; it is even possible that more people read him than read Herodotos. Even men like Plato and Aristotle were acquainted with him, and we may assume that Aristotle's most famous pupil, Alexander the Great, read him too. Indeed, Alexander's admiral, Nearchos (IV-2 B.C.) tells us that the king was deeply fascinated by the stories concerning Semiramis and Cyros.[127] The imagination of men of action is more likely to be fired by myths than by scientific accounts; it may be that Herodotos was too scientific for the great king and that Ctesias was more attractive to him. Thus, Ctesias had a share of responsibility in the Asiatic campaigns of Alexander.

[125] In Ionian the word *elleboros* is spelled with smooth breathing, in Attic with rough breathing. This explains the two English orthographies, ellebore and hellebore, the first of which is now obsolete. The dried rhizome and roots of various species of hellebore were much used by the Greeks and Romans as drugs; they contain various alkaloids acting as sedatives and repressants; also externally as insecticides. There are a good many references to hellebore in the Hippocratic corpus; see Littré, *Oeuvres complètes d'Hippocrate*, vol. 10, pp. 628–630; they are much fewer in Galen; see K. G. Kühn, *Galeni opera omnia* (20 vols.; Leipzig, 1821–1833); vol. 20, p. 296. Hippocratic physicians used it for a great variety of purposes.

[126] Oribasios of Pergamon (IV-2), physician to Julian the Apostate. The text occurs in the *Iatricai synagōgai*, VIII, 8. See the excellent edition of Bussemaker and Daremberg (6 vols.; Paris, 1851–1876), vol. 2 (1854), p. 182.

[127] According to Strabon, XV, 1, 5; 2, 5.

XIII

GREEK MEDICINE OF THE FIFTH CENTURY, CHIEFLY HIPPOCRATIC

Though this book is not a history of medicine, many references have already been made to medical topics. Strangely enough, the peak of ancient medicine before this time had been reached more than a millennium earlier, by the Egyptians in the seventeenth century and before. The fame of Egyptian medicine had reached Greece: witness the Odyssey [1] and Herodotos [2] and the Hippocratic writings.[3] True, by the time of Darios (king of Persia *and* Egypt from 521 to 485), the Egyptian physicians were no longer what they had been in their golden age and those attending to him would have been impaled but for Democedes' intercession.[4] In spite of that, we hear that Darios restored the Egyptian college of medicine at Sais.[5] It is possible that the Greeks derived a certain amount of medical knowledge also from Babylonia. At any rate, since the days of Homer they had obtained much knowledge of their own, and by the second half or the end of the fifth century medicine had been raised to a new level, a much higher level than had ever been attained in Egypt or Mesopotamia. In order to explain that revolution, the Hippocratic revolution, we must briefly relate the long evolution that led up to it.

FROM HOMER TO HIPPOCRATES

The *Iliad* reveals a good deal of medical (chiefly surgical) knowledge, and two early physicians are named,[6] Podaleirios and Machaon, two good doctors, two sons of Asclepios, himself son of Apollo. This brings us back to the religious origin of medical teaching. In Homer's time Asclepios was not a god, but a blameless physician; later the cult of Asclepios flourished in a large number of temples.[7] Some 320 places have been listed in the Greek world where the cult was celebrated. The rites included lustral bathing and incubation during which the

[1] *Odyssey*, IV, 227–232.

[2] Herodotos, II, 84.

[3] There are a good many references to Egypt in the Hippocratic corpus; see Littré, *Oeuvres complètes d'Hippocrate* (10 vols.; Paris, 1839–1961), vol. 10, p. 572.

[4] Herodotos, III, 129, 132.

[5] Heinrich Schäfer, "Die Widereinrichtung einer Ärzteschule in Sais unter König Darius I," *Z. aegyptische Sprache* 37, 72–74 (1899), quoting the text on the "naophore statue" in the Vatican, the only inscription of its kind in Egyptian archaeology.

[6] *Iliad*, II, 731–732.

[7] Emma J. Edelstein and Ludwig Edelstein, *Asclepius, a collection and interpretation of the testimonies* (2 vols.; Baltimore: Johns Hopkins University Press, 1945) [*Isis* 37, 98 (1947)].

patient experienced dreams, which, being interpreted, helped their healing. The patients who had been cured made gifts to the temple (*ex voto*), many of which have been preserved. After his apotheosis, Asclepios was represented with a head like Zeus, bearing a staff with a single serpent twisted around it. The serpent is a symbol and witness of the ancient chthonic worship with which Asclepios himself was associated.[8]

The rite of incubation was already practiced in Egypt and the Greeks may have derived it from there, but they might have developed it independently, for it is natural enough. Patients all over the world would pray to their gods for health and fertility. In warm climates they would be tempted to sleep in the temple court. Whenever wise priests were in charge, they would do their best to make the circumstances of incubation as favorable as possible: sufficient rest and enthusiasm, complete peace and trust. In the following morning the patients would love to speak of their experience and to narrate the events of that wonderful night which they had been privileged to spend in the sanctuary. The main events were dreams, which the priests would explain and from which they would obtain a better knowledge of the patient's needs. The details of the cult would vary from place to place, and its application to healing purposes would depend upon the wisdom of its attendants. The practice might be grossly superstitious in some temples,[9] and almost scientific in others, for it is certain that the practice of incubation, at its best, was excellent: all the resources of suggestion and autosuggestion could be enlisted; one could hardly devise a better method to restore the patient's morale and to invigorate his soul.

The cult was relatively late in developing. It began perhaps in Epidauros[10] (hardly before 500), which remained the main sanctuary for the worship of Asclepios. In addition to Epidauros, the outstanding temples were eventually those of Cnidos, Cos, Rhodes, and Cyrene. The importance of those temples for the development of early Greek medicine can hardly be exaggerated, for even if there were no medical attendants, intelligent priests would accumulate case histories and perhaps keep records of them. They might even begin to classify these cases, more or less consciously, and build up gradually a treasury of medical experience. The interpretation of dreams might be the opportunity for a heart-to-heart talk between priest and patient, something comparable to modern consultation with a religious or medical adviser, or with a psychoanalyst. We should never forget, however, that a rational treatment could be combined, and probably was, with a certain amount of irrational practices. Many patients need such practices, ask for them, and get them.

Moreover the temple treatment, however rational, was largely restricted to psychologic means. Some drugs might occasionally be prescribed, but no surgical or obstetric operations would be attempted, and minor physical means, such as

[8] For serpent worship in general see *Encyclopedia of Religion and Ethics*, vol. 11 (1921), pp. 396–423. M. Oldfield Howey, *The encircled serpent. A study of serpent symbolism in all countries and ages* (422 pp., ill.; London, 1926). J. P Vogel, *Indian serpent lore or the Nagas in Hindu legend and art* (quarto, 332 pp., 30 pls.; London, 1926) [*Isis 10*, 234 (1928)].

[9] The more superstitious people would not go to the Asclepieia but to the places where mysteries were celebrated or to such places as the temple of Amphiáraos near Oropos (near the frontier of Boeotia and Attica, close to the sea, facing Euboea) or the oracle of Trophonios in a cave of Lebadeia (in Boeotia).

[10] Epidauros is on the shore of the Saronic gulf, northeast of the Peloponnesos.

bleeding, scarification, and massage, would be abandoned to lay hands working in other places. Therefore, the medical experience that might accumulate in certain temples would be almost exclusively in the psychologic field, an immense field, to which the Greek physicians always gave proper attention.

The medical teaching that has come down to us may have been influenced at first by temple practice, but it should be emphasized that the Hippocratic writings are almost exclusively lay and rational, with few traces of superstitions and hardly any reference to religion.[11]

The main information concerning drugs had been accumulated for centuries by the collectors of herbs and the root diggers (*rhizotomoi*). Judging from the great mass of empirical knowledge available and the extreme slowness of empirical methods, the work must have been continued for untold generations. A great many plants had been tried, some of their virtues or powers (*dynamis*) recognized, and elaborate means devised for the collection of the most useful herbs. No rational explanation of their virtues could be given, and therefore this part of folklore was deeply impregnated with magic. We could not go into that without losing ourselves in the jungle growth of superstitions. It must suffice to state the fact that the pharmacodynamic properies of many plants were known to root diggers long before the beginning of scientific medicine. The Hippocratic physicians had received from their anonymous ancestors a treasury of drugs. The herbs that they needed were gathered for them by professional herbalists who observed all kinds of superstitious practices in their work. For example, they would have to be in a state of ceremonial purity, otherwise the plants collected by them would fail; some plants had to be plucked only in darkness, or at the time of the waxing (or waning) moon; various spells had to be intoned during the gathering, definite tools used, and the plants dealt with according to a ritual. The possibilities of variations are innumerable. Every part of the work was dominated by magical concepts. As Conway Zirkle amusingly put it, "The gathering of herbs or the digging of roots from the bosom of Mother Earth was considered vaguely analogous to pulling hairs from the back of a sleeping tiger, a dangerous occupation unless proper precautions were observed." [12] At any rate, the new physicians did not have to discover the plants or the roots; they had them and their task was simply to reinvestigate their properties and to determine the proper use and dosage of each of them in a more scientific manner.

While the guardians of the Asclepieia were obtaining a better knowledge of men's psychologic defense against disease, and while the rhizotomists were gathering and testing roots, stems, leaves, flowers, and fruits, various schools of philosophers were devising theories. Let us recall briefly the philosophic influences that might come and did come from four parts of the Greek world: South Italy (Magna Graecia), Sicily, Ionia, and Thrace.

From South Italy came the mystical teachings of Pythagoras and his school. The main physician of that school was Alcmaion of Croton who had some visions of genius; for example, he realized the importance of the brain as the center

[11] The only reference I can think of is in the *De decenti habitu*, VI; Littré, vol. 9, p. 235.

[12] This subject has been very beautifully investigated by Armand Delatte, *Herbarius. Recherches sur le cérémonial usité chez les anciens pour la cueillette des simples et des plantes magiques* (Brussels: Académie royale de Belgique, 1936) [*Isis* 27, 531–532 (1937)]; rev. ed. (180 pp., 4 pls.; Liége: Université de Liége, 1938) [*Isis* 30, 395 (1939)].

of sensations and that health was due to an equilibrium of forces. Democedes carried the experience of Croton to the court of Persia in Susa. Philolaos, though primarily concerned with astronomy, had also some physiologic ideas; he was the first to distinguish between sensory, animal, and vegetative functions, centralized respectively in the brain, the heart, and the navel (not so bad except for the third one!). More influential than the medical particularities were the general ideas, which have never completely ceased to flow and have colored more or less the thinking of physicians as well as of philosophers.

The Sicilian prophet was Empedocles, deeply interested in physiology and medicine, but too much of a poet and a seer (a kind of Greek Paracelsus). His main disciples were Acron of Agrigentum [13] (V B.C.) and a little later Philistion of Locroi (IV–1 B.C.). Both attached special importance to air, inside the body and out. Acron distinguished between different currents of air, useful to man or not. According to Suidas he wrote a regimen for healthy people (*peri trophēs hygieinōn*); according to Plutarch he ordered fires to be lighted to purify the air during the plague of Athens. This seems doubtful because Thucydides does not refer either to the practice or to Acron. The idea that the plague was carried by the air and might be averted by purifying the air was a plausible one, however, and it reoccurred periodically apropos of each epidemic until the nineteenth century.

The third cradle of medical theory was Ionia (or Asia Minor). It must suffice to recall the names of Anaximenes of Miletos, Heracleitos of Ephesos, Anaxagoras of Clazomenai, Archelaos of Miletos (?), and finally, perhaps, Diogenes of Apollonia.[14] These men were physiologists in the old sense, some of them in the new sense. Their cosmologic theories had physiologic applications. Anaxagoras and Diogenes made anatomic dissections.[15] Diogenes accentuated the pneumatic tendencies of Anaximenes and of the Sicilians.

Finally, there were the Thracian influences, Democritos of Abdera, whom Hippocrates knew personally, and Herodicos of Selymbria,[16] who is said to have been his teacher. Herodicos attached great importance to gymnastics, and claimed that physical activity and diet must complete and balance each other (this was one of the central Hippocratic doctrines). As to Democritos, we have some curious letters exchanged by Hippocrates and him; [17] these letters are apocryphal, yet they testify to the popularity of both men and are documents for the study of the Hippocratic legend, the formulation of which began very early. These letters deal with insanity and the use of hellebore, and it is a fact that Democritos was deeply interested in what might be called psychomedical problems or, to use an ugly modern term, psychosomatic medicine. The best Greek medicine was decidedly psychosomatic, which is not surprising considering the origins which we have described (incubation, philosophy). The encyclopedic tendencies of Democritos are illustrated by the scope of his medical studies, for various anatomic investigations are

[13] Agrigentum (Acragas in Greek, Girgenti in Italian), close to the middle of the southern coast of Sicily.

[14] In my *Introduction*, vol. 1, p. 96, I wrote "Apollonia in Crete." There were many places called Apollonia, and this one is more probably Apollonia in Phrygia. Crete was Dorian, and Diogenes wrote in Ionian; this does not prove that he was not a Cretan, but the Phrygian origin is more plausible. Or is it? I am beginning to doubt. See Pauly Wissowa, vol. 9 (1903), p. 763. At any rate, Diogenes is generally quoted as the last representative of Ionian philosophy.

[15] None is mentioned in the Hippocratic corpus (Littré, index).

[16] Selymbria is on the north shore of the sea of Marmara.

[17] Littré, vol. 9, pp. 381–399.

ascribed to him, he tried to account for inflammation, hydrophobia, epidemic contagion, and he approached many difficult questions, such as the nature of enthusiasm, artistic creation, genius, and folly. Apparently at that time efforts had already been made (perhaps in the healing temples) to cure patients by musical means, for Democritos tried to explain the cures that were thus obtained. Music was used mainly for the healing of psychologic troubles, but it was also used in other cases, such as intoxication caused by snake bites. It is probable that the psychologic symptoms accompanying such intoxication suggested the use of musical therapy.[18] Democritos' attempts to explain all the modalities and mysteries of psychologic life were perhaps premature – our ignorance of these matters is still deep today – but so were all the Greek scientific efforts of his time; it was easier to ask questions than to solve them, yet an unusual amount of wisdom and imagination was already required for the asking of them; it is typical of the Greek genius that it was ready and eager to ask difficult questions and did so.

We may now speak of the two places where medical thought matured, Cnidos and Cos, both in the same region, Caria, in the southwest corner of Asia Minor.[19] The occurrence of the two main medical schools in that little corner is not accidental. A look at the map will show that if one were to sail northwest from Cos, one would review the Ionian islands, or southward a short run would take one to Rhodes. From Rhodes one can sail in a circle to Cypros, Phoenicia, Egypt, Cyrenaica, and back to Crete. The Cyclades take one step by step to Greece. One can sail across the Aegean almost without ever losing sight of land. The main point is that Caria, with its back to Asia, was relatively near to Crete, Cypros, and Egypt, and hence was a strategic location for intellectual exchanges. Of course, Cnidos and Cos need not have been as close to each other as they were, and this we cannot explain. It may be that one was the offshoot of the other. We cannot know; both schools appear over the horizon at about the same time, after a period of obscure preparation, which may have taken two or three generations in each case but which we have no means of measuring.

As this and the following chapters will be chiefly devoted to the school of Cos, let us speak first of its contemporary rival.

THE SCHOOL OF CNIDOS

The main difference between the school of Cnidos and that of Cos lies in the fact that the latter was more interested in disease in general and the former in particular diseases. To use modern terms, we might say that the doctors of Cos dealt chiefly with general pathology and their Cnidian colleagues with special pathology. Both tendencies were justifiable, and one might argue that the second was at least as necessary as the former, but even so it was premature. According to Galen, Cnidian doctors recognized seven diseases of the bile and twelve of the bladder; that was obviously artificial. The means of exact diagnosis were utterly insufficient to discover characteristic symptoms, that is, to distinguish between symptoms that have differential value and those that have not. The Cnidians were not able to make

[18] Armand Delatte, "Les conceptions de l'enthousiasme chez les philosophes présocratiques (80 pp.; Paris, 1934), reprinted from *L'antiquité Classique* 3. There are no references to musical therapy in the Hippocratic corpus (Littré, index).

[19] Cos is an island, while Cnidos is at the end of a very long promontory, which is almost the same for practical purposes as an island.

such distinctions; they attached too much importance to unessential details and thus created nosologic ghosts (this summarizes the Coan criticism of them).

We have already become acquainted with one Cnidian physician, the historian Ctesias, who flourished at the court of Persia. Their main physician, however, was Euryphon of Cnidos, who may have been the author or editor of a collection of aphorisms, the *Cnidian sentences* (*Cnidiai gnōmai*), and of other Cnidian treatises which are preserved in the Hippocratic corpus.[20] The *Sentences* are unfortunately lost and thus we lack what could have been an excellent tool for distinguishing between the two schools. The distinction is not easy, because it is quantitative rather than qualitative. Rival medical schools can never be mutually exclusive and their points of agreement must necessarily be far more numerous than their points of disagreement. It would seem, for example, that the Cnidians paid more attention than the Coans to obstetrics and gynecology, yet it is obvious that the latter could not completely abandon the care of women.[21]

Euryphon had made anatomic studies and had written a book on the "livid fever" (*peliē nosos*); he explained pleurisy as an affection of the lungs and treated consumption with milk and a red-hot iron. A third Cnidian physician, Chrysippos of Cnidos, flourished somewhat later; as he was a pupil of Philistion as well as of Eudoxos,[22] he combined in himself the doctrines of Cos and of Sicily.

Cnidos gave birth not only to the physicians Euryphon, Ctesias, and Chrysippos, but also to the architect Sostratos (III–1 B.C.), builder of the lighthouse of Alexandria, and to the geographer Agatharchides (II–1 B.C.). Her most illustrious son, however, was Eudoxos (IV–1 B.C.). In the second half of the fourth century, many pilgrims thronged to the temple of Cnidos to see the statue of Aphrodite, one of Praxiteles' masterpieces.

THE SCHOOL OF COS

While the Cnidian physicians were practicing and meditating on their promontory, another school was developing in an island of their immediate neighborhood. Another look at the map will show that the island of Cos is situated at the entrance of a gulf (Ceramicus Sinus), and the sailor entering that gulf or bay has Halicarnassos to his left and Cnidos to his right. Thus Herodotos, Euryphon, and Hippocrates were at some time close neighbors. Cos is a small island (111 mi^2), but fertile and beautiful, admirably situated; it produced vines, ointments, and silk. The *bombyx* of Cos lived on the leaves of oak, ash, and cypress, not on mulberry like the true silkworms. The silk obtained was different from the Chinese silk. A Coan woman called Pamphila, daughter of Plateus,[23] invented means of producing and weaving the local silk, out of which were made very thin and almost transparent fabrics, which became one of the great luxuries of the Augustan age.[24] Cos, rich in

[20] The following treatises may be considered to be Cnidian in various degrees: *Diseases II, III, IV, Affections, Internal affections, Generation, Nature of the child, Diseases of women, Barrenness*. The list is not exclusive. The text of all these treatises may be found in Littré, vols. 6 to 8.

[21] A good many of the Hippocratic aphorisms deal with gynecology, obstetrics, and pediatrics. There are abundant references to those subjects in other Hippocratic writings.

[22] The mention of Eudoxos here is somewhat unexpected, for he was a mathematician and astronomer, whose fundamental work will be discussed in another chapter. Yet he had received some medical training from Philistion.

[23] Aristotle speaks of her, *Historia animalium*, v, 15, p. 551B, but does not say when she lived.

[24] The Coan dresses (*Coae vestes*) were famous in antiquity, but distinguished from the Chinese ones (*vestes sinicae*) made of Chinese

grapes and silk, was blessed also in her men, for she was the birthplace (or main residence) of three poets of the third century B.C., Philetas, Herodas, and Theocritos, and of the great artist Apelles (fl. 336–306), who had painted for her temple a famous picture of Aphrodite rising out of the sea (*hē anadyomenē Aphroditē*). It is pleasant to think of Hippocrates and his disciples in the middle of vineyards and mulberry groves, and to associate with his memory that of an illustrious painter and of poets; it is pleasant to think of Asclepios vying with Aphrodite to attract pilgrims to the island.[25] For us, of course, Cos is primarily the home of the greatest medical school of antiquity. Hippocrates was not its founder, but he towered so high above the other physicians of that island that Coan medicine and Hippocratic medicine are now interchangeable expressions. Who was Hippocrates?

HIPPOCRATES OF COS

It does not take very long to recite all that we know about Hippocrates. He was born in Cos *c.* 460 and was taught the art by his father Heraclides and by Herodicos of Selymbria. He traveled considerably in Greece. For example, the cases related in *Epidemics I, III* refer to the island Thasos, to Larissa in Thessaly, Abdera in Thrace (it was probably there, or in Athens (?) that he became acquainted with Democritos), Maliboea in Magnesia (east of Thessaly), Cyzicos, south of the sea of Marmara, and elsewhere. He was consulted by Perdicas II (king of Macedonia *c.* 450–413) and by Artaxerxes II Mnemon (king of Persia 405–359) and lived to a very old age; he died in Larissa. If the date of birth c. 460 is correct, and he lived until the age of almost eighty-five, his date of death would be c. 375, that is, well into the fourth century.[26]

There are three ancient biographies of Hippocrates, the earliest being that of Soranos (II–1), but his existence was attested long before that, in the first place by his younger contemporary Plato. In the *Protagoras*[27] Plato speaks of a young man who goes to visit Hippocrates the Asclepiad of Cos to learn medicine, and in *Phaidros*[28] he discusses one aspect of Hippocratic doctrine, the need to understand nature if we would understand the body and the soul of man. We may conclude from these two references that Hippocrates of Cos belonged to a family of Asclepiads (we shall explain the meaning of this presently), that he taught the medical art, and that he was already enjoying some amount of fame within his own lifetime.

In the *Politica*[29] Aristotle speaks of Hippocrates' greatness as a physician. Why should one want other testimonies than those of Plato and Aristotle?

It is paradoxical, however, that no early reference is made to his writings,[30] so

silk. The difference between real silk, *nēma sēricon, metaxa* (of Chinese origin) and wild silk (Indian origin?, Coan) is difficult to state. See F. Warre Cornish, ed., *Concise dictionary of Greek and Roman antiquities* (London, 1898), p. 574. Albert Neuburger, *The technical arts and sciences of the ancients* (London, 1930), pp. 165–167.

[25] There is some piquancy in the fact that Cos and Cnidos were rivals in the worship of Aphrodite as well as in that of Asclepios: while the former boasted a painting of the goddess by Apelles, the second owned a statue by Praxiteles. Would that our American cities could foster rivalries of that kind.

[26] It is perhaps safer to say that he died between 380 and 370. Sudhoff states that Hippocrates died in 390 at 70. All guesswork. See *Ann. Medical History* 2, 18 (1930).

[27] Plato, *Protagoras*, 311B.

[28] Plato, *Phaidros*, 270C–E.

[29] Aristotle, *Politica*, 1326A.

[30] Aristotle referred to the *Nature of man* but ascribed it to Polybos (IV–1 B.C.). The *Phaidros* might refer implicitly to the same work or to *Ancient medicine*. It is not possible to know which particular books Menon (IV–2 B.C.) had in mind.

much so that Wilamowitz Moellendorff could speak of him as "a name without writings." Nevertheless, there is no doubt about the existence of a good many Hippocratic writings. Their genuineness will be discussed in the next chapter.

Hippocrates belonged to an Asclepiad (medical) family; both his grandfather, Hippocrates, and his father, Heraclides, had practiced medicine before him; the latter was naturally his first teacher. Hippocrates II was succeeded by his sons, Thessalos and Dracon of Cos, and by his son-in-law, Polybos of Cos.

The surgical treatises on *Fractures* and *Joints* which are one of the glories of Hippocratic medicine were once ascribed to his grandfather, Hippocrates son of Gnosidicos;[31] this ascription is generally rejected, but it proves that the grandfather was a physician of some distinction.

Thessalos flourished at the court of Archelaos, king of Macedonia from 413 to 399; he was one of the founders of the Dogmatic school. The edition of *Epidemics II*, *VI*, and even *IV* was ascribed to him without any proof. Galen called him the most eminent of Hippocrates' sons.

Polybos (IV–1 B.C.) was the greatest of Hippocrates' successors; he may be the author of the treatise on the *Nature of Man*, as was suggested by Aristotle.

The only thing we know about Hippocrates' physical appearance is that, like many other great men, he was of small stature.

HIPPOCRATIC MEDICINE

It is best to proceed with the Hippocratic writings as we did with the *Iliad* and the *Odyssey*, that is, to study their contents and tendencies, and postpone the consideration of their authorship. Indeed, the main reality for us is constituted by these writings, which are of their nature eternal while the authors, whoever they were, have passed away like shadows.

For the sake of clearness, let us consider the Hippocratic ideas under a series of definite topics.

1. *Anatomy and physiology*. Anatomy was rudimentary. The Hippocratic physicians might have had a sufficient knowledge of the bones, especially if they were surgeons, but their knowledge of the internal organs, of the vessels, sinews, and nerves, was exceedingly vague. Yet they needed some anatomic and physiologic guidance; therefore, they did what other learned physicians did under the same circumstances – they invented or postulated a system of general physiology. Happily, their path along that dangerous slope was braked by a few careful observations and their fantasies restrained by Greek common sense and moderation. What would have happened without such restraints is well illustrated by the development of Hindu and Chinese medicine.[32]

Their general physiology was the so-called theory of humors, which had been adumbrated many centuries before. It is obvious that the bodies of men (or of other animals, which it was easier to observe) include various liquids of considerable importance, such as blood, watery phlegms, biles. Some distempers are evidenced by the appearance of liquid excretions, for example, the slimy mucus

[31] Galen, xv, 456.

[32] For Hindu medicine, see *Isis 34*, 174–177 (1942–43); *41*, 120–123 (1950); for Chinese medicine, *Isis 20*, 480–482 (1933–34); *22*, 267–272 (1934-35); *27*, 341–343 (1937); *33*, 277–278 (1941–42); *41*, 230 (1950); *42*, 265–266 (1951).

running from one's nose when one has a cold in the head, expectorations, diarrhetic stools. The Pythagorean Alcmaion of Croton (VI B.C.) was the first to think of health as an equilibrium of the body, and of disease as an upsetting of that equilibrium (*isonomia* as against *monarchia*). Now, such thinking would naturally be centered on the fluid and variable parts of the body rather than on the fixed organs. These views were repeated by Empedocles, who made them more precise by stating that health (or disease) is conditional on the equilibrium (or disequilibrium) of the four elements (fire, air, water, earth) of which human bodies (as well as everything else) are composed. The theory of four elements evoked the complementary theory of four qualities [33] (dry and moist, hot and cold) which is referred to in *Ancient medicine* [34] and *Sacred disease*.[35] Later still, it evoked the theory of four humors (phlegm, blood, black bile, yellow bile). The first explanation of the theory of four humors (implying the four elements, the four qualities, and even the four seasons) is found in the treatise on the *Nature of man*, which Aristotle ascribed to Polybos. Curiously enough, the theory of humors is not explained in the Hippocratic treatise on humors (*Peri chymōn*). To complete this pyramid of quaternions, the theory of four temperaments was developed and explained for the first time by Galen (II-2); [36] it remained the central theory of post-Galenic medicine until the nineteenth century; it is still alive today, at least, in the nonmedical world, as is witnessed by many expressions in almost every language.

There is a fundamental difference, however, between the late theory of four temperaments and the previous theories. The four elements, the four qualities, and the four humors are present in every body, and health implies their equilibrium in each of them. On the other hand, the theory of temperaments is an anthropologic theory, a means of classifying people: each man is characterized by a single temperament, and one cannot speak of an equilibrium of temperaments except in a social, political, sense.[37]

It would be interesting to compare these quaternions with other physiologic theories, the tridosha (three humors) or pañcabhūta (five elements) of Āyurveda, the Buddhist theory of four elements, the Chinese concept of *yin* and *yang*, all of which illustrate the intellectual need of symmetry, which has guided men of science all over the world and sometimes misguided them.

2. *Prognosis versus diagnosis.* As indicated above, while the Cnidian doctors tried to distinguish (or diagnose) special diseases, their rivals of Cos were more inter-

[33] The elements were called by Empedocles *rhizōmata* and later, by Plato, *stoicheia*; this second term prevailed and is preserved in our own terminology (stoichiology, stoichiometry).

The qualities (properties or powers) were called by Hippocrates or before him *dynameis*. That word remained popular for a long time in Greek and Latin (dinamidia); our word pharmacodynamics is a reminiscence of it.

The four qualities were discussed by the anatomist Quintos (*Coïntos*), who flourished in Rome under Hadrian (117-138) and founded a medical school to which Galen's teachers belonged; he was banished and died in Pergamon in 148(?). Galen wrote a book criticizing Quintos' view on the four qualities. See *Isis 8*, 699, no. 105 (1926); *Introduction*, vol. 1, 281.

[34] *Ancient medicine*, XIV.

[35] *Sacred disease*, XXI.

[36] Sarton, "Remarks on the theory of temperaments," *Isis 34*, 205-207 (1942-43).

[37] Differences of temperament or constitution due to climate or race were clearly recognized in the Hippocratic treatise *Airs waters places*, but there was no question of four temperaments. The Greek word for temperament is *crasis* (mixing), for any temperament is caused by a particular blending of the four elements, qualities, and humors. Galen's treatise is entitled *Peri craseōn, De temperamentis*; see K. G. Kühn, *Galeni opera omnia*, vol. 1, pp. 509-694.

ested in general pathology. The tendency of the latter was to consider all diseases as belonging to one of two groups (see section 4 below), or even to one group. The important matter, then, is prognosis, the ability to foretell the development of the disease, and whether the issue is likely to be fatal or not. We must bear in mind that few, if any, diagnoses were possible in the fifth century and that the patients were more concerned with health than with medical labels. They appealed to the physician somewhat in the same spirit as they consulted the oracles. Would they live and be well? How long would the ailment last? These were the questions.

Prognosis enabled the doctor to recognize, and eventually, as his experience increased, to foretell, different stages that occurred in every disease. During the initial stage (what we would call today the period of incubation), the humors were gradually disturbed and their equilibrium upset. Hippocrates called that stage "coction" (*pepsis*), a homely metaphor derived from the experience of cooking food or brewing drinks. After a definite number of days the cooking was done and the *crisis*, that is, the determination or judgment, was manifested. That judgment was not always final; even when the crisis was favorable, it might be followed by a relapse (*hypostrophē*) or by the ejection or abscession (*apostasis*) of peccant matters (in the form of abscess or tumor). Moreover, many diseases that fell under the observation of the Greek physicians, being malarial fevers, had a rhythmic development that must have been recognized very early. That is, new crises recurred periodically on "critical days" (*crisimos hēmera*).[38] In *Prognostic*, the series of critical days is 4, 7, 11, 14, 17, 20, 34, 40, 60; in *Epidemics*,[39] 4, 6, 8, 10, 14, 20, 24, 30, 40, 60, 80, 120 (all even days) or 3, 5, 7, 9, 11, 17, 21, 27, 31 (all odd days).

The good physician is one who can form a general opinion of the disease in its early stage and can foretell the dangers ahead (critical days) and thus strengthen the patient's will against them.

3. *What diseases did the Hippocratic physicians know?* In the first place, they recognized the fundamental symptom of disequilibrium in human bodies: a state of fever. They could not measure the temperature as we do, but they could appreciate it and may have been subtler in that than we could be today. They could observe the skin, tongue, eyes, sweating, urine, and stool, and make many distinctions between various kinds of fevers. It is possible that some of these distinctions were artificial; it is probable that many had a real differential value. Did they notice the acceleration of the pulse? Apparently not yet, or not clearly; this is one of the main puzzles of the Hippocratic writings – there is hardly any mention of the pulse in them. We find it hard to believe that the early Greek physicians did not feel the pulse of their patients, for the observations of pulsations (in arm or leg) is one that an intelligent man could not help making sooner or later.

This matter is so strange that we must stop a moment to examine it more closely. The early Egyptian physicians were well aware of the pulse.[40] How did that knowledge slip away? Democritos, it is true, refers to the beating of the pulse (*phlebopalia*), but there is only one mention of the pulse in the Hippocratic

[38] *Aphorisms*, VII, 85.

[39] *Prognostic*, XX. *Epidemics I*, XXVI.

[40] James Henry Breasted, *The Edwin Smith surgical papyrus* (Chicago: University of Chicago Press, 1930), vol. 1 [*Isis 15*, 355–367 (1931)].

corpus, to wit in *Nutriment*: [41] "Pulsations of veins and breathing of the lungs according to age, harmonious and unharmonious, signs of disease and of health, and of health more than of disease, and of disease more than of health." This is meager enough; the mixing of pulsation with respiration is confusing and the cryptic tone unpleasant.[42] A study of the pulse was ascribed to a shadowy physician of the Hippocratic age, Aigimios of Elis,[43] and to Praxagoras of Cos (IV–2 B.C.), but we are on solid ground only with the great Hellenistic anatomist, Herophilos of Chalcedon (III–1 B.C.). From that time on (but we are now in an altogether different world, the Hellenistic age focused on Alexandria), the Greek knowledge of the pulse advanced considerably. Its results as published by Galen (II–2), *Synopsis peri sphygmōn*, were the basis of sphygmology until the modern age.[44]

To return to the Hippocratic physicians: they were aware of the occurrence of various fevers, even if they could not measure temperature and count pulses as we do. The fevers varied greatly from the point of view of prognosis, for each had its own evolution, its own rhythm and critical days. Consider this enumeration in *Epidemics*:

> Some fevers are continuous, some have an access during the day and an intermission during the night, or an access during the night and an intermission during the day; there are semitertians, tertians, quartans, septans, nonans. The most acute diseases, the most severe, difficult and fatal, belong to the continuous fevers. The least fatal and least difficult of all, but the longest of all, is the quartan. Not only is it such in itself, but it also ends other, and serious, diseases. In the fever called semitertian, which is more fatal than any other, there occur also acute diseases, while it especially precedes the illness of consumptives, and of those who suffer from other and longer diseases. The nocturnal is not very fatal, but it is long. The diurnal is longer still and to some it also brings a tendency to consumption. The septan is long but not fatal. The nonan is longer still but not fatal. The exact tertian has a speedy crisis and is not fatal. But the quintan is the worst of all. For if it comes on before consumption or during consumption the patient dies.[45]

The meaning of all this has been very well explained by W. H. S. Jones in his books on malaria and Greek history.[46] The most common diseases in the Hippocratic times and places were chest troubles and malaria. In both cases the humors were much in evidence: phlegm (mucus, expectorations), blood (in hemor-

[41] *Nutriment*, xlviii.

[42] There is no mention of "pouls" or "sphygmologie" in Littré's index, but look under "battements" (violent pulsations of the temples, etc). On the other hand, the index to Kühn's edition of Galen devotes a very large amount of space (20, 506–516) to pulses and all their varieties. This helps us to measure the medical progress made between the fifth century B.C. and the second after Christ.

[43] Aegimios wrote a book on palpitations or pulsations, *Peri palmōn*, to which Galen refers; he is otherwise unknown. See *Biographisches Lexikon der hervorragenden Aerzte aller Zeiten und Völker* (ed. 2, 6 vols.; Berlin, 1929–1935), vol. 1, p. 37.

[44] Emmet Field Horine, "Epitome of ancient pulse lore," *Bull. History of Medicine 10*, 209–249 (1941).

[45] *Epidemics I*, xxiv. Chapters xxv and xxvi, which I have no space to quote, give additional information on the evolution of various fevers, such as the critical days.

[46] W. H. S. Jones, *Malaria, a neglected factor in the history of Greece and Rome* (114 pp.; Cambridge, 1907); *Malaria and Greek history* (184 pp.; Manchester, 1909) [*Isis 6*, 48 (1924–25)]. Jones claims that the decadence and fall of Greece and later of Rome were largely due to malaria. His thesis cannot be completely proved, but he certainly helped us to realize the tremendous importance of malaria in ancient history. That disease is still dominating the stage in many parts of the world and is the main cause of the backwardness of some Oriental countries; see *Isis 41*, 380 (1950). A short and good account of the history of malaria and of the ominous nature of that disease to this day may be found in Norman Taylor, *Cinchona in Java* (New York: Greenberg, 1945) [*Isis 36*, 230 (1946)].

rhages), black and yellow bile (in the vomiting fits of remittent malaria). Malaria was the dominating factor. As Jones remarks,

> In malarious countries, all diseases, and not malaria only, tend to grow more severe periodically; latent malaria, in fact, colors all other complaints.[47]

This helps to account for Hippocrates' interest in prognosis (as against diagnosis), for an experienced physician would have recognized the essential alikeness of most ailments in spite of changing rhythms and other differences. Hippocrates was thus led to attach more importance to disease in general (*versus* health) than to varieties of it.

The fevers dealt with in the Hippocratic corpus were all of them malarial fevers,[48] or such as are concomitant with pneumonia, pleurisy, consumption. There is no mention of smallpox, measles, scarlet fever, diphtheria, bubonic plague, syphilis. We are almost certain that syphilis was imported from America only at the end of the fifteenth century, but what about the other diseases: did they not exist in ancient days? And if they did, how is it that the old physicians failed to recognize some of their clear symptoms? It is very baffling, as is always the mixture of knowledge and wisdom with deep ignorance.

Another puzzle: considering the immense catastrophe caused by the plague of Athens, how is it that we find no clear description or even mention of it in any of the medical writings, and would not even know of its existence but for the book of a layman, Thucydides?

There are many references to ophthalmias, and that is not surprising, for various eye diseases have always prevailed in the Near East, but scarcely any technical knowledge of them appears. The malarial fevers are tolerably well described, as well as the state of general ill-health and despondency to which they occasionally lead, the so-called malarial cachexia, characterized by weakness, anemia, darkened complexion, and enlarged spleen. Cases of delirium and various mental troubles are also described; such diseases could not be ignored, for they advertise themselves.

4. *Hygiene and therapeutics.* The scientific nature of Hippocratic efforts appears clearly in their therapeutics. The main difference between a scientist and a nonscientist is often this: the former is fully aware of his ignorance while the latter "knows" (in that respect Socrates was a man of science). "Je sais tout" is the motto of crass ignorance. In the same way we might say that the main difference between an honest physician and a quacksalver is that the latter promises a cure while the former is more prudent. It is not true that all quacks are crooks who care for nothing but money, and some expert physicians are as greedy as quacks; the difference

[47] Jones, *Hippocrates* (Loeb Classical Library), vol. 1, p. lv.

[48] Of course, the Hippocratic physicians could not understand the essential nature of malarial diseases, nor could they know the specific drug, cinchona, a South American plant, the marvelous efficacy of which was revealed to the world by Peruvian Indians in the seventeenth century. Quinine was extracted from cinchona by Pelletier and Caventou in 1820. The beginnings of the scientific knowledge of malaria are summarized by the following facts. In 1880, Laveran found protozoans of the genus *Plasmodium* in red corpuscles of malarial patients; in 1897 Sir Ronald Ross found *Plasmodium* in the stomachs of mosquitos; in 1898 Giovanni Battista Grassi showed that only the *Anopheles* mosquito carries the parasite of malaria. Note that these fundamental discoveries were made, Laveran's in Constantine, Algeria; Ross's at Begumpet, Secunderabad, near Hyderabad; Grassi's in Rome. The story of quinine, on the other hand, ranges from Peru to Java. All of which is very far away from Cos in space and time.

between them lies not so much in greediness as in lack of criticism. Quacks are often good-natured and benevolent people who wish to help as many of their neighbors as possible; their anxiety to cure is equal to the average man's anxiety to know; in both cases the wish is father to the thought. Hippocrates was very prudent and very humble. The therapeutic means at his disposal were very few and weak and he was aware of it. He used purgatives, emetics, cordials, emmenagogues, enemata and clysters, bloodletting,[49] starvation diets in order to evacuate the body, fomentations and baths, frictions and massage, barley water and barley gruel (*ptisanē*, hence the English ptisan and the French tisane, designating all kinds of infusions), wine, hydromel (honey with water), oxymel (honey with vinegar). Remember that the Greeks had no sugar but honey.[50] The best thing that the physician could hope to do was to assuage pain whenever possible and to strengthen the patient's body and soul.

The main Hippocratic idea is neatly expressed by the Latin words *vis medicatrix naturae* (the healing power of nature).[51] To use the physical language of today, health is a condition of stable equilibrium, disease is a break in that equilibrium; if the break is not too deep, the equilibrium tends to reëstablish itself automatically. The bodily and spiritual peace of the patient must be guarded in such a way that the healing power of nature may assert itself and function without hindrance and that health (the state of equilibrium) may be promptly regained. The physician's main duty is to stand by and to help nature.

Therefore, therapeutics is much less a matter of drugs than of diet. The main guarantee of health is a proper regimen combining a moderate amount of nourishment with a moderate amount of exercise. Walking is one of the best forms of exercise for sedentary men. These views are explained in *Regimen III–IV* and *passim* in other Hippocratic writings.

5. *Medical climatology*. One of the Hippocratic treatises, the genuineness of which has hardly ever been questioned, is the one entitled *Airs waters places* (*Peri aerōn hydatōn topōn*). This is certainly the first treatise on medical climatology. It describes the effects of topography and climate upon health, and upon character.

With the exception of balneologists and other physicians attached to watering places, modern physicians do not bother so much about the climate as did their

[49] For bloodletting, see *Introduction*, vol. 2, p. 76. Hippocrates practiced venesection and cupping, but did not use leeches. The only mention of leeches (*bdella*) in the Hippocratic corpus is *Prorrhetic*, II, 17, and it is casual; it is remarked that when the throat often fills with blood this may be due to a hidden leech. This suggests that the early physicians did not discover leeches; it was rather the leeches that discovered them. In places where leeches occur naturally, they are a great nuisance; some physicians of genius realized that the nuisance might be turned to advantage. There are many references to leeches in Galen; see index to Kühn's edition *s.v.* hirudines.

[50] Edmund O. von Lippmann, *Geschichte des Zuckers* (Berlin, 1929) [*Isis 13*, 393–395 (1929–30)]. Sugar cane was hardly known west of India before the early Islamic conquests (VII-1); see *Introduction*, vol. 1, p. 465. It appeared in Egypt in 643, in Syria (Damascus) in 680, in Cypros in 700, in Spain in 714, in Provence in 750, in Crete in 818, in Sicily in 827.

[51] For a history of that idea see Max Neuburger, "The doctrine of the healing power of nature throughout the course of time" (184 pp.), *J. Am. Inst. Homeopathy* (New York, 1932). The *vis medicatrix naturae* might be considered the first example of the idea of automatic regulation in living organisms. Cf. the *milieu intérieur* of Claude Bernard and the wider concept of homeostasis of Walter Bradford Cannon (1871–1945), *Isis 36*, 258–260 (1946). It might even be connected with the general law stated by Henri Le Châtelier (1850–1936) in 1887: the equilibrium of a system, when displaced by a stress, is displaced in such a way as to tend to relieve the stress.

ancient and medieval colleagues. This is partly because our distant ancestors were more completely at the mercy of the weather than we are, especially in the cities, where one lives, as it were, in an artificial climate. It may be due also to gradual neglect and ignorance of climatic factors because of the overwhelming attraction of other factors. We should perhaps attach more importance to climate than we do: it is highly probable that some patients would be more easily cured in certain localities than in others.[52]

The study of the relations of climate to health has always been of special concern to historians of medicine, partly because of the Hippocratic example, partly because of balneologic tradition,[53] and chiefly because of the bearing of climatic and topographic factors on the spread of epidemics. On the other hand, the pedagogues of Europe considered history and geography two parallel disciplines, and continued to do so until yesterday. Therefore, it is not surprising that the same scholars were tempted to study the "geography of medicine" just as well as the history of medicine.[54]

6. *Scientific aspects of Hippocratism.* Some scientific aspects have already been evidenced in the previous sections, but we must come back to that because it is really the central point. If one were asked to characterize Hippocratic medicine in the briefest way, one would say: it is scientific medicine, the first in Greece, if not in the world.[55]

Hippocrates set himself the task of solving medical problems in a rational way. In fact, he has sometimes laid himself open to an accusation often made against modern experts, that he cared less for individual cures than for knowledge. There is nothing to prove his indifference to patients, except that his clinical stories are dispassionate, as they should be. The fact that these stories do not betray any irrelevant feelings does not prove that he did not have feelings and that he did not suffer when patients died. Examples of such clinical stories will be given in the next chapter. They are astounding. In *Epidemics I* and *III* Hippocrates describes cases just as one of our own physicians would do, giving what he considers the essential, nothing more and nothing less. Forty-two cases are described, twenty-five of which ended in death. Hippocrates, like a true scientist, realized that truth matters above everything else, and therefore he recorded his failures as accurately as his successes (a quacksalver would have hidden his failures, not necessarily because he was dishonest, but because the whole business of medical charlatanry implies overconfidence).

The scientific nature of Hippocrates' genius appears in his careful observations, moderate judgments, love of truth, and also indirectly, in his rejection of superstitions, irrelevant philosophy, and rhetoric.[56]

7. *Psychologic healing.* When Hippocrates explained the prime duty of favoring the healing power of nature, he was conscious that the means of favoring it were psychologic as well as physical. It is not enough that the body be allowed to rest

[52] This is fully recognized, of course, in the case of at least one disease, tuberculosis.

[53] *Introduction*, vol. 3, pp. 286, 1240.

[54] It is typical that the third Janus (1896–1941) bore as subtitle Archives internationales pour l'histoire de la médecine et la géographie médicale.

[55] This qualification is added in remembrance of the best Egyptian medicine which has been described in Chapter II.

[56] *Ancient medicine.*

as completely as possible (confinement to bed, very light diet), but the soul must rest (quietness) and be invigorated by good cheer and hope. The physician must deal very gently with his patients.

Here is a typical passage of the *Precepts*, a relatively late cento but derived from good Hippocratic sources:

> I urge you not to be too unkind, but to consider carefully your patient's superabundance or means. Sometimes give your services for nothing, calling to mind a previous benefaction or present satisfaction. And if there be an opportunity of serving one who is a stranger in financial straits, give full assistance to all such. For where there is love of man, there is also love of the art. For some patients, though conscious that their condition is perilous, recover their health simply through their contentment with the goodness of the physician. And it is well to superintend the sick to make them well, to care for the healthy to keep them well, but also care for one's own self, so as to observe what is seemly.

Hippocrates' interest in psychologic healing was natural enough if he had witnessed (as he probably had) the practice of incubation in the Asclepieia or other temples. If so, he had heard of the miraculous cures which the priests and pilgrims would be sure to advertise, and he could appreciate the therapeutic value of such methods. Body and soul are very closely interrelated; neither can be healthy if the other is not and the physician cannot heal the one while neglecting the other; therefore he must try to strengthen both.

It is very tempting to corroborate these views with an extract from Plato's *Charmides*, wherein Socrates refers to one of the Thracian physicians of Zalmoxis

> who are said even to make one immortal. This Thracian said that the Greeks were right in advising as I told you just now: "but Zalmoxis," he said, "our king, who is a god, says that as you ought not to attempt to cure eyes without head, or head without body, so you should not treat body without soul"; and this was the reason why most maladies evaded the physicians of Greece – that they neglected *the whole*, on which they ought to spend their pains, for if this were out of order it was impossible for *the part* to be in order. For all that was good and evil, he said, in the body and in man altogether was sprung from the soul, and flowed along from thence as it did from the head into the eyes. Wherefore that part was to be treated first and foremost, if all was to be well with the head and the rest of the body.[57]

The Zalmoxian criticism reported by Socrates might apply to certain Greek physicians; it certainly did not apply to Hippocrates.

THE HIPPOCRATIC ACHIEVEMENTS

The main achievement of Hippocrates was the introduction of a scientific point of view and scientific method in the cure of diseases, and the beginning of scientific medical literature and clinical archives. The importance of this can hardly be exaggerated. Hippocrates' personality, however shadowy, symbolizes one of the greatest initiatives in the history of mankind. In praise of him it may suffice to say that he did whatever could be done in his time, with genius alone, without the drugs and instruments of a later age. It is remarkable that the idea of writing and collecting clinical cases, as he did in *Epidemics*, was not continued after him. The stories told by Galen are very inferior in spirit; they are more in the nature of self-advertisement than plain and honest reports in the Hippocratic manner. Galen was

[57] Plato, *Charmides*, 156.

more interested in puffing his reputation than in publishing the truth. After Galen there are no clinical cases on record until the time of al-Rāzī (IX-2), and after that I can think of nothing else but a few of the medieval *regimina* and *consilia* and the post-mortem analyses of the Florentine Antonio Benivieni (d. 1502) – but between Hippocrates and Benivieni almost two millennia have elapsed.[58]

Though Hippocrates was more concerned with general than with special pathology, he has left clinical pictures of phthisis, puerperal convulsions, and epilepsy, and has described the typical countenance of a dying or dead man, or of a man whose body has been exhausted by starvation, excessive evacuations, long sickness and suffering. This aspect is still called *facies Hippocratica*. One also speaks of Hippocratic fingers, which are symptomatic of some chronic heart diseases; on account of insufficient oxygenation the terminal joints are enlarged and clublike. Or consider this case, described in *Epidemics*:

> In Thasos the wife of Delearces, who lay sick on the plain, was seized after a grief with an acute fever with shivering. From the beginning she would wrap herself up, and throughout, without speaking a word, she would fumble, pluck, scratch, pick hairs, weep and then laugh, but she did not sleep; though stimulated, the bowels passed nothing. She drank a little when the attendants suggested it. Urine thin and scanty; fever slight to the touch; coldness of the extremities.
>
> Ninth day. Much wandering followed by return to reason; silent.
>
> Fourteenth day. Respiration rare and large with long intervals becoming afterwards short.[59]

The breathing described in the last lines is now generally called Cheyne-Stokes breathing, after two Dublin physicians (1818), or "changed-stroke" by medical students.[60]

Hippocrates' common sense, wisdom, and modesty were sometimes forgotten and overshadowed by the excessive rationalization and the immoderate pride of the Galenic and Arabic physicians, but men of genius were always ready to pay tribute to the father of medicine and to try to imitate him. I am not thinking of the medical philologists like Anuce Foes (1528–1591) of Metz or the Dutchman Van der Linden, whose editions of Hippocrates (published respectively in 1595 and 1665) were much used by students and physicians, but rather of clinicians like Thomas Sydenham (1624–1689). A new wave of medical self-conceit was created at the end of the last century by the triumphs of bacteriology, and for a time many physicians were hypnotized by microbes to such an extent that they failed to consider the patient in his wholeness. In combination with other factors, this has caused a Hippocratic revival which has sometimes been carried too far.[61] Yet intelligent physicians make a distinction between knowledge and wisdom, and recognize that, in spite of the immense and almost incredible medical progress, there is something in the Hippocratic achievement that cannot be superseded.

58 Max Meyerhof, "Thirty-three clinical observations by Rhazes, *c.* 900 A.D.," *Isis 23*, 321–372 (1935), including Arabic text, 14 pp. Meyerhof published separately two pages of errata to that text, copies of which may be obtained from G. Sarton. For *regimina* and *consilia*, see *Introduction*, vol. 3, pp. 285–286, 1238–1240. As to Benivieni, his little but famous book *De abditis nonnullis ac mirandis morborum et sanationum causis* (Florence, 1507; other editions, 1521, 1528, 1529, 1581) contains a record of 20 autopsies and a number of clinical cases.

59 *Epidemics III*, case 15.

60 John Cheyne (1777–1836) described that kind of breathing in the *Dublin Hospital Reports* 2, 216 (1818). William Stokes (1804–78) described more cases in 1846.

61 *Isis 34*, 206 (1942–43).

THE ASCLEPIADAI

One of the very few things that we know about Hippocrates is that he was an Asclepiad (Plato tells it), and on the other hand we know that there were temples dedicated to Asclepios, patron and god of medicine. Who were the Asclepiads? The first idea was that they were priests of such temples. The intelligent priests of healing temples might accumulate medical experience, without too much trouble, almost unconsciously. It is probable, however, that in addition to such men, half priest, half physician, there were in such centers as Cnidos and Cos professional physicians who were called Asclepiadai, either because they were supposed to be descendants of the god or hero Asclepios, or because their duties were inspired by that god.

Such a profession tended to be restricted to certain families, for it was natural for a father to train his son and to bequeath to him his experience and his practice. We have already become acquainted with two medical families – the family of Ctesias in Cnidos and that of Hippocrates in Cos. Hippocrates had been trained by his father Heraclides and his work was continued by his sons and his son-in-law.

These medical families were united by common interests and it is possible that in some places, at least, their union was expressed in the form of rules and regulations, written or unwritten. The Asclepiadai of a district may have formed a kind of guild,[62] that is, a professional association, the structure of which might be as weak or as strong as you please, and the spirit of which might be purely economic, social, scientific, or religious, or it might be colored by various combinations of these influences.

The presence in the Hippocratic corpus of a number of deontologic books does not prove the existence of medical guilds, but if such guilds existed they would have favored the composition of books wherein the duties of physicians, their manners, and their customs would be defined and explained. The deontologic books are first of all the *Oath*, then the *Law*, *Decorum*, *Precepts*, and the *Physician*, chapter 1. Some of them are late, yet they embody earlier traditions and it is with such traditions that we are at present concerned.

The short text entitled the *Oath* includes a professional oath and a kind of indenture (*syngraphē*) binding medical students to their teachers. The constitution of every guild would imply both things, for it must hold the members together, and provide for the education and admission of new ones, and for the defense and continuation of their traditions. The guild might be secret; it would certainly be private; its regulations bind only the members and facilitate their protection against other bodies or against incompetent outsiders. We should beware, however, of thinking too much in terms of modern experience; all of the activities of a modern guild would exist in an ancient one *in potentia* yet unformulated and unformalized. For example, the guild might have a ritual and liturgy to be employed on special occasions, such as the admission of members or their funerals.

We know nothing definite; the absence of documents would seem to prove that even if the Asclepiadai were organized the importance of their guilds cannot have been considerable; if medical guilds existed in certain places, such as Cos, their importance was restricted to a small district and a short period.[63]

[62] See articles Guilds by A. E. Crawley and J. S. Reid, *Encyclopedia of Religion and Ethics*, vol. 6 (1914), pp. 214–221; see also *Introduction*, vol. 3, pp. 152–156.

[63] W. H. S. Jones, "Secret societies and the Hippocratic writings," *Hippocrates* (Loeb Classical Library), vol. 2 (1923), pp. 333–336.

XIV

THE HIPPOCRATIC CORPUS

The Hippocratic tradition will be briefly discussed later in this chapter, but I feel bound to declare at the outset that until recently my knowledge of Hippocratic writings was derived mainly from the splendid edition prepared by Emile Littré, the tenth volume of which contains an elaborate index.[1] Philologists laboriously preparing the *n*th edition of a Hippocratic text may speak ill of Littré, but such criticisms do not decrease his gigantic stature nor increase their own dwarfish one by a single inch. During the last thirty years many Hippocratic editions, translations, and monographs have passed through my hands and some of them have been analyzed in *Isis*. As I was plotting out this chapter, in order to refresh my knowledge I examined very carefully the selected writings edited in Greek and English for the Loeb Classical Library by William Henry Samuel Jones and Edward Theodore Withington.[2] Littré was not a pedantic philologist, but he knew Greek and medicine very well and at his best he is an excellent guide. Jones and Withington had the advantage of doing their bit of work three-quarters of a century later. I am much in sympathy with them and generally willing to accept their guidance in controversial questions, for example, Jones's theory on the deleterious and far-reaching effects of malaria in the ancient world. As to Withington, I am indebted to him directly for many of his studies on the history of medicine and indirectly for his medical contributions to the revision of Liddell and Scott.[3]

COMPLETE OR PARTIAL GENUINENESS OF THE HIPPOCRATIC WRITINGS

The books to which Plato and Menon refer cannot be identified with certainty and hence skeptics may claim that "Hippocrates" is a "name without writings" and that no Hippocratic work can be considered absolutely genuine. The general question of Hippocratic genuineness is thus essentially different from that concerning Platonic or Aristotelian genuineness, for there are enough books of Plato

[1] Emile Littré (1801–1881), *Oeuvres complètes d'Hippocrate* (10 vols.; Paris, 1839-1861). Léon Guinet, "Emile Littré," *Isis 8*, 77–102 (1926), with portrait; p. 87 enumerates the contents of Littré's edition of Hippocrates, volume by volume.

[2] Jones edited vols. 1–2 (1923) [*Isis 6*, 47 (1923-24); 7, 175 (1925)] and vol. 4 (1931). Withington edited the surgical writings in vol. 3 (1927) [*Isis 11*, 406 (1928)].

[3] *A Greek-English lexicon*, by Henry George Liddell (1811-1898) and Robert Scott (1811–1887); new edition revised by Sir Henry Stuart Jones (2160 pp.; Oxford: Clarendon Press, 1925–1940). Withington read for lexicographic purposes the whole of the extant remains of Greek medical literature; see *Isis 8*, 200-202 (1926).

or Aristotle the genuineness of which is as certain as can be and which can be used as standards; it is more like the question concerning the authorship of the *Iliad* and the *Odyssey*. We can accept the authenticity of many of the Hippocratic writings in the same spirit and with the same reservations as we accept the authenticity of the Homeric poems; the personality of Hippocrates, however, is far more tangible than that of Homer.

For practical purposes that is sufficient; yet we should be careful. The Hippocratic spirit and method are defined on the basis of a group of writings; we cannot claim then that certain of these writings are necessarily genuine, because they reflect Hippocratic qualities, without arguing in a circle. The statements made by Plato and Menon are sufficient, however, to define the essential characteristics of Hippocratism and may help to put the Hippocratic writings in a certain order of probable genuineness. We cannot do more, but that is enough for our main purpose.

Independently of the probability of their genuineness, the Hippocratic works available to us are in various stages of composition and preservation. Some are well written, others not so well; others are in the form of a sketch or rough notes that have not been properly edited. The composition of certain books (for example, the one on *Humors*) is exceedingly casual. Moreover, some writings have not reached us in their original integrity. The earliest books were in the shape of rolls (*volumina*), more fragile than those whose shape we are familiar with; the end parts of a roll were especially frail and were easily broken off. This explains why so many of the old manuscripts (not simply Hippocratic) are either headless or tailless. In the case of literary texts, that condition was recognized and generally respected; in the case of medical writings, of which the librarians or editors did not always understand the meaning and the structure, the missing parts were sometimes replaced by another text; a *volumen* might be cut into two or more parts, or the fragments of diverse *volumina* might be fastened together. The composition of certain Hippocratic writings cannot be explained otherwise. In short, some texts were badly composed; in the case of other texts the original composition, whether good or bad, was not transmitted to us; the *volumina* were accidentally torn to pieces, then the odds and ends were put together by careless people.

The contents of the Hippocratic books vary as much as their form. Some of the books were written for physicians or medical students, others for laymen; others are notes scribbled by teachers to guide their lecturing or by apprentices to strengthen their memory; some are notebooks wherein a physician jotted down the results of his experience, others are essays carefully written for polemical or rhetorical purposes. Most of the books represent the doctrines of the school of Cos, but some reflect those of the neighboring and rival school of Cnidos, others still contain the views of outsiders. This is easy enough to understand if we assume that the collection that has come down to us was originally the library of Cos (or part of it, with possible accretions from outside). The temple, school, or guild of Cos would have a library and that library would include not only the local writings, but others given to it by the authors, or obtained for it by the Coan doctors for the sake of study or curiosity.

In the presence of such a variety of form and content one realizes the enormous difficulty, or rather the impossibility, of establishing the genuineness of each item.

Can this or that one be credited to Hippocrates, or to an immediate disciple or a later one, or was it written by a sophist interested in medicine, or by a philosopher who cared less for medicine than for general ideas? In the last case, a definite coloration, say Epicurean or Stoic, may prove the lateness of its composition. The question of personal genuineness is relatively unimportant. We are more concerned in distinguishing the writings that are Hippocratic from others that represent other schools, and then in placing them in a rough chronologic sequence. Some of the writings are obviously ancient, pre-Hippocratic; others, whether written by Hippocrates himself or not, belong to his age and school; others still are definitely post-Hippocratic, yet continue the Hippocratic teachings. The problem is aggravated by the fact that a late writing may very well incorporate a nucleus of early doctrine. Many ancient books are like monuments various parts of which were built and rebuilt at different periods. The question, "What is the date of that monument?" becomes almost meaningless; one must determine as far as possible the dates of various layers. Even so, when one tries to date Hippocratic books complete and accurate solutions are out of reach; we should not attempt the impossible but do our best and be satisfied with it.

Philologists hope to solve such questions by the methods of textual criticism, that is, by investigations of the language, but this involves uncertainties of the same kind, for how can we be sure that the language that has come down to us is the original language? The idea of reproducing exactly every linguistic peculiarity of a text is a modern conceit; the early (say the Hellenistic) editors were more concerned with the substance than with the form of medical texts [4] and did not hesitate to modernize them if they felt like it. Happily, they were often too lazy or too busy to do so, and they reproduced more or less the original text because they followed the path of least effort.

There is one peculiarity that has been preserved in every ancient medical text: they are all written in the Ionian dialect. This is very remarkable, because Cos (and also Cnidos) had been invaded and governed by Dorians, yet the intellectual prestige of the Ionian colonies nearby was so great that the Ionian dialect symbolized learning and elegance. Remember that Herodotos, who was not more of an Ionian than Hippocrates, also wrote in Ionian. This is of some help, but not very much. The fact that a medical book is written in Ionian does not necessarily prove that it belongs to the Hippocratic age, for the language, having been associated with a certain genre, continued to be used for other writings of the same class.[5] The Ionian language of various Hippocratic texts is not by any means the same; there are variations from the Ionian norm, even as there are variations in Herodotos, because that language was somewhat artificial to the writers, being different from the language that they actually spoke.[6] The writers living in that southwest corner

[4] As distinguished from a purely literary text, the form of which, whether in prose or verse, would be appreciated and respected.

[5] Compare the use of Galician by Spanish poets of the Middle Ages (*Introduction*, vol. 3, pp. 337, 344); the use of Latin by French doctors of the seventeenth century; and the use of Anglo-Norman words in the legal jargon to this very day.

[6] Says W. H. S. Jones, *Hippocrates* (Loeb Classical Library), vol. 2, p. liv: "We cannot hope to restore the text beyond reaching the best textual tradition current in the time of Galen. Occasionally even this aim cannot be reached. It is futile to attempt to restore the exact dialect actually written by the authors. They probably did not all write exactly the same kind of Ionic, as it was a literary and not a spoken dialect as far as medicine and science generally are concerned. It is more than futile to think that we know whether the author wrote, e.g., *tois, toisi* or *toisin*."

of Asia Minor were submitted to so many influences (Dorian, Cretan, Carian, Ionian, Attic) that their dialect might easily take various colorations.

EARLY COMMENTARIES

Our study of the Hippocratic writings is facilitated by that of the early commentators, but unfortunately the earliest of them all, Herophilos of Chalcedon (III-1 B.C.), is already late, too late to enable us to separate the fourth-century writings from those of the preceding century. Moreover, Herophilos was not a commentator pure and simple, but an anatomist, the greatest of antiquity. Next to him came two of his pupils, Bacchios of Tanagra [7] and Philinos of Cos.[8] Bacchios edited *Epidemics III*, annotated three other Hippocratic writings, and compiled a glossary; Philinos (considered to be the founder of the Empirical school of medicine) is said to have written Hippocratic commentaries and six books directed against Bacchios. It would be instructive to read the divergent views of Hippocratic commentators of the third century, but those texts are lost.

Three distinguished commentators flourished in the first half of the first century B.C. – Heraclides of Tarentum, Glaucias of Tarentum, and Apollonios of Cition. In the first century of our era, the Hippocratic writings were much used by Celsus (I-1),[9] and glossaries were collected by Erotianos (I-2) and Herodotos of Rome (I-2).[10] The most important as well as the most copious of the ancient commentators was Galen (II-2). Galen wrote so many commentaries on Hippocrates that their names are united, and many scholars (unfamiliar with the history of medicine) speak of them together – Hippocrates-Galen – as if they were twin brothers, representatives of a single period and a common school. That is silly; six centuries are stretched between them. Galen was about as close to the father of medicine as we are to the father of English poetry, Geoffrey Chaucer.

One of the writings of Galen is a study of the books of Hippocrates that are genuine and those that are not, *De genuinis scriptis Hippocratis*. That text is lost, but we know from the catalogue of Ḥunain ibn Isḥāq (IX-2),[11] that Ḥunain had a manuscript of it and had prepared a Syriac translation and summary of it for 'Īsā ibn Yaḥyā. The Syriac text was translated into Arabic by Ḥunain's son, Isḥāq ibn Ḥunain (IX-2) for 'Alī ibn Yaḥyā.[12] The Arabic title was Kitāb fī kutub

[7] Tanagra in Boeotia, a place famous for its business, its fighting cocks, and above all the lovely terra-cotta figurines excavated from its necropolis in 1873 and later.

[8] Bacchios and Philinos were not included in my *Introduction*, because of the loss of their works and of uncertainties concerning their personalities. For Bacchios, see M. Wellmann, Pauly-Wissowa, vol. 4 (1896), p. 2790; for Philinos, see Diller, *ibid.*, vol. 38 (1938), pp. 2193–94, and K. Deichgräber, *Die griechische Empirikerschule* (Berlin, 1930). Jones has compiled a very instructive list of the Hippocratic writings known respectively to Bacchios, Celsus, and Erotianos; *Hippocrates* (Loeb Classical Library), vol. 1, pp. xxxviii–xxxix.

[9] Celsus is not a commentator, but his Latin treatise *De re medicina* is full of Hippocratic memories. See in W. G. Spencer's edition a list of the parallel passages in Hippocrates and Celsus; (Loeb Classical Library), vol. 3 (1938), pp. 624–627. Celsus appeared in print as early as 1478, before Hippocrates and Galen.

[10] Erotianos compiled a Hippocratic glossary which is very precious; other glosses were collected by Herodotos or can be deduced from Galen's commentaries.

J. G. F. Franz, ed., *Erotiani Galeni et Herodoti glossaria in Hippocratem ex recensione Henrici Stephani* (Leipzig, 1780); modern edition of Erotianos' glossary by Ernst Nachmanson (Uppsala, 1918).

[11] Is the *Peri tōn gnēsiōn Hippocratis syngrammatōn* really lost? It is not included in Kühn's edition. In Ḥunain's list it is No. 104. See Bergsträsser's edition (1925) or Meyerhof, *Isis 8*, 699 (1926).

[12] Abū-l-Ḥasan 'Alī ibn Yaḥyā (d. 888) was the son of Yaḥyā al-munajjim (= the astrologer). Yaḥyā had been converted to Islām and was in the

Buqrāṭ al-ṣahīḥa wa ghair al-ṣahīḥa; the recovery and edition of that text in Arabic or in translation is highly desirable.

Some twenty-three of the Hippocratic writings were known to Bacchios and forty-nine to Erotianos; the Littré edition contains seventy items. If Erotianos knew as many as forty-nine works, this suggests that there was already in his time a kind of Hippocratic canon. The word canon is perhaps a little too strong, for there can hardly be a canon without a canonizing authority. It is probable that the Hippocratic collection was hardly more in ancient times than a group of volumes such as would be found in a library, where all the books had been classified roughly by subject. Some such collection was known to the Byzantine scholars of the seventh century, or perhaps long before,[13] and the whole, or a part of it, was eventually translated into Syriac and into Arabic.

To return to the Greek tradition, the manuscripts should give us the best kind of information, but unfortunately those extant are very late, none prior to the tenth century. The early manuscripts contain lists of Hippocratic writings; the earliest one, Vindobensis med. iv, tenth century, contains only a dozen writings; the Marcianus Venetus 269, eleventh century, lists fifty-eight items; the Vaticanus Graecus 276, twelfth century, lists sixty-two.[14]

Printed editions. The first printed editions of Hippocrates were Latin translations of separate treatises, or a few treatises, the best example being the editions of the *Articella* (1476–1500).[15] For other incunabula see Klebs or some of my notes below. The Hippocratic writings were among the most popular scientific incunabula, "Hippocrates" being the third author in popularity; the first two, far ahead of him, were Albert the Great and Aristotle.[16]

The first general editions of Hippocrates were the Latin one prepared by Fabius Calvus (723 pp.; Rome, 1525) and the Greek Aldine (233 pp.; Venice, 1526), both folio volumes, the second being the true *princeps* (Fig. 70). This was the beginning of a very long series. The most important of the early editions are the second Greek one by Janus Cornarius (Basel, 1538) (Fig. 71), the Greek-Latin one by Anuce Foes (folio; Frankfurt, 1595; very often reprinted), to be used together with Foes' dictionary *Oeconomia Hippocratis alphabeti serie distincta* (folio; Frankfurt, 1588) (Fig. 72), and the Greek-Latin one by Joan. Antonides Van der Linden (2 vols., octavo; Leiden, 1665).[17] Among later editions it will suffice to mention the Greek-French one by E. Littré (10 vols.; Paris 1839–1861) (Fig. 73), the Greek one by Franciscus Zacharias Ermerins (3 vols.; Utrecht, 1859–1864), and the Greek one by Hugo Kühlewein (2 vols.; 1894–1902).

service of the caliph al-Ma'mūn. The son 'Alī was secretary to the caliph al-Mutawakkil and was a great collector of books and a lover of science; many of the Arabic translations of Galen were made for him or under his patronage; see *Isis 8*, 714 (1926). 'Īsā ibn Yaḥyā was presumably a brother of 'Alī.

[13] *Introduction*, vol. 1, p. 480. This section must be corrected in two ways. John the Grammarian (VII-1) should be identified with John Philoponos (VI-1), and the second date (VI-1) is the correct one. The medical writings ascribed to John are apocryphal. The Byzantine Hippocratic collection cannot be dated, for no early manuscript is extant; it may be that the earliest Byzantine collection was simply a copy of the Alexandrian one.

[14] Lists edited by I. L. Heiberg in "Hippocratis indices librorum," *Corpus medicorum graecorum*, vol. 1 (1927), 1, pp. 1–3 [*Isis 11*, 154 (1928)].

[15] Klebs, 116. This refers to No. 116 in Arnold C. Klebs, "Incunabula scientifica et medica," *Osiris 4*, 1–359 (1938), a carefully ordered list of all the fifteenth-century printed books dealing with science or medicine. The same abbreviation will be used repeatedly without further explanation.

[16] The numbers of incunabula ascribed to each are 151 for Albert the Great (XIII-2), 98 for Aristotle, 52 for Hippocrates, apocryphal items being counted together with the genuine ones; see *Osiris* 5, 183, 186 (1938).

[17] G. Sarton, "J. A. Van der Linden," *Singer Festschrift* (Oxford: Clarendon Press, 1952). The early editions including Van der Linden's (1665) and perhaps even later ones were not prepared for philologists and historians but for physicians and medical students.

ἍΠΑΝΤΑ ΤᾺ ΤΟΥ͂
ΙΠΠΟΚΡΆΤΟΥΣ.
OMNIA OPERA
HIPPOCRATIS.
AL DVS.
Ne quis alius impune, aut Venetiis, aut uſquam locorum hos Hippocratis libros imprimat, & Clementis VII. Pont. Max. & Senatus Veneti decreto cautum eſt.

Fig. 70. Title page of the Greek *princeps* of the *Omnia opera Hippocratis,* containing the Greek text of 59 Hippocratic writings without Latin translation, edited by Franciscus Asulanus, and printed by the illustrious Aldine firm Aldus and Andreas Asulanus (Aldo Manuzio and Andrea Torresani of Asola) in Venice, 1526. This splendid folio volume begins with a letter addressed by Clement VII (Giulio de' Medici, pope, 1523–1534) to the sons of Andrea Torresani and the heirs of Aldo Manuzio (1449–1515). [From the copy in the Harvard College Library.]

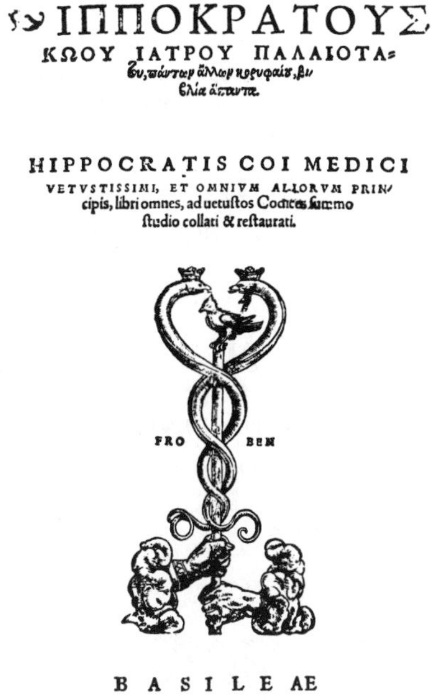

ΙΠΠΟΚΡΑΤΟΥΣ
ΚΩΟΥ ΙΑΤΡΟΥ ΠΑΛΑΙΟΤΑ-
HIPPOCRATIS COI MEDICI
VETVSTISSIMI, ET OMNIVM ALIORVM PRINcipis, libri omnes, ad uetuſtos Codices ſummo ſtudio collati & reſtaurati.
FRO BEN
BASILEAE
M D XXXVIII

Fig. 71. Title page of the second Greek edition (folio) of Hippocrates by Janus Cornarius (Johann Hagenblut of Zwickau), printed by Frobenius (Johann Froben) in Basel, 1538. The humanists of Basel were always competing with their Venetian rivals. [From the copy in the Harvard College Library.]

The *Corpus medicorum graecorum* sponsored by the German academies naturally includes Hippocrates, but only one part of the Hippocratic edition has appeared, twelve works, edited by Hermann Diels and J. L. Heiberg, vol. 1, 1 (158 pp.; Leipzig, 1927).[18] The *Corpus medicorum graecorum* reproduces Littré's pagination and thus pays to it a fine homage.

The two outstanding English translations are the one by Francis Adams (2 vols.; London: Sydenham Society, 1849) and the recent one by W. H. S. Jones and E. T. Withington (4 vols.; Loeb Classical Library, 1923–1931), to which reference has already been made.

In short, there is no Hippocratic canon, only collections the composition of which varies from manuscript to manuscript and from one edition to another. The gen-

[18] *Isis 11,* 154 (1928).

uineness of each work must be discussed separately; it is never certain, but many writings are certainly apocryphal; the probability of a work's being genuine thus varies from zero to less than 100 percent.

When we were dealing with men like Herodotos and Thucydides, who have each but one book to his credit, all the generalities applied necessarily to that single book. The case of Hippocrates is very different; a great many books are ascribed, rightly or wrongly, to him and to his school, and these books vary in so many ways that we must consider them separately – not all, for that would be too long and unnecessary, but some thirty of them. The reader who follows me in the brief

OECONOMIA
HIPPOCRATIS,
ALPHABETI SE-
RIE DISTINCTA.

*IN QVA DICTIONVM APVD HIP-
pocratem omnium, præsertim obscuriorum, vsus explicatur, &
velut ex amplissimo penu depromitur: ita vt* LEXI-
CON HIPPOCRATEVM *merito
dici possit.*

ANVTIO FOESIO MEDIOMATRICO
MEDICO, AVTHORE.

FRANCOFVRDI,
Apud Andreæ Wecheli heredes,
Claudium Marnium, & Io. Aubrium,
ANNO S. MDLXXXVIII.
Cum Priuilegio S. Cæsareæ Maiestatis.

OEUVRES
COMPLETES
D'HIPPOCRATE,
TRADUCTION NOUVELLE
AVEC LE TEXTE GREC EN REGARD,
COLLATIONNÉ SUR LES MANUSCRITS ET TOUTES LES ÉDITIONS;
ACCOMPAGNÉE D'UNE INTRODUCTION,
DE COMMENTAIRES MÉDICAUX, DE VARIANTES ET DE NOTES PHILOLOGIQUES;
Suivie d'une table générale des matières.

PAR É. LITTRÉ.

Τοῖς τῶν παλαιῶν ἀνδρῶν
ὁμιλῆσαι γράμμασι.
GAL.

TOME PREMIER.

A PARIS,
CHEZ J. B. BAILLIERE,
LIBRAIRE DE L'ACADÉMIE ROYALE DE MÉDECINE,
RUE DE L'ÉCOLE DE MÉDECINE, 17;
A LONDRES, CHEZ H. BAILLIÈRE, 219 REGENT-STREET.
1839.

Fig. 72. Title page of the first Hippocratic encyclopedia and dictionary by Anuce Foes of Metz (1528–1595), a monument of medical learning, which is still a valuable tool for the study of Greek medicine (folio, 33 cm, 700 pp., printed in small type in 2 cols.; Frankfort, 1588). In spite of its size the work was reprinted in Geneva, 1662. [From the copy in the Harvard College Library.]

Fig. 73. Title page of the first volume of Littré's Greek-French edition of Hippocrates (10 vols.; Paris, 1839–1861). [From the copy in the Harvard College Library.]

analysis of these books will have a sounder idea of the Hippocratic corpus than could be conveyed in a general way.

One should not attach too much importance to the order in which they are dealt with. A chronologic sequence, which would be the most natural one, is impossible to establish. Some writings are probably pre-Hippocratic, namely *De hebdomadibus* (see p. 215), *Prorrhetic* (*Praedicta*), *Coan prenotions*, and the substance of the *Oath*. We shall consider thirty works, divided roughly as follows: 1–6, Main medical writings; 7–11, Surgical; 12–20, Medical philosophy and essays; 21–24, Aphoristic; 25–29, Deontology; 30, Epistles.

MAIN MEDICAL WRITINGS[19]

1. *The sacred disease; De morbo sacro; Peri hierēs nosu.*[20] This is not by any means the most popular of the Hippocratic writings, but it is one of the outstanding ones from the point of view of medical historians. It is probably genuine, and it is certainly of Hippocrates' time. The sacred disease is epilepsy (the falling sickness), but the treatise deals also with other seizures and other forms of mental illness. The disease is said to originate in the brain, the immediate cause of the seizures being the stoppage of air in the blood vessels by phlegm coming from the head; this pneumatic explanation was perhaps derived from Hippocrates' contemporary, Diogenes of Apollonia. The brain (and not the heart or midriff) is considered to be the seat of consciousness; this may have been derived from Alcmaion (VI B.C.); it was accepted by Plato but rejected by Aristotle (this was one of Aristotle's worst errors) and therefore it took considerable time to rediscover it.

The most startling feature of this book is the rejection of the name commonly given to epilepsy, the "sacred disease." Hippocrates[21] claims that there are no two kinds of diseases, natural and sacred, or human and divine; all are natural and in a sense all are divine. Here are his own amazing words:

> I am about to discuss the disease called "sacred." It is not, in my opinion, any more divine or more sacred than other diseases, but has a natural cause, and its supposed divine origin is due to men's inexperience, and to their wonder at its peculiar character. Now while men continue to believe in its divine origin because they are at a loss to understand it, they really disprove its divinity by the facile method of healing which they adopt, consisting as it does of purifications and incantations. But if it is to be considered divine just because it is wonderful, there will be not one sacred disease but many, for I will show that other diseases are no less wonderful and portentous, and yet nobody considers them sacred. For instance, quotidian fevers, tertians, and quartans seem to me to be no less sacred and god-sent than this disease, but nobody wonders at them. Then again one can see men who are mad and delirious from no obvious cause, and committing many strange acts; while in their sleep, to my knowledge,

[19] Hippocratic works are best known under their Latin titles, which have international currency. For each work reference is made to the Littré, Loeb, and *corpus medicorum graecorum* (*CMG*) editions, as far as these are available. In the study of any work one should pay very special attention to the Galenic commentary. If such commentary was made by Galen and transmitted to us, its text will be found in the Greek-Latin edition by Karl Gottlob Kühn, *Galeni opera omnia* (20 vols.; Leipzig, 1821–1833); vol. 20 is the general index.

[20] Littré, vol. 6, pp. 350–397; Loeb, vol. 2, pp. 129–183.

[21] For the sake of convenience I shall often use the word "Hippocrates" in these notes, meaning the author, whoever he was. We cannot reopen that discussion apropos of each item.

many groan and shriek, others choke, others dart up and rush out of doors, being delirious until they wake, when they become as healthy and rational as they were before, though pale and weak; and this happens not once but many times. Many other instances, of various kinds, could be given, but time does not permit us to speak of each separately.

My own view is that those who first attributed a sacred character to this malady were like the magicians, purifiers, charlatans, and quacks of our own day, men who claim great piety and superior knowledge. Being at a loss, and having no treatment which would help, they concealed and sheltered themselves behind superstition, and called this illness sacred, in order that their utter ignorance might not be manifest.[22]

The anatomy of the blood vessels is very poor; there are good clinical observations, but the definition of epilepsy is insufficient. We should be indulgent however, for in spite of very sophisticated methods of approach (electroencephalography), we have not yet succeeded in explaining the "sacred disease"; nor are we yet able to cure its victims or to give them much help.

We seldom forget our first impressions. This treatise was the first Greek scientific treatise that I read and the spirit animating it moved me deeply. This was my initiation as a historian of science. My fellow students and I at the University of Ghent read it in the (partial) edition included in the *Griechisches Lesebuch* of Wilamowitz, under the wise direction of Joseph Bidez.[23]

2. *Prognostic; Prognostica sive praenotiones; Prognōsticon.*[24] This is traditionally ascribed to Hippocrates without dissenting voice. The development of acute diseases is described in order to enable the physician to foretell it once it has begun. This work remained in practical use until the middle of the seventeenth century, and therefore is represented by a large number of manuscripts and editions in many languages.

Latin editions of the *Prognostica* appeared very early, in the six editions of the *Articella* (1476 to 1500) and separately by Henri Estienne (Paris, 1516). I am not sure whether the Latin-German edition of the *Prognostica Ypocratis cum aliis notatis* (Memmingen, 1496?; Klebs, 521) is really the same text.

The first chapter reads:

I hold that it is an excellent thing for a physician to practice forecasting. For if he discover and declare unaided by the side of his patients the present, the past, and the future, and fill in the gaps in the account given by the sick, he will be the more believed to understand the cases, so that men will confidently entrust themselves to him for treatment. Furthermore, he will carry out the treatment best if he know beforehand from the present symptoms what will take place later. Now to restore every patient to health is impossible. To do so indeed would have been better even than forecasting the future. But as a matter of fact men do die, some owing to the severity of the disease before they summon the physician, others expiring immedi-

[22] Similar ideas occur in ch. 21 of the same work and in ch. 22 of *Airs waters places* apropos of the Scythian disease, effeminacy of certain men. "But the truth is, as I said before, those affections are neither more nor less divine than any others, and all and each are natural." This would suggest that the same man wrote *The sacred disease* and *Airs waters places*.

[23] Ulrich von Wilamowitz-Moellendorff (1848–1931), *Griechisches Lesebuch* (2 vols. in 4; Berlin, 1902–1906); vol. 1, pp. 269–277; vol. 2, pp. 168–172. For Joseph Bidez (1867–1945), see *Osiris 6* (1939). For a fuller treatment, see Oswei Temkin, *The falling sickness. A history of epilepsy from the Greeks to the beginnings of modern neurology* (359 pp.; Baltimore: Johns Hopkins University Press, 1945) [*Isis 36*, 275–278 (1946)].

[24] Littré, vol. 2, pp. 110–191; Loeb, vol. 2, pp. 1–56.

ately after calling him in – living one day or a little longer – before the physician by his art can combat each disease. It is necessary, therefore, to learn the natures of such diseases, how much they exceed the strength of men's bodies, and to learn how to forecast them. For in this way you will justly win respect and be an able physician. For the longer time you plan to meet each emergency the greater your power to save those who have a chance of recovery, while you will be blameless if you learn and declare beforehand those who will die and those who will get better.

The very last sentence seems to have been written in opposition to the Cnidian physicians:

Do not regret the omission from my account of the name of any disease. For it is by the same symptoms in all cases that you will know the diseases that come to a crisis at the times I have stated.

3. *Regimen in acute diseases; De diaeta* (or *De ratione victus in acutis*); *Peri diaitēs oxeōn nosēmatōn.*[25] The genuineness of this treatise has never been doubted. It is a kind of supplement to *Prognostic.* The acute diseases dealt with, characterized by high fever, are chest complaints and remittent malaria. The treatment prescribed is very mild, with insistence on diet (as the title indicates). Hippocrates recommends the use of a gruel or ptisan of barley, hot fomentations, baths and rubbings, various wines and honey drinks, and so on; very few drugs are mentioned.[26]

I should most commend a physician who in acute diseases, which kill the great majority of patients, shows some superiority. Now the acute diseases are those to which the ancients have given the names of pleurisy, pneumonia, phrenitis, and ardent fever, and such as are akin to these, the fever of which is on the whole continuous. For whenever there is no general type of pestilence prevalent, but diseases are sporadic, acute diseases cause many times more deaths than all others put together.[27]

The Latin text was included in every incunabula edition of the *Articella,* six of them (from before 1476 to 1500; Klebs 116). The first separate edition of the Greek text was that of Haller (Paris, 1530). There were many other editions, chiefly in Latin.

The work was known also under other titles, *On the ptisan* (*De ptisana*), because of the importance attached to ptisan, and *Against the Cnidian sentences,* because Cnidian teachings are criticized in the first three chapters.

4. *Prorrhetic II*; *Praedicta II*; *Prorrhēticon b′.*[28] We place this book here in spite of the fact that ancient critics like Erotianos and Galen did not consider it genuine. It has all the appearances, however, of belonging to the early Hippocratic period. We place it here because it is in some respects comparable to *Regimen in acute diseases*; it might have been entitled *Regimen in chronic diseases.*

It is very different from *Prorrhetic I* which is a collection of 170 aphorisms. *Prorrhetic II* is divided into forty-three chapters, some of which are fairly long. It contains a good many medical observations, and two curious statements. In chap. 3 one reads, "Touching the belly and veins with one's hands one is less likely to be deceived than by not touching them;" this must be a reference to pulsations. The Hippocratic physicians did not know much about the pulse, but pulsations had been observed by them (how could they have failed to observe them?). In chap. 17

[25] Littré, vol. 2, pp. 224–377; Loeb, vol. 2, pp. 59–125.

[26] Chap. XXIII.

[27] Chap. V.

[28] Littré, vol. 9, pp. 1–75.

there is a reference to a leech (*bdella*) hidden in the throat, which may be the cause of bleeding. The Hippocratic physicians did not use leeches, but they had recognized the harm that leeches could cause accidentally; this was a correct observation in a country where these animals occur.[29]

5. *Epidemics I and III; Epidemiorum libri I et III; Epidēmiōn biblia a′, g′.*[30] This work is one of the masterpieces of Greek science; it is not well written, for the author was hardly thinking of form. It is a collection of "constitutions" (*catastasis*), and of particular clinical histories. The "constitutions" describe the general circumstances of climate and disease in definite places; three of them refer to the island of Thasos, with which we must assume that the author (Hippocrates?) was very familiar. The clinical cases are forty-two in number, twenty-five of them ending in death.

The scientific nature and dispassionate tone of these medical notes are marvelous. Here are a few examples.

Epidemics I. First constitution. This is a description of epidemic parotitis (mumps); it is particularly interesting because it mentions the orchitis which may be one of the complications of mumps (orchitis parotidea).

In Thasos during autumn, about the time of the equinox to near the setting of the Pleiades, there were many rains, gently continuous, with southerly winds. Winter southerly, north winds light, droughts; on the whole, the winter was like a spring. Spring southerly and chilly; slight showers. Summer in general cloudy. No rain. Etesian winds few, light and irregular.

The whole weather proved southerly, with droughts, but early in the spring, as the previous constitution had proved the opposite and northerly, a few patients suffered from ardent fevers, and these very mild, causing hemorrhage in few cases and no deaths. Many had swellings beside one ear, or both ears, in most cases unattended with fever, so that confinement to bed was unnecessary. In some cases there was slight heat, but in all the swellings subsided without causing harm; in no case was there suppuration such as attends swellings of other origin. This was the character of them: — flabby, big, spreading, with neither inflammation nor pain; in every case they disappeared without a sign. The sufferers were youths, young men, and men in their prime, usually those who frequented the wrestling school and gymnasia. Few women were attacked. Many had dry coughs which brought up nothing when they coughed, but their voices were hoarse. Soon after, though in some cases after some time, painful inflammations occurred either in one testicle or in both, sometimes accompanied with fever, in other cases not. Usually they caused much suffering. In other respects the people had no ailments requiring medical assistance.

Epidemics I. End of second constitution:

Pains about the head and neck, and heaviness combined with pain, occur both without and with fever. Sufferers from phrenitis have convulsions, and eject verdigris-colored vomit; some die very quickly. But in ardent and the other fevers, those with pain in the neck, heaviness of the temples, dimness of sight, and painless tension of the hypochondrium, bleed from the nose; those with a general heaviness of the head, cardialgia, and nausea, vomit afterwards bile and phlegm. Children for the most part in such cases suffer chiefly from the convulsions. Women have both these symptoms and pains in the womb. Older people, and those whose natural heat is failing, have paralysis or raving or blindness.[31]

[29] *Introduction*, vol. 2, p. 76.

[30] Littré, vol. 2, pp. 598–717; 24–149; Loeb, vol. 1, pp. 141–287.

[31] Chap. XII.

Epidemics I ends with fourteen cases (*arrōstoi tessarescaideca*). We quote the second *in extenso*:

Silenus lived on Broadway near the place of Eualcidas. After overexertion, drinking, and exercises at the wrong time he was attacked by fever. He began by having pains in the loins, with heaviness in the head and tightness of the neck. From the bowels on the first day there passed copious discharges of bilious matter, unmixed, frothy, and highly colored. Urine black, with a black sediment; thirst; tongue dry; no sleep at night.

Second day. Acute fever, stools more copious, thinner, frothy; urine black; uncomfortable night; slightly out of his mind.

Third day. General exacerbation; oblong tightness of the hypochondrium, soft underneath, extending on both sides to the navel; stools thin, blackish; urine turbid, blackish; no sleep at night; much rambling, laughter, singing; no power of restraining himself.

Fourth day. Same symptoms.

Fifth day. Stools unmixed, bilious, smooth, greasy; urine thin, transparent; lucid intervals.

Sixth day. Slight sweats about the head; extremities cold and livid; much tossing; nothing passed from the bowels; urine suppressed; acute fever.

Seventh day. Speechless; extremities would no longer get warm; no urine.

Eighth day. Cold sweat all over; red spots with sweat, round, small like acne, which persisted without subsiding. From the bowels with slight stimulus there came a copious discharge of solid stools, thin, as it were unconcocted, painful. Urine painful and irritating. Extremities grow a little warmer; fitful sleep; coma; speechlessness; thin, transparent urine.

Ninth day. Same symptoms.

Tenth day. Took no drink; coma; fitful sleep. Discharges from the bowels similar; had a copious discharge of thickish urine, which on standing left a farinaceous, white deposit; extremities again cold.

Eleventh day. Death.

From the beginning the breath in this case was throughout rare and large. Continuous throbbing of the hypochondrium; age about twenty years.

Epidemics I, case 6:

Cleanactides, who lay sick above the temple of Heracles, was seized by an irregular fever. He had at the beginning pains in the head and the left side, and in the other parts pains like those caused by fatigue. The exacerbations of the fever were varied and irregular; sometimes there were sweats, sometimes there were not. Generally the exacerbations manifested themselves most on the critical days.

About the twenty-fourth day. Pain in the hands; bilious, yellow vomits, fairly frequent, becoming after a while like verdigris; general relief.

About the thirtieth day. Epistaxis from both nostrils began, and continued, irregular and slight, until the crisis. All the time he suffered no thirst, nor lack of appetite or sleep. Urine thin, and not colorless.

About the fortieth day. Urine reddish, and with an abundant, red deposit. Was eased. Afterwards the urine varied, sometimes having, sometimes not having a sediment.

Sixtieth day. Urine had an abundant sediment, white and smooth; general improvement; fever intermitted; urine again thin but of good color.

Seventieth day. Fever, which intermitted for ten days.

Eightieth day. Rigor; attacked by acute fever; much sweat; in the urine a red, smooth sediment. A complete crisis.

Epidemics I, case 11:

The wife of Dromeades, after giving birth to a daughter, when everything had gone normally, on the second day was seized with rigor; acute fever. On the first day she began to feel pain in the region of the hypochondrium; nausea; shivering, restless; and on the following days did not sleep. Respiration rare, large, interrupted at once as by an inspiration.

Second day from rigor. Healthy action of the bowels. Urine thick, white, turbid, like urine which has settled, stood a long time, and then been stirred up. It did not settle. No sleep at night.

Third day. At about midday rigor; acute fever; urine similar; pain in the hypochondrium; nausea; an uncomfortable night without sleep; a cold sweat all over the body, but the patient quickly recovered heat.

Fourth day. Slight relief of the pains about the hypochondrium; painful heaviness of the head; somewhat comatose; slight epistaxis; tongue dry; thirst; scanty urine, thin and oily; snatches of sleep.

Fifth day. Thirst; nausea; urine similar; no movement of the bowels; about midday much delirium, followed quickly by lucid intervals; rose, but grew somewhat comatose; slight chilliness; slept at night; was delirious.

Sixth day. In the morning had a rigor; quickly recovered heat; sweated all over; extremities cold; was delirious; respiration large and rare. After a while convulsions began from the head, quickly followed by death.

It is clear that the book was not ready for publication; it is doubtful whether it was ever meant to be published or to be used outside of the medical school. It may have been written for Hippocrates' personal use, except that it is too well composed for that purpose.

The doctrine of temperaments is adumbrated in *Epidemics III*:

The physical characteristics of the consumptives were: – skin smooth, whitish, lentil-colored, reddish; bright eyes; a leucophlegmatic condition; shoulder blades projecting like wings. Women too so. As to those with a melancholic or a rather sanguine complexion, they were attacked by ardent fevers, phrenitis and dysenteric troubles. Tenesmus affected young, phlegmatic people; the chronic diarrhoea and acrid, greasy stools affected persons of a bilious temperament.[32]

6. *Epidemics II, IV–VII; Epidemiorum libri II, IV, V, VI, VII; Epidēmiōn biblia b′, d′–z′.*[33] We separate these five books of *Epidemics* from the other two (*I*, *III*) in agreement with an old tradition. They are supposed to be less genuine; the ancients ascribed *I* and *III* to the master himself and the other books to other Hippocratic physicians. *Epidemics II*, *VI*, and *IV* (?) were occasionally ascribed to Hippocrates' son Thessalos; *Epidemics VI* was commented upon by one of the ancient physicians, Glaucias of Tarentum (I–1 B.C.).

In one essential respect, the five books that we are going to consider are similar to the other two: they all are collections of clinical cases and medical notes, in various degrees of elaboration. *Epidemics I* and *III* are closer to perfection; *V* and and *VII* less close; *II*, *IV*, and *VI* much more distant; but the general purpose of all books is the same.

The five books are a jungle of clinical notes of many kinds, some of them well written (at their best, like the cases in *Epidemics I* and *III*), others jotted down more rapidly; some notes were written down immediately after some observations had been made without waiting for the continuation and ending of the case; some items are ungrammatical and unclear, others are completely obscure. Some cases are recognizable to the modern physician (and Littré did recognize them), others are mysterious.

One's impression is that these notes were originally the archives of a single physician or of many. They were written on separate pieces of papyrus. At some unknown time all these pieces were put together, and "edited" – if the word "edited" may be used for such careless work. My guess is that the edition was made relatively late (say in the third century), when the Hippocratic school had

[32] Chap. XIV.

[33] Littré, vol. 5, pp. 3–429.

already attained considerable fame. The "editor" was too respectful of those fragments to modify them in any way and he published them just as they were. In that he was right; but he was wrong in leaving them in great disorder and in perpetuating such blunders as inserting *Epidemics VI* between *V* and *VII*, which obviously belong together.

It is perhaps a blessing that those rude notes were permitted to come down to us, because the study of them enables us to recreate the life and experience of the Hippocratic physicians. We watch these men at their work, and are given glimpses of their meditations. In *Epidemics V* there are many examples of self-correction; the physician concludes that his former judgment on this or that case and the treatment which he prescribed were mistaken. In *Epidemics IV*, 6, relating a case of abortion the physician adds, "Did the woman speak the truth, I wonder."

Three physicians are named, Herodicos,[34] whose methods are blamed, Pythocles,[35] who gave his patients milk diluted in much water, and the consultant Mnesimachos.[36] Many other physicians are referred to anonymously.

The randomness of the collection is illustrated by the abundance of repetitions occurring in them, especially within the groups *II*, *IV*, *VI* and *V*, *VII*. It would seem that some notes were written more than once and were thus represented by various pieces of papyrus; each piece was left where it was and the scribe who copied them all on a single roll did not notice the duplications or did not bother about them.

The repetitions extend not only to the collection itself but to many other works of the Hippocratic corpus. Littré has carefully indicated all the fragments that are identical with, or very close to, passages that can be read in the *Aphorisms*, *Prorrhetic I*, *Prognostic*, *Airs waters places*, *Regimen in acute diseases*, *The physician's office*, etc. This is extremely instructive. It shows that part of the Hippocratic corpus was available for reference when the physicians wrote these clinical notes, or that the same men wrote those notes and various other Hippocratic books. In other words, *Epidemics* helps us to realize the integrity of a good part of the Hippocratic corpus. This has been very well shown by Littré in the footnotes and in his introduction to *Epidemics* in general and to each book in particular. Littré's argument has been restated with greater detail by Deichgräber,[37] who confirmed Littré's classification and ventured to give definite dates to each group. According to him, the dating could be represented roughly by the following scheme: *I*, *III*, *c.* 410; *II*, *IV*, *VI*, *c.* 399–395; *V*, *VII*, *c.* 360.

We need not discuss that precise dating, which seems a little bold to me, considering the disorder and heterogeneity of each book; it will suffice to accept the general conclusion that *Epidemics* represent in their totality the medical experience of a definite group of physicians, the Coan school, during a relatively short period, say half a century.

Deichgräber had the advantage of working almost a century later than Littré, but the disadvantage of being less able than Littré to deal with the *realia*, for he is simply a philologist, not a physician.

In order to discuss more thoroughly various features of Hippocratic medicine it

[34] *Epidemics VI*, 3, 18.

[35] *Epidemics V*, 56.

[36] *Epidemics VII*, 112.

[37] Karl Deichgräber, "Die Epidemien und das Corpus Hippocraticum. Voruntersuchungen zu einer Geschichte der Koischen Ärzteschule," *Abhandl. Preuss. Akad.*, Philos. Kl., nr. 3 (172 pp., quarto; Berlin, 1933).

would be worth while to prepare a new edition of *Epidemics*, arranged as much as possible by topics. Inasmuch as the present text is accidental, one would have the right to disregard the casual sequences that it has crystallized and to assume that the original collection of pieces of papyrus has been given back to us in its initial disorder. We would proceed to reëdit it, but intelligently; that is, we would begin by classifying the fragments as well as possible, putting together those that belong together, for example, all those that deal with the curious epidemic which obtained in Perinthos [38] in the winter of an unnamed year: [39] coughing combined with many other troubles such as angina, night blindness, paralysis of various members; the disease took different complexions according to the profession and experience of each patient, for example, town criers and singers suffered from angina, laborers using their arms had pains in the arms, and so forth. Compare this with a statement in *Aphorisms*: "If previous to an illness a part of the body has suffered pains, it is in that part that the disease will settle." [40] The method of editing that I have suggested might be extended to other parts of the corpus; it should be undertaken not by a pedantic philologist but by an experienced physician who was also, but secondarily, a good Hellenist, an editor like Littré or Petrequin. The *realia*, we should always remember, cannot be learned in books but only in the practice of a scientific profession.

Many of the observations recorded in *Epidemics* are singular and yet have all the appearance of being genuine. Here is one, perhaps the most singular of all:

> In Abdera, Phaitusa, housekeeper of Pytheas, had had children, but her husband having abandoned her, she stopped menstruating for a long time, then she had pain in the joints and red patches upon them. These things being so, her body took on the appearance of a man's body, and became covered with hair, she grew a beard, her voice became harsh. In spite of all our efforts to reëstablish her menstruation it did not reappear, and she died not long afterward. The same thing happened to Nanno, the wife of Gorgippos, in Thasos. According to all the physicians whom I talked with, the only hope of restoring her female nature lay in the resumption of her courses; but in this case, also, their efforts failed, and she soon died.[41]

In spite of its singularity, this is a good example of the medical stories told in *Epidemics* (some 567 of them); [42] some are much longer, many much shorter, reduced to aphorisms. The tone is strictly medical, scientific, irrelevant details being avoided, without nonsense.

THE SURGICAL BOOKS

The surgical books are quite as important for the formulation of Hippocratic medicine as the medical books that we have been thus far discussing, but they are a bit too technical for the general reader, and we cannot devote much space to them. Every intelligent person can appreciate the wisdom of Hippocratic medicine

[38] Perinthos, on the north shore of the Propontis, in Thrace, near Selymbria. In the fourth century it was a more important place than Byzantium.

[39] *Epidemics VI*, 7, 1, etc.; also *Epidemics II*, *IV*.

[40] *Aphorisms*, 4, 33.

[41] *Epidemics VI*, 32; Littré, vol. 5, p. 357.

[42] *Epidemics II* is divided into 6 sections, a total of 116 items; *Epidemics VI* into 8 sections, a total of 160 items; *Epidemics IV, V, VII* contain respectively 61, 106, and 124 items; grand total, 567. Each item deals generally with one story or one medical note or aphorism. Some items deal with more than one, such as the one just quoted, which combines two cases of the same kind.

as revealed in the treatise *Regimen in acute diseases*, but it takes a surgeon to appreciate the fine points of Hippocratic surgery, and no amount of explanation would help other readers to judge them correctly.

In spite of their relative excellence, the surgical treatises are less astonishing than some of the other medical treatises, because we know that the surgical profession was very old in Greece (not to mention the Egyptian tradition, which was many centuries older). The Homeric poems reveal already the existence of much surgical knowledge. It is very interesting to compare them with the medieval romances of chivalry, "where there is no end of wounds and violence but an almost complete absence of definiteness or surgical interest." [43] In the *Iliad* some 147 wounds are described so clearly that a surgeon can identify each of them. The Greeks obtained much surgical experience from the accidents not only of war but also of gymnastics and sport. For example, the shoulder was frequently dislocated in wrestling and a good surgeon should know all the ways of putting it in again. Surgical knowledge included not only the setting of broken and dislocated bones, but various forms of bandaging, the application of splints, chiropractic, the use of massage and ointments. The Hippocratic surgeons did whatever could be done with the means available to them; there was, of course, no antisepsis and no anesthesia, except the most rudimentary. The reputation of Greek surgeons had spread abroad as far as Persia before the end of the sixth century; witness the story of Democedes of Croton, who was called to the court of Darios (p. 215). The Hippocratic treatises mark the climax of a long tradition.

An admirable Greek-French edition of the surgical writings was prepared by the French surgeon Joseph Eleonore Petrequin (1810–1876), who devoted to it the leisure time of thirty years, *Chirurgie d'Hippocrate* (2 vols., 1222 pp.; Paris: Imprimerie nationale, 1877–78). Both volumes include very elaborate notes but the long introductions prepared for each treatise of vol. 1 are missing in vol. 2, which the author could not complete and which was edited posthumously by Emile Jullien.

7. *Wounds in the head; De capitis vulneribus; Peri tōn en cephalē trōmatōn.*[44] This is one of the greatest Hippocratic treatises, dating probably from the end of the fifth century and generally ascribed to Hippocrates himself. It contains descriptions of various kinds of skulls (variations in the sutures) and the theory of fracture by contrecoup. There is a remarkably modern method of trephining and discussion of the cases when trephining is recommended and of those when it is better to abstain.

8. *In the surgery; De officina medici; Cat' iētreion.*[45] This is a collection of notes dealing mainly with bandaging, and explaining how the surgeon should behave, which instruments to use, and so on. It may have been a notebook prepared by a teacher or by a student; there are many repetitions, but good teaching implies frequent rehearsing. The following extracts will give a better idea of it than any description.

2. Operative requisites in the surgery; the patient, the operator, assistants, instruments, the light, where and how placed; their number, which he uses, how and when; the [patient's?] person and the apparatus; time, manner, and place.

[43] Withington, in Loeb, vol. 3, p. xii.

[44] Littré, vol. 3, pp. 182–261; Loeb, vol. 3, pp. 2–51.

[45] Littré, vol. 3, pp. 262–337; Loeb, vol. 3, pp. 54–81.

3. The operator whether seated or standing should be placed conveniently to himself, to the part being operated upon, and to the light.

Now, there are two kinds of light, the ordinary and the artificial, and while the ordinary is not in our power the artificial is in our power. Each may be used in two ways, as direct light and as oblique light. Oblique light is rarely used, and the suitable amount is obvious. With direct light, so far as available and beneficial, turn the part operated upon toward the brightest light – except such parts as should be unexposed and are indecent to look at – thus while the part operated upon faces the light, the surgeon faces the part, but not so as to overshadow it. For the operator will in this way get a good view and the part treated not be exposed to view. . .

4. The nails neither to exceed nor come short of the finger tips. Practice using the finger ends especially with the forefinger opposed to the thumb, with the whole hand held palm downward, and with both hands opposed. Good formation of fingers: one with wide intervals and with the thumb opposed to the forefinger, but there is obviously a harmful disorder in those who, either congenitally or through nurture, habitually hold down the thumb under the fingers. Practice all the operations, performing them with each hand and with both together – for they are both alike – your object being to attain ability, grace, speed, painlessness, elegance, and readiness. . .

6. Let those who look after the patient present the part for operation as you want it, and hold fast the rest of the body so as to be all steady, keeping silence and obeying their superior.

This little treatise is certainly Hippocratic and relatively early. Thessalos, Hippocrates' son, has been named as the author. Irrespective of its genuineness, there is perceptible in it the influence of a great, original teacher.

9–11. *Fractures, Joints, Instruments of reduction; De fractis, De articulis reponendis, Vectiarius; Peri agmōn, Peri arthrōn, Mochlicon.*[46] These three treatises may be considered together; the first two, which were certainly written by the same physician, once formed a single work; the third (*Mochlicon*) is an abbreviation of the parts in 1 and 2 dealing with dislocations. All of them are too technical for the general reader.

The genuineness of *Fractures* and *Joints* has never been doubted, and Galen put them in his first group of Hippocratic writings, the most genuine. Curiously enough, some ancient commentators ascribed them not to Hippocrates himself but to his grandfather Hippocrates, son of Gnosidicos.[47] This confirms the view that the surgical tradition was old and that Hippocrates did not originate it; at best he (if not his grandfather before him) standardized it. The two main works, *Fractures* and *Joints*, are not clearly separated from each other; the first contains a good deal (one quarter of it) about dislocations, while the second has some chapters on fractures. What is more surprising, both treatises contain rhetorical passages, such as are not found in the best Hippocratic works, but these passages may be due to the editorial care of a pedantic disciple.

In the treatise on *Joints* (chap. 9) the author discusses massage in surgical cases and announces his intention of devoting a special book to the subject (*anatripsis*); that book was not written, however, and no reference was made to it, except in that single passage.[48]

A commentary on the treatise on *Joints* was written by Apollonios of Cition

[46] Littré, vol. 3, pp. 338–563; vol. 4, pp. i–xx, 1–395; Loeb, vol. 3, pp. 84–455.

[47] Galen, xv, 456.

[48] For the history of massage, see *Introduction*, vol. 3, p. 288.

(I–1 B.C.).[49] That commentary has obtained a great importance, because of an accident in its transmission. A manuscript of it in Florence [50] is a Byzantine copy of the ninth century, including surgical illustrations (for example, with reference to reposition methods), which might go back to the time of Apollonios and even of Hippocrates. Iconographic traditions of this kind are very rare, because the copying of figures was far more difficult than the writing of the text and was often abandoned. Thanks to Apollonios, we have very clear notions of the ancient practice of surgery.

MEDICAL PHILOSOPHY AND ESSAYS

12. *Ancient medicine; De prisca medicina; Peri archaiēs iētricēs.*[51] This treatise is ancient, say the end of the fifth century, but could not have been written by the author of *The sacred disease*, *Regimen in acute diseases*, *Epidemics*, because its form is too literary. It was probably composed by an early disciple of the master who was a physician and also a sophist or rhetorician and who felt the necessity and duty of defending the medical art in a form acceptable to his colleagues.

It begins with a protest against philosophic speculation in medicine, a defense of the "ancient medicine," that is, scientific (as against philosophic) medicine.

Long experience was needed to discover what kind of food was wholesome and what was not, how it should be prepared and how much of it should be taken to preserve the health of strong people or to increase the strength of weaker ones. Now the medical art is but the refinement of the art of nourishing oneself well. The discoveries of the good physicians were of the same kind as those of the early dietists ("My own view is that their reasoning was identical and the discovery one and the same." [52]). They had to find the kind of nourishment that would be suitable for sick people (diluted foods, paps or gruels, *rhophēmata*) and would restore their health instead of destroying what remained of it.

The four qualities (wet and dry, hot and cold) are comparatively unimportant; other qualities or virtues (*dynameis*), not limited to four, are probably more important – virtues such as strength, saltness, bitterness, sweetness, sharpness, sourness, moistness, and many others, plus their innumerable combinations. This was a very remarkable outburst of medical common sense against premature classification.

The burden of the polemical part of the treatise was a denunciation of irresponsible hypotheses; [53] the physician must restrict himself to available and controllable evidence; he must be rational and modest; we would simply say scientific.

The author was acquainted with Alcmaion, Empedocles, Anaxagoras, but his main interest was technical.[54] His good appreciation of ancient medicine is some-

[49] Cition was one of the nine chief towns in Cypros. Apollonios flourished in Alexandria. For the story of the illustrations to Apollonios' commentary see *Introduction*, vol. 1, p. 216. These illustrations were beautifully reproduced by Hermann Schöne, *Illustrierter Kommentar zu* peri arthrōn (75 pp., 31 pls.; Leipzig, 1896).

[50] Codex Laurentianus, LXXIV, 7.

[51] Littré, vol. 1, pp. 557–637; Loeb, vol. 1, pp. 3–64; *CMG*, vol. 1, pp. 36–55.

[52] Chap. VIII.

[53] The author was the first to use the Greek word *hypothesis*, not as we use it, however, but with the meaning of unverifiable and irresponsible assumption. The theory of four qualities was such an assumption.

[54] "Technical" is derived from the Greek word *technē*, which means art but also method and thus comes sometimes close to "science," even as the English words technical and scientific may assume comparable meanings. The difference between the Greek words *technē* and *epistēmē*

what misleading, for there was empirical medicine (and surgery) but little scientific medicine before Hippocrates, and pioneers like Alcmaion were led astray by Pythagorean hypotheses. He was perhaps too modest for his older contemporaries and too generous to their predecessors. He attacked the philosophers, the premature rationalists, but had nothing to say of the charlatanism that was flourishing in the sanctuaries. It may be that he did not discuss superstitions (even as our own physicians do not speak of them) because he considered them irrelevant and below contempt. His reference to bad physicians, "who comprise the great majority,"[55] implies not charlatanry but incompetence.

Read the significant beginning:

All who, on attempting to speak or to write on medicine, have assumed for themselves a postulate (*hypothesis*) as a basis for their discussion – heat, cold, moisture, dryness, or anything else that they may fancy – who narrow down the causal principle of diseases and of death among men, and make it the same in all cases, postulating one thing or two, all these obviously blunder in many points even of their statements, but they are most open to censure because they blunder in what is an art, and one which all men use on the most important occasions, and give the greatest honors to the good craftsmen and practitioners in it. Some practitioners are poor, others very excellent; this would not be the case if an art of medicine did not exist at all, and had not been the subject of any research and discovery, but all would be equally inexperienced and unlearned therein, and the treatment of the sick would be in all respects haphazard. But it is not so; just as in all other arts the workers vary much in skill and in knowledge, so also is it in the case of medicine. Wherefore I have deemed that it has no need of an empty postulate, as do insoluble mysteries, about which any exponent must use a postulate, for example, things in the sky or below the earth. If a man were to learn and declare the state of these, neither to the speaker himself nor to his audience would it be clear whether his statements were true or not. For there is no test the application of which would give certainty.

But medicine has long had all its means to hand, and has discovered both a principle and a method, through which the discoveries made during a long period are many and excellent, while full discovery will be made, if the inquirer be competent, conduct his researches with knowledge of the discoveries already made, and make them his starting point. But anyone who, casting aside and rejecting all these means, attempts to conduct research in any other way or after another fashion, and asserts that he has found out anything; is and has been the victim of deception.

And chapter xx:

Certain physicians and philosophers assert that nobody can know medicine who is ignorant what man is; he who would treat patients properly must, they say, learn this. But the question they raise is one for philosophy; it is the province of those who, like Empedocles, have written on natural science, what man is from the beginning, how he came into being at the first, and from what elements he was originally constructed. But my view is, first, that all that philosophers or physicians have said or written on natural science no more pertains to medicine than to painting. I also hold that clear knowledge about natural science can be acquired from medicine and from no other source, and that one can attain this knowledge when medicine itself has been properly comprehended, but till then it is quite impossible – I mean to possess this information, what man is, by what causes he is made, and similar points accurately. Since this at least I think a physician must know, and be at great pains to know, about natural science, if he is going to perform aught of his duty, what man is in relation to foods and drinks, and to habits generally, and what will be the ef-

or *mathēma* may be nothing more than the difference between practical and theoretical knowledge.

[55] Chap. IX.

fects of each on each individual. It is not sufficient to learn simply that cheese is a bad food, as it gives a pain to one who eats a surfeit of it; we must know what the pain is, the reasons for it, and which constituent of man is harmfully affected. For there are many other bad foods and bad drinks, which affect a man in different ways. I would therefore have the point put thus: — "Undiluted wine, drunk in large quantity, produces a certain effect upon a man." All who know this would realize that this is a power of wine, and that wine itself is to blame, and we know through what parts of a man it chiefly exerts this power. Such nicety of truth I wish to be manifest in all other instances. To take my former example, cheese does not harm all men alike; some can eat their fill of it without the slightest hurt, nay, those it agrees with are wonderfully strengthened thereby. Others come off badly. So the constitutions of these men differ, and the difference lies in the constituent of the body which is hostile to cheese, and is roused and stirred to action under its influence. Those in whom a humor of such a kind is present in greater quantity, and with greater control over the body, naturally suffer more severely. But if cheese were bad for the human constitution without exception, it would have hurt all.[56]

There are two recent editions: by W. H. S. Jones, "Philosophy and medicine in ancient Greece," Supplement 8 to the *Bulletin of the History of Medicine* (100 pp.; Baltimore, 1946) [*Isis* 37, 233 (1947)], including new edition of the text and English translation; and by A. J. Festugière, *L'ancienne médecine* (136 pp., Paris: Klincksieck, 1948), including Heiberg's Greek text and a French translation. Both editors provide abundant notes and elaborate introductions.

13. *The art; De arte; Peri technēs.*[57] This short treatise of the early Hippocratic age was written to prove that there is such a thing as the medical art, and to defend its practitioners against various kinds of detractors. The author may have been a layman; some scholars have tried to identify him with Protagoras or with Hippias; such attempts, proceeding from the common wish of finding an author for an anonymous work, are futile when there is little more than the wish to support them.

We gather from it that in Hippocrates' time, as in our own, there were people who spoke ill of physicians, saying that cures were due to luck, that patients often recovered without medical help, that some died in the doctor's hands, and that doctors refused to treat some diseases. The first three objections contained enough truth to be impressive. The fourth one would not be used today; physicians do not refuse any more to treat certain hopeless patients, though they sometimes wish they did not have to treat them.

14. *Nature of man; De natura hominis; Peri physios anthrōpu;*[58] and *Regimen in health; De salubri victus ratione; Peri diaitēs hygieinēs.*[59] These two works are put together, because they formed a single work in ancient times and are joined together in manuscripts. Aristotle quoted a fragment from the *Nature of man*, which he introduced with the words "Polybos writes to the following effect." On that basis

[56] Jones's translation in Loeb *Hippocrates*, vol. 1, pp. 13, 53. He uses the word "postulate" for the Greek word *hypothesis* in order to avoid misunderstandings, for we now restrict the meaning to "good, valuable, hypotheses" as distinguished from the unwarranted one. The tone of both extracts is astonishingly modern. The author speaks like a man of science of today, who repeats, "Do not generalize a priori; do not use concepts until their operational value has been tested."

[57] Littré, vol. 6, pp. 1–27; Loeb, vol. 2, pp. 186–217; *CMG*, vol. 1, pp. 9–19.

[58] Littré, vol. 6, pp. 29–69; Loeb, vol. 4, pp. 1–41.

[59] Littré, vol. 6, pp. 70–87; Loeb, vol. 4, pp. 44–59.

that treatise has been ascribed to Polybos, son-in-law of Hippocrates, an ascription that is plausible [60] and is partly confirmed by Menon.[61]

Taking both works together, they are not a well-organized whole but rather a collection of fragments, put together arbitrarily. Hence, the discussion of the authorship is somewhat futile; there may be many authors. Menon, ascribing chapter 9 to Aristotle and chapter 3 to Polybos, may be right in both cases. The beginning of the *Nature of man* is reminiscent of *Ancient medicine*, and there are points of contact with other books of the corpus.

The most important part of the *Nature of man* is the discussion of the theory of humors. It is the only Hippocratic work discussing seriously that theory, while the one ostensibly devoted to it (*Peri chymōn*) does not deal with it. The author argues against philosophers who think that the universe is made of a single substance and extend that theory to medicine; if that were the case there would be but one disease and one remedy. The human body is constituted of four separate humors whose balance is the condition of health; yet different humors predominate in each season. There follow therapeutic rules derived from those premises. Chapter II contains a confused account of the vascular system (the oldest Greek descriptions of it were those of Syennesis of Cypros, Diogenes of Apollonia, and this one).

Regimen in health gives rules for diet and exercise according to the seasons, complexion, and age; how to become leaner or fatter; [62] when to use emetics and clysters; regimen of children, women, and athletes.

There are six incunabula of the Latin text (Klebs, 519, 644, 826), the earliest in Milan, 1481. The latest edition of the Greek text is by Oskar Villaret (88 pp.; Berlin, 1911).

15. *Humors; De humoribus; Peri chymōn.*[63] This is perhaps the most chaotic and puzzling book of the corpus; Littré said that it deserved to be called *Epidemics VIII* (and he printed it immediately after *Epidemics II, IV–VII*) and Jones, going him one better, says, "It is obviously a scrapbook of the crudest sort; it has no literary qualities and it is obscure to a degree." Yet it is a genuine Hippocratic scrapbook, known to the early commentators. Is it a collection of teacher's or of student's notes. Every guess is permitted and none can be substantiated.

It is full of puzzles, beginning with its very title, for it hardly deals with humors. The only Hippocratic work dealing with that is the *Nature of man*.

In spite (or because) of its obscurity it was frequently copied and printed.

16. *Airs waters places; De aere locis aquis; Peri aerōn hydatōn topōn.*[64] Undoubtedly genuine (meaning ancient Hippocratic), this treatise is also one of the most astonishing fruits of the Hippocratic (or call it the Greek) genius. It is the first treatise in world literature on medical climatology (see our discussion of that in the preceding chapter) and it is also the first treatise on anthropology.

[60] The quotation occurs in *Historia animalium* (3, 3, p. 512 *b*), and the passage quoted is taken from chapter XI of *Nature of man*, a confused description of the veins.

[61] W. H. S. Jones, *The medical writings of Anonymus Londinensis* (Cambridge: University Press, 1947), p. 75 [*Isis* 39, 73 (1948)].

[62] This form of speech is used rather than the one more natural to us, "how to lose or gain weight," because there is no mention of weight. Nobody was ever weighed in antiquity.

[63] Littré, vol. 5, pp. 470–503; Loeb, vol. 4, pp. 62–95.

[64] Littré, vol. 2, pp. 12–93; Loeb, vol. 1, pp. 66–137; *CMG*, vol. 1, part 1, pp. 56–78.

Hippocrates explains that the physician should pay full attention to the climate of each locality, and to the variations of that climate caused by changeable seasons, by different exposures, by the nature of the available water and food, and so on. Each medical case must be considered in its own geographic and anthropologic background. Diseases vary from place to place according to the difference in topography, climate, and human nature. The explanation is supported by a great many examples which the author had collected in his travels.

The second part of the book (chaps. 12–24) deals with the effect of climate upon character, and is a kind of anthropologic discussion of history. What is the difference between Europe and Asia, or between the Hellenes and the barbarians? Hippocrates ascribes those differences chiefly to physical (geographic) causes. So did his contemporary Herodotos, who put that teaching in the mouth of Cyros, the king of Persia, and thus gave to his *History* the most significant ending.

One of the most remarkable chapters in the Hippocratic anthropology is the twenty-second, discussing the case of the Scythian eunuchs or androgynes.[65] We can hardly expect the author's physical explanation of that mysterious situation to be correct, but it is very astonishing that he did try to give such an explanation, especially when we remember that the impartial discussion of sexual abnormalities is often believed to be a conquest of our own times.

The popularity of this treatise is witnessed by the number of manuscripts and editions. There are four incunabula of the Latin version, the first in 1481 (Klebs, 644.2, 826.1–3). Among modern editions of the Greek text, special mention should be made of the one prepared by the Greek scholar and patriot, Adamantos Coraes ("Coray," 1748–1833) with a French translation (2 vols.; Paris, 1800). There are at least five English translations, the first by Peter Low (London, 1597). See also Ludwig Edelstein, *Peri aerōn und die Sammlung der Hippokratischen Schriften* (196 pp.; Berlin, 1931) [*Isis 21,* 341 (1934)], and Arne Barkhuus, "Medical surveys from Hippocrates to the world travelers, medical geography, geomedicine," *Ciba Symposia* 6, 1986–2020 (1945).

For additional discussion of this treatise, see Chapter XIII.

17. *Nutriment; De alimento; Peri trophēs.*[66] *Nutriment* might be considered one of the aphoristic works, for it is divided into fifty-five chapters, eighteen of which are two lines long or less in the Greek text, twenty-nine covering from three to five lines, and only eight a little longer, though less than ten lines; thirty-five chapters out of the fifty-five are less than four lines long. It is unique in the Hippocratic corpus, because of its strong Heracleitean coloration. The dating of it is post-Heracleitos, probably prior to the fourth century, say the end of the fifth century.

The author was trying to explain the infinitely complex process of nutrition; as there could be no real understanding of that before the development of modern chemistry, it is not surprising that he was baffled and took refuge in obscure, sibylline utterances. In many chapters two opposite meanings are conveyed — take your choice. One thing he understood clearly, that food must be fluid to be assimilated,[67] also the obvious fact, that food is essential to life (the *dynamis* of food

[65] The chapter begins, "Moreover the great majority among the Scythians become impotent, do women's work, live like women and converse accordingly. Such men they call Anaries (*Anarieis*)." Herodotos refers to the same people, giving them almost the same name, *Enarees* (I, 105; IV, 67). That was probably a Scythian word, equivalent to androgyne or homosexual.

[66] Littré, vol. 9, pp. 94–121; Loeb, vol. 1, pp. 337–361; *CMG*, vol. 1, part 1, pp. 79–84.

[67] LV. Moisture the vehicle of nourishment.

replaces the Heracleitean fire). But again, how could anybody understand in the fifth century the mysterious chemistry of food transformed into flesh and bones, with blood and milk as "excess" (*pleonasmos*).[68] No food is good absolutely, but only with regard to a definite person and a definite purpose; "all things are good or bad relatively." [69]

Let us consider a few other examples: [70] (four chapters are quoted, each complete):

Nutriment and form of nutriment, one and many. One, inasmuch as its kind is one; form varies with moistness or dryness. These foods too have their forms and quantities; they are for certain things, and for a certain number of things.

This is a form of the puzzle that exercised the mind of early Greek philosophers: the one *versus* the many. Many kinds of foods produce the same result, organic growth.

To illustrate the Heracleitean type of obscurity:

Nutriment is that which is nourishing; nutriment is that which is fit to nourish; nutriment is that which is about to nourish. The beginning of all things is one and the end of all things is one, and the end and beginning are the same.

The best chapter is:

Pulsations of veins and breathing of the lungs according to age, harmonious and unharmonious, signs of disease and of health, and of health more than of disease, and of disease more than of health. For breath too is nutriment.

This is valuable not only because it is more concrete than the rest, but also because it is the earliest mention of pulse in Greek literature, and refers to air as food. The absence of other references to simple pulsations is one of the curiosities of the Hippocratic corpus.[71] As to air, it was clearly indispensable to life, but to recognize it as food could then be only a guess or a metaphor.

18. *The use of liquids; De liquidorum usu; Peri hygrōn chrēsios.*[72] This is a collection of notes concerning sweet and salt water, vinegar, wine, and the use of warm and cold liquids. It was possibly an abridgment of a larger treatise that is lost. Our only reason for listing it here is the fact that it is available in the *Corpus medicorum graecorum.*

19. *Regimen I–IV; De victu* (Book IV is often called *Dreams, De insomniis,* or *De somniis*); *Peri diaitēs, Peri enypniōn.*[73] This work has been ascribed to Herodicos of Selymbria, to Hippocrates, to Philistion of Locroi, and to others. It dates probably from the Hippocratic age but is decidedly not Hippocratic in the good sense, for it is full of philosophic fancies and arbitrary "hypotheses." One finds traces in it of the teachings of Heracleitos, Empedocles, Anaxagoras, and the Pythagoreans.

[68] Chap. XXXVI.

[69] End of chap. XLIV.

[70] Chaps. I, VIII, IX, and XLVIII are quoted, each complete.

[71] The first Greek studies of pulse were made by Praxagoras of Cos (IV-2 B.C.) and by Herophilos of Chalcedon (III-1 B.C.), which takes us into the Hellenistic age. Hippocratic physicians recognized the excited palpitations occurring in fevers (throbs; see Littré's index, *s.v.* "battements"). See section 4 above.

[72] Littré, vol. 6, pp. 116–137; *CMG*, vol. 1, part 1, pp. 85–90.

[73] Littré, vol. 6, pp. 462–663; Loeb, vol. 4, pp. 224–447.

Modern editions include four books, the fourth having an alternate title, *Dreams.* Some of the early editions began with Book II; in Galen's time the work was divided into three parts, Book IV being simply the end of Book III. At any rate, the four books are held together by what the author calls his "discovery" (*heurēma*): the two main factors of health are food and exercise; these two factors must be well balanced; if one of them predominates, precautions must be taken to reëstablish the equilibrium. This gives the physician a method for treating his patients.

The author accepts the existence of the four elements but tries to reduce them to two – fire and water – and his physiology is then derived from the conflict between these two, which produces endless changes. The general conception is not clear, and its applications (for example, to embryology) are highly artificial and nebulous. In the first book such fancies are used to explain the composition of living bodies, the differences between ages and sexes, the nature of physical and mental health. The second deals with the properties of different countries, winds, foods and drinks, exercises. The third describes the signs revealing the imbalance of food and exercise and the onset of disease. The fourth explains how dreams may help to indicate the disorders that are brewing.

Embryologic problems are discussed in Book I, VI–XXXI. The author shows that the fetus is developed from the sperm, identified with the soul. The sperm-soul is a mixture of fire and water and is constituted of parts (*merea*) issuing from the bodies of both parents. Fetal development is compared with the execution of a piece of music, the fetus itself with a musical instrument. These musical-embryologic fantasies are obviously of Pythagorean origin. The obscurity of these ideas is increased by the corruption of the text.[74]

One of the most interesting parts for the modern reader is the description and comparison of different kinds of exercises (natural ones such as walking, and violent ones such as racing and wrestling) and their methods and results.[75] Book IV, on dreams, is also very instructive; there are two kinds of dreams, those of divine origin, which concern oneiromancers, and those of physiologic origin, which give clues to the physicians. When the diviners venture to interpret the dreams of the second kind they are likely to fail.

> They recommend precautions to be taken to prevent harm, yet they give no instruction how to take precautions, but only recommend prayers to the gods. Prayer indeed is good, and while calling on the gods a man should himself lend a hand.[76]

The four books combine fantastic notions with good observations. They illustrate the confusion that obtained even in the best minds when they tried to explain physical and physiologic complexities that were still hopelessly beyond their reach. Hippocratic common sense emerges here and there, in spite of the premature theories.

The book on dreams was the first "scientific" treatment of a subject that fascinated the people of antiquity and of the Middle Ages, and indeed the people of all ages. However strange and inadequate it may seem to the modern scientist, it

[74] Armand Delatte, *Les harmonies dans l'embryologie hippocratique* (Mélanges Paul Thomas, pp. 160–171, Bruges, 1930). Joseph Needham, *A history of embryology* (Cambridge: University Press, 1934), pp. 13–19 [*Isis* 27, 98–102 (1937)].

[75] Book II, LXI–LXVI.

[76] End of LXXXVII.

represents the first attempt to explain rationally the mysteries of dreamland and to apply them to healing purposes. The author of that book was a distant ancestor of Freud.

Some of the dreams considered are relative to celestial phenomena (one may see the Sun or the Moon in one's dream). It is striking that the author does not classify such dreams with those of divine origin, but with the physiologic ones. From that point of view alone it is not correct to suggest (as Jones did)[77] that *De insomniis* is the first occurrence in classical literature "of a supposed connection between the heavenly bodies and the fates of individual human lives." Moreover, it is not certain that that treatise is older than the *Epinomis* of Plato, or even than its posthumous publication by Philip of Opus.

The *De insomniis* was one of the earliest Hippocratic books to appear in printed form, its Latin version being printed separately in Rome in 1481, then added to the early editions of the *Aphorismi* of Maimonides and of the *Liber Almansoris* of al-Rāzī (Klebs, 517, 644.2, 826.2–3); in all, four incunabula, ranging from 1481 to 1500.

20. *On winds* or *Breaths; De flatibus; Peri physōn.*[78] This book, which dates probably from the early Hippocratic age, helps us to realize the high complexity of medical thought in that age. It is for that very reason that it is so useful to consider separately so many writings. The complexity of medical thought is not astonishing if one remembers that it was an age of great curiosity and intellectual effervescence. Medical observations were accumulating in certain favored places and intelligent physicians tried to put them in order on the basis of their philosophic conceptions. Their philosophic background was rarely, if ever, homogeneous, for by the end of the fifth century they had been submitted to many different influences. The thoughtful physician, facing insoluble problems, tried to solve them from the point of view that seemed the most promising to him.

Anaximenes had reached the conclusion that air (*pneuma*) was the original principle; this point of view had been applied to physiology by Diogenes of Apollonia. Of course the importance of air was obvious enough. Think of wind in all its varieties, the gentle breezes of spring, the sudden squalls of summer, the biting gales of winter, the deadly storms, think of earthquakes;[79] in the human body, a need of free air was as evident as the danger of a lack of it or of an imperfect circulation of it. The physician could observe the normal breathing of healthy people, the difficult breathing of sick ones, the agonies of incipient suffocation, he could observe also eructations, flatuses, borborygmus, crepitus ventris; he was familiar with the pains of flatulency. Truly air (*pneuma*) was one of the conditions of life, and when a man had given out his last breath, he died. Perhaps the soul (*anima*) was a kind of air?

The author of the book on *Breaths* was not a Hippocratic physician and perhaps he was not a physician at all; he was certainly a sophist, one especially interested in the facts of life and health. His book is a kind of discourse, the burden of which

[77] Jones, Loeb *Hippocrates*, vol. 4, p. lii.

[78] Littré, vol. 6, pp. 88–113; Loeb, vol. 2, pp. 221–253; *CMG*, vol. 1, part 1, pp. 91–101.

[79] Earthquakes being frequent in the Mediterranean area, early philosophers like Anaximenes, Anaxagoras, Democritos had tried to give a rational explanation of them. According to Aristotle (*Meteorologica*), who discussed their views, earthquakes and volcanic phenomena are caused by underground winds. Archibald Geikie, *Founders of geology* (London, 1905), pp. 13–14.

is that all the diseases are caused by air, and more particularly by the kind of air that is in living bodies (*physa*). It is possible that other Hippocratic treatises like the *Nature of man* and *Ancient medicine* were written partly to refute him (and his kind).

It is worth while to compare the pneumatic ideas represented by the *De flatibus* with similar ideas in early Sanskrit literature. The comparison has been made by Jean Filliozat,[80] who quotes and translates relevant texts taken from Caraka, Bhela, and Suśruta. These texts establish the Hindu theory of pneumatism, the essential virtue of "winds" in the whole of nature as well as in living bodies, in short the same general conception as is crystallized in the various meanings of the words *pneuma*, *anima*, and *spiritus*. It is impossible, however, to prove any derivation from Sanskrit into Greek or vice versa. The main ideas are common, but many others are different; there are no textual identities. The alikeness of the Greek and Hindu traditions may be explained by a vague diffusion of ideas, for there were many contacts between India and Greece before Alexander, but it may be explained also by independent cogitations on facts of common experience: the need of "winds" in nature and in our own bodies, and the troubles that such "winds" may occasionally cause, are too obvious to escape observation.

The *De flatibus* was often published in Greek and Latin in the sixteenth century. The most recent edition of the Greek text, in addition to the Loeb and *Corpus medicorum graecorum* editions, is the one by Axel Nelson, *Die Hippokratische Schrift* Peri physōn (Uppsala, 1909). This includes two Renaissance Latin translations by Francesco Filelfo (1398–1481) and Janus Lascaris (1445–1535).

APHORISTIC WRITINGS

A number of books of the Hippocratic corpus may be grouped together because of their composition in the form of brief aphorisms that have been collected under a single title with very little or no order. We have already come across one of them, *Nutriment*.

The oldest of these writings was probably the *Cnidian sentences*, which is lost but the very title of which suggests that it was a collection of aphorisms, summarizing the wisdom of the Cnidian physicians (there are other Cnidian writings in the Hippocratic corpus, for the two schools of Cos and Cnidos were very close to each other, and thus Cnidian books would naturally be found in the Coan library). One might claim that aphoristic books must be early because the use of proverbs is a primitive form of expression. It is almost certain that some such collections are early but one must beware of generalization; the love of proverbs and aphorisms is common to all peoples and ages, with ups and downs perhaps, but with no stops. Jones [81] would place all the aphoristic books of the Hippocratic corpus in the second half of the fifth century, approximately in this order: *Prorrhetic I*, 440; *Aphorisms*, 415; *Coan predictions*, 410; *Nutriment*, 400; *Dentition*, later(?). With the exception of *Nutriment* which has already been dealt with, I shall consider them in that order.

Poetry and proverbs are the earliest forms of literature in every nation. Aphoristic statements have the advantage of being easy to memorize, and the people

[80] J. Filliozat, *La doctrine classique de la medecine indienne* (Paris: Imprimerie Nationale, 1949), pp. 161–190 [*Isis 42*, 353 (1951)].

[81] Jones, Loeb *Hippocrates*, vol. 2, p. xxviii.

repeating them give themselves without trouble an air of knowledgeableness and wisdom. The success of the medical aphorisms of the fifth century was caused not only by the popular love of proverbs but also by the aphorisms of Heracleitos and other philosophers, and the poems of Pindar and of other interpreters of Greek ideals. It was tempting to quote the most significant lines of a great poem, and such lines, often repeated, became aphorisms. Even so today, many people express their feelings by means of proverbs, or they quote a line of the Bible or of Shakespeare. That is easy enough, and it is pleasant.

21. *Prorrhetic I; De praedictionibus; Prorrhēticon a'.*[82] This is a collection of medical aphorisms arranged without any order. It includes 170 short aphorisms, out of which only seventeen (one-tenth) are exclusive to it. Indeed, most of the collection was incorporated into the Coan prenotions.

One of the aphorisms[83] has been the origin of considerable discussion: "Phrenetics drink little, are disturbed by noise and are quivering." The word "drink little" (*brachypotai*) is the bone of contention. If it is understood as a reference to hydrophobia (rabies), then that disease is not new but very ancient. A passage of Aristotle refers clearly to hydrophobia, though it ends with an error.[84]

Prorrhetic I is very different from *Prorrhetic II*, the writing of the latter being as good as that of the former is poor. See sec. 4.

22. *Aphorisms; Aphorismi sive sententiae; Aphorismoi.*[85] This is the most popular book of the whole corpus, its popularity being partly due to the love of all peoples for "compressed wisdom," wisdom in tablets that can be easily swallowed. Its popularity is attested by the abundance of manuscripts in many languages,[86] and the number of commentaries, supercommentaries, and imitations. The most famous of the imitations was the *Kitāb al-fuṣūl fī-l-ṭibb* of Maimonides (XII-2), which was itself the beginning of a new tradition.

The *Aphorisms* were first printed (in Latin) in 1476, and innumerable editions have appeared ever since in many languages. Until the eighteenth century, every educated physician owned a copy of the *Aphorisms* and used it as a kind of medical breviary.

The collection as we have it is divided into seven sections containing a total of 412 aphorisms, irregularly distributed[87] among them and without order, except that one sometimes comes across a small group related to the same subject. They deal with almost every medical subject except surgery. Some of the *Aphorisms* occur in other Hippocratic books; for example, sixty-eight of them may be read also in the *Coan prenotions*.

A work of this kind defies analysis and the best we can do is to quote a few specimens.

The first aphorism is generally known, not only to physicians but to educated people in general; most people know only the first sentence however. They do not

[82] Littré, vol. 5, pp. 504–573.

[83] *Prorrhetic I*, 16 = *Coan prenotions*, 95.

[84] Aristotle, *Historia animalia*, VIII, 22, 604A, "Dogs suffer from three diseases: rabies, quinsy, and sore feet. Rabies drives the animal mad, and any animal whatever, excepting man, will take the disease if bitten by a dog so afflicted; the disease is fatal to the dog itself, and to any animal it may bite, man excepted."

[85] Littré, vol. 4, pp. 450–609; Loeb, vol. 4, pp. 98–221.

[86] At least 140 manuscripts in Greek, 232 in Latin, 70 in Arabic, 40 in Hebrew; these total 482, and there are many others in other languages.

[87] Section 1 contains the fewest (25), section 7 the most (87).

know the second, which is independent of the first (perhaps two different aphorisms got stuck together in the manuscript tradition), and expressed one of the fundamental tenets of Hippocratic medicine.

Life is short, the Art long, opportunity fleeting, experience treacherous, judgment difficult. The physician must be ready, not only to do his duty himself, but also to secure the co-operation of the patient, of the attendants and of externals.[88]

The following aphorism deals with the regimen of athletes; it is not quoted completely.

In athletes a perfect condition that is at its highest pitch is treacherous. Such conditions cannot remain the same or be at rest, and, change for the better being impossible, the only possible change is for the worse. For this reason it is an advantage to reduce the fine condition quickly, in order that the body may make a fresh beginning of growth. But reduction of flesh must not be carried to extremes, as such action is treacherous; it should be carried to a point compatible with the constitution of the patient . . .[89]

Here are a few others, taken almost at random:

Old men endure fasting most easily, then men of middle age, youths very badly, and worst of all children, especially those of a liveliness greater than the ordinary.

Bodies that are not clean, the more you nourish the more you harm.

It is a good thing when an ophthalmic patient is attacked by diarrhoea.

Such as become hump-backed before puberty from asthma or cough, do not recover.[90]

Such a collection is like a building the stones of which have not been cemented together. There are many variations in the editions and translations, because it was easy enough to interpolate new aphorisms or to leave out those for which the editor did not care.

See the last section of this chapter, on the medieval tradition of Hippocrates.

23. *Coan prenotions; Praenotiones Coacae; Cōacai prognōseis.*[91] This work is divided into seven sections, like *Aphorisms*, and contains 640 aphorisms arranged without any order. Many of them invite medical commentary, and Littré quotes medical cases of his own time to illustrate those referred to by the Coan physician.

24. *Dentition; De dentitione; Peri odontophyiēs.*[92] This collection of thirty-two aphorisms deals with the hygiene and treatment of infants and especially with teething. It may be divided into two parts, the first (1–17) concerned with dentition (*odontophyia*), the second (18–32) with ulceration of the tonsils (*paristhmia*), uvula, and throat. It may be that *Dentition* has been extracted from a larger collection by an editor whose interest was narrowed to pediatrics. As such, it is the earliest treatise restricted to that branch of medicine, though there are of course pediatric remarks in many other books of the corpus.

88 *Aphorisms*, I, 1.

89 *Ibid.*, I, 3.

90 *Ibid.*, I, 13; II, 10; VI, 17; VI, 46. The last is a short description of Pott's disease, named after the English surgeon Percival Pott (1714–1788).

91 Littré, vol. 5, pp. 574–733.

92 Littré, vol. 8, pp. 542–549; Loeb, vol. 2, pp. 317–329.

DEONTOLOGY

It is natural to group together a number of texts concerned with the duties of physicians and the proper way of dealing with patients. The composition of those books seems to suggest that the physicians were beginning to organize themselves into a professional body having definite obligations and privileges. We have no other proof of the existence of such a body and hence it is impossible to say how far the organization went. It may have been a guild, or more probably an informal group, of the elder physicians, their younger associates, and apprentices. The earliest and most significant of these texts is the famous *Oath of Hippocrates*.

25. *The oath; Iusiurandum; Horcos.*[93] This is the oath that was taken by the apprentices before they were accepted as members in the guild or society of Coan physicians. According to the first sentence, it was not only an oath but an indenture (*syngraphē*); the apprentice undertook to treat the children of his master as if they were his brothers, to share his livelihood with his teacher and help him in case of need, to teach his master's children without fee or indenture, to give full instruction to his own children, his master's children, to a few other students who had taken the oath and signed the indenture, and to no others. That is, not only was the profession organized but also its continual monopoly was protected. Thus was medical teaching established on a guild basis.

It is impossible to determine the date of the *Oath*, but it was probably administered from the golden days of the Coan school.

One passage is very puzzling: "I will not use the knife, not even, verily, on sufferers from stone, but I will give place to such as are craftsmen therein." It has been suggested that it was not lithotomy that was forbidden but castration; Greek physicians were not afraid, however, of using the proper word. The idea that surgery should not be permitted to the physician, but abandoned to inferior assistants, does not tally with what we know of Hippocratic surgery. The prejudice against surgery is not ancient, but medieval. In modernized editions this passage is generally left out.

The *Oath* is the fundamental document of medical deontology. Its popularity was enormous, for it always was an intrinsic part of the corpus, and what is more, the ideals that it defended were accepted by almost all the medical schools in the Greco-Arabic-Latin tradition down to our own day. For the history of it, see W. H. S. Jones, *The doctor's oath* (61 pp.; Cambridge, 1924) [*Isis 11*, 154 (1928)]; Ludwig Edelstein, *The Hippocratic oath. Text, translation, and interpretation* (70 pp.; Baltimore: Johns Hopkins University Press, 1943) [*Isis 35*, 53 (1944)], and various queries in *Isis*: *20*, 262 (1933–34); *22*, 222 (1934–35); *32*, 116 (1947–49); *38*, 94 (1947–48). As to the perpetuation of the oath, with necessary modifications, to our own day, see *Isis 40*, 350 (1949). There are some nine incunabula of the Latin text (see Klebs), and the first edition of the Greek text appeared in 1524, together with the text of Aisopos[94] and with Latin translations by Niccolò Perotti of Sassoferrato (1430–1480).

[93] Littré, vol. 4, pp. 628–633; Loeb, vol. 1, pp. 291–301; *CMG*, vol. 1, part 1, pp. 4–6.

[94] Aesōpos, traditional author of the Greek fables, the history of which is inextricable. According to Herodotos (II, 134), Aesopos the story writer (*ho logopoios*) was a slave in Samos during the reign of Amasis (king of Egypt, 569–525). A biography of him was written by Maximos Planudes (XIII-2). Ben Edwin Perry, *Studies in the text history of the life and fables of Aesop* (256 pp., 6 pls.; Haverford, Pennsylvania: American Philological Association, 1936). Article "Fable," *Oxford classical dictionary*, p. 355.

26. *Law; Lex; Nomos.*[95] This text, which is not much longer than the *Oath* (less than two pages in Greek), is much younger than the latter, for it shows traces of Stoic influence. It was known to Erotianos. It is less matter-of-fact or businesslike than the *Oath*, but more philosophic, and it is elegantly written. It aims to delineate the education of a good physician, and suggests that by the time it was written the medical guild had become a kind of secret brotherhood.

Let us quote the first and the two final sections:

Medicine is the most distinguished of all the arts, but through the ignorance of those who practice it, and of those who casually judge such practitioners, it is now of all the arts by far the least esteemed. The chief reason for this error seems to me to be this: medicine is the only art which our states have made subject to no penalty save that of dishonor, and dishonor does not wound those who are compacted of it. Such men in fact are very like the supernumeraries in tragedies. Just as these have the appearance, dress and mask of an actor without being actors, so too with physicians; many are physicians by repute, very few are such in reality . . .

These are the conditions that we must allow the art of medicine, and we must acquire of it a real knowledge before we travel from city to city and win the reputation of being physicians not only in word but also in deed. Inexperience on the other hand is a cursed treasure and store for those that have it, whether asleep or awake; it is a stranger to confidence and joy, and a nurse of cowardice and of rashness. Cowardice indicates powerlessness; rashness indicates want of art. There are in fact two things, science and opinion; the former begets knowledge, the latter ignorance.

Things however that are holy are revealed only to men who are holy. The profane may not learn them until they have been initiated into the mysteries of science.

There are eight incunabula of the Latin version (Klebs).

27. *The physician; De medico; Peri iētru.*[96] This book was not mentioned by the ancients, such as Erotianos and Galen, but it has many affinities with writings of the corpus. Only the first chapter is deontologic; it describes the character of a good physician for the body and the soul. There are fourteen chapters in all, explaining principles of medical practice, how to arrange the surgery and all instruments and other objects that are needed in it, how to dress and bandage wounds, how to cup patients, and so on; the last chapter is devoted to military surgery, which can be learned only in the field. It is very practical. The anatomic basis is very poor, which suggests an early Hippocratic date.

28. *Decorum; De decenti habitu; Peri euschēmosynēs.*[97] The poor language of this text as well as its affectations (use of rare words) suggests that it is relatively late. Moreover, it is colored with Stoic ideas, and some chapters (out of eighteen) are artificial and (intentionally?) obscure, all of which is not Hippocratic in the good sense. Nevertheless, the subject is interesting. The author explains how the physician should behave at the bedside for the patient's good and for his own reputation. The physician should be not a sophist but a wise man, gracious and truthful. "A physician loving wisdom is godlike" (*iētros gar philosophos isotheos*).[98] In chapter vi, which is unfortunately marred with obscurities, the author insists on the importance of religion; this passage is unique in the corpus. Many practical details

[95] Littré, vol. 4, pp. 638–643; Loeb, vol. 2, pp. 257–265; *CMG*, vol. 1, part 1, pp. 7–8.

[96] Littré, vol. 9, pp. 198–221; Loeb, vol. 2, pp. 305–313, chap. i only; *CMG*, vol. 1, part 1, pp. 20–24.

[97] Littré, vol. 9, pp. 222–245; Loeb, vol. 2, pp. 269–301; *CMG*, vol. 1, part 1, pp. 25–29.

[98] Chap. v.

are given concerning the observations to be made in the dispensary or at the bedside, the preparation of drugs, and so forth. It is necessary to visit patients frequently, and sometimes to leave an apprentice in charge during the physician's absence.

29. *Precepts; Praecepta; Parangeliai.*[99] This seems to be a late compilation, as late perhaps as the Roman age, though pre-Galenic. It is full of obscurities and the style is at once poor and pretentious; the first two chapters have an Epicurean coloration.

The largest part (chaps. 3–13, out of fourteen) is deontologic, dealing with medical etiquette, the avoidance of charlatanry and of quack's patter. (Maybe the itinerant quacks had already learned the art of haranguing the people and puffing their wares when they reached a village.) Chapters 1 and 2 constitute a kind of introduction: the medical art must be based on observations, not "hypotheses." The last chapter is a collection of unrelated sentences; these may be notes which the author had no opportunity of working out.

Chapter 6 of *Precepts* has been quoted in full in Chapter XIII (see p. 345).

LETTERS

30. *Apocryphal letters.* The ninth volume of Littré's edition (pp. 308–466) contains letters and other documents that are apocryphal yet interesting for the study of the development of the Hippocratic legend. Some of the letters tend to show that Hippocrates saved Athens and Greece from the plague, and this would have been known otherwise if it had been true. Among the correspondents are the great King Artaxerxes, Hystanes, Persian governor of the Hellespont, the citizens of Cos and of Abdera, Hippocrates' son Thessalos, King Demetrios. Long letters between Hippocrates and Democritos deal with the alleged madness of the latter.

It is noteworthy that the ancient scholars wanted to complete the *opera omnia* of great men with "authentic" letters (cf. Plato, Aristotle); they could not have the documents that modern editors can so easily collect, but they found it permissible to "create" the letters that they needed. After all, writing a plausible letter, or one that seemed plausible to them, could not be much worse than writing "speeches," as was the accepted habit of the ancient historians, including so truthful a man as Thucydides.

Latin translations of some of the letters were printed as early as 1487 and 1492 (Klebs, 337) together with letters of Diogenes of Sinope, the Dog (*c.* 400–325), founder of the Cynic sect.

The reader who has been patient enough to follow me in this examination of the most significant Hippocratic writings will realize the richness and complexity of their contents. The mass of them was written in the fifth century; others were later by a century or more, yet continued a great tradition, one of the noblest traditions in the history of mankind.

THE MEDIEVAL TRADITION OF HIPPOCRATES

The greatness of a man can be deduced from the size of the shadow that he casts ahead of him throughout the ages. In order to understand the greatness of Hippoc-

[99] Littré, vol. 9, pp. 246–273; Loeb, vol. 1, pp. 305–333; *CMG*, vol. 1, part 1, pp. 30–35.

rates it is necessary to appreciate the influence that he exerted upon his posterity. We attempt to give an account of the events in their chronologic order, and in that order "Hippocrates" appears in the second half of the fifth century, but we should realize that what Hippocrates, whoever he was, did in that period was only the beginning of a very long story. If that story were written it might be entitled "The life of Hippocrates from the fifth century B.C. until to-day," and if it were told with any completeness it would fill an enormous book. Great men are truly immortal; they may be more alive after their death than before.[100]

The study of Hippocratic traditions is peculiarly complex, because the Hippocratic writings do not form a single solid block, like those of Herodotos and Thucydides, or like the *Iliad* or the *Odyssey*. The many writings, genuine or not, are not closely integrated by a rigid canon, as is the case for the Bible. One has to consider the tradition of each item, or of each group of items. Some items were brought together by the care of early librarians, copyists, and editors, and also by the curricula of medical schools. For example, the *Aphorismi*, *Prognosticum*, and *Regimen acutorum* (*De diaeta in acutis*) were often combined, as was the case in the school of Montpellier in 1309 and 1340.[101]

For the sake of illustration we shall outline the tradition of a single book, the most popular of all, the *Aphorisms*.

Galen wrote commentaries on some seventeen Hippocratic books,[102] and *Aphorisms* is one of them; thus in this case, as in many others, the Galenic tradition is combined with the Hippocratic and reinforces it. The early medieval tradition of Galen is fortunately well known because of a treatise written by one of the greatest philologists of the Middle Ages, Ḥunain ibn Isḥāq al-'Ibādī (IX-2), called in Latin Joannitius, who flourished in Jundīshāpūr, then in Baghdād, and died in 877. Ḥunain was a Nestorian, a physician, a translator from Greek into Syriac and into Arabic; he himself translated many of the scientific classics written by Hippocrates, Plato, Aristotle, Dioscorides, Ptolemy, and Galen, directed a school of translators, and transmitted to them an admirable discipline. The treatise of his to which I just referred is a survey of the Syriac and Arabic translations of Galen, wherein he appreciates the comparative value of those translations and does not hesitate to criticize severely some of his own.[103]

Here is what he says of the *Aphorisms*:

> Hippocrates' explanation of the book of Aphorisms (Tafsīr li kitāb al-fuṣūl). This book is divided into seven parts.[104] It was poorly translated [into Syriac] by Ayyūb; Jibrīl ibn Bakhtyashū' tried to improve the translation but made it worse. Therefore, I have collated it with the Greek text, corrected it to the extent of making a new

[100] *Introduction*, vol. 3, p. 10.

[101] *Introduction*, vol. 3, pp. 247–248.

[102] These 17 works constitute, if not a canon, at least a definite group, each item of which might fall under the notice of any student of Galen. These works are: *De officina medici, Prognosticum (Praenotiones), De diaeta in acutis, Prorrhetic (Praedicta), Epidemiorum libri, De fracturis, De articulis, De natura hominis, De humoribus, De alimento, Aphorismi, De salubri victus ratione* (all these are in Kühn's edition of Galen; all of them except the last are in Hunain's list), *De capitus vulneribus, De aëre aquis locis, Iusiurandum, De ulceribus, De natura pueri.*

[103] It was published in Arabic and German by Gotthelf Bergsträsser (1886–1933), *Ḥunain ibn Isḥāq über die syrischen und arabischen Galen-Uebersetzungen* (Leipzig, 1925), and summarized by Max Meyerhof (1874–1945) in *Isis* 8, 685–724 (1926). My references to either edition are indicated thus: Ḥunain, No. *x*.

[104] The Arabic word *maqāla* is used to translate the Greek *tmēma* (section); in Latin one used the word *liber*. These three words are equivalent yet different metaphors.

[Syriac] translation possible, and added the text of Hippocrates' own words. Aḥmad ibn Muḥammad al-Mudabbir had asked me to translate it for him. I translated a single part into Arabic. He then proposed to me not to begin the translation of another part, before having read to him the part already translated; he was kept busy otherwise, however, and therefore my translation was interrupted. Yet as Muḥammad ibn Mūsā examined each part he begged me to continue my work and thus have I completed my translation of the whole.[105]

Ḥunain does not refer to a translation by Sergios of Resaina (VI–1), who was one of the earliest and greatest translators from Greek into Syriac. Sergios had studied in Alexandria and died at Constantinople in 536; he was not a Nestorian like Ḥunain but a Monophysite.[106] He may have translated the *Aphorisms* (not Galen's explanation of them), but that is doubtful.[107]

Curiously enough, I find no trace of special interest in the *Aphorisms* for the period of almost one and one-half centuries from Ḥunain's death in 877 to about 1025. By the middle of the eleventh century, at least two Arabic commentaries were written, the first by the Egyptian 'Alī ibn Ridwān (XI–1) and the second by the Persian 'Abd al-Raḥmān ibn 'Alī ibn abī Ṣādiq,[108] both of whom died about 1067.

A century later the Spaniard Yūsuf Ibn Ḥasdai (XII–1) wrote another Arabic commentary entitled *Sharḥ al-fuṣūl*. After this the translations and commentaries increase in number, so much so that it will be convenient to deal with them in successive half century periods.

Second half of the twelfth century. One of the dominating personalities of this age is another Spaniard, the Jew Maimonides (XII–2). The most important as well as the most famous of his medical works is another collection of aphorisms, generally called *Fuṣūl Mūsā*, derived almost exclusively from Galen.[109] His commentary on the *Aphorisms* of Hippocrates is a different work, much less known. Though the *Fuṣūl Mūsā* are derived from Galen, it is probable that they include here and there remarks concerning directly or indirectly the *Aphorisms* of Hippocrates.

Both Ibn Ḥasdai and Maimonides spent the best part of their lives not in Spain

[105] Translated from Bergsträsser's Arabic text (Ḥunain, No. 88). Ayyūb al Ruhāwī al-Abrash (IX–1), Job of Edessa the Spotted, was a translator from Greek into Syriac; Jibrīl ibn Bakhtyashū' (IX–1), another translator from Greek into Syriac; Aḥmad ibn Muḥammad al-Mudabbir, a great administrator and patron of science; see *Isis* 8, 715 (1926). Muḥammad ibn Mūsā was one of the Banū Mūsā, that is, one of the three sons of Mūsā ibn Shākir (IX–1); they patronized the translations into Arabic; Muḥammad lived until 872/3.

[106] The orthodox point of view concerning Christology is that there are in Christ two natures (human and divine) but one person. The Nestorians claimed that there are two natures and two persons; they were condemned by the council of Ephesos in 431. The Monophysites went to the other extreme and claimed there is in Christ but one nature and one person; they were condemned by the council of Chalcedon in 451. The transmission of Greek science to the Islamic world was largely effected by these two (opposite) groups of Christian heretics, Nestorians and Monophysites. Those of Asia used the same language, Syriac, but two different scripts; *Introduction*, vol. 2, p. 501. There are thus *two* Greek-Syriac-Arabic traditions, duplicating or completing each other. We cannot go into the details of that; this is done in my *Introduction*.

[107] Henri Pognon, *Une version syriaque des Aphorismes d'Hippocrate* (2 vols.; Leipzig, 1903), Syriac-French edition. Pognon suggests that the Syriac text might have been written by Sergios and even earlier (vol. 1, p. xxx) but he does not prove it.

[108] Not dealt with in my *Introduction*. There is a copy of 'Abd al-Raḥmān's commentary on the *Aphorisms* in the Escorial. See H. P. J. Renaud's catalogue (Paris, 1941) No. 877 [*Isis* 34, 34–35 (1942–43)].

[109] So much so that Latin writers like Jean de Tournemire (XIV–2) called it *Flores Galieni*. For various editions in Arabic, Hebrew, Latin of the *Fuṣūl Mūsā* see *Introduction*, vol. 2, p. 377, No. 8, and *Osiris* 5, 109 (1938), Figs. 28–29. Maimonides' collection was much larger than Hippocrates', containing some 1500 aphorisms as compared with 412.

but in Egypt. A third Spaniard, or more exactly a Catalan, Joseph ben Meïr ibn Zabara (XII–2), who studied in Narbonne but resided mostly in his native city, Barcelona, may be the author of a satiric parody of the *Aphorisms* in Hebrew, *Momeri ha-rofe'im.*

In the meanwhile Burgundio of Pisa (XII–2) translated the *Aphorisms* directly from Greek into Latin, and the anatomist Maurus of Salerno (XII–2) wrote a Latin commentary on them. As Maurus died some 20 years after Burgundio (1214 compared with 1193), he might have used Burgundio's translation instead of previous ones made from the Arabic, but this is not clear from his text without deeper study of it than was possible to me.[110]

First half of the thirteenth century. My notes on the first half of this century concern only Arabic writings, written in Damascus, or at least by physicians who flourished in that city.

Three commentaries on the *Aphorisms* were written by two Muslim physicians, Ibn al-Dakhwār, who died in Damascus in 1230,[111] and Ibn al Lubūdī of Aleppo (XIII–1), who was educated in Damascus and died after 1267, and by the Samaritan physician Ṣadaqa ben Munaja' al-Dimishqī (XIII–1); Ṣadaqa's commentary is entitled *Sharḥ fuṣūl Buqrāṭ.*

Second half of the thirteenth century. In the second half of the century, the *Aphorisms* attract the attention of every physician west of India and are discussed in Arabic, Hebrew and Latin.

Arabic commentaries were written in Arabic by two Eastern physicians, the Christian, Abū-l-Faraj, called Barhebraeus (XIII–2),[112] and the Muslim, Ibn al-Nafīs (XIII–2).

In Latin, commentaries were given to us by the Portuguese Peter of Spain of Lisbon (XIII–2), who died as Pope John XXI in 1277, and by the Italian Taddeo Alderotti of Florence (XIII–2), who lived until 1303.

There are at least five Hebrew translations of the *Aphorisms.*[113] The most interesting is the one that was completed in Tarascon in 1267 by Shem-ṭob ben Isaac of Tortosa (XIII–2). This affords a good illustration of the vicissitudes of literary tradition. Shem-ṭob's Hebrew text includes commentaries by Palladios the Iatrosophist (V–1), which are unknown in the Greek original. Moses ibn Tibbon of Marseille (XIII–2), one of the greatest medieval translators, translated Maimonides' commentary from Arabic into Hebrew in 1257 or 1267. Nathan ha-me'ati of Cento (XIII–2), who flourished in Rome *c.* 1279–1283, translated the *Aphorisms* from Arabic into Hebrew, together with Galen's commentary.

First half of the fourteenth century. The last Arabic commentaries known to me date from this period and we owe them, curiously enough, to two Turkish physicians, 'Abdallāh ibn 'Abd al-'Azīz of Sīwās (XIV–1) and Aḥmad ibn Muḥammad al-Kīlānī (XIV–1). 'Abdallāh's commentary, dating from the beginning of the century, is entitled 'Umdat al-fuḥūl fī sharḥ al-fuṣūl. Aḥmad's was written

110 The text of the Glosule amphorismorum secundum magistrum Maurum was edited by Salvatore de Renzi, *Collectio salernitana* (Naples, 1856), vol. 4, pp. 513–557.

111 *Introduction,* vol. 2, p. 1099, note.

112 There is a possibility that the commentary ascribed to Barhebraeus was written by another Christian, also called Abū-l-Faraj but less famous, Abū-l-Faraj Ya'qūb Ibn al-Quff of Karak (XIII–2). In Renaud's catalogue of Escorial manuscripts, No. 878 is tentatively ascribed to Ibn al-Quff. Of course, it is possible that both men wrote a commentary.

113 *Introduction,* vol. 2, p. 846.

somewhat later, because it was dedicated to Jānī Beg Maḥmūd, who was khān of the Blue Horde of western Qipčāq in 1340–1357.

The production of Latin editions and commentaries was naturally increased by the growing needs of medical schools, especially the most important one of that time, the school of Montpellier in Aragon. The *Aphorisms* was one of the texts that medical students were expected to con.[114] Thus, we have Latin commentaries by Bartholomew of Bruges (XIV–1) who obtained his M.D. in Montpellier before 1315, by Berenger of Thumba (XIV–1), who was in Montpellier in 1332, and (perhaps) by Gerald de Solo (XIV–1), who was professor there and died *c.* 1360.

The medical school of Bologna was almost as important as that of its Aragonese rival, and we have two commentaries prepared by Bolognese professors, Niccolò Bertuccio (XIV–1) and Alberto de' Zancari (XIV–1). Alberto's commentary was rather a new edition, wherein the aphorisms were arranged for the first time in logical order: *Anforismi Ypocratis per ordinem collecti.*

Second half of the fourteenth century. The activities of Hebrew commentators seem to peter out like those of their Arabic rivals. I can quote only one Jewish commentator, the Catalan Abraham Cabret (XIV–2).

For the sake of curiosity we may mention also the summary of the Aristotelian Organon, *Minḥat Judah*, written by the Judeo-Greek philosopher and mathematician, Joseph ben Moses ha-Kilti (XIV–2), in the form of aphorisms, this being almost certainly a conscious or unconscious imitation of the Hippocratic work. Joseph flourished at the end of the fourteenth century or in the beginning of the fifteenth.

Martin de Saint Gilles (XIV–2), who was flourishing in Avignon in 1362, made a French translation of the *Aphorisms* together with Galen's commentary.[115] This introduces a new tradition and suggests that we might explore all the vernaculars of Europe into which the *Aphorisms* were translated sooner or later, but that would take us too far out of our own field. None of those vernacular traditions concerns the average historian of science, though they may be of very great interest to particular ones. For example, the story of the Polish translations is significant for the students of Polish science and of Polish letters.

The learned people of Western Europe did not need vernacular translations and looked down on them; the Latin text was more desirable to them and remained so for many centuries.

Marsiglio of Sancta Sophia (XIV–2), professor at Padua, wrote *Quaestiones in aphorismos* which were printed in Padua in 1485 and many times afterward.[116] Marsiglio died *c.* 1405.

This brings us into the fifteenth century, of which I have not made a sufficient study. Yet two commentators of the early part of that century may be quoted, Giacomo della Torre, and Ugo Benzi.[117] Both are children of the fourteenth century and their commentaries were very influential, because they were frequently printed.

The commentary of Giacomo della Torre, alias Jacopo da Forlì (*c.* 1350–1413), was first printed in Venice in 1473 and there are six incunabula editions of it; [118] the one by Ugo Benzi of Siena (*c.* 1370–1439) was first printed in Ferrara in 1493 and re-

[114] *Ibid.*, vol. 3, p. 248.

[115] Germaine Lafeuille is preparing a study of that French translation to appear in 1953–54.

[116] Klebs, 546.3–6.

[117] *Introduction,* vol. 3, p. 1195.

[118] Klebs, 476.

Fig. 74. Hippocrates' *Aphorisms*, the first independent edition, a Latin translation of the *Aphorisms* and of Galen's commentary by Laurentius Laurentianus of Florence, printed by Antonio Miscomini in Florence, 1494. It is a volume of 98 leaves without title page; we reproduce the colophon [*Osiris* 5, 100 (1938)]. [Courtesy of the British Museum.]

FINIVNT
Sententiæ Hippocratis Et Item Commentationes
Galeni In Eas Ipſas Sententias Editæ Laurentio
Laurentiano Florentino Interprete Viro Cla
riſſimo Quas Antonius Miſcominus
Ex Archetypo Laurentii Diligenter
Auſcultauit & Formulis Imprimi Curauit.
FLORENTIAE
Anno Salutis .M.CCCCLXXXXIIII.
Decimoſeptimo.kal.Nouembris

printed only once before the sixteenth century.[119]

Independently of the commentaries by Marsiglio da Sancta Sophia, Giacomo della Torre, and Ugo Benzi, the Latin text of the *Aphorisms* was printed at least eight times before the sixteenth century, six times in the *Articella*, 1476 to 1500, and in two other editions, 1494, 1496 (Fig. 74).[120]

Later editions in many languages are innumerable. Very long yet incomplete lists of them may be found in Littré [121] and in the catalogues of the British Museum and of the Bibliothèque Nationale of Paris.

Our own account of the tradition of the *Aphorisms* is also very incomplete for many reasons. To begin with, we could speak only of the Hippocratic commentators, of whom we know definitely that they have either translated the *Aphorisms* or commented upon them. The translations and commentaries here mentioned should be considered simply as specimens of a large class. A deeper cause of error lies in the fact that the indirect and hidden commentators were probably more numerous than the direct and obvious ones. To put it otherwise, many so-called commentaries or supercommentaries are more original than books that are supposed to be independent. That is true in all ages: the tradition of *X* cannot be deducted from the books definitely devoted to *X*, or even from those that quote *X*. Not only the plagiarists but mediocre minds in general are often as anxious to hide their sources as the river Nile; the more they steal the less are they inclined to confess their indebtedness.

A similar article could be written concerning the tradition of the other Hippocratic books, and indeed concerning the tradition of any scientific book of antiquity. One would discover great differences in popularity. The *Aphorisms* were one of the most popular; other books, which were early lost or forgotten, represent the other extreme. The pattern of each story would be the same, though the names of the actors would vary considerably. The tradition was international, interracial, interreligious. The main linguistic steps were Greek, Syriac, Arabic, Latin, Hebrew, vernaculars; the main religious steps were pagan, Muslim, Christian, Jewish.

119 Klebs, 1002. Dean Putnam Lockwood, *Ugo Benzi* (Chicago: University of Chicago Press, 1951) [*Isis 43*, 60–62 (1952)].

120 Klebs, 116.1–6, 520.1–2.

121 Littré, vol. 4, pp. 446–457.

XV

COAN ARCHAEOLOGY

The personality of Hippocrates dominates so completely the development of ancient Greek medicine, and it is so intimately connected with the island of Cos, that it is worth while to approach the subject from the archaeologic end.

In spite of its smallness, Cos was the cradle of many physicians,[1] and yet this is very puzzling. Hippocrates and his brethren seem to have practiced their art not so much in Cos as in other parts of Greece very distant from it. If we define Hellas, *stricto sensu*, as the islands of the Aegean sea and the lands surrounding it – Greece proper in the west, the Balkans in the north, Ionia in the east, Crete in the south – we find that Cos was near to the southeast corner of that area, and the Hippocratic physicians were exercising their skill in the northern part of it, in Thessaly, Macedonia, Thrace. Should one compile a list of the patients named in the clinical stories and of the places where cases were observed, one would find that Hippocratic experience was very largely gathered in the north (as defined above) and hardly at all in Cos. There are but two references to Coan patients in the corpus, the first to "the sister of the man of Cos" who suffered from an enlargement of the liver,[2] the second to Didymarchos in Cos.[3] The second case was observed in Cos, but we cannot say where the first was, for the "sister of the man of Cos" may have roamed far away from home. In another book[4] the wine of Cos, "astringent and very black," is twice recommended,[5] but wine could be very easily exported and if it was good we may assume that it was drunk outside of the island as much as inside. Hence, we have to face a paradox: the Hippocratic physicians are spoken of as representing the school or guild of Cos, yet as far as we can locate their activities they practiced elsewhere.

In order to solve that paradox, let us consider briefly the history of Cos. We have already indicated (p. 336) that the island was rich in produce, chiefly grapes and silk, but it is well to realize that its prosperity in Hippocratic days and later was not a novelty. Cos was not an upstart among the islands of that wonderful sea. Thanks to large deposits of obsidian, it was already a commercial center in the Stone Age.[6] Much of the obsidian was quarried in Cos itself, much also of a purer

[1] See index, under "Cos."

[2] *Epidemics II*, XXIII.

[3] *Prorrhetic I*, XXXIV.

[4] *De morbis internis*, XXV and XXX.

[5] The wine of Cos was famous. Says Strabon, XIV, 2, 19, "Cos is everywhere well supplied with fruits, but like Chios and Lesbos it is best in respect to its wine."

[6] Obsidian is a volcanic glass, very hard and very sharp, an excellent material for (Stone Age) tools.

quality in the little island of Hyali,[7] placed between Cos and the Cnidian chersonese. The obsidian trade gave that district (Cos and Cnidos) a kind of supremacy; it created wealth, and made possible the efflorescence of culture and learning. We may be sure that there were already practicing physicians in Cos long before the Dorian invasion.

The Dorians came probably from Crete about the ninth century, and displaced or dispossessed the native Carians. It may be that it was the Dorians who introduced the cult of Asclepios and thus gave a new prestige to the art of healing. On the other hand, Cos was magnificently located at the crossroads of many nations, so that its commercial importance was necessarily international. The Coan merchants had business dealings with Greece and Crete, with Caria and Ionia, with Asia and Europe. Their trade relations with the Ionian cities were so intense that in spite of the Dorian overlordship Cos became to some extent an Ionian city itself. At any rate, its higher culture was Ionian, not Dorian, and the Ionian dialect was considered to be the polite language.

The prosperity of the island and the international intercourse that it enjoyed were excellent conditions for the success of any kind of scientific effort. All that was needed was such an initial fermentation as is provided by the intervention of a man of genius. One of the Asclepiad families, the family of Hippocrates, provided that opportunity. It is not surprising then that the medical school which they created, or recreated, flourished as well as it did, and it would have continued to flourish but for the calamities of war.

The Ionization of the island had probably been facilitated by the Persian conquest. Under Darios (king of Persia, 521–485) Cos was part of a Persian satrapy, and the educated people, loving their Greek brethren and hating their Persian lords, would naturally rally around the Ionian teachers and affect Ionian speech and manners, which represented then and there the highest ideals of Hellas. After the naval victory of Mycale in 479 they threw off the Persian yoke and sooner or later were persuaded by the Ionians to join an Athenian confederation against Persia. As a result, they were involved in the Peloponnesian War, on the Athenian side. Indeed, Thessalos, son of Hippocrates, took part in the ill-fated Sicilian expedition (415–413). That period was a tragic one for Cos, for the island was devastated by an earthquake[8] and a little later invaded by the Spartans.

We may assume that the youth of the Hippocratic school in Cos coincided with that half century of peace between Mycale and the beginning of the Peloponnesian War. Hippocrates was educated during that period and revealed his genius, but his work and that of his disciples had to be continued elsewhere. The tumults[9]

[7] Hyali, from *hyalos* meaning rock crystal, glass; the island derived its name from its main source of wealth. The island is now called Istros.

[8] The earthquake of 413–12 was certainly not the first, and, as we shall see, it was not the last. The island's bad reputation as an earthquake center is confirmed by mythology. Polybotes, one of the giants who fought against the gods, was pursued by Poseidon (Neptunus) across the sea as far as Cos. The god of the sea was infuriated, broke off part of the island, threw it at Polybotes, and buried him under it! The popular inventors of that myth did not choose Cos at random; they chose Cos because of its known instability.

[9] The tumults were much aggravated by the heterogeneity of the Coan people. They were philhellenes, but in a distant way, and we may be sure that Dorian sympathies were not extinguished and that many of them were pro-Spartan. This was fully proved by the Social War, which began in 357 and was mainly directed against the Athenian protectorate. Cos allied itself with Mausolos, king of Caria, 377–353, who was anti-Athenian as well as anti-Persian. They concluded a peace with

caused by the war were not auspicious for scientific research and it is not surprising that Hippocrates and the other Asclepiads left their island home and began the life of wandering exiles. This explains the paradox that the Hippocratic teachings were formulated largely outside of Cos. It may explain another paradox, the persistence of Hippocratic positivism in spite of the Asclepiadic heritage. No matter how strong and pervasive the influence of Asclepios was, the Hippocratic doctors escaped it; instead of their allowing themselves to be subverted by magic rites the opposite happened, and the temple of Asclepios in Cos eventually capitalized on the fame of Hippocrates for its own religious purposes.

We cannot tell when the cult of Asclepios began in Cos, but the remains of the oldest temple there date back only to the third century, or say the end of the fourth. The site was thoroughly excavated in 1898 and the following years by members of the Archaeological Institute of Germany; after the first World War, when the Dodecanese was in Italian hands, new excavations were made by Italian archaeologists (Fig. 75). The sanctuary was not in the walled city of Cos, but about 1½ miles west of it on the slope of a hill. It was laid out on three artificial terraces. On the top, one may still see the remains of the Doric temple of Asclepios, with six columns on its short and eleven columns on its long sides. On the middle terrace there were smaller temples. The lower terrace was a promenade bounded by porticoes and there was a sacred well. Near the well was a little temple dedicated to Nero (emperor, 54–68), in the character of the god Asclepios, by a physician, C. Stertinius Xenophon.[10]

The earliest mention of the temple is relatively late, in the *Geographica*[11] of Strabon (I–2 B.C.). It reads: "In the suburb [of Cos] is the Asclepieion, a temple exceedingly famous and full of numerous votive offerings among which is the Antigonos of Apelles." Many of the inscriptions, of which the temple was full, have been preserved; they memorialize rites of purification, invitations to the festivals, decrees in honor of physicians of Cos, many of whom had obtained distinction in foreign service, and so forth. The "votive offerings" to which Strabon refers and which were probably far more abundant than the other inscriptions, represent another group of monuments, common enough in the sanctuaries of all countries and ages. The people who were sorrow-laden because of disease, infirmities, or other calamities appealed to the god and made vows; if they were healed and their troubles were removed they expressed their gratitude by means of an ex-voto. These monuments vary considerably in size, value, and contents. They may represent the god Asclepios, the snakes that were his attributes and the instruments of his grace, or the patient or, more specifically, the part of his body that had been cured. Among the old medical offerings there are some that

Athens in 355. Cos remained in Carian power until 346. Soon afterward, it fell under the control of Alexander the Great. After Alexander's death, Coan sympathies oscillated between Macedonia, Syria, and Egypt. The island obtained her main glory under the Ptolemies. In the first half of the third century she was graced by the presence of two poets, Philetas of Cos and his pupil, Theocritos of Syracuse. During the Roman period Cos enjoyed a kind of limited autonomy, being a *libera civitas* in the province of Asia. Claudius, emperor, 41–54, influenced by his physician Xenophon of Cos, granted various privileges to the island.

[10] C. Stertinius Xenophon is the same physician mentioned in footnote 9. He was archiater to Claudius and Agrippina and belonged to an old Asclepiad family. The first Xenophon of Cos was a pupil of Praxagoras of Cos (IV–2 B.C.); A. N. Modona, *L'isola di Cos*, p. 128. The stele bearing Xenophon's dedication is reproduced in Modona, pl. 8.

[11] Strabon, *Geographica*, XIV, 2, 19.

represent a pregnant woman, babies, eyes, womb and bladder, cancer of the breast, a dropsical body, hernia of the bowels.[12] One of the most beautiful medical ex-votos known to me is here reproduced (Fig. 76). It shows an old man holding in his arms an enormous leg with varicose veins. Votive offerings are so common everywhere that we may consider them characteristic of human nature; they are especially abundant in the Catholic churches, and the pilgrims to Lourdes can easily imagine how the Asclepieion of Cos looked, say, in Strabon's time. I call them characteristic of human nature, for imitation is almost certainly excluded: a grateful patient gives a pair of crutches to the sanctuary of Lourdes in the same spirit as he would have given them to the temple of Cos or Epidauros (Fig. 77).

We have definite ideas concerning the methods of treatment followed by the Hippocratic physicians; as shown in previous chapters, those methods were astoundingly rational. On the other hand, we know nothing except what votive

[12] Many are reproduced in T. Meyer-Steineg und Karl Sudhoff, *Geschichte der Medizin im Ueberblick* (Jena, 1921; [*Isis 4*, 368 (1921–22)]; ed. 2, 1922) [*Isis* 5, 188 (1923)]. William Henry Denham Rouse, *Greek votive offerings* (480 pp., ill.; Cambridge, 1902), or Rouse's article in *Encyclopedia of Religion and Ethics*, vol. 12 (1922), p. 641.

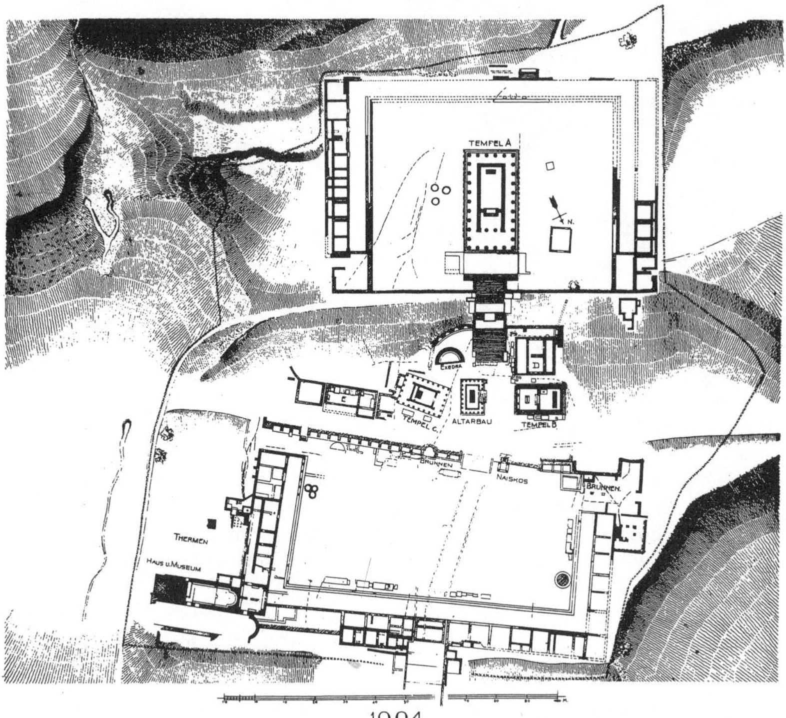

Fig. 75. Plan of the Asclepieion made by the German archaeologists in 1904. The three successive terraces are shown, the highest one being at the top of the figure. Italian archaeologists have excavated later a fourth terrace, which would be represented below the bottom part of this figure. [From Schazmann, *Asclepieion* (Berlin, 1932), pl. 37.]

Fig. 76. Ex voto. A man holding in front of him a gigantic leg with varicose vein. [*Mitt. krl. deut. Archaeol. Inst., Athenische Abt., 18* (Athens, 1893), pl. 11.] The original is preserved in the National Museum, Athens.

offerings tell us (and that is negligible) about the medical cures effected in the Asclepieion of Cos. It is probable, however, that that Asclepieion was somewhat controlled and its priests restrained by the lay practice flourishing in their neighborhood and by the Hippocratic ideals, that their methods were more rational (or less irrational) than those obtaining in other asclepieia, that they used more common sense and less magic, or that they used the latter less blatantly.[13] One cannot repeat too often that the essence of the temple practice (incubation, rest, and confidence) was rational and excellent; the irrationalities of Epidauros and other places were accretions caused by the credulity of the people and the greed of the priests.

All we can say is that no votive tablets have been excavated in Cos that are comparable to those that have been discovered in Epidauros. Here are three of these Epidauros inscriptions:

[13] I assume that the Asclepieion of Cos was guided and restrained by the Asclepiads. Ancient but late witnesses made the opposite assumption: the physicians had obtained their initial knowledge from the temple. Thus, Strabon (I–2 B.C.): "It is said that the dietetics practiced by Hippocrates were derived mostly from the cures recorded on the votive tablets in Cos (*Geography,* XIV, 2, 19). Pliny (I–2) makes a similar statement in *Natural history,* XXIX, 1(2), 4. They were very probably wrong, but I do not exclude the possibility of an exchange of good influences between temple and medical office.

Cleo was with child for five years. After she had been pregnant for five years she came as a suppliant to the god and slept in the Abaton.[14] As soon as she left it and got outside the temple precincts she bore a son who, immediately after birth, washed himself at the fountain and walked about with his mother. In return for this favor she inscribed on her offering: "Admirable is not the greatness of the tablet, but the Divinity, in that Cleo carried the burden in her womb for five years, until she slept in the Temple and He made her sound."

A man of Torone with leeches. In his sleep he saw a dream. It seemed to him that the god cut open his chest with a knife and took out the leeches, which he gave him into his hands, and then he stitched up his chest again. At daybreak he departed with the leeches in his hands and he had become well. He had swallowed them, having been tricked by his stepmother who had thrown them into a potion which he drank.

A man had his toe healed by a serpent. He, suffering dreadfully from a malignant sore in his toe, during the daytime was taken outside by the servants of the Temple and set upon a seat. When sleep came upon him, then a snake issued from the Abaton and healed the toe with its tongue, and thereafter went back again to the Abaton. When the patient woke up and was healed he said that he had seen a vision: it seemed to him that a youth with a beautiful appearance had put a drug upon his toe.[15]

[14] *Abaton* means the untrodden, inviolable (place), the holy of holies.

[15] The text is taken from Edelstein, *Asclepius* (vol. 1), § 423, Stele 1 of Epidauros, Nos. 1, 13, 17. That stele describes 20 cases; at the top of it is written: "God and Good Fortune. Cures of Apollo and Asclepios."

Fig. 77. Ex voto to Amphiaraos. Healing scene (National Museum, Athens). [From Maxime Gorce and Raoul Mortier, eds., *Histoire générale des religions* (Paris: Quillet, 1944), vol. 2, p. 137.]

Fig. 78. Asclepios with his main attribute, a snake entwined around his staff; bronze in Berlin Museum. [From W. H. Roscher, *Ausführliches Lexikon der griechischen und römischen Mythologie* (Leipzig, 1884–1890), vol. 1, p. 636.]

The snakes that were kept in the Asclepieia have already been mentioned thrice (chiefly p. 332). The presence of snakes and their medical use is a proof of the antiquity of the cult. The main attributes of the god Asclepios were a staff and a serpent, the latter generally entwined around the former. We need not worry about the exact meaning of those attributes, because the ancients did not agree in their explanations, and modern scholars cannot do better than pile up a series of guesses. It just happened that way. A dignified old man, with a full beard, holding a heavy staff around which a serpent seems to be gliding – that is unmistakably Asclepios, and don't ask any more questions (Figs. 78 and 79).[16]

The Asclepieion of Cos was famous in Hellenistic and Roman days, but it suffered at the hands of Christian iconoclasts in the fourth century and was destroyed in the earthquake of 554.

To the archaeologic evidence may be added two local traditions, which we are willing to accept, if not literally, at least as symbols of the gratitude and devotion of the sons of Cos to the most illustrious of their countrymen.

The first concerns the ancient plane tree that stands in the market place of the chief town of the island.[17] It is claimed that Hippocrates taught under its shade. The tree is certainly very old and its branches spread over the whole agora; it is supported by marble columns taken from the Asclepieion. It may be contemporary with Hippocrates, or it may be the offshoot of another tree that existed at the same place in Hippocrates' time. Who could tell? Remember the old trees in the Garden

[16] When the medical corps of the U. S. Army was organized, it chose as its emblem (embroidered upon uniforms, etc.) a wand with *two* serpents entwined around it. That was a mistake, for the caduceus was the staff of office not of Asclepios, the god of medicine, but of Hermes (Mercury), god of business and communications.

[17] A beautiful photograph of the tree forms the frontispiece to vol. 4 of the Loeb *Hippocrates*; it is described on p. lix.

Fig. 79. Homage to Asclepios' snake (Berlin Museum). [From Gorce and Mortier, *Histoire générale des religions*, vol. 2, p. 135.]

of Gethsemane which the Franciscan fathers say were contemporary with Christ. It is true, the Cos tree would have to be at least four centuries older than the olive trees of Jerusalem.

There is an islet off the southeast coast of Cos called Palaionisi; it is told that Hippocrates composed some of his writings in its seclusion.[18]

In short, the neighboring places, Cos and Cnidos, were the cradles of scientific medicine; thanks to the fact that the Asclepiad family of Hippocrates belonged to Cos, the island became more famous than its continental neighbor and almost eclipsed it. Hippocratic medicine began in Cos but developed chiefly in the north of the Greek area. It is possible that members of the family remained in Cos and continued the glorious tradition begun by Hippocrates. In the third century, the building of an Asclepieion (or of a new one vaster than the preceding) increased the prestige of religious healing. Scientific medicine and religious healing may have coexisted in Cos as they do in Boston.

The students of Greek medicine are more fortunate than those of Greek poetry, for they can see the place where Hippocrates grew up and dreamed, they can sit in the shade of an old plane tree and fancy that the master sat there twenty-five centuries ago, while it is impossible to visualize Homer's immediate surroundings.

For the study of Coan history and archaeology I have made use of the following publications:

F. H. Marshall, *Discovery in Greek lands* (Cambridge, 1920), pp. 82–84 [*Isis 4*, 59 (1921–22)];

Karl Sudhoff, "Cos and Cnidos," *Ann. Medical History* 2, 13–19 (1930) [*Isis 15*, 199 (1931)];

Archäologisches Institut des deutschen Reiches, *Kos. Ergebnisse der deutschen Ausgrabungen und Forschungen*, vol. 1, Paul Schazmann, *Asklepieion* (folio, 110 p., 57 pl., 1 map; Berlin, 1932);

Aldo Neppi Modona, *L'isola di Cos nell'antichità classica* (Rhodes: Memorie dell'Istituto storico di Rodi, 1933), vol. 1 (folio, 240 pp., 18 pls., 2 maps);

Emma J. and Ludwig Edelstein, *Asclepius. A collection and interpretation of the testimonies* (2 vols.; Baltimore: Johns Hopkins University Press, 1945) [*Isis* 37, 98 (1947)].

[18] The story was told in 1844 by natives to the German archaeologist Ludwig Ross (1806–1859).

PART THREE

THE FOURTH CENTURY

XVI

PLATO AND THE ACADEMY

POLITICAL BACKGROUND

The beginning of the new (fourth) century was dismal. The Peloponnesian Wars had ended in 404 with the surrender of Athens. Sparta had won, but it could not rule Greece without establishing garrisons in many cities and obtaining the help of oligarchies, powerful little groups of local "collaborators." Athens was prostrate; the Spartan domination was very hard to endure, not only in Attica but everywhere.

In the meanwhile, economic conditions had changed as fast and deeply as the political ones. The farms of Attica had been devastated during the war; the little farmers were the main victims; there appeared a new class of large landowners, manufacturers, bankers. Let us stop a moment to evoke one of these, Pasion, who had been a slave employed by other bankers, but manumitted by them as a reward for his zeal and fidelity. Pasion started a banking business of his own, together with a manufactory for the making of shields, and became the wealthiest man of his time; his benefactions to Athens were rewarded with the freedom of the city. When he died in 370 his freedman Phormion married his widow and took charge of his business and of his sons, Apollodoros and Pasicles. The former of these dissipated a good part of his patrimony. We are pretty well informed about Pasion, his business, and his family because of the law suits in which they were involved and the speeches of Isocrates and Demosthenes. The life of Pasion is very much like that of a self-made millionaire of today, and it throws light upon the capitalism that was growing in Athens while the government of the city and of other parts of Greece was festering.

Another result of the long wars was the existence of a relatively large body of veterans who had lost interest in peaceful arts and could not be easily reassimilated. Many of them became mercenaries, ready to take part in other people's wars in Egypt, Asia Minor, Persia. We shall come across later a body of them abandoned in the Tigris valley and obliged to fight their way back home under Xenophon's leadership.

The Spartans accumulated more hatred against them in a shorter time than the Athenians had done before and their supremacy did not last much more than thirty years (404–371). The common hostility was capitalized and organized by the Thebans, led by Epaminondas, the greatest tactician and one of the noblest men of his time, who caused the creation (in 370) of the Arcadian League against Sparta. Epaminondas invaded the Peloponnesos four times, and died in his last victorious battle at Mantineia (in Arcadia) in 362. In spite of her defeat, Sparta refused to accept the conditions of peace and more trouble followed, but Greek independence was almost over and the Greek cities fell now in the orbit of the growing Macedonian power.

This outline is restricted to the main facts, and leaves out the many little wars, the political intrigues, the alliances made and broken, the heroic deeds of courageous men and the crimes of greedy cowards and traitors. The warp and woof of the political life of Greece was so complex that a clear account would require considerable space; one would have to explain the troubles occurring within each city and the endless vicissitudes of their mutual relations. The main point is that the political web was disintegrating and breaking to pieces; the running down had become irremediable and irreversible.

And yet the spiritual life went on, though here too one could detect symptoms of disease. The mysteries, especially those of Eleusis, were flourishing; Orphism became almost a national religion. Foreign gods, imported from Egypt and Asia, were more welcome than ever. In spite of the efforts of Isocrates of Athens (436–338), national unity could not be realized and the Greeks were united only in their superstitions.

SCOPAS AND PRAXITELES

The earlier Attic school of sculpture, represented by Pheidias, serene and restrained, was followed by the school of Scopas and Praxiteles, whose works showed more individuality, sensibility, and emotion. The activity of Scopas of Paros lasted at least from 394 to 351 (it covers almost exactly the Platonic age); one of his last works was the frieze of the Mausoleum in Halicarnassos.

Praxiteles of Athens was a generation younger, for he was born *c.* 390, when Scopas had already completed his decoration of the temple of Tegea (in Arcadia). As far as can be judged from his dated works, he flourished about the middle of the century (356–346). His art was exceedingly gracious. His statue of Aphrodite (in Cnidos), an idealization of Phryne's[1] body, became the symbol of perfect beauty. His masterpiece, however, was the Hermes of Olympia. It must suffice to recall these glorious deeds in the briefest manner, and one should remember that the creation of beauty is not incompatible with political chaos.

We may now introduce Plato in those surroundings of confusion, terror, and beauty. We cannot understand him well unless we see him in the midst of them.

PLATO'S LIFE

Plato was born in Athens in 428; his father, Ariston, and his mother, Perictione, were members of aristocratic families and he was always very conscious of his noble ancestry. He received as good an education as a rich Athenian boy could obtain, and when he was about twenty met Socrates and was one of the latter's pupils for eight years. At the time of the master's execution (399), Plato and other disciples took refuge at Megara (about halfway between Athens and

[1] Phryne, the most famous of the Athenian *hetairai* (*courtesans*), was born in Thespiai, Boeotia. She inspired not only the sculptor Praxiteles but also the painter Apelles. It is said that after Alexander had destroyed Thebes in 336, she offered to rebuild the walls on condition that an inscription would record the deed: "Alexander destroyed the walls but Phryne, the *hetaira*, rebuilt them."

Corinth); one of those disciples was Euclid, who founded the school of Megara.[2] Plato did not stay there very long, and during the next dozen years (398–386) he traveled extensively in Greece, Egypt, Italy, Sicily. In 387 he was welcomed in Syracuse by the tyrant Dionysios (*c.* 430–367), who had pretensions to literary taste and claimed to be a philosopher. During his stay, he became very friendly with Dion of Syracuse and Archytas of Tarentum.[3] On his way back, Plato was captured by pirates, enslaved, and ransomed in Aegina. Soon afterward, in 387 – being then a man of forty – he began his teaching at the Academy. But for short absences (two visits to Syracuse, in 367 and 361) Plato spent the rest – the second half – of his life at the Academy. He died in Athens in 347, at the age of eighty-one.

THE ACADEMY (387 B.C. TO A.D. 529)

When Plato completed his *Wanderjahre*, he felt in him the vocation of teaching, but did not answer the call in the casual manner of Socrates; he realized the need of a school established in a definite place; he did not want to teach in the streets and markets, but on the contrary in a place that was sufficiently distant from the madding crowd and secluded. He chose a piece of land on the Cephissos, some 6 stadia from the Dipylon, the western gate of Athens.[4] The land originally belonged to the hero Academos[5] and the school was therefore called Academia. It is because of that accident, the use made of Academos' land by Plato, that the word "academy" has been included in almost every language of Europe; the fortune of that word would be a fair subject for a semantic study.[6]

The place was very wisely chosen by Plato, for it had been a kind of sacred place for a long time. Hipparchos, patron of letters (murdered in 514), younger

[2] Euclid's teaching combined Eleatic philosophy with Socratic dialectics and ethics. The Megaric or Dialectic school existed rather ingloriously until the end of the fourth century.

[3] The fruits of his acquaintance with Archytas will be considered in the next chapter; those of his friendship with Dion must be tasted right now. That friendship was ominous, for himself, for Dion, and for Syracuse. Dion was a relative and minister of Dionysios I; being influenced by Plato and probably full of hopes and good intentions, he tried to educate the king and the latter's son. When the son (Dionysios II) succeeded his father in 367, at the age of thirty, being like him a dilettante but weaker and irresolute, he played the part of a patron of letters and philosophy. Dion invited Plato to return to Syracuse; Dionysios II banished him, confiscated his property, and tried in vain to retain Plato.

Dion lived for a time in Athens, attending the Academy. In 357, helped by other members of that school, he reëntered Syracuse by force and expelled Dionysios II. He became of necessity a tyrant in his turn and was murdered a few years later. Many of these facts are known from Plato's letter 7 (the genuineness of which is uncertain), addressed in his old age to Dion's partisans after the latter's death, and urging them to be moderate. The letter proves that Plato himself and other members of the Academy had been deeply mixed up in the intrigues and crimes of Syracusan politics. With regard to the letter ascribed to Plato, see *Isis 43*, 68 (1952).

[4] According to a letter kindly sent to me by Prof. Michael Stephanides (Athens 23 July 1950), the place is now a popular quarter of Athens, vulgarly called *Astryphos* (*Hagios Tryphōn*) but also called *Acadēmia*. The location is open to visitors, but there is no memorial monument.

[5] Academos it was who revealed to the Dioscuroi (Castor and Pollux) the place where their sister, Helen of Sparta, had been hidden away. Therefore, when the Lacedaemonians invaded Attica, they spared the Academy.

[6] The successive meanings of academy (and its variants in other European languages) are, briefly: (1) the school founded by Plato; (2) college of higher learning; (3) secondary school; (4) special school (academy of music, naval academy, etc.); (5) place of education or training in general; (6) society of learned men.

It was felt early that "academy" was an honorable term, a "glamour" word; its further use increased the glamour (Académie des sciences); it was also misused. There are in the world too many worthless academies. For any humanist who remembers Plato, "academy" is a sacred word.

son of Pisistratos of Athens, had walled it in. It was dedicated to Athena and contained a grove of olive trees, the oil from which was given to the victors of the Panathenaian games. At the time of the great Dionysia, the statue of Dionysos Eleutherios was brought to it in great pomp. It included a park, a grove, and an athletic field, and the famous Athenian soldier and statesman Cimon (*c.* 512–449) had embellished it. Plato used it as a regular meeting place for his disciples, and he owned property in the neighborhood.

We may assume that in his time it already included some buildings, for example, a chapel or museum (a temple to the Muses), perhaps a few chambers for teachers and disciples, and halls for assembly, for lecturing, and for eating together, if only on formal occasions. Considering the climate of Athens, it is possible that much of the teaching took place either in the grove or in a portico, wherein one could be protected from the sun yet enjoy the open air.

We do not know more about the teaching itself than we do about the material organization except as far as can be judged from the writings of Plato, his disciples, and his successors. It is possible that the dialectic method of Socrates was largely used, especially at the beginning, and that there was less lecturing than discussion, somewhat as in the so-called seminar meetings of our own universities. Everything was informal and tentative. The center of attraction was Plato's own personality; students came to him from far and near, as they had come to Socrates before and to other famous teachers; but for the first time they came to a definite place. Plato was the chief attraction, but they went to the Academy as the students of today go to the university.

The Academy was not a novelty as a school, for there had been schools many centuries before its foundation, not only in Greece but in Babylonia, Egypt, Crete. Indeed, wherever a government exists it is necessary to train clerks for its service; wherever there is a church it is necessary to train priests and acolytes; wherever there are business houses and banks it is necessary to train accountants. The novelty consisted in the kind of teaching that was provided. Continuing the tradition of the sophists and of Socrates, Plato was not interested in teaching reading, writing, and arithmetic, even less in teaching methods of business. His aim was considerably higher, he wanted to enlighten the students, to give them the love of knowledge and of wisdom, to make philosophers of them and perhaps statesmen; he was teaching not any special knowledge, except perhaps logic and mathematics, but the principles of knowledge, of education, of ethics and politics. The Academy was not a school created by the government for its own administrative needs; it was a higher school of philosophy and politics, independent of the government and more often than not hostile to it. The Academy may be called the first institution for higher learning; it was a private institution.[7]

The students of various ages who flocked to it did not come to obtain degrees or certificates that would give them a right to a job; they passed no examination and obtained no credit of any kind, except such as was implicit in the good will of teachers and classmates. That was the best feature of the Academy. The teachers and disciples were disinterested, as disinterested as scholars can be; their ideal

[7] The privacy was perhaps a necessity suggested by the condemnation of Socrates. Such teaching as Plato had in mind could not be made in public without danger; it was more prudent to teach privately, if not secretly, in a secluded place.

was the old Pythagorean one — the search for knowledge is the greatest purification. We shall see presently that Plato did not remain faithful to that ideal, and that political passion led him to betray his master Socrates.

LATER HISTORY OF THE ACADEMY (347 B.C. TO A.D. 529)

We shall be better able to appreciate Plato's foundation if we turn aside for a moment from our main subject to outline the history of the Academy. Shortly after Plato's death in 347, he was succeeded by the son of his sister, Speusippos, who completed the organization of the school. The following successors were Xenocrates of Chalcedon, master or director of the Academy from 339 to 315, Polemon of Athens from 315, Crates of Athens from *c.* 270. With Crates the Old Academy came to an end. Its fame is due not only to the five masters who have just been named, but also to disciples or assistant teachers such as Philip of Opus, Eudoxos of Cnidos, Heracleides of Pontos, Crantor of Soli (in Cilicia). We shall have much to say later about the first three, and it will suffice now to give a brief account of the last one. Crantor studied under Xenocrates and Polemon and was the first to write commentaries on Plato's works. Of his own works the most famous was the one on grief, *Peri tu penthus*, which is lost but fragments of which survive in the *Tusculan disputations* and in the *Consolation* which Cicero was moved to write after the loss of his daughter Tullia.[8]

After Crates the Academy continued to function but took a different (skeptical) color under the direction of Arcesilaos of Pitane (in Aeolis, *c.* 315–241), who is sometimes called the founder of the second or Middle Academy. Arcesilaos was followed by Carneades of Cyrene (213–129), who increased the skeptical tendencies, and is called the founder of the Third Academy. Carneades was sent by the Athenians as their ambassador to Rome, where he obtained so much success that Cato the Censor (II-1 B.C.) took fright, denounced him, and caused the Senate to drive him out. A Fourth Academy was created by Philon of Larissa, who leaned toward Stoicism. Finally, a Fifth Academy was begun by Antiochos of Ascalon (d. 68 B.C.), who tried to reconcile the teachings of Plato, Aristotle, and the Stoa. This Fifth Academy is generally called the New Academy. Both Philon and Antiochos visited Rome, and Cicero listened to the former in 88 and to the latter ten years later. Thanks to Carneades, Philon, and Antiochos, the various teachings of the Academy reached the Roman world, Cicero (I-1 B.C.) and Varro (I-2 B.C.) being the outstanding interpreters.

During the siege of Athens by Sulla (in 86 B.C.) the latter, needing timber, cut down the trees of the Academy. It has been claimed that the Academy was then moved inside the town and remained there until the end, but if that were true, its town location would be known and no such location has ever been mentioned. Hence, we must assume that in spite of the damage caused by Sulla's soldiery the Academy remained where it was. Its further history is very obscure, however, until the fifth century, when it was given a new fame as a center of neo-Platonic teaching, chiefly under Proclos (V-2). The last seven directors of the Academy were Plutarchos of Athens, or Plutarchos the Great, who died very old in 431, Syrianos of Alexandria (V-1), who died in 450, Domninos of Larissa (V-2), Proclos, who died in 485, Marinos of Sichem (V-2), Isidoros of Miletos, one of the architects

[8] Long article on Crantor by von Arnim in Pauly-Wissowa, vol. 22 (1922), pp. 1585–1588.

of Hagia Sophia *c.* 532, and Damascios of Damascus (VI–1), director from *c.* 510 to 529, when the Academy was closed by Justinian as a school of pagan and perverse learning.

Justinian closed the Academy but he did not kill the teachers, and some of these escaped to the court of the king of Persia, Chosroes (Nūshīrwān the Just, ruled 531–579), probably in Jundīshāpūr, Khūzistān, where the king established a famous medical school. This is exceedingly important, for the exiles, philosophers and physicians, brought with them the seeds of Greek science and wisdom that were to develop a few centuries later under Muslim patronage. Justinian closed a door, Chosroes opened another, and thus science continued its march from Athens to Baghdād.

Among the philosophers welcomed by Chosroes, the most prominent were Simplicios of Cilicia (VI–1) and Priscianos of Lydia (VI–1), who might be said to represent the Academy in exile, the Athenian academy of Persia!

It is significant that out of the nine academicians just named, the last seven directors and the two exiles, only two were Greeks of Greece (Plutarchos and Domninos); the seven others were Egyptian or Asiatics.[9]

The Academy had lasted many centuries. At the time when Justinian closed its doors it might have celebrated its 916th anniversary. Whether that would have been completely justified, I do not know, for we have no proof that there was no solution of continuity in that long existence. Institutions are not like single men, whose age at any time of their life can be obtained by subtraction of their date of birth from the current date; they can die and disappear for many years, or many centuries, and then reappear again. Moreover, the Academy changed considerably in the course of centuries; it is only the Old Academy that may be considered as Plato's Academy, and it lasted a century and a half or less. To this one might reply that every institution is bound to change with the vicissitudes of time and that the longer it lives the more it must be expected to change. Bearing these remarks in mind, we may put it this way: the Academy of Athens, the Academy founded by Plato, lasted more than nine centuries.

ORIENTAL INFLUENCES

The temptation to narrate the vicissitudes of the Academy could not be resisted, though it took us far out of our immediate subject. It is a history of the Hellenic impregnation of the East which began with Alexander, a generation later than Plato, continued with ups and downs for a thousand years, and reached a new climax when Justinian closed the Academy. Justinian's purpose was to defend Christianity against paganism, but the chief result of his decision was to nourish and strengthen the Eastern peoples who would become under Islamic guidance the main challengers of Christian culture.

This history becomes even more startling when one takes into account, as one should, its counterpart, the Orientalization of Greece. The origin and development of Greek culture was stimulated by Oriental influences; Greek wisdom was nursed in an Oriental cradle and throughout its growth it was excited over and over again by the examples of barbarian friends or enemies. The reader has already been

[9] Proclos of Byzantion is counted among the Asiatics, though Byzantion is on the western (European) side of the Bosporos.

prepared for this in the previous chapters dealing with pre-Hellenic civilization, or with the Oriental sources of Pythagoras and Democritos. It is clear that Plato was also submitted to Oriental influences, but in his case those influences were liminal rather than continuous; moreover, it is not possible to separate his immediate borrowings from those that he made unwittingly through the intermediacy of Pythagoras, Archytas, Democritos, or his own disciples, Eudoxos and Philip of Opus.

Though Plato was not as friendly to the barbarians as Herodotos was, he was more friendly than his disciple Aristotle. He had explored Egypt, visited her wonderful monuments, and obtained some knowledge of her science and religion, rites and manners. He realized that the Egyptian civilization was considerably older than that of Greece. This is neatly expressed in *Timaios*,[10] in the form of a talk between Solon [11] and an Egyptian priest who was exceedingly old. Said the priest of Sais, "O Solon, Solon, you Greeks are always children: there is not such a thing as an old Greek." On hearing this Solon asked, "What mean you by this saying?" And the priest replied, "You are young in soul, every one of you. For therein you possess not a single belief that is ancient and derived from old tradition, nor yet one science that is hoary with age." The old priest treated his illustrious Greek visitor in about the same way as American visitors have often been treated by their European hosts; he then kindly explained the beautiful features of Egyptian society, its division into castes, and so forth. Solon marveled, and Plato marveled even more.

He had no immediate experience of Mesopotamia, but he referred to the laws of the Assyrians (the empire of Ninos). His astral mysticism was very probably of Chaldean provenience. As to Persia, the traditional enemy of his people, every educated Greek knew something about her. Stimulated by Democritos and Eudoxos, Plato knew much more than most; he had read the accounts of Ctesias and Herodotos and perhaps of other historians and their revelations of the Achaemenidian empire pleased him very much. Persian autocracy and order seemed to him far superior to Athenian democracy and chaos. The myth of Er the Pamphylian in the *Republic* [12] is of Chaldeo-Iranian origin.

The myth of the Earthborn is called in the text [13] a sort of Phoenician tale (*Phoinicicon ti*), and it might well be, like the Cadmos tradition and various others.

The dualistic ideas latent in the last dialogues of Plato may also have been derived from the Iranian religion, though we must admit that the derivation was indirect and tenuous. Zoroaster is named but once in Plato's writings.[14]

According to ancient tradition, when Plato was very old he received the visit of a Chaldean guest, but became feverish, and a Thracian flutist was invited to play in order to soothe him. He died soon afterward. Others would have it that many Magians were present at the time of the master's death. Realizing that he had died on a day sacred to Apollo and had lived nine times nine years, they concluded that

[10] *Timaios*, 22B.

[11] Solon (*c.* 638-*c.* 558), the illustrious Athenian lawgiver, one of the Seven Wise Men. After he had completed his code of laws, he absented himself from Athens for ten years, visiting Egypt, Cypros, and Lydia, where he had his famous interview with Croesos. Shortly after his return, supreme power was seized by Peisistratos, his constitution was revoked and he died two years later, *c.* 558.

[12] *Republic*, x, 616.

[13] *Ibid.*, 414.

[14] In *Alcibiades I* (121E–122A), of doubtful genuineness. At the age of 14 the young Persian was taught the magianism of "Zoroaster son of Horomazos."

Plato must have been a hero (a superhuman being) and they made a sacrifice to his memory.

There are many analogies between Platonic philosophy on the one hand and Sāmkhya and Vedanta philosophy on the other, but there is no proof that Plato was ever submitted to Hindu influences.

See Richard Reitzenstein and H. H. Schaeder, *Studien zum antiken Synkretismus aus Iran und Griechenland* (355 pp.; Studien der Bibliothek Warburg 7; Leipzig, 1926); Joseph Bidez and Franz Cumont, *Les mages hellénisés* (2 vols.; Paris: Les belles lettres, 1938) [*Isis 31*, 458–462 (1939–40)]; J. Bidez, *Eos ou Platon et l'Orient* (256 pp.; Brussels: Hayez, 1945) [*Isis 37*, 185 (1947)]; Simone Pétrement, *Le dualisme chez Platon, les Gnostiques et les Manichéens* (354 pp.; Paris: Presses Universitaires de France, 1947); Franz Cumont, *Lux perpetua* (558 pp.; Paris: Geuthner, 1949) [*Isis 41*, 371 (1950)].

THE THEORY OF IDEAS [15]

We have no intention of describing the details of Plato's philosophy, but we must discuss the theory of Ideas, which is the core of it and dominates Plato's thought on every subject.

The objects that we see with our own eyes are only appearances, like the shadows in the cave.[16] If there is any truth, there must be things that really exist. These things are the "Ideas" or "Forms." [17] To each kind of being or object there corresponds an Idea, which is as it were its womb and its cause. For example, we see "horses" all of which are different and imperfect; however good they may seem to be, they are bound to weaken and sooner or later to pass away. The Idea of the horse, however, or let us call it the "ideal horse," is perfect and eternal. The ideal horse cannot be seen or touched, but while the horses of sense are as ephemeral and nonexistent as shadows, it truly exists; it is the archetype of all possible horses, born or unborn.

This theory enables one to classify all the objects in their reality, instead of having to consider only their evanescent appearances. It helps us to understand the law of change and decay, which seems to be universal, and it gives us new principles of thought and conduct. The sensible world is submitted to corruption and death, but the Ideas, being immaterial, are incorruptible and ageless; the world of Ideas is real and permanent. The Idea is not only the essential reality of a thing, it is also its definition and its name; hence, we are given at one and the same time the tools of knowledge and its valid elements. The Ideas are not fancies, but beings, living and eternal; they are Forms, patterns, wombs, standards; at the same time they are like magical names.

The Ideas lend themselves easily to classification and hierarchy. The supreme Idea is the Idea of Good, which comes very close to God.

We may have opinions concerning the material objects, but real knowledge can be built only upon the basis of the immaterial Ideas. The aim of science is thus to

[15] When Platonic Ideas are meant the word is capitalized to distinguish this very particular and exalted acceptation from the common ones.

[16] As in the parable at the beginning of *Republic* vii, 514 ff. We are like prisoners in a cave who are aware of outside events only because of the shadows projected upon the inside walls.

[17] The terms used by Plato are *hē idea* (idea) and *to eidos* (form, shape). The second term is semantically curious, for its original meaning is "that which is seen" and the Idea cannot be seen. All our abstract terms have necessarily concrete origins.

investigate, understand, and know those Ideas. The real philosopher is the man whose soul can grasp them beyond the fleeting and deceiving appearances, and he derives his greatest reward from the contemplation of the purest and highest Ideas. Let us listen to the wise woman of Mantineia, Diotima:

> Such a life as this, my dear Socrates, spent in the contemplation of the beautiful, is the life for men to live; which if you chance ever to experience, you will esteem far beyond gold and rich garments, and even those lovely persons [18] whom you and many others now gaze on with astonishment, and are prepared neither to eat nor drink so that you may behold and live for ever with these objects of your love! What then shall we imagine to be the aspect of the supreme beauty itself, simple, pure, uncontaminated with the intermixture of human flesh and colors, and all the other idle and unreal shapes attendant on mortality; the divine, the original, the supreme, the monoeidic beautiful itself? What must be the life of him who dwells with and gazes on that which it becomes us all to seek? Think you not that to him alone is accorded the prerogative of bringing forth, not images and shadows of virtue, for he is in contact not with a shadow but with reality; with virtue itself, in the production and nourishment of which he becomes dear to the Gods, and if such a privilege is conceded to any human being, himself immortal.[19]

If a man has a real knowledge of virtue, that is, if he truly sees the Idea of virtue, he is virtuous, for nobody having attained such pure knowledge can do wrong readily.[20]

One of the most beautiful dialogues, the *Phaidon*, has already been referred to, for we borrowed from it Plato's moving account of Socrates' death (see pp. 266–270). The purpose of that dialogue is to show that the philosopher is happy to die. The Idea of the soul implies its immortality. The discussion leads to the conclusion that the Ideas are the sole causes of all things and the sole objects of knowledge. The theory of Ideas helps to prove the immortality of the soul, and vice versa.

The two conceptions that there are entities intermediary between the Ideas (or Forms) and things and that Ideas are numbers, which are ascribed to Plato in Aristotle's *Metaphysics*,[21] are not found in his dialogues, yet the ascription may be correct, for we may assume that Plato's teaching, which Aristotle had received from his own lips, is not completely represented by his writings. Every great teacher teaches much more than he can possibly write down.

The theory of Ideas is the source of logical realism and of the problem of universals, which, being formulated by Boetius (VI–1) and again more explicitly by St. Anselm (XI–2) (*universalia ante rem*), dominated medieval thought. The opposite theory, nominalism (*universalia post rem*), was explained by St. Anselm's contemporary, Roscelin of Compiègne (XI–2), but did not make much headway until its revival by William of Occam (XIV–1).[22] The Platonic point of view allured poets and metaphysicians, who fancied that it made divine knowledge possible; unfortunately, it made the more earthbound scientific knowledge impossible. The

[18] Like many other translators of Plato, Shelley, from whose translation this passage is taken, hides the fact, clear enough in the Greek text, that the "lovely persons" are not women, but beautiful boys and striplings. Platonism leads easily to hypocrisy.

[19] From the translation of the *Symposium* (211) by Shelley reprinted in *Five dialogues of Plato bearing on poetic inspiration* (Everyman's Library).

[20] Virtue is the condition of happiness, wickedness or sin is a wrong calculation. The truly virtuous man, in the Platonic sense, is the dialectician who knows the Idea of good.

[21] Aristotle, *Metaphysics*, 991.

[22] *Introduction*, vol. 3, pp. 81–83, 549–557.

Platonic method of leading from the general to the particular, from the abstract to the concrete, is intuitive, swift, and sterile. It is sterile because it is unworkable, or, to use our modern terminology, it is not "operational": [23] abstract good is no good and one cannot ride an ideal horse. The opposite method (nominalism, the *via moderna*), leading from known particulars to abstract notions of increasing generality, is slow but fruitful; it prepared very gradually the way for modern science. In spite of the incredible fertility and power of science, Platonism is not dead and will never die, for there will always be impatient metaphysicians wanting universal and immediate answers to their queries, and there will always be (let us hope) poets electing dreams instead of realities.

Strangely enough, those metaphysicians and poets are often called "realists." It is perhaps a little less ambiguous to call them idealists.[24] Yet this is a cause of new misunderstandings, for there are plenty of simple-minded people who believe that idealists have the monopoly of ideals. "Idealists" prefer ideals to realities, and try to explain the latter in terms of the former. In that sense, Plato was their archetype. Men of science have ideals of their own, but they do not subordinate them to realities; their ideals grow out of the realities and are limits of them which man can hope to approach asymptotically. We cannot give people credit for their passive and uncontrollable ideals but only for their active thoughts and tangible deeds. Gratuitous ideals can lead only to hypocrisy, cynicism, and skepticism.

The analogies between Platonic philosophy and various forms of Hindu wisdom are numerous and obvious, but it does not follow that definite borrowings were made from either side by the other. It will suffice to remember the indefinite contact that had existed for centuries between Greece and the East, and the unity of the human mind. Given certain premises, such as the illusions of the visible world and the greater reality of an invisible one, men are bound to draw similar conclusions.

PLATO'S WRITINGS

BIBLIOGRAPHIC SUMMARY

In this summary we enumerate only a few general editions of all of the works, or of most of them.

The first printed edition was the Latin translation by Marsiglio Ficino (folio; Florence, 1483–84). The Greek princeps edited by A. P. Manutius and M. Musurus was printed by the Aldine press thirty years later (Venice, 1513) (Fig. 80). A Greek-Latin edition, with the new Latin version by J. Serranus, was printed by Henricus Stephanus (Henri Estienne) (3 vols., folio; Paris, 1578) (Fig. 81). This edition is very important because its pagination has been repeated in every scientific edition. The best way to refer to a Platonic passage is to quote the title of the work and the Stephanus volume and page (the title being given, the volume number is superfluous).

The best Greek edition is the one by John Burnet (5 vols. in 6; Oxford: Clarendon Press, 1899–1906).

The first French translation was by André Dacier (1651–1722), *Les oeuvres de Platon* (2 vols.; Paris, 1699). A Greek-French edition is being published by the Association Guillaume Budé (Paris, 1920 ff.).

The first English translation was made from the French version of Dacier (2 vols.;

[23] For a definition of operationism see Dagobert D. Runes, *Dictionary of philosophy* (New York: Philosophical Library, 1942), p. 219 [*Isis* 39, 128 (1948)].

[24] The weasel word idealist is sometimes understood as the opposite of realist.

London, 1701). The first English translation from the Greek was by Floyer Sydenham and Thomas Taylor (quarto, 5 vols.; London, 1804). The most famous English translation is the one by Benjamin Jowett (1817–1893), Master of Balliol (4 vols.; Oxford, 1871; 5 vols.; 1875; etc.). There are Greek-English editions in the Loeb Classical Library (1914 ff.).

See also Friedrich Ast, *Lexicon platonicum* (3 vols.; Leipzig, 1835–1838; anastatic reprint, Berlin, 1908). There is an English index in vol. 5 of Jowett's translation. Ast's very full glossary and Jowett's index refer to the Stephanus page numbers and hence can be used with any edition of Plato that quotes these numbers.

PLATO'S WORKS AND THEIR CHRONOLOGIC ORDER

The list of works varies because the genuineness of some is doubted. It includes the *Apology of Socrates*, plus twenty-five to twenty-eight dialogues, and thirteen letters (letter seven is probably genuine).

There are apocryphal works but (this is very remarkable) no lost ones. This implies early and continuous appreciation of Platonic writings.

Endless controversies have been and will be devoted to the chronology of Plato's writings, but there is a general agreement *grosso modo* on the following basis.

1. The Socratic dialogues – *Euthyphron, Charmides, Laches, Lysis, Criton* – as well as the *Apology* were Plato's first writings, when he was completely under Socrates' influence and tried to reproduce faithfully the latter's thought.

2. Second group. Educational dialogues. Dialogues criticizing sophistry: *Protagoras, Euthydemos, Gorgias, Phaidros; Menon, Symposium; The Republic; Phaidon, Cratylos.*

3. Third group. *Parmenides, Philebos, Theaitetos, Sophist, Statesman.*

4. Final group (old age). *Timaios. Laws* (this was his last as well as his longest work).

This list is not complete but is sufficient for a rough chronology. It would be wiser perhaps to simplify it even more and say that Plato wrote the Socratic dialogues at the beginning of his career, *Timaios* and *Laws* at the end, and the rest in the middle.

It is remarkable that all the works except the *Apology* and the dubious letters are in the form of dialogues, which we remember as the Platonic form par excellence. It enables the writer to illustrate various sides of a question, and even to suspend his own judgment, or at least to hide it from the reader. Thus, we are given inconclusive dialogues, like *Protagoras.*

Socrates is one of the dramatis personae in all the dialogues, except *Laws*; in *Parmenides, Sophist, Statesman, Timaios* he appears, but only in a subordinate capacity. In the early "Socratic" dialogues, he is the chief speaker, and we feel more confident that we are listening to the real Socrates. In later dialogues we are invited to listen to what the commentators have been pleased to call a "Platonizing" or "idealized" Socrates, but what seems to be more often a debased and degraded one.

The dialogues are sometimes interrupted by myths, such as the myth of Atlantis at the beginning of the *Timaios*, that of Er at the end of the *Republic*, and that in the *Statesman*, and more often by statements which are so long that they read like lectures, and that the other speakers are almost forgotten. The dialogic form enables us to see the argument from many angles, we are permitted

142 ΠΛΑΤΩΝΟΣ ΤΙΜΑΙΟΣ, Η ΠΕΡΙ ΦΥΣΕΩΣ.

ΤΑ ΤΟΥ ΔΙΑΛΟΓΟΥ, ΠΡΟΣΩΠΑ.

Σωκράτης. Κριτίας. Τίμαιος. Ἑρμοκράτης.

Χ ΣΩ. Εἷς, δύο, τρεῖς· ὁ δὲ δὴ τέταρτος ἡμῖν ὦ φίλε Τίμαιε, ποῦ τῶν χθὲς μὲν δαιτυμόνων, τανῦν δ' ἑστιατόρων. ΤΙ. Ἀσθένειά τις αὐτῷ συνέπεσεν ὦ Σώκρατες· οὐ γὰρ ἂν ἑκὼν τῆσδε ἀπελείπετο τῆς ξυνουσίας. ΣΩ. Οὐκοῦν σὸν τῶνδέ τε ἔργον, καὶ τὸ ὑπὲρ τοῦ ἀπόντος ἀναπληροῦν μέρος. ΤΙ. Πάνυ μὲν οὖν, καὶ κατὰ δύναμίν γε οὐδὲν ἐλλείψομεν· οὐδὲ γὰρ ἂν εἴη δίκαιον, χθὲς ὑπὸ σοῦ ξενισθέντας, οἷς ἦν πρέπον ξενίοις, μὴ οὐ προθύμως σε τοὺς λοιποὺς ἡμῶν ἀντεφεστιᾶν. ΣΩ. Ἆρ' οὖν μέμνησθε, ὅσα ὑμῖν καὶ περὶ ὧν ἐπέταξα εἰπεῖν; ΤΙ. Τὰ μὲν, μεμνήμεθα· ὅσα δὲ μή, σὺ παρὼν ὑπομνήσεις. μᾶλλον δέ, εἰ μή τί σοι χαλεπόν, ἐξ ἀρχῆς διὰ βραχέων πάλιν ἐπάνελθε αὐτά, ἵνα βεβαιωθῇ μᾶλλον παρ' ἡμῖν. ΣΩ. Ταῦτ' ἔσται. χθές που τῶν ὑπ' ἐμοῦ ῥηθέντων λόγων, περὶ πολιτείας ἦν τὸ κεφάλαιον, οἵα τε καὶ ἐξ οἵων ἀνδρῶν, ἀρίστη κατεφαίνετ' ἄν μοι γενέσθαι. ΤΙ. Καὶ μάλα γε ἡμῖν ὦ Σώκρατες ῥηθεῖσα πᾶσι κατὰ νοῦν. ΣΩ. Ἆρ' οὖν οὐ τὸ τῶν γεωργῶν, ὅσαι τε ἄλλαι τέχναι, πρῶτον ἐν αὐτῇ χωρὶς διειλόμεθα, ἀπὸ τοῦ γένους τοῦ τῶν προπολεμησόντων. ΤΙ. Ναί. ΣΩ. Καὶ κατὰ φύσιν δὴ δόντες τὸ καθ' αὑτὸν ἑκάστῳ πρόσφορον, ἓν μόνον ἐπιτήδευμα, μίαν ἑκάστῳ τέχνην, τούτους οὓς πρὸ πάντων ἔδει πολεμεῖν, εἴπομεν· ὡς ἄρα αὐτοὺς δέοι φύλακας εἶναι μόνον τῆς πόλεως. εἴ τις ἔξωθεν ἢ καὶ τῶν ἔνδον ἴοι κακουργήσων, δικάζοντας μὲν πρᾴως τοῖς ἀρχομένοις ὑπ' αὐτῶν, ἅτε φύσει φίλοις οὖσι· χαλεποὺς δὲ ἐν ταῖς μάχαις τοῖς ἐντυγχάνουσι τῶν ἐχθρῶν γιγνομένους. ΤΙ. Παντάπασι μὲν οὖν. ΣΩ. Φύσιν γάρ, οἶμαί τινα τῶν φυλάκων τῆς ψυχῆς ἐλέγομεν· ἅμα μὲν θυμοειδῆ, ἅμα δὲ φιλόσοφον δεῖν εἶναι διαφερόντως, ἵνα πρὸς ἑκατέρους δύναιντο ὀρθῶς πρᾶοι καὶ χαλεποὶ γίγνεσθαι. ΤΙ. Ναί. ΣΩ. Τί δαὶ τροφήν; ἆρ' οὐ γυμναστικῇ καὶ μουσικῇ μαθήμασί τε ὅσοις προσήκει, τούτοις ἐν ἅπασι τεθράφθαι; ΤΙ. Πάνυ μὲν οὖν. ΣΩ. Τοὺς δέ γε οὕτω τραφέντας, ἐλέχθη που, μήτε χρυσὸν μήτε ἄργυρον, μήτε ἄλλο ποτὲ μηδὲν κτῆμα, ἑαυτῶν ἴδιον νομίζειν δεῖν. ἀλλ' ὡς ἐπικούρους μισθὸν λαμβάνοντας τῆς φυλακῆς παρὰ τῶν σωζομένων ὑπ' αὐτῶν, ὅσος σώφροσι μέτριος. ἀναλίσκειν τε δὴ κοινῇ, καὶ ξυνδιαιτωμένους μετ' ἀλλήλων ζῆν· ἐπιμέλειαν ἔχοντας ἀρετῆς διὰ παντός· τῶν ἄλλων ἐπιτηδευμάτων ἄγοντας σχολήν. ΤΙ. Ἐλέχθη καὶ ταῦτα ταύτῃ. ΣΩ. Καὶ μὲν δὴ καὶ περὶ γυναικῶν ἐπεμνήσθημεν, ὡς τὰς φύσεις τοῖς ἀνδράσι παραπλησίας εἴη ξυναρμοστέον. καὶ τὰ ἐπιτηδεύματα πάντα κοινὰ κατά τε πόλεμον, καὶ κατὰ τὴν ἄλλην δίαιταν δοτέον πάσαις. ΤΙ. Ταύτῃ καὶ ταῦτα ἐλέγετο. ΣΩ. Τί δὲ δὴ τὸ περὶ τῆς παιδοποιίας; ἢ τοῦτο μὲν διὰ τὴν ἀήθειαν τῶν λεχθέντων εὐμνημόνευτον, ὅτι κοινὰ τὰ τῶν γάμων καὶ τὰ τῶν παίδων, πᾶσιν ἁπάντων ἐτίθεμεν, μηχανωμένους ὅπως μηδείς ποτε τὸ γεγενημένον αὐτῷ ἰδίᾳ γνώσοιτο· νομιοῦσι δὲ πάντες πάντας αὑτοὺς ὁμογενεῖς· ἀδελφὰς μὲν καὶ ἀδελφούς, ὅσοιπερ ἂν τῆς πρεπούσης ἐντὸς ἡλικίας γίγνωνται· τοὺς δ' ἔμπροσθεν καὶ ἄνωθεν, γονέας τε καὶ γονέων προγόνους· τοὺς δ' εἰς τὸ κάτωθεν, ἐκγόνους τε καὶ παῖδας ἐκγόνων. ΤΙ. Ναὶ καὶ ταῦτα εὐμνημόνευτα ᾗ λέγεις. ΣΩ. Ὅπως δὲ δὴ κατὰ δύναμιν εὐθὺς γίγνοιντο ὡς ἄριστοι τὰς φύσεις, ἆρ' οὐ μεμνήμεθα, ὡς τοὺς ἄρχοντας ἔφαμεν καὶ τὰς ἀρχούσας δεῖν εἰς τὴν τῶν γάμων σύνερξιν λάθρᾳ μηχανᾶσθαι κλήροις τισίν, ὅπως οἱ κακοὶ χωρίς, οἵ τ' ἀγαθοὶ ταῖς ὁμοίαις ἑκάτεροι ξυλλήξονται. καὶ μή τις αὐτοῖς ἔχθρα διὰ ταῦτα γίγνηται, τύχην ἡγουμένοις αἰτίαν τῆς ξυλλήξεως. ΤΙ. Μεμνήμεθα. ΣΩ. Καὶ μὴν ὅτι γε τὰ μὲν τῶν ἀγαθῶν θρεπτέον, ἔφαμεν εἶναι· τὰ δὲ τῶν φαύλων, εἰς τὴν ἄλλην λάθρᾳ διαδοτέον πόλιν. ἐπαυξανομένων δέ, σκοποῦντας ἀεὶ τοὺς ἀξίους πάλιν ἀνάγειν δεῖν. τοὺς δὲ παρὰ σφίσιν ἀναξίους, εἰς τὴν τῶν ἐπανιόντων χώραν μεταλλάττειν. ΤΙ. Οὕτω. ΣΩ. Ἆρ' οὖν διεληλύθαμεν ἤδη καθάπερ χθές, ὡς ἐν κεφαλαίοις πάλιν ἐπανελθεῖν, ἢ ποθοῦμεν ἔτι τῶν ῥηθέντων ὦ φίλε Τίμαιε ἀπολειπόμενον. ΤΙ. Οὐδαμῶς. ἀλλ' αὐτὰ ταῦτ' ἦν τὰ λεχθέντα ὦ Σώκρατες. ΣΩ. Ἀκούοιτ' ἂν ἤδη τὰ μετὰ ταῦτα περὶ τῆς πολιτείας ἣν διήλθομεν, οἷόν τι πρὸς αὐτὴν πεπονθὼς τυγχάνω. προσέοικε δὲ δή τινί μοι τοιῷδε τὸ πάθος· οἷον εἴ τις ζῷα καλά που θεασάμενος, εἴτε ὑπὸ γραφῆς εἰργασμένα, εἴτε καὶ ζῶντα ἀληθινῶς, ἡσυχίαν δὲ ἄγοντα, εἰς ἐπιθυμίαν ἀφίκοιτο θεά-

Fig. 80. A page of the Greek *princeps* of Plato's works (Venice, 1513), edited by Aldo Manuzio (Aldo il Vecchio, 1449–1515) and by the Cretan Marco Musurus (1470–1517). This page is the beginning of the *Timaios* (17A to 19B). Compare with Fig. 60. [From the copy in the Harvard College Library.]

ΠΛΑΤΩΝΟΣ

ΑΠΑΝΤΑ ΤΑ ΣΩΖΟΜΕΝΑ.

PLATONIS

opera quæ extant omnia.

EX NOVA IOANNIS SERRANI INterpretatione, perpetuis eiusdẽ notis illustrata: quibus & methodus & doctrinæ summa breuiter & perspicuè indicatur.

EIVSDEM Annotationes in quosdam suæ illius interpretationis locos.

HENR. STEPHANI de quorundam locorum interpretatione iudicium, & multorum contextus Græci emendatio.

EXCVDEBAT HENR. STEPHANVS,

CVM PRIVILEGIO CÆS. MAIEST.

Fig. 81. Title page of the Greek-Latin edition of Plato published by Henri Estienne (3 vols., folio; Paris, 1578). The pagination of that edition is repeated in every scientific edition, and the best way of referring to a Platonic text is to quote the Stephanus page. [From the copy in the Harvard College Library.]

to turn around it, as it were, but this is often more illusory than real. Many dialogues, especially the political ones, are as dogmatic as it is possible to be, and the objections of various interlocutors seem to be introduced only in order to illuminate the same dogmas from another side. Another disadvantage of that form is that it is a cause of repetition and prolixity, and jeopardizes the unity of the subject.

Plato's style is the perfection of Attic prose of the golden age, when the Greek language was still pure. It is easy, yet elegant, sometimes humorous, sometimes poetic, rich in metaphors, very flexible, and full of surprises. In spite of the dryness of many of the arguments, Plato often manages to astonish his reader and to charm him. This is especially true if one is privileged to read him in the original Greek, and to read with sufficient fluency.

It must be confessed that many words written in praise of Plato's charm are disingenuous, because they were written by people whose knowledge of Greek was utterly insufficient. In order to appreciate the literary merit of any text, the subtleties of the author's thought and tongue, one must know his language exceedingly well. One must know the dictionary and the grammar so deeply that one does not think of them any more, but only of the living flow, the rhythm, the images, the intriguing associations between the ideas and their wording. The admiration of Plato's style by incompetent people is a curious form of snobbishness; and one should never underestimate the strength of that weakness; it has helped to nourish the love of Greek ideals and to keep alive teachers of Greek.

POLITICS. THE GREAT BETRAYAL [25]

As far as we can judge from his writings, Plato's teaching in the Academy must have been very largely devoted to political questions, or let us say to politics and ethics, two subjects that were (and will always be) very intimately connected. The good citizen, not to mention the good statesman, must be a good man to begin with. Only three of Plato's works deal chiefly with politics, but their total length is considerable. His political ideals were explained by him in his middle age in the *Republic*; later some ideas were made more precise in the *Statesman*, and at the end of his life he wrote the largest of his books, the *Laws*.[26] The *Laws* was a practical adaptation of his political dreams to human weakness. It included a great wealth of material regulating every aspect of public and private life, and as such had a great influence upon Hellenistic and Roman legislation. Many codes of law had been drafted and enacted before Plato's, but one could hardly speak of legal philosophy before his time, and therefore he may be called the founder of jurisprudence.

In order to understand Plato's meditations one must bear in mind the political circumstances in the midst of which his mind had grown. He was a child of the Peloponnesian War; he had witnessed not only the utter defeat of Athens but also the downfall of democracy; during the most sensitive years of his adolescence he

[25] In my appreciation of Plato's politics, I have been much helped by Warner Fite, *The Platonic legend* (340 pp.; New York: Scribner, 1934); Benjamin Farrington, *Science and politics in the ancient world* (243 pp.; New York: Oxford University Press, 1940) [*Isis* 33, 270-273 (1941–42)], and above all by Karl R. Popper, *The open society and its enemies* (2 vols.; London: Routledge, 1945; new ed. in 1 vol., 744 pp.; Princeton: Princeton University Press, 1950). My references are to the first edition.

[26] In Jowett's translation the *Republic* covers 338 pp., the *Statesman* 68, the *Laws* 361, total 767 pp. No other work covers as much as 100 pp.

had seen crimes perpetrated first by mobs, then by aristocrats; he was twenty-four when the Thirty Tyrants were functioning (404–403), whose exactions were such that the worst deeds of the democrats were condoned and forgiven. After that, things went from bad to worse. In 399 his teacher Socrates was condemned to death and Plato was obliged to leave the city. Plato was a man of substance, connected with some of the oligarchs; the political chaos was very painful to him, the condemnation of his friends and of his revered teacher impossible to bear. The Athens of his day was not pleasant to contemplate; Sparta and Crete in the distance seemed better. When he was writing the *Republic*, he was already disillusioned and was escaping from reality into utopian dreams. Political despair was his motive power. We know well enough from our own experience how strong that power can be; political passions are often so intense and so bitter that they can fill a man's heart with anguish and hatred and drive him to the commission of outrageous deeds. Plato observed evil and chaos around him, and he himself suffered all the pains of hopelessness and frustration. Matters became worse and worse, and we may assume that the Academy, which could be attended only by men of leisure, was a cradle of discontent. The author of the *Laws* was a disgruntled old man, full of political rancor, fearing and hating the crowd and above all their demagogues; his prejudices had crystallized and he had become an old doctrinaire, unable to see anything but the reflections of his own personality and to hear anything but the echoes of his own thoughts. The worst of it is that he, a noble Athenian, admired the very Spartans who had defeated and humiliated his fatherland. Plato was witnessing a social revolution (even as we are) and he could not bear it at all. His main concern was: how could one stop it?

It is especially hard for us to understand his admiration of Sparta, for we are able to compare Athens with Sparta from such a distance that our judgment is naturally impartial and objective, and if we ask ourselves what each of them has given to the world, the answer is peremptory. Our debt to Athens is immense, our debt to Sparta negligible. This was not as obvious to Plato's contemporaries as it is to us. In the first place, they were suffering the evils of war and chaos, military defeat and misgovernment; we do not have to bear that terrible burden, and we can concentrate our minds upon the literary and scientific legacy of Athens and the spiritual ineptitude of Sparta. This great Athenian praising the virtues of Sparta reminds us of the disgruntled Americans (not great men in any sense) who carried their hatred of their own government so far that they were ready to admire the Fascists and the Nazis.[27] The puzzle remains, for Plato was a philosopher and they were not, yet political passion may make fools of the best men.

The madness of a philosopher, however, is likely to take a special philosophic color. We have seen that Plato's conception of the universe was dominated by the theory of Ideas: the changing visible world is only a poor copy of the unchanging invisible one. That scheme extended itself most naturally to the political events, which illustrated the process of decay and corruption more shockingly than all others. Athenian politics was a mess of stinking fermentation. Plato invented a

[27] To make the comparison more precise, imagine that we had been beaten by the Germans, because we had begun our preparations too late or because they had built an atomic bomb ahead of us, and that a professor of government in Harvard University started to praise and preach Nazi doctrines . . .

political utopia and took refuge in it. His Republic, as the utopia was called,[28] was supposed to describe an ideal city, a city that would be perfect by definition and changeless. The heavenly city would escape the law of increasing corruption and increasing corruptibility. How could Plato devise such a heavenly city, one wonders, and make the invisible, visible and tangible? How could he flatter himself that the city born in his own brain would be identical with the divine city, and that it could ever be accepted without criticism, as a pattern of final perfection?

At any rate, in his own mind, change and corruption were equated. The same might be said of every conservative, but in the case of Plato the equation was proved by his theory of Ideas. Such a metaphysical proof was conclusive, was it not? What is more remarkable, Plato seems to have believed that it would be possible to establish a perfect, ideal, state, that such a state would be viable and might continue to exist, and that political change might be arrested. He might just as well have tried to arrest the rotation of the heavenly spheres.

Let us consider his utopia more closely. The Republic that Plato has created to serve as an ideal pattern is small, as small as or smaller than Athens. How could it isolate itself from the rest of the world, so as to escape the contagion of their wickedness?

The inhabitants are divided into three classes: the rulers, the soldiers or guardians, and the rest. The rest was at least 80 percent of the population; it is not clear to me whether it included the slaves or not.[29] The three classes are natural, not artificial; they are compared in the *Republic* with the three souls animating man's body – reason, spirit, and appetite; [30] the rulers are the "reason" of the state, the lower classes have nothing but gross appetites. It would perhaps be more correct to say that the citizens of Plato's Republic were divided into only two classes – the rulers and their auxiliaries, on the one hand, and the ruled on the other. Indeed, the differences between the first two classes are not great and can be bridged without difficulty; for example, as the auxiliaries grow older, less fit for military labor and more fit for reflection, they may rise to the top; between the rulers and the masses, however, there is an unbridgeable abyss. They are separated not by a temporary difference of class or function, but by a permanent difference of race or caste. (The comparison of Platonic classification with Hindu castes is substantially correct, but it is not necessary to assume that Plato was aware of their existence.) [31]

In the *Statesman*, the rulers of the state are likened to shepherds of men. That comparison and similar ones occur many times in Plato: the rulers are shepherds,

[28] The original title is *Politeia ē peri dicaiu*, Polity or concerning justice. The first word might be englished "polity"; the translation "republic" is somewhat misleading but too well established to be modified. "Republic" is to be taken in its original sense – *res publica*.

[29] The question of slavery need not be discussed here. Slaves were originally prisoners of war whose life was forfeited, and who chose servitude as a lesser evil than death. Slavery was accepted as a natural institution, not only by Plato and Aristotle, but also sixteen centuries later by such a man as St. Thomas Aquinas (XIII-2). See *Introduction*, vol. 2, p. 916. From Plato's point of view, the masses were on the same spiritual level as the slaves.

[30] *Nus, thymos, epithymia.* These three "souls" correspond, respectively, to the three *pneumata* of Galenic physiology – psychic, vital, and natural spirits – which formed the basis of physiology until the time of Harvey and even later. The comparison of the whole state with the body of a single person is typical of Platonic philosophy.

[31] Plato describes Egyptian castes in *Timaios* 24. For comparison with Hindu castes, see E. Senart (1847–1928), *Les castes dans l'Inde* (Paris, 1896, 1927) [*Isis 11*, 505 (1928)] and J. H. Hutton, *Caste in India* (Cambridge: University Press, 1946) [*Isis* 39, 107 (1948)].

the guardians are the dogs, the masses are the herd. The art of ruling men is not essentially different from that of managing and breeding cattle.

The rulers might properly claim: 'L'Etat c'est nous." They are the state, indeed, and therefore their class as a body cannot be controlled except by itself. Out of its own wisdom it knows what is best for the other people, that is, for the great majority of the population.

In order to insure the self-control of that hereditary oligarchy and its undivided devotion to the state (that is, to itself), it must be protected against disrupting and corrupting influences, the main ones of which are financial and sexual greed. Therefore, the elite of the Republic is obliged to accept communism, and its communism covers not only property but wives and children. This does not imply debauchery or promiscuity, but no man can claim a woman as his own; all upper-class citizens are brethren; the children are common children, their family is the state.

In a golden age when so many wonderful things were created not only by architects and sculptors but also by artisans, these artisans had no standing in Plato's eyes. The laborers of any kind are members of the herd; they are by definition low-minded brutes who want to fill their bellies; they have no ideals, only desires.

It is strange that Plato realized the disintegrating nature of human passions, such as the love of money or the love of family, but that he did not realize that other passions might be equally dangerous. One of the fundamental human passions is the love of power; love of money is only an aspect of it; men love money only because of the power that money gives them. Plato was very much afraid of property, chiefly in the form of money, gold and silver, but if money were demonetized, if it lost its power of purchase, would men lose their greed? Of course not; their greed would adapt itself to the new circumstances. The greed of power could not be eradicated. Even when the elite was definitely master of the masses, there could still be (and there certainly were) conflicts of power between themselves. Plato must have witnessed many illustrations of the statement often ascribed to Lord Acton: "Power corrupts; absolute power corrupts absolutely"; yet there is no evidence that he ever drew that conclusion.

Plato's rejection of property and family for the sake of fortifying the elite has been compared with the poverty and chastity that are enforced upon the Catholic clergy and monastic orders. That comparison is fallacious in many respects. It may be true that clerical asceticism is not only a matter of discipline and self-control, but also a means of separation from the laity and of a more complete control of it. The purpose, however, is purely religious and fraternal; it is not, or should not be, mixed with any velleity of political and economic control. The clergy and monks are not rulers of the state but its servants.

It is necessary to insist that Plato's integral communism concerned only the upper classes; the lower classes did not need any superior morality and therefore might indulge their appetites as much as they pleased, provided they kept quiet and obedient, and had good opinions.[32]

Plato's communism cannot be understood except as an aristocratic reaction

[32] *Doxa alēthēs*. Having good opinions, that is, being orthodox, right-minded, is simply a deeper form of obedience.

against the growing capitalism of his time. It was hard for the old aristocrats to be challenged and supplanted by the *nouveaux riches*, who were as often as not people of low manners and low caste, even slaves.[33] It is very hard for any elite to feel itself pushed out by a new one. If money could break the natural distinction between gentle people and the others, well, money must go. It is much more difficult to understand Plato's communism of women and children, that is, his virtual destruction of the family spirit of the "best" people. The *Republic* is the work of a disgruntled fanatic, yet it is hard to believe that he could carry his fanaticism and heartlessness to that extreme. Plato was never married, but he had a mother and father, and a family of his own. Did his parents mistreat him, one cannot help wondering? The fanaticism of a good man has generally a definite cause. Plato's communism of property could be explained in terms of his disillusionment and of his disgust with financial excesses; his communism of women and children could not be explained in the same way. I cannot explain it at all except in terms of sexual aberration.

Is there any generous man who has not suffered deep in his heart from the curse of money and wished that he could eradicate it? Is there any generous man who has not been comforted in his anguish by the love of family? How could a man destroy at the same time the greatest evil of life and its greatest blessing? Yet Plato did it, or at least he tried to do it.

PLATO'S POLITICAL PROBLEM

It was all very well to outline an ideal republic, but a self-respecting philosopher and dialectician had to prove that such a republic could really exist and continue. Where would one find an elite worthy of such an exalted position and one that would not abuse it? Inasmuch as the elite was very small (say one fifth of the population or even less), it could not retain its enormous privileges unless it was strong enough to defend them against the overwhelming majority of the people.

The elite was a natural elite, it existed, all that was needed was to strengthen and unite it. Plato was the earliest eugenist.[34] One should begin with a good stock and breed men as one breeds cattle. The leading aristocratic families provided a blooded stock. Again one can but wonder at Plato's naïveté. However high the correlation between good birth and good character may be, we can never be sure that a man of good birth is *ipso facto* a good man. Plato himself could easily have named many aristocrats who were utterly unreliable and contemptible.

But let us assume that we had a good stock to begin with; the eugenic problem would be to keep it as pure as possible. The best families would create as many children as the state needed, no more. The good birth of those children would not suffice, however; they would have to be very carefully and strictly educated. Plato was so convinced of the formative value of education that a great part of the *Republic* is devoted to it: the *Republic* is very largely a treatise on political education, an education meant exclusively for the ruling class.

[33] Cf. the story of Pasion (p. 395), the ex-slave who had become the richest man of Athens and had bought high honors with his benefactions.

[34] It would be better to say that Plato was the first theorist of eugenics. Eugenic views had been expressed two centuries before by the aristocratic poet Theognis (fl. 544–541). M. F. Ashley Montagu, "Theognis, Darwin and social selection," *Isis* 37, 24–26 (1947).

The future rulers must be at the same time strong and gentle, and this double aim must always be kept in mind. The two corresponding parts of education are gymnastics and music. The former includes all the physical exercises that help to make vigorous men and good warriors; the latter means not simply music as we understand it but the *bonae litterae*, the humanities in general.[35] Music was for the soul what gymnastics was for the body. It was excessively regulated. There was to be no jazz in the Republic, but only music in certain definite moods that would induce vigor and virtue. The same applies to belles-lettres and poetry; only certain poems would be accepted, and Homer himself, "the educator of Hellas," would be banished from the city.[36] The Greek classics would be given to the youths only after censorship and adaptation to the needs of good communists. Indeed, poetry, art, and music must be subordinated to political needs. Almost the whole of Greek literature would have been driven out or mutilated by the "divine" Plato! He would have forbidden the very things (except mathematics) that we have in mind when we speak of the glory of Greece. In that respect he was about on the same level as those great literary and artistic critics, Thomas Bowdler and Adolf Hitler.

Though deeply immersed in politics, Plato gave but little thought to economics. Business and trade would be left to the lower classes. How would the upper classes live? Well, they would be landowners and slaveowners, and would not the lower classes work for them? It was hardly worth while to worry about such base matters. Aristotle remarks, however,[37] that 5000 warriors[38] of the *Laws* "will require a territory as large as Babylon . . . if so many persons are to be supported in idleness together with their women and attendants. . . In framing an ideal," concluded Aristotle, "we may assume what we wish but should avoid impossibilities."

How could Plato imagine for a moment the practical possibility of such a state as is described in the *Republic* (or even the more moderate one of the *Laws*), and if it could ever have been established, how could it have been preserved? We shall come back to the question of leadership presently, but we may remark at once that if the first rulers of that crazy commonwealth were sufficiently wise and capable of keeping it going, how could one be sure of the wisdom of their successors? One may object that the Republic was a utopia, a dreamland, yet we must expect the dreams of a philosopher to have some sort of consistency and logic. Plato's idea was one of stability and changelessness, yet the Republic conceived by him was essentially unstable.

We may pause a moment here and ask ourselves where he got his inspiration. The primary sources were his hatred of Athenian politics and his approval of the Dorian institutions of Crete and Sparta. His idealization of the latter was as unfair and as ardent as his hostility to the former. It is not the case that his knowledge

[35] The "musical man" (*musicos anēr*) was what we would call a humanist, but Plato's humanist was belittled and debased, for his freedom of thought was very restricted.

[36] *Republic*, 398A.

[37] Aristotle, *Politics*, 1265A, 14.

[38] *Laws*, 737 limits the number of citizens (i.e., the whole elite) to 5,040 (not 5,000). The number was to be kept constant, and children would be produced only in sufficient quantity to keep the population stationary! The limit was determined by one of Plato's numerologic fantasies: $5040 = 21 \times 20 \times 12 = 35 \times 12 \times 12$. The number 5040 has as many as 59 divisors, including all the numbers from 1 to 12, except 11; it is almost divisible by 11 (*Laws*, 738, 771). Had Plato known that $5040 = 7!$ his enthusiasm for that number would have been carried even higher.

of politics had remained exclusively theoretical. During his travels and the vicissitudes of his political life, he had observed innumerable variations between the perfect city, which did not exist except in his brain, and the real cities of the world. He had classified the latter into six groups: absolute monarchy, constitutional monarchy, oligarchy, democracy, chaos, and tyranny. These forms might succeed each other, and then perhaps the whole cycle might begin again. This was a remarkable sociologic investigation, because of which Plato might be called the first sociologist as well as the first student of constitutional history. In the *Laws* [39] he gives us a history of the decline and fall of Persia that is the first analysis of its kind. Moreover, he had obtained plenty of experimental knowledge when he had been Dionysios' adviser, but one must admit that his meddling with Syracusan politics had been unfortunate for all concerned.

Plato did not lack political experience – far from it – but he was too much of a doctrinaire to profit from it. His political rancor was too bitter and his dreams were too strong to be influenced by changing and ephemeral realities.

The fundamental dogma of Platonic politics is the absolute supremacy of the state. Only the state can be perfect and self-sufficient; individuals are only imperfect, incomplete copies of it. Only the state can be changeless and enduring; individuals pass away in quick succession. Hence, the individual must be submitted and if necessary sacrificed to the state; that is good communist and totalitarian doctrine.

But how could the state be perfect unless created by God himself? Imperfect as it must be when created by Plato, how could it be brought nearer to perfection if no criticism and no change were ever allowed?

The main weakness of the totalitarian (as compared with the democratic) state is the difficulty, if not the impossibility, of obtaining independent, sincere criticism. We may excuse Plato for not having realized that as clearly and strongly as we realize it today.[40] Let us pay homage in passing to St. Thomas Aquinas (XIII-2) who was the first, in his commentary on Aristotle's *Politics*, to assert vigorously the subordination of every group to its members, and of every government to its dependents. Plato is the more excusable when we remember the totalitarian evils that were perpetrated even after the time of St. Thomas, and never on so large a stage, nor in so cruel a manner (a "scientific" manner), as in our own day.

LEADERSHIP

Plato saw clearly that it was not enough to have a ruling class; that class must have a chief, an absolute leader. Without a leader it could not subsist. The next problem then is: Who shall be the leader? The conclusion he had reached was that the philosophers should become kings or the kings philosophers, or else "there can be no cessation of troubles for our states nor, I fancy, for the human race either." [41] Had Plato learned nothing in Syracuse? And how could he imagine the possibility of such a conjunction? What philosopher, except perhaps Plato, would

[39] *Laws*, 694–698.

[40] Cf. the views of Walter Bradford Cannon (1871–1945) on homeostatic control of social conditions, *Isis* 36, 260 (1946). In a way these views would have pleased Plato, for they introduced a new analogy between physiology and politics, between the microcosm and the macrocosm.

[41] *Republic*, 473. The same idea is expressed again and again, half a dozen times, in the *Republic*.

ever wish to become a king? And how could a king extricate himself sufficiently from his innate passions and his daily worries to become a philosopher? The simultaneous existence of such different vocations in a single person would be nothing short of a miracle.

He thought that the problem could be solved by designing institutions for the education of future leaders. Much of the *Republic* is devoted to that, and the result was a radical corruption of the theory and practice of education.

At any rate, once the leader is chosen he should be obeyed implicitly even in the smallest matter.

> Now for expeditions of war much consideration and many laws are required; the great principle of all is that no one of either sex should be without a commander; nor should the mind of any one be accustomed to do anything, either in jest or earnest, of his own motion, but in war and in peace he should look to and follow his leader, even in the least things being under his guidance; for example, he should stand or move, or exercise, or wash, or take his meals, or get up in the night to keep guard and deliver messages when he is bidden; and in the hour of danger he should not pursue and not retreat except by order of his superior; and in a word, not teach the soul or accustom her to know or understand how to do anything apart from others.[42]

Absolute leaders are in constant danger of violent suppression, and they cannot brook independence and originality in their immediate entourage. They are surrounded by hypocrites and flatterers, men of a mediocre and cowardly nature. Where will their successors be found? That is an insoluble riddle. The best practical solution is to trust heredity and determine the succession by an organic law of the state, as was done for the absolute monarchs by divine right, but even that is an awful gamble.

There is no safe way of selecting a ruler, and in practice the ruler, if not hereditary, has always selected himself, seized the power, and cowed the opposition by the charm of his personality and the implacability of his defense.

One of the best parts of Popper's excellent book[43] is the one wherein he shows that by expressing the problem of politics in the form of that ominous question, "Who shall rule the state?" Plato has created a lasting confusion in political philosophy. The intelligent creative question is rather, "How can we organize political institutions so that bad and incompetent rulers can be prevented from doing too much damage?"

Here again, as always, one must fall back on education, for the best institutions do not suffice. They must be manned, and the good men who are needed can be produced only by appropriate training. The purpose of education is no longer the perverse Platonic purpose of creating leaders but the honest one of creating good men. In the course of time the best of these men, or rather the fittest for statesmanship, will or may become rulers, but even so their power must be restricted by constitutional checks.

[42] *Laws,* 942. The quotation is from the standard English translation of Benjamin Jowett (1817–1893), master of Balliol, hereafter referred to as "Jowett." This passage occurs in Stephanus' edition (3. vols.; Paris: Henri Estienne, 1578), vol. 2, p. 942, and in Jowett (ed. 3), vol. 4, p. 330.

[43] Popper, *The open society*, vol. 1, chap. 7. John Stuart Mill had explained the necessity of constitutional checks in his *System of logic* (1843), and in *The subjection of women* (1869) he had remarked: "Who doubts that there may be great goodness and great happiness and great affection under the absolute government of a good man? Meanwhile, laws and institutions require to be adapted, not to good men but to bad." "Who doubts?" asked Mill. Popper does, with good reason.

POLITICS AND MATHEMATICS

Plato's mathematics will be discussed in the next chapter, but this discussion may be anticipated by a few remarks concerning his mathematization of political thought. The mathematicians of today who deal with political problems approach them from the statistical or the economic side, but that could not have been Plato's approach, for he had no inkling of statistics, and no interest whatsoever in economic matters. It does not seem to have ever occurred to him that economic factors might influence private and public life, and yet he cannot have overlooked the troubles caused in families or nations by financial difficulties; when such troubles arise they are too obvious and too strident to be ignored. Did he never have to meet financial obligations, I wonder, either his own or those of other people to him? Were these obligations meaningless to him?

His approach was not arithmetic (in our sense) but geometric. The secret of the universe (cosmos) is order and measure. Plato extended that conception to everything domestic and political and he did it without moderation. Everything in the perfect city must be regulated; no change is foreseen, therefore there is no opportunity, no choice, no fancy. The city will function like a machine. Some chapters of *Laws* regulate private life with so much detail and so little restraint that they are to the modern mind repulsive and obscene.

Plato sometimes argues with words as if they were geometric symbols; in that respect he may be considered the earliest ancestor of the symbolic (or mathematical) logicians of today.

NEITHER FREEDOM NOR TRUTH IN THE REPUBLIC

Considering the mathematical pattern of the Republic, it is clear that there is little room for freedom. Freedom is the negation of virtue. Each man must know his station and stay there. He must know his duty and do it. He cannot choose his station and duty. The ruler himself has no freedom; though nobody can control him, he is self-controlled. Each man must mind his own business. Social conformity is carried to the limit.

In the *Laws*[44] young people are forbidden to criticize the city regulations. An old man may do this, but only when no youth is present.

Education is dominated by censorship. The citizens, young and old, must be given no opportunity to read anything that is not approved by the state, to listen to unorthodox sayings, or to hear improper music.

When Waldo Frank visited Moscow, he talked with a young mechanician and explained to him "that in the New York papers every morning one might find every possible shade of judgment on all possible subjects."[45] "I don't see the use of it," answered the young man. "Every problem has a right answer. It seems to me the press would be serving the people a lot better if they found each day the right opinion on each important subject and printed only that. What is the sense in printing a lot of different points of view, when only one can be right?" This would have pleased Plato, and if one had asked him which is the right point of view, he would have said without hesitation, "The point of view of the state."

[44] *Laws*, 634D.

[45] Waldo Frank, *Dawn in Russia* (New York: Scribner, 1932), p. 163.

Happily, Plato was not a dictator, except in his own dreams, and if he had been one his dictatorship would have been tempered by technical inefficiency. What the French call *bourrage de crâne* has been made very easy whenever modern governments have managed to control the press, the telegraph, the telephone, radio, television, but it was not so easy in ancient times. Plato's censorship would have been of necessity very incomplete and the nets of his inquisitors full of holes.

The destruction of freedom implies unavoidably the destruction of truth. If it becomes the ruler's duty to provide citizens only with wholesome ideas, the ideas must be sieved and graded. When the people are told only a part of the news, the lying is on, but Plato did not stop there. "Opportune falsehoods" and noble lies [46] may be necessary to deceive not only the people, but the elite itself. There is no doubt about it, the dictator must lie, or his assistants must lie for him (which amounts to the same thing). How could Plato reconcile that conclusion with the theory of the philosopher-king? For the philosopher is a man who loves the truth and if the king must lie, even if it be only occasionally, how will the philosopher in him take it? The search for the truth and the exercise of absolute power are utterly incompatible.

As Popper remarks,

> Socrates had only one worthy successor, his old friend Antisthenes, the last of the Great Generation. Plato, his most gifted disciple, was soon to prove the least faithful. He betrayed Socrates, just as his uncles had done. These, besides betraying Socrates, had also tried to implicate him in their terrorist acts, but they did not succeed, since he resisted. Plato tried to implicate Socrates in his grandiose attempt to construct the theory of the arrested society; and he had no difficulty in succeeding for Socrates was dead.
>
> I know of course that this judgment will seem outrageously harsh, even to those who are critical of Plato. But if we look upon the *Apology* and the *Crito* as Socrates' last will, and if we compare these testaments of his old age with Plato's testament, the *Laws*, then it is difficult to judge otherwise. Socrates had been condemned, but his death was not intended by the initiators of the trial. Plato's *Laws* remedy this lack of intention. Coolly and carefully they elaborate the theory of inquisition. Free thought, criticism of political institutions, teaching new ideas to the young, attempts to introduce new religious practices or even opinions, are all pronounced capital crimes. In Plato's state, Socrates would never have been given the opportunity of defending himself publicly; he would have been handed over to the secret Nocturnal Council for the "treatment," and finally for the punishment, of his diseased soul.[47]

Starting from the concept of transcendental truth as represented by the eternal Ideas, Plato has gradually fallen to the level of propaganda, inquisition, and beneficent lying. At first view, there would seem to be an abyss between absolute truth and definite lying, but Plato bridged it without seeming to realize his disingenuousness. Compare his aberration with the views of men of science: we do our best to approach the truth by successive approximations; we do not claim to have the truth, but we are reaching for it and coming gradually closer to it; we do not start with the whole truth, but we obtain more and more of it. This is impossible without freedom. Truth is not, as Plato thought, an ideal from which we have fallen off; it is an ideal toward which we are steadily progressing. Truth is a goal and a limit, and so is pure democracy.

[46] *Republic,* 414B, 389B.

[47] Popper, *The open society*, vol. 1, p. 171.

PLATO'S RELIGION

In his ideal state, Plato "instituted a religion considerably different from the current religion, and proposed to compel all the citizens to believe in his gods on pain of death or imprisonment. All freedom of discussion was excluded under the cast-iron system which he conceived. But the point of interest in his attitude is that he did not care much whether a religion was true, but only whether it was morally useful; he was prepared to promote morality by edifying fables; and he condemned the popular mythology not because it was false, but because it did not make for righteousness."[48]

PLATO'S LACK OF HUMANITY

Plato's ideal of frozen perfection and his communism entailed a hatred not only of freedom but of individualism in all its forms. His attack against individualism is strangely tortuous and disingenuous. To put it as briefly as possible, one may oppose individualism to collectivism and egoism to altruism.[49] Plato's jugglery (which, let us hope, was inconscient) consists in equating the first and the last terms of each pair (individualism with egoism, collectivism with altruism); one may then conclude that individualism is incompatible wth altruism. Q.E.D. A man must be a communist or else he is a selfish brute! The whole trend of political progress from St. Thomas on to our own day has been, on the contrary, to combine individualism (freedom of conscience) with altruism.

Not only did Plato reject individualism but he had no respect for personality. This was made clear in the passage from the *Laws* quoted above,[50] and in many others. Let me quote one more, taken from the same book:

> The first and highest form of the state and of the government and of the law is that in which there prevails most widely the ancient saying, that "Friends have all things in common." Whether there is anywhere now, or will ever be, this communion of women and children and of property, in which the private and individual is altogether banished from life, and things which are by nature private, such as eyes and ears and hands, have become common, and in some way see and hear and act in common, and all men express praise and blame and feel joy and sorrow on the same occasions, and whatever laws there are unite the city to the utmost, – whether all this is possible or not, I say that no man, acting upon any other principle, will ever constitute a state which will be truer or better or more exalted in virtue.[51]

It is paradoxical that the writer who hated individualism was canonized as a master of humanism, and some enthusiasts went so far as to consider him a kind of proto-Christian. Plato's subordination of the individual to the state was so complete that his philosophy became almost inhuman. And yet his self-delusion was so deep that he gave to the *Republic* the alternative title of Justice, and a good part of the book is devoted to the discussion of abstract justice.

What is justice? That which is in the interest of the state. The city is just when the castes are determined and invariable and everybody remains in his proper place, when the principle of class rule and class privilege is meekly accepted by all the people. The well-ordered and unchangeable city is the symbol of eternal justice.

[48] John Bagnell Bury (1861–1927), *History of the freedom of thought* (New York, 1913), p. 35.

[49] For details, see Popper, vol. 1, p. 87.

[50] *Laws*, 942, quoted above.

[51] *Laws*, 739; Jowett, vol. 5, p. 121. See also *Republic*, 462, and *passim* by means of Jowett's index.

Plato's definition of justice was chosen to strengthen his totalitarian conceptions, while the popular and common-sense idea of justice was exactly the opposite; hence we continue to turn in the same circle.

There are occasionally feelings of humanity in Plato's writings, especially in the early Socratic dialogues, for example, in *Gorgias* when he argues that it is better to suffer injustice than to perpetuate it, but he never grasped the idea of humanity as transcending that of the crystallized city of his dreams. Let humanity be sacrificed to that city or else the latter would degenerate and fall.

He could not understand that justice should never be divorced from love. Love without justice is erratic and dangerous, but justice without love is inhuman. Abstract justice is dangerously near to injustice.

We cannot blame Plato for being un-Christian, but he deserves blame for having sacrificed to his political dogmatism the generous ideals of Pericles, Democritos, Socrates, and of Gorgias' disciples Alcidamas, Lycophron, and Antisthenes.[52] It is because of that wanton sacrifice that I entitled this section "The great betrayal." It was a betrayal not only of Athenian democracy but also of the master who had been his first guide and whom he had loved. Inded, many of the arguments against democracy were put in the mouth of Socrates; Plato made his old master say the very opposite of what he had taught. Was Plato's power of self-illusion so deep that he could no longer distinguish between the real Socrates and the Socrates created by his own fancy? [53]

Can there be a deeper betrayal than that? Plato did not deny his master; what he actually did was considerably worse; in his later works he presented a caricature of him, a monstrous deformation. Let us repeat that Socrates was a democrat, an individualist, an equalitarian. Plato gradually became the opposite of all this; Socrates' main purpose was to teach self-criticism, and he was always ready to recognize his ignorance; Plato, on the contrary, was the master who knew, the philosopher-king who must be obeyed implicitly, the creator of a Republic that is perfect by definition and therefore cannot change without disgrace.

There is another kind of betrayal, however, for which Plato is not responsible and which comes closer to the one described by the French author, Julien Benda (1867–), in his *Trahison des clercs*.[54] The clerks who betrayed us, I would say, are the many commentators who explained Plato's political thought but gave us an entirely false picture of it, because they glossed over his totalitarianism and his views on the communism of property, women, and children.

I cannot do better once more than to quote Popper's words:

> What a monument of human smallness is this idea of the philosopher king. How far removed it is from the simple humaneness of Socrates, from the Socratic demand that the responsible statesman should not be dazzled by his own excellence, power, or

[52] Alcidamas of Elaia (Aeolis) condemned slavery as contrary to natural law. According to Lycophron, "law is only a convention, a surety to one another of justice and has no power to make the citizens good and just." Aristotle, *Politics*, 1280B 10.

Antisthenes of Athens, founder of the Cynic school, was a disciple of Socrates, at whose death he was present. He taught at the Cynosarges, a gymnasium outside of the walls of Athens for the use of those who were not of pure Athenian blood. He was an "impure" Athenian himself, his mother being a Thracian. He died in Athens at the age of 70.

[53] The reader may object: How do you know that? Well, the real Socrates is the one about whom Plato and Xenophon agree, and whose genius is defined in the early Socratic dialogues of the former.

[54] J. Benda, *Trahison des clercs* (Paris, 1927).

wisdom, but that he should know what matters most: that we are all frail human beings. What a distance from this world of irony and truthfulness and reason, to Plato's kingdom of the sage, whose magical powers raise him high above ordinary men; but not high enough to forego the use of lies, nor to neglect the sorry game of all shamans, the sale of taboos – of breeding taboos – for power over his fellow-men.[55]

THE TIMAIOS

The scientific ideas of Plato will be analyzed later, but it is proper to speak now of the book that most scholars think of as his main scientific book. This is the *Timaios*, dealing not with science in the restricted sense but with cosmology, that is, the study of the universe in its fullness, order, and beauty. Science, as we understand it, concerns itself with limited objects, and it owes its success and immense fertility to its deliberate and severe restraint. Cosmology is the opposite: it deals with the whole universe, and therefore, irrespective of the amount of scientific ingredients which he may include in his survey, the cosmologist is to be judged as a metaphysician rather than as a man of science.

This is especially true of the *Timaios*, which many commentators have considered for thousands of years as the climax of Platonic wisdom, but which modern men of science can only regard as a monument of unwisdom and recklessness.[56]

Toward the end of his life, say during the last two decades, Plato began the writing of a trilogy – *Timaios*, *Critias*, and *Hermocrates*. He finished the *Timaios*, but interrupted *Critias* abruptly (in the middle of a sentence) and did not even begin the third piece. The whole was to be a story of the world, from prehistoric times to the dawn of the future. The two last parts were political, and as his notes accumulated he probably realized that the original frame was too narrow. He then abandoned it and started the composition of the *Laws*, his last and longest work. It is clear that if one begins to legislate for the future, and wants to do it with sufficient detail, the composition must increase enormously, far beyond the scope of the original dialogues.

The *Timaios* is so called after the main speaker, Timaios of Locris, who cannot be identified with any actual person, and who was perhaps a poetic creation.[57] It is the cosmologic basis of the trilogy. The argument may be divided into three parts (the numbers in parentheses indicate their relative lengths): (i) Introduction (8) including the myth of Atlantis as told by Solon, the wisest of the Seven Wise Men;[58] (ii) cosmology proper (42), the making of the soul of the world, theory of the elements, theory of matter and of sense objects; (iii) physiology, normal and pathologic (23), the making of man's soul and body. The second part is the

55 Popper, *The open society*, vol. 1, p. 137.

56 There are innumerable editions of it. The English readers may use Jowett, vol. 3, or R. G. Bury's edition (Greek-English) in the Loeb *Plato*, vol. 7 (1929), pp. 3–253, or the translation with running commentary by Francis Macdonald Cornford (1874–1943), *Plato's Cosmology* (394 pp.; London: Kegan Paul, 1937) [*Isis 34*, 239 (1942–43)]. Cornford's is the most convenient for the historian of science. Heinrich Otto Schröder, *Galeni in Platonis Timaeum commentarii fragmenta. Appendix II. Mosis Maimonidas Aphorismorum praefatio et excerpta a Paulo Kahle tractata* (140 pp.; *Corpus medicorum graecorum, Suppl.* 1; Leipzig, 1934).

57 Attempts were made to identify Timaios of Locris (Locri Epizephyrii, southeast Bruttium, Italy) with an old Pythagorean who would have been Plato's teacher and who wrote in Doric dialect a treatise (*Peri psychas cosmu cai physios*) on the soul of the world and nature. This was taken by the Neoplatonists as a genuine work, but has been shown to be apocryphal, not earlier than the first Christian century. Far from being prior to the *Timaios*, it is a late summary of it.

58 (*Timaios*, 20E). The myth was related to Solon by an old priest of Sais in the Delta. We have already referred to their conversation (p. 401).

main one, much longer than the two others combined. It discusses the essence of physics, being and becoming, model and copy (the visible world is only a likeness of the real world); the creation, the body of the world and its soul, the coöperation of reason and necessity; and so on. A fuller analysis would take considerable space and would only serve to sidetrack the reader.

The ideal city described in the *Republic* is supposed to have really existed in the remote past, in prehistoric Athens. The purpose of *Timaios*, however, is to link the ideal Republic with the organization of the whole universe; the Republic is only the political aspect of the universe; human morality is only a reflection of cosmic intelligence.

The world artificer (*dēmiurgos*) is not a creator but, like the *nus* of Anaxagoras, an ordinator. Let us call him the divine reason; the ordered world is divine itself to the extent of its reasonableness. The distinction between matter and mind is not quite clear, for both can be expressed in terms of the universal intelligence. There is another form of dualism in the *Timaios*, however, the distinction between the greater world (*macrocosmos*) and the smaller one (*microcosmos*). That distinction had already been made by Democritos (p. 251), but Plato developed it considerably.

The universe is like a single living body, the rationality of which is evidenced by the regularity of astral motions. The soul of the universe is comparable to the soul of man; both are divine and immortal.

The planets and stars are the most sublime representations of the Ideas; one might call them gods. Astronomy is the basic knowledge for wisdom, health, and happiness. The divine mathematics represented by the movements of the stars can be traced also in music and in the theory of numbers. When men die, their souls return to their native stars.[59]

The astrologic nonsense that has done so much harm in the Western world and is still poisoning weak-minded people today was derived from the *Timaios*, and Plato's astrology was itself an offshoot of the Babylonian one. In justice to Plato it must be added that his own astrology remained serene and spiritual and did not degenerate into petty fortunetelling. To his contemplative mind the planets are like perfect clocks which reveal the march of time, the rhythms of the universal soul.

On account of the number of planets these rhythms are very complex, but given a certain grouping of them, after a certain interval of time the same grouping will recur. That interval is the Great Year,[60] measured by the perfect number (36,000 years, 760,000 years?).

The poetic analogy between the little world and the big world (microcosm and macrocosm), between our body and the universal body, can be carried very far.[61] It guided Plato's thought and, largely because of him, dominated the minds of many medieval thinkers, and even of such a "modern" man as Leonardo da Vinci. The particular aspect of that analogy which interested Plato most was of course this one: the perfect city of his dreams is an image of the divine city. *Timaios* is a cosmologic justification of the *Republic*.

The cosmos is made up of four elements: earth, water, air, fire, of which the

[59] *Timaios*, 42B.

[60] *Republic*, 546B; *Timaios*, 39D.

[61] The concepts of "great year" and of "microcosm vs. macrocosm" are probably also of Oriental, Babylonian, origin.

second and third are mean proportionals between the first and the last.[62] These elements are solid bodies, however, can be resolved into geometric parts, and are related to four regular solids.[63]

Plato met Philistion of Locroi (see p. 334) in Syracuse and may have been influenced by him, or he might have been influenced if he had been less impervious to experimental science. Philistion was not simply a theorist, following Empedocles; he was a distinguished anatomist, who had made dissections and even vivisections. He considered the heart as the main regulator of life, and his observations of the living heart were very keen. He discovered that the ventricles die before the auricles (we know indeed that the right auricle is the last part of the heart to die, *ultimum moriens*) and that the sigmoid (or semilunar) valve of the pulmonary artery is weaker than the sigmoid valve of the aorta (quite so, for the pressure in the pulmonary circulation is but one-third of the pressure in the systemic circulation). Philistion's observations are astounding, for they imply a certain amount of experimentation, but our ascription of them to him is based on the assumption that he was the author of the Hippocratic treatise on the heart.[64]

The circulation of food and blood in the body is like the circulation of water in the earth[65] or "like that movement of all things in the universe which carries each thing toward its own kind."[66]

Plato recognized three groups of diseases. The first group was caused by alterations of the four elements; the second by the corruption of humors derived from those elements; the third by pneuma, phlegm, and bile.[67] Now, the third group suggests comparison with the *tridosa* in *Āyurveda*. The ideas of Plato and of the Hindu physicians being equally vague, such a comparison leads nowhere.[68]

The lost island, Atlantis,[69] which was somewhere west of Gibraltar, has caused considerable speculation of a peculiarly irrational kind. For example, when the hypsometry of the Atlantic Ocean was better known and geologists began to formulate the hypothesis of lost islands or continents on a solid basis of observations, it was suggested that Plato had anticipated their discovery! Many geologists have wasted their time in trying to give some appearance of reality to Plato's dream.

Such aberrations have been carried to the limit by a Polish logician, Wincenty Lutoslawski, in his remarkable book on the *Origin and growth of Plato's logic*.[70] Lutoslawski found in Plato's writings anticipations of spermatozoa[71] and of the true constitution of water, three atoms, two of one gas, and one of another.[72] *Risum*

[62] *Timaios*, 31B ff.

[63] *Timaios*, 53C ff. The fantastic comparison of the elements with the Platonic solids was not translated by Chalcidius, whose version and commentary stopped just short of it.

[64] Littré's edition of that treatise, *Peri cardiēs*, vol. 9, pp. 76–93, is very insufficient. Better edition by Friedrich Karl Unger (Utrecht thesis, 1923). G. Leboucq, "Une anatomie antique du coeur humain. Philistion de Locres et le Timée," *Revue des études grecques* 57, 7–40 (1944). This includes a new edition of the *Peri cardiēs* by Joseph Bidez.

[65] For the circulation of water in the earth (*perirrhoē*) see *Phaidon*, 111 D–E.

[66] *Timaios*, 81.

[67] *Timaios*, 82–84.

[68] Dhirendra Nath Ray, *The principle of tridoṣa in Āyurveda* (376 pp.; Calcutta: Banerjee, 1937) [*Isis* 34, 174–177 (1942–43)]. Jean Filliozat: *La doctrine classique de la médecine indienne* (Paris: Imprimerie nationale, 1949) [*Isis* 42, 353, (1951)].

[69] *Timaios*, 24E.

[70] Wincenty Lutoslawski, *Origin and growth of Plato's logic, with an account of Plato's style and of the chronology of his writings* (565 pp.; London, 1897) see p. 484. In this book an effort was made to put Plato's writings in chronologic order, on the basis of a systematic investigation of 500 peculiarities of his style.

[71] *Timaios*, 91C.

[72] *Timaios*, 56D.

teneatis? This shows to what extent the cult of Plato can go. For if Plato could have anticipated Leeuwenhoek and Lavoisier without instruments he would have been not a man of science but a magician or wonderworker. Lutoslawski reminds me of the people who read scientific anticipations in the Bible or the Qur'ān; at any rate, there is more logic in their efforts than in his, for if those sacred books were directly inspired by God, well, God knows the future. To make the same claim for Plato without asserting his divinity involves fundamental contradictions.

If one of our own contemporaries, a well-trained and distinguished philosopher like Lutoslawski, is able to read such things in *Timaios*, we should not be surprised by the fantastic interpretations of ancient and medieval scholars. Thanks to the prodigious fame of the divine Plato, his *Timaios* was taken not as a poetical fancy, but as a kind of cosmologic gospel. Its very obscurity allured many people; it may have been partly deliberate, but it was largely due to the confusion existing in Plato's own thoughts; it is the kind of obscurity that one calls oracular and that weak-minded people take for a proof of divinity and certainty. The skeptical philosopher and poet Timon of Phlius[73] coined a new verb, *timaiographein*, to mean writing in the oracular style of *Timaios*, to vaticinate. Julian the Apostate (IV–2) opposed the *Timaios* to Genesis, and Proclos (V–2), one of the last directors of the Academy, would have liked to destroy all the books except *Timaios* and the Chaldean Oracles.[74]

The influence of *Timaios* upon later times was enormous and essentially evil. A large portion of *Timaios* had been translated into Latin by Chalcidius (IV–1), and that translation remained for over eight centuries the only Platonic text known to the Latin West.[75] Yet the fame of Plato had reached them, and thus the Latin *Timaios* became a kind of Platonic evangel which many scholars were ready to interpret literally.[76] The scientific perversities of *Timaios* were mistaken for scientific truth. I cannot mention any other work whose influence was more mischievous, except perhaps the Revelation of St. John the Divine. The apocalypse, however, was accepted as a religious book, the *Timaios* as a scientific one; errors and superstitions are never more dangerous than when they are offered to us under the cloak of science.

PLATONIC LOVE

We read in *Laws*[77] that "among men all things depend upon three wants and desires, of which the end is virtue, if they are rightly led by them, or the opposite if wrongly." These three desires are hunger and thirst, which begin at birth, and sexual lust, which sets in later. In *Timaios*[78] Plato states that "since human nature is two-fold, the superior sex is that which hereafter should be designated man."

[73] Timon of Phlius (northeast Peloponnesos) had studied philosophy at the school founded by Euclid in Megara; after many years of wandering, he spent the remainder of his life in Athens, where he died in very old age. He wrote mock poems called silli (*silloi*) and is therefore called Sillographos.

[74] It is significant that the only two works which Proclos would have preserved were both Oriental. Indeed, there is more Oriental lore in the *Timaios* than Greek wisdom.

[75] More exactly, Chalcidius's incomplete translation of *Timaios* remained the only Platonic text available in Latin until the translation of *Menon* and *Phaidon* c. 1156. In Henri Etienne's edition the *Timaios* covers pp. 17 to 92 of vol. 3; Chalcidius' translation and commentary stopped at 53B.

[76] See the last section of this chapter, outlining the medieval tradition of the *Timaios*.

[77] *Laws*, 782D.

[78] *Timaios*, 42.

At the very end of the same book he introduced a fantastic theory of sex. His embryology appears in a kind of appendix and sex itself as a kind of afterthought of creation, a disturbing factor:

> Wherefore in men the nature of the genital organs is disobedient and self-willed, like a creature that is deaf to reason, and it attempts to dominate all because of its frenzied lusts. And in women again, owing to the same causes, whenever the matrix or womb, as it is called, — which is an indwelling creature desirous of childbearing, — remains without fruit long beyond the due season, it is vexed and takes it ill; and by straying all ways through the body and blocking up the passages of the breath and preventing respiration it casts the body into the uttermost distress, and causes, moreover, all kinds of maladies; until the desire and love of the two sexes unite them.[79]

In another part of the same work, after having referred to the sexual passions he says,

> And if they shall master these they will live justly, but if they are mastered, unjustly. And he that has lived his appointed time well shall return again to his abode in his native star, and shall gain a life that is blessed and congenial; but whoso has failed therein shall be changed into woman's nature at the second birth; and if, in that shape, he still refraineth not from wickedness he shall be changed every time, according to the nature of his wickedness, into some bestial form after the similitude of his own nature; nor in his changings shall he cease from woes until he yields himself to the revolution of the Same and Similar that is within him, and dominating by force of reason that burdensome mass which afterwards adhered to him of fire and water and earth and air, a mass tumultuous and irrational, returns again to the semblance of his first and best state.[80]

In the speech of Diotima in *Symposium*, it is explained that sexual desire is the lowest form of our passion for immortality. Plato realized the need of marriage and the begetting of children. In the perfect Republic the sexual relations of the best people would be reserved for certain solemn occasions, and regulated according to demographic needs. Plato does not seem to have realized that married love involved a peculiarly intimate relation between two persons, required much kindness and sweetness from both of them, and if fortunate brought great rewards. He thought of short marriages somewhat in the mood of a stockbreeder. It does not seem to have occurred to him that marriage is not simply a matter of sexual convenience and eugenics, but is a relation between persons, a communion of hearts; that for the development of rich personalities and harmonious couples it is the long marriages that count, the longer the better; and that happy and enduring unions are the greatest blessings of life.

How is it that the idealist Plato did not think of such matters? The reason is simply that whenever he idealized sexual desires — and he did so frequently, whenever he thought of a struggle between the spirit and the flesh, whenever he took a romantic view of love, his background was not heterosexual but homosexual. "Platonic love" has for us two meanings; the first is an urge to union with the beautiful and contemplation of the ideal (as expressed by Diotima), the second is a spiritual friendship without sexual desire. When we think of Platonic love in the second sense, we think of spiritual friendship between a man and a woman;

[79] *Timaios*, 91; Loeb, vol. 7, p. 249.

[80] *Timaios*, 42B; Loeb, vol. 7, p. 91. Similar views concerning the transformation of men into women or into animals are again expressed at the end of the *Timaios* (91–92).

Plato, however, thought of a spiritual friendship between a man and a boy. Platonic love for him was the sublimation of pederasty; true love is called in *Symposium*[81] the right method of boy loving (*to orthōs paiderastein*).

Plato was not necessarily a pederast in the physical sense, but he was almost certainly homosexual. He never married, and though he speaks occasionally of sexual relations between men and women he does so without emotion; his tender feelings were reserved for homosexual relations. He was somewhat of a woman hater. That is revealed to us many times in his writings. Compare, for example, Xenophon's gentle account of Xanthippe in the *Memorabilia*[82] with Plato's brutal one in *Phaidon*. Xenophon spoke like a *paterfamilias*, Plato described the scene like the misogynist he was.[83] How could one believe that Plato, gentle and gracious as he could be in other respects, would have sacrificed women and the sanctity of marriage as he did in the *Republic*? It was relatively easy, however, for a homosexual to accept the community of wives and children.

In fairness to Plato it must be added that in his last work, the *Laws*, pederasty was condemned.[84] It may be argued also in his defense that the practice of pederasty was fairly common in Athens and even more so in the countries that he admired so much, Crete and Lacedaimon. According to him, the story of Zeus and Ganymedes,[85] which was the divine exemplar of pederasty, had been invented in Crete. It is probable that pederasty was more common in Athens among the aristocrats, the idle rich, and the sophisticated than among the simpler people, and in any case, heterosexuality must have been the rule, not the exception, otherwise the race would have perished. The Greeks honored marriage and wanted children even as we do, and perhaps even more, because male descendants were needed to continue the domestic cult and to perform religious rites when their father died. The atmosphere of Plato's writings is homosexual, but the same is not true of other Greek writers of his time, such as Xenophon. We must assume that the average normal man in Greece, as in our own day, was inclined to love women and to beget children.

It was necessary to clarify these matters, though they have no immediate bearing on the history of science, because we should be able to appreciate Plato's personality and to measure the hypocrisy of his commentators. Most of them have preferred to draw a veil over Plato's homosexuality, as they did over his integral communism. English translators found it easy to disguise homosexual references, for an adjective like "beloved" may refer to a woman as well as to a man, while the Greeks used a masculine participle leaving no room for ambiguity. The translators may try to justify their prudishness by the need of reverence for youth. It would be better, however, to avoid a text than to misinterpret it deliberately;

[81] *Symposium*, 211B.

[82] *Memorabilia*, 2, 2. Socrates reproves his eldest son, Lamprocles, for being out of humor with his mother and ungrateful to her.

[83] Just before Socrates drank the hemlock, his wife Xanthippe came in. "She cried out and said the kind of thing that women always do say: 'Oh Socrates, this is the last time now that your friends will speak to you or you to them.' And Socrates glanced at Crito and said, 'Crito, let somebody take her home.' And some of Crito's people took her away wailing and beating her breast" (*Phaidon*, 60). Then Socrates talks of something else. The whole story was quoted above. Socrates' dismissal of his poor wife is unbelievably churlish and cruel in this account.

[84] *Laws*, 636C, 836C.

[85] Ganymedes became the nickname of boys prostituted to men. The word must have been used frequently in Roman times, because it wore out and was corrupted in Latin to catamitus (hence the English catamite).

there is no excuse for lies, and the lies used to illustrate a false idealism are of all lies the worst ones.

See David Moore Robinson and Edward James Fluck, *Study of the Greek love names including a discussion of paederasty* (210 pp.; Baltimore: Johns Hopkins University Press, 1937); Warner Fite, *The Platonic legend* (New York: Scribner, 1934). Hans Kelsen, "Platonic love," *The American Imago* 3, 110 pp. (Boston, 1942).

CONCLUSION

Plato was a poet and a metaphysician, a craftsman who used wonderfully well a literary instrument of almost unbelievable exquisiteness, the Greek prose of the golden age. His scientific activities will be discussed in the next chapter, but we may remark here that he was not a man of science; he was a cosmologist, a metaphysician, a seer. The history of Platonism is the history of a long series of ambiguities, misunderstandings, prevarications.

Our own discussion of his political and sexual fancies might seem out of place in a book devoted to the history of science; yet the subterfuges and evasions of commentators concerning his aberrations deserve to arrest our attention. There is nothing comparable to it in world literature, except perhaps the general blindness to some obscene verses of the Old Testament. It seemed as if the divine Plato could do no wrong, and one could not suspect him without becoming oneself an object of suspicion and a stumbling block. The story of Averroism is also a sequence of misunderstandings, but of a very different kind. While Plato was generally praised to the skies, and his faults hidden or glossed over, Ibn Rushd was painted blacker than he was. The two cases had this in common, however, that the judgment of scholars was governed and colored by the popular verdict. That verdict was generally in favor of Plato, while Averroes was condemned; or, to put it more clearly, it had become a matter of good breeding and good usage to pay homage to Plato, while if one spoke at all of Averroes, it was to blame him. A gentleman was naturally a Platonist, while every Averroist was somewhat of a radical and troublemaker.

Such uncritical praise implied hypocrisy and falseness. One cannot extol a man for his divine wisdom, and yet condone his nonsense; that is not honest.

The matter is not so bad as it looks if one bears in mind that the Platonic legend is largely due to literary prejudice. The language of Plato was so beautiful and so difficult to understand that the contents were overlooked, beauty was mistaken for justice, and obscurity for depth. Plato ended by occupying in Greek culture a place almost as high as Homer, and like the latter he dominated Greek education.

This was the supreme misunderstanding. He was not interested in individuality or personality, and hence we cannot call him a true humanist, and yet the humanists of Byzantion and of Florence considered him their master. They were so sure of that and so anxious to protect their faith that they always refused to read in his writings the obvious proofs of his lack of humanism.

Plato had a right to his own opinions, and we should not blame him for having expressed them, but the commentators who glossed over everything that was objectionable in his thoughts deserve to be castigated. Their attitude is very puzzling. That schoolmasters entrusted with the education of the future rulers of their country should be pleased with Plato's aristocratic assumptions, and even with his totalitarian methods, we can understand; but how could they be blind to his ideas on communism and catamiting, to his lack of respect and tenderness for women, and other things that were completely at variance with their own predilections? How could Plato get away with murder? [86]

Plato was a great poet and had glimpses of wisdom, but he was not always a safe guide, and in many cases he was very unsafe and would have led us to the abyss. Happily, the men who praised him so much did not follow him. It would have been best, perhaps, to deal with him as he did with Homer – to crown him with flowers and drive him out of the city. No, that would not do at all; we should not imitate his worst manners. He should be allowed to remain and to have his say. Let him stay and let us see him and show him to others as he was – sometimes great, and sometimes not.

Theologians and philosophers may gloss over his aberrations, but for men of science that is the unpardonable sin. A *paideia* based on lies is bad, and the better it looks on the surface the more seductive it is, and the more pernicious.

The cult of Plato being part and parcel of Western humanities, it required considerable courage to criticize him. One of the first to do so was Charles Crawford, in his dissertation on the *Phaido* (London, 1773); Crawford was a young rebel of Queen's in Cambridge and his book is marred by petulance and verbosity (Fig. 82). We must pay tribute to George Grote (1794–1871), who wrote a large work on *Plato and the other companions of Sokrates*,[87] intended as a sequel and supplement to his *History of Greece*; Grote admired Plato but was not afraid of criticizing him.

More recent books revealing Plato as he was, by quoting his own words, have already been mentioned, the most important being those of Fite (1934), Farrington (1940) and Popper (1945).

Warner Fite (1867–   ) was professor of ethics in Princeton University. In a long letter which he did me the honor of writing to me (dated Hopewell, N. J., 1 July 1944) he gave me an outline of the criticism to which his book *The Platonic legend* had been subjected. Some of the critics condemned the book for maligning Plato, others for saying irreverently what everyone knew to be true (but what no one, except Grote, had said in print). At the end of his letter he observed: "If I were rewriting the *Legend* I should try to make some changes in the emphasis. After all the 'animus' was not so much directed against Plato as against his interpreters. And after chapter viii, but especially in ix to xi, I was rather more interested in developing the picture of a scientific theorist than in negative criticism. But at seventy-seven and nine years emeritus, I have to leave it as it stands."

The same remark applies to this chapter, the purpose of which was to destroy the false image of Plato that many generations of adulators have created. *Amicus quidem Plato sed magis amica veritas.*[88]

[86] In 1950, politicians wishing to discredit the U. S. Department of State insinuated that many officers of that department were communists or homosexuals. Is it possible that those officers were simply Platonic gentlemen?

[87] 3 vols.; London, 1865.

[88] That sentence is often quoted but few people could trace it to its source. It is taken from the life of Aristotle by Ammonios Saccas (III–1) edited in Greek and Latin by Ant. Westermann in *Diogenis Laërtii vitae philosophorum* (Paris: Didot, 1862), part 2, p. 10. Ammonios applied it to Socrates, not to Plato; yet the numerous quotations always read *Amicus Plato.*

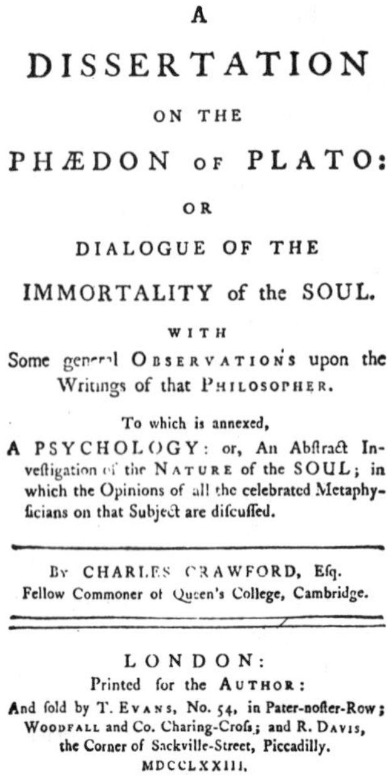

A

DISSERTATION

ON THE

PHÆDON OF PLATO:

OR

DIALOGUE OF THE

IMMORTALITY of the SOUL.

WITH

Some general OBSERVATIONS upon the Writings of that PHILOSOPHER.

To which is annexed,

A PSYCHOLOGY: or, An Abſtract Inveſtigation of the NATURE of the SOUL; in which the Opinions of all the celebrated Metaphyſicians on that Subject are diſcuſſed.

By CHARLES CRAWFORD, Eſq.
Fellow Commoner of Queen's College, Cambridge.

LONDON:
Printed for the AUTHOR:
And ſold by T. EVANS, No. 54, in Pater-noſter-Row; WOODFALL and Co. Charing-Croſs; and R. DAVIS, the Corner of Sackville-Street, Piccadilly.
MDCCLXXIII.

Fig. 82. A curiosity of English literature. The first attack on Platonic philosophy, by Charles Crawford, 1773. [From the copy in the Harvard College Library.]

A NOTE ON THE ANCIENT AND MEDIEVAL TRADITION OF THE TIMAIOS

Until the middle of the twelfth century the learned people of the West knew only the *Timaios* out of all the books of Plato, and therefore Plato to them was simply or chiefly the author of the *Timaios*. It is worth while to retrace briefly the tradition of that fateful book.

The *Timaios* had been also one of the first books to attract the attention of commentators. The first Platonic commentary was devoted to it by Crantor of Soli (Cilicia, fl. *c.* 300 B.C.), and extracts from it have been preserved by Plutarch and Proclos. Other Greek commentaries on the *Timaios* were composed by Posidonios of Apamea (I-1 B.C.), Adrastos of Aphrodisias (Caria, II-1) Galen (II-2),[89] Proclos of Byzantion (V-2), and the latter's pupil, Asclepiodotos of Alexandria (V-2). The Neoplatonic philosophers were acquainted with it. So much for the Greek tradition.

The Latin one began with Chalcidius (IV-1), who translated the *Timaios* into Latin down to the middle of 53 C. The next Platonic dialogues to be Latinized were *Menon* and *Phaidon*, and that was not done until *c.* 1156. The main events in that tradition are represented by the names of John Scot Erigena (IX-2), William of Conches (XII-1), Bernard Silvester (XII-1), Albert the Great (XIII-2), William of Moerbeke (XIII-2) and St. Thomas Aquinas (XIII-2). I find nothing in the fourteenth century except that the dialogue composed by Jean

[89] Galen devoted two commentaries to the *Timaios*, the second of which, lost in Greek but preserved in Arabic, has recently been edited by Paul Kraus and Richard Walzer, *Galeni compendium Timaei Platonis aliorumque dialogorum synopsis quae extant fragmenta* (130 pp. + 67 pp. in Arabic; London: Warburg Institute, 1951) [*Isis 43*, 57 (1952)].

Bonnet (XIV–1) of Paris, *Les secrets aux philosophes*, is probably a reflection of the *Timaios*; the two interlocutors are called Placides and Timeo. I dealt with it in the part of my *Introduction* devoted to the first half of the fourteenth century, but it may have been written at the end of the thirteenth; it is certainly prior to 1304. The Latin tradition of the *Timaios* is not always easy to disentangle from Neoplatonic traditions.

The Arabic tradition overlaps the Latin, even as the Latin overlapped the Greek. It begins with Yaḥyā ibn Baṭriq (IX–1) who translated the *Timaios* into Arabic; another translation is said to have been made by Ḥunain ibn Isḥāq (IX–2); the translation (whichever it was) was corrected by Yaḥyā ibn 'Alī (X–1).

The ascription of a translation to Ḥunain ibn Isḥāq may be due to a misunderstanding. Ḥunain did translate into Syriac, and also partly into Arabic, Galen's commentary on the medical section of the *Timaios*. Ḥunain's Arabic translation was completed by his nephew Ḥubaish ibn al-Ḥasan (IX–2).[90] This translation was probably the source of another error committed by al-Mas'ūdī (X–1) in his Kitāb al-tanbīh, where he ascribed to Plato a medical *Timaios* distinct from the *Timaios* itself. We may safely assume that that medical *Timaios* is simply the medical part of the *Timaios*, which had been distinguished and separated from the rest in Galen's commentary translated by Ḥunain.[91]

Irrespective of the Arabic translation of the *Timaios*,[92] the essence of it was known to Arabic philosophers through the *Theology of Aristotle* (V–2) and other Neoplatonic writings. That tradition was rather confused, the views of Plato being mixed with those of Plotinos and others.

Ḥunain ibn Isḥāq wrote a treatise entitled "That which ought to be read before Plato's works." [93] This title suggests the one used by Theon of Smyrna (II–1), but Theon's introduction to Plato was restricted to mathematics.

Brief as it is, this outline is sufficient to illustrate the vicissitudes of Platonic traditions before the printing of the Greek and Latin editions.

The study of the other Platonic works would lead to similar conclusions. For example, the *Republic* was commented upon in Greek by Proclos (V–2); it was translated into Arabic by Ḥunain ibn Isḥāq (IX–2) and commented upon in that language by Ibn Rushd (XII–2) and in Hebrew by Samuel ben Judah of Marseille (XIV–1) and by Joseph Kaspi (XIV–1). The Greek text was Latinized by Manuel Chrysoloras (XIV–2), and Gemistos Plethon (*c.* 1356–1450) must have spoken of it when he explained to the Florentine scholars the difference between Plato and Aristotle.

The medieval tradition of Plato (in Greek, Arabic, Latin, and Hebrew) is extremely complex, and each work introduces a few novelties and new names.

Plato's prestige had grown by leaps and bounds, first during the Byzantine renaissance of the ninth and tenth centuries, then under the patronage of the

[90] No. 122 in G. Bergsträsser's edition of the catalogue of Ḥunain's translations (1925) [*Isis* 8, 701 (1926)].

[91] See Carra de Vaux's translation of al-Mas'ūdī, *Le livre de l'avertissement* (Paris, 1897), p. 223, and his article on Aflāṭūn, *Encyclopedia of Islam,* vol. 1 (1908), pp. 173–175.

[92] There is a manuscript of the Arabic version of the *Timaios* in Aya Sofia, No. 2410. As far as I know, that text is still unpublished.

[93] So says Carra de Vaux, *Encyclopedia of Islam,* vol. 1, p. 174, but this is not confirmed by Giuseppe Gabrieli, "Ḥunáyn ibn Isḥāq," *Isis* 6, 282–292 (1924).

School of Chartres (XI, XII-1), finally under that of the Platonic Academy of Florence. The prestige of the *Timaios* grew in proportion, and many scholars were deceived into accepting the fantasies of that book as gospel truths. That delusion hindered the progress of science; and the *Timaios* has remained to this day a source of obscurity and superstition.

XVII

MATHEMATICS AND ASTRONOMY IN PLATO'S TIME

Now that we have made the acquaintance of Plato the man, the philosopher, the politician, the moralist, it is time to ask ourselves what kind of man of science he was.

There is a great contrast between his manner of thinking and that of such men as Hippocrates and Thucydides, and even Herodotos. We already realize that Plato is the typical and "ideal" philosopher, whose knowledge or wisdom is supposed to come from above, and to stoop like an eagle on the objects below. The knowledge of a worthy metaphysician is complete to begin with and proceeds from heaven downward; the knowledge of the man of science, on the contrary, begins with homely things on the face of the earth, then soars slowly heavenward. The two points of view are fundamentally different. Indeed, Plato would have gone so far as to say that men of science have only opinions, no substantial knowledge, for knowledge can be derived only from absolute ideals while material objects can yield nothing more valuable than doubtful and precarious opinions.

His philosophy was colored with mathematical ideas, which he obtained from his Pythagorean friends, and especially from Theodoros of Cyrene and from Archytas of Tarentum. We have already spoken of Theodoros, who was a much older man (p. 282), and we shall come back to Archytas presently. We may assume that Plato had received a good mathematical training; though Socrates did not care for mathematics, he was fond of using forms of argument that could be easily applied to mathematical questions. Hence, the paradox that Plato had received an essential part of his mathematical training from Socrates, who was definitely not a mathematician.

MATHEMATICS

Plato's general attitude to mathematics is well explained in the *Republic*:

"It is befitting, then, Glaucon, that this branch of learning should be prescribed by our law and that we should induce those who are to share the highest functions of state to enter upon that study of calculation and take hold of it, not as amateurs, but to follow it up until they attain to the contemplation of the nature of number, by pure thought, not for the purpose of buying and selling, as if they were preparing to be merchants or hucksters, but for the uses of war and for facilitating the conversion of the soul itself from the world of generation to essence and truth." "Excellently said," he replied. "And, further," I said, "it occurs to me, now that the study of reckoning has been mentioned, that there is something fine in it, and that it is useful for our purpose in many ways, provided it is pursued for the sake of knowledge and not for huckstering." "In what respect?" he said. "Why, in respect of the very point of which we were speaking, that it strongly directs the soul upward and compels it to discourse about pure numbers, never acquiescing if anyone proffers to it in the discussion numbers attached to visible and tangible bodies." [1]

Irrespective of its mathematical interest, this extract is typically Platonic because of its juristic slant. Mathematics in Plato's eyes is very important, so much so that "there ought to be a law" making the study of it obligatory for would-be statesmen (I wonder how our own statesmen would take to that).

When he speaks of mathematics, Plato is thinking, of course, of pure mathematics which gives us a vision of eternal truth and affords the best means of raising one's soul to the Idea of Good, and to God. Plato carried his dislike of "applied mathematics" to the extreme of deprecating the use of instruments, except perhaps the ruler and the compass.[2]

His general point of view is beautifully expressed in the statement that "God is always geometrizing" (God is primarily a mathematician!).[3] It is illustrated by the traditional inscription upon the door of the Academy: "Nobody should enter who is not a mathematician." [4]

The Platonic Idea is perfectly understood in the mathematical field, and it was possibly from his conception of it in that field that he ventured to extend it to the whole universe of thought. If we define the circle as a closed plane curve every point of which is equidistant from a point within, we create an Idea, the ideal or essential circle (*autos ho cyclos*), which no drawn circle could possibly emulate. The same applies to every mathematical definition; we can define a tangent, but it is impossible even with the finest instruments to draw a line and a circle having only one point in common. The ideal circle makes sense, while the ideal horse does not. And yet, according to Aristotle, Plato placed things mathematical (*ta mathēmatica*) somewhat below the pure Ideas, and considered them intermediate between the latter and tangible things, because the Idea of triangle is one, while there are many "ideal triangles." [5] This seems farfetched. In spite of that quibbling, we may safely assume that the Platonic theory of Ideas had a mathematical origin, and take its formulation as one proof among others of Plato's immoderate and irrational mathematization of everything.

Plato's contributions to mathematical knowledge were chiefly of the philosophic kind; he improved the definitions and increased the logical tightness of the elements. It is not possible to measure the extent of those contributions and their originality. The Academy attached much importance to mathematical discussions; an increase in mathematical rigor was the main result, which cannot be ascribed definitely to the master or to any other member of the school but was to some extent a collective achievement.

[1] Plato, *Republic*, 525C–D; Paul Shorey's translation in the Loeb Classical Library.

[2] For discussion of this see Heath: Greek mathematics (1, 287–88, 1921).

[3] According to Plutarch (I–2), who discussed the statement in his *Quaestiones convivales, lib.* VIII, 2: *Pōs Platōn elege ton theon aei geōmetrein.*

[4] For the history of this tradition in Byzantine and Arabic letters see *Introduction*, vol. 3, facing p. 1019.

[5] Heath, *History of Greek mathematics* (Oxford, 1921), vol. 1, p. 288; *Mathematics in Aristotle* (Oxford: Clarendon Press, 1949) [*Isis 41*, 329 (1950)].

Fig. 83. The locus of points equidistant from two intersecting lines

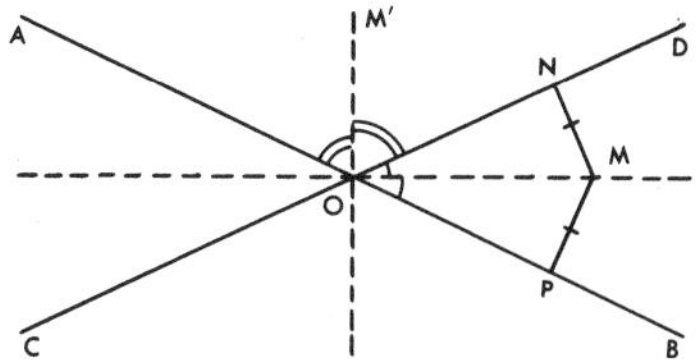

Did Plato invent geometric analysis? It is highly probable that the invention was made by Hippocrates of Chios (p. 277). Yet Plato may have improved it, or explained it more clearly (classroom discussion would easily lead to that), or he may have been the first to realize the need of completing the analysis with a synthesis.

Example of analysis. Suppose we have to prove that *A* is *B*. Let us assume that *A* is *B*, then *B* is *C*, *C* is *D*, *D* is *E*; therefore *A* is *E*. If that is untrue, then the theorem is disproved by *reductio ad absurdum.*

But if *A* is *E*, the theorem is not yet proved, and the analysis must be completed by the reverse process, called synthesis.

Synthesis. If *A* is *E*, *E* is *D*, *D* is *C*, *C* is *B*; therefore *A* is *B*.

It is also possible that Plato was the inventor (or developer) of problematic analysis.

Suppose one has to find the locus of all points at equal distances from two intersecting lines. Consider the two lines *AB* and *CD* crossing in *O* (Fig. 83), and suppose we have found one point *M* at equal distances from both lines. This means that if we draw perpendiculars from *M* to both lines, the segments *MN* and *MP* are equal. Let us draw the line *OM* and compare the triangles *OMN* and *OMP*; these triangles are equal; hence, the angles *NOM* and *MOP* are equal. Hence, *OM* is the bisector of the acute angle. A similar result would be obtained if one considered a point *M′* in the obtuse angle.

The next step is the construction of the locus, that is, the drawing of the two bisectors.

The last step is the synthesis, which consists in proving (1) that any point on the bisectors is at equal distances from both lines, (2) that any other point is not at equal distances from both lines.

Or suppose we are asked to draw a tangent from a point *A* to a circle *C* (the circle and point being in the same plane) (Fig. 84). Let us assume that the tangent is *AT*; then the radius *CT* is the shortest distance from *C* to *AT* and the angle *ATC* is a right angle. The locus of the vertexes of right angles subtended by *AC* is the circle of which *AC* is a diameter. Let us construct that circle. It cuts the circle *C* in two points, *T* and *T′*, and therefore we can draw two tangents, *AT* and *AT′*. Synthesis: We must now prove that *AT* and *AT′* are really tangents, and that there are no others.

Did Plato develop these methods? Or were they developed by his disciples, with him or without him, in the Academy or outside? It is impossible to tell, but the Platonic or Academic invention or rigorous formulation of the invention is very plausible.

We have already explained that Plato was profoundly impressed by the mathematical regularities that the Pythagoreans had discovered in the musical intervals. Thus, mathematics was connected with music on the one hand and with astronomy on the other. Might one not conclude then that there was music in astronomy? This was an intoxicating thought, which led Plato to his conception of the harmony of heavens, or the harmony of the world soul.[6]

[6] *Timaios*, 35–36.

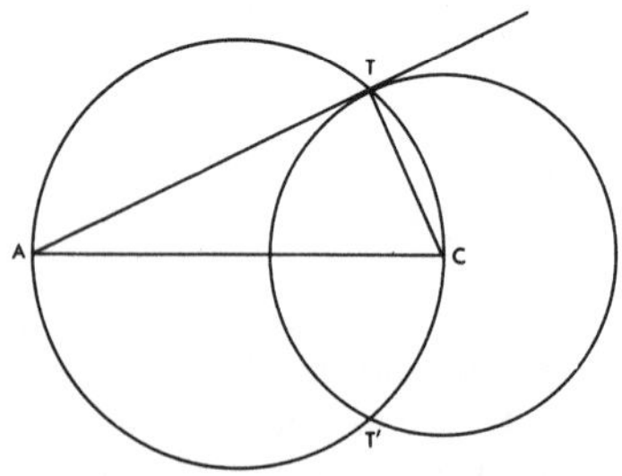

Fig. 84. Construction of a tangent to a circle from a point.

The reader is familiar with the medieval conception of the seven liberal arts, which is generally traced back to Boetius (VI-1) but can already be found in St. Augustine (V-1).[7] In reality, the idea is more ancient (as far as the *quadrivium* is concerned). The liberal arts constituted (and still do) a kind of general education (*encyclios paideia*).[8] In the course of time their number and contents varied. According to the medieval combination with which we are most familiar, the seven arts were divided into two groups, the *trivium* (grammar, logic, rhetoric) and the *quadrivium* (arithmetic, geometry, music, astronomy). This means that the second or higher level of general education was wholly mathematical.[9] That idea is often ascribed to Plato, but it is more correct to call it Pythagorean, though we cannot trace it further back than Plato's time. Plato conceived a kind of mathematical *quadrivium*, but curiously enough it did not include music. It included arithmetic, geometry, stereometry, astronomy; the distinction between planimetry and stereometry or between plane and solid geometry betrays the immaturity of contemporary mathematics. The familiar division of the *quadrivium* (with music and without stereometry) was adumbrated by Archytas (text quoted below), but then disappeared; it reappeared only in the first century of our era, in the *Pinax* of pseudo-Cebes and in Seneca (I-2), then in Sextos Empiricos (II-2) and in Porphyry (III-2), later in St. Augustine (V-1), Martianus Capella (V-2), Boetius (VI-1), Cassiodorus (VI-1), Isidore of Seville (VII-1), and others. Plato did not introduce the medieval *quadrivium*, but it was he who caused the higher general studies to be mathematical.

The discovery of the regular solids has sometimes been ascribed to Plato. What does that mean? The regular solids were certainly known before him; the simplest had been known from time immemorial. The most difficult to recognize, the dodecahedron, was already known to Hippasos of Metapontum (p. 283) or to other Pythagoreans who liked to play with pentagrams and pentagons. We may thus assume that the Pythagoreans knew the five regular solids. They could construct them by putting together 4, 8, or 20 equilateral triangles, 6 squares, or 12 pentagons. That was not very difficult. But did they realize that there could not be more than five regular solids? That realization was the crux of the discovery, which was

[7] Henri Irénée Marrou, *Saint Augustin et la fin de la culture antique* (Paris: Boccard, 1938) [*Isis 41*, 202-204 (1950)], chiefly pp. 211-275. According to a fragment of a lost treatise of Archytas of Tarentum (IV-1 B.C.), quoted below, Pythagorean mathematics were already divided into four branches: arithmetic, geometry, astronomy and music. That division is exactly the same as that of the quadrivium.

[8] The term is Hellenistic. It is used by Dionysios of Halicarnassos (I-2 B.C.), also by Plutarch (I-2), etc.

[9] We call it the higher level of *general* education; in medieval times, the whole of general education was an introduction to professional studies, such as medicine and law, or to the highest studies, philosophy and theology.

probably made by Theaitetos, and communicated to his friend Plato. The latter's original contribution to the theory is of very doubtful quality. Considering on the one hand the four elements and on the other the five solids, these two groups must be somehow related, must they not? Plato associated the tetrahedron (pyramid) with fire, the hexahedron (cube) with earth, the octahedron with air, the icosahedron with water. What then shall be done with the fifth solid? That is easy. Plato associated it with the whole universe.[10]

It has been argued that because Plato assumed that the particles of earth were cubic, the particles of fire pyramidal, and so on, he was an atomist. That is quibbling. Plato was definitely on the side of the antiatomists with Anaxagoras and with Aristotle. He rejected the possibility of a vacuum.[11] He was not interested in the regular solids as "atoms" but simply as means for cosmologic analogies. The theory of the four elements was nonsensical and the attempt to dovetail the four elements with the five solids, doubly so.

Another mystical fancy to which innumerable commentaries have been devoted is the geometric or nuptial number of the *Republic*.[12] The number is called "nuptial" because it is related by Plato, in a language that is suitably obscure, to the time required for the breeding of perfect rulers. "Now for divine begettings there is a period comprehended by a perfect number" (*arithmos teleios*) and the perfect number is determined in such a sibylline manner that the interpretations of it vary considerably. There are really two numbers to be determined, not one, and Hultsch and Adam have arrived at the same two numbers in different ways. We quote their solution for the sake of example, and without attaching too much importance to it, for whether we know these numbers or do not is immaterial. Their two numbers are $216 = 2^3 + 3^3 + 4^3 = 2^3 \times 3^3$, and $12{,}960{,}000 = 60^4 = 3600^2 = 4800 \times 2700$.

The first number, 216, may express the shortest period of human gestation in days. As to the larger one, 12,960,000, it "represents two current aeons in the life of the universe, in which the world waxes and wanes alternately, the harmony 3600^2 meaning the cycle of uniformity and the harmony 4800×2700 the cycle of dissimilarity described by Plato in the *Statesman*." [13]

Let us approach the matter from another angle; the number 3600, being one of the units of the sexagesimal system, suggests a Babylonian origin: $12{,}960{,}000 = 3600 \times 3600 = 360 \times 36{,}000$, that is, 36,000 years of 360 days; [14] according to Berossos (III–1 B.C.), the period of 36,000 years was the duration of a Babylonian cycle, it was called later *magnus platonicus annus*. Moreover,

> all the multiplication and division tables from the temple libraries of Nippur and Sippar and from the library of Ashurbânapal are based upon 12,960,000. This coincidence can scarcely be accidental. We must necessarily come to the conclusion that Plato, or rather Pythagoras, whom he closely followed, borrowed his famous number and the whole idea of a decisive influence exercised by it upon the life of man directly from Babylonia.[15]

[10] *Timaios*, 55–56.

[11] *Timaios*, 80c. *Introduction*, vol. 3, p. 148. Paul Friedländer, *Structure and destruction of the atom according to Plato's Timaeus* (University of California publications in philosophy 16, 4 fig.; 1949), pp. 225–248 [*Isis 41*, 58 (1950)].

[12] *Republic*, VIII, 546B–D.

[13] *Statesman*, 270. James Adam, *The Republic of Plato* (Cambridge, 1902), vol. 2, pp. 201–209, 264–312. For the geometric number see also *Introduction*, vol. 1, p. 115; Heath, *History of Greek mathematics*, vol. 1, pp. 305–308.

[14] A year of 360 days was shockingly obsolete in Plato's time.

[15] H. V. Hilprecht, *Mathematical, metrological and chronological tablets from the temple library of Nippur* (Philadelphia, 1906), p. 31.

This much is clear: the geometric number is almost certainly of Babylonian origin, but we need not worry concerning Plato's interpretation of it or modern interpretations of Plato's. It is typical of the harm done by the *Timaios* that many scholars have racked their poor brains and perhaps been driven to distraction and insanity by the puzzle offered to them, in such solemn terms, by the divine Plato. We should not imitate them, and abondon the solution of Platonic riddles to wits, or better still, to half-wits.[16]

Even if Plato made no mathematical discoveries (and there is no evidence that he made any), he was probably an up-to-date mathematician, but very definitely an amateur.[17] Yet his influence upon the progress of mathematics was immense. This was very neatly expressed by Proclos (V-2) in his commentary on the first book of Euclid:

> He caused mathematics in general and geometry in particular to make a very great advance by reason of his enthusiasm for them, which of course is obvious from the way in which he filled his books with mathematical illustrations and everywhere tries to kindle admiration for these subjects in those who make a pursuit of philosophy.[18]

One could not put it better. It was because of Plato that the higher level of the liberal arts was mathematical. His mathematical enthusiasm was contagious. One must love mathematics before one knows it; otherwise, one would never learn it; it is a kind of faith which Plato communicated to others. He did not create mathematics, but he created mathematicians.

He intimated again and again that a gentleman should know mathematics and for that reason mathematics was an essential part of the classical tradition nourished by the public schools of England. Most of the boys took their mathematics as they took cod-liver oil; it was a painful process but one had to submit to it; some of them, however, took to it in great earnest. Plato was their initiator and their leader – in this respect, at least, a good leader.

Unfortunately, Plato was not unlike other amateurs, even amateurs of genius, in that his enthusiasm betrayed him and caused him to misuse mathematics grievously. We have given enough examples of his misapplications in this chapter and in a previous one. He was a mathematician with a vengeance.

The mathematical tradition that Plato initiated in the Academy was continued by his successors, and the Academy remained through the centuries a cradle of mathematicians. Let us speak now of those who were his contemporaries, were influenced by him and influenced him in return. It is a curious situation: they were genuine mathematicians, which he was not, yet they owed perhaps their vocation to him, and in any case he fostered it.

For the practical student of the history of mathematics, it is a great relief to pass from Plato to the real mathematicians, from hot air to golden fruits. We shall

[16] Plato had the impudence to make a distinction between real knowledge (derived from Ideals) and opinions (what we would call scientific knowledge). The real distinction to be made is that between rational, demonstrable knowledge and pseudoknowledge (magic and nonsense). The geometric number, the obtention of which seemed to many infatuated Platonists the acme of wisdom, is absolutely meaningless and worthless.

[17] Julian Lowell Coolidge, *The mathematics of great amateurs* (Oxford: Clarendon Press, 1949) [*Isis 41*, 234-236 (1950)]. The first chapter of that delightful book is rightly devoted to Plato (pp. 1-18).

[18] G. Friedlein, *Procli in primum Euclidis elementorum commentarii* (Greek text; Leipzig, 1873), p. 66, lines 8-14; Heath, *History of Greek mathematics*, vol. 1, p. 308.

restrict ourselves to Theaitetos, Leodamas, Neocleides, Leon, Archytas, and the greatest of them all, Eudoxos.

THEAITETOS

We do not know much about the life of Theaitetos (*c.* 415–*c.* 369), not even his father's name, but we know that he was an Athenian, a disciple of Socrates and Theodoros of Cyrene and a contemporary of Plato and Archytas.

One of the best of Plato's dialogues, entitled *Theaitetos*, is a conversation between the mathematician in his youth, Theodoros of Cyrene, and Socrates, shortly before the latter's death. Their conversation is not reported directly. A prelude describes another conversation, occurring in 369 in Megara, between Euclid and Terpsion in front of the former's house. Euclid tells Terpsion[19] that as he was going to the harbor he met Theaitetos, who had been wounded in a battle fought for the Athenians near Corinth, and was carried back to Athens, suffering from his wounds and from dysentery, more dead than alive. They praise him for his courage and genius and Euclid recalls the dialogue proper which he has written down and which is now being read to them by his servant. The *Theaitetos* is thus a dialogue within a dialogue. As Plato knew the man, we may trust the physical portrait that he gives of him. Theodoros introduces him to Socrates with the following words,

> Yes, Socrates, I have become acquainted with one very remarkable Athenian youth, whom I commend to you as well worthy of your attention. If he had been a beauty I should have been afraid to praise him, lest you should suppose that I was in love with him; but he is no beauty, and you must not be offended if I say that he is very like you; for he has a snub nose and projecting eyes, although these features are less marked in him than in you.[20]

And at the end of the dialogue, Socrates remarks to Theaitetos that his "snubnosedness" is characteristic of him. Thus, if we do not know Theaitetos very well, yet we can see him with the eyes of our imagination.

From that dialogue we also gather that Theaitetos was not only a mathematician but also a philosopher who distinguished between numbers perceived by the senses and numbers conceived by the mind. This is not very surprising, because every mathematician of that time was a philosopher.

Moreover, we can be sure that he was a Pythagorean, because the two subjects to which he owes his fame are both Pythagorean – the theory of irrationals and the theory of the regular solids.

The early history of irrationals has been told above apropos of Theaitetos's master, Theodoros of Cyrene (pp. 282–285). Theaitetos continued the elaboration of that theory; he introduced distinctions between various kinds of irrationals (medial, binomial, apotome) which are described in Book x of the *Elements*.[21]

[19] Terpsion of Megara, one of the disciples who attended Socrates' death (*Phaidon*, 59c).

[20] Jowett, vol. 4, p. 195; *Theaitetos*, 143.

[21] The elaborate classification of irrationals in Euclid x, of which Theaitetos laid the foundation, is difficult and, in spite of its accuracy, obsolete. According to Eudemos (IV–2 B.C.), Theaitetos associated these three particular kinds of irrationals, medial, binomial, and apotome, respectively with the arithmetic, the geometric, and the harmonic means. As I do not like to use undefined terms, here are the definitions of these three kinds of irrationals (out of many more) according to Euclid x: *Prop.* 21. The rectangle contained by rational straight lines commensurable in square only is irrational, and the side of the square equal to it is irrational. Let the latter be called *medial* (*mesē*). *Prop.* 36. If two rational straight lines

In particular, proposition 9 of that book is definitely ascribed to him (that the sides of squares that have not to one another the ratio of a square number to a square number are incommensurable). In short, he laid the foundation of the knowledge included in *Euclid*, x.[22]

As to the theory of regular solids, it was said that Theaitetos discovered the octahedron and the icosahedron, and was the first to write on the five regular solids. The first part of that statement cannot be correct as it stands. Earlier Pythagoreans knew these two solids and were probably able to build them by means of 8 or 20 equilateral triangles (cut in leather, wood, or stone). That is, by holding 3, 4, or 5 equilateral triangles (of equal size) around a common vertex, they could form solid angles, and by combining 4, 6, or 12 such solid angles they could construct regular solids of 4, 8, and 20 faces. That was one thing, but the geometric construction was quite another. Still another was the realization that there were five regular solids and that there could not be more.

Theaitetos was the first to write on the five regular solids.[23] How much did he write? In the case of the irrationals, we ascribed to him an indefinite part of Book x of the *Elements*; in the regular solids we can ascribe to him in the same imprecise way a part of Book xiii. It was natural for him to study the regular solids, the mathematical construction of which implied irrationals. If he wrote of the five solids, this implies that he knew that there could not be more. Could he know that? Why not? After all, the proof as given in Euclid[24] is simple enough, so simple indeed that we can afford to insert it here (but I give it in my own way, for greater clearness).

There can be only five regular convex solids.

1. In every solid angle the sum of the plane angles is smaller than four right angles. The maximum (four right angles) would be attained only if the solid angle were completely flattened out around its vertex; that is, the solid angle would cease to exist.

2. If the faces are triangles, there can be around a point:

(i) three triangles, and the solid will be a tetrahedron or pyramid (4 faces);

(ii) four triangles, and the solid will be an octahedron (8 faces);

(iii) five triangles, and the solid will be an icosahedron (20 faces).

(There cannot be six, for six such angles would equal 4 right angles.)

3. If the faces are squares, there can be only three around a point, and the resulting solid will be a hexahedron (cube, 6 faces).

4. If the faces are pentagons, there can be only three around a point (for the angle of a pentagon equals 6/5 of a right angle) and this leads to a dodecahedron (12 faces).

5. No others are possible, for the angle of the hexagon equals 4/3 of a right angle, and three of these equal 4 right angles.

6. Thus there can be only five regular solids, having respectively 4, 6, 8, 12, 20 equal faces.

---

commensurable in square only be added together the whole is irrational; and let it be called *binomial* (*ec duo onomatōn*). *Prop.* 73. If from a rational straight line there be subtracted a rational straight line commensurable with the whole in square only, the remainder is irrational; and let it be called an *apotomē*.

Example. The two segments of the "golden section" are apotomes (Euclid, *Elements*, Book xiii, prop. 6).

[22] For discussion see Heath, *History of Greek mathematics*, vol. 1, pp. 209–212; *Euclid* (Cambridge, ed. 2, 1926), vol. 3. A commentary of Pappos (111-2) on Book x of the *Elements* has come down to us only in the Arabic translation of Abū 'Uthmān al-Dimishqī (X-1); the Arabic text was edited and translated by William Thomson (Cambridge, 1930) [*Isis 16*, 132–136 (1931)]. Gustav Junge added to that book a history (in German) of the theory of irrationals.

[23] This is told by a late witness, Suidas (X-2), but the tradition is plausible.

[24] *Elements*, Book xiii, prop. 18.

It was necessary to add the word "convex" at the head of the demonstration, because it was discovered much later that there were other regular solids which were not convex; these are called stellated polyhedrons and are to the convex polyhedrons somewhat as pentagrams are to pentagons. In 1810, Louis Poinsot (1777–1859) discovered four stellated polyhedrons, to wit, three dodecahedrons and one icosahedron; in 1813, Augustin Cauchy (1789–1857) proved that these nine solids exhaust the series of regular solids; his proof was rigorous but difficult. It was simplified by Joseph Bertrand (1822–1900), who showed that the vertexes of every stellated polyhedron must be the vertexes of a convex concentric one. It suffices then to take the five Pythagorean solids and to examine which other regular solids could be obtained by grouping their vertexes differently.[25]

To return to the five convex solids, the discovery that there could only be five of them, whether it was made by Theaitetos or not, must have been a great surprise and a shock. The investigation of polygons had not prepared one for that, because the regular polygons are infinite in number. If there is a regular polygon with n sides, one can easily construct others with $2n$ sides, $4n$ sides, etc. It is strange to pass from an infinity of polygons to the very small group of five polyhedrons. That extraordinary and sudden restriction appeared to Plato a mathematical mystery, requiring some kind of philosophic interpretation. If the regular solids are restricted to five, those five bodies (later called the Platonic bodies) must each have some definite meaning. They could not be connected with the planets, of which there were seven. Plato bethought himself of the four elements; the fifth solid would then represent the whole universe. This patching up of the theory, plus the finding of a meaning for the superfluous solid, is typical of the analogies invented by numerologists and other mystical mathematicians; they change the rules of their game as often as necessary and thus prove what they want to prove. In his interpretation of the regular solids, Plato stooped to the level of the Chinese cosmologists.

LEODAMAS, NEOCLEIDES, AND LEON

The progress of geometric discovery and organization under the influence of the Academy is symbolized by these three mathematicians, about whom we know only what Proclos tells us in his commentary on Euclid, Book I, and that is more tantalizing than sufficient.

Said Proclos:

> At about the same time Leodamas of Thasos, Archytas of Tarentum and Theaitetos of Athens increased the number of theorems and inserted them in a more scientific context; then Neocleides who was younger than Leodamas and his disciple Leon [IV-1 B.C.] produced many more things than their forerunners, so much so that Leon was able to assemble *Elements* (*ta stoicheia*) which are of great interest because of their number and their usefulness, and to invent distinctions (*diorismoi*) which show when a problem is solvable and when it is not.[26]

[25] Gaston Darboux, *Eloges académiques* (Paris, 1912), p. 33. Another extension of the idea of regular solid leads to the conception of the so-called Archimedean solids. There are 13 such solids, each of which has its solid angles equal; the faces are regular polygons but not all of the same species.

[26] G. Friedlein, *Procli in primum Euclidis elementorum commentarii* (Leipzig, 1873), pp. 66, 211. Ver Eecke, *Commentaires de Proclus sur le premier livre d'Euclide* (Bruges: Desclée De Brouwer, 1948).

That is all he has to say about Neoclides and Leon, but he adds this about Leodamas: "Plato had explained to him the analytical method which enabled him (Leodamas), it is reported, to invent many things in geometry." That information is meager and vague, but helps us to realize that much geometric research was done by younger contemporaries of Plato. There was an emulation among them for the discovery of new theorems and, what is even more important, for a better dovetailing of all of them in a single synthesis. Proclos has nothing more to say about Archytas but happily we know much about the latter from various other sources.

ARCHYTAS OF TARENTUM

At the time of Plato's first visit to Sicily in 388 he made the acquaintance of the Pythagorean Archytas, who was a very important man in Tarentum, not simply as a philosopher and mathematician, but also as a politician (or statesman) and as a general. He is said to have saved Plato's life, through his influence with Dionysios. At the time of Plato's last visit to Sicily (in 361–60) Archytas was still alive.

As far as one can judge from the fragments of his lost writings, his was a rich and complex personality. One of them shows that the classification of mathematical subjects that was later crystallized in the *quadrivium* existed already in the minds of the early Pythagoreans, or at least in his own.

> The mathematicians (*hoi peri ta mathēmata*) seem to me to have arrived at correct conclusions, and it is not therefore surprising that they have a true conception of the nature of each individual thing: for, having reached such correct conclusions regarding the nature of the universe, they were bound to see in its true light the nature of particular things as well. Thus, they have handed down to us clear knowledge about the speed of the stars, their risings and settings, and about geometry, arithmetic, and sphaeric, and last, not least, about music; for these branches of knowledge (*mathēmata*) seem to be sisters.[27]

Archytas was an astronomer whose memory as such was still alive in the time of the poet Horace (65–8 B.C.), who celebrated it in one of his odes.[28] He speculated concerning the finitude or infinity of the universe and concluded that it must be unlimited. The most astonishing of his mathematical achievements is his solution of the famous problem of the duplication of the cube. That problem had been reduced by Hippocrates of Chios to the finding of two mean proportionals in continued proportion between two given straight lines. Archytas determined these two mean proportionals by means of the intersection of three surfaces of revolution. The intersection of two of these, a cylinder and a tore (or anchor ring) with inner diameter zero, is a curve of double curvature. The point where that curve pierces the third surface, a right cone, gives the solution. This is the first example in history of the use of a curve of double curvature. Archytas' boldness is amazing.

His was an inventive mind, mechanical. It is said that he had devised a flying toy, a wooden dove, which did not continue its flight, however, after it had settled down. In Aristotle's *Politics* we find an amusing reference to another toy:

[27] Archytas' fragments in Diels, *Vorsokratiker*, vol. 1[4], pp. 330–331; English translation by Heath, *History of Greek mathematics*, vol. 1, p. 11.

[28] Horace, *Odes*, I, 28.

> Children should have something to do, and the rattle of Archytas, which people give to their children in order to amuse them and prevent them from breaking anything in the house, was a capital invention, for a young thing cannot be quiet.[29]

The anecdote is charming, but assuming that it refers to our Archytas, it does not help us to appreciate his mechanical ingenuity. The creation of a flying dove would have been a very remarkable achievement, but no mechanical genius was needed to make a good rattle.

Did Archytas write a book on mechanics, which would be, of course, the first of its kind? We do not know. Was he the founder of theoretical mechanics? [30] We have no right to make such a statement. All we can say is that he was interested in mechanics (in the crude sense of that word); he may have seen the possibility of relations between mechanics and mathematics even as he improved the mathematical study of music; [31] he had found a mechanical solution of a mathematical problem,[32] and he may have thought of applying mathematics to mechanics. We cannot say more than that. At any rate, that Sicilian philosopher and mathematician is a kind of prototype of a greater countryman of his, Archimedes of Syracuse (III-2 B.C.).

EUDOXOS OF CNIDOS

As far as we can trust Diogenes Laërtios (III-1), and there is no reason to distrust the substance of his account, Eudoxos' life is fairly well known and of unusual interest to the student of international relations. The dates of his birth and death are uncertain, but we may assume them to be *c.* 408 and 355.[33] Eudoxos, son of Aischines, was born in Cnidos; he learned geometry from Archytas and medicine from Philistion of Locroi. At the age of 23 (*c.* 385) he traveled to Athens and became a pupil of Plato (the Academy had been opened in 387); his traveling expenses were defrayed by Theomedon the physician. He was so poor that he remained in the Peiraieus, where he had disembarked, and walked to Athens every day. After spending two months in that way he returned to Cnidos; later, he traveled to Egypt with the physician Chrysippos of Cnidos, bearing a letter of introduction from Agesilaos to Nectanabis,[34] who recommended him to the priests (learned men). He remained in Egypt sixteen months, conforming to the customs of his hosts (he had his beard and eyebrows shaved), and there he wrote his *Octaëtēris*. From Egypt he went to Cyzicos, on the southern shore of the

[29] Aristotle, *Politics*, 1340B; Jowett's translation in the Oxford English Aristotle. This passage occurs in a discussion of the musical education of children. We cannot guarantee that the Archytas referred to by Aristotle is our Archytas of Tarentum. The name was not uncommon.

[30] Imprudent statement made in my *Introduction*, vol. 1, p. 116.

[31] He gave the numerical ratios representing the intervals of the tetrachord on three scales, the anharmonic, the chromatic and the diatonic; see Heath, *History of Greek mathematics*, vol. 1, p. 214.

[32] The duplication of the cube mentioned above. To understand his discovery of that extraordinary solution we must think of it very concretely, in mechanical terms.

[33] The climax of his activity is assumed to fall *c.* 367. George de Santillana, "Eudoxus and Plato, a study in chronology," *Isis* 32, 248–262 (1940–49), would place it 10 years later. Students having no access to Diogenes Laërtios, VIII, 86–91, will find the relevant text in Santillana, p. 251.

[34] Agesilaos, king of Sparta from 398 to 361, Xenophon's friend. Nectanabis (Nekht-har-hebi) was the first king of the Sebennite dynasty (*c.* 378–350), one of the native dynasties which re-established in Egypt a precarious independence after the Persian conquest in 525 and before Alexander's in 332. Nectanabis ruled from *c.* 378 to 364. Putting these facts together indicates that Eudoxos went to Egypt during the period 378–364, but remained there only 16 months.

Propontis (Sea of Marmara) and to other places in that neighborhood, earning his living as a teacher (*sophisteuonta*), then returned to his native region and attended the court of Mausolos in Halicarnassos.[85] He then visited Athens, but this time he came not as a poor young student, but as a master accompanied by his own disciples. Plato gave a banquet in his honor. After his return to Cnidos he helped to write laws for his fellow citizens and was very much honored by them.

Apollodoros of Athens (II–2 B.C.) said that he died in his fifty-third year (which fixes 355 if we accept 408). According to Favorinos of Arles (*fl.* under Hadrian, emperor 117–138), when Eudoxos was in Egypt with Chonuphis of Heliopolis, the sacred bull Apis licked his cloak and the priests augured that he would become famous but would not live very long. (The statements of Apollodoros and Favorinos are reported by Diogenes.)

The prediction of the Egyptian priests was confirmed, imperfectly with regard to his age (for to attain his fifty-third year was not so bad), completely with regard to his fame. He is considered to be the greatest mathematician and astronomer of his age, and one would have to speak of him even in the briefest outline of the whole history of science. Plato is known to a much larger public, but from the point of view of science, the time of Plato should be called the time of Eudoxos.

His well-deserved mathematical fame rests on three grounds – his general theory of proportion, the golden section, and the method of exhaustion. On that triple basis he deserves to be called one of the greatest mathematicians of all time.

A new theory of proportion had become necessary because of the revolutionary discovery of irrationals by Theodoros of Cyrene and Theaitetos of Athens. The Pythagoreans had observed parallelisms between numbers and lines (such as triangular or square numbers, and the theorem of Pythagoras). A ratio between lines might be represented by the ratio between two integers m and n, and conversely m/n might represent the ratio of two lines m and n units long. Now, the newly discovered lines or numbers, the irrationals (*alogos*)[86] were not integers, and could not be represented by any ratio of integers. Thus the structure of Pythagorean mathematics was breaking down. There were only two ways out, either to reject the parallelism between geometry and arithmetic, or to recognize a new kind of number, the irrationals. The second alternative was more complex than the nonmathematician would imagine, for it implied not only the definition of those numbers and the proof of their existence but also the proof that they could be handled in the same way as other numbers and the validation of geometric demonstrations that included or might include irrational elements. To put it otherwise, it was necessary to extend the idea of number, so that irrationals would be included, and to extend the idea of length so that theorems concerning any lines would still be correct if some of the lines were irrational. This extension was accomplished by Eudoxos in his general theory of proportions, which was to be developed later in Books V and VI of Euclid's *Elements*. How much of that was done by Theaitetos and how much by Eudoxos it is impossible to state with precision, but it is traditionally assumed that the latter's contribution was decisive.

What is the golden section? According to Proclus,[87] the theorems concerning "the section" (*ta peri tēn tomēn*) originated with Plato, and Theaitetos applied to

[85] Mausolos was king of Caria from 377 to 353.

[86] The diagonal of a square was an irrational line; the diagonal of the square of side 1 was an irrational number $\sqrt{2}$.

[87] Friedlein's edition, p. 67, 6.

them the method of analysis. It is more likely that the theorems were discovered by Theaitetos or other mathematicians and that Plato applied them to his own fancy. The curious use of the definite article "the section" (*hē tomē*) must refer to a very exceptional section, almost certainly to the division of a straight line in extreme and mean ratio,[38] which obtruded itself in the construction of the pentagon and of the dodecahedron. At a later time, that remarkable section was called divine (by Luca Pacioli, 1509), and much later still, golden.[39] The term "golden section" was enormously successful, and a number of artists and mystics played with the idea that that particular section was one of the secrets of beauty.[40]

Eudoxos' share in the theory of the golden section gives him a modicum of popularity and glamour, but his outstanding mathematical achievements are the general theory of proportion and the method of exhaustion.

The method of exhaustion is a true infinitesimal method, the first of its kind, and it was based upon a rigorous notion of limit. By his invention of it Eudoxos became one of the distant ancestors of the integral calculus. Integration of simple areas had been made before him, and such results as that circles are to one another as the squares of their diameters[41] had no doubt been obtained. Indeed, Hippocrates is said to have proved that theorem. How did he do it?

The demonstration given by Euclid is based on the use of the method of exhaustion invented by Eudoxos, and hence we may assume that it is essentially Eudoxos' demonstration.

Given two circles, of areas A and B and radii a and b; we claim that $A/B = a^2/b^2$.

It has been proved before that similar polygons inscribed in circles are to one another as the squares of the diameters.[42] That was easy; the difficulty was to pass to the limit.

(1) Let us inscribe in the circles A and B regular polygons of areas A' and B', having so many sides that the differences $A - A'$ and $B - B'$ are arbitrarily small.

(2) We have to prove that

$$a^2/b^2 = A/B.$$

Let us suppose that that is not true, and that

$$a^2/b^2 = A/C.$$

Can C be smaller than B?

We can reduce the difference $B - B'$ so much that

$$B - B' < B - C \text{ or } B' > C.$$

The equalities

$$a^2/b^2 = A/C = A'/B'$$

are irreconcilable because

$$A > A', C < B'.$$

One shows in the same way that C cannot be greater than B. If C can be neither smaller nor greater than B, then $C = B$ and the theorem is proved.

[38] Euclid, II, 11; VI, 30. To refresh the reader's memory, here is the problem as expressed by Euclid, II, 11: "To cut a given straight line so that the rectangle contained by the whole and one of the segments is equal to the square on the remaining segment." Or in algebraic terms, given the line a, to divide it into two segments x and $a-x$ in such a way that

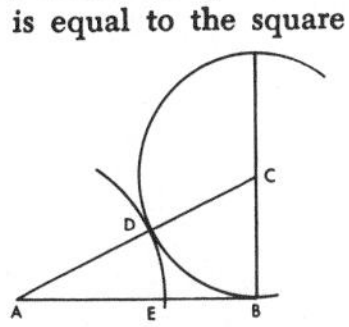

Fig. 85

$$a/x = x/(a - x).$$

The solution is easy (Fig. 85). Given the line AB equal to a, draw a perpendicular at B, equal to a, and use it as diameter of the circle C. Join AC, which cuts the circumference in D. The circle of radius AD cuts the line in E and divides the line AB in extreme and mean ratio. The demonstration is so simple that we need not quote it.

[39] G. Sarton, "Query no. 130. When did the term golden section or its equivalent in other languages originate?" *Isis* 42, 47 (1951).

[40] For a general discussion, see G. Sarton, "The principle of symmetry and its applications to science and to art," *Isis* 4, 32–38 (1921).

[41] Euclid, XII, 2.

[42] Euclid, XII, 1.

This might be generalized, but the ancients failed to do so. The method of exhaustion was rigorous but particular; a special demonstration had to be elaborated in each case. Its use enabled Eudoxos to prove rigorously the formulas concerning the volumes of the pyramid and the cone that Democritos had discovered.[43]

By the middle of the fourth century, thanks chiefly to the efforts of Theaitetos and Eudoxos, geometry had been raised to a much higher plane, approaching the Euclidean one. We are now past the stage of intuitive discovery, and mathematicians well trained in logic are no longer satisfied with partial results; they need rigor. What had been the share of Plato in that development? It is impossible to say. He may have insisted on clearness and good logic, but the main achievements, the purely mathematical ones, were not his. He may have helped the mathematicians; they could have done without him; he could not have done without them.

ASTRONOMY

The astronomic deeds of the Platonic age are as brilliant as the mathematical ones, and they were accomplished chiefly by the same man, Eudoxos of Cnidos. The history that we have to tell is rather complex. We shall deal first with the achievements ascribed to Babylonian astronomers. The Greek story must be divided into three sections: The Precursors; Eudoxos; Plato and Philip of Opus.

KIDINNU

In order to explain the part that Babylonian astronomers may have played in the development of Greek astronomy, we must anticipate a little. According to Ptolemy (II-1),[44] Hipparchos of Nicaea (II-2 B.C.), comparing his own observations of the fixed stars with other observations made a century earlier in Alexandria by Aristyllos and Timocharis (III-1 B.C.), concluded that all the stars had moved a little toward the east, that is, he discovered the precession of the equinoxes. Hipparchos assumed that the displacement of the stars in longitude, that is, the precession, amounted to 45″ or 46″ a year, which would be only 1°10′ in a century. (Ptolemy corrected that to 36″ a year or exactly 1° in a century, but Hipparchos was nearer to the truth, the real value being 50″.26.) Could Hipparchos have noticed a difference of the order of 1°? Yes, that was not impossible, but it would have been easier for him to discover the precession if older observations had been available to him,[45] and accurate observations made

[43] Said Archimedes in his *Method* (discovered only in 1906 by Heiberg), "It is of course easier, when we have previously acquired, by the method, some knowledge of the questions, to supply the proof than it is to find it without any previous knowledge. This is a reason why, in the case of the theorems the proof of which Eudoxos was the first to discover, namely that the cone is a third part of the cylinder, and the pyramid of the prism, having the same base and equal height, we should give no small share of the credit to Democritos who was the first to make the assertion with regard to the said figure though he did not prove it." Translated by T. L. Heath, *The Method of Archimedes*, 152 pp.; Cambridge, 1912, p. 13.

[44] *Almagest*, VII, 1-2.

[45] It is shocking to realize, however, that Ptolemy's determination of the equinoxes was 26 percent worse than Hipparchos', though his basis was three centuries longer. Hipparchos was an observer of astounding precision, Ptolemy was a very poor one. Worse than that the "Catalogue of stars" of the *Almagest* was not based upon new observations, but was derived from the catalogue of Hipparchos, the longitudes being increased by the same constant. Thanks to Ptolemy's wrong estimate of the precession, the true epoch of his "Catalogue" is A.D. 58, while his own observations extended from 127 to 151. Christian H. F. Peters and Edward Ball Knobel, *Ptolemy's Catalogue of stars* (Washington, 1915) [*Isis* 2, 401 (1914-19)].

by Babylonians may have been. Ptolemy refers to Chaldean observations made in 244, 236, 229 B.C.[46] The theory has been advanced that Hipparchos not only had earlier Oriental observations at his disposal (which is very likely), but that the precession had been discovered as early as 379 by the Babylonian astronomer Kidinnu.[47]

It is certain that Chaldean astronomers had gathered a large number of observations of astounding precision. The earliest of them known to us by name were Naburianos (Naburimannu son of Balatu), who flourished at Babylon in 491, and Kidinnu, who flourished *c.* 379; they devised lunar tables according to two different systems; then came the astronomers responsible for the Chaldean observations recorded in the *Almagest.* It is almost certain that Hipparchos was acquainted with these observations, which facilitated his own work, in particular his discovery of the precession.[48]

The discovery, it should be noted, was unavoidable just as soon as lists of stars sufficiently distant in time were compared. The astronomers making those comparisons could not fail to notice that all the longitudes were increased by the same quantity; that quantity was very small, about 1°24′ in a century, 4°12′ in three centuries, 5°36′ in four centuries. No matter how gross the observations, a time must come when the precession is noticed (I do not say explained; that is another story).

We cannot abandon this subject without introducing another remark even if this obliges us to make more anticipations. The precession being finally recognized, as it had been by Hipparchos, and published by Ptolemy,[49] additional observations of stellar longitudes would confirm it, and therefore one would expect such a fundamental discovery to be firmly established. Not at all! Most of the followers of Ptolemy let it drop; the only ones who refer to it are Theon of Alexandria (IV-2) and Proclos (V-2), but the latter denied it and Theon, while accepting the Ptolemaic value (1° per century), suggests that it is restricted to an oscillation along an arc of 8°, which means that the precession would accumulate for some eight centuries and then be reversed. Proclos made a similar admission; the tropical points do not move in a whole circle but some degrees to and fro.

Theon was thus the originator of the theory of the "trepidation of the equinoxes," which enjoyed a long popularity in spite of its wrongness. The theory of continuous precession as discovered by Hipparchos and explained by Ptolemy and the theory of trepidation were contradictory, though many astronomers tried to

[46] *Almagest,* IX, 7; XI, 7; Heiberg's edition, vol. 1, part 2, pp. 268, 267, 419; Halma's edition, vol. 2, pp. 171, 170, 288.

[47] The theory was defended by Paul Schnabel, "Kidenas, Hipparch und die Entdeckung der Präzession," *Z. Assyriologie* 3, 1-60 (1926) [*Isis 10,* 107 (1928)]. For Kidinnu or Cidenas (*Cidēnas*), see Wilhelm Kroll, *Catalogus codicum astrologorum graecorum,* vol. 5, part 2, p. 128; Joseph Heeg, *Ibid.,* vol. 8, part 2, pp. 125-134; W. Kroll, Pauly-Wissowa, vol. 21 (1921), p. 379; according to this article, Cidenas, who flourished at the latest in the second century B.C., found the equation 251 synodic months = 269 anomalistic months. There are in the British Museum lunar tablets written in cuneiform on 22 December 103 B.C. after Cidenas. Kroll concludes that Cidenas may have been one of the Chaldean astronomers referred to by Ptolemy, but that would bring him only as early as 244 B.C., and the Cidenas who "discovered" the precession in 379 must then be another man.

[48] J. K. Fotheringham, "The indebtness of Greek to Chaldaean astronomy," *The Observatory 51,* No. 653 (1928); reprinted in *Quellen und Studien* [*B*] 2, 28-44 (1932). A. T. Olmstead, *History of the Persian empire* (Chicago: University of Chicago Press, 1948), pp. 453-457. Otto Neugebauer, "The alleged Babylonian discovery of the precession," *J. Am. Oriental Soc. 70,* 1-8 (1950).

[49] *Almagest,* VII, 1-2.

find a compromise. Trepidation was accepted by the Hindu astronomer Āryabhaṭa (V–2), who was perhaps the link between Theon and Proclos on one hand and Thābit ibn Qurra (IX–2), the first Arabic writer to speak of it, on the other. It must be said to the credit of the Muslim astronomers that most of them rejected the idea of trepidation; that is the case for al-Farghānī (IX–1), al-Battānī (IX–2), 'Abd al-Rahmān al-Ṣūfī (X–2), and Ibn Yūnus (XI–1). But unfortunately al-Zarqālī (IX–2) al-Biṭrūjī (XII–2) sponsored that wrong idea, and as their influence was considerable, they were largely responsible for its diffusion among the Muslim, Jewish, and Christian astronomers, so much so that Johann Werner (1522) and Copernicus himself (1543) were still accepting it; Tycho Brahe and Kepler had doubts concerning the continuity and regularity of the precession, but they finally rejected the trepidation.[50] The matter could not be completely elucidated until the precession of the equinoxes had been explained in Newton's *Principia* (1687).

The persistence of the false theory of trepidation is difficult to understand. At the very beginning of our era, the time span of the observations was still too small to measure the precession with precision and without ambiguity, but as the centuries passed there could not remain any ambiguity. Between the stellar observations registered in the *Almagest*[51] and those that could be made by Copernicus almost fifteen centuries had elapsed, and the difference of longitudes would amount to 21°.[52] How could such a difference be explained in terms of trepidation? How could it be explained otherwise than by a steady accumulation of differences of the same sign?[53]

The vicissitudes of the theories of precession and trepidation, the true versus the false, is one of the best examples of human inertia. It helps us not to be too optimistic, and to remain modest. If it is so difficult to establish scientific facts, which are relatively tangible and unambiguous, we should not expect much progress in other fields and be very humble and very patient.

THE PRECURSORS OF SCIENTIFIC ASTRONOMY: PHILOLAOS, HICETAS, AND ECPHANTOS

Philolaos was a contemporary of Socrates; Hicetas and Ecphantos, both of Syracuse, were younger; the first flourished perhaps and the second certainly in the fourth century. Their views were explained in a previous chapter (pp. 288–291), as it was more convenient not to separate Hicetas and Ecphantos from Philolaos, but we should remember that the fruition of their ideas occurred definitely in Plato's age. These ideas may be summarized as follows. The universe is spherical and limited; the earth is not necessarily in the center, it is a planet like the others, and turns eastward around its axis.[54] Did Plato know of them? He mentions Philolaos in *Phaidon*;[55] considering his Pythagorean and Sicilian connections, he probably heard of the other men, yet he does not refer to them.

[50] For more details on the history of trepidation see my *Introduction, passim*; summaries in vol. 2, pp. 18, 295, 749, 758; vol. 3, p. 1846.

[51] The true epoch of Ptolemy's "Catalogue" is A.D. 58; *Isis* 2, 401 (1914-19); see footnote 45.

[52] The correct value of the precession is about 50".26. Thus, 5026" = 84' = 1°24' per century; and this amounts to 21° in 15 centuries.

[53] It might be objected to this that as long as precession was not explained (as Newton did) one could not be certain of its indefinite continuation in the same direction: it might pile up for, say, 8°, or 80°, or 150° and then possibly stop or be reversed.

[54] Counterclockwise for an observer above the North Pole.

[55] *Phaidon*, 61D.

THE FOUNDER OF SCIENTIFIC ASTRONOMY, EUDOXOS OF CNIDOS, AND HIS THEORY OF HOMOCENTRIC SPHERES

We have given above an outline of Eudoxos' biography and told that he spent sixteen months in Egypt (sometime in the period 378–364) and was admitted to the company of learned priests. Before that he had studied at the Academy and had become familiar with Pythagorean astronomy. Yet that did not suffice him and, being of a rigorous cast of mind, he may have been especially dissatisfied with the lack of observations. Not only did he obtain communication of Egyptian observations, but he made new ones and the observatory used by him, between Heliopolis and Cercesura,[56] was still pointed out in the time of Augustus (emperor, 27 B.C.–A.D. 14). Later, he built another observatory in his native land, Cnidos, and there he observed the star Canopus, which was not then visible in higher latitudes.

Eudoxos' relatively long stay in Egypt accounts for his knowledge of Egyptian astronomy, but was he familiar with Babylonian astronomy, which was so much richer? There is no evidence of his having traveled in Mesopotamia and Persia, but he had a deep knowledge of the ancient world and wrote an elaborate description of it (*Periodos gēs*) which was the first of its kind and scope. As far as can be judged from the fragments that have come down to us, Eudoxos' geography contained a vast body of geodesic and topographic data, plus information on natural history, medicine, ethnology, religion. For example, he had noted the importance of Zoroastrianism, and Plutarch's knowledge of Isis and Osiris was partly derived from his.[57] Far superior to the geographers of the fifth century, he is in this respect the forerunner of Eratosthenes of Cyrene (III–2 B.C.).

Even if Eudoxos did not travel to Mesopotamia, his residence in Cnidos enabled him to drink from Asiatic springs, whether Persian or Chaldean, for Cnidos (as well as its neighbors, Halicarnassos and Cos) was a cosmopolitan center of the first order. There is some probability that he is the author of bad-weather predictions (*cheimōnos prognōstica*),[58] which are definitely of Babylonian origin. He had identified the twelve main gods of the Greeks with the zodiacal signs. This is interesting, but we shall not insist upon it, for no matter how well he was acquainted with Egyptian and Babylonian astronomy his merit lies in another direction. No doubt he had profited by his study of Oriental methods of observation and he had played with Chaldean astrology, but no Oriental astronomer could have suggested to him his main achievement: the theory of homocentric spheres.[59]

The purpose of that theory was to give a mathematical account of the positions of the heavenly bodies at any time, or, if we may use the strong Greek phrase, "to save the phenomena" (*sōzein ta phainomena*). That was easy enough as far as the stars were concerned, but how could one account for the planets, the

[56] Cercesura on the west bank of the Nile, at the point where the river is divided into its three principal branches, the east one, or Pelusiac, the central one, and the west one, or Canopic.

[57] The fragments of Eudoxos' *Periodos* have been edited and explained by Friedrich Gisinger, *Die Erdbeschreibung des Eudoxos von Knidos* (Stoicheia 6, 142 pp.; Leipzig, 1921). For the oriental sources of Eudoxos, see J. Bidez, *Eos* Brussels: Hayez, 1945), pp. 24–37 [*Isis* 37, 185 (1947)].

[58] Text in *Catalogus codicum astrologorum graecorum*, vol. 7 (1908), pp. 183–187; see also vol. 8, part 3, p. 95.

[59] We know that theory and of its ascription to Eudoxos from Aristotle's Metaphysics (1073 B 17–1074 A 15) and from the commentary of Simplicios (VI–1) on the *De caelo*.

trajectories of which were extremely puzzling? At times they seem to stop, to retrograde, and to trace a singular curve such as the one that Eudoxos investigated and that he called the *hippopedē* or horsefetter, a spherical lemniscate, looking like a figure eight. This was a difficult geometric or kinematic problem. He had to discover a combination of movements, circular or spherical, of such a nature that a single planet, say Mercury or Venus, would seem to describe in the heavens a *hippopedē*.

Eudoxos' solution of that problem was typical of the Greek mathematical genius and of his own. Mercury is supposed to be placed on the equator of a sphere centered upon the earth and revolving with constant speed around one of its diameters (on account of an old Pythagorean prejudice all the movements must be circular and uniform). Let us call that diameter after its two poles AA'. If that diameter did not change its position, Mercury (M) would describe a circle around the earth, but let us suppose that the diameter AA', instead of being fixed, is carried by another concentric sphere turning with constant speed around the diameter BB'; the apparent motion of M will then be the combination of the two rotations, of speed ω around AA' and ω' around BB'. If that is not enough to "save the phenomena" we may suppose that the diameter BB' is not fixed but is carried by a third concentric sphere rotating with speed ω'' around the axis CC'; the apparent motion of M will then be the kinematic resultant of the three rotations of speeds ω, ω', ω'' around the axis AA', BB', CC'. There is no need to stop at this third sphere; once the principle is accepted one may use as many auxiliary or starless (*anastroi*) spheres as are needed. And the problem can thus be stated in these terms: given the apparent trajectory of any celestial body, to find enough spheres concentric with the earth of speeds ω, ω', ω'', . . . and axes AA', BB', CC', . . . to account for it. When a solution has been found, it can be checked as frequently as one pleases and in fact it is checked every time one compares the calculated position of that body with the observed one. If these two positions do not coincide, the solution may be improved either by changing the speeds and axes of the auxiliary spheres or by adding a new one.

In order to explain the motions of all the celestial bodies, Eudoxos was forced to postulate the existence of no fewer than 27 concentric spheres,[60] each of which turned around a definite axis with a definite speed. The boldness of that conception is staggering. It was the first attempt to explain astronomic phenomena in mathematical terms; the explanation was very complicated (inasmuch as it required the simultaneous motion of 27 spheres rotating with different speeds around different axes), but it was adequate and elegant. It did "save the phenomena" with sufficient approximation. The working out of that solution implied an advanced knowledge of spherical geometry; it is probable that Eudoxos himself contributed to its advancement, for he needed it badly.

The theory of homocentric spheres is a magnificent example of Greek rationalism. Eudoxos introduced as many spheres as were needed for his kinematic purpose; he did not speculate on the real existence of those spheres or on the cause of their motions. It matters not, he might have said, whether the spheres exist or not, or why they move as they do; the only thing that matters is that

[60] To explain the apparent trajectories of the fixed stars one sphere was needed; of the Sun and Moon, three spheres each; of the five Planets, four each; total: 27 spheres.

their imaginary functioning together "saves the phenomena." The theory provides a kinematic re-creation and verification of the observations.

Remarkable as it was, that theory was unavoidably imperfect; the observations available to Eudoxos were themselves insufficient in number and precision. His ideas on the sizes and distances of the celestial bodies were very crude. For example, we know from Aristarchos of Samos (III–1 B.C.) that he believed the diameter of the Sun to be nine times that of the Moon.

Eudoxos wrote two astronomic books, entitled the *Mirror* (*Enoptron*) and the *Phainomena*, a description of the heavens, which was the source of a famous astronomic poem by Aratos of Soli (III–1 B.C.).[61] The *Phainomena* of Eudoxos and of Aratos were commented upon by Hipparchos (II–2 B.C.) in his youth, and this commentary is, strangely enough, the only text of Hipparchos that has been transmitted to us in its integrity. Hipparchos corrected some of Eudoxos' errors, for example, the latter's belief that the north pole was occupied by a particular star; the north pole, said Hipparchos, is empty but there are three stars close to it [α and κ of Draco and β of Ursa minor] with which the point at the pole forms a square.

According to Diogenes Laërtios,[62] Eudoxos wrote the *Octaëtēris* while he was in Egypt. This may refer to an attempt made by Eudoxos to discuss or correct the eight-year cycle introduced by Cleostratos (p. 179), but we do not know the nature of his correction.

These are secondary matters, however. The fame of Eudoxos rests upon his invention and development of the theory of homocentric spheres, thanks to which he must be considered the founder of scientific astronomy and one of the greatest astronomers of all ages.

THE ASTRONOMIC FANCIES OF PLATO AND PHILIP OF OPUS
THE INTRODUCTION OF SIDEREAL RELIGION INTO THE WESTERN WORLD

After having breathed the pure air of Eudoxean rationalism, it is a terrible shock to come down again to the low level of Platonic sublimities. Plato insists [63] that "each planet moves in the same path, not in many paths, but in one only, which is circular, and the varieties are only apparent. Nor are we right in supposing that the swiftest of them is the slowest, nor conversely, that the slowest is the quickest," and also [64] that "the movements can be apprehended only by reason and thought, not by sight." That is, he realized that the universe is a cosmos, but that its order and regularity could not be immediately deduced from the appearances.[65] Eudoxos had proved that, for if it were possible to represent the motions of the celestial bodies by a kinematic system, these motions were well ordered; one might not know their causes or the rules producing them, but one could be certain that there were rules (natural laws).

[61] The *Phainomena* as preserved by Aratos is the earliest Greek treatise on astronomy extant. Eudoxos' description was partly derived from Democritos and, directly or indirectly, from the Babylonian astronomers.

[62] Diogenes Laërtios, VIII, 87.

[63] *Laws*, VII, 822.

[64] *Republic*, VII, 529.

[65] According to Sosigenes (Julius Caesar's astronomer), Eudemos of Rhodes (IV–2 B.C.) declared that Plato set it as a problem for astronomers to find "which uniform and ordered movements must be assumed to account for the apparent movements of the planets" (Simplicios on *De caelo*, 488, 20–31 in Heiberg's edition). Eudoxos solved that problem; it is highly probable that it was he, not Plato, who stated it.

The relation between Plato and Eudoxos is not clear. The latter was a younger contemporary of the former, and was for a time his pupil but left, whether rejected by the master or disgusted with his philosophy. There were certainly exchanges of influence between Eudoxos and the Academy. Eudoxos is nowhere mentioned by Plato. I suspect that they could not understand each other; they spoke different languages.

The astronomic views of Eudoxos have been explained in the previous section; these views were scientific views of the highest order. The observations at Eudoxos' disposal were insufficient in number and precision, but his method was excellent. On the other hand, the Platonic views expressed in *Timaios* (and also in *Phaidon*, *Republic*, and *Laws*) are unscientific; Plato asserts this and that but he proves nothing, and his language is often as unclear as that of any soothsayer. His astronomic knowledge was of Pythagorean origin; it was far from up-to-date, being inferior not only to that of Eudoxos, but also to that of the later Pythagoreans like Philolaos and Hicetas. Let us give a brief outline of it.

The universe is spherical; in the center of it lies the Earth, which is also spherical and immovable, and stays there (in the center) for reasons of symmetry. The axis of the universe and of the Earth passes through their common center. The outer sphere of the universe rotates around that axis with constant speed in 24 hours, as is witnessed by the motion of the fixed stars. The Sun, Moon, and other planets are also carried round by the motion of the outer sphere, but they have other circular motions of their own. On account of these independent motions the planet's real trajectory is a spiral in the zodiacal zone. The angular speeds of planets decrease in the following order: Moon, Sun, Venus and Mercury traveling with the Sun, Mars, Jupiter, Saturn. The order of distances from the Earth is the same, and the distances are deduced from two geometric progressions: 1, 2, 4, 8; 1, 3, 9, 12, the distances being: Moon, 1; Sun, 2; Venus, 3; Mercury, 4; Mars, 8; Jupiter, 9; Saturn, 12.

In the *Timaios*[66] it is suggested that Venus and Mercury move in a direction opposite to that of the Sun.[67] Plato knew the periods of Moon, Sun, Venus, and Mercury (the periods of the three last named he believed to be the same, 1 year),[68] but not those of the other planets, yet he speaks of the Great Year[69] when the eight revolutions (the seven planetary ones plus that of the outer sphere) are brought back to their starting point. That Great Year is equal to 36,000 years.[70] How did he measure it? He measured nothing, but copied a Babylonian tradition (see p. 71).

We leave out other fancies connecting the planets with the regular solids or with musical notes, the harmony of the spheres. The music of the heavens referred to in *Timaios* is not one audible to human ears, however. It may be caused by the

[66] *Timaios*, 38D.

[67] This may have suggested to Heracleides of Pontos (IV-2 B.C.), the theory according to which Venus and Mercury rotate around the Sun.

[68] We may recall that the true periods are, in terms of the Earth's period (that is of the Sun's period if the Earth is in the center of the world): Mercury, 0.24; Venus, 0.62; Earth (Sun), 1; Mars, 1.88; Jupiter, 11.86.

[69] *Timaios*, 39.

[70] 36,000 years is also the time of completion of the precession cycle on the (wrong) Ptolemaic assumption that the precession amounts to 1° in a century (*Almagest*, VII, 2). This is a curious coincidence, for Plato had no knowledge of precession. 36,000 is a factor of the Geometric number.

relative speeds of the planets, yet it exists only in the world soul. Do not expect me to explain these mysteries.

According to Aristotle, Plato believed that the Earth rotates around its axis; according to Theophrastos, "in his old age Plato repented of having given the Earth the central place in the universe, to which it had no right." These two sayings have caused polemics, but we are justified in rejecting them, because they contradict Plato's own writings, all of which have come down to us.

The success of Plato's astronomy, like that of his mathematics, was due to a series of misunderstandings: the philosophers believed that he had obtained his results by the aid of his mathematical genius; the mathematicians did not like to discuss the same results because they ascribed them to his metaphysical genius. He was speaking in riddles, and nobody dared to admit that he did not understand him for fear of being considered a poor mathematician or a poor metaphysician. Almost everybody was deceived, either by his own ignorance and conceit or by his subservience to fatuous authorities. The Platonic tradition is very largely a chain of prevarications.

THE EPINOMIS

We must still speak of a short dialogue called *Epinomis* (or *Nocturnal council*, or *Philosopher*). As its main title indicates, it is an appendix to the *Laws*.[71] The Nocturnal Council named in the second title was a secret body of inquisitors who had to make sure that the laws were obeyed. The *Epinomis* might be described as a discussion concerning the education of the members of that Council; but because this purpose appears only in the first and final paragraphs, the reader is likely to forget it. According to Diogenes Laërtios and Suidas, the *Epinomis* was written or published posthumously by one of Plato's disciples, Philip of Opus.[72] Philip had been Plato's secretary (*anagrapheus*) in the latter's old age; he edited the *Laws*, divided them into twelve books, and added the *Epinomis*. Many other writings are ascribed to him, dealing with mathematics (for example, on polygonal numbers and on means), astronomy (on the distances of the planets, *parapegma*, which are astronomic tables or almanacs), optics, meteorology, ethics. Was he the author or simply the editor of the *Epinomis*? And if he was the editor, how far did his editorship extend? It is impossible to answer such questions. We must take the *Epinomis* as we find it (the text itself gives no clue as to authorship or editorship). It is Platonic in form and contents, though more Pythagorean than the other Platonic writings. The astronomy of *Epinomis* is essentially the same as that of *Timaios*, except for the stronger Pythagorean note, which concerns the metaphysics of astronomy rather than astronomy proper.

The main purpose of the *Epinomis* is to emphasize the importance of astronomy for the attainment of true wisdom. As expressed by a master of the history of ancient religion, the late Franz Cumont, the *Epinomis* was "the first Gospel

[71] *Laws*, XII, 966–967. The speakers are the same, Megillos the Spartan, Cleinias the Cretan and an Athenian stranger who does almost all of the talking. Their conversation occurs in Crete, they had begun it on the previous day (*Laws*, 1, 625) while walking from Cnossos to the temple of Zeus beneath Mount Ida in the center of the island.

[72] *Philippos ho Opuntios*. *Opus* is probably the place of that name in Locris Opuntia on the Euboic gulf. Philip of Opus has been identified with Philip of Mende, Mende being on the west coast of the Thermaic Gulf, Macedonia. He was probably born in Mende and later moved to Opus and to Athens. Elaborate article in Pauly-Wissowa, vol. 38 (1938), pp. 2351–2366.

preached to the Hellenes of the stellar religion of Asia." [73] That religion was born in Babylonia, where priests were astronomers and where an amazingly clear sky invited observations. At the beginning of the Achaimenian empire (Cyros the Great, ruling 559–529), which included Babylonia, these ideas were diffused by Persian priests called magi and by native ones called Chaldeans. It reached Greek-speaking people from both sources (Persian and Chaldean) and the *Epinomis* was its first evangel in the Greek language.

The argument of the *Epinomis* is far from clear, but we shall indicate some of the salient ideas. Number is extremely important, and nowhere more obviously so than in the regular motions of celestial bodies, the stars, Sun and Moon, and the planets. The five regular solids are equated to the five elements, but the fifth element is called aether.[74] The soul is older and more divine than the body. Order is equated to intelligence, and disorder to nonintelligence; the supreme order of the celestial motions represents supreme intelligence. There are eight "powers in heaven" (the seven planets and the eighth sphere), which are equally divine. The planets must be gods. That is what the Egyptians and Syrians (meaning Babylonians) have known for thousands of years; we should accept their knowledge and their religion, after having improved it, as the Hellenes always do when they borrow anything from the barbarians. While paying due reverence to the ancient gods, according to venerable traditions, the cult of the visible gods, the celestial bodies, should become a state religion. That religion would give the Hellenes a vision of divine unity, and would provide for them a bond (*desmos*) universal and immaterial. Note that many of the astrologic fancies were already expressed in other Platonic writings (*Phaidros*, *Timaios*, *Laws*). The novelty lies in the religious accent, the equation of astronomy and piety, the idea of an astrologic state religion.

The aim of wisdom is to contemplate numbers, and especially celestial numbers. The most beautiful things are those revealed to our understanding by our own souls, or the cosmic soul, and the heavenly regularities.[75] The cult of the stars must be introduced into the laws.

Astronomy is not only the climax of scientific knowledge, it is the rational theology. Members of the Nocturnal Council should receive a mathematical education, leading them to astronomy and religion. The supreme magistrates of the city are not to be philosophers, but rather astronomers, that is, theologians.

There are so many irrational statements in *Epinomis* (parading under the garb of supreme rationalism) that discussion of them would be as endless as it would be futile. There is one point, however, that I would like to touch, because it has puzzled me more than the others and has received but little attention. The author criticizes foolish people for associating initiative (freedom) with intelligence; [76]

[73] Franz Cumont (1868–1947), *Astrology and religion among the Greeks and the Romans* (New York, 1912), p. 51. A much amplified rewriting in French was posthumously published under the title *Lux perpetua* (Paris: Geuthner, 1949) [*Isis 41*, 371 (1950)].

[74] In *Timaios* the fifth solid was equated to the whole universe. In the *Epinomis* the elements are first named in the order fire, water, air, earth, aether, but later in a more "logical" order (from spirituality to grossness), fire, aether, air, water, earth (981, 984). It is curious that aether is given the second, not the first, place.

[75] One cannot help thinking of Kant's statement, "Two things fill one's conscience with ever-increasing wonder and awe, the stars in heaven and the moral law in oneself," *Critique of practical reason* (Riga., 1788) [*Isis 6*, 479 (1924)], but the words of a rational mystic like Kant are incomparably more impressive than those of the irrational author of the *Epinomis*.

[76] *Epinomis*, 982.

real intelligence, however, is characterized by repetitious order; the planets reveal supreme divine intelligence by the eternal accuracy of their motion. Now we might admit with Plato that the planetary motions reveal God, but not that the planets themselves are gods. Think of the popular argument of the clock. Its mechanism and regular motions reveal the existence of a clockmaker. Nobody ever said that the clockmaker was in the clock, or that the clock was itself the clockmaker. Yet, according to the new astrologic religion, the planets did not simply reveal God, they were themselves gods, each planet regulating its own motion with divine intelligence and repeating it eternally to evidence its own wisdom. Does that make sense? Yet the argument was accepted by the New Academy and by the Stoics and we find it very clearly stated by Cicero.[77] The confusion of thought was probably caused by a wrong generalization: the soul or intelligence of an animal is within itself; we may say that the animal has intelligence or that he is an intelligent being; his intelligence, however, is revealed not by the regularity and precision of his motions, but rather by their unexpectedness.

It is significant that the *Epinomis*, first Greek gospel of the astrologic religion, does not include any astrology in the vulgar sense. There is a passing reference[78] to the divinity of birth (*to theion tēs geneseōs*), but it is not clear and does not prove that the author had accepted the fundamental assumption of practical astrology: that a man's fate is determined at the time of his birth (or conception?) and can be deduced from the study of his horoscope.[79] Yet judicial or mundane astrology had been practiced in Babylonia from time immemorial, and as to the Greeks, if they took astronomy and divination with equal seriousness, the step to astrologic divination was unavoidable.

If we believe that the stars or planets are gods and that there are communications between them and us, we cannot but conclude that they must influence us. The communication is sufficiently proved by the fact that we see them, for this implies that something passes from them to us.[80] Astrologic divination becomes possible only after certain assumptions had been made, such as the one referred to above, and after the establishment of "scientific" horoscopes had been determined by the acceptance of a series of conventions.[81]

[77] *De natura deorum,* II, 16. Cicero put it in the mouth of Gaius Aurelius Cotta, the Academic, at whose house the dialogue *De natura deorum* was supposed to have taken place, *c.* 77 B.C. (the dialogue was written *c.* 45 B.C.). "It is therefore likely that the stars possess surpassing intelligence, since they inhabit the ethereal region of the world and also are nourished by the moist vapours of sea and earth, rarefied in their passage through the wide intervening space. Again, the consciousness and intelligence of the stars is most clearly evinced by their order and regularity; for regular and rhythmical motion is impossible without design, which contains no trace of casual or accidental variation; now the order and eternal regularity of the constellations indicate neither a process of nature, for it is highly rational, nor chance, for chance loves variation and abhors regularity; it follows therefore that the stars move of their own free-will and because of their intelligence and divinity." quoted from H. Rackham's translation in the Loeb Classical Library.

[78] *Epinomis,* end of 977.

[79] The word *hōroscopos* and cognate words were coined very late. It is used by Manilius (I–1), Sextos Empiricos (II–2), Clement of Alexandria (*c.* 150–220), but I cannot find any earlier use.

[80] The transmission of stellar power to the Earth was demonstrated by a dramatic experiment staged in Chicago on 27 May 1933. The illumination of the Century of Progress Exposition was turned on by light that had left the star Arcturus 40 years before, at the time of the Columbian World Fair! The light of Arcturus was picked up by telescopes at the Yerkes Observatory, Williams Bay, Wisconsin, and focused on photoelectric tubes; the current thus obtained was much amplified and directed to Chicago, *Science News Letter* 23, 307 (1933).

[81] The early history of that is obscure. Eudoxos of Cnidos declared that the Chaldeans who foretold the life of a man from his birthday should not be given any credence (Cicero, *De divinatione,*

In any case, the astrologic religion first explained in the *Epinomis* became gradually the highest religion of the pagan world, Greek and Latin. The old gods were still worshiped and the old myths were celebrated by the poets and the artists, but thinking people could no longer acquiesce in them except in a traditional mood and with poetic ambiguities. As compared with the mythologic childishness and immorality, the cult of the stars seemed highly rational. Not only the *Epinomis*, but Pythagorean and Platonic ideas in general, provided a philosophic substructure upon which the new religion could be established so solidly that most members of the intellectual elite took it as a kind of science. The influence of that "Pagan science" upon the best minds of the Roman empire was so deep that Christianity itself could not completely erase it. Indeed, it is represented to this day by one of our oldest and most ubiquitous institutions, one that controls the labor and rest of every man, the week. The number of days in the week is of astrologic origin, and their names in most European languages are planetary names.[82]

II, 42, 87), but we cannot conclude from that the Chaldean rules had been already Hellenicized. We are told that Panaitios of Rhodes (II-2 B.C.) rejected astrology, and we may assume that his contemporary Hipparchos (II-2 B.C.) did the same, but who invented the horoscopic rules? The earliest treatise on astrology extant is the *Tetrabiblos* ascribed to Ptolemy (II-1); that treatise is still used by astrologers of our own time! *Isis* 35, 181 (1944).

[82] Francis Henry Colson, *The week, an essay on the origin and development of the seven-day cycle* (133 pp.; Cambridge, 1926). The seven-day cycle was diffused unofficially through the Roman empire not long before Christ; its gradual diffusion throughout the world is the most remarkable instance of cultural convergence next to the decimal system. Nobody planned it; it just happened. See also Solomon Gandz, "Origins of the planetary week or the planetary week in Hebrew literature," *Proc. Am. Acad. Jewish Research 18*, 213-254 (1949).

XVIII

XENOPHON

This short chapter was written as a kind of intermezzo and should be read as such. A scholar dealing with the history of science *stricto sensu* might omit Xenophon or dismiss him in a single paragraph, but if we take general education into account (and we certainly wish to do so), we must give him a much larger place. Indeed, Xenophon not only tried to improve the education of his own time, he exerted a strong influence upon the education of later generations, much later ones, such as the Elizabethan or even our own. Moreover, he continued Thucydides and was one of the outstanding disciples of Socrates. A wonderful literary instrument had been perfected in his time – the Attic prose of the golden age – and Xenophon wielded it with consummate art.

Xenophon, son of Gryllos, an Athenian of the countryside, was born *c.* 430 and died in Corinth about the middle of the fourth century. Diogenes Laërtios said of him,[1] "Xenophon was a remarkable man in many respects, notably his love of horses, hunting and the military art, he was a pious man who loved to offer sacrifices, was familiar with religious rites and was a faithful disciple of Socrates." This brief description is excellent, and it is completed with anecdotes that help us to realize what a kind of man he was. For example, Diogenes tells us how Socrates and he came together. "It is said that Xenophon having come across Socrates in the street, the latter closed the way with his stick and asked him where one could buy the necessities of life. Xenophon gave him the information. 'And to become an honest man,' asked Socrates, 'where should one go?' Xenophon could not answer. 'Follow me then,' said Socrates, 'and I will show you.'" Is not that a pretty story? It suggests that Socrates had enough perspicacity to recognize a good man when he saw one. The story impresses us the more deeply, because we cannot help remembering how Christ called Peter and Andrew, James and John, and how they obeyed his call and followed him.

Xenophon was a man of means who could indulge his taste for riding and hunting and may have been employed in the Athenian cavalry, but had no definite profession. He was thus free in 401 to join the army of Greek mercenaries that Cyros the Lesser was levying to march against his brother and sovereign, Artaxerxes. Cyros was defeated and killed at the battle of Cunaxa, and the Greek army had to find its way home as best it could. After the murder of the generals, Xenophon was elected leader and he succeeded in guiding the Ten Thousand to Trebizond. Early in 399 he handed over what remained of the Greek army to

[1] Diogenes Laërtios, II, 56.

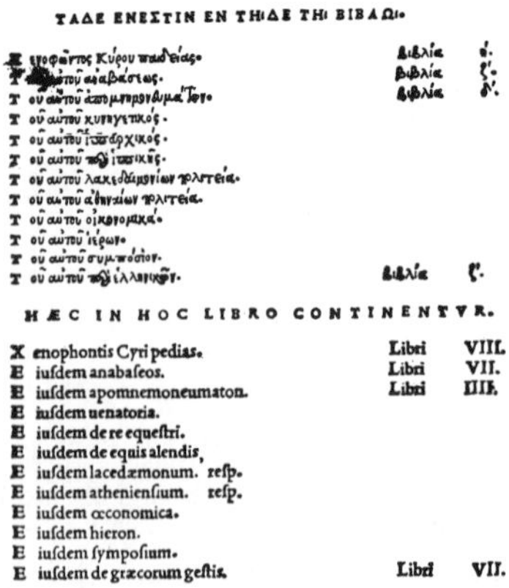

Fig. 86. There is no title page to the Greek *princeps* of Xenophon's *Opera* (folio; Florence: Giunta, 1516), except this table of contents. The titles (but not the works themselves) are translated into Latin. [From the copy in the Harvard College Library.]

a Spartan general who was then stationed in Asia. It was at about that time that he was banished from his native city, and he deserved his banishment. He continued in the Spartan service and became a friendly admirer of Agesilaos (king of Sparta, 399–360), one of the best generals and noblest men of his nation. He fought under him against the Persians, came back with him to Greece, and was present (in the Spartan contingent) at the battle of Coroneia.[2] In the meanwhile, Xenophon had married, and by 394 his sons were old enough to be educated in Sparta. Later, the Spartans gave him a large estate at Scillus, near Olympia, where he lived like a squire, managing his property, riding, hunting, and writing. Many of his books were composed during the twenty years of his life at Scillus. It was certainly there that he wrote the best-known of them, the *Anabasis*, between 379 and 371, when the vicissitudes of war caused him to lose his estate and forced him to begin a new life in Corinth. In 369, the Athenians made peace with Sparta and readmitted Xenophon to their fellowship; his sons served eventually in the Athenian cavalry.[3]

We have not given the whole of Xenophon's military record, but it is clear that he had obtained considerable experience as a cavalry and military man, and that he had obtained it not only in the famous march of his youth from Cunaxa to the Black Sea, but also in the service of the enemies of his country. He was a great admirer of Spartan education and discipline, and shortly after Agesilaos' death in 360 he wrote a eulogy of him.

XENOPHON'S WRITINGS

Xenophon's writings[4] are various and abundant (Fig. 86). One or two excepted, they can hardly be prior to his warlike activities (401–394), and thus

[2] The Spartans, led by Agesilaos, defeated at Coroneia (Western Boeotia) in 394 a Greek coalition (Thebes, Corinth, Argos, and Athens) subsidized by Persian gold.

[3] His elder son Gryllos fell at the battle of Mantineia in 362; according to a legend, it was he who had given to Epaminondas, the Theban general, his mortal wound. In that battle the Athenians were allied with the Spartans and many other Greeks against Thebes. Epaminondas' victory was indecisive.

[4] Xenophon has always been a favorite author and the manuscripts, editions, and translations of his works are very numerous. The princeps of the Greek opera was published by Luca Antonio Giunta in Venice, 1516 (again, 1527); complete Latin edition, Basel, 1534; Greek edition by Edgar Cardew Marchant (5 vols.; Oxford, 1900–1910).

they belong definitely to the fourth century. Many were composed at Scillus (394–371), but he continued to write until the last years of his life. His works will now be rapidly listed, with a few remarks added for the sake of historians of science.

We began with a group of three books (I–III) dealing with hunting and horsemanship, because the first of them is supposed to have been written in his youth, before his departure from Athens for Asia.

1. *Cynēgeticos* is a treatise on hunting, especially of the hare, including the breeding of dogs; it is the first treatise of its kind known to us.

2. *On horsemanship* (*Peri hippicēs*) was thought to be the first treatise on the subject in any literature, until in 1931 Hrozny published a Hittite treatise on horsemanship.[5] It was written by a man who was a lover of horseflesh and a horseman of long experience.

3. *Hipparchicos*, on the duties of a cavalry commander, is a continuation of the preceding item, on the application of horsemanship in all its aspects to warlike purposes.

Xenophon's books on horsemanship (2 and 3) are well known to French readers because of a masterly translation by Paul Louis Courier (1772–1825). Courier had a racy use of the French language, and he was both a horseman and a Hellenist.

The most famous of Xenophon's writings were the two devoted to Asiatic matters (4 and 5).

4. *Anabasis* (*Cyru Anabasis*) (Fig. 87). This account of the greatest adventure of his life, the share of the Ten Thousand Greek mercenaries in the revolt of Cyros the Lesser, their defeat at Cunaxa, and their retreat to Trebizond, was the first chronicle of its kind and has remained one of the outstanding military memoirs; it is also the first description of the country that they traversed, the highlands of Armenia. The book is full of curious details, references to ostriches,[6] to army surgeons,[7] to poisonous honey,[8] to tatooing,[9] to the iron work of the Chalybes,[10] to bookselling.[11] He illustrates with his own example, the need of a military officer to be just, generous, pious, to love the soldiers and win their devotion. The difficulties of leadership were particularly great in his case, because the Ten Thousand were a highly heterogeneous group, adventurers recruited from every Greek land, a kind of human flotsam, who had no quality in common except a rudimentary Hellenism, exacerbated by their loneliness in the midst of barbar-

---

Xenophon's works are included in all the collections of classics such as the Budé and the Loeb series. Gustav Sauppe, *Lexicologus Xenophonteus* (156 pp.; Leipzig, 1869).

[5] See Chap. III, note 66. Courier's translation of the *Hippichē* has been improved by that of another French Hellenist and horseman, Edouard Delebecque, *Xenophon. De l'art équestre,* (195 pp.: Paris: Les belles lettres, 1950).

[6] *Struthoi ai megalai; Anabasis,* I, 5, 2. The existence of ostriches in ancient Mesopotamia and ancient China is confirmed by monuments, and therefore their presence in Asia Minor is not surprising. When did they disappear? Their (natural) habitat is restricted to Arabia and Africa. Berthold Laufer, *Ostrich egg-shell cups of Mesopotamia and the ostrich in ancient and modern times* (51 pp., ill.; Chicago, 1926) [*Isis 10*, 278 (1928)].

[7] *Anabasis,* III, 4, 30.

[8] *Ibid.,* IV, 8, 20–21.

[9] *Ibid.,* V, 4, 32.

[10] *Ibid.,* V, 5, 1.

[11] *Ibid.,* VII, 5, 14. Xenophon refers to many written papyri found on the Thracian shore of the Black Sea ready to be shipped. Compare with another curious statement in Plato's *Apology*, 26E. Books containing Anaxagoras' views could be bought for a drachma in the orchestra.

ΠΑΙΔΕΙΑΣ, ΟΓΔΟΟΝ.

καὶ ἐκ τούτου φαίνεται. οὐ μέντοι ὄφελός τι οὐδὲν αὐτῶν εἰς πόλεμον. δῆλοι δὲ καὶ αὐτὰ τὰ γιγνόμενα. κατὰ τὴν χώραν γὰρ αὐτῶν ῥᾴδιον οἱ πολέμιοι ἢ οἱ φίλοι ἀναστρέφονται. καὶ γὰρ δὴ ὁ Κῦρος, τῶν μὲν ἀκροβολίζεσθαι ἀποπαύσας, θωρακίσας δὲ καὶ αὐτοὺς καὶ ἵππους, καὶ ἓν παλτὸν ἑκάστῳ δοὺς εἰς χεῖρα, ὁμόθεν τὴν μάχην ἐποιεῖτο. νῦν δὲ οὔτε ἀκροβολίζονται ἔτι, οὔτ' εἰς χεῖρας συνιόντες μάχονται. καὶ οἱ πεζοί, ἔχουσι μὲν γέρρα καὶ κοπίδας καὶ σαγάρεις, ὥσπερ οἱ ἐπὶ Κύρου τὴν μάχην ποιησόμενοι. εἰς χεῖρας δὲ ἰέναι, οὐδ' οὗτοι ἐθέλουσιν. οὐδέ γε τοῖς δρεπανηφόροις ἅρμασιν ἔτι χρῶνται ἐφ' ᾧ Κῦρος αὐτὰ ἐποιήσατο. ὁ μὲν γάρ, τιμαῖς αὐξήσας τοὺς ἡνιόχους, καὶ ἀγαθοὺς ποιήσας, εἶχε τοὺς εἰς τὰ ὅπλα ἐμβαλοῦντας. οἱ δὲ νῦν οὐδὲ γιγνώσκοντες τοὺς ἐπὶ τοῖς ἅρμασιν, οἴονται σφίσιν ὁμοίους, τοὺς ἀνασκήτους τοῖς ἠσκηκόσιν ἔσεσθαι. οἱ δέ, ὁρμῶσι μέν. πρὶν δ' ἐν τοῖς πολεμίοις εἶναι, οἱ μέν, ἑκόντες ἐκπίπτουσιν, οἱ δ' ἐξάλλονται. ὥστε ἄνευ ἡνιόχων γιγνόμενα τὰ ζεύγη πολλάκις, πλείω κακὰ τοὺς φίλους ἢ τοὺς πολεμίους ποιεῖ. Ἐπεὶ μέντοι καὶ αὐτοὶ γιγνώσκουσιν, οἷα σφίσι τὰ πολεμιστήρια ὑπάρχει, ὑφίενται. καὶ οὐδεὶς ἔτι ἄνευ Ἑλλήνων εἰς πόλεμον καθίσταται. οὔτε ὅταν ἀλλήλοις πολεμῶσιν, οὔτε ὅταν οἱ Ἕλληνες αὐτοῖς ἀντιστρατεύωνται. ἀλλὰ καὶ πρὸς τούτους ἐγνώκασι μεθ' Ἑλλήνων τοὺς πολέμους ποιεῖσθαι. ἐγὼ μὲν δὴ οἶμαι, ἅπερ ὑπεθέμην ἀπειργάσθαι μοι. φημὶ γὰρ Πέρσας, καὶ τοὺς σὺν αὐτοῖς, καὶ ἀσεβεστέρους περὶ θεούς, καὶ ἀνοσιωτέρους περὶ συγγενεῖς, καὶ ἀδικωτέρους περὶ τοὺς ἄλλους, καὶ ἀνανδροτέρους τὰ εἰς τὸν πόλεμον, νῦν ἢ πρόσθεν ἀποδεδεῖχθαι. εἰ δέ τις τἀναντία ἐμοὶ γιγνώσκει, τὰ ἔργα αὐτῶν ἐπισκοπῶν, εὑρήσει αὐτὰ μαρτυροῦντα τοῖς ἐμοῖς λόγοις.

ΞΕΝΟΦΩΝΤΟΣ ΚΥΡΟΥ ΑΝΑΒΑΣΕΩΣ ΠΡΩΤΟΝ.

δ

Αρείου καὶ Παρυσάτιδος γίγνονται παῖδες δύο, πρεσβύτερος μὲν Ἀρταξέρξης, νεώτερος δὲ Κῦρος. ἐπεὶ δὲ ἠσθένει Δαρεῖος καὶ ὑπώπτευε τελευτὴν τοῦ βίου, ἐβούλετο τὼ παῖδε ἀμφοτέρω παρεῖναι. ὁ μὲν οὖν πρεσβύτερος παρὼν ἐτύγχανε. Κῦρον δὲ μεταπέμπεται ἀπὸ τῆς ἀρχῆς, ἧς αὐτὸν σατράπην ἐποίησε. καὶ στρατηγὸν δὲ αὐτὸν ἀπέδειξε πάντων, ὅσοι εἰς Καστωλοῦ πεδίον ἁθροίζονται. ἀναβαίνει οὖν ὁ Κῦρος λαβὼν Τισσαφέρνην ὡς φίλον. καὶ τῶν Ἑλλήνων δὲ ἔχων ὁπλίτας ἀνέβη τριακοσίους. ἄρχοντα δὲ αὐτῶν, Ξενίαν Παρράσιον. ἐπειδὴ δὲ ἐτελεύτησε Δαρεῖος, καὶ κατέστη εἰς τὴν βασιλείαν Ἀρταξέρξης, Τισσαφέρνης διαβάλλει τὸν Κῦρον πρὸς τὸν ἀδελφόν, ὡς ἐπιβουλεύοι αὐτῷ. ὁ δέ, πείθεταί τε καὶ συλλαμβάνει Κῦρον, ὡς ἀποκτενῶν. ἡ δὲ μήτηρ ἐξαιτησαμένη αὐτόν, ἀποπέμπει πάλιν ἐπὶ τὴν ἀρχήν. ὁ δ' ὡς ἀπῆλθε κινδυνεύσας καὶ ἀτιμασθείς, βουλεύεται ὅπως μήποτε ἔτι ἔσται ἐπὶ τῷ ἀδελφῷ. ἀλλ' ἢν δύνηται, βασιλεύσει ἀντ' ἐκείνου. Παρύσατις μὲν δὴ ἡ μήτηρ, ὑπῆρχε τῷ Κύρῳ φιλοῦσα αὐτὸν μᾶλλον, ἢ τὸν βασιλεύοντα Ἀρταξέρξην. ὅστις δ' ἀφικνεῖτο τῶν παρὰ βασιλέως πρὸς αὐτόν, πάντας οὕτω διατιθεὶς ἀπεπέμπετο, ὥσθ' ἑαυτῷ μᾶλλον φίλους εἶναι ἢ βασιλεῖ. καὶ τῶν παρ' ἑαυτῷ δὲ βαρβάρων ἐπεμελεῖτο, ὡς πολεμεῖν τε ἱκανοὶ εἴησαν, καὶ εὐνοϊκῶς ἔχοιεν αὐτῷ. τὴν δὲ Ἑλληνικὴν δύναμιν ἤθροιζεν, ὡς μάλιστα ἐδύνατο ἐπικρυπτόμενος, ὅπως ὅτι ἀπαρασκευαστότατον λάβοι βασιλέα. ὧδε οὖν ἐποιεῖτο τὴν συλλογήν. ὁπόσας εἶχε φυλακὰς ἐν ταῖς πόλεσι, παρήγγελλε τοῖς φρουράρχοις ἑκάστοις λαμβάνειν ἄνδρας Πελοποννησίους, ὅτι πλείστους καὶ βελτίστους, ὡς ἐπιβουλεύοντος Τισσαφέρνους ταῖς πόλεσι. καὶ γὰρ ἦσαν αἱ Ἰωνικαὶ πόλεις Τισσαφέρνους τὸ ἀρχαῖον, ἐκ βασιλέως δεδομέναι. τότε δ' ἀφειστήκεσαν πρὸς Κῦρον πᾶσαι πλὴν Μιλήτου. ἐν Μιλήτῳ δὲ ὁ Τισσαφέρνης προαισθόμενος τὰ αὐτὰ ταῦτα βουλευομένους ἀποστῆναι πρὸς Κῦρον, τοὺς μὲν ἀπέκτεινε, τοὺς δ' ἐξέβαλεν. ὁ δὲ Κῦρος ὑπολαβὼν τοὺς φεύγοντας, συλλέξας στράτευμα, ἐπολιόρκει Μίλητον καὶ κατὰ γῆν καὶ κατὰ θάλατταν. καὶ ἐπειρᾶτο κατάγειν τοὺς ἐκπεπτωκότας. αὕτη οὖν ἄλλη πρόφασις ἦν αὐτῷ τοῦ ἁθροίζειν στράτευμα. πρὸς δὲ βασιλέα πέμπων, ἠξίου ἀδελφὸς ὢν αὐτοῦ, δοθῆναί οἱ ταύτας τὰς πόλεις μᾶλλον, ἢ Τισσαφέρνην ἄρχειν αὐτῶν. καὶ ἡ μήτηρ συνέπραττεν αὐτῷ ταῦτα. ὥστε βασιλεὺς τῆς πρὸς ἑαυτὸν ἐπιβουλῆς οὐκ ᾐσθάνετο. Τισσαφέρνει δὲ ἐνόμιζε πολεμοῦντα αὐτὸν ἀμφὶ τὰ στρατεύματα δαπανᾶν. ὥστε οὐδὲν ἤχθετο αὐτῶν πολεμούντων. καὶ γὰρ ὁ Κῦρος ἀπέπεμπε τοὺς γιγνομένους δασμοὺς βασιλεῖ ἐκ τῶν πόλεων, ὧν Τισσαφέρνης ἐτύγχανεν ἔχων. ἄλλο δὲ στράτευμα συνελέγετο αὐτῷ ἐν Χερρονήσῳ τῇ καταντιπέρας Ἀβύδου, τόνδε τὸν τρόπον. Κλέαρχος Λακεδαιμόνιος φυγὰς ἦν. τούτῳ συγγενόμενος ὁ Κῦρος, ἠγάσθη τε αὐτόν, καὶ δίδωσιν αὐτῷ μυρίους δαρεικούς. ὁ δέ, λαβὼν τὸ χρυσίον, στράτευμα συνέλεξεν ἀπὸ τούτων τῶν χρημάτων. καὶ ἐπολέμει ἐκ Χερρονήσου ὁρμώμενος, τοῖς Θρᾳξὶ τοῖς ὑπὲρ Ἑλλήσποντον οἰκοῦσι. καὶ ὠφέλει τοὺς Ἕλληνας. ὥστε καὶ χρήματα συνεβάλλοντο αὐτῷ εἰς τὴν τροφὴν τῶν στρατιωτῶν, αἱ Ἑλλησποντιακαὶ

g iiii

Fig. 87. Beginning of the *Anabasis* taken from the Greek *princeps*. A little delta has been printed in the empty square to guide the limner who was expected to draw a beautiful capital. [From the copy in the Harvard College Library.]

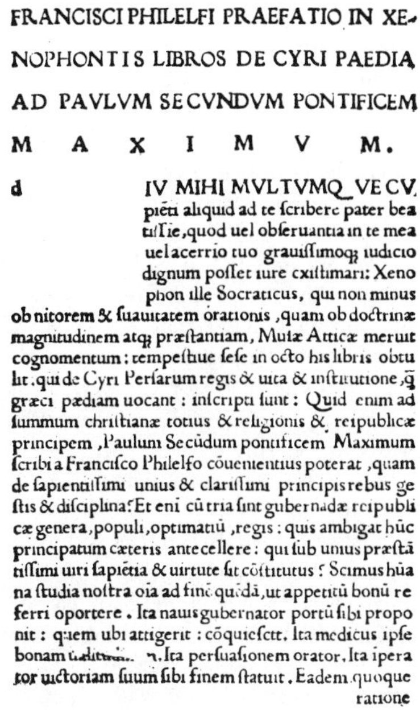
FRANCISCI PHILELFI PRAEFATIO IN XE-
NOPHONTIS LIBROS DE CYRI PAEDIA
AD PAVLVM SECVNDVM PONTIFICEM
M A X I M V M.

d IV MIHI MVLTVMQVE CV
pieti aliquid ad te scribere pater bea
tissime, quod uel obseruantia in te mea
uel acerrio tuo grauissimoq; iudicio
dignum posset iure existimari: Xeno
phon ille Socraticus, qui non minus
ob nitorem & suauitatem orationis, quam ob doctrinæ
magnitudinem atq; præstantiam, Musæ Atticæ meruit
cognomentum: tempestiue sese in octo his libris obtu
lit. qui de Cyri Persarum regis & uita & institutione, q̄
græci pædiam uocant: inscripti sunt: Quid enim ad
summum christianæ totius & religionis & reipublicæ
principem, Paulum Secūdum pontificem Maximum
scribi a Francisco Philelfo conuenientius poterat, quam
de sapientissimi unius & clarissimi principis rebus ge
stis & disciplina? Et eni cū tria sint gubernādæ reipubli
cæ genera, populi, optimatiū, regis: quis ambigat hūc
principatum cæteris antecellere: qui sub unius præstā
tissimi uiri sapiētia & uirtute sit cōstitutus? Scimus hūa
na studia nostra oia ad finē quēdā, ut appetitū bonū re
ferri oportere. Ita nauis gubernator portū sibi propo
nit: quem ubi attigerit: cōquiescit. Ita medicus ipse
bonam [illegible]. Ita persuasionem orator. Ita ipera
tor uictoriam suum sibi finem statuit. Eadem quoque
ratione

Fig. 88. Franciscus Philelphus translated the *Cyropaedia* into Latin in 1467; this was printed by Arnoldus de Villa in Rome, 1474, a volume of 146 leaves, the only one that can be ascribed to that printer. Some copies contain his name and the date Rome 10 March 1474. There is no title page; the first page, here reproduced, is Philelphus' dedication to Paul II (pope, 1464–1471). [From the copy in the Harvard College Library.]

ians. It required military genius to make these desperate men pull through together.[12]

The *Anabasis* is a literary masterpiece, which would suffice to immortalize its author. The most popular of his works, however, remained for centuries, not the *Anabasis*, but the *Cyropaedia*.

5. *Cyropaedia* (*Cyru paideia*) (Fig. 88) is a romanticized biography of Cyros the Great. The constitution and the manners of Persia which it is supposed to describe are much closer to the constitution and the manners of Sparta, or rather to an idealization of them by a man who admired the Spartans as much as he despised Attic indiscipline.

It is one of the golden books of world literature. We may call it the prototype of a type that obtained some popularity in the Middle Ages, the *Regimina principum* (*De eruditione filiorum nobilium*, etc.), books written in order to educate the sons of kings and noblemen, and to teach future governors their duties and privileges.[13]

The *Cyropaedia* should not be taken literally (as it was in the past), for it is full of historical errors mixed with truth. The main purpose is aristocratic, yet Xenophon had not forgotten his teacher Socrates (he never did), and therefore it includes Socratic methods and notions, and even the delightful portrait of an

[12] Considering the great importance of the role played by Xenophon in that epic retreat (according to his own account of it), it is very puzzling that Diodoros of Sicily (I-2 B.C.) could describe that event (*Bibl. hist.*, XIV, 25–30) without even mentioning Xenophon!

[13] Examples in many languages are given in my *Introduction, passim*. The *Cyropaedia* was the prototype for Western peoples; earlier examples exist in Egyptian literature (*Introduction*, vol. 3, p. 314) but remained unknown in the West until our own time.

Armenian Socrates.[14] It even includes some glimpses of democratic ideas. For example, he refers (ironically it is true) to equal freedom of speech (*isēgoria*) and more seriously to the fact that "in Persia equality of rights is considered justice." [15] These inconsistencies exist because Xenophon's goodheartedness was stronger than his prejudices. There are many delightful stories or pictures — on the goodness of bread (as compared with meats, etc.), since one does not need to clean one's hand after eating it; [16] the "junior republic"; [17] the zoölogic parks or menageries; [18] the danger of wealth; [19] the postal system; [20] and wise sayings: "battles are decided more by men's souls than by the strength of their bodies," [21] "the considerate people are those who avoid what is offensive when seen; the self-controlled avoid that which is offensive, even when unseen"; [22] perhaps an interpolation. The most moving part of the whole book is the final chapter,[23] describing the death of Cyros and his last recommendations, and discussing the immortality of the soul. A comparison with the *Phaidon* of Plato is not by any means to Xenophon's discredit.

This Greek educational romance (a distant ancestor of *Télémaque*) is full of life and good humor and this helps to explain its popularity. It is a bit long, yet it blends pleasantly all the subjects that awakened the author's curiosity or passions at different periods of his life (the Asiatic country that he had explored, the barbarians whom he had learned to know, methods of education, military service and technique, hunting, politics, Socratism). If Xenophon wrote it relatively early it was prophetic of his other writings; if he wrote it late in life, as seems more probable, it was a summary of their main messages in romantic garb, a gentle valediction.

We might now examine Xenophon's Socratic writings (6–9) which were probably written at Scillus.

6. *Memorabilia* (*Apomnēmoneumata*). This is a defense of Socrates and reminiscences of his conversations. We cannot accept them literally, yet they give us a general account of Socrates' habits, which is probably true and serves to complete and correct the Platonic account. In both cases we have to do with reminiscences, but Xenophon's inspire more confidence than Plato's.

7. *Apology* (*Apologia*). This also completes the account that Plato published under the same title.[24] Some parts repeat the *Memorabilia*.

8. *Symposium* (*Symposion*). This new duplication of a Platonic dialogue cannot be accidental. We must assume that it is later than Plato's *Symposium*; it is inferior in style to it.

[14] *Cyropaedia*, III, 1, 38.

[15] *Ibid.*, I, 3, 10; I, 3, 18.

[16] *Ibid.*, I, 3, 5.

[17] *Ibid.*, I, 2, 6.

[18] *Ibid.*, I, 3, 14; I, 4, 5. Compare with their equivalents in ancient and medieval times, *Introduction*, vol. 3, pp. 1189, 1470, 1859.

[19] *Cyropaedia*, VIII, 2, 20; VIII, 3, 46–47.

[20] *Ibid.*, VIII, 6, 17.

[21] *Ibid.*, III, 3, 19; VIII, 1, 2.

[22] *Ibid.*, VIII, 1, 31.

[23] *Ibid.*, VIII, 7, the final chapter of the original text. Chapter 8, describing the degeneracy of the "modern" Persians, i.e., those who were Xenophon's contemporaries, seems to be a later addition.

[24] Xenophon's *Apology* is much shorter than Plato's (in the proportion 6:17) and less lofty. At the beginning Xenophon refers to other apologies (possibly by Lysias and Theodectes, not necessarily by Plato, for Plato's might be later). The existence of so many apologies proves that Socrates' condemnation to death was a great scandal. According to Xenophon, Socrates insisted upon the argument that it is better to die before the miseries and humiliations of old age. He reports Socrates' answer to Apollodoros, who was shocked by the unjust condemnation: "Would you prefer to see me die guilty?"

9. *Oeconomicos* (*Oiconomicos*), is a dialogue between Socrates and Critobulos concerning estate management and domestic economy. Socrates, who had no interest in farming and country life, reports his conversation with a gentleman farmer, Ischomachos. The latter's views are obviously Xenophon's; they are typical of his frame of mind – earthbound, practical, good natured, essentially kind.

The only work of Xenophon's which professes to be straight history is *Hellenica.*

10. *Hellenica* comprises two distinct parts. The first continues Thucydides' history from 411 to the end of the Peloponnesian War in 404. The second part is another continuation, to the battle of Mantineia (362), but in a different manner. Xenophon's prejudices for Sparta and against Thebes appear many times on the surface. Though he worked at this second part until 358, it is not quite finished. He lived probably a few more years but was obliged to drop his pen.

Xenophon's political books constitute a final group (our order is not chronologic; the exact sequence of his writings cannot be established).

11. *Agesilaos.* This is the biography of the king of Sparta whom Xenophon had served and whom he admired; it was composed shortly after Agesilaos' death in 360.

12. *Polity of the Lacedaimonians* (*Lacedaimoniōn politeia*). This eulogy of Spartan institutions as established by the legendary Lycurgos was probably written before 369; he added a palinode sometime later.

A similar work on the *Polity of the Athenians* (*Athēnaiōn politeia*), formerly ascribed to Xenophon, is almost certainly an earlier work written by an oligarch before 423.[25]

Both works are entitled *Politeia,* as was Plato's work the title of which is generally translated *The Republic.*

13. *Hieron.* This is an imaginary dialogue between Hieron the Elder, tyrant of Syracuse from 478 to 467, and the lyric poet Simonides of Ceos (*c.* 556–468) on the double subject, Is a tyrant happier than the people he rules? How can he win their regard and affection? Xenophon may have been inspired to compose this piece at the time of the accession of Dionysios II (367), of whom Plato had hoped to make a philosopher-king.

14. *Ways and means* (*Peri porōn*) contains practical suggestions for improving Athenian finances. It was written at the very end of his life, long after he had made his peace with his native country.

None of his writings is lost, but some that bear his name may be apocryphal.

PLATO AND XENOPHON

The reader of the preceding list will have noticed many titles similar to the titles of Plato's works or reminiscent of them, such as the *Apology, Symposium,*

[25] *The Polity of the Lacedaimonians* is possibly apocryphal; it may have been written by Antisthenes the Cynic. K. M. T. Chrimes, *The Respublica Lacedaemoniorum ascribed to Xenophon* (Manchester: Manchester University Press, 1948) [*Isis 42,* 310 (1951)].

The *Polity of the Athenians* is certainly apocryphal. It was written during Xenophon's childhood in the period 430–424, and is thus the oldest extant book in Attic prose. It is also the oldest treatise on political theory, or rather the oldest political pamphlet. The author cannot be named. It has been suggested that he might be identified with Critias, one of Socrates' unworthy pupils, one of the Thirty Tyrants established by the Spartans in Athens in 404. Critias was a distinguished orator, but his authorship of this particular book cannot be proved. All that one can say is that the author was an Athenian oligarch.

Ernst Kalinka, *Die pseudoxenophontische Athēnaiōn politeia* (330 p.; Leipzig, 1913), Greek text, German translation, commentary.

Politeia, and of course many others that are entirely original such as the Asiatic and cynegetic writings. Plato and Xenophon were almost exact contemporaries (Plato was born a couple of years earlier and died a few years later; he reached the age of eighty while Xenophon died at about seventy-five). Both men were friends of Socrates and enemies of Athens. They have many points in common and the use of identical titles for three of their works is strange. The differences between them, however, are far deeper than the resemblances. As both are great men, fully representative of their age and climate, it is tempting to pursue the comparison: it is a study in contrast that may help us to understand both of them better.

Both had received the same general education, completed with higher studies in Socrates' open-air seminar. They were both men of letters, knowing to perfection, by instinct as well as by training, the purest Attic Greek. Both received the political education that every Athenian partook of as a matter of course, and in addition they had equal, though very different, opportunities of learning practical politics, first both of them in Athens, then Plato at the court of Syracuse and Xenophon in his familiarity with the king of Sparta. They lived in very different parts of the Greek world, but their paths crossed frequently, and they were bound to discuss the same political and ethical problems. Both became enemies of democracy, Plato increasingly so to the end of his life; Xenophon, with more moderation always, was reconciled to his native country in his old age.

Both were aristocrats, but in very different ways, Xenophon as a gentleman farmer, a conservative, reminiscent of the good old days, Plato as a proud intellectual, laying down the law for others to obey. Both explained their ideal of aristocratic government, but what a difference between the monarch of the *Cyropaedia* and the dictator of the *Republic!* Both were moralists as well as politicians, Plato like a professor, Xenophon like a paterfamilias. This suggests another difference, cutting very deep; Plato was a confirmed bachelor, Xenophon a loving husband and father. We may assume that in the *Oeconomicos* he described his own experience as a husband and father managing his estate at Scillus. The farmer Ischomachos in that dialogue was no doubt Xenophon himself, and the farmer's wife, unnamed but charming, was Xenophon's wife, Philesia. It is a great tribute to the writer's honesty that the wife is so much more attractive than the husband (that is, himself).

Plato is generally conceived as a sublime idealist, while Xenophon is looked down upon because of his simple piety, and because he was too practical and earthbound, more interested in good recipes than in principles. Yet the latter was affectionate and goodhearted, while the former was a doctrinaire, dogmatic to the point of inhumanity.

If we try to visualize them in their usual surroundings, the contrast is even greater, for Xenophon was a soldier and a farmer, while Plato was a professor. We see the former among his comrades in the mountains of Anatolia, or on his estate, riding, hunting, inspecting his fields, vines, and stables, directing the husbandry; while we can hardly imagine Plato except in the gardens of the Academy discussing philosophy and mathematics, quarreling with his colleagues and disciples.

The best in both of them they owed to their old master Socrates, and Xenophon was grateful to him until the end, while Plato in his pride betrayed him.

XENOPHON AS EDUCATOR

In spite of their great variety, Xenophon's books have much in common, not only in their style (which is natural enough) but in their contents. The dominant note is didactic. Xenophon was not a philosopher, but he was, like his master Socrates, a born and irrepressible educator. He believed in the power of education and in his own ability to educate other people. His vision was never sublime, but it was clear and honest. He tried to understand the little world around him (not the universe) and to explain it as lucidly and simply as possible. His theory and practice of education are set forth in the *Memorabilia* (especially in Book IV) and indirectly in the *Cyropaedia.* It had been influenced not only by Socrates but also by Democritos and by the Pythagoreans. There was a colony of these not far from Scillus where he spent twenty years of his life, the happiest ones and the most fruitful. It was probably from those Pythagorean neighbors that he learned the need of a good diet combined with exercise, the value of moral and religious traditions, the importance also of mathematics (though he had little taste for mathematics himself). A good training is necessary for all men, but especially so for those boys who have the richest natural endowment. He realized keenly the three fundamental elements of any education, the natural gifts (*physis*), learning (*mathēsis*) and exercise (*ascēsis*). His criticism of booklearning[26] helps us to realize that in his time books were already abundant, and this implies not only the creation of such books but the existence of a regular book trade. Young men must be trained to express themselves, to increase their self-control, to act as needed by circumstances, to be resourceful and independent; they must be prepared to take part in political discussions and in administration.

His main purpose was the same as Socrates'; in fact, his recommendations are put in the latter's mouth. He was continuing or trying to continue his master's teaching, interpreting it and adding to it the fruits of his rich experience. He was primarily interested in general education, which every gentleman required in order to fulfill his destiny, yet he realized the need of adapting that education to the special qualities of each student. Men have different qualities, each of which can be improved by adequate training; it is the educators duty to be on the lookout for good aptitudes and to develop them. In any case, moral and religious education is fundamental. The teacher should not try to teach too much; it is more to the point to strengthen the student's spirit, to form his character.

All of which does not seem very original today, but Socrates and Xenophon were the first to explain it. Remember that Xenophon was writing in the first half of the fourth century B.C. and that some of our own educators have not understood it yet.[27]

FUNCTIONAL ARCHITECTURE

One of the most curious sections of the *Memorabilia* is Socrates' dialogue with Aristippos. Beauty, the master explains, is a function of the purpose that one wishes to attain. Said Aristippos:

[26] Well illustrated in Socrates' dialogue with the handsome Euthydemos (*Memorabilia,* IV). As to the book trade, see the *Anabasis,* VII, 5, 14.

[27] Armand Delatte, "La formation humaniste chez Xénophon," *Bull. Acad. Belgique* (lettres, 35, fas. 10, 20 pp.; Brussels, 1949).

"Do you mean that the same things are both beautiful and ugly?"

"Of course – and both good and bad. For what is good for hunger is often bad for fever, and what is good for fever bad for hunger; what is beautiful for running is often ugly for wrestling, and what is beautiful for wrestling ugly for running. For all things are good and beautiful in relation to those purposes for which they are well adapted, bad and ugly in relation to those for which they are ill adapted."

Again, his dictum about houses, that the same house is both beautiful and useful, was a lesson in the art of building houses as they ought to be.

He approached the problem thus:

"When one means to have the right sort of house, must he contrive to make it as pleasant to live in and as useful as can be?"

And this being admitted, "Is it pleasant," he asked, "to have it cool in summer and warm in winter?"

And when they agreed with this also, "Now in houses with a south aspect, the sun's rays penetrate into the porticoes in winter, but in summer the path of the sun is right over our heads and above the roof, so that there is shade. If, then, this is the best arrangement, we should build the south side loftier to get the winter sun and the north side lower to keep out the cold winds. To put it shortly, the house in which the owner can find a pleasant retreat at all seasons and can store his belongings safely is presumably at once the pleasantest and the most beautiful. As for paintings and decorations, they rob one of more delights than they give."

For temples and altars the most suitable position, he said, was a conspicuous site remote from traffic; for it is pleasant to breathe a prayer at the sight of them, and pleasant to approach them filled with holy thoughts.[28]

XENOPHON'S VIEWS ON DIVINATION

We have already drawn attention to the outstanding superstition of classical antiquity, the firm belief in divination. At the risk of repetition we must come back to it, for we cannot have a balanced view of the Greek spiritual life if we leave out an aspect of it that was as significant to them as it is unpleasant to us.

The Greeks (and the Romans after them) believed that it was possible to interpret the meaning of past and future events from the observation of natural phenomena of many kinds.[29] In the *Anabasis*[30] we are given many examples of Xenophon's trust in divination, and of the necessity when in trouble of interpreting omens not only for his own sake but also for the sake of his soldiers. This is not exceptional in ancient literature, but normal.

In the *Memorabilia* Xenophon was very anxious to prove that the accusations leveled against his master Socrates were unfounded and his condemnation unjust. In particular, he wanted to show that Socrates had always been a religious and pious man, sharing the beliefs of his people and practicing the accepted rites. The most popular rites were those concerning divination, the traditional interpretation of sacred omens. Therefore, Xenophon quoted examples of Socrates' firm belief in divination.

He offered sacrifices constantly, and made no secret of it, now in his home, now at the altars of the state temples, and he made use of divination with as little secrecy. Indeed it had become notorious that Socrates claimed to be guided by "the deity": it was out of this claim, I think, that the charge of bringing in strange dei-

[28] *Memorabilia*, III 8. Translation by E. C. Marchant in the Loeb Classical Library.

[29] The best account of divination by an ancient writer is relatively late, Cicero's *De divinatione*, but one may find the equivalent of it scattered in many Greek writings of much earlier date. There is a good introduction to that large field by Arthur Stanley Pease, *Oxford Classical dictionary*, pp. 292–293. For comparative studies of divination, see *Encyclopedia of Religion and Ethics*, vol. 4 (1912), pp. 771–830.

[30] *Anabasis*, VI, 4; also VII, 8, 20.

ties arose. He was no more bringing in anything strange than are other believers in divination, who rely on augury, oracles, coincidences and sacrifices. For these men's belief is not that the birds or the folk met by accident know what profits the inquirer, but that they are the instruments by which the gods make this known; and that was Socrates' belief too . . .

In so far as we are powerless of ourselves to foresee what is expedient for the future, the gods lend us their aid, revealing the issues by divination to inquirers, and teaching them how to obtain the best results . . .

When anyone was in need of help that human wisdom was unable to give he advised him to resort to divination; for he who knew the means whereby the gods give guidance to men concerning their affairs never lacked divine counsel.[31]

The best explanation of the importance of divine omens is given by Cambyses to his son Cyros the Great.[32] It is every man's duty, and especially the king's, to obey the divine guidance, but how shall he know it? Cambyses warns his son that he must not be at the mercy of the soothsayers and should be able to interpret the omens himself. But how could one check the interpretation? It is astonishing that the intelligent Greeks never asked themselves that question, or at any rate never gave a good answer to it. For granted that the divine will may be implied in any event, how shall we discover that will and be sure that we have understood it? How can one obey an order that is not clear?

We should remember, however, that wise men were not at the mercy of soothsayers, who might be stupid as well as dishonest, but interpreted the omens in their own way. The main warning was one of gravity and fatality; a decision had to be taken and that decision should be as wise as possible: the signs could be interpreted *ad hoc* and generally were. The omens were symbols of divine immanence and of general guidance; special guidance had to be determined for each man by his own conscience.[33]

XENOPHON'S HUMOR

Xenophon, like Plato, and like their teacher Socrates, could be very humorous in a simple way. A good example is put in Socrates' mouth in *Memorabilia.* In order to ridicule the stupidity and conceit of men who are candidates for public office without having any of the needed qualifications, he suggests that those wretched candidates might address their constituency as follows:

"Men of Athens, I have never yet learnt anything from anyone, nor when I have been told of any man's ability in speech and in action, have I sought to meet him, nor have I been at pains to find a teacher among the men who know. On the contrary, I have constantly avoided learning anything of anyone, and even the appearance of it. Nevertheless, I shall recommend to your consideration anything that comes into my head."

This exordium might be adapted so as to suit candidates for the office of public physician. They might begin their speeches in this strain:

"Men of Athens, I have not yet studied medicine, nor sought to find a teacher among our physicians; for I have constantly avoided learning anything from the physicians, and even the appearance of having studied their art. Nevertheless I ask you to appoint me to the office of a physician, and I will endeavor to learn by experimenting on you."

The exordium set all the company laughing.[34]

[31] *Memorabilia* I, 1; IV, 3, 12; IV, 7, 10.

[32] *Cyropaedia,* I, 6, 1; XVI, 44–46.

[33] A good example of rational understanding of omens was given by Homer (*Iliad,* XII, 243): *eis oiōnos aristos, amynesthai peri patrēs,* the one best omen is to fight for one's country. Every educated Greek would remember that. The omen is up to himself.

[34] *Memorabilia,* IV, 2, 4–5.

The second example shows incidentally that the office of public or city physician already existed in those days,[35] and this is the more remarkable, because that office vanished later and only reappeared relatively late in medieval times, in the thirteenth century.[36]

XENOPHON'S INFLUENCE

The influence of Xenophon has been very great, partly because of his didactic intentions, partly because of the good stories that he had to tell, and told very well, partly because of his humanity and the purity of his style. He was good-natured and his prose was so easy and pleasant that he was nicknamed the Attic bee. Quintilian found a good phrase to define his style, *jucunditas inaffectata* (unaffected cheerfulness), and because of that quality Xenophon became for many generations a master of language. This had a serious drawback, for many students have suffered so much plodding without sufficient preparation through the *Anabasis* that their memory of it is rather painful. Their judgment of the *Anabasis* and of Xenophon is irrelevant, however. All the classics have become instruments of torture in the same way, and this condemns bad students and bad teachers, but nothing else.

Xenophon's influence was already considerable in ancient times. It has been argued that his Asiatic books, chiefly the *Anabasis*, illustrated the relative facility of dealing with Asiatics and gave the Macedonian kings the ambition of conquering Asia. We may be certain that these books were studied by the young Alexander; on the other hand, Xenophon's description of an ideal Asiatic monarchy was a bewitching prefiguration of the Hellenistic kingdoms. Roman gentlemen studied hunting, domestic economy, ethics, and government in Xenophon's books. They found in them clear answers, in simple and concrete language, to most of their questions.

During the Byzantine Renaissance, Xenophon's works were studied and their Atticism imitated. The main literary models of Ioannes Cinnamos (XII–2) were Herodotos and Xenophon. His works were translated into Latin by the early Hellenists: Poggio of Florence, Leonardo Bruni of Arezzo, Francesco Filelfo of Tolentino (Ancona). English gentlemen of the period 1530–1630 read the *Cyropaedia* and tried to find in it the solution of their own problems. This was the first historical novel in world literature, and it entertained and educated not only the English but also the French, and indeed the gentlemen of every civilized part of Europe. The *Cyropaedia* was a kind of vade mecum of Socratism and politics, as well as an Oriental introduction. Later, the *Anabasis* was preferred to it (I do not know exactly why), but Xenophon remained one of the outstanding teachers of Greek and of Hellenism. As an educator, he has done more good and less harm than Plato.

[35] *Ho tēs poleōs iatricos*. Compare references to a board of health, public physicians, and dispensary in the *Cyropaedia*, I, 6, 15; VIII, 2, 24. The necessary use of military surgeons may have inspired the appointment of town physicians.

[36] *Introduction*, vol. 3, p. 1244, 1861.

XIX

ARISTOTLE AND ALEXANDER. THE LYCEUM

THE GROWTH OF MACEDONIAN POWER

We now approach a new age, the Age of Aristotle, which is essentially different in many respects from the preceding one, the Age of Plato, in spite of its closeness to it and of their mutual interpenetration. The political background is not Hellenic in the old way, but Macedonian. This requires a few words of explanation.

If one looks at a map, one will see that Macedonia is a Balkan country, north of Thessaly, east of Illyria, and west of Thrace. The map does not show boundaries between these countries, but their names in large letters indicate their general location. One could not do better. Moreover, the boundaries, wherever they existed, were not permanent, and the kings of Macedonia extended their territory from time to time. For example, it finally included the Chalcidice, the three-legged peninsula (a miniature Peloponnesos) at the northwest end of the Aegean Sea, a country more closely related to the Aegean islands than to Macedonia proper. The inhabitants of Macedonia were not of a special type; there was not a Macedonian type but rather a mixture of Thracian and Illyrian (Albanian) types. They did not speak Greek, but it is difficult to define their own languages. These belonged to the Indo-European group, but were probably different as well from the Hellenic subgroup as from the Slavonic one. The Thracian dialects were connected with the Phrygian ones spoken in the northwest part of Asia Minor (south of the Propontis); the Illyrian ones are represented today by the Albanian language.[1] However, as southern Macedonia was so close to Thessaly and Epeiros, Greek refugees arrived in early days and there was a large movement of emigration from Argos (in the Peloponnesos). The Doric dialect of the Greek emigrants was soon permeated with barbarian words. We may assume that a Macedonian Greek arriving in Athens was easily detected as a stranger by the very women of the market place, even if he was himself well educated.

The Macedonians were ruled by a dynasty of kings that began, according to some, with Caranos of Argos (*c.* 750), or according to others, with Perdiccas I, also of Argive descent (700–652). Little is known of their history until the reign

[1] A. Meillet and Marcel Cohen, *Les langues du monde* (Paris, 1924), pp. 47, 52 [*Isis 10*, 298 (1928)].

of the sixth[2] king, Amyntas I (540–498), but even he, an ally of the Persians, did not attract much attention. Still king followed king and it was only when the twenty-second began his rule that the picture changed. That twenty-second king was Philip II (ruled from 359 to 336). These Macedonian kings were Greeks, yet they married native women and the Greek strain in them was repeatedly mixed with native ones. Thus, we hear that Philip's mother learned the Greek language only in her old age; Philip, however, received a Greek education. When he gained power in 360, he understood very well the Greek situation: political chaos interrupted by precarious armistices; alliances made, broken, and replaced by new combinations; no hope of peace except upon the order of a ruler of overwhelming power. Philip determined to be that ruler. During his detention in Thebes, he had observed new military methods; he not only mastered them but improved upon them. He created a professional army and taught it to move and fight in a new formation, the Macedonian phalanx. It was a combination of infantry and cavalry, foot in the center and horse in the wings, trained to work together. The Macedonian tactics were generally irresistible; it remained the best military method for centuries; it was simple enough, yet its realization required unusual gifts of generalship, and its value depended very largely upon the genius that the general officer could evidence, slowly and steadily on the training fields, and more rapidly, with as many improvisations as might be required, on the field of battle. Philip managed to put an end to the tribal feuds of his mountaineers and created a national union. He had plenty of opportunities for the exercising of his army in his own region, south of the Danube and west of the Black Sea, and he gradually increased the area and solidarity of his native realm. After that, he was ready to tackle the disorganized Greeks. We need not tell the history of his campaigns.

What was the reaction of the Greeks and especially of the Athenians to the rise of Macedonia? We must bear in mind that in their eyes Philip, in spite of his education, was not a pure Hellene; he was not a barbarian, yet he was a stranger. His ambitions became more transparent every year. Would the Greeks, who had been thus far impatient with every leader, submit to the domination of an outsider?[3] There were two great parties in Athens. The first, led by the old Isocrates (436–338), might be called in modern phraseology the party of collaborators. The second was inspired by the greatest of all Attic orators, Demosthenes (385–322). He delivered virulent orations wherein he denounced Philip's ominous designs and defended Greek freedom against him.[4] In the fourth of these great orations he suggested that Persia might be asked to help protect Greek freedom against Macedonian imperialism. The incessant civil wars that had desolated the Greek world for more than a century were poisoned by Persian interventions, for Persia was always ready to interfere and either the one or the other of the groups

[2] Sixth, if Perdiccas I was the first, ninth if Caranos was. I follow the list given by A. M. H. J. Stokvis, *Manuel d'histoire, de généalogie et de chronologie de tous les états du globe* (3 vols.; Leiden, 1888–1893), vol. 2, pp. 448–450.

[3] It is curious, by the way, that dictators have frequently been outsiders, foreigners. Think of Philip the Macedonian, Napoleon the Corsican, Mehemet Ali the Albanian, Hitler the Austrian, Stalin the Georgian.

[4] These orations were called *Philippica* and the word "philippic" in various European languages is a reminiscence of them. That word is used to designate political discourses denouncing a particular leader and generally full of invective. In particular, it was used to designate Cicero's many orations against Mark Antony. Lincoln, Woodrow Wilson, and Franklin Roosevelt have been the targets of many philippics.

fighting for hegemony would be susceptible to Persian gold and ready to ally itself with the national enemy in order to attain its own purpose. Hence, the Greek civil wars were always to some extent internationalized. From the time of Philip's accession, the situation had changed; there were now two foreign powers in the offing, Persia and Macedonia, and the fighting between Greeks was preceded and accompanied by a vast amount of propaganda, diplomatic intrigue, and treacherous infiltration. The Greeks were unable to stand together without a foreign armature. The question was, which of the two enemies and prospective guardians was the more dangerous: Macedonia, half Greek, or Persia, wholly Oriental?

Demosthenes and his partisans considered themselves more patriotic than the others, and maybe they were. Both parties realized the imperative need of a national union; the "collaborators" claimed that the union was impossible or would not be viable except under Macedonian hegemony, the other party fought for national independence as well as for union. Looking at it from a long distance, it would seem that the collaborators were right; there was no hope or possibility of reconciling national freedom with national union. Needless to say, Philip did not consider himself a conqueror but rather the champion of Greek union and Greek culture against anarchy.

Thanks to his well-trained armies, he defeated his adversaries in many battles, the final one being that of Chaironeia (Boeotia) in 338. The last composition of Isocrates was his letter of congratulation to Philip for having gained that victory, a victory which he was sharing, for he himself had won it against Demosthenes; he died happy a few days later, almost a centenarian. Demosthenes had been present at the battle of Chaironeia; he survived it sixteen years, suffering many vicissitudes; he finally took refuge in the temple of Poseidon on the island of Calaureia (in the Saronic Gulf off the coast of Argolis) and committed suicide in 322.

To return to Chaironeia, 338, the peace that followed it established the Hellenic League, wherein all the Greek states (except Sparta) were represented and of which Philip was the head and protector. Soon afterward, Philip began operations in Asia Minor for the liberation of Greek colonies from the Persian yoke, but these operations were stopped by his murder in 336, at the age of forty-seven, in the twenty-fourth year of his rule. He was succeeded by his son, Alexandros III, whom posterity remembers under the name Alexander the Great. Philip was the creator of Macedonian power and the man who made the Alexandrian adventures and achievements possible. Many of Alexander's qualities (such as the love of science and letters) were already in him, but they were smothered by sensuality and unscrupulousness. His murder was probably a fruit of the corruption that surrounded him.[5]

[5] We would know more about him if the *Philippica* of his contemporary Theopompos of Chios (IV-2 B.C.) had survived. The *Philippica* (not to be confused with Demosthenes' famous orations) was a history of Philip II and in fact of the whole of Greece, continuing Xenophon's from 362 to 336. It was a monument of literary vanity, but Theopompos was well informed and outspoken. He was one of the founders of psychologic history, the forerunner of Tacitus (I-2). Though he considered Philip to be the greatest man the world had known, he did not flatter him but, on the contrary, gave a terrible picture of his weaknesses, of the dissolute living of his boon companions. R. H. Eyssonius Wichers, *Theopompi Chii fragmenta* (308 pp.; Leiden, 1829). For example, see fragment 249 describing in avenging terms the corruption of Philip's court.

Chaironeia marked the end of Greek independence, and thus the background of this period – the age of Aristotle – is the decadence and fall of political Greece. We attend the agony of the great nation that gave the world one of its most precious possessions, democratic ideals. She died in the travail of realizing them. Yet the Greek spirit is immortal, and even when national freedom was gone and lost, it produced wonderful achievements.

THE LIFE OF ARISTOTLE

Chalcidice is much more like an island in the north of the Aegean Sea than a part of Macedonia. The main lines of communication are sea lines, just as in the case of the other isles. The peninsula was colonized very early by Greek emigrants who came from Chalcis [6] (hence its name Chalcidice); its early Greek culture was Ionian, and its natural relations were with the other Ionian colonies of the Aegean Sea and the Asiatic coast. Chalcidice was a member of various Greek leagues established for mutual defense. Its main enemies were Persia and Macedonia; it was so close to the latter and was so obviously a part of its natural territory that it was bound to arouse Macedonian cupidity. To make a long story short, it was finally conquered and annexed by Philip, who is said to have replaced the Greek colonists by Macedonian veterans.

It was in that region that Aristotle was born, in 384. His birthplace was the city of Stageira, situated just north of the easternmost leg, the Mount Athos peninsula, the Holy Mountain. At the time of his birth, Chalcidice, or at least the easternmost part of it, was still independent and Ionian; in any case, the higher culture remained Ionian even after the Macedonian conquest. Aristotle may thus be called an Ionian philosopher; as we shall see in a moment, the appellation Macedonian would be equally proper.

We know nothing of his mother, except her name, Phaistis. His father, Nicomachos, belonged to an Asclepiad family; he became physician to Amyntas II (king of Macedonia, 393–370), and then moved from Stageira to the Macedonian capital of that time. The boy, Aristotle, was thus educated in Macedonia and must have obtained some familiarity with court life. His youth was dominated by three kinds of influence: Ionian, Macedonian, and medical. The first and the third were excellent for the shaping of a future man of science.

At the age of seventeen he was sent to Athens to complete his education (this was a normal procedure which would appeal equally to the philhellenes of Macedonia and to the Ionians of Chalcidice). Aristotle spent the following twenty years in Athens (367–347), and this is often stated in the following terms: he joined the Academy in 367 and was Plato's disciple for twenty years, until the latter's death. Such a statement is certainly wrong. Aristotle was Plato's pupil at the beginning of his stay, and Plato appreciated his precocity and his youthful energy; he called him the reader or the mind (*anagnōstēs*, *nus*). Considering Aristotle's intellectual curiosity, it is probable that he listened to other teachers, such as Isocrates, and he certainly shared with the Athenians the lessons of eloquence and politics that could be obtained in the agora or the Areiopagos. He

[6] Chalcis is the main city of the island Euboea; it is located at the narrowest point of the Euripos, the strait that separates Euboea from Boeotia on the mainland. The strait is so narrow at Chalcis that it was bridged as early as 411 B.C.

must have listened to some of the orations of Demosthenes.[7] A man of his originality and zeal would not have remained Plato's disciple for twenty years, but he was a member of the Academy and visited it from time to time; as far as we can judge from the fragments of his lost writings, he was a Platonist, at least until Plato's death, but with increasing reservations. The time of his membership was the second half of the Academy's existence; it had already abandoned its Socratic features and had become excessively Platonic, un-Socratic. There were occasional conflicts between the old master and his outstanding disciple; note that there was a difference of forty-four years between them; that is an enormous difference; Plato was not one generation but two generations back of him. According to Diogenes Laërtios,[8] Aristotle seceded from the Academy while Plato was still alive; hence the remark attributed to the latter: "Aristotle spurns me as colts kick out at the mother who bore them"; the situation as well as the remark itself are plausible.[9] It is, of course, impossible to say when Aristotle ceased to be a Platonist; it would be impossible even if we had all his early writings, and if these were dated; the boundaries between Platonism and non-Platonism are not sufficiently determined.

The way I would put it is this: Aristotle spent twenty years of study in Athens; during the first years, he was a regular student of the Academy, later he came back as a postgraduate student or alumnus, as a friend of the master and of other Academicians. The Academy was his main center of reunion, for there he would find not only the old master but a number of congenial men with whom he could discuss philosophic and scientific questions. Membership in the Academy was not a formal matter (as it would be today); it was informal; an old student, already distinguished, would always be sure of a welcome.

At the time of Plato's death, his nephew Speusippos was elected head of the school (*scholarchēs*), and he directed it for eight years (348–47 to 339). Did that choice annoy other members of the Academy? At any rate, Aristotle and his friend Xenocrates decided to leave; they accepted the invitation of a fellow student, Hermeias, ruler of Atarneus.

We must tell the story of Hermeias, for it illustrates the variety, the complexity, the unexpectedness of life in that age (as in all ages). The eunuch, Hermeias, who began his career as a money-changer, was a kind of financial wizard and became very wealthy and powerful. He obtained possession of a vast domain in Troas (northwest part of Mysia), and was known as the tyrant of Atarneus (opposite Lesbos). So far, his story is not very unusual; similar things have happened everywhere. The following is more typical of his own setting. He had been a student in the Academy (was this compatible with money brokerage? Why not? Many financiers are Harvard men), had remained a great admirer of Plato and had probably asked for his advice and help in government. Was not Plato the greatest master of politics? Two other alumni of the Academy, Erastos

[7] Demosthenes delivered his first Philippic in 351 and in 349–48 the three Olynthiac orations, in defense of Olynthos (in Chalcidice), threatened by Philip. Aristotle could not help being concerned with the fate of a city so close to his own birthplace, but his upbringing put him in the Macedonian party. Demosthenes and Aristotle were exact contemporaries (384–322).

[8] Diogenes Laërtios, v, 2; translation of R. D. Hicks in Loeb Classical Library.

[9] Another story, more difficult to believe, yet possible, was told by me in *Lychnos* (Uppsala, 1945), p. 253.

and Coriscos, both of Scepsis (in Troas), were Hermeias' assistants and were trying to establish a better government under Plato's patronage.[10] They had actually founded a new school (call it a branch of the Academy), in Assos.[11] after Speusippos' election as *scholarchēs* of the Academy, Aristotle and Xenocrates joined the school of Assos and they were followed later by Callisthenes and Theophrastos. Aristotle spent three years in Assos (347–344) and was one of Hermeias' familiars. He married Pythias, who was Hermeias' niece and adopted daughter. It was probably through Aristotle's instrumentality that Hermeias negotiated with Philip to obtain Macedonia's alliance. As Mysia was more or less under Persian suzerainty, Hermeias' negotiations were from the Persian point of view treasonable. Mentor, a Rhodian condottiere in Persian service, invited Hermeias to a conference, had him seized, and delivered him to the Great King. Hermeias was questioned and tortured concerning his relations with Philip, but instead of revealing his secrets and the names of his associates to Artaxerxes Ochos (ruled 359–338), as Demosthenes had predicted, he refused to speak. The Great King was moved by Hermeias' gallantry, and wanted to reprieve and befriend him, but his counselors discouraged such magnanimity. He then asked Hermeias what was his last request. "I wish my friends to know," answered Hermeias "that I have done nothing to be ashamed of or unworthy of philosophy." Hermeias was crucified; this occurred in Susa in 344. Aristotle dedicated at Delphi a monument to celebrate the heroic death of his friend, and wrote for it an inscription in two distichs. He also composed a longer poem in his honor, in the form of an ode to virtue, a paean, that is, a liturgic hymn, the purpose of which was to worship Hermeias. This poem (16 lines) and the inscription have both been preserved and give us a fair idea of Aristotle as a poet.

During his stay in Assos, Aristotle sailed from time to time to Mytilene (Lesbos) nearby, which was the native place of his new friend, Theophrastos. These three years in Assos and Mytilene were exceedingly fruitful; they enabled Aristotle to make many observations (for example, in the field of zoölogy) and to develop his own philosophy. Aristotle found himself in Assos.

Philip needed a tutor for his son, Alexander. It is possible that Aristotle was recommended to him by Hermeias; at any rate, Aristotle was known to him, and his merits as a negotiator and as a leader of the school of Assos must have been recognized. The royal offer was accepted, and Aristotle proceeded to Pella, which was Philip's residence. Aristotle's tutorship of Alexander lasted from 343 to 340, when the young boy (he was then only sixteen) was already obliged to replace his father (absent on military duties) as regent of the kingdom. It is not clear where Aristotle lived from 340 to 335, except that it was in Macedonian territory. Perhaps he remained in Pella, where he must have been an honored guest, or he may have returned to Stageira. He had good opportunities in any case to think out his new conceptions. His tutorship had compelled him to formulate his knowledge and wisdom as clearly and simply as possible; when the young prince lacked time for additional lessons, his tutor had more time for deeper meditations.

When the prince succeeded Philip, Aristotle remained his counselor and friend,

[10] The sixth of Plato's epistles is addressed to Hermeias, Erastos, and Coriscos. Edited and translated by R. G. Bury in Loeb Classical Library; Plato, vol. VIII (1929), pp. 456–461.

[11] Assos was in Hermeias' territory. It was an impregnable citadel and harbor opposite Lesbos. Assos was the birthplace of Cleanthes the Stoic (III-1 B.C.).

at least until the imprisonment and death of Callisthenes. Soon after Alexander's accession to the throne and while he was putting down rebellions in the Balkans and in Greece, Aristotle returned to Athens for the consummation of his great purpose: the creation of a new school and center of research, the Lyceum (335).

When the Alexandrian meteor had ended its astounding but brief trajectory in 323, the anti-Alexandrian factions of Athens resumed their vigor and their virulence. The king's patronage of the Lyceum and his benevolence to Aristotle compromised the latter. It was remembered by his enemies that Aristotle had written a paean in Hermeias' honor, and he was accused of impiety. Aristotle did not want Athens to repeat the unpardonable crime it had perpetrated when it had condemned Socrates to death and he preferred to take refuge in Chalcis (the mother city of his own native Chalcidice). He died there of disease within a few months, in 322 (Demosthenes committed suicide in the same year).

Aristotle had been married twice, the first time to Pythias of Assos, by whom he had a daughter bearing the same name. His second wife, Herpyllis, gave him a son who was named Nicomachos after her father-in-law, the Asclepiad, and was immortalized by the *Ethics* dedicated to him (the only ethical treatise of Aristotle the authenticity of which is unquestionable).

According to Diogenes Laërtios, Aristotle "spoke with a lisp. . . , his calves were slender, his eyes small and he was conspicuous by his attire, his rings and the cut of his hair." [12] We must be satisfied with these meager indications, because no plastic representation of him has come down to us. It is true the Austrian philologist, Franz Studniczka, claimed that the marble head preserved in the Kunsthistorisches Museum in Vienna was an authentic portrait, but his argumentation is unconvincing and worthless.[13] He remarks that the Vienna head suggests comparisons with Melanchthon and Helmholtz, but even that does not prove that it represents Aristotle!

We have a better knowledge of Aristotle's spiritual personality than of his physical appearance, through his abundant writings and also through his will, published by Diogenes Laërtios.[14] This will shows that Aristotle was a good paterfamilias, grateful to his wives, thoughtful of his children and of his servants. It is a document full of simple humanity.

THE LOST ARISTOTLE. HIS EARLY, PLATONIC WRITINGS

Aristotle's writings may be divided into three groups: (1) the early ones dating from the time [15] of his membership in the Academy, (2) erudite compilations dating probably from the Lyceum days, (3) a series of treatises prepared during his teaching years in Assos, Pella, and Athens.

All the works that have come down to us in their integrity belong to the third group, except for a single representative of the second group, the *Athenian Constitution*.

[12] Diogenes Laërtios, v, 1.

[13] F. Studniczka, *Das Bildnis des Aristoteles* (35 pp., 3 pls.; Leipzig, 1908). Discussed by me in *Lychnos* (1945), pp. 249–256. Studniczka's memoir is a monument of pedantic stupidity, but it fooled many philologists, including Jaeger, *Aristotle*, p. 322 (see note 16).

[14] Diogenes Laërtios, v, 11–16.

[15] We must repeat that the length of that time is difficult to determine. A man belongs to a school; for a time he is a devoted member of it; after a while his enthusiasm cools down and he attends fewer and fewer of its meetings, then ceases to come, and finally declares his opposition. How many stages of affection and disaffection must we recognize and when exactly does one pass from the one to the other?

Though the works of the first group are lost, there are enough fragments of them and references to them in ancient literature to enable us to appreciate their contents.[16] Indeed, these lost works were not immediately lost, far from it, and for many centuries Aristotle's fame was largely based upon them. These early works were written for the educated public in general, not for special students.[17] They were in the form of dialogues, which was the Platonic form of predilection, and reflected more or less faithfully the teachings of the Academy. Some of these works were related not only to Plato in general but also to definite Platonic writings; for example, the *Eudemos* of Aristotle is derived from *Phaidon*, his *Gryllos* from *Gorgias*, his books on *Justice* from the *Republic*, his *Protrepticos* from the *Euthydemos*.

Let us examine three of them: *Eudemos*, *Protrepticos*, and *Philosophy*.

The *Eudemos* is a dialogue on the immortality of the soul, so named after Aristotle's friend, Eudemos of Cypros,[18] who had been killed in 354. When we bewail the death of a person whom we loved, we cannot help asking ourselves anxiously whether the physical death is final. Aristotle accepts the Platonic theory that the soul of man comes from heaven and returns thither when released from its bondage.

The *Protrepticos*[19] (hortatory) is a treatise (not a dialogue) addressed to Themison, a prince of Cypros, exhorting him to study philosophy and to take a philosophic view of life. All the imperfections of mortal life are perfected in a transcendental world; death is the escape into a higher life. The imprisonment of the soul in the body is the cause of all our troubles and sufferings. The philosopher must keep himself as free as possible from worldly entanglements, which can only hinder his return to God. There are so many points of agreement between *Protrepticos* and *Epinomis* that the authors must have drunk from the same Platonic fountain, or else one of them copied the other.[20] *Protrepticos* interests us specially because of its extraordinary fame. Cicero prepared a Latin adaptation of it under the title *Hortensius*.[21] It influenced Iamblichos (IV-1) and Julian the Apostate (IV-2), and Cicero's version made a very deep impression on St. Augustine (V-1). St. Augustine was nineteen years of age when he read the *Hortensius*, and it was that very book which stirred him to the study of wisdom.[22] Is not that a singular fortune? Young Aristotle was the instrument of St. Augustine's awakening. Note

[16] The pioneer student of the early Aristotle was Werner Jaeger whose *bahnbrechend* work appeared in Berlin 1923. When we refer to it, we refer to the English translation, *Aristotle. Fundamentals of the history of his development* (410 pp.; Oxford: Clarendon Press, 1934). Ettore Bignone, *L'Aristotele perduto e la formazione filosofica di Epicuro* (2 vols.; Florence: Nuova Italia, 1936). Joseph Bidez, *Un singulier naufrage littéraire dans l'antiquité* (70 pp.; Brussels: Office de Publicité, 1943) [*Isis* 36, 172 (1946)].

The fragments were edited by Valentin Rose: Aristotelis fragmenta qui ferebantur librorum (Leipzig 1886) and by Richard Walzer, Aristotelis dialogorum fragmenta in usum scholarum selegit (Florence: Sansoni, 1934).

[17] Cicero (I-1 B.C.) called them *exōtericos* in the letter to Atticus written in 700 U.C. = 54 B.C. *Epistolae ad Atticum*, IV, 16.

[18] Eudemos of Cypros, Plato's disciple, was one of the Academicians whom Dion had enlisted for the insurrection against Dionysios the Younger. Eudemos was killed in one of the battles around Syracuse in 354. He is not to be confused with a younger man, Eudemos of Rhodes (IV-2 B.C.), who flourished *c.* 320, was Aristotle's disciple and probably the editor of the *Ethica Eudemia*.

[19] *Protrepticos eis philosophian*, exhortation to (the study of) philosophy.

[20] Philip of Opus (IV-1 B.C.) was, like Aristotle, a pupil of Plato. He may have been younger than Aristotle or older. The *Epinomis* may have been composed soon after *Protrepticos*, or before.

[21] *Hortensius* is a Latin equivalent of *protrepticos*, but not used elsewhere; the common form is *hortatorius*.

[22] St. Augustine, *Confessions*, III, 4; VIII, 7.

the temporal distance between them, almost eight centuries, and their very different orientations, Aristotle toward science, Augustine toward Christ.

The longest of the lost Aristotelian books, as far as we can judge from the fragments, was the treatise on *Philosophy* in three books. Going back to the speculations of the Seven Wise Men and early Delphic inscriptions (for example, *gnōthi sauton*), Aristotle explained in Book I his conception of the eternal return of doctrines,[23] in Book II he criticized the Platonic Ideas, in Book III, he outlined an astral theology. In that third book, he conceived the soul to be endowed with a spontaneous and eternal motion,[24] like the celestial bodies, each of which has a will of its own. He thus continued the strange aberration of the *Timaios* and of the *Epinomis*, according to which the regular periodicity of celestial bodies is considered to be a proof of their intelligence and divinity. It would seem that at the time of his writing this dialogue, Aristotle already thought of the fifth essence of heaven (or aether) as the very substance of which the souls are made.[25] This I find very difficult to grasp. After having ascribed divinity to the planets because of their regularity, how could he assimilate to them the souls of men, the motions of which are unpredictable? Perhaps he was led astray by the idea of automaticity which he ascribed to the stars as well as to the souls. His cosmology in this treatise is comparable to that of the *Timaios*, with an important difference: divinity is not transcendental, as Plato understood it, but tangible in the celestial bodies. The main source of wisdom is the contemplation, no longer of abstract Ideas, but of the perfect motions of stars and planets.

Aristotle's conviction of God's existence was derived from two sources — the prophetic abilities of the soul (as revealed in dreams) and the sight of the starry heavens.[26] His endorsement of the astral theology must have contributed powerfully to its general acceptance in the Hellenistic age. Jaeger summed up that situation very aptly and beautifully when he wrote:

> The establishment of the worship of the stars, which are confined to no land or nation but shine on all the peoples of the earth, and of the transcendental God who is enthroned above them, inaugurates the era of religious and philosophic universalism. On the crest of this last wave Attic culture streams out into the Hellenistic sea of peoples.[27]

This early philosophy of Aristotle is already independent of Plato's, but not much; the essential of his metaphysics is still Platonic (except for the rejection of the Ideas) and permeated with the Chaldean and Iranian thoughts that were circulating in the Academy. This is not astonishing. His teaching at Assos and Pella had already oriented his mind in a new direction; he was putting in order his

[23] We, the moderns of today, who can read the history of the philosophic vicissitudes of three millennia, cannot help thinking, as Aristotle did, when he was a modern, of the same cyclic repetitions.

[24] The technical term is *endelecheia*, meaning continuity, persistency, which has been confused by all the editors with *entelecheia*, full, complete reality. It does not occur in Bonitz' *Index aristotelicus!* Bidez, *Un singulier naufrage littéraire*, pp. 33–37.

[25] Aristotle's later views on the soul were very different from his early, Platonic ones. The souls, he finally concluded, being the "forms" of the material bodies, do not survive the latter any more than vision survives the loss of an eye. There is something in the individual soul, however, that comes from the outside and is a part of pure reason. When a man dies, that part of his soul goes back to the universal reason (God) which absorbs it. There is thus a kind of impersonal immortality.

[26] Fragment 10. Transmitted by Sextos Empiricos (II–2).

[27] Jaeger, *Aristotle*, p. 166.

knowledge of logic, mathematics, astronomy, natural history, and for the time being was willing to accept Platonic metaphysics much as he found it. His position was very similar to that of a modern man of science who carries on his investigations without trying to inquire into the religious ideas and the religious practice that are essential parts of his family tradition.

The creation of these early writings is even less astonishing. The fact that they are essentially different from those of his maturity requires no explanation; he was a man of extraordinary genius, but genius itself must grow; it is foolish to expect it to be prematurely mature. The infant prodigies more often than not reach very early a low level of maturity and then fail to soar much higher. A real man of genius, on the contrary, is likely to develop more slowly than other men. Many men of science began their career with poetic or philosophic publications which they repudiated later on, or abandoned to oblivion.[28] Such a course is natural enough; it was very much so in the case of a man submitted for twenty years to the fatuous speculations of the Academy. Aristotle was saved from the voodoo of the *Timaios* by his scientific curiosity, his growing habit of careful investigations, the practical duties put on his shoulders in Assos and Pella, and above all, by his own rationality and independence.

All things considered, the evolution of Aristotle was not exceptional, but normal. His mind was cleared of Platonic phantasies in proportion to the growth of his scientific knowledge.

We would not bother much about his early writings except for the importance attached to them during three or four centuries, and their subsequent mysterious disappearance. It is as if one Aristotle had been known for centuries and then suddenly replaced by another one, very different. What puzzles me most is the eclipse of the old Aristotle. As his writings enjoyed some fame, each must have been represented by many copies; how is it that all have disappeared and that we do not have the complete text of a single one? This illustrates once more the precariousness of manuscript tradition. Yet why was it more precarious in the case of Aristotle's popular writings than in that of Archimedes' very technical ones? We cannot answer that question. The preservation of manuscripts was exceedingly capricious and hazardous.

THE LIVING ARISTOTLE. HIS PERMANENT WRITINGS

A historian of science would be tempted to say that the Platonic writings of Aristotle were lost because they were displaced and obliterated by his later writings, but if he did say that he would give a bad example of self-centeredness. We must bear in mind that Platonic illusions were for many centuries (and are even today) more agreeable to the majority of people than scientific matter-of-factness. The final loss of the old Aristotelian writings is very mysterious, and the temporary loss and rediscovery of his later ones romantic to a degree.

Here is what happened. After Aristotle's death, his papers became the property of his friend and successor, Theophrastos. The latter bequeathed them, not to his own successor or to the Lyceum, as we would have expected, but to his nephew,

[28] For example, Claude Bernard. Until very recently the education that future men of science received in the secondary schools was very largely humanistic. Hence, their juvenile ambition was fired by literary models and their scientific genius did not find its direction until later. They were very much in the same situation as Aristotle.

Neleus of Scepsis.[29] Neleus does not seem to have cared for them in any way, but his heirs sold some items to Ptolemy Philadelphos (ruled 285–247), who was building up the library of Alexandria. Being afraid that the rest might be seized by Attalos of Pergamon (ruled 269–197), who was their own king and was building up the rival library of Pergamon, they hid all the manuscripts in a cave. Some time later, Apellicon of Teos, who was passing through Scepsis, heard of that treasure and obtained it for the private library that he was gathering in Athens. This Apellicon was a Peripatetic and a rich collector of books; we do not know much about him except that he died soon before Sulla's siege and sack of Athens (84 B.C.). Sulla bought or took the Aristotelian manuscripts and carried them to Rome. A Greek grammarian nicknamed Tyrannion was taken captive by Lucullus soon afterward (in 72 B.C.) and brought to Rome, and the arrangement of Apellicon's books was intrusted to him. Tyrannion was a capable scholar, praised by Cicero and Strabon, but he does not seem to have done more than catalogue or describe the Aristotelian manuscripts, or if he began an edition of them his work was insufficient. The first edition was prepared at about the same time by Andronicos of Rhodes (I–1 B.C.), and Andronicos' edition is fundamental; all other editions are derived from it, directly or indirectly. We should not conclude that the Aristotelian writings remained unknown until Andronicos' publication *c.* 70 B.C., because there must have been an oral and a written tradition of them in the Lyceum. Andronicos' edition, I take it, was the first that reached outsiders.

This story throws an interesting sidelight upon the cultural progress of the Hellenistic age, for example, upon the growth of libraries in Alexandria, Pergamon, Athens, and Rome.

The writings preserved by Andronicos were presumably the same as those that have reached us today. It will suffice at present to give a short list of them, with few remarks; we shall discuss some of them at greater length later on. We list them in the order that has become traditional, as represented, for example, in the Bekker text (1831) and in the English *Aristotle*.[30]

Vol. 1 (pp. 1–184). The Organon: *Categoriae. De interpretatione. Analytica priora et posteriora. Topica. De sophistis elenchis.*

Vol. 2 (pp. 184–338). *Physica. De caelo. De generatione et corruptione.*

Vol. 3 (pp. 338–486). *Meteorologica. De mundo. De anima. Parva naturalia.*[31] *De spiritu.*

Vol. 4 (pp. 486–633). *Historia animalium.*

Vol. 5 (pp. 639–789). *De partibus, motu, incessu, et generatione animalium.*

Vol. 6 (pp. 791–858). *De coloribus. De audibilibus. Physiognomonica. De plantis. De mirabilibus auscultationibus. Mechanica.* (pp. 968–980). *De lineis insecabilibus. Ventorum situs et cognomina. De Melisso, Xenophane, Gorgia.*

Vol. 7 (pp. 859–967). *Problemata.*

Vol. 8 (pp. 980–1093). *Metaphysica.*

Vol. 9 (pp. 1094–1251). *Ethica Nicomachea. Magna moralia. Ethica Eudemia.*

Vol. 10 (pp. 1252–1353). *Politica. Oeconomica.* (pp. 1*–69*, text Berlin Academy, 1903). *Atheniensium res publica.*

Vol. 11 (pp. 1354–1462). *Rhetorica. De rhetorica ad Alexandrum. De poetica.*

[29] Neleus was a disciple of Aristotle and Theophrastos. He was the son of Coriscos, who had been Aristotle's friend and associate in Assos.

[30] The indications of volumes refer to the English edition; those of pages, to Bekker.

[31] *De sensu et sensibili. De memoria et reminiscentia. De somno et vigilia. De somniis. De divinatione per somnum. De longitudine et brevitate vitae. De iuventute et senectute. De vita et morte. De respiratione.*

All these writings except one belong to the third group, that is, they are textbooks representing lectures given in the Lyceum either by Aristotle or by others. The exception is the *Athenian republic* (in vol. 10), which is the only extant representative of the second group, surveys prepared for the Lyceum. Aristotle had made a comparative study of 158 Greek constitutions, the most important being presumably the Athenian, which alone has come down to us. It includes two main parts: (1) the constitutional history of Athens from the origins to Aristotle's time, each stage being intelligently and clearly described; (2) analysis of the Athenian constitution and government as it existed *c.* 330.

The tradition of that text is curious. Until 1891, only fragments of Aristotle's constitutional studies were known. In that year a papyrus found in Egypt, preserved in the British Museum, was edited by Kenyon: this was the editio princeps of the *Athenian Constitution.*[32]

The corpus of Aristotelian writings is encyclopedic in scope. It covers logic, mechanics, physics, astronomy, meteorology, botany, zoölogy, psychology, ethics, economics, politics, metaphysics, literature, etc. No large treatise is devoted to mathematics, but there are a good many discussions of mathematical topics, scattered in various books.

Are these writings authentic? The question is more complex than it seems at first sight, and it cannot be answered *in toto.* The authenticity of each separate writing has been discussed by its editors, who have not always reached the same conclusions. If the question concerns the literal authorship — the actual writing of each text — it is probable that few were written by Aristotle himself. We cannot even say that all of them represent indirectly his own teaching; some may represent the teaching of Theophrastos or of other members of the Lyceum. The texts available to us may represent Aristotle's ideas or the ideas of other Peripatetics; if they represent his own ideas, it does not follow that they represent his own words, except in so far as a good student would have taken pains to reproduce the master's *ipsissima verba,* at least with regard to essentials.

With the exception of a few writings that are generally conceded to be apocryphal, the consensus of opinion seems to be that the books bearing Aristotle's name may be taken to be the substance of his lectures; the original manuscripts (as edited by Andronicos) were written down on the basis of his own lecture notes (in various stages of development) or on the basis of the notes taken by auditors, revised (or not) by himself. The possible variations of that hypothesis are endless.

The documentation of some books, especially the zoölogic ones, may have been collected partly by the master, partly by his assistants and students. This would not diminish his authorship, however, for in such cases the author is not so much, or exclusively, the man who discovers the individual facts, but rather he who puts them together and explains them.

The chronology of Aristotle's writings is very uncertain. Some of them were drafted, if not composed, in Assos or Macedonia; others owe their origin to the Lyceum. Most of them are the fruit of a long evolution, and were probably planned, written, and rewritten in many instalments. Professor Jaeger has proved

[32] Frederic George Kenyon (1863–), *Aristotle on the Constitution of Athens* (British Museum, 1891, third and rev. ed., 295 pp., British Museum, 1892). A facsimile of the whole papyrus was published by the British Museum in 1891. Elaborate bibliography in Kenyon's edition.

that this was the case for the *Metaphysics*, the *Ethics*, and the *Politics*. Every author, and especially every teacher of long experience, will understand this easily. One may be able to date the completion of a book, and sometimes its inception, but it is difficult, if not impossible, to date its various parts. If two books have been completed in the years t and $t+l$, it does not follow that the second is wholly later than the first; in fact, the first may include references to the second.

Many of the traditional views concerning the composition or style of Aristotle are as arbitrary and legendary as those concerning Plato, except that the legends developed in opposite directions. The same pedants who admired Plato's style (often without a sufficiently fluid knowledge of the Greek language to appreciate such subtleties), were agreed that the Aristotelian writings were poorly written, that Aristotle had *no* style, and so on. We recognize a fallacy that often dominates the mind of literary critics when they have to judge scientific writings. The main difference between scientific writings and imaginative ones concerns the relation of content to form. The man of science worries much more about what he has to say than about the way of saying it; he is satisfied when he has succeeded in explaining clearly his ideas and in describing exactly the results that he has obtained. His effort is likely to stop at that moment, for he has no patience with literary vanities, while the man of letters would take any amount of additional effort to express his thoughts more beautifully, with more wit and grace and a better rhythm. There is a subtle polarity between the form or style and the content of a book. In scientific books, style is subordinated to content; in poetic compositions, the opposite relation is more natural. When the critic is aware that the content of a book is intrinsically important, and that the form is severe and terse, he jumps easily to the conclusion that the author cannot write properly. His conclusion may be sometimes justified, because many scientific books are poorly written, but it is quite as often wrong and unfair. Being unable to appreciate the beauty of the content and chilled by the uncompromising austerity of the language, he decides that the book is "not" written, that it is "not" literature. The Aristotelian books are scientific books and their content mattered infinitely more than their form; the latter is sometimes a bit careless; on the other hand, we find in them occasional lapidary expressions that reveal the master's genius (*ex ungue leonem*). Aristotle, I would say, was anxious to write as well as possible, because he was a poet [33] and never forgot his Platonic education; if some of his books are imperfect and slipshod, it is not owing to his own carelessness but simply to the fact that he lacked the opportunity of finishing them as he would have liked to do.

The form of many of the books bearing his name might possibly have been improved by a stylist, but could the latter have done that without sacrificing some of the ideas, taking their edge off and putting on another, less worth while? We

[33] Michael Stephanides, "Aristotle as a poet," *Practica of the Academy of Athens* (1950), pp. 249–253, in Greek. Professor Stephanides says that the *De mundo* was written with particular elegance, and that the Alexander to whom it is addressed was probably Aristotle's pupil, Alexander the Great. This does not tally with the conclusion of Wilhelm Capelle, *Neue Jahrbücher* *15*, 529–568 (1905), that the *De mundo* is founded on two works of Poseidonios (I–1 B.C.). It is amusing to reflect that the clearness of the *De mundo* was used as an argument against its authenticity by the famous Dutch philologist, Daniel Heinsius (1580–1655), "Le Traité en question n'offre nulle part cette majestueuse obscurité qui dans les ouvrages d'Aristote, repousse les ignorants" (as quoted by François Arago in his eulogy of Gay-Lussac, *Oeuvres*, vol. 3, p. 53).

all agree that form and content are as inseparable as body and soul, but literary critics often behave as if the form were the soul, while the soul of a book is in its ideas, that is, in its content. That is certainly true of scientific books.

The language of the Aristotelian writings, it must be admitted, is no longer the Attic language of the golden age; it is mixed up not only with technical terms but also with common terms of various origins. Aristotle may be considered one of the founders of the new common language (*hē coinē dialectos*). His terminology is remarkable; it includes superfluities, but that was unavoidable in his time; the elimination of unnecessary terms, as well as the creation of new ones, is one of the aspects of scientific development. The marvel is not that many Aristotelian terms have been dropped but rather that so many have been preserved in our own languages.

EDITIONS, TRANSLATIONS, INDEXES

The purpose of this book is not bibliographic, yet it is necessary to speak of some of the early editions – for these are historical landmarks – and to advertise the modern editions that are the most convenient for reference.

For incunabula editions, most of them in Latin, with or without the Averroës commentaries, see Klebs, Nos. 82–97 (Fig. 89).

One of the greatest of all incunabula is the Greek princeps of Aristotle published by Aldus Manucius in Venice, 1495–1498, in five folio volumes (Fig. 90). The printers of Basel were always competing with those of Venice, and thus a new edition of all the Aristotelian works was prepared by Erasmus of Rotterdam and Symon Grynaeus (2 vols., folio; Basel, 1531) (Figs. 91 and 92). The Greek text was edited again by Friedrich Sylburg (1536–1596) and printed in Frankfurt (11 vols.; 1584–1587). The first collected edition with Latin translation appeared in Lyons in 1590.

The most important of the modern editions is the one prepared by Immanuel Bekker (1785–1871) and published under the auspices of the Academy of Berlin with Latin translations (5 vols., quarto; Berlin, 1831–1870).[34] Bekker's pagination has been preserved in almost all the later editions. The Bekker Greek text was reprinted in Oxford,[35] with the addition of the "indices Sylburgiani" (11 vols.; Oxford, 1837). The Didot edition, Greek-Latin, is by F. Dübner, U. C. Bussemaker, and E. Heitz (5 vols.; Paris, 1848–1874).

Jules Barthélemy-Saint-Hilaire (1805–1895) devoted the best part of his life to a French translation of the Aristotelian corpus (1839 ff.). Though his translations are no longer up-to-date, it is often very worth while to refer to them.

The works of Aristotle have been translated into English under the editorship of W. D. Ross (11 vols.; Oxford: Clarendon Press, 1908–1931). The contents of those volumes has been indicated above, p. 477.

Many Aristotelian works are available in Greek and English in the Loeb Classical Library, for example, *Parts, movements and progression of animals* (1937) [*Isis* 29, 205 (1938); *30*, 322 (1939)]; *On the heavens* (1939) [*Isis* 32, 136 (1947–49)]; *Generation of animals* (1943) [*Isis* 35, 181 (1944)].

The English translation included in the Oxford volumes and in the Loeb series are up-to-date and convenient, but their annotations are insufficient. There is real need of new translations (preferably with the Greek text) fully explained by a scholar who knows the history of science as well as the history of philosophy, acquainted not only with philologic details but with all the realia expressed or implied.

Indexes. Marco Antonio Zimara, *Tabula dilucidationum in dictis Aristotelis et Averrois* (folio; Venice, 1537) [*Isis 41*, 106 (1950)]. The indexes relative to each separate work compiled by Friedrich Sylburg (1584–87) are reprinted in the Oxford Bekker edition (1837). Very elaborate index by Hermann Bonitz, *Index Aristotelicus*

[34] Vols. 1–2 (1831), Greek text; vol. 3 (1831), Latin translations; vol. 4 (1846), Greek scholia; vol. 5 (1870), index.

[35] Unfortunately, the Oxford reprint of Bekker's edition does not include the Bekker pagination. That is an almost incredible aberration.

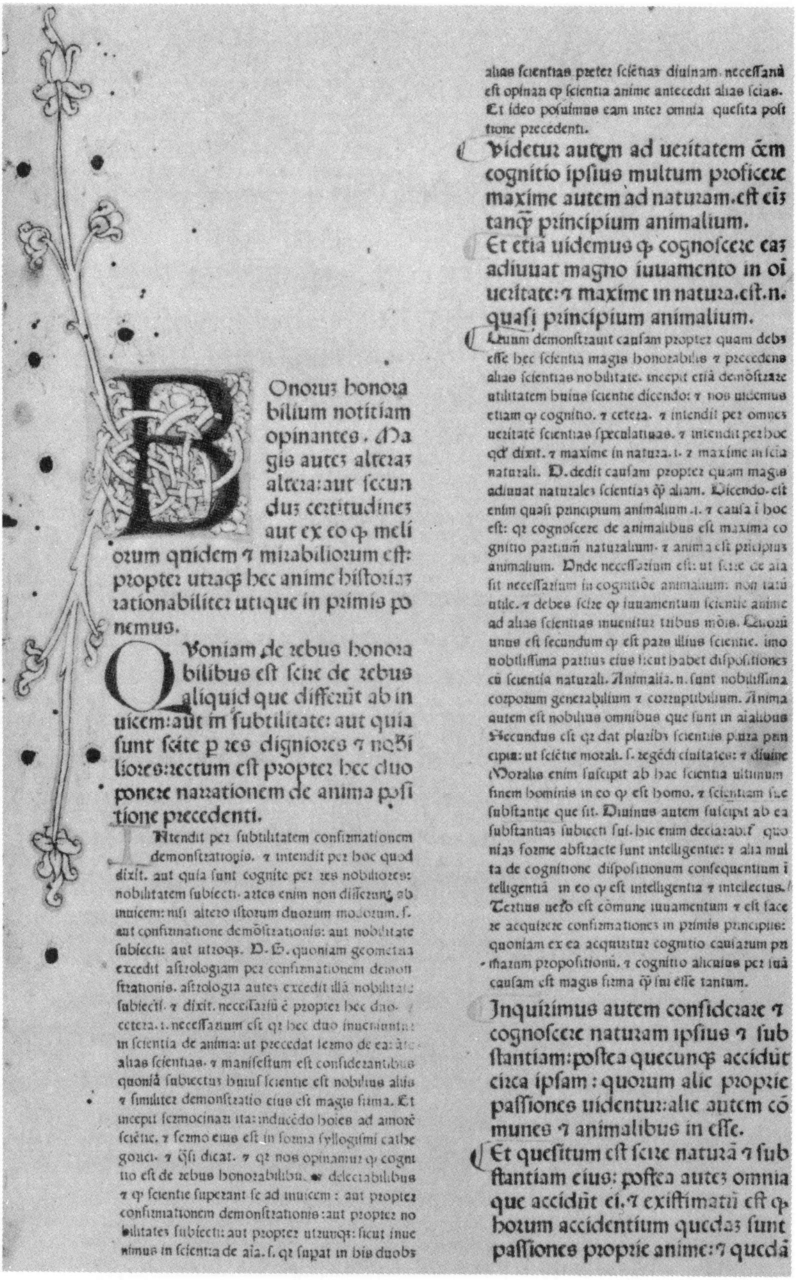

Bonorum honorabilium notitiam opinantes. Magis autem alteram altera: aut secundum certitudinem aut ex eo quod meliorum quidem et mirabiliorum est: propter utraque hec anime historiam rationabiliter utique in primis ponemus.

Quoniam de rebus honorabilibus est scire de rebus aliquid que differunt ab invicem: aut in subtilitate: aut quia sunt scire per res digniores et nobiliores: rectum est propter hec duo ponere narrationem de anima positione precedenti.

Intendit per subtilitatem confirmationem demonstrationis. et intendit per hoc quod dixit. aut quia sunt cognite per res nobiliores: nobilitatem subiecti. artes enim non differunt ab invicem: nisi altero istorum duorum modorum. s. aut confirmatione demonstrationis: aut nobilitate subiecti: aut utroque. D. G. quoniam geometria excedit astrologiam per confirmationem demonstrationis. astrologia autem excedit illam nobilitate subiecti. et dixit. necessarium est propter hec duo. cetera. i. necessarium est quod hec duo inveniuntur in scientia de anima: ut precedat sermo de ea ante alias scientias. et manifestum est considerantibus quoniam subiectum huius scientie est nobilius aliis et similiter demonstratio eius est magis firma. Et incepit sermocinari ita: inducendo homines ad amorem scientie. et sermo eius est in forma syllogismi cathegorici. et quasi dicat. et quia nos opinamur quod cognitio est de rebus honorabilibus et delectabilibus et quod scientie superant se ad invicem: aut propter confirmationem demonstrationis: aut propter nobilitatem subiecti: aut propter utrumque: sicut invenimus in scientia de aia. s. quia superat in his duobus alias scientias preter scientiam divinam. necessarium est opinari quod scientia anime antecedit alias scias. Et ideo posuimus eam inter omnia quesita positione precedenti.

Videtur autem ad veritatem omnem cognitio ipsius multum proficere maxime autem ad naturam. est enim tanquam principium animalium.

Et etiam videmus quod cognoscere eam adiuvat magno iuvamento in omni veritate: et maxime in natura. est. n. quasi principium animalium.

Cum demonstravit causam propter quam debet esse hec scientia magis honorabilis et precedens alias scientias nobilitate. incepit etiam demonstrare utilitatem huius scientie dicendo: et nos videmus etiam quod cognitio. et cetera. et intendit per omnes veritates scientias speculativas. et intendit per hoc quod dixit. et maxime in natura. i. et maxime in scientia naturali. D. dedit causam propter quam magis adiuvat naturales scientias quam aliam. Dicendo. est enim quasi principium animalium .i. et causa in hoc est: quia cognoscere de animalibus est maxima cognitio partium naturalium. et anima est principium animalium. Unde necessarium est: ut scire de aia sit necessarium in cognitione animalium: non tamen utile. et debes scire quod iuvamentum scientie anime ad alias scientias invenitur tribus modis. Quorum unus est secundum quod est pars illius scientie. immo nobilissima partium eius sicut habet dispositiones cum scientia naturali. Animalia. n. sunt nobilissima corporum generabilium et corruptibilium. Anima autem est nobilius omnibus que sunt in aialibus. Secundus est quia dat pluribus scientiis prima principia: ut scientie morali. s. regendi civitates: et divine. Moralis enim suscipit ab hac scientia ultimum finem hominis in eo quod est homo. et scientiam sue substantie que sit. Divinus autem suscipit ab ea substantias subiecti sui. hic enim declarabitur quoniam forme abstracte sunt intelligentie: et alia multa de cognitione dispositionum consequentium intelligentiam in eo quod est intelligentia et intellectus. Tertius vero est commune iuvamentum et est facere acquirere confirmationes in primis principiis: quoniam ex ea acquiritur cognitio causarum primarum propositionum. et cognitio alicuius per suam causam est magis firma quam sui esse tantum.

Inquirimus autem considerare et cognoscere naturam ipsius et substantiam: postea quecunque accidunt circa ipsam: quorum alie proprie passiones videntur: alie autem communes et animalibus in esse.

Et quesitum est scire naturam et substantiam eius: postea autem omnia que accidunt ei. et existimatum est quod horum accidentium quedam sunt passiones proprie anime: et quedam

Fig. 89. First Latin edition of *De anima* (Padua, 1472), printed by Laurentius Canozius of Lendenaria, who worked in Padua from 1472 to 1475 (British Museum *Catalogue*, vol. 7, p. 907; Klebs, 84.1). The printing of this book was completed on 22 November 1472, 90 leaves folio, two columns. Each paragraph of Aristotle's text is given twice, in a new and an old Latin version, and is followed by Ibn Rushd's commentary on the latter. We reproduce the first page. [From the copy in the Harvard College Library.]

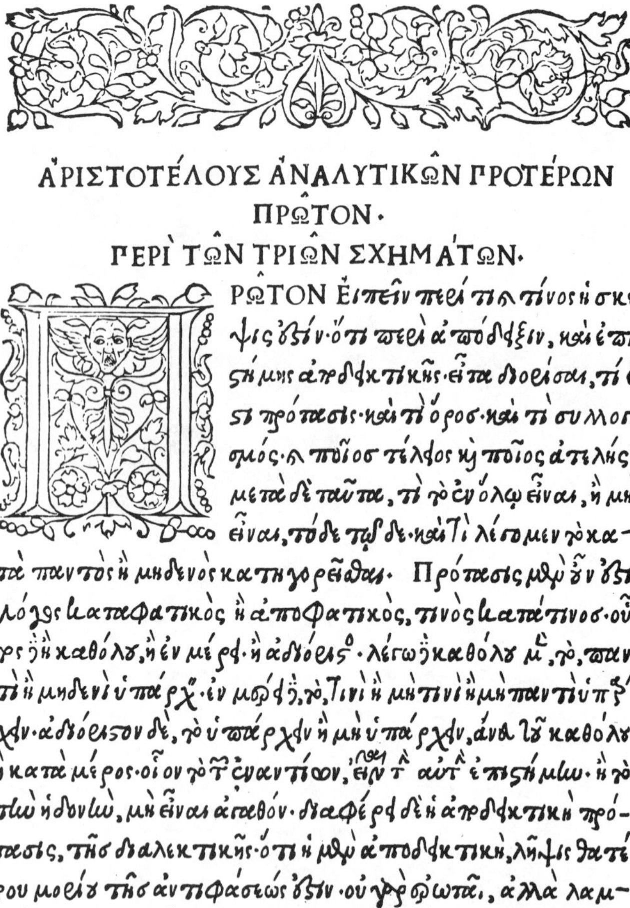

ἈΡΙΣΤΟΤΈΛΟΥΣ ἈΝΑΛΥΤΙΚΩ͂Ν ΠΡΟΤΈΡΩΝ ΠΡΩ͂ΤΟΝ.

ΠΕΡῚ ΤΩ͂Ν ΤΡΙΩ͂Ν ΣΧΗΜΆΤΩΝ.

ΠΡΩ͂ΤΟΝ εἰπεῖν περὶ τί καὶ τίνος ἡ σκέψις ἐστίν· ὅτι περὶ ἀπόδειξιν, καὶ ἐπιστήμης ἀποδεικτικῆς· εἶτα διορίσαι, τί ἐστι πρότασις· καὶ τί ὅρος· καὶ τί συλλογισμός· καὶ ποῖος τέλειος καὶ ποῖος ἀτελής· μετὰ δὲ ταῦτα, τί τὸ ἐν ὅλῳ εἶναι, ἢ μὴ εἶναι, τόδε τῷδε· καὶ τί λέγομεν τὸ κατὰ παντὸς ἢ μηδενὸς κατηγορεῖσθαι. Πρότασις μὲν οὖν ἐστὶ λόγος καταφατικὸς ἢ ἀποφατικός, τινὸς κατά τινος· οὗτος δὲ ἢ καθόλου, ἢ ἐν μέρει, ἢ ἀδιόριστος· λέγω δὲ καθόλου μὲν, τὸ παντὶ ἢ μηδενὶ ὑπάρχειν· ἐν μέρει δὲ, τὸ τινὶ ἢ μὴ τινὶ ἢ μὴ παντὶ ὑπάρχειν· ἀδιόριστον δὲ, τὸ ὑπάρχειν ἢ μὴ ὑπάρχειν, ἄνευ τοῦ καθόλου ἢ κατὰ μέρος· οἷον τὸ τῶν ἐναντίων εἶναι τὴν αὐτὴν ἐπιστήμην· ἢ τὸ τὴν ἡδονὴν, μὴ εἶναι ἀγαθόν· διαφέρει δὲ ἡ ἀποδεικτικὴ πρότασις, τῆς διαλεκτικῆς· ὅτι ἡ μὲν ἀποδεικτική, λῆψις θατέρου μορίου τῆς ἀντιφάσεώς ἐστιν· οὐ γὰρ ἐρωτᾷ, ἀλλὰ λαμβάνει ὁ ἀποδεικνύων· ἡ δὲ διαλεκτική, ἐρώτησις τῆς ἀντιφάσεώς ἐστιν· οὐδὲν δὲ διοίσει πρὸς τὸ γενέσθαι τὸν ἑκατέρου συλλογισμόν· καὶ γὰρ ὁ ἀποδεικνύων καὶ ὁ ἐρωτῶν, συλλογίζεται, λαβών τι κατά τινος ὑπάρχειν ἢ μὴ ὑπάρχειν· ὥστε ἔσται συλλογιστικὴ μὲν πρότασις, ἁπλῶς κατάφασις ἢ ἀπόφασίς τινος κατά τινος, κατὰ τὸν εἰρημένον τρόπον· ἀποδεικτικὴ δὲ, ἐὰν ἀληθὴς ᾖ, καὶ διὰ τῶν ἐξ ἀρχῆς ὑποθέσεων εἰλημμένη· δια-

Fig. 90. A page of the first Greek edition of Aristotle's works, 5 folio volumes in 6 printed by Aldus Manutius in Venice, 1495–1498 (Klebs, 83.1). This page is taken from vol. 1, containing the Organon, November 1495. It is the beginning of the *Prior analytics*. Note the beautiful printing full of ligatures; except for the top lines it looks like a manuscript page. The colophon includes a privilege granted by the Venetian Senate, forbidding other printers to publish the same texts. [From the copy in the Harvard College Library.]

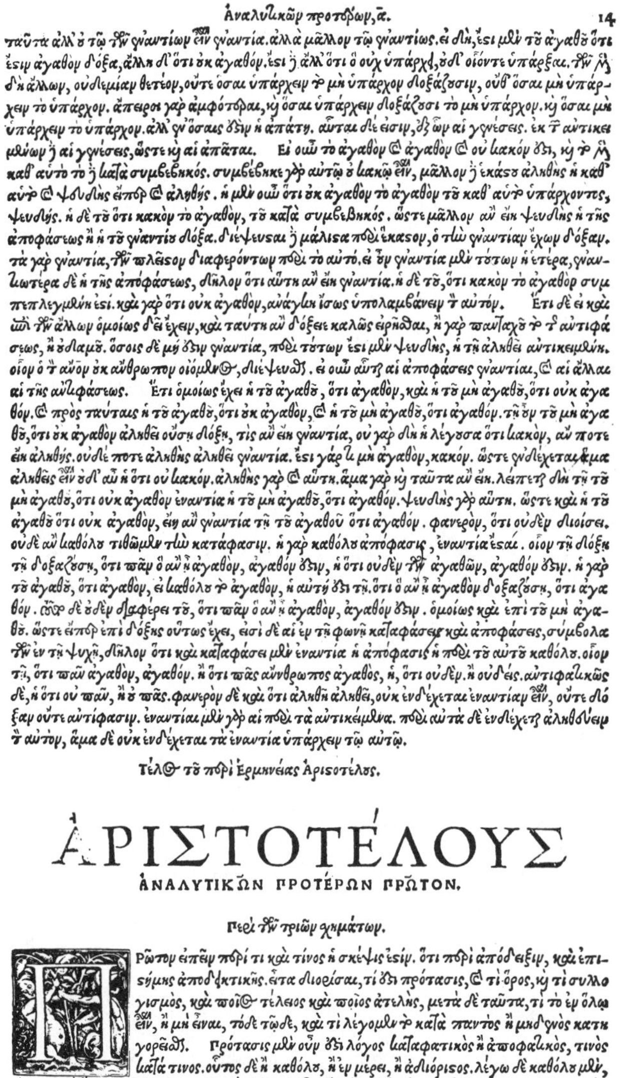

Ἀναλυτικῶν προτέρων, α. 14

ταῦτα ἀλλ' ὁ τῶν ἐναντίων εἶν' ἐναντία. ἀλλὰ μᾶλλον τῶν ἐναντίως. εἰ δή, ἐστι μὲν τοῦ ἀγαθοῦ ὅτι ἐστὶν ἀγαθὸν δόξα, ἄλλη δ' ὅτι οὐκ ἀγαθόν. ἔστι δ' ἀλλ' ὅτι ὃ οὐχ ὑπάρχει, οὐδ' οἷόν τε ὑπάρξαι. τῶν μὲν δὴ ἄλλων, οὐδεμίαν θετέον, οὔτε ὅσαι ὑπάρχειν τὸ μὴ ὑπάρχον δοξάζουσιν, οὔθ' ὅσαι μὴ ὑπάρχειν τὸ ὑπάρχον. ἄπειροι γὰρ ἀμφότεραι, καὶ ὅσαι ὑπάρχειν δοξάζουσι τὸ μὴ ὑπάρχον, καὶ ὅσαι μὴ ὑπάρχειν τὸ ὑπάρχον. ἀλλ' ἐν ὅσαις ἐστὶν ἡ ἀπάτη. αὗται δέ εἰσιν, ἐξ ὧν αἱ γενέσεις. ἐκ τῶν ἀντικειμένων δὲ αἱ γενέσεις, ὥστε καὶ αἱ ἀπάται. Εἰ οὖν τὸ ἀγαθὸν καὶ ἀγαθὸν καὶ οὐ κακόν ἐστι, καὶ τὸ μὲν καθ' αὑτὸ, τὸ δὲ κατὰ συμβεβηκός. συμβέβηκε γὰρ αὐτῷ οὐ κακῷ εἶναι, μᾶλλον δὲ ἑκάστου ἀληθὴς ἡ καθ' αὑτὸ καὶ ψευδὴς εἴπερ καὶ ἀληθής. ἡ μὲν οὖν ὅτι οὐκ ἀγαθὸν τὸ ἀγαθὸν τοῦ καθ' αὑτὸ ὑπάρχοντος, ψευδής. ἡ δὲ τοῦ ὅτι κακὸν τὸ ἀγαθὸν, τοῦ κατὰ συμβεβηκός. ὥστε μᾶλλον ἂν εἴη ψευδὴς ἡ τῆς ἀποφάσεως ἢ ἡ τοῦ ἐναντίου δόξα. διέψευσται δὲ μάλιστα περὶ ἕκαστον, ὁ τὴν ἐναντίαν ἔχων δόξαν. τὰ γὰρ ἐναντία, τῶν πλεῖστον διαφερόντων περὶ τὸ αὐτό. εἰ οὖν ἐναντία μὲν τούτων ἡ ἑτέρα, ἐναντιωτέρα δὲ ἡ τῆς ἀποφάσεως, δῆλον ὅτι αὕτη ἂν εἴη ἐναντία. ἡ δὲ τοῦ, ὅτι κακὸν τὸ ἀγαθὸν συμπεπλεγμένη ἐστί. καὶ γὰρ ὅτι οὐκ ἀγαθὸν, ἀνάγκη ἴσως ὑπολαμβάνειν τὸν αὐτόν. Ἔτι δὲ εἰ καὶ ἐπὶ τῶν ἄλλων ὁμοίως δεῖ ἔχειν, καὶ ταύτῃ ἂν δόξειε καλῶς εἰρῆσθαι, ἢ γὰρ πανταχοῦ τὸ τῆς ἀντιφάσεως, ἢ οὐδαμοῦ. ὅσοις δὲ μή ἐστιν ἐναντία, περὶ τούτων ἐστὶ μὲν ψευδὴς, ἡ τῇ ἀληθεῖ ἀντικειμένη. οἷον ὁ τὸν ἄνθρωπον οὐκ ἄνθρωπον οἰόμενος, διέψευσται. εἰ οὖν αὗται αἱ ἀποφάσεις ἐναντίαι, καὶ αἱ ἄλλαι αἱ τῆς ἀντιφάσεως. Ἔτι ὁμοίως ἔχει ἡ τοῦ ἀγαθοῦ, ὅτι ἀγαθὸν, καὶ ἡ τοῦ μὴ ἀγαθοῦ, ὅτι οὐκ ἀγαθόν. καὶ πρὸς ταύταις ἡ τοῦ ἀγαθοῦ, ὅτι οὐκ ἀγαθὸν, καὶ ἡ τοῦ μὴ ἀγαθοῦ, ὅτι ἀγαθόν. τῇ οὖν τοῦ μὴ ἀγαθοῦ, ὅτι οὐκ ἀγαθὸν ἀληθεῖ οὔσῃ δόξῃ, τίς ἂν εἴη ἐναντία, οὐ γὰρ δὴ ἡ λέγουσα ὅτι κακὸν, ἅμα ποτὲ ἂν εἴη ἀληθής. οὐδέποτε ἀληθὴς ἀληθεῖ ἐναντία. ἔστι γάρ τι μὴ ἀγαθὸν, κακόν. ὥστε ἐνδέχεται ἅμα ἀληθεῖς εἶναι. οὐδ' αὖ ἡ ὅτι οὐ κακόν. ἀληθὴς γὰρ καὶ αὕτη. ἅμα γὰρ καὶ ταῦτα ἂν εἴη. λείπεται δὴ τῇ τοῦ μὴ ἀγαθοῦ, ὅτι οὐκ ἀγαθὸν ἐναντία ἡ τοῦ μὴ ἀγαθοῦ, ὅτι ἀγαθόν. ψευδὴς γὰρ αὕτη. ὥστε καὶ ἡ τοῦ ἀγαθοῦ ὅτι οὐκ ἀγαθὸν, εἴη ἂν ἐναντία τῇ τοῦ ἀγαθοῦ ὅτι ἀγαθόν. φανερὸν, ὅτι οὐδὲν διοίσει. οὐδὲ ἂν καθόλου τιθῶμεν τὴν κατάφασιν. ἡ γὰρ καθόλου ἀπόφασις, ἐναντία ἔσται. οἷον τῇ δόξῃ τῇ δοξαζούσῃ, ὅτι πᾶν ὃ ἂν ᾖ ἀγαθὸν, ἀγαθόν ἐστιν, ἡ ὅτι οὐδὲν τῶν ἀγαθῶν, ἀγαθόν ἐστιν. ἡ γὰρ τοῦ ἀγαθοῦ, ὅτι ἀγαθὸν, εἰ καθόλου τὸ ἀγαθὸν, ἡ αὐτή ἐστι τῇ, ὅτι ὃ ἂν ᾖ ἀγαθὸν δοξαζούσῃ, ὅτι ἀγαθόν. τοῦτο δὲ οὐδὲν διαφέρει τοῦ, ὅτι πᾶν ὃ ἂν ᾖ ἀγαθὸν, ἀγαθόν ἐστιν. ὁμοίως καὶ ἐπὶ τοῦ μὴ ἀγαθοῦ. ὥστε εἴπερ ἐπὶ δόξης οὕτως ἔχει, εἰσὶ δὲ αἱ ἐν τῇ φωνῇ καταφάσεις καὶ ἀποφάσεις, σύμβολα τῶν ἐν τῇ ψυχῇ, δῆλον ὅτι καὶ καταφάσει μὲν ἐναντία ἡ ἀπόφασις ἡ περὶ τοῦ αὐτοῦ καθόλου. οἷον τῇ, ὅτι πᾶν ἀγαθὸν, ἀγαθόν. ἢ ὅτι πᾶς ἄνθρωπος ἀγαθὸς, ἡ, ὅτι οὐδέν. ἢ οὐδείς. ἀντιφατικῶς δὲ, ἢ ὅτι οὐ πᾶν, ἢ οὐ πᾶς. φανερὸν δὲ καὶ ὅτι ἀληθῆ ἀληθεῖ, οὐκ ἐνδέχεται ἐναντίαν εἶναι, οὔτε δόξαν οὔτε ἀντίφασιν. ἐναντίαι μὲν γὰρ αἱ περὶ τὰ ἀντικείμενα. περὶ αὐτὰ δὲ ἐνδέχεται ἀληθεύειν τὸν αὐτόν, ἅμα δὲ οὐκ ἐνδέχεται τὰ ἐναντία ὑπάρχειν τῷ αὐτῷ.

Τέλος τοῦ περὶ ἑρμηνείας Ἀριστοτέλους.

ΑΡΙΣΤΟΤΕΛΟΥΣ

ΑΝΑΛΥΤΙΚΩΝ ΠΡΟΤΕΡΩΝ ΠΡΩΤΟΝ.

Περὶ τῶν τριῶν σχημάτων.

Πρῶτον εἰπεῖν περὶ τί καὶ τίνος ἡ σκέψις ἐστίν. ὅτι περὶ ἀπόδειξιν, καὶ ἐπιστήμης ἀποδεικτικῆς. εἶτα διορίσαι, τί ἐστι πρότασις, καὶ τί ὅρος, καὶ τί συλλογισμός, καὶ ποῖος τέλειος καὶ ποῖος ἀτελής, μετὰ δὲ ταῦτα, τί τὸ ἐν ὅλῳ εἶναι, ἢ μὴ εἶναι, τόδε τῷδε, καὶ τί λέγομεν τὸ κατὰ παντὸς ἢ μηδενὸς κατηγορεῖσθαι. Πρότασις μὲν οὖν ἐστὶ λόγος καταφατικὸς ἢ ἀποφατικός, τινὸς κατά τινος. οὗτος δὲ ἢ καθόλου, ἢ ἐν μέρει, ἢ ἀδιόριστος. λέγω δὲ καθόλου μὲν, τὸ, παντὶ ἢ μηδενὶ ὑπάρχειν. ἐν μέρει δὲ, τὸ, τινὶ ἢ μὴ τινὶ ἢ μὴ παντὶ ὑπάρχειν. ἀδιόριστον δὲ, τὸ ὑπάρχειν ἢ μὴ ὑπάρχειν, ἄνευ τοῦ καθόλου ἢ κατὰ μέρος. οἷον τὸ τῶν ἐναντίων, εἶναι τὴν αὐτὴν ἐπιστήμην. ἢ τὸ τὴν ἡδονὴν, μὴ εἶναι ἀγαθόν. διαφέρει δὲ ἡ ἀποδεικτικὴ πρότασις, τῆς διαλεκτικῆς, ὅτι ἡ μὲν ἀποδεικτικὴ, λῆψις θατέρου μορίου τῆς ἀντιφάσεώς ἐστιν. οὐ γὰρ ἐρωτᾷ, ἀλλὰ λαμβάνει ὁ ἀποδεικνύων, ἡ δὲ διαλεκτικὴ, ἐρώτησις τῆς ἀντιφάσεώς ἐστιν. οὐδὲν δὲ διοίσει πρὸς τὸ γενέσθαι τὸν ἑκατέρου συλλογισμόν. καὶ γὰρ ὁ ἀποδεικνύων καὶ ὁ ἐρωτῶν, συλλογίζεται, λαβών τι κα

Fig. 91. A page of second Greek edition of Aristotle's works prepared by Erasmus of Rotterdam and Grynaeus of Heidelberg and printed by Bebel in Basel, 1531; two volumes, folio, generally bound in one. The printing is far less beautiful than that of the *princeps*. For the sake of comparison we have chosen the same text, the beginning of *Prior analytics*, which is preceded by the end of *De interpretatione*. [From the copy in the Harvard College Library.]

LIBRORVM OMNIVM QVI HOC OPERE CON tinentur, & quos uidere nobis hactenus græce impressos con tigit, catalogus. Extant enim latine quidam, qui nusquam dum impressi fuerunt.

IN PRIMO TOMO SVNT

IN SECVNDO TOMO SVNT

Fig. 92. Another page of the second Greek edition (Basel, 1531). The main text is preceded by eight introductory leaves, including Erasmus' Latin dedication to John More dated Freiburg im Breisgau 1531, the short life of Aristotle by Guarino da Verona, and the table of contents (folio of last leaf), which we reproduce. Note that the text proper begins with the *Introduction* of Porphyry (III–2), very often added to the Organon. [From the copy in the Harvard College Library.]

(896 pp., quarto; Berlin, 1870). This is the last volume of the Bekker edition, vols. 1–4 of which appeared in 1831–1846. There is also an elaborate index by Emile Heitz in vol. 5 of the Didot edition (932 pp.; Paris, 1874), vols. 1–4 of which appeared in 1848–1869. There are indexes to each separate work in the Oxford English *Aristotle*.

Troy Wilson Organ, *Index to Aristotle in English translation* (183 pp.; Princeton: Princeton University Press, 1949) [*Isis 40*, 357 (1949)].

The Berlin Academy has published a large series of commentaries. *Commentaria in Aristotelem graeca* (23 vols., 1882–1909); *Supplementum Aristotelicum* (3 vols., 1885–1903).

For special investigations one must always refer to the latest critical editions of the text required. These editions are too numerous to be listed here. Most needs will be answered, however, by referring to the general editions listed above.

ALEXANDER THE GREAT (356–323) AND THE MACEDONIAN EMPIRE[36]

Alexander was born in Pella twenty-eight years later than Aristotle, in the summer of 356, being the son of Philip II by the Epirote princess, Olympias, a passionate and superstitious woman. We do not know how he was educated in his younger years, but when he was thirteen Aristotle was invited to be his tutor. The tutorship lasted only three years, because at the age of sixteen Alexander was forced to act as regent of Macedonia during his father's absence, and he took part very early in military affairs; at eighteen, he commanded his father's left wing at Chaironeia. The following year, his father having married Cleopatra, palace intrigues obliged Alexander and his mother to flee to Illyria. What would have happened to the young man if he had remained in exile? The wheel of fortune turned very fast for him. Another year passed; Philip was assassinated,[37] and Alexander became king of Macedonia at the age of twenty (336).

Let us return for a moment to Aristotle's tutorship. Though it did not last very long, it exerted a very deep influence upon his pupil. What did Aristotle teach him? Poetry, especially the *Iliad* (Alexander kept under his pillow a copy of the *Iliad* that his tutor had revised for him), the history of Greece and Persia, the geography of Asia Minor, ethics and politics. The content of Aristotle's lessons does not matter so much as the spirit that informed them. We may be sure that the teaching was sensible, practical, moderate, and yet high-minded and generous; Aristotle could easily have been the best of tutors, even as Plato would have been the worst. When Alexander had to assume administrative and military duties, the tutorship came naturally to an end, but Aristotle remained an honored friend and a trusted counsel.[38] Friendly relations continued between them at least until the murder of Callisthenes in 327.[39]

There are many proofs of Alexander's kindness to his former tutor. As soon as he was in power, he ordered the restoration of the latter's native place, Stageira, which had been destroyed by Philip; when he conquered Lesbos, he protected it

[36] The best account is William Woodthorpe Tarn, *Alexander the Great* (2 vols.; Cambridge: University Press, 1948), based upon a minute and wise study of all the sources.

[37] The murder was ascribed to Persian intrigue; it was also ascribed to Olympias' jealousy. Neither assumption can be proved; either or both may be true.

[38] The situation has often occurred, the tutor of a royal prince becoming in the course of the time the friend and counselor of a king. For example, Nicole Oresme (XIV-2), tutor of the dauphin Charles, became the adviser of Charles V (*Introduction*, vol. 3, p. 1486).

[39] Callisthenes of Olynthos (in the Chalcidic peninsula), who accompanied Alexander's expedition as historian, and extolled Alexander's Panhellenism. They fell out and Callisthenes was executed on a charge of treason.

from pillage for the sake of Aristotle's friend, Theophrastos; when he visited the tomb of Achilles in Troas, he had with him Aristotle's nephew, Callisthenes. He gave much assistance to the Lyceum, to Aristotle personally, and to his assistants in their scientific undertakings.

Though our readers have but little interest in military conquests, we must give a brief outline of Alexander's campaigns in order to illustrate his astounding genius.

His campaigns began in Greece, for he had to subdue the rebellions that had occurred in various places after his father's death. In order to illustrate his ruthlessness and to discourage further revolts, he destroyed Thebes, sparing only (and this was typical of him) the house of Pindar. In spite of Athens' submission and allegiance, Demosthenes, whose efforts were financed by Persia, created new trouble. Alexander forgave them; and the Hellenic League was reëstablished (still without Sparta) and Alexander elected its leader. He was now able to resume Philip's plan of Asiatic conquest; the champion of Hellenism could do no less, and it was clear enough that Greek unity would remain precarious as long as Persia was capable of stirring up bad blood between the Greek states and of abetting insurrections.

Alexander had a strong dramatic sense and knew how to excite the loyalty and wonder of his soldiers and to feed the superstitions that were favorable to him. Having levied a Macedonian army including contingents from all the Greek states (except Sparta), he began his conquests in the northwest corner of Asia Minor, camped in the plain of Troas, and worshiped in the temple of Athena, evoking the ancient glories that every Greek had learned from the *Iliad*. To his soldiers he thus appeared in the form of a new Achilles. His first great battle was won in 334 near the river Granicos (in Mysia); the Persian satraps were unable to resist the Macedonian phalanx and were utterly defeated. After that Alexander was free to proceed southward, liberating one Greek colony after another. His position was endangered, however, by the existence of a strong Persian navy that might cut his communications with Macedonia and Greece. Therefore he resolved to make himself master of all the harbors (in Asia Minor, Syria, or Egypt) without which the Persian navy could not operate. This was done with astonishing speed. Alexander led his armies across Asia Minor, then across the Cilician gates, and fought in 333 another great battle at Issos,[40] defeating the main Persian army led by the Great King himself, Darios III, the last of his line. Darios begged for peace, offering to abandon to the Greeks all the region west of the Euphrates, but by this time Alexander knew his power and could no longer restrain his ambition. Before completing his conquest of the Persian empire, he captured the Phoenician harbors and Egypt. The Persian fleet could not function any longer and was scattered or destroyed. Alexander then resumed his conquest of the East, crossed the Euphrates and the Tigris, and defeated Darios III again at Arbela (331). Darios was assassinated by one of his own men and Alexander showed great generosity to his family. There was nothing now to hinder him from taking the Persian cities, Babylon, Susa, Pasagarda (where he visited the tomb of Cyros), Persepolis (the marvelous palaces of which were set afire), Ecbatana. Alexander could not stop

[40] Issos in Cilicia at the end of the Gulf of Issos, which is the northeast corner of the Mediterranean. At Issos one is out of Asia Minor, and approaches Syria from its northern end.

himself any more; he obliged his armies to march through the Iranian Plateau, to cross the Oxos and Jaxartes rivers, then to turn southward toward India. He would have continued indefinitely but for the despair and anger of his soldiers. They sailed down the Indus on 800 ships, and when they reached the Indian Ocean, the Greeks were amazed by the spectacle of the tides, which was new to them. The return journey to Babylon was effected partly afoot along the Persian deserts, partly on a fleet of ships following the shores of the Indan Ocean and sailing up the Persian Gulf and the Shatt-al Arab. The survivors of that incredible journey reached Babylon in 323.

These prodigious conquests had gradually changed Alexander's character. His natural disposition was generous and he proved his magnanimity on many occasions. On the other hand, he could not help feeling exalted; if not a god, he felt himself to be more than a man, a superman, in the Greek meaning, a hero. During his stay in Egypt, he had spent three weeks of his time visiting the temple of Amon in the western desert, and there he had been proclaimed a son of Zeus-Amon. To the Egyptians he was a living god; to the Asiatics he was the successor of the Great King, an absolute ruler, whom nobody could gainsay; to the Greeks, he was the head and protector of the Hellenic League, a conquering hero, a dictator. Like every other dictator of any time, he was the main victim of his irresponsible and unrestrainable power. Those who dared oppose him, either in the government, in a debate, or simply in an orgy, must die, and he was the direct or indirect cause of many murders: the execution of Philotas, son of his best general Parmenion, in 330, the murder of Parmenion himself, the killing by his own hand of his best friend, Cleitos, who had saved his life at the Granicos, the execution of Callisthenes in 327, and others. This was the price that Alexander paid for his glory, the infamous deeds that no amount of victory and majesty could ever compensate.

There was but one friend left to him, the Macedonian Hephaistion, son of Amyntor, but Hephaistion died of fever in 324 and the king mourned him extravagantly. While he was making new plans for the conquest of Arabia and perhaps of the Western Mediterranean (for the making of such plans was part of his nemesis), he fell sick with fever and died at Babylon in June 323, at the age of thirty-three. His phenomenal career had lasted thirteen years, during which he had conquered a large part of the world and in spite of his generosity caused the death and misery of innumerable people.

Thus lived and died Alexander the Great, a man whose deeds could never be forgotten, nor forgiven.

Alexander was fortunate in his death, more fortunate than other conquerors, because he did not witness the disintegration of his empire. Great as his achievements were, he had done only the beginning of his task, the easiest part. An immense amount of work remained to be accomplished in order to consolidate his victories, to organize the empire, and to stop the innumerable causes of conflict and of weakness. It had been possible to steal the world from feeble hands; it would have been impossible, even with the strongest hands, to keep it sound and whole. The gods were more generous to Alexander than he deserved and permitted him to die while he was at the acme of his glory. He was like a gambler who has collected all the chips on his side of the table and expires suddenly before losing them.

The Alexandrian empire did not survive him. For the next fifty years his main officers fought one another for as much power as they could get. By 275, three new dynasties had emerged: the Antigonids, holding Macedonia and Greece; the Seleucids, holding Western Asia; the Ptolemies, holding Southern Syria, Egypt, Cyrenaica, Cypros. Greece disintegrated into its old elements, some of which would sometimes ally themselves against the others. Not only did the empire vanish but Greece and Macedonia were gradually absorbed into the new Roman world. By 200, Greek and Macedonian freedom were approaching the end. Macedonia had enjoyed a long existence before Alexander; it did not live two centuries after him; it collapsed in 167 and became a Roman province in 146.[41] Alexander did not found a durable empire, but he helped to destroy his own country, his own patrimony.

Did Alexander mistake himself for a god? If he had a grain of intelligence, how could he? Do the gods suffer pains and illusions? Did he dream of a world empire? Probably not in a conscious way, but his nemesis obliged him to conquer more and more. Such as it was, his empire was unwieldy, heterogeneous, weakened by all kinds of external and internal tensions. The only way of relieving such tensions is domestic or foreign war; thus, the expansion continued while the domestic repressions were postponed. If Alexander had lived longer, his remaining years would have been wasted in continuous and fruitless strife.

It is possible that other men fancied him to be a god, for his power was incalculable. The Egyptians accepted his divinity and maybe some Asiatics did; the Greeks took it with a grain of salt. The superstitious reverence paid to dictators in our own enlightened time helps us to understand the situation that shaped itself twenty-four centuries ago.

Alexander was very generous but impulsive. In one essential respect he was more generous than Aristotle, not to mention Plato. Both philosophers had considered the barbarians, that is, the non-Greeks, as inferiors by nature. It was meet to wage war against them, to extirpate or enslave them. The Greeks were born freemen, and the barbarians were born slaves. It is much to Alexander's credit that he was able to raise himself to a higher plane than his tutor.[42]

Alexander perceived the unity of mankind, which they did not. The explication of his moral superiority in this respect over theirs lies in his greater experience of men. He had known the seamy side of Greek and Macedonian life from his very childhood. As the intelligent boy grew up, the corruption of his father's court cannot have been completely hidden from him; his mother, Olympias, may have opened his eyes if his father did not. On the other hand, he must have known many Orientals who were good men. He must have discovered early that the opposition of Greek to barbarian was false and wrong. In the course of his short but crowded life his experience of men must have increased by leaps and bounds; as he himself was idolized he was exalted to so high a plane that all men appeared equal in their inequality to him. He was so high above them that it was easy

[41] Perseus (who ruled from 179 to 168) was the forty-third and last king of Macedonia. He was not an undistinguished king, but the situation that he was facing was hopeless. The Macedonian kingdom had lasted 532 years.

[42] I borrow this remark from Tarn, who amplifies it (vol. 1, p. 9).

for him to be tolerant of their diversities and to recognize their fundamental brotherhood.

It is not probable that Alexander dreamed of a world empire but it is almost certain that he dreamed of world *homonoia* (*concordia*). He realized that men should not be classified blindly according to their races, but intelligently and kindly according to their merits. One may object that other conquerors may have had the same idea: the only defense of their deeds would be to claim that they had come not to conquer, but to unite, not to enslave but to liberate.[43] True, but Alexander was the first, and his merit is greater, because it would have been natural for him to continue the evil tendencies of Plato and Aristotle. The fact that he was able to overcome unaided those tendencies is the best proof of his genius.

His ideas on the necessity of the fusion of races for the common tasks of humanity may have been facilitated by his own ancestry; he was not a pure Greek like Plato, but half a barbarian.[44] At any rate, he did his best to realize his new political ideal by appointing Orientals to satrapies and to other high offices, mixing soldiers of various races in his armies, mixing populations in the new cities, marrying the Bactrian princess Roxane and encouraging mixed marriages around him. It is probable that all these measures were very insufficient, yet they prove his good will, and the beginning of a new policy, radically different from the earlier one. As Tarn put it: "Aristotle's State had still cared nothing for humanity outside its own borders; the stranger must still be a serf or an enemy. Alexander changed all that. When he declared that all men were alike sons of one Father, and when at Opis he prayed that Macedonians and Persians might be partners in the commonwealth and that the peoples of his world might live in harmony and in unity of heart and mind, he proclaimed for the first time the unity and brotherhood of mankind." [45]

That idea – the brotherhood of mankind – has often been ascribed to the Cynics, to the Stoics, to the Christians, but Alexander was ahead of them all.[46] Zeno the Stoic, we should remember, was born about the time when Alexander began his conquests, and was only twelve years old when the conqueror died.

Diogenes of Sinope (*c.* 400–325), who is often considered the founder of the Cynic school, was older than Alexander and, if a famous anecdote is true, he met the latter at the general assembly of the Greeks held at the Isthmus of Corinth. Alexander having been appointed leader of the expedition against Persia, many people came to congratulate him. Not so, however, Diogenes, who was living in Corinth but took not the slightest notice of the king. "Alexander went in person to see him; and he found him lying in the sun. Diogenes raised himself up a little when he saw so many persons coming towards him, and fixed his eyes upon Alexander. And when that monarch addressed him with greetings, and asked if

[43] Napoleon took that attitude, but not Hitler, whose purpose was definitely to enslave the non-Germans or to extirpate them.

[44] What about Aristotle? How Greek was he? And how much of a barbarian? It is impossible to know.

[45] Opis on the Tigris. When Alexander reached that place, his army was rebellious; he made a speech to explain his policy and succeeded in restoring their confidence in him; Tarn, *Alexander the Great*, vol. 1, p. 115. The capital of the Seleucid empire, Seleuceia, was built *c.* 312 near Opis, and became a great commercial center, being connected with the Euphrates by a canal, and an outpost of Greek culture in the East.

[46] It has been suggested by Jaeger, *Aristotle*, p. 24 that Aristotle's early dialogue on Alexander or Colonization may have dealt with Alexander's racial policy and condemned it.

he wanted anything, 'Yes,' said Diogenes, 'stand a little out of my sun.' It is said that Alexander was so struck by this, and admired so much the haughtiness and grandeur of the man who had nothing but scorn for him, that he said to his followers, who were laughing and jesting about the philosopher as they went away, 'But verily, if I were not Alexander, I would be Diogenes.' "[47]

Diogenes may have inspired Alexander, but the cosmopolitanism of the Cynics (if it ever existed) was a much later creation.[48]

Thanks to his genius and to Aristotle's tuition, Alexander was not a vulgar conqueror; he might have become a greater man than he was if unfortunate circumstances had not forced him to conquer the world. He was interested in Aristotle's undertakings and ready to support the Lyceum and to obtain for it all the specimens that were needed.[49] His expeditions into Asia might be called the first scientific expeditions. Not only did he take with him engineers able to construct military machines or to deal with water works or mining, architects, geographers, and surveyers, but there were a secretarial or historical department headed by Eumenes of Cardia, philosophers and literary men such as Callisthenes of Olynthos, Anaxarchos the Democritean and his pupil Pyrrhon who founded the Skeptic school, Onesicritos, seaman and romancer, naturalists collecting specimens for the Lyceum, and the future king Ptolemy, son of Lagos (Ptolemy I Soter, *c.* 367–282, king of Egypt), to whom we owe the most reliable information concerning Alexander's career. In all this Alexander showed the same kind of intellectual ambition that would redeem Bonaparte's fame twenty-one centuries later.

Alexander's dream of a world united under Greek hegemony was too premature to be realized, but he accomplished a certain community of culture that superficial though it was, was never obliterated. That is what is often called the Hellenization of the East. Thanks to his efforts, Hellenic ideals were spread over Western Asia, and reached India and even China. The most striking illustration of that Hellenization is given to us by the beginnings of Buddhist iconography as it developed under Greek influence in the Gandhāra.[50] It was chiefly in Western Asia, however, that the Hellenization was felt (it had been felt before Alexander and continued after him) and because of it that part of the world was knit more closely to Europe than to the rest of Asia. The Hellenization of the East cannot be denied, but one should not forget that it was accompanied by another movement in the opposite direction, which we may call the Orientalization of the West.[51] It was because of the example given by Alexander in Babylon and by his successors in Egypt and Asia that new ideas of sovereignty, politics, and gov-

[47] Plutarch, *Lives, Alexander,* 14; translation by Bernadotte Perrin, Loeb Classical Library, vol. 7, p. 259.

[48] According to Tarn, vol. 2, p. 409, "there was no such thing."

[49] According to Pliny (I-2), whose account of Alexander's help (*Natural history,* VIII, 17) seems exaggerated; we shall quote from it in another chapter. Athenaios of Naucratis (III-1) writes that "the Stagirite received 800 talents from Alexander to further his research on animals" (*Deipnosophistai,* IX, 398E).

[50] A. Foucher, *L'art gréco-bouddhique du Gandhāra* (2 vols.; Paris, 1905–1918); *The beginnings of Buddhist art* (Paris, 1917). J. P Vogel, *Buddhist art* (Oxford: Clarendon Press, 1936).

[51] To the Westernization of Oriental art that occurred in Gandhāra corresponded many centuries later in the Near East the Orientalization of Western art of which the late Josef Strzygowski (1862–1941) has given many examples. The parallelism is curious: early Buddhist art influenced by Western artists, early Christian art influenced by Eastern ones.

ernment were introduced into the West. The Hellenization of the East had begun long before Alexander, and it was continued throughout the Hellenistic and the Roman ages, and even to some extent by the Byzantine autocrats; in the same way, the Orientalization of the West was not by any means a novelty in Alexander's age, but both movements reached a climax in that age.

When that is said it must be emphasized again that both the Hellenization and the Orientalization were exceedingly superficial. That is the way most cultural influences spread, like oil upon the face of the waters. The waters are not changed. There were Greek manners in the East, but Greek ideals could not be understood, and hence could not serve as a bond of union. It is for that reason above all others that the Macedonian empire was essentially unstable: there was no cement to hold it together, nothing whatever except Alexander's personal might.

The Greek culture that developed in Western Asia was definitely post-Alexandrian; it developed under the Roman patronage and was given a certain stability because of the continuity and the relative length of the Pax Romana. We may assume that in many cases seeds of the Alexandrian age did not fructify until Roman peace gave them a chance. The best example of this is the astrologic religion and all that belongs to it (such as the seven-day cycle) which dates back to Plato and Philip of Opus, yet did not really flourish except in Roman times.

The Alexandrian influence was felt in another way in the form of legends. Such influence should not be despised; the legends were crude travesties of reality but they were accepted by the majority of the people as truthful. The world knew Alexander through those legends, even as it knew Helen and Achilles through the *Iliad*. For the great mass of the people, East and West, the legendary Alexander was the real one. The Alexander romance spread everywhere; more than eighty versions of it have been collected in twenty-four languages. When the Muslims conquered the world a millennium after him, they helped to advertise the story of the great hero, Alexander Dhūl-qarnain (the two-horned king), and the Arabic story was retranslated into other languages.[52]

Some of the earliest legends, spread by Peripatetics, who could not forgive the murder of Callisthenes, were very unfavorable to Alexander. He was represented as a very good pupil of Aristotle, who had been ruined by his extraordinary fortune and had degenerated into a cruel tyrant. The legends of a later time abandoned political implications and made of Alexander a supernatural hero and a wizard to whom all the conceivable *mirabilia* could be ascribed. All of which is popular literature and folklore without scientific value of any kind, yet full of humanity.

Among the creatures immortalized by the history of Alexander and the Alexandrian folklore, let me select for mention a single one, Bucephalos, the hero's favorite horse, which was killed at the battle of Hydaspes in 326.[53] Bucephalos is the most illustrious representative of his species.

[52] Iskandar-nāma, *Encyclopedia of Islam*, vol. 2 (1921), p. 535. For the early legend of the pseudo-Callisthenes, see Tarn, *Alexander the Great*, vol. 2, by index.

[53] The river Hydaspes (or Jhelum), one of the tributaries of the Indus in the Punjab. Alexander founded the city of Bucephala in memory of his horse, near the spot where it died. According to Plutarchos (Life of Alexander 32), Alexander, "as long as he was riding about and marshalling some part of his phalanx, or exhorting or instructing or reviewing his men, spared Bucephalos, who was now past his prime, and used another horse; but whenever he was going into action, Bucephalos would be led up, and he would mount him and at once begin the attack."

Alexander the Great will continue to ride his faithful steed Bucephalos as long as men exist.

THE LYCEUM, 335. ITS FOUNDATION AND EARLY HISTORY

Though Aristotle must have ceased his tutorship of Alexander when the latter assumed administrative and military duties, he remained at Pella (or perhaps in Stageira) a few more years. In 336, Alexander succeeded Philip as king and soon afterward began his campaignings first in Thrace and Illyria, then in Greece. By 335, he was the master of Greece and was preparing his conquest of Asia, to which the rest of his short life would be devoted. By 335, Macedonia was on a war footing, not very comfortable for a scholar, and Aristotle returned to Athens. What was his situation there? He had spent twenty years of his youth (aet. 18–38) as a student, associate, and friend of the Academy; now, after an interval of twelve years, he was coming back in the train of the Macedonian army. He could not be welcome to all the Athenians, but only to the collaborators.

At any rate, he could not go back to his old school, and he founded a new one in another part of the city. The Academy was northwest of the walls, outside of the Dipylon gate; the Lyceum was east of the walls, near the road to Marathon.[54] From the gardens of the Lyceum one could see Mount Lycabettos northward and the river Ilissos in the south. It was a grove sacred to Apollon Lyceios (the wolf god) and its name, Lyceum, was derived from that dedication. In the warm climate of Athens much of the teaching took place in the open, under the trees or under a portico. The teacher and students might be sitting for a while, then walking up and down, hence their nickname, the Peripatetics.

There are great differences between Plato's foundation and Aristotle's. Half of Plato's life was spent as director and oracle of the Academy; Aristotle founded the Lyceum in the opposite direction from Athens fifty-two years later, and he remained its head for thirteen years only (not forty, like Plato). Plato's foundation was a great innovation and his teaching experience was relatively small; when Aristotle started the Lyceum, he was fifty, and he had obtained considerable experience of men and students in Assos and Pella. Plato had always dreamed of the intimate association between a great king and a leading philosopher, but his dreams had not been realized. On the contrary, Aristotle was supported by Alexander, the most powerful king of antiquity; Alexander gave him money (perhaps this came under the head of Macedonian propaganda) and, what is almost equally important, he provided for the museum that was a part of the new school specimens of natural objects of many kinds. If there were anything needed to make the teaching more concrete and more effective, Aristotle could always obtain it from his patron.

This fact underlines the fundamental difference between the Lyceum and the Academy. It is not so much that Aristotle could obtain exhibits if he needed them; it is that he did need them, while Plato would have spurned them. Plato was satisfied with eternal and immortal Ideas; Aristotle wanted tangible objects. We know little about details of teaching. Aulus Gellius (II–2) tells us that Aristotle offered two kinds of lessons, in the morning for the initiated (*esōterica*,

[54] The road to Kephissia and to Marathon. The present Byzantine Museum is close to the place where the Lyceum was.

acroamatica) and in the evening for a larger public (*exōterica*). He is a late witness, but what he tells is plausible enough. There are open and closed courses in almost every school, for both kinds answer natural purposes.

Both schools were philosophical, but the Academy tended towards metaphysics or transcendentalism, even when it dealt with practical subjects such as education or politics. The Lyceum was philosophical in another sense which will be defined presently; Aristotle was interested in logic and science; under his direction it became an organization for personal and even for collective research. Our "Academies of Science" are misnamed; the name "Lyceum" would fit them better. Languages are very capricious, however, and no one could safely foretell how given words, whether domestic or foreign, will be eventually accepted. The word "lyceum" has become as popular as "academy" in almost every Western language: in France it is used to designate all the state high schools (secondary schools); in the United States it enjoyed some popularity to designate free associations for lectures, debates, concerts, and "improving" entertainments of sundry kinds.

In spite of the fact that the Academy and the Lyceum were as different as the spirits of their founders, one should not exaggerate their differences, or rather one should not forget that there were between them similarities as well. Both were institutions of higher and disinterested learning; the master of the second was an illustrious alumnus of the former. We may imagine that students passed from one to the other, or, if they were sufficiently wide awake, attended lectures in both. The history of the two institutions reveals many examples of interplay. There was no reason for not discussing the writings of Plato in the Lyceum or those of Aristotle in the Academy. Many commentators of a later time commented on Plato as well as on Aristotle.

These two men represented antipodal types, however, the alternance of which seems to exhaust possibilities, so much so that it has been claimed that "every thinking man is either a Platonist or an Aristotelian." That claim cannot be completely substantiated, but it is remarkable that it could be made at all.

We shall now tell the early history of the Lyceum, even as we told that of the Academy, and for the same reason. One cannot know a living thing except while it is alive and changing; one cannot know what the Lyceum was without contemplating its evolution. This is somewhat paradoxical, for Aristotle could not foresee the vicissitudes of the Lyceum any more than a father can foresee those of his children, let alone his more distant progeny.

Aristotle's headmastership of the Lyceum lasted only thirteen years. Toward the end of his life two men were considered fit to succeed him, Eudemos of Rhodes and Theophrastos of Eresos. Aulus Gellius tells us[55] that Aristotle expressed his preference for the latter by comparing the wine of Rhodes with that of Lesbos. "Both are good but the Lesbian wine is sweeter" (*hēdiōn ho Lesbios*). Theophrastos succeeded him and may be called the second founder of the Lyceum, for he was the leader for thirty-eight years (323–286), and he completed its organization. He bequeathed some of his property to the trustees of the Lyceum with definite indications as to its use; his library, however, he gave to Neleus. Theophrastos was

[55] Aulus Gellius, XIII, 5.

succeeded by Straton of Lampsacos (III–1 B.C.), who was the head for nineteen years (286–268). This completed the golden age of the Lyceum. The fourth *scholarchēs*, Lycon of Troas, ruled for forty-four years (268–225), but this was a period of relative decadence. He had no interest in science and restricted himself to ethics and rhetoric. Curious information on the first four masters of the Lyceum is provided by Diogenes Laërtios, who gives us[56] the full text of their wills; he must have obtained those four remarkable documents from a single source. After Lycon's time the history of that famous school is full of holes, but a few names emerge, chiefly that of Andronicos of Rhodes (I-1 B.C.), who flourished in Athens *c.* 80 B.C. and was the tenth successor of Aristotle.

A complete history of the Lyceum should not be restricted to the activities of the headmasters. One should speak of some of their collaborators, and not forget the interplay and occasional collaboration with Academicians. During Aristotle's headmastership, the head of the Academy, the third head, was his friend Xenocrates of Chalcedon, and among his pupils were Theophrastos and Eudemos, Aristoxenos of Tarentum, Dicaiarchos of Messina, Clearchos of Soloi. Among Theophrastos' pupils was Demetrios of Phaleron, who was to be the founder of the Alexandrian library.

After Andronicos' time, the Peripatetic school loses its identity, and its members are no longer pure Peripatetics but Stoics, Academicians, Neoplatonists. Great leaders of thought like Panaitios of Rhodes (II–2 B.C.), Posidonios of Apamea (I–1 B.C.), Ptolemy (II–1), Galen (II–2) are Peripatetics only in a measure; they have studied some of Aristotle's books, and they continue some of his tendencies.

Beginning with the third century, one speaks no longer of headmasters but of commentators, one of the first and greatest being Alexander of Aphrodisias (III–1), the commentator par excellence (*ho exēgētēs*), who was actually the head of the Lyceum from 198 to 211. By this time it had already become necessary to free Aristotelian thought from Platonic or Neoplatonic interpretations. The Lyceum becomes relatively insignificant; the main philosophic school in Athens during the first five centuries of our era (or rather until 529) is the Academy. It alone continues to exist as an administrative entity; it loses its philosophical entity; its main tendency is Neoplatonic but that is combined with much else. The Lyceum has gone and the Academy has become a school of pagan philosophy.

EARLY COMMENTATORS

A history of Aristotelianism would be a cross section not only through the history of philosophy but also through the history of science, at least as far as the eighteenth century. We cannot go into that without inserting too long a digression. Incidentally, we might remark that what makes the history of science so difficult is that the significance of each stage cannot be appreciated except in the light of everything that happened before and everything that happened later, and that is a pretty big order. The whole ancient and medieval tradition of Aristotle is explained implicitly in my *Introduction*. We must satisfy ourselves at present with a very brief sketch. The influence of Aristotle was continued not

[56] Diogenes Laërtios, v.

only by translators and commentators but also by philosophers, theologians, men of science who could not help meeting Aristotle at every step and were obliged to bow to his superiority or to fight him.

We have already referred to Alexander of Aphrodisias, the Commentator, but he was not by any means the first one. The first editor, Andronicos of Rhodes (I-1 B.C.), had been naturally the initiator; he was followed in the second half of the same century by Boethos of Sidon, Ariston of Alexandria, Xenarchos of Seleucia (Cilicia), Nicolaos of Damascos (I-2 B.C.). In the first century after Christ there was Alexandros of Aigai, tutor of Nero (emperor 54–68). In the second century, the number of commentators is singularly large: Ptolemaios Chennos of Alexandria (flourished under Trajan and Hadrian, emperors, 98–138), the author of the *De mundo*,[57] Aspasios, Adrastos of Aphrodisias (II-1), Ptolemy (II-1), Galen (II-2), Aristocles of Messina in Sicily, Herminos. The last named was the teacher of the illustrious Alexander of Aphrodisias (III-1), whose very elaborate commentaries have come to us in the original Greek or in Arabic translation.

With Alexander of Aphrodisias begins a new era in Aristotelian scholarship which is represented by the names of the Syrian Porphyrios (III-2), Anatolios of Alexandria (III-2), Themistios of Paphlagonia (IV-2), Syrianos of Alexandria (V-1), head of the Academy, and in the sixth century by Damascios of Damascus (VI-1), the Arab Doros, Ammonios son of Hermias (VI-1) and his pupil Asclepios of Tralles (VI-1), Simplicios of Cilicia (VI-1), who flourished in Athens and Persia, and the greatest of all, John Philoponos of Alexandria (VI-1). To the same century belonged also the earliest Latin translator and commentator, Boetius of Rome (VI-1).[58] The history of the Greek tradition includes quite a few distinguished names of later times, such as Stephanos of Alexandria (VII-1), who flourished in Constantinople, Eustratios of Nicaea (*c.* 1050–1120), Michael of Ephesos, pupil of Michael Psellos (XI-2) Sophonias (XIII-2).

In the meanwhile, the Aristotelian tradition was continued by means of the Arabic detour, and the Arabic leaders were the Arab al-Kindī (IX-1), the Persians or Turks al-Fārābī (X-1) and Ibn Sīnā (XI-1), and above all Ibn Rushd of Cordova (XII-2), known to the Latin world as Averroës. The Aristotle reëxplained by Averroës influenced St. Thomas Aquinas (XIII-2) and the other Catholic schoolmen, and their Christian interpretation dominated medieval thought. We need not continue that story, which from this point on is sufficiently well known.

The main thing to remember is that the Aristotelian thought had been commented upon by a large number of scholars, first in Greek, then in Arabic, finally in Latin and the Western vernaculars. It was explained by pagan interpreters, then by Muslim, Jewish, and Christian ones. The Christian Aristotle was the great master, "il Maestro di color che sanno,"[59] the "Magister dixit" whose authority became so overwhelming and paralyzing that it impeded further progress. The history of modern science begins with the rebellion against Aristotle.

[57] Wilhelm Capelle would place the *De mundo* in the first half of the second century; *Neue Jahrbücher 15*, 529–568 (1905).

[58] For the Latin tradition see Amable Jourdain, *Recherches critiques sur l'âge et l'origine des traductions latines d'Aristote* (Paris, 1819; ed. 2, 1843). Alexandre Birkenmajer, "Classement des ouvrages attribués à Aristote par le Moyen âge latin" (Prolegomena in Aristotelem latinum consilio et impensis Academiae Polonae litterarum et scientiarum edita, 1, 21 pp.; Cracovie, 1932).

[59] Dante, *Inferno*, IV, 131.

SOME ASPECTS OF ARISTOTLE'S PHILOSOPHY

The study of so long a tradition offers many difficulties, the greatest of which are caused by the fact that the subject changes with the passing of time. The Aristotle that Cicero knew, was not the same as the one explained by Alexander of Aphrodisias; al-Kindī in the ninth century and Ibn Rushd in the twelfth had not read the same Aristotelian books, or they had read them in different moods; the Aristotle praised by St. Thomas in the thirteenth century is not the same as the one blamed by Ramus in the sixteenth century or by Gassendi in the seventeenth. There were times when the passions for or against Aristotle ran so high that objective appraisals became almost impossible. Now that these passions are dead and cannot be revived, even in the institutes devoted to scholastic philosophy, we are able to rediscover the real Aristotle, who never was as omniscient and all-wise as some people believed, nor as dogmatic and preposterous as his enemies accused him of being.

Aristotle's scientific views and achievements will be discussed in the following pages, but we must try to show him at once in his wholeness. Perhaps the simplest way of doing that is to compare him with his old master, Plato. The latter's scientific training had been restricted to mathematics and astronomy; Aristotle's was primarily medical. His father, Nicomachos, was an Asclepiad, and the Asclepiadean tradition was handed down directly from father to son. The young Aristotle may have visited patients with his father or assisted him in his treatment of them in the surgery; at any rate, he could not help, wide-awake as he was, learning much from his father's lips and above all absorbing the empirical point of view. A mathematician, especially one like Plato, inspired by Pythagorean arithmology, would satisfy himself with a priori conceptions of the universe; a physician soon realized that one should assume and foretell as little as possible, but observe, take notes, induce and deduce prudently. Plato was imaginative and tender-minded; [60] Aristotle, experimental and tough-minded; yet, one should remember that Aristotle had begun his intellectual life as a Platonist and never shook off completely some of his Platonic fantasies. To my mind, this illustrates his greatness; he never was as dogmatic as his master had been and was so keenly aware of the mysteries of life that he remained somewhat Platonic in his growing resistance to Plato.

Aristotle had some experience of the occult practices of Greek religion and like Plato he compared intuitive knowledge with the initiation into the mysteries, yet he evaded mystical exaggerations. He appreciated the value of enthusiasm and of mystical and healing cults, but tried to build up a rational and transmissible system of thought. He fully realized the existence of two kinds of knowledge (intuitive and discursive) and of two modes of psychologic life (intellectual and emotive), but emotive life, important as it is, should be regulated by self-restraint instead of being exacerbated by Corybantic rituals. One of his disciples, Clearchos of Soloi, tells us that Aristotle, having attended a seance of hypnotism, was convinced that the soul could be separated from the body; [61] this illustrated his

[60] Under the influence of the Platonic mirage Plato's tenderness is generally exaggerated. Some of the extravagances of the *Republic* and of the *Laws* show that he could be exceedingly cruel. His tenderness was of the dubious kind that dictators advertise.

[61] See the excellent study by Jeanne Croissant, *Aristotle et les mystères* (228 pp.; Liége: Université de Liége, 1932) [*Isis 34*, 239 (1942–43)].

open-mindedness; yet he was always anxious to explain things in a scientific way. His residual mysticism was very much like that of the great scientists of every age who are humble and prudent and never unmindful of the infinite complexity of the universe.[62]

One of his fundamental ideas, expressed by the word teleology, may be called mystical, for its validity cannot be completely proved. The idea is very typical of the Plato-Aristotle relation, for it was derived from the Platonic conceptions of Idea or Form as prior to the existing object, as their metaphysical womb, as it were; for Aristotle the Idea is an unattainable goal. Plato tended to assimilate change with corruption; Aristotle, on the contrary, conceives change as a motion toward an ideal. Plato rejected the possibility of progress, while Aristotle accepted it. Things change because of the potentialities inherent in them; they change in order to attain or to approach their perfection. The Idea or Form is *in* the thing (like the adult in the embryo), not outside. The destiny of a thing is foretold by its hidden unrealized essence. Evolution proceeds as it does, not because of material causes producing natural consequences, pushing them on by a *vis a tergo*, but by final causes pulling them ahead by a *vis a fronte*. All the things that exist are directed toward an end (which is potentially inside of them); their development is shaped by a purpose. The world is gradually realized because of a transcendental Design, or call it Divine Providence.

Aristotle realized that mechanism and purpose are complementary and inseparable aspects; in the study of nature one must seek for a mechanical explanation or for the leading reason; sometimes the mechanism is clearer, sometimes the reason. In his time practically no mechanism (for example, a physiologic mechanism) was conceivable; hence, there remained only the teleologic explanation.

To a hard-boiled man of science of today such an explanation is mere verbiage. It is futile to ask the "why" of things, he would say; it suffices to answer as carefully as possible the question "how?" Aristotle was trying prematurely to answer the question "why?" and was giving that question first place. Was he all wrong? The question might be premature, but it was not futile; it had in a first approximation a guiding value. To his credit we should bear in mind also: (1) that his conception of terminal ideas (teleology) was an enormous improvement upon Plato's conception of germinal ideas: (2) that the teleologic explanations, even if insufficient, are yet very useful; every man of science uses them wittingly or unwittingly; the purpose of an organ helps us to understand and to remember its anatomy and physiology; (3) that the vitalists use teleologic language, and there are still many of them among us; it is impossible to suppress the vitalist point of view; it dodges every blow and reappears under a new form; (4) finally, if one accepts Divine Providence one cannot reject teleology.

The teleologic appearances of nature are obvious enough; do they correspond to an inner reality or are they simply illusions? The question can be put this way: Is the argument of design valid or is it a paradox? Aristotle was the first to use that argument and to attach considerable importance to it. Who will be the last?[63] Aristotelian teleology is one of the proofs of his genius.

[62] Compare Einstein's saying quoted in my *Introduction*, vol. 3, p. v.

[63] Excellent discussion of teleology from the point of view of modern chemistry and physiology

The teleologic point of view implied the concept of evolution, evolution toward an ideal, progress. To understand things we must penetrate their purpose, their genesis and growth. Aristotle applied these ideas to natural history, rather than to human history; otherwise he would have been one of the ancestors of the historians of science.

Aristotle was primarily an encyclopedist and with the partial exception of Democritos he was the first one. Earlier philosophers had tried to explain the universe, but Aristotle, who shared their ambition, was the first to realize that such an explanation should be preceded by as complete an inventory and description of it as possible. He did not simply understand that need but, what is more remarkable, he satisfied it. The totality of his works represents an encyclopedia of the available knowledge, much of which was obtained by himself or because of his leadership. It is easy to find holes or errors in that encyclopedia, but the amazing thing is that it was as good, as comprehensive and durable, as it was.

The encyclopedic purpose implies the belief that there is some unity and order in the universe and the conviction that the same unity and order should be transparent in our knowledge of it. The unity is proved by one's study of first principles (philosophy, theology), the order by proper classification and description.

As to first principles, there is a soul in every living thing and there is in each soul something divine, something connected with pure reason. God exists, for it is the necessary principle and end of everything, the first motor. All motion and all life symbolize an immense and universal impulse to perfection, to God; that impulse is obscure in the lower forms of existence, but it becomes clearer and clearer in men according to their degree of intelligence. Much of this could and did eventually lead to scholasticism and to mysticism, but in Aristotle's mind these sublime thoughts were restrained by his matter-of-factness and moderation. Aristotle's classification made a first distinction between the various branches of science, theoretical, productive, and practical. The theoretical ones have no aim but the apprehension and contemplation of truth; they are mathematics, physics, metaphysics (first philosophy, theology). The productive branches concern the arts. Practical philosophy seeks to regulate human actions; its two chief branches are ethics and politics. In spite of its insufficiency, Aristotle's classification exerted a very strong influence upon the whole development of philosophy and science down to our own day.[64]

His encyclopedic ambition was a very elementary one as compared with ours. He could not help believing that it could be achieved by an accumulation of definitions (that is why I used the word "inventory" above), and his definitions were verbal, not truly explicative. To the modern mind this is very insufficient indeed,

---

by Lawrence J. Henderson, *The order of nature* (240 pp.; Cambridge: Harvard University Press, 1917) [*Isis* 3, 152 (1920–21)]. "As the German physiologist, Ernst Wilhelm von Bruecke, remarked, 'Teleology is a lady without whom no biologist can live. Yet he is ashamed to show himself with her in public.'" Walter Bradford Cannon, *The way of an investigator* (New York: Norton, 1945), p. 108 [*Isis* 36, 259 (1946)].

[64] For further study of the classification of science and bibliography see *Introduction*, vol. 3, pp. 76–77.

but one had to begin with such inventories and fill them in gradually with more and deeper meaning.[65]

Scientific knowledge of a thing is possible when we know its causes, and the main cause is the essence.[66] We should know the varieties of each kind of thing, and this means enumeration and description. Ideas of increasing generality are not established a priori but are derived from the observation of increasing numbers of things. Aristotle, his colleagues, and his disciples accumulated a large number of observations; they provided good analyses and descriptions, and intelligent interpretations. Their terminology was often artificial, but much of it was apposite and has survived in modern languages. Unfortunately, the search for the essence of things opened the door to metaphysics; explanations were often wordy, and the enumerations incomplete. Aristotle did not realize their incompleteness and often concluded an enumeration with the words *cai para tauta uden* (and beyond that nothing else); he believed himself nearer to the goal than he was, or than he could possibly be. That was natural enough. His school had done so much that its illusions were pardonable; illusions of complete knowledge are far less pardonable today.

That philosophy was satisfying, because it is full of common sense and is moderate. Aristotle's love of order, of clearness and tidiness, of the *via media*, appealed to the Greek mind. After the days of paganism, when religious fervor increased, all that was necessary to preserve the popularity of his philosophy was to harmonize it with the dogmatic theology of other nations, and this was done by various doctors, for example, Ibn Rushd for the Muslims, Maimonides for the Jews, St. Thomas Aquinas for the Christians.

It has sometimes been said that Aristotelianism as compared with mystical deviations lacked humanity, tenderness, and even ideals. This was very misleading. Its main ideal was the scientific ideal, the discovery of the truth, an ideal that is always far ahead of men, yet a guide in the darkness. Aristotle's conception of science was very insufficient as compared with ours, but that was unavoidable. Because of his willingness to compromise he has been accused of mediocrity; that is another way of saying that he lacked ideals. That seems very unfair to me. He was trying hard to reach the truth; he could not realize as clearly and strongly as we do that truth (scientific truth) is not attainable, though we may approach it indefinitely.

THE ORGANON

Strangely enough, logic was not included in the Aristotelian classification; it was a kind of external introduction to philosophy and to science. Yet Aristotle devoted to it a series of books that constitute magnificent propylaea to the rest of his work.

[65] Popper remarks in *The open society*, vol. 2, p. 11: "Science does not develop by a gradual encyclopaedic accumulation of information, as Aristotle thought, but by a much more revolutionary method; it progresses by bold ideas, by the advancement of new and very strange theories (such as the theory that the earth is not flat, or that 'metrical space' is not flat) and by the overthrow of the old ones." True enough, but one had to begin as Aristotle did, and the encyclopedic approach was perfectible in many ways, in depth as well as in extension.

[66] To put it otherwise, the (Aristotelian) essence of a thing is the same as the final stage toward which it is developing. It is to be realized in the distant future, while the Platonic Form or Idea had been realized in the distant past.

These books, no fewer than six of them (*Categoriae, De interpretatione, Analytica priora, Analytica posteriora, Topica, De sophisticis elenchis*) are designated in their totality by the word Organon, meaning instrument. This is the instrument for philosophic intercourse, the instrument par excellence. Nobody would study logic today in the Aristotelian Organon, and it is easy enough to discover weaknesses in it, the main one being excessive verbalism. Yet it was an astounding creation, the greatest perhaps of the many creations that we owe him, and such as it is the most enduring. Aristotle invented logic and wrote its very first treatises, treatises of astounding complexity and richness.

These treatises examine and analyze such things as the ten categories or heads of predication (substance, quantity, quality, relation, place, time, position, possession, action, affection), the quality, quantity, and conversion of propositions, the syllogism and its correct figures, demonstrations by deduction (*apodeixis*) and induction (*epagōgē*), classification of fallacies, the art of reasoning correctly *versus* the art of disputation (dialectics), and so forth. All of which had been debated by sophists before Aristotle, and then more systematically in the Academy and the Lyceum, but Aristotle was the first to put all these things together, to make other people realize their propaedeutic importance, and to give the Western world its fundamental instrument, the universal key to philosophic and scientific discussion.

Such an instrument could easily be abused and was abused by the schoolmen; it was and is still abused by logicians who love logic for its own sake, but we cannot blame Aristotle for that. On the other hand, there is no doubt that his immense prestige and excessive authority in medieval times and later was largely due to his creation of the Organon. The abstraction of that work had the dual effect of discouraging some readers and of increasing their superstitious reverence for the author. In modern times we have frequently witnessed the same kind of paradoxical spectacle. People who were flabbergasted by the (to them) incomprehensible writings of a mathematician accepted the soundness of his philosophic views with incredible passivity.[67] They seemed to think that because they could not understand his mathematics they need not try to understand his philosophy, and might accept the latter as well as the former. The creator of the Organon became naturally in the popular opinion the master of all knowledge.

[67] I am thinking of Alfred North Whitehead, whose fame as a philosopher was derived partly from his authorship (with Bertrand Russell) of the *Principia mathematica* (1910 ff.) [*Isis* 8, 226–231 (1926); *10*, 513–519 (1928)] and the profound esoterism of that work.

XX

MATHEMATICS, ASTRONOMY, AND PHYSICS IN ARISTOTLE'S TIME

I. MATHEMATICS

ARISTOTLE THE MATHEMATICIAN

Aristotle, having spent twenty years of his life in or near the Academy, was necessarily a mathematician. He was not a professional mathematician like Eudoxos, Menaichmos, or Theudios, but he was less of an amateur than Plato. This is proved positively by the mass of his mathematical disquisitions[1] and negatively by his lack of interest in the mathematical occultism and nonsense that disgraced Platonic thought. He was well trained but not quite up-to-date, and inclined to avoid technical difficulties. He was probably well acquainted with Eudoxos' ideas, but not so well with those of other contemporaries like Menaichmos. His references to incommensurable quantities are frequent, but the only example quoted by him is the simplest of all, the irrationality of the diagonal of a square in relation to its side. He was primarily a philosopher, and his mathematical knowledge was sufficient for his purpose. All considered, he is one of the greatest mathematicians among philosophers, being surpassed in this respect only by Descartes and Leibniz. Most of his examples of scientific method were taken from his mathematical experience.

In his classification of sciences he considered most exact those that are most concerned with first principles. On that basis, mathematics came first, arithmetic being ahead of geometry.[2] Like Plato he was interested in knowledge for its own sake, for the contemplation of truth, rather than for its applications. Moreover, he was more interested in generalities than in particularities, and more interested in the determination of general causes than in the multiplicity of consequences.

He made a distinction between axioms (common to all sciences) and postulates (relative to each science). Examples of axioms or common notions (*coinai ennoiai*) are the "law of excluded middle" (everything must be either affirmed or

[1] English versions of all the mathematical texts have been put together by Sir Thomas Heath, *Mathematics in Aristotle* (305 pp.; Oxford: Clarendon Press, 1949) [*Isis 41*, 329 (1950)]. Heath's posthumous book is disappointing in that all the texts are published in the order of the books in which they appeared (*Organon*, *Physics*, *De caelo*, etc.), that is, helter-skelter instead of being classified by topics. The book is handy, however, and it illustrates the continuity of Aristotle's mathematical thought throughout his life.

[2] *Metaphysics*, 982A, 25–28.

denied), the "law of contradiction" (a thing cannot at the same time both be and not be), and "if equals are subtracted from equals the remainders are equal." As to definitions, they must be understood; they do not necessarily assert the existence or nonexistence of the object defined. We must assume in arithmetic the existence of the unit or monad, and in geometry of points and lines. More complex things, like triangles or tangents, must be proved to exist, and the best proof is the actual construction.

Aristotle's greatest service to mathematics lies in his cautious discussion of continuity and infinity. The latter, he remarked, exists only potentially, not in actuality. His views on those fundamental questions, as developed and illustrated by Archimedes and Apollonios, were the basis of the calculus invented in the seventeenth century by Fermat, John Wallis, Leibniz, and the two Isaacs, Barrow and Newton (as opposed to the lax handling of pseudo infinitesimals by Kepler and Cavalieri).[3] This statement, which cannot be amplified in a book meant for nonmathematical readers, is very high praise indeed, but justice obliged us to make it, the more so because Plato is more famous as a mathematician than Aristotle, and that is exceedingly unfair. Aristotle was sound but dull; Plato was more attractive but as unsound as could be. Aristotle and his contemporaries built the best foundation for the magnificent achievements of Euclid, Archimedes, and Apollonios, while Plato's seductive example encouraged all the follies of arithmology and gematria and induced other superstitions. Aristotle was the honest teacher, Plato the magician, the Pied Piper; it is not surprising that the followers of the latter were far more numerous than those of the former. But we should always remember with gratitude that many great mathematicians owed their vocation to Plato; they obtained from him the love of mathematics, but they did not otherwise follow him and their own genius was their salvation.

SPEUSIPPOS OF ATHENS

Let us now leave Aristotle and the Lyceum and return to the Academy. We should always bear in mind that mathematical studies were then fashionable in Athens and were conducted in both schools, probably in friendly emulation. Most of the mathematical work was probably done in the Academy; Speusippos and Xenocrates were Plato's successors at the head of it; the brothers Menaichmos and Deinostratos were both mentioned by Proclos[4] as friends of Plato and pupils of Eudoxos; Theudios of Magnesia wrote the textbook of the Academy; on the other hand, Eudemos of Rhodes, quoted as a pupil of Aristotle and Theophrastos, must be assigned to the Lyceum. These matters cannot be settled with any certainty, for we know the headmasters of both schools (some of them at least), but there never were any lists of students, and it is possible that attendance was informal. So-and-so are named disciples of Plato or of Aristotle, not members of the Academy or the Lyceum.

Speusippos, nephew of Plato, succeeded him in 348/47 as master of the Academy. Judging from the fragments, his lost work "On the Pythagorean numbers" was derived from Philolaos and dealt with polygonal numbers, primes *versus* composite numbers, and the five regular solids.

[3] Carl B. Boyer, *The concepts of the calculus* (352 pp.; New York: Columbia University Press, 1939; reprinted, Hafner, 1949) [*Isis 32*, 205–210 (1947–1949); *40*, 87 (1949)].

[4] G. Friedlein, *Procli in primum Euclidis elementorum commentarii* (Leipzig, 1873), p. 67.

XENOCRATES OF CHALCEDON[5]

At the time of Speusippos's death there was an election for a new master and the votes were almost equally divided between Heracleides of Pontos and Xenocrates of Chalcedon, but the latter won and was the head of the Academy for twenty-five years (339–315). Note that Aristotle, Heracleides, Xenocrates were all "northerners," and that the new master was an old friend of Aristotle (who referred many times to him in his writings). Hence, we must assume that Xenocrates was as familiar with Aristotle's mathematical views as with Plato's. He continued Plato's policy of excluding from the Academy the applicants who lacked geometric knowledge and said to one of them, "Go thy way for thou hast not the means of getting a grip of philosophy." [6] The story is plausible.

Xenocrates wrote a great many treatises, all of which are lost, but judging from the titles [7] some of them dealt with numbers and with geometry. The perennial controversy on geometric continuity which had been dramatized by Zeno's paradoxes led him to the conception of indivisible lines. He calculated the number of syllables that could be formed with the letters of the alphabet (according to Plutarch that number was 1,002,000,000,000); this is the earliest problem of combinatorial analysis on record.[8] Unfortunately, we know nothing about his activities but the meager information just given.

MENAICHMOS

Menaichmos and Deinostratos were two brothers, about whose circumstances we know only what Proclos told us in a short paragraph of his commentary on Book I of the *Elements* of Euclid: "Amyclas of Heraclea, one of Plato's friends, Menaichmos, a pupil of Eudoxos who had also studied with Plato, and Deinostratos his brother made the whole of geometry more nearly perfect." [9]

We do not know when and where these brothers were born, but they lived in Athens, attended the Academy, and sat at the feet of Plato and later of Eudoxos. We may conclude that they flourished about the middle of the century.

Both brothers were concerned with the building up of a geometric synthesis. Menaichmos was especially interested in the old problem of the duplication of the cube. That problem had been reduced by Hippocrates of Chios (V B.C.) to the finding of two mean proportionals between one straight line and another twice as long. In modern language we would say that Hippocrates had reduced the solution of a cubic equation to that of two quadratic equations. How would these be solved? Menaichmos found two ways of solving them by determining the intersection of two conics — two parabolas in the first case, a parabola and a rectangular hyperbola in the second.

This marks the appearance of conics in world literature, and the discovery of those curves is ascribed to Menaichmos. His construction of them seems very peculiar to us; he imagined that a plane cuts a right circular cone, the plane being

[5] Chalcedon in Bithynia at the entrance of the Bosporos, almost opposite Byzantion. It is thus on the Asiatic side of the strait, where Kadiköy, a suburb of Istanbul, is now.

[6] Iamblichos (IV-1), *Life of Pythagoras*, as translated by T. L. Heath, *History of Greek mathematics* (Oxford, 1921), vol. 1, p. 24.

[7] As quoted by Diogenes Laërtios, IV, 11–15.

[8] Plutarch, *Quaestiones convivales*, VIII, 9, 13, 733A.

[9] Friedlein, *Procli in primum Euclidis*, p. 67. Amyclas is probably a mistake for Amyntas; he came from Heraclea in Pontos. Nothing is known about him but what has just been quoted.

always perpendicular to the generating line of that cone. The three different conics (which he seems to have differentiated) were obtained by increasing the cone's angle;[10] as long as the angle is acute, the section is an ellipse; when the angle is right the section is a parabola; when the angle is obtuse one obtains the two branches of a hyperbola. Neugebauer has surmised that Menaichmos may have been led to his discovery by the use of sundials.[11] If he is right (and his argument is very plausible to me), it is strange to think that those curves, of astronomic origin, were not introduced into astronomic theory until almost two millennia later. Menaichmos discovered them (*c.* 350 B.C.) because of his solar observations, but not until Kepler (1609) were they used for the explanation of the solar system.

Alexander the Great asked Menaichmos whether there was not a short cut to geometric knowledge and Menaichmos answered, "O King, for traveling over the country there are royal roads and roads for common citizens, but in geometry there is one road for all."[12] The story has become a commonplace, and it has been ascribed to Euclid and Ptolemy, as well as to Menaichmos. It fits the last best because he is the most ancient and because Alexander, whose intellectual ambitions had been fanned by Aristotle, might well have asked such a question. The great king was naturally impatient, but he had to find out that it might take longer to acquire sound knowledge than to conquer the world.

DEINOSTRATOS

We have explained above (p. 278) that geometric thinking was activated in the fifth century by the emergence of three problems: (1) the squaring of the circle, (2) the trisection of the angle, (3) the duplication of the cube. Hippocrates of Chios and Menaichmos were especially interested in the third of those problems; Hippias of Elis found an ingenious solution of the second by means of the curve invented by him, the quadratrix. That name was given to it because Deinostratos, Menaichmos' brother, applied it to the solution of the first problem. We thus see that the three famous problems were still exercising the minds of the geometers of the Academy in the fourth century and helping them to extend the frontiers of their knowledge.

THEUDIOS OF MAGNESIA

Said Proclos: "Theudios of Magnesia distinguished himself in mathematics and in other branches of philosophy; he arranged beautifully the Elements (*ta stoicheia*) and made many partial theorems more general."[13]

This statement is very significant in spite of its concision. It reveals the existence of a book which might be called "The geometric textbook" (or the "Elements") of the Academy. The mathematicians of that time were interested, some in discovery, others in synthesis and logical consistency; the former were like

[10] The cone's angle here is the total angle $2a$, twice the angle a whose rotation generates the cone.

[11] Otto Neugebauer, "The astronomical origin of the theory of conic sections," *Proc. Am. Philosophical Soc.* 92, 136–138 (1948) [*Isis 40*, 124 (1949)].

[12] Stobaios (V–2), *Anthologion*, II, 13, 115; Englished by Heath, *History of Greek mathematics*, vol. 1, p. 252.

[13] The Greek text (Friedlein, *Procli in primum Euclidis*, p. 67) is not quite clear, but there is no doubt as to the general meaning.

adventurers or conquerors, the latter like colonizers. The two tendencies have always coexisted in times of healthy mathematical development, and they are equally necessary. There must be continual pressure on the frontiers and better organization within. As far as we can guess from Proclos' laconic account, Theudios' task was to put the geometric knowledge already obtained by the pioneers into as strong and beautiful a logical order as possible. Theudios was the forerunner of Euclid, and made the latter's achievement easier.

EUDEMOS OF RHODES

Eudemos was a pupil of Aristotle and a friend of Theophrastos. We may thus conclude that he flourished in the third quarter of the century and that he was a member of the Lyceum. In fact, Proclos, who quotes him four times in his commentary on Euclid I, calls him Eudemos the Peripatetic.[14] Among the writings ascribed to him, but lost, were histories of arithmetic, geometry, and astronomy. He is the first historian of science on record,[15] and, though only fragments have come to us, we have good reason to assume that his work was the main source out of which whatever knowledge we possess of pre-Euclidean mathematics has trickled down. One of the most important fragments is the one concerning the quadrature of the lunes by Hippocrates of Chios, of which we have already spoken.

The appearance at this time of a historian of mathematics and astronomy is very significant, for it proves that so much work had already been accomplished in these two fields that a historical survey had become necessary. Let us remember with gratitude the name of the first historian of mathematics and consider his presence in Athens around the year 325 as a new illustration of the glory of Hellenism.[16]

ARISTAIOS THE ELDER[17]

The last mathematician of this century marks the transition between the age of Aristotle and the age of Euclid. Two treatises of great originality are ascribed to him. One of them was devoted to solid loci connected with conics, that is, it was a treatise on conics regarded as loci, and was prior to Euclid's book on the same subject.[18] He defined the different kinds of conics in the same way as Menaichmos, as sections of cones with acute, right, and obtuse angles. The other book was entitled *Comparison of the five figures*, meaning the five regular solids, and among other things it proved the remarkable proposition that "the same circle circum-

[14] Proclos; see Friedlein, *Procli in primum Euclidis*, p. 379; Ver Eecke, *Commentaires de Proclus sur le premier livre d'Euclide* (Bruges: Desclée De Brouwer, 1948), p. 324.

[15] With the possible exception of the historian of medicine Menon, another Peripatetic, of whom we shall speak later.

[16] When Otto Neugebauer and Raymond Clare Archibald founded a journal devoted to the history of mathematics and astronomy, they called it *Eudemus*, in homage to their earliest spiritual ancestor; only one number was published (Copenhagen, 1941) [*Isis 34*, 74 (1942–43)].

[17] I call him Aristaios the Elder after Pappos' *Collection*, ed. by F. Hultsch (Berlin, 1876–78), beginning of VII, vol. 2, p. 634, but there was an older mathematician of the same name, to wit, Aristaios of Croton, son of Damophon, son-in-law of Pythagoras and his immediate successor (Pauly-Wissowa, vol. 2, p. 859). Pappos of Alexandria (III-2) flourished probably under Diocletian (emperor, 284–305), but his *Mathematical collection* was probably written by him late in life, after 320 [*Isis 19*, 382 (1933)].

[18] Pappos' *Collection*, VII; Hultsch, vol. 2, pp. 674–679; Heath, *History of Greek mathematics*, vol. 2, pp. 116–119.

scribes both the pentagon of the dodecahedron and the triangle of the icosahedron when both solids are inscribed in the same sphere." [19]

How beautiful a result this was, and how unexpected! For who could have foreseen that the faces of two different regular solids are equally distant from the center of the sphere enveloping them? These two solids, the icosahedron and the dodecahedron, had thus a special relation which the three other solids did not have. How much more beautiful indeed in its truth and honesty than the Platonic illusions on the same "figures."

MATHEMATICS IN THE SECOND HALF OF THE FOURTH CENTURY

The second half of the century did not witness the renewal of revolutionary efforts comparable in their pregnancy to those of Eudoxos of Cnidos, yet the total amount of new mathematics was splendid. The members of the Lyceum headed by Aristotle improved the definitions and axioms and more generally the philosophic substructure; Eudemos facilitated the needed synthesis by his historical surveys. Under the guidance of Speusippos and Xenocrates the Academy continued geometric investigations of various kinds which led to the composition of the "Elements" by Theudios. The brothers Menaichmos and Deinostratos, and Aristaios were creative geometers of the first order. We owe to Menaichmos and to Aristaios the first study of conics.

ASTRONOMY

HERACLEIDES OF PONTOS

Pride of place in our astronomic section must be given to Heracleides not only because of his age but also because of his singular greatness. He was born in Heracleia Pontica [20] *c.* 388, before Aristotle, and he lived until the ninth decade of the century (*c.* 315–310). His singularity was such that he has been called "the Paracelsus of antiquity," a silly nickname, yet meaningful, whether it is taken as praise or blame. To compare him with a man who appeared nineteen centuries later is to invite unnecessary trouble; it is more helpful to compare him with his predecessor, Empedocles, a man whom he greatly admired and tried to emulate.

We know little of his life except that he was wealthy, emigrated to Athens, and was a pupil of Plato and Speusippos, perhaps also of Aristotle. When Speusippos died in 339 and was replaced by Xenocrates (Aristotle's friend), Heracleides returned to his country. He wrote many books on philosophy and mythology which obtained some popularity not only among the Greeks but also among the Romans of the last century B.C. For example, Cicero admired him and one can detect traces of Heracleides' influence in "Scipio's dream." [21] Even as Plato had written

[19] Pappos' *Collection*, Hultsch, vol. 1, p. 435. It was Hypsicles (II-1 B.C.) who ascribed that discovery to Aristaios in the so-called XIVth book of Euclid (prop. 2).

[20] One must add Pontica, because many Greek cities had been named after the most popular hero of antiquity, Heracles (Hercules). Heracleia Pontica is on the southern shore of the Black Sea, in the western part, the coast of Bithynia. Its modern Turkish name is Ereğli.

[21] "Somnium Scipionis" in book VI of Cicero's *De republica*; the "Somnium" was often printed with the commentary by Macrobius (V-1), which was the main source of Platonism in the Latin west outside of the partial translation of Timaios by Chalcidius (IV-1).

a revelation of other-world mysteries in his myth of Er, Heracleides wrote a similar revelation in his myth of Empedotimos:[22] his Hades where the disincarnated souls found their last refuge was located in the Milky Way; the souls were illuminated!

Such poetic fancies explain his popularity but would not justify our own praise in this volume. Yet to be a spiritual descendant of Empedocles was a remarkable thing and we must pause a moment to consider it: there was an irrational trend in Greek thought cutting through the centuries via the Pythagoreans, Empedocles, Plato, Heracleides, and their epigoni. Heracleides, however, combined his apocalyptic with scientific tendencies, and we must speak of him at greater length because of his astronomic theories, which make him one of the forerunners of modern science.

One more word, however, concerning his relation with Empedocles. The latter's view of the universe included the four elements and the two antagonistic forces (love and strife). Heracleides conceived the world as made up of jointless particles (*anarmoi oncoi*), as opposed probably to the Democritean atoms, which had various shapes and could cling to one another. The Heracleidean particles might hold together by some kind of Empedoclean attraction.[23]

Heracleides' astronomy was more rational, as we would expect, than his cosmology. He had probably heard of the views expressed by Hicetas and Ecphantos and agreed with them. On the basis of those views and of other Pythagorean-Platonic ideas he explained his own theory, which can be summarized as follows. The universe is infinite. The Earth is in the center of the solar system; the Sun, Moon, and superior planets revolve around the Earth; Venus and Mercury (the inferior planets) revolve around the Sun; the Earth rotates daily on its own axis (this rotation replaces the daily rotation of all the stars around the Earth).[24] This geoheliocentric system had an astounding fortune. It was not sufficiently bolstered up with observations to deserve the acceptance of the practical astronomers of Heracleides' time; yet the hypotheses that it included were never forgotten. They reappeared in Chalcidius (IV-1), Macrobius (V-1), Martianus Capella (V-2), John Scotus Erigena (IX-2), William of Conches (XII-1).[25]

Looked at from the modern point of view, Heracleides' system is a compromise between the Ptolemaic (centered upon the Earth) and the Copernican (centered upon the Sun), but this should not be exaggerated as is done by the historians who call Heracleides the Greek Tycho![26] The compromise suggested by Tycho Brahe (1588; regular publication, 1603) and by Nicholas Reymers (1588) was deeper: all the planets, not two only, were supposed to revolve around the Sun. Strangely enough, the Jesuit, Giovanni Battista Riccioli, in his *Almagestum novum* published

[22] Empedotimos of Syracuse. Note that the name Empedotimos is etymologically equivalent to Empedocles. J. Bidez, *Eos* (Brussels: Hayez, 1945), pp. 52–59 [*Isis* 37, 185 (1947)].

[23] The comparison of this with molecular attraction, suggested by Gomperz and later by Bidez, *Eos*, p. 56, is unwarranted.

[24] Heracleides' views on the rotation of the Earth on its axis are reported by Aëtios and by Simplicios (VI-1), those on the motion of Mercury and Venus round the Sun by Vitruvius (I-2 B.C.), Chalcidius (IV-1), and Martianus Capella (V-2). English versions of their statements in Heath, *Greek astronomy* (London: Dent, 1932), pp. 93–95 [*Isis* 22, 585 (1934–35)].

[25] Charles W. Jones, "A note on concepts of the inferior planets in the early Middle Ages," *Isis 24*, 397–399 (1936).

[26] The Italian astronomer, Giovanni Virginio Schiaparelli (1835–1910), claimed that Heracleides anticipated not only Tycho Brahe but also Copernicus. Such claims cannot be sustained; *Introduction*, vol. 1, p. 141.

half a century later (Bologna, 1651), came back somewhat closer to Heracleides, for he accepted the rotation of three planets around the Sun, the two most remote ones (Jupiter and Saturn) moving around the Earth.[27]

Heracleides was not a Copernicus, nor even a Brahe, yet his conception of the solar system, imperfect as it was, was astoundingly good for its time.

CALLIPPOS OF CYZICOS

In the meanwhile, the work of Eudoxos was being continued by Aristotle and Callippos. They worked together at the Lyceum; though Callippos was somewhat younger than his chief, he seems to have been the originator in astronomic research. That would be natural enough, for Aristotle was obliged to busy himself with the whole institution and with the logical and philosophic teaching. If he had been tempted to make special investigations on his own account, he would probably have made them in the field of zoölogy, or he would have devoted more time to zoölogy than he was able to do.

After his return from Egypt, Eudoxos had spent some time in Cyzicos (Sea of Marmara), where he started a school of his own. Now, Callippos was born in that very place *c.* 370 and he may have known Eudoxos in his youth. In any case, he must have heard of Eudoxos' mathematical and astronomic teaching, either directly or from a disciple such as his countryman, Polemarchos of Cyzicos, who is quoted as one of the first critics of the theory of homocentric spheres.[28] Indeed, he was Polemarchos' pupil and followed him to Athens, where "he stayed with Aristotle helping the latter to correct and complete the discoveries of Eudoxos." [29] The date of Callippos' arrival in Athens was probably after the beginning of Alexander's rule (336), and before the beginning of Callippos' cycle (330). According to Aristotle,[30] Callippos realized the imperfections of Eudoxos' system and tried to remove them by adding seven more spheres, that is, two each for the Sun and the Moon, and one more for each of the other planets, except Jupiter and Saturn. The theory as improved by Callippos thus required a total of 33 concentric spheres rotating simultaneously each on its own axis and with its own speed.

Callippos concerned himself also with the reform of the calendar, the last establishment of which had been made in Athens in 432 by Meton and Euctemon. Better solstitial and equinoctial observations enabled him to determine more exactly the lengths of the seasons (beginning with the spring, 94, 92, 89, 90 days, the errors ranging from 0.08 to 0.44 day). He improved the Metonic cycle of 19 years by dropping 1 day out of each period of [$19 \times 4 =$] 76 years. The epoch of the new era was possibly 29 June 330.[31] The comparison of Callippos' calendar with Meton's gives us a measure of the progress in astronomic observation that had been achieved in a century.

[27] In short, according to Heracleides (*c.* 350 B.C.), two planets revolve around the Sun; according to Tycho Brahe (1588), five; according to Riccioli (1651), three.

[28] By Simplicios (VI-1), in his *Commentary on Aristotle's De caelo* (Heiberg ed., 1894), p. 505. Polemarchos wondered how the variations in brightness of the planets could be reconciled with the theory of homocentric spheres, for according to that theory the distance between the Earth and the planets is invariable; he seems to have rejected his own objection on the ground that the changes in brightness were too small to be taken into account.

[29] Simplicios' *Commentary on De caelo* (Heiberg ed.), p. 493.

[30] *Metaphysics*, 1073B.

[31] For Callippos' calendar see Geminos of Rhodes (I-1 B.C.), Greek edition with German translation by Karl Manitius (Leipzig, 1898), pp. 120–122.

ARISTOTLE THE ASTRONOMER

Aristotle's views on astronomy are explained in *Metaphysics lambda*, in *Physics*, in *De caelo*,[32] and in Simplicios' *Commentary*. He was not satisfied with the theory of homocentric spheres, even as perfected by Callippos. As Heath puts it,

> In his matter-of-fact way, he thought it necessary to transform the system into a mechanical one, with material spherical shells one inside the other and mechanically acting on one another. The object was to substitute one system of spheres for the Sun, Moon, and planets together, instead of a separate system for each heavenly body. For this purpose he assumed sets of *reacting* spheres between successive sets of the original spheres. Saturn being, for instance, moved by a set of four spheres, he had three reacting spheres to neutralize the last three, in order to restore the outermost sphere to act as the first of the four spheres producing the motion of the next lower planet, Jupiter, and so on. In Callippos' system there were thirty-three spheres in all; Aristotle added twenty-two reacting spheres making fifty-five. The change was not an improvement.[33]

This is typical of Aristotle's mind; in his anxiety to give a mechanical and tangible explanation of planetary movements, he introduced unnecessary complications. Did Aristotle believe in the physical reality of the homocentric spheres? We cannot be sure; yet his transformation of the geometric concept into a mechanical one suggests such a belief. It is a good example of the eternal conflict between the explanation that satisfies the mathematician and the one that the practical man requires. The practical man is often defeated by his very practicality, and so was Aristotle in this case.

We cannot dissociate his astronomic views from the physical ones. Let us describe them rapidly together. There are three kinds of motion in space: (1) rectilinear, (2) circular, (3) mixed. The bodies of the sublunar world are made out of the four elements. These elements tend to move along straight lines, earth downward, fire upward; water and air, being relatively heavy and relatively light, fall in between. Hence, the natural order of the elements, starting from the Earth, is: earth, water, air, fire. Celestial bodies are made out of another substance, not earthly, but divine or transcendent, the fifth element or aether, whose natural motion is circular, changeless, and eternal.

The universe is spherical and finite; it is spherical, because the sphere is the most perfect shape; it is finite, because it has a center, the center of the earth, and an infinite body cannot have a center.[34] There is but one universe and that universe is complete; there can be nothing (not even space) outside of it.

Is there a transcendent mover of the spheres (that is, a superior and unmoved mover of the spheres and of everything else)? Aristotle could not reach a certain answer on that fundamental question.[35] His final conclusion in *De caelo* was that

[32] The *Metaphysics* is certainly Aristotle's work; we are less certain with regard to the *Physics* and the *De caelo*. The *De caelo*, as we have it, is a text prepared by Aristotle for teaching, and possibly amended by himself or disciples; its state of incompleteness is proved by many contradictions [*Isis* 32, 136 (1947–49)].

[33] Heath, *Greek astronomy*, p. xlviii [*Isis* 22, 585 (1934–35)].

[34] That argument was curiously reversed later, e.g., by Plutarch (I–2). The universe is infinite, hence it has no center and one cannot say that the Earth is in the center of it. This was repeated by all the medieval philosophers who believed in the infinity of the universe, for example, Nicolaus Cusanus (1401–1464).

[35] In the Loeb Classical Library edition and translation of the *De caelo* (1939) [*Isis* 32, 136 (1947–49)], W. K. C. Guthrie gives a list of Aristotelian passages (*a*) that exclude the transcendent mover, (*b*) that imply it.

the sphere of the fixed stars was the prime mover (though itself moving) and hence the foremost and highest god; [36] but in the *Metaphysics lambda*, his conclusion is that there is behind the fixed stars an unmoved mover influencing all the celestial motions as the Beloved influences the Lover. This implies that the celestial bodies are not only divine but alive, sensitive, and makes us realize once more, and more deeply, that ancient physics and ancient astronomy were very close to metaphysics, so close that one could not know any more where one was. Is this astronomy or metaphysics or theology?

We come closer to reality in Aristotle's discussion of the shape of the Earth and estimate of its size. The Earth must be spherical for reasons of symmetry and equilibrium; the elements that fall upon it fall from every direction and the final result of all the deposits can only be a sphere. Moreover, during lunar eclipses the edge of the shadow is always circular, and when one travels northward (or southward) the general layout of the starry heavens changes; one sees new stars or ceases to see familiar ones. The fact that a small change in our position (along a meridian) makes so much difference is a proof that the Earth is relatively small. Here is the relevant text:

> There is much change, I mean, in the stars which are overhead, and the stars seen are different, as one moves northward or southward. Indeed there are some stars seen in Egypt and in the neighborhood of Cyprus which are not seen in the northerly regions; and stars, which in the north are never beyond the range of observation, in those regions rise and set. All of which goes to show not only that the earth is circular in shape, but also that it is a sphere of no great size: for otherwise the effect of so slight a change of place would not be so quickly apparent. Hence one should not be too sure of the incredibility of the view of those who conceive that there is continuity between the parts about the Pillars of Hercules and the parts about India, and that in this way the ocean is one. As further evidence in favor of this they quote the case of elephants, a species occurring in each of these extreme regions, suggesting that the common characteristic of these extremes is explained by their continuity. Also, those mathematicians who try to calculate the size of the earth's circumference arrive at the figure 400,000 stades. This indicates not only that the earth's mass is spherical in shape, but also that as compared with the stars it is not of great size.[37]

The mathematicians referred to are probably Eudoxos and Callippos. Their estimate of the size of the Earth as quoted by Aristotle is the earliest of its kind; it was too large yet very remarkable.[38] This fragment of Aristotle was the first seed out of which grew eventually in 1492 the heroic experiments of Christopher Columbus.

The main achievement of the astronomers of this period, if not of Aristotle himself, was the completion of the theory of homocentric spheres. This achievement

[36] *De caelo*, 279A.

[37] *De caelo*, 298A, following J. L. Stocks' translation in the Oxford Aristotle (1922).

[38] It is impossible to appreciate its precision without knowing the length of the stadium. Aubrey Diller, "Ancient measurements of the earth," *Isis 40*, 6–9 (1949). The circumference of the Earth in thousands of stadia was for Aristotle 400; for Archimedes (III-2 B.C.), 300; for Eratosthenes (III-2 B.C.), 252; for Poseidonios (I-1 B.C.), 240, but also 180; for Ptolemy (II-1), 180. The trouble is that the stadia varied in length from place to place and from time to time. It is possible that the two Poseidonian values are really the same adjusted to two different lengths of stadia. The ratio of 10 stadia/mile to 7.5 stadia/mile = 4:3 = 240:180.

The Erastosthenian value was supposed to be the best in antiquity; *Introduction*, vol. 1, p. 172. If both Eratosthenes and Poseidonios used stadia of length ten to a mile, then their results came very close, for 252:240 = 21:20.

implied the availability of a fairly large number of solar, lunar, and planetary observations. Where did Eudoxos, Callippos, and Aristotle obtain them? In Egypt and Babylonia.

According to Simplicios' commentary on the *De caelo*, the Egyptians possessed a treasure of observations extending over 630,000 years, and the Babylonians had accumulated observations for 1,440,000 years.[39] A more modest estimate was quoted by Simplicios from Porphyry, according to which the observations sent from Babylon by Callisthenes, at Aristotle's request, covered a period of 31,000 years. All that is fantastic, but Oriental observations covering many centuries were actually available to the Greek theorists and were sufficient for their purpose. The Greeks obtained them in Egypt and Babylonia; they could not have obtained them in Greece, where men of science had preferred to philosophize each in his own way and where no institution had ever been ready to continue astronomic observations throughout the centuries. Simplicios' exaggerations are simply a tribute to the antiquity and the admirable continuity of Oriental astronomy.

To return to Aristotle, though he was acquainted in a general way with Egyptian and Babylonian astronomy, he did not need their observations as keenly as did professionals like Eudoxos and Callippos. Being primarily a philosopher, he was more interested in questions of such generality that observations were of little help. For example, in the *De caelo* we find discussions concerning the general shape of the heavens, the shape of the stars, the substance of the stars and planets (which he assumed to be "aether"), the musical harmony caused by their motions. This may seem very foolish, but in justice to Aristotle and his contemporaries we should remember that many irrelevant and futile questions had to be asked and discussed before the pertinent ones were disentangled from the rest. In science immense progress is made whenever the right question is asked, the asking in proper form is almost half of the solution, but we can hardly expect these right questions to be discovered at the beginning.

The fortune of Aristotelian astronomy was singular. The theory of homocentric spheres was eventually displaced by the theories of eccentrics and epicycles, which was eventually crystallized in the *Almagest* of Ptolemy (II–1). Later, as the weaknesses of the *Almagest* appeared more clearly, some astronomers went back to Aristotle. The history of medieval astronomy is largely a history of the conflict between Ptolemaic and Aristotelian ideas; the latter were relatively backward and hence the growth of Aristotelianism retarded the progress of astronomy.[40]

AUTOLYCOS OF PITANE

In order to complete our survey of mathematics and astronomy in this golden age we must still speak of one great person, whose appearance ends it beautifully. Autolycos was born in Pitane[41] in the second half of the century, and he flourished probably in the last decade. He was an older contemporary of Euclid.[42] Hence,

[39] *Simplicios' commentary* (Heiberg ed.), p. 117, 25. With regard to Greek astronomy (Eudoxos, Callippos) Simplicios quoted many times Sosigenes the Peripatetic (Caesar's astronomer), who had been able to use the lost history of astronomy by Eudemos of Rhodes; see Heiberg, p. 488, 20.

[40] This has been discussed repeatedly in my *Introduction*, see, e.g., vol. 2, p. 16; vol. 3, p. 484.

[41] Pitane was on the coast of Aeolis (Mysia, Asia Minor).

[42] Euclid made use of Autolycos' work in his *Phainomena*, though without naming him.

he represents the transition between the great Hellenic school of mathematics and the Alexandrian age.

We know almost nothing about him, not even the place where he flourished. Did he go to Athens? That would have been natural enough. Yet Pitane was a civilized and sophisticated place, a well-located harbor facing Lesbos, not very far from Assos where Aristotle had taught. We know that Autolycos was the teacher of a fellow citizen of his, Arcesilaos of Pitane (315–240), founder of the Middle Academy. This suggests that he resided in Pitane and fixes the date approximately, the turn of the century.

Our ignorance concerning his personality is in paradoxical contrast with the fact that he wrote two important mathematical treatises, which are the earliest Greek books of their kind transmitted to us in their integrity. We know his works exceedingly well, but nothing of himself, except that he was the author of them.

Before speaking of these two books we must refer briefly to a third one which is lost and wherein he criticized the theory of homocentric spheres. He wondered how that theory could be reconciled with the changes of relative size of Sun and Moon and with the variations in the brightness of the planets, especially Mars and Venus. Judging from his controversy with Aristotheros, he could not solve that difficulty.[43]

The two books that have come down to us deal with the geometry of the sphere.[44] As all the stars were supposed to be on a single sphere (and in any case one might always consider their central projections on that sphere), mathematical problems concerning their relations were problems of spherical geometry. For example, any three stars are the vertexes of a spherical triangle, the sides of which are great circles. When we try to measure the distance between two stars on that sphere (one side of the triangle), what we measure really is the angle which that side subtends at the center of the earth or as seen by a terrestrial observer. All such problems are solved now by means of spherical trigonometry, but trigonometry had not yet been invented in Autolycos's time and he tried to obtain geometric solutions.

Irrespective of their practical value, which was considerable, these books are of great interest to us because of their Euclidean form, before Euclid. That is, the propositions follow one another in logical order; each proposition is clearly enunciated with reference to *lettered* figures, then proved. Some propositions, however, are not proved; that is, they are taken for granted, and this suggests that Autolycos' books were not the first treatises on spherical geometry, but had been preceded by at least one other now lost. The substance of the lost treatise is somewhat preserved in the *Sphaerics* of Theodosios of Bithynia (I-1 B.C.), which gives the proofs of theorems unproved by Autolycos.

The first of Autolycos' treatises, entitled *On the moving sphere*, deals with spherical geometry proper; the second, *On risings and settings* [of stars], is more

[43] Aristotheros was the teacher of Aratos of Soloi (III-1 B.C.), but is otherwise unknown. He is referred to by Simplicios, Heiberg, p. 504, 25. The same argument had been used also (independently?) by Polemarchos of Cyzicos (p. 508).

[44] Greek edition with Latin translation by Friedrich Hultsch (Leipzig, 1885). New Greek edition without translation by Joseph Mogenet, *Autolycus de Pitane. Histoire du texte suivie de l'édition critique des traité de la sphère en mouvement et des levers et couchers* (336 pp.; Louvain: Université de Louvain, 1950) [*Isis 42*, 147 (1951)].

astronomic, that is, it implies observations. Both treatises are too technical to be analyzed here.

How did it happen that such books were preserved? Their practical value was immediately realized by mathematical astronomers, who transmitted them from generation to generation with special care. Their preservation was facilitated and insured by the fact that they were eventually included in a collection called "Little astronomy" (in opposition to the "Great collection," Ptolemy's *Almagest*). The "Little astronomy" was transmitted in its integrity to the Arabic astronomers, and became in Arabic translation a substantial part of what they called the "Intermediate books." [45] The maxim "l'union fait la force" (part of the heraldic achievement of Belgium) applies to books as well as to men: when books become parts of homogeneous collections, each helps the other to survive.

ASTRONOMY IN ARISTOTLE'S TIME

The main achievement is the completion of the theory of homocentric spheres by Callippos; this may be put to the credit of the Lyceum. The Greeks were theorists rather than observers, but they were fortunate in that a treasure of Egyptian and Babylonian observations was available to them. It is almost impossible to determine their use of it except in a very general way. We can see only the fruits of that use, the main one being the theory of homocentric spheres. Heracleides was the first to propose a kind of geoheliocentric system, that is, to postulate the rotation of some planets around the Sun. He may be called the first Greek forerunner of the Copernican astronomy. At the end of the century Autolycos was building the geometric foundation of astronomy. Aristotle helped to state astronomic problems and to explain their relation to the rest of knowledge.

Note that none of these men was a Greek of Greece proper; their birthplaces were in Macedonia (Stageira) or in Asia Minor (Heracleia Pontica, Cyzicos, and Pitane).

PHYSICS

PHYSICS IN THE EARLY LYCEUM

Aristotle, his colleagues, and his younger disciples must have devoted much time to the discussion of physical questions; it was the old Ionian tradition of research *de natura rerum*, though already much better focused. A part of that was astronomic, but astronomy was always mixed with physics. The great advantage of astronomy proper, and the main cause of its early progress, was that some problems at least were very definite, and could be isolated with relative ease – such problems as how to account for the regular irregularities of planetary motions, or what are the shapes of the Earth and the planets, their mutual distances, their sizes. Not only was it possible to state these problems, but solutions were offered, some of which were sufficient at least as first approximations.

The universe was divided into two parts, essentially different – the sublunar world and the rest. Physical questions applied mainly to the sublunar world, astronomical ones to the Moon and beyond.

[45] *Kitāb al-mutawassiṭāt,* about which see *Introduction,* vol. 2, p. 1001. Mogenet's edition of 1950 includes an elaborate study of the tradition of Autolycos, in Greek, Arabic, Latin and Hebrew. For the "Little astronomy," see Mogenet, pp. 166, 172.

Auerrois comment. in libros aristotelis de physico auditu.

Intentio mea in hoc sermone est glosare librum Arist. qui dicitur auditus naturalis. Et secundum usum glosatorum dicetur in principio libri octo capitula: scilicet intentionem libri: utilitates: ordines: divisionem: proportionem: uiam doctrine. nomen libri: nomen auctoris: dicam & ego licet plura ex istis possint intelligi ex proemio libri: sicut in subsequentibus ostendemus. Dico quod intentio scientie naturalis in summa cuius intentionis iste liber est pars: est cognoscere causas specierum sensibilium & causas uisibiles in eis speciebus accidentium: quia subiectum istius artis non sunt nisi res sensibiles: subiectum dico secundum quod transmutantur de se idest secundum quod habent ex se principium motus & quietis: & intentio huius libri est cognoscere de istis causas communes omnibus rebus naturalibus. s. materiam: formam finem: & agentem: et cognoscere causas primas ex hoc quod est possibile scire in hac scientia: sicut materia primam & mouens primum. forma autem prima & mouens primum a metaphysico considerantur: & consideratio in consequentibus communibus omnibus rebus naturalibus sicut tempore loco & eis similia est inquisitio naturalis. Utilitas eius est pars utilitatis scientie speculatiue: & declaratum est in scientia considerante in operationibus uoluntariis Quod esse hominis secundum ultimam perfectionem ipsius & substantia eius perfecta est ipsum esse perfectum per scientiam speculatiuam: & ista dispositio est sibi felicitas ultima & sempiterna uita: et in hac scientia manifestius est quod predicatio nominis hominis perfecti a scientia speculatiua: & non perfecti sine non habentis aptitudinem quod perfici possit: est equiuoca: sicut nomen hominis predicatur de homine uiuo & homine mortuo: siue predicatio hominis de rationali & lapideo: & cum hoc consequitur cognitionem scientie speculatiue de moralitate uirtuosa: quoniam scientes istam scientiam quum erunt secundum ordinem naturalem oportet eos de necessitate esse uirtuosos in omnibus speciebus uirtutum moralium: que sunt iustitia: abstinentia: audacia: sublimitas: bonitas: ueritas: fiducia: & alie de uirtutibus hominum. & declarauit Alexander in proemio huius libri quomodo sequitur scientia harum uirtutum ad scientiam speculatiuam quia quum nouerit sapiens paruitatem uite sue & perpetuitatem sempiternam & motum continuum: & quod proportio sue uite ad tempus sempiternum est sicut puncti ad lineam. uel sicut finiti ad infinitum non procurabit propriam uitam: & tunc erit audax ex necessitate: & similiter quum nouit quod mors est ex necessitate hyle siue materie & propter quam habet perfectionem humanam & dubitat quod a perfectione expolietur tunc non est mirum si aliquando eligit mortem pre uita: sicut fecit Socrates cum athenensibus: quamuis opinetur de Socrate quod non fuerunt ei declarate uie discipline communes & propter quod uoluit ambulare in uia propria populorum & similiter quum sciuerit quod perfectionem habere pot erit ei mors preeligenda super uitam: & manifestum est quod quum sapiens dubitauerit non esse perfectus quod audacia eius non erit nisi in acquirendo hanc perfectionem Iterum oportet sapientes esse iustos & abstinentes insuper quum sciuerint naturam iustitie existentem in substantia rerum: tunc amant similari illi nature & acquirere illam formam. Abstinentes quum sciuerint indignitatem uoluptatum: quia non sunt de dispositionibus necessariis in esse hominum nedum de perfectionibus ipsorum: sed sunt consequentia ex necessitate materie de se ipsis eiciunt uoluptates omnino: & erunt abstinentes & manutenebunt leges diuinas & imitabuntur ordines naturales: & manifestum est quod debent esse boni & liberi quia uilipendunt pecuniam quum sciant essentie pecunie uilitatem: & quod studiositas in ea est unum de desideriis preter naturam. Similiter debent esse sublimes & amantes ueritatem in operationibus & dictionibus ipsorum cum aliis consequentibus eis ex uirtute humana. Si autem dicitur quod esse non testatur super hoc: quia plures sapientes huius artis huius temporis non sunt tales: dicitur quod hoc accidit eis propter dispositionem in naturalem in hac communitate. & ista dispositio est declarata exercitantibus in disciplina practice. & signum super hoc quod accidit eis res extra naturam: est quod inuenimus Alex. & alios dicentes: quod sapientes uirtuosi fuerunt in tempore suo: sapientes uero temporis huius quia a communitate gentium uiles habentur et principes existimant quod non sunt digni regere ciuitatem: neque digni sint esse eorum secretarii & proximi: sapientes tunc existimant quod hec uilitas ex scientia accidet eis: & propter hoc declinabit dispositionem quam naturaliter a scientia habuerunt a dispositione opinionis istius. & declinant uniuersaliter ad uoluptates: & accidit eis dispositio extra naturam: in tantum quod non sunt digni esse pars ciuitatis et hec dispositio plures dialecticis aduenit: ergo secundum naturam scientie sequuntur eos omnes uirtutes & secundum accidens sequuntur eos uicia propter societatem hominum temporis huius. Ordo autem huius libri est primus librorum positorum ab Arist. in scientia naturali quia res naturales diuiduntur ad diuersas naturas: & quum diuersificantur in eo quod proprium est unicuique eorum coicant etiam in rebus communibus existentibus in eis: & propter hoc diuiditur consideratio huius artis in duas partes. prima est consideratio in rebus communibus omnibus rebus naturalibus: secunda est consideratio

Fig. 93. Beginning of the Aristotelian physics in Latin translation, *Physica sive De physico auditu* (Padua, 1472–1475; Klebs, 93.1). First edition of the *Physics* in any language. It contains the double text in Latin with commentary by Ibn Rushd (XII-2). The anonymous printer was Laurenzius Canozius, in Padua. [Courtesy of the Bibliothèque Nationale, Paris.]

Aristotelian physics, or more correctly Peripatetic physics, is found in many books, such as *Physica* (Fig. 93), *Meteorologica, Mechanica, De caelo, De generatione et corruptione*, and even in *Metaphysica*, and the dating of some of these works is very uncertain. For example, the *Mechanica* has been ascribed not only to Aristotle, but also to Straton of Lampsacos (III-1 B.C.), who was Euclid's contemporary. The fourth book of the *Meteorology* is also ascribed to Straton. Let us forget for a moment these differences and try to describe the physical ideas that were explained in the Lyceum in the fourth and third centuries.

In order to avoid confusion we must try to forget another thing, our present conception of physics, which is relatively recent. In ancient and medieval times, and even down to the seventeenth century, physics concerned the study of nature in general, inorganic and organic.

The center of Aristotelian[46] physics is the theory of motion or of change. Aristotle distinguished four kinds of motion:

(1) Local motion, that is, our kind, translation of an object from one place to another. Such local motion, Aristotle recognized, is fundamental; it may and does occur in the other kinds.

(2) Creation and destruction; metamorphoses. As such changes are eternal, they imply compensations, or some kind of cyclic return. If they proceeded only in one direction they could not continue eternally. Creation is the passage from a lesser to a higher perfection (say the birth of a living being); destruction is the passage from a higher form to a lower (say the passage from life to death). There is neither absolute creation nor absolute destruction.

(3) Alterations, which do not affect the substance. Objects may receive another shape yet remain substantially alike. A man's body may be altered by injury or by disease.

(4) Increase and decrease.

Everything that happens, happens because of some kind of motion as defined above. The physicist studies these "motions" for their own sake but also better to understand the substance undergoing them.

It is impossible, however, to explain nature only in terms of "material motions" or mechanism. One has to take into account some general ideas, such as that of universal economy: God (or nature) does nothing in vain. Every motion has a direction and a purpose. The direction is toward something better or more beautiful. The purpose of a being is revealed by the study of its genesis and evolution. We are falling back upon the theory of finalism (or teleology) which has been discussed in the previous chapter.

Everything in nature has a double aspect: material and formal. The form expresses the aim, which cannot be accomplished, however, except through some kind of matter. The weaknesses, imperfections, monstrosities that occur in nature are caused by the blind inertia of matter, defeating the purpose.

Aristotle had inherited and accepted the theory of four elements, at least to account for the changes that occur in the sublunar world. (For the changeless

[46] In what follows I shall often use the adjective Aristotelian in a loose sense, for the sake of simplicity. Every statement of mine can be justified with quotations taken from the Aristotelian corpus, but one might claim that this or that quotation represents the thought not of Aristotle himself, but of Straton or of another philosopher, known or unknown. Thus, every statement might necessitate long discussions, out of place here.

world above the Moon it was necessary to postulate a fifth, incorruptible, element, the aether.) He had also accepted the four qualities; at least, he considered them (wet and dry, hot and cold) the fundamental ones, to which others (for example, soft and hard) could be reduced. Only the necessities are formal; individual objects are contingent. It is the forms that the scientist must try to understand, but he cannot understand them except through individual (accidental) examples. We are thinking of Plato, and in some way, Aristotle is as idealistic as his predecessor, yet with a difference: Plato passes from the Form (the Idea) to the object, Aristotle does the reverse. That difference is simple but immense.

Aristotle made an exception, however, for some fundamental beings, such as the Prime Mover or the Elements, beings whose essence implies existence, and which cannot be known except a priori. All the rest can be known only empirically, by gradual induction, from individual cases to more general ones, and from inferior forms to superior ones. Mechanism alone can never explain the universe, yet analyses, descriptions, and inductions must precede every synthesis. That procedure is essentially the procedure of modern science.

Though he often quoted Democritos and praised him repeatedly, Aristotle rejected the atomic theory and what might be called Democritian materialism. He rejected the concept of vacuum,[47] because he could not conceive motion except in a definite medium, and was not everything that happened due to a kind of motion? It is possible that Aristotle rejected the atomic theory only because of the wrong use that Democritos (or his disciples) had made of it. It was claimed that Democritos tried to explain everything in mechanical terms, while the Aristotelian explanations were partly material and partly formal.

Celestial bodies move eternally, with constant speed, along circles. Sublunar bodies do not move if they are in their natural places; if they are removed from those places they tend to return to them along a straight line. There are two possible motions along a straight line, upward and downward.[48] Heavy bodies like earth move downward; light ones like fire, upward. Between these two elements, which are absolutely heavy and absolutely light, occur the two others, water and air, which are respectively less heavy than earth and less light than fire.

Aristotelian mechanics includes adumbrations of the principle of the lever, of the principle of virtual velocities, of the parallelogram of forces, of the concept of center of gravity, and of the concept of density. Some of these ideas were to be given explicit and quantitative formulation by Archimedes of Syracuse (III-1 B.C.), others would be developed later, but the germs were already in the Aristotelian corpus.

Most discussions of Aristotelian mechanics center upon his dynamics. The genesis of Aristotle's ideas on this subject is extremely instructive. We have seen that he did not accept the concept of vacuum.[49] Motion is inconceivable in empti-

[47] His most definite statement on that subject occurs, curiously enough, in the *De respiratione*, 471A, apropos of the breathing of fishes: "Anaxagoras says that when fishes discharge water though their gills, air is formed in the mouth, for there can be no vacuum."

[48] Aristotle recognized two possible directions in a straight line, but only one in a circle. All the celestial motions known to him had the same direction; yet were motions in the opposite direction inconceivable?

[49] That Aristotelian prejudice is generally expressed in this form (Natura abhorret a vacuo, Nature abhors a vacuum), the exact origin of which I do not know; it is a medieval saying. For the history of the vacuum see Cornelis De Waard, *L'expérience barométrique* (Thouars: Imprimerie nouvelle, 1936) [*Isis 26*, 212 (1936)].

ness; hence, when he considered the movement of bodies it was always in a resisting medium. On the basis of gross observations he concluded that the speed of a body is proportional to the force pushing (or pulling) it and inversely proportional to the resistance of the medium. Any object moving in a resisting medium is bound to come to a standstill unless a force continues to push it. (In a vacuum, the resistance would be zero and the speed infinite.) He also remarked that the speed of a falling body would be proportional to its weight, and that it would increase as the body was further removed from its point of release and came closer to its natural place. Hence the velocity would be proportional to the distance fallen.

The discovery of the true laws of motion became possible only when the Aristotelian prejudice against a vacuum was removed. Instead of rejecting motion in a vacuum as absurd, one assumed its possibility and considered what would happen if resistances were eliminated. Thanks to that happy abstraction, Galileo found that the speed was independent of the weight or mass of the falling body. He first thought that the speed would be proportional to the distance fallen but then realized that it was proportional to the time elapsed. The final laws of motion were discovered by Newton, chiefly the one that motive forces are proportional not to the speed of the body moved but to its acceleration. In fairness to Aristotle, however, one must remember that his conclusions were not unreasonable within the frame of his experimental knowledge. Mach was unjust to him and Duhem perhaps too generous. It is just as unfair to condemn Aristotle for not accepting what the invention of the air pump would prove, as for not seeing what could be seen only after the invention of the telescope.

The great difficulty of terrestrial (as compared with celestial) mechanics consisted in the extreme complexity of natural events. These could not become understandable without abstractions of great boldness. Aristotle's imagination was not equal to that, not because it was inferior to Galileo's or Newton's, but because it could not depend upon the same mass of experience and could not soar off from the same altitude.

The *Meteorologica* ascribed to Aristotle contains meteorology in our sense, plus much else that we would classify under physics, astronomy, geology, even chemistry.[50] The astronomic part came in because Aristotle considered such phenomena as comets and the Milky Way as originating below the Moon; these phenomena were thus for him meteorologic rather than astronomic. Such errors were natural and pardonable in his time, and indeed until the end of the sixteenth and the seventeenth centuries. The unpredictable behavior of comets seemed absolutely different from the complex and solemn regularities of planetary motions. The planets suggest eternity and divinity; on the contrary, what better examples of capriciousness and evanescence could one adduce than the comets, which appear in the sky and after a relatively short time dissolve and disappear? Moreover, comets were generally seen outside of the zodiac. That Aristotelian prejudice was not shaken until the publication by Tycho Brahe, in 1588, of his observations of the comet of 1577. Brahe proved that its parallax was so small that the comet could not be sublunar; its orbit exceeded that of Venus.[51]

[50] Brief analysis in *Isis 6*, 138 (1924).

[51] Tycho Brahe, *De mundi aetherii recentioribus phaenomenis liber secundus qui est de illustri stella caudata* . . . (Uraniborg, 1588). Though it is irrelevant to my immediate purpose, I cannot resist mentioning that Brahe concluded in that

As to the Milky Way, which divides the heavens as a great circle along the solstitial colure, it also was supposed to be a meteorologic phenomenon, formed by dry and hot exhalations, similar to those that cause the meteors. A better understanding of the Milky Way was hardly possible without a telescope. Aristotle's views were finally disproved by Kepler, according to whom the Milky Way was concentric with the Sun, on the inner surface of the starry sphere.

A great many other phenomena are described and discussed in the *Meteorology*, such as meteors, rain, dew, hail, snow, winds, rivers and springs, the saltness of the sea, thunder and lightning, earthquakes. The consideration of each of them would require at least a page, and space is lacking, the patience of our readers limited. Let us restrict ourselves to a few remarks concerning Aristotle's optical theories. He rejected the view that light is material, being due to corpuscles emitted by the luminous object or emanating from the eye; on the contrary, he suggested that it was a kind of aetherial phenomenon. (Please do not call this an anticipation of the wave theory of light.) He was aware of the repercussions of sound (echo) and of light, and offered a theory of the rainbow, based upon the reflection of light in water drops and thus incomplete, yet very remarkable. His theory of colors has been compared to Goethe's, a comparison that is not very complimentary to the latter but is very much to Aristotle's credit.[52]

It is right to marvel at the endless number of physical questions in the Aristotelian corpus, but one should resist the temptation of reading into them too many ideas that are comparable to modern ideas yet could not possibly have had in their author's mind the meaning and pregnancy that they have in ours. One should never forget that the authority of a statement is a direct function of the knowledge and experience upon which it is based; many Aristotelian statements are brilliant, yet as irresponsible as the queries of an intelligent child.

The fourth book of the *Meteorology* is probably the work of Straton.[53] As it has come to us, it might be called the first textbook of chemistry. It discusses the constitution of bodies, the elements and qualities, generation and putrefaction, concoction and inconcoction (indigestion), solidification and solution, properties of composite bodies, what can and what cannot be solidified and melted, homoiomerous bodies.[54] The final conclusion is that end and function are more evident in nonhomoiomerous bodies than in the homoiomerous bodies that compose them, and in these than in the elements. Aristotle (or Straton) had been thinking hard on the differences that may or may not occur when two different bodies are mixed together; they may remain separate or separable, or they may be combined into something essentially new; their two forms may disappear or exist only *in potentia*, while a new form is created.[55]

---

same treatise of 1588 that the orbit of the comet of 1577 was not circular but elliptic. This was the first time that an astronomer referred to an orbit which was neither a circle nor a combination of circles. Kepler's discovery of elliptic trajectories was published in 1609.

C. Doris Hellman, *The comet of 1577: its place in the history of astronomy* (New York: Columbia University Press, 1944) [*Isis 36*, 266–270 (1946)].

[52] Aydin M. Sayili, "The Aristotelian explanation of the rainbow," *Isis 30*, 65–83 (1939). Carl B. Boyer, "Aristotle's physics," *Scientific American* (May 1950), pp. 48–51.

[53] *Isis 3*, 279 (1920–21).

[54] Homoiomerous: consisting of similar parts, homogeneous. The contrary is heteromerous, heterogeneous. Aristotle uses the words *homoiomerēs, anhomoiomerēs*.

[55] To use a modern comparison, the forms of hydrogen and oxygen disappear when a proper number of molecules of these gases unite to become water. There is no more hydrogen in that water, except *in potentia*.

All of which is again very impressive, especially when we bear in mind the impenetrability of the chemical jungle until the end of the eighteenth century. Aristotle and Straton went as far as it was possible to go in their time, or more exactly, their thinking far exceeded their experimental reach, and more than two thousand years would be needed to bring it to maturity and to fruitage.

We have given a few examples of the long acceptance of Aristotelian ideas and prejudices. One might say in a general way that Aristotelian physics dominated European thought until the sixteenth century. Then the revolt that had been gathering strength for centuries became more articulate, more intense, and better organized. In the middle of that century Ramus [56] went to the extreme of proclaiming that everything that Aristotle had said was false. The foundations of Aristotelian physics were undermined in the following century by Gassendi, who revived atomism, and by Descartes,[57] who accepted some of Aristotle's prejudices yet built up an entirely new structure. Yet even then the general conception of physics remained as broad as ever. Knowledge was hardly strong and sharp enough in any part of the immense field to separate that part from the rest, or to create physics as we understand it now.[58]

Aristotle's views were rejected, but they were not forgotten or overlooked, and there remained an active Scholastic and Peripatetic opposition. Aristotle was still very much alive, though on the defensive, as late as the eighteenth century.

GREEK MUSIC. ARISTOXENOS OF TARENTUM

One disciple of Aristotle must still be introduced before we close this chapter, not the least of them, the musician, or rather the theorist of music, Aristoxenos. Aristotle himself was much interested in music, not only in the ethical value of it, somewhat in the Platonic manner,[59] but also in the more technical sense. He was familiar with the Pythagorean discovery, the numerical aspect of musical harmony. Pythagoras or one of his early disciples had observed that when the vibrating string of a musical instrument was divided in simple ratios ($1{:}\frac{3}{4}{:}\frac{2}{3}{:}\frac{1}{2}$) one obtained very pleasant accords. Aristotle [60] extended the same operation to reed pipes.[61] He realized the importance of frequency of vibration, yet confused it with speed of transmission, and wrongly believed with Archytas that the speed of sound increased with the pitch. He asked the question, Why is the voice higher when it echoes back?[62] The question was curious and pertinent, but it was not answered until 1873 by Lord Rayleigh's theory of harmonic echoes.[63]

[56] Pierre La Ramée (1512–72), one of the victims of the St. Bartholomew purge.

[57] Gassendi (1592–1655) and Descartes (1596–1650) were almost exact contemporaries. They were antagonists and between them dominated the second quarter of their century.

[58] Consider the famous *Traité de physique* by Jacques Rohault (Paris, 1671), which remained for half a century the main textbook of Cartesian physics. It includes not only physics *stricto sensu* but cosmology, astronomy, meteorology, geography, physiology and medicine. G. Sarton, "The study of early scientific textbooks," *Isis 38*, 137–148 (1947–48).

[59] For the ethical aspect of music in ancient Greece (and in ancient China), see *Introduction*, vol. 3, pp. 161–162.

[60] More exactly, the unknown author of the *Problemata*. The *Problemata* contain probably Aristotelian ingredients, to which other Peripatetic ones were gradually added. The work as we have it may be relatively late, say fifth or sixth century; *Isis 11*, 155 (1928).

[61] *Problemata*, 919B, 5.

[62] *Problemata*, 918A, 35.

[63] Rayleigh, *Nature 8*, 319 (1873); *Theory of sound* (London: Macmillan, 1878; ed. 2, 1896; reprinted, 1926), vol. 2, p. 152.

It is probable that other members of the Lyceum discussed questions concerning acoustics and music, because the books of Aristoxenos, which we shall examine presently, contain a body of knowledge on that subject that is remarkable alike because of its relative depth, extent, and complexity.

Most of what we know concerning Aristoxenos is derived from Suidas (X–2), but Suidas used ancient books that are lost to us, and whatever he tells us is sufficiently confirmed from various other sources to be reliable. Aristoxenos was born in Tarentum, close to the country where Pythagorean fancies had matured; he was educated by his father, Spintharos, who was a musician, by Lampros of Erythrai and Xenophilos the Pythagorean,[64] finally by Aristotle. After the master's death the election of Theophrastos instead of himself as head of the Lyceum infuriated him. Suidas says that he flourished in the 111th Olympiad (336–333)[65] and that he was a contemporary of Dicaiarchos of Messina; he adds that Aristoxenos' writings dealt with music, philosophy, history, and all the problems of education, and that he wrote altogether 453 books!

The only work of his that has come down to us is his *Elements of harmony* (*Harmonica stoicheia*), which is the most significant treatise of its kind in ancient literature. As we have it, it seems to be an artificial recombination of two separate works. It covers (in Macran's edition) 70 pages or some 1610 lines.[66] It is a tedious book wherein Aristoxenos applied the logical methods of the Lyceum to the exposition of the knowledge that had been transmitted to him by Spintharos, Lampros, and Xenophilos or that he had obtained by his own experiments. It is divided into three parts, treating (1) generalities, pitch, notes, intervals, scales; (2) *idem*, plus keys, modulation, melody (the polemical tone of this suggests the existence of other writings now lost); (3) some twenty-six theorems on the combination of intervals and tetrachords in scales.

The most original part of Aristoxenos' work is the theoretical determinations of the intervals. Starting from the three Pythagorean intervals (2/1, 3/2, 4/3; octave, fifth, and fourth) he takes as unit the difference between the fifth and the fourth (the tone). That unit is too large, however; in order to obtain subunits he divides the interval arithmetically (not by extraction of roots). For example, in the descending fourth *la–mi* he inserts two tones, which gives the notes *sol*, *fa*. The new interval between *fa* and *mi* is the semitone. If this new interval is really a semitone, there are 5 semitones in the fourth, 7 in the fifth, and 12 in the octave. Aristoxenos went even further and considered not only semitones but also thirds, fourths, and

[64] Lampros and Xenophilos are otherwise unknown. Their names are mentioned, because it is notable that Aristoxenos was educated by at least one Pythagorean. Lampros hailed from Erythrai, but there were many places of that name; this was possibly the Ionian one opposite Chios (one of the twelve Greek cities of Asia Minor), for many Ionians had taken refuge in Magna Graecia. He is not the Lampros mentioned by Plato, also a musician but of an earlier time (first half of the fifth century).

[65] This may be taken to mean that Aristoxenos came to Athens *c.* 336–333; it may also mean that he was about 40 in 336; and this would make him a little older than Theophrastos. Whether he was 40 years old or 50 at the time of Aristotle's death (322), he had had time enough to prove his value and to be a worthy candidate for the mastership.

[66] *Introduction,* vol. 1, p. 142. Henry S. Macran, *The Harmonics of Aristoxenos* (Greek and English with notes, 303 pp.; Oxford, 1902). Louis Laloy, *Aristoxène de Tarente et la musique de l'antiquité,* (418 pp.; Paris, 1904), includes Aristoxenian lexicon; reprinted in 1924 [*Isis* 8, 530 (1926)].

Léon Boutroux, "Sur l'harmonique aristoxénienne," *Revue générale des sci.* *30*, 265–274 (1919) [*Isis* 3, 317 (1920–21)]; mathematical comparison of the Pythagorean and Aristoxenian ideas derived from Ptolemy's *Harmonics.*

even eights of the tone; these smaller divisions fell into abeyance. The empirical confusion between a leimma [67] and a semitone led Aristoxenos to a calculus comparable to the calculus by logarithms: the intervals (which are ratios) are calculated by means of additive units. This is extremely interesting, yet it would be foolish to conclude that Aristoxenos was a forerunner of Napier! There's many a slip 'twixt the cup and the lip, and there are many more between an idea and the theory eventually built upon it.[68]

The treatise of Aristoxenos is nevertheless highly significant, one of the masterpieces of Hellenic thought. Its influence was considerable, either directly or through the intermediary of the *Harmonics* of Ptolemy (II–1). The higher learning of late antiquity and of the medieval period included four main subjects (hence the name quadrivium),[69] and those four subjects were arithmetic, music, geometry, astronomy. Music, not physics! Thanks to Pythagoras and Aristoxenos, music was a mathematical science, while physics remained in a qualitative stage, close to philosophy.

Aristoxenos was less influential in the West, because the first great teacher of music in the Latin language was Boetius (VI–1), whose handbook was based chiefly upon the Pythagorean tradition rather than on the Aristoxenian one. The Byzantine musicologists, on the contrary, followed Aristoxenos. For Manuel Bryennios (XIV–1), who composed the latest Byzantine *Harmonics*, the history of music was divided into three periods – pre-Pythagorean, Pythagorean, and post-Pythagorean. The third of these periods was the one initiated by Aristoxenos and continued by the other musicologists of classical and Byzantine times; Manuel himself was still in that third and last age, the age of Aristoxenos. Indeed, Greek musical theory never surpassed Aristoxenos' exposition; nor did the practice of music (composition, playing, singing, teaching) change materially after him.[70]

Ancient music included not only music as we understand it but also metrics, poetry, for Greek poetry was composed to be chanted. Moreover, it had an ethical and cosmologic aspect; the theory of harmony in music was a part of the theory of harmony in the whole cosmos or in the soul of man. Thus music was a branch of philosophy as well as a branch of mathematics. It brought the humanities into the quadrivium.

[67] The word *leimma*, meaning remnant, residue, serves in music to designate the interval 256/243 left over when two tones (*tonoi*) of 9/8 are measured off from the fourth or tetrachord (*dia tessarōn*): 9/8) × (9/8) × (256/243) = 4/3. Plutarch misunderstood the interval 256/243 to mean 256 – 243, or 13, in *De animae procreatione in Timaeo Platonis*, 1017F.

[68] A claim to the invention of logarithms was made by modern Arabs on a similar ground, the musical theory of al-Fārābī; *Isis 26*, 552 (1936). This is even less justifiable, for the Arabic idea was borrowed from the Greek, and the Greek idea itself was a curious coincidence, not an invention.

[69] The quadrivium originated in Greek lands, but its success was more complete in the West, beginning with Boetius (VI–1). There is no single Greek word corresponding to quadrivium. The treatise of Georgios Pachymeres (XIII–2) is entitled *Syntagma tōn tessarōn mathēmatōn* (Stephanou edition; Rome, 1940) [*Isis 34*, 218–219 (1942–43)].

[70] Paul Henry Láng, *Music in Western civilization* (1124 pp., ill.; New York: Norton, 1941) [*Isis 34*, 182–186 (1942–43)]. Gustave Reese, *Music in the Middle Ages, with an introduction on the music of ancient times* (520 pp., 8 pls.; New York: Norton, 1940) [*Isis 34*, 182–186 (1942–43)].

XXI

THE NATURAL SCIENCES AND MEDICINE IN ARISTOTLE'S TIME

For the sake of greater clarity we shall divide this chapter into four main sections – geography; zoology and biology; botany; geology and mineralogy; medicine – in spite of the fact that this entails a few repetitions, chiefly in the case of Aristotle, who will naturally reappear in every section. This is another way of appreciating the comprehensiveness of his mind, the universality of his genius. One cannot deal with any science, or with any branch of science, without having to drag him in.

GEOGRAPHY

ARISTOTLE THE GEOGRAPHER

The most fundamental set of queries on natural history would naturally deal with the Earth itself, its shape, size, and surface. The shape and size have already been discussed in the astronomic section, and we have seen that Aristotle's estimate of the size was too large, but not shockingly so.[1] His knowledge of the size of the whole Earth was based upon calculations, which could be gradually improved without going out of a relatively small region, but his knowledge of the part of the Earth that was inhabited (the *oicumenē*) was derived from the reports of explorers and travelers. At the very best it was guesswork, for however well one might know certain regions this did not enlighten one at all with regard to the others. By the middle of the century many explorations had already been conducted (we have briefly described them in previous chapters), but if these were outlined on a sphere, one would realize at once that they covered only a very small portion of it. Alexander's expeditions improved considerably the knowledge of the Middle East and of the region west of the Indus and Jaxartes,[2] but their results were not completely available to Aristotle. The latter could avail himself, however, of the information collected by Scylax of Caryanda, whose *Periplus* was

[1] Roughly, Aristotle's estimate of the circumference of the Earth was to the correct measurement in the ratio 8 to 5. In bulk, the Aristotelian Earth was about four times the size of the real one.

[2] The Jaxartes is one of the two rivers, the eastern one, emptying into the Aral Sea, the Oxus being the other. Many cities (at least nine) were called Alexandria to honor the conqueror; one of them, called Alexandria Ultima, established on the Jaxartes, marked the limit of Alexander's efforts in Sogdiana.

published *c.* 360–347 (p. 299). It is highly problematic how much Aristotle knew of descriptive geography,[3] yet he was bold enough to postulate the extension of the inhabited world in the temperate zone "round the whole circle." [4] If that habitable region did not extend beyond the Pillars of Hercules in the west and India in the east, this was caused by the presence of the Ocean, not by climatic difficulties. On the other hand, he assumed a priori that the inhabited world was limited in breadth, because the cold would be too severe for human life in higher latitudes. If he had heard of Pytheas' voyages, he would have been more prudent.

The idea of zones goes back to Parmenides. He it was who conceived the spherical earth as divided into five parallel zones: a broad equatorial one, torrid; two polar ones, frigid; and in between two zones of moderate climate. The Greek *oicumenē* was included in the north temperate zone. This was made a little more precise by Aristotle (or rather by the author of the *Meteorology* [5]) but he was still unable to determine the limits of each zone. More precision would be introduced a century later by Eratosthenes of Cyrene (III-2 B.C.), and it is he, not Aristotle, who deserves to be called the founder of mathematical geography.[6]

PYTHEAS OF MASSILIA

If one means by "Italian" a man who was born and lived in the territory at present governed by the Italian Republic, then we have already come across many "Italians." Indeed, Magna Graecia [7] (*hē megalē Hellas*) was one of the cradles of Greek science. If Zenon of Elea was an "Italian," then Pytheas was a "Frenchman." It is better, however, not to mix old history with modern geography. Pytheas was born at Massilia (modern Marseille), in Gaul, and was thus the earliest representative of Western Europe in the history of science. He was probably one of Aristotle's younger contemporaries, for the latter did not know of his achievements, but they were quoted by Dicaiarchos.

Pytheas was one of the greatest navigators of antiquity. It is possible that his voyages were undertaken by order and at the expense of the Massilian colony, which was competing bitterly with its Carthaginian rivals and was anxious to outdo them in foreign trade, especially the rich trade in amber and tin.[8] It is equally possible that he was driven by his own eagerness and scientific curiosity. In the history of geographic discoveries, both motives, the personal and the social, are generally combined. Great deeds can be done only by great men, but however great, these men need help in order to accomplish their bold designs.

[3] The author of the *Meteorologica* gives much geographic information in Books I and II, information that seems to be derived from a geographic treatise or even from an atlas. One could put all that information on a map, but the result would be very unsatisfactory, full of holes. Moreover, we are always subdued by the same doubt; how much Aristotelian knowledge is there in the *Meteorologica?*

[4] *Meteorology,* 362–363.

[5] *Meteorology,* 363.

[6] For the history of zones, see Ernst Honigmann, *Die sieben Klimata* (Heidelberg: Winter, 1929) [*Isis 14*, 270-276 (1930)].

[7] The term Magna Graecia referred loosely to South Italy; it might or might not include Sicily. Greek colonies were restricted to a limited number of cities along the coasts. T. J. Dunbabin, *The Western Greeks* (518 pp.; Oxford: Clarendon Press, 1948) [*Isis 40*, 154 (1949)]; a very elaborate story but, unfortunately, it stops at about 480 B.C.

[8] Phocaia, the northernmost of the Ionian cities on the west coast of Asia minor, between Lesbos and Chios, distinguished itself among other Ionian cities by founding the westernmost colonies, Massilia in Gaul and Mainaca in Andalusia (east of Malaga). Such colonies were challenging Phoenician efforts in the Western Mediterranean. When the Phocaians were colonizing Massilia (*c.* 600), they defeated the Carthaginians in a sea fight (Thucydides, I, 13). The naval and commercial rivalry between Massilia and Carthage remained intense.

Pytheas was a scientific navigator; he was able to determine exactly the latitude of Massilia by means of a gnomon, and was one of the first Greeks to establish a relation between the Moon and the tides. This was due, of course, not so much to special intelligence as to the fact that he navigated outside of the Mediterranean where the tides were too small to attract attention. On the Atlantic shores the tides were high and as ancient people (not only the educated ones but also farmers and shepherds) observed the Moon carefully, they could not have failed to notice any relation that might exist between the lunar cycle and the tidal one.

Our knowledge of Pytheas' navigations is second hand [9] and he reported so many "marvels" that some of the ancient historians, like Polybios (II–1 B.C.) and Strabon (I–2 B.C.) distrusted him. His fate was comparable to that of Marco Polo in later times; some of the things that they told were so extraordinary, so contrary to common experience, that wise and prudent men could not believe them and concluded that they were fables. In both cases the stories that had been disbelieved were vindicated by later observations.

While certainty is out of the question, historians of ancient geography are now agreed that the achievements ascribed to Pytheas are real and occurred in Aristotle's time or very soon afterward (say in the period 330–300). There are, of course, unavoidable divergencies concerning localities and other details, but the general account as summarized below may be accepted as truthful.[10]

Pytheas and his companions sailed from Massilia, passed the Pillars of Hercules, visited Gades just west of them, then followed the Spanish and French coast northward. They were aware of the immense depth of the Bay of Biscay and of the enormous size of the Armorican peninsula (Brittany). Reaching the British Isles, they visited tin mines and the island Ictis,[11] connected to the shore at low tide, which was their commercial center. Pytheas gave a rough description of Britain as seen by a circumnavigator; yet he had made excursions inland, he noticed the native use of mead, the use of barns for threshing in bad weather, and the decrease of cultivation as one goes farther north. The general shape of Great Britain is a triangle of which the three vertexes are Orcas in the north (*Orcades insulae*, the Orkney and Shetland Islands), Belerion in the southwest (Land's End), and Cantion (Kent) in the southeast.

According to Polybios,[12] Pytheas followed the European coast all the way from Gades to Tanais. What is Tanais? [13] Two very different conjectures have been made; the Tanais would be a Baltic river, either the Vistula, which enters the Baltic at Danzig, or the Dvina, which enters it further east in Courland. The Tanais is more generally identified, however, with the Don, flowing into the Sea of Azov (Maeotis Palus). Pytheas visited places where amber was found, and the most

[9] It is derived mainly from Geminos of Rhodes (I–1 B.C.), Strabon (I–2 B.C.), Diodoros of Sicily (I–2 B.C.), and Pliny (I–2).

[10] It would take considerable space to explain how the summary was pieced up from many sources. See H. F. Tozer, *History of ancient geography* (ed. 2, Cambridge: University Press, 1935), pp. 152–164, xx [*Isis 26*, 537 (1936)]. Gaston E. Broche, *Pythéas le Massaliote, découvreur de l'extrême occident et du nord de l'Europe* (266 pp.; Paris: Société française d'imprimerie, 1935), with a map of Pytheas' navigations. J. Oliver Thomson, *History of ancient geography* (Cambridge: University Press, 1948) [*Isis 41*, 244 (1950)].

[11] Ictis was almost certainly Saint Michael's Mount, in the Bay of Penzance, Cornwall.

[12] In Strabon, II, 4, 1.

[13] The translators write the Tanais, but there is no article in Greek (*apo Gadeirōn heōs Tanaidos*). However, in another passage, Strabon, II, 4, 5, used the article: "The Tanais flows from the summer rising of the sun" (*ho de Tanaïs rhei apo therinēs anatolēs*).

famous of such places were along the southern shore of the Baltic. It may be that he sailed in the Baltic as far east as the longitude of the Maeotis Palus (strictly speaking that would be impossible, but the determination of longitudes was very vague).

We are on safer ground concerning the North Sea. He navigated it very far to the north, witnessed (or heard about) the extraordinary inrush of the sea at Pentland Firth, and may have sailed as far as the island Thule, to which he gave that name. Was Thule Iceland or northern Norway?[14] It was six days' voyage north of Britain and close to the frozen ocean. Did he go there or hear of it? Every traveler is tempted to extend his voyages beyond the limits that he has actually reached, by referring to countries beyond that are known to him by hearsay. It is clear that wherever one goes, one is likely to meet natives who have been farther.

At any rate, among Pytheas' incredible stories were the first reports of arctic conditions. He spoke of regions where the nights were exceedingly short, and of the "sleeping place of the sun," meaning perhaps the Arctic Circle, where during one day in the year the sun does not appear above the horizon. He described the inextricable mixture of air, sea, and land obtaining in these regions, and the frozen sea which can neither be traversed on foot nor by boat. Arctic travelers of our own time vindicate Pytheas, saying that his descriptions contain many details that could not have been invented. Says Fridtjof Nansen:

> What Pytheas himself saw may have been the ice sludge in the sea which is formed over a great extent along the edge of the drift ice, when this has been ground to a pulp by the action of waves. The expression 'can neither be traversed on foot nor by boat' is exactly applicable to this ice sludge. If we add to this the thick fog, which is often found near drift ice, then the description that the air is also involved in the mixture, and that land and sea and everything is merged in it, will appear very graphic.[15]

It is certain that arctic travelers are more competent to appreciate the verisimilitude of the statements ascribed to Pytheas than are armchair philologists, and their verdict is in his favor. This ought to satisfy us.

We owe to Pytheas not only the first account of northwest Europe and particularly of Great Britain, but also our first vision of the arctic world. This was an enormous increase of the knowledge available to Greek geographers.

NEARCHOS THE CRETAN

After this entirely unexpected trip to the arctic world, let us return to more familiar regions, the Mediterranean Sea and the Near East. When we gave a brief description of Alexander's conquests, we remarked that they increased considerably the geographic knowledge available to the Greeks. Much of our knowledge of the world originated in that very way: the *terrae incognitae* were not gently unveiled by lovers of science; they were brutally raped by conquistadors and their

[14] Later geographers identified Thule or Ultima Thule with Iceland, but that does not prove that the first user of the name Thule, Pytheas, meant that island and naught else. See Ferrari and Baudrand, *Novum lexicum geographicum* (Patavii, 1697), vol. 2, p. 228.

[15] Fridtjof Nansen (1861–1930), *In northern mists* (2 vols.; London, 1911). This contains an enthusiastic chapter on Pytheas (vol. 1, pp. 43–73); the passage quoted above appears on p. 67. Vilhjalmur Stefansson in his book *Greenland* (New York: Doubleday, Doran, 1942), pp. 28–41 [*Isis 34*, 379 (1942–43)] is even more enthusiastic.

followers, men of prey, whose only interest was power and wealth, yet who could not help increase geographic knowledge. Even if there had been no geographers attached to Alexander's army or detached by him for exploration, if there had been no scholars around him but historians without special interest in geographic facts, even then they could not have described intelligibly their master's razzias without explaining as clearly as possible where these took place. Historical events happen in a definite geographic background, and the geography that is inseparable from historiography, the geography of history, includes valuable fragments of the history of geography.

As a matter of fact, Alexander, who was a scientific organizer as well as a conqueror, had in his train not only secretaries, men of letters, historians, but also explorers, pathfinders,[16] surveyors, some of whom are known by name: one Heracleides, Archias, Androsthenes, Hieron of Soloi, Diognetos, Baiton; by far the most important was Nearchos, whose account has been preserved for us in the *Indica* of Arrian.[17]

A fleet having been rigged in 327 to convey Alexander's army from the Hydaspes (one of the tributaries of the Indus) to Persia, Nearchos was put in charge as admiral, while Onesicritos was the helmsman of Alexander's own ship.[18] Nearchos was a Cretan, but he flourished in Amphipolis;[19] he had been employed by Philip, then disgraced, but Alexander had realized his merit and taken him back in the Macedonian service. Nearchos acquitted himself very well in a highly difficult and dangerous undertaking. He sailed his fleet down the Hydaspes and the Indus, then to the Persian Gulf, the Shatt al-'Arab, the Tigris, the Pasitigris, and the Choaspes to Susa. His voyage occupied five months. He observed tidal phenomena (unknown to the Mediterranean sailors); he could not help observing them, of course, even as Pytheas could not help observing them along the Atlantic shores at about the same time. The fact that the tides existed in the Atlantic and in the Arabian Sea encouraged Eratosthenes (III–2 B.C.) to conclude that the whole of the outer ocean was a single mass of water.[20]

Nearchos made other observations; he was aware of the immense size of India (as compared with the Mediterranean scale of countries) and the fabulous length of her rivers. After passing Karachi they sailed along the country of the *ichthyophagi* (fish-eating people). They encountered a shoal of whales, and Nearchos (or Arrian) gives a vivid account of that astonishing and frightening spectacle. In the Persian Gulf he observed the pearl fisheries, which have continued to be operated there until our own day.[21]

As far as can be judged from many comparisons and verifications, Arrian's account is faithful and reliable.

[16] It is clear that he could not venture his army into unknown regions without preliminary scouting. Otherwise, the army might have perished in deserts, marshes, or inaccessible mountains.

[17] Arrianos (II–1) of Nicomedia in Bithynia, better known as the editor of Epictetos (II–1).

[18] Alexander himself did not sail farther than the mouth of the Indus, whence he continued by land; Nearchos remained in charge of the fleet

[19] Amphipolis in Macedonia, so called because the river Strymon (Struma), which separates Macedonia from Thrace, flows around the town, encircling it almost completely. Amphipolis is on the lower course of the river, near the sea, just east of the Chalcidice.

[20] "According to Eratosthenes, the whole of the outer sea is confluent, so that both the Western Sea and the Red Sea are one" (Strabon, I, 3, 13).

[21] They are now in relative decadence because of the competition of the native pearls produced by Japanese methods, of artificial pearls, and even more because of the increasing importance of oil in the Persian Gulf area and the growing industrialization of that area. There is more wealth in oil than in oysters, more in the earth than in the sea.

DICAIARCHOS OF MESSINA

The men of whom we have spoken thus far were explorers, travelers, and though their activities extended geographic knowledge considerably they were not professional geographers. Dicaiarchos, who now engages our attention, was a historian and geographer. His abundant writings dealt with history, politics, literature, philosophy, and geography proper, but only fragments of them remain.[22] He was born in Messina, Sicily, but flourished on the Greek continent, especially in the Peloponnesos and in Athens. He was a disciple of Aristotle, a friend of Theophrastos and Aristoxenos; thus we may place his acme (his flowering age) in the last quarter of the century.

His main work seems to have been a kind of cultural history of Greece, significantly called *The life of Hellas* (*Bīos Hellados*) of which nineteen fragments have come to us. We are more directly interested, however, in his geographic books, one of which was a description of the world, possibly accompanied with maps (*Periodos gēs*), and the other a treatise on the measurement of mountains; the fragment of it that has come to us deals with mountains in the Peloponnesos.

The reason why we suggest that his description of the world was illustrated with maps, or that he used maps, is a statement made by Agathemeros:

> Dicaiarchos divides the earth . . . by a completely straight line from the Pillars through Sardinia, Sicily, Peloponnese, Caria, Lycia, Pamphylia, Cilicia, Tauros, and on to Imaos. Of the regions thus formed he names one part the northern, the other the southern.[23]

Another novelty to the credit of Dicaiarchos is his attempt to measure the heights of mountains.[24] His estimates were generally too high, and yet he concluded that these mountains were as nothing compared with the size of the Earth. This was a bold conclusion; it required imagination and courage to declare that the gigantic mountains, the ascension of which may tax our strength to the limit, are but wrinkles of the Earth's surface. He influenced Eratosthenes and later geographers, like Strabon (I–2 B.C.), who admired him, but also philosophically minded writers like Cicero. The latter, who knew Dicaiarchos better than we can know him today, took his life as a model of the practical life, while Theophrastos' was a model of the theoretical life. This opinion was perhaps based upon Dicaiarchos' interest in measurements.[25] Aristotle's estimate of the size of the Earth was probably derived from that of his disciple. Dicaiarchos realized that the tides were influenced not only by the Moon but also by the Sun.

Thanks to the Alexandrian epic and to the commercial rivalries between the Greek and the Phoenican colonies, knowledge of geography, climatology, and anthropology had grown so much that the scholars who flourished during the last quarter of the century had a vision of the inhabited world that was at the same

[22] Carolus Müller, *Fragmenta historicorum graecorum* (Paris, 1848), vol. 2, pp. 225–268; *Geographi graeci minores* (Paris, 1882), vol. 1, pp. 97–110, 238–243. All the fragments are given in Greek with Latin translation and notes.

[23] Agathemeros wrote a geographic epitome the date of which is unknown except that it is later than Ptolemy (II-1).

[24] A passage of Theon of Smyrna (II-1) suggests that Dicaiarchos may have used a diopter (Hiller's ed., pp. 124–125). That is not at all impossible; any intelligent person wishing to determine azimuths or other angles with precision would be bound to invent a kind of diopter or theodolite (without lenses, of course); a simple type would be easy to construct and to use.

[25] Florian Cajori, "History of determinations of the heights of mountains," *Isis 12*, 482–514 (1929).

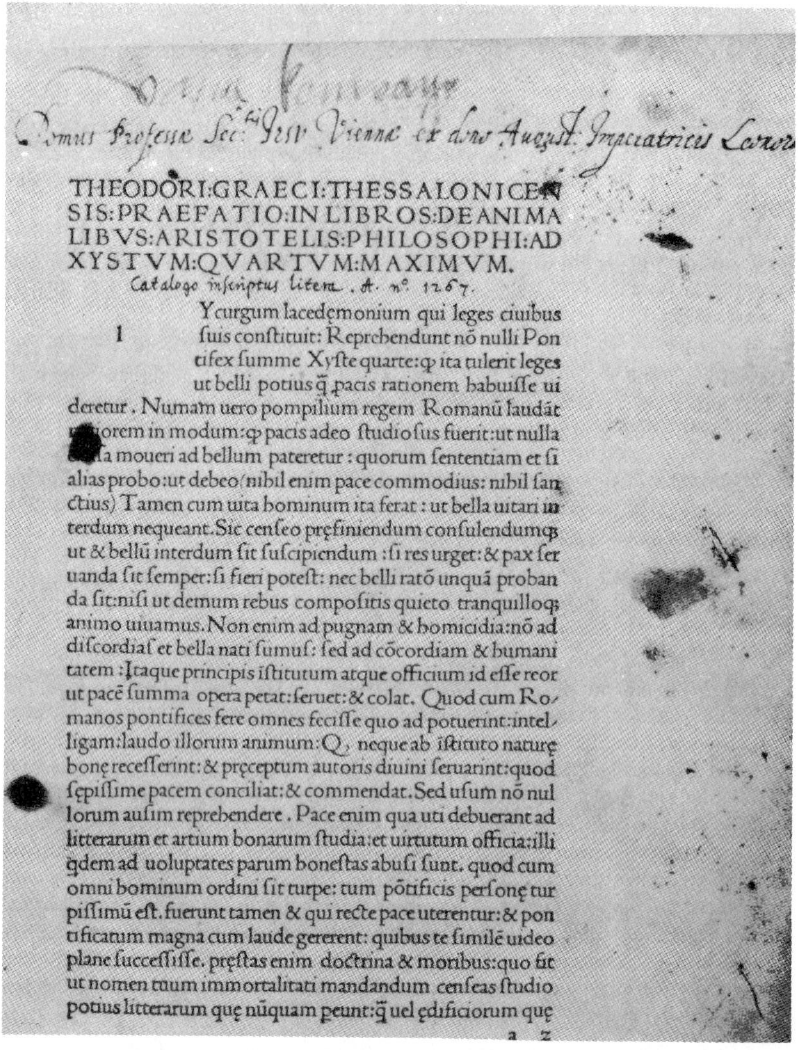

Domus Professae Soc.tis Jesu Viennae ex dono August. Imperatricis Leonor

THEODORI:GRAECI:THESSALONICEN
SIS:PRAEFATIO:IN LIBROS:DE ANIMA
LIBVS:ARISTOTELIS:PHILOSOPHI:AD
XYSTVM:QVARTVM:MAXIMVM.

Catalogo inscriptus litera. A. nº. 1267.

Ycurgum lacedęmonium qui leges ciuibus
suis constituit: Reprehendunt nō nulli Pon
tifex summe Xyste quarte: ꝙ ita tulerit leges
ut belli potius q̄ pacis rationem habuisse ui
deretur. Numam uero pompilium regem Romanū laudāt
[...]orem in modum: ꝙ pacis adeo studiosus fuerit: ut nulla
[...]a moueri ad bellum pateretur: quorum sententiam et si
alias probo: ut debeo (nihil enim pace commodius: nihil san
ctius) Tamen cum uita hominum ita ferat: ut bella uitari in
terdum nequeant. Sic censeo pręfiniendum consulendumq;
ut & bellū interdum sit suscipiendum: si res urget: & pax ser
uanda sit semper: si fieri potest: nec belli ratō unquā proban
da sit: nisi ut demum rebus compositis quieto tranquilloq;
animo uiuamus. Non enim ad pugnam & homicidia: nō ad
discordias et bella nati sumus: sed ad cōcordiam & humani
tatem: Itaque principis īstitutum atque officium id esse reor
ut pacē summa opera petat: seruet: & colat. Quod cum Ro-
manos pontifices fere omnes fecisse quo ad potuerint: intel-
ligam: laudo illorum animum: Q, neque ab īstituto naturę
bonę recesserint: & pręceptum autoris diuini seruarint: quod
sępissime pacem conciliat: & commendat. Sed usum nō nul
lorum ausim reprehendere. Pace enim qua uti debuerant ad
litterarum et artium bonarum studia: et uirtutum officia: illi
q̄dem ad uoluptates parum honestas abusi sunt. quod cum
omni hominum ordini sit turpe: tum pōtificis personę tur
pissimū est. fuerunt tamen & qui recte pace uterentur: & pon
tificatum magna cum laude gererent: quibus te similē uideo
plane successisse. pręstas enim doctrina & moribus: quo fit
ut nomen tuum immortalitati mandandum censeas studio
potius litterarum quę nūquam peunt: q̄ uel ędificiorum quę

a 2

Fig. 94. Title page of the *Liber de animalibus*, as Latinized by Theodoros Gaza (*c.* 1400–1475) of Thessalonica; first edition, folio, 30 cm (Venice: John of Cologne and John Manthen de Gherretzen, 1476; Klebs, 85.1). Theodoros was one of the collaborators of Vittorino da Feltre in Mantua; he translated many books from Greek into Latin, and also from Latin into Greek. [From the copy in the Harvard College Library.]

time broader and more detailed. We may assume that the efforts made by Dicaiarchos were not unrelated to that new vision. As one's knowledge increases and becomes more precise, it invites new surveys. Dicaiarchos prepared such a survey and began a new series of measurements which would make possible eventually the creation of scientific geography by Eratosthenes.

ZOÖLOGY AND BIOLOGY

ARISTOTLE, THE ZOÖLOGIST, THE BIOLOGIST

The main texts for the study of Aristotelian biology (Figs. 94, 95, 96) are *De anima, Historia animalium, De partibus animalium, De motu animalium, De incessu animalium, De generatione animalium.* These books touch some of the fundamental problems of biology and contain an almost incredible wealth of information on innumerable subjects. Much of that information has naturally lost its validity, but the surprising thing is that so much of it is still valid today, with relatively few qualifications. The abundance of facts mentioned in the zoölogic treatises is such that it would have been impossible for a single man to collect them. We must assume that Aristotle was helped by many colleagues and disciples. This assumption entails a relatively late date of composition[26] even if Aristotle's own investigations began very early. His interest in natural history may have originated in his boyhood when his father took him with him on his

[26] Say the second Athenian residence, or Lyceum period (335–322).

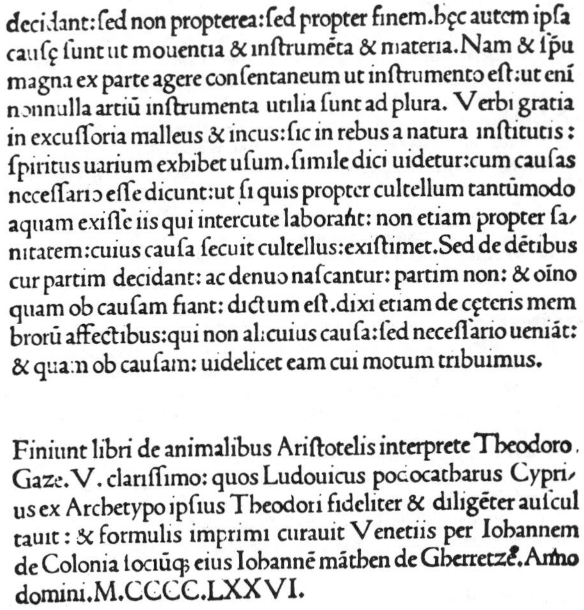
decidant:ſed non propterea:ſed propter finem.hęc autem ipſa
cauſę ſunt ut mouentia & inſtrumēta & materia.Nam & ſpū
magna ex parte agere conſentaneum ut inſtrumento eſt:ut enī
nonnulla artiū inſtrumenta utilia ſunt ad plura. Verbi gratia
in excuſſoria malleus & incus:ſic in rebus a natura inſtitutis:
ſpiritus uarium exhibet uſum.ſimile dici uidetur:cum cauſas
neceſſario eſſe dicunt:ut ſi quis propter cultellum tantūmodo
aquam exiſſe iis qui intercute laborāt: non etiam propter ſa-
nitatem:cuius cauſa ſecuit cultellus:exiſtimet.Sed de dētibus
cur partim decidant: ac denuo naſcantur: partim non: & oīno
quam ob cauſam fiant: dictum eſt.dixi etiam de cęteris mem
brorū affectibus:qui non alicuius cauſa:ſed neceſſario ueniāt:
& quam ob cauſam: uidelicet eam cui motum tribuimus.

Finiunt libri de animalibus Ariſtotelis interprete Theodoro.
Gaze.V.clariſſimo: quos Ludouicus podocatharus Cypri-
us ex Archetypo ipſius Theodori fideliter & diligēter auſcul
tauit: & formulis imprimi curauit Venetiis per Iohannem
de Colonia ſociūq; eius Iohannē māthen de Gherretzē.Anno
domini.M.CCCC.LXXVI.

Fig. 95. Colophon of the *Liber de animalibus.* [From the copy in the Harvard College Library.]

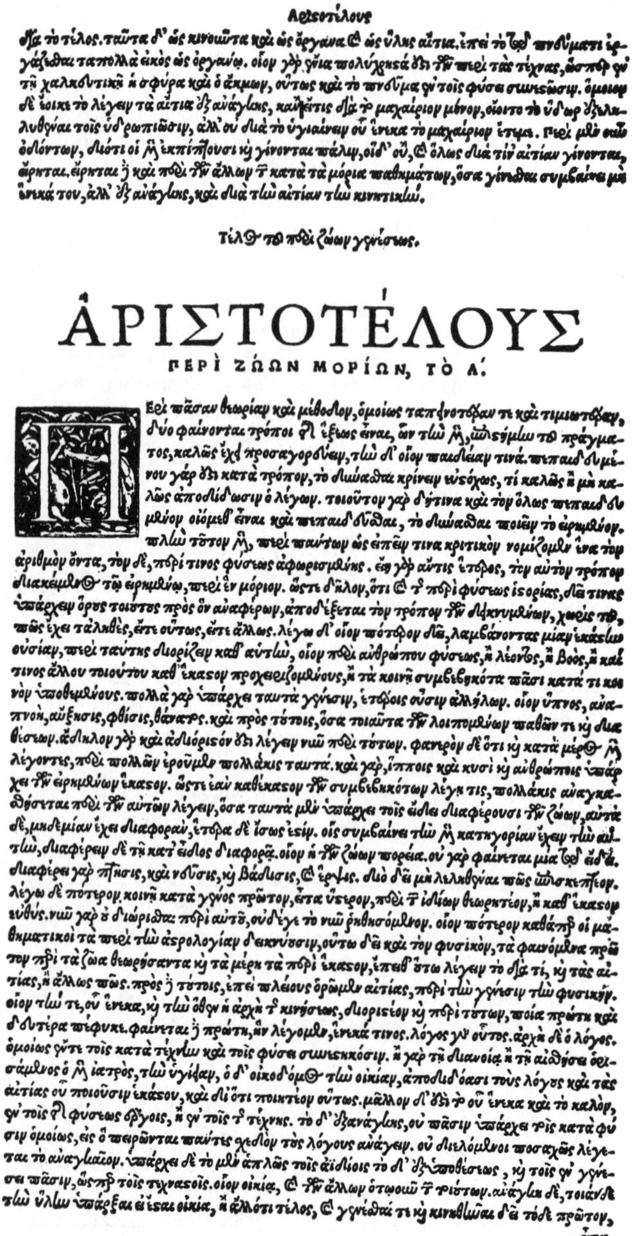

Ἀριστοτέλους

Τέλος τοῦ περὶ ζώων γενέσεως.

ἈΡΙΣΤΟΤΈΛΟΥΣ

ΠΕΡῚ ΖΏΩΝ ΜΟΡΊΩΝ, ΤῸ Α.

Fig. 96. Page 232 verso of volume 1 of the second edition of the Greek Aristotle, prepared by Erasmus of Rotterdam with emendations by Simon Grynaeus (folio; Basel: Bebel, 1531). This page shows the end of the *De generatione animalium* and the beginning of the *De partibus*. [From the copy in the Harvard College Library.]

The Greek *princeps* of Aristotelian zoölogy was in vol. 3 (1497) of the Aristotle-Theophrastos princeps (5 vols.; Venice: Manutius, 1495–1498), represented in Fig. 102.

medical rounds; it continued in Athens and was probably stimulated during the years spent at the seashore in Assos and Lesbos. Among Aristotle's helpers was Alexander the Great, who provided information and obtained specimens in distant countries. No matter how many men collaborated with him, the zoölogic books were very probably written by himself; the style is uniform in its scientific austerity, and the teleologic point of view permeating the whole is typical of Aristotelian thought.[27]

English readers may easily consult these texts in the Oxford English *Aristotle* and in the Loeb Classical Library. The Loeb volumes are especially convenient, because they contain the Greek original on opposite pages to the translation.

Among recent studies of Aristotelian biology we may mention Thomas East Lones, *Aristotle's researches in natural science* (302 pp., ill.; London, 1912) [*Isis 1*, 505–509 (1913)], and above all the many publications of my good friends, the late D'Arcy W. Thompson[28] and Charles Singer. Of Sir D'Arcy's, it will suffice to recall his *Glossary of Greek birds* (London: Oxford University Press, 1895, 1936) [*Isis 29*, 135–138 (1938)], his translation of *Historia animalium* (Oxford, 1910), his *Aristotle as a biologist* (London, 1913),[29] and his *Glossary of Greek fishes* (London: Oxford University Press, 1947) [*Isis 38*, 254 (1947–48)]; of Singer's, his *Greek biology* (Studies in the history and method of science, vol. 2, pp. 1-101; (Oxford: Clarendon Press, 1921) and his *Short history of biology* (London: Harper, 1931; often reprinted).

The strange vicissitudes of Aristotle's fame in antiquity have been related in the previous chapter. To Cicero and the latter's contemporaries he was known as a Platonist; later, his early Platonic writings vanished, and he became known by the works of his maturity. Not by all of these, however; for centuries, attention was centered upon the Organon; then his other works were gradually appreciated, those dealing with astronomy, physics, ethics, government. The books on natural history were read also, but modern biologists gradually lost their respect for them, as their own ideas became more "scientific." It was only in the second half of the nineteenth century that the best parts of Aristotelian biology were properly appreciated, and from that time on, Aristotle the zoölogist, Aristotle the biologist was the subject of increasing praise and wonder. Some enthusiasts went so far as to say that Aristotle's genuine fame rested on his biology alone, the rest of Aristotelian writings might be abandoned,[30] but those dealing with natural history were truly astounding.

It is my hope that the four chapters devoted to Aristotle in this book will make a better-balanced judgment of him possible. He was certainly one of the very greatest men in the whole of our past, but greatness is never absolute. Aristotle's

[27] Many errors due to carelessness and ill arrangement tend to support the "notebook" hypothesis: the biologic books have come down to us not as the master wrote them himself but as students took them down. To this we offer two answers: (1) Aristotle remains the author, if not the editor, of his thoughts; (2) we must always bear in mind that ancient books were not submitted, like ours, to the long ordeal of final revision and proofreading. Any author knows the many differences obtaining between his final "personal" manuscript and the text that finally appears in print.

[28] G. Sarton, "D'Arcy Wentworth Thompson, 1860–1948," *Isis 41*, 3–8 (1950), with portrait.

[29] Herbert Spencer lecture, Oxford, 1913; reprinted in *Science and the classics* (London: Oxford University Press, 1940) [*Isis* 33, 269–270 (1941–42)].

[30] There was even a reaction against Aristotelian logic, the essence of which had been accepted for more than 22 centuries! The attack against Aristotelian logic was led by the Polish philosopher, Alfred Habdank Korzybski (1879–1950); *Isis 30*, 517 (1939); *41*, 202 (1950).

encyclopedic knowledge was astounding indeed; yet it was very imperfect; it could not be otherwise.

Modern biologists reading the Aristotelian books concerning their own studies are astonished by the wealth of detail and even more so by the broadness and complexity of his outlook. He opened up the main fields of inquiry – comparative anatomy and physiology, embryology, customs of animals (ethology), geographic distribution, ecology – and in each field he assembled the relevant facts, described and discussed them, and drew philosophic conclusions. The facts have been gradually corrected because of the availability of better observational and experimental methods, yet many of the conclusions have reappeared periodically in sundry dresses; they are still acceptable in our own day to a number of well-informed biologists.

The writings enumerated above may be classified as follows.

The *Historia animalium* contains all the zoölogic observations collected under Aristotle's guidance.

In spite of its title, *De partibus animalium* is physiologic rather than anatomic. That title (who is responsible for it?) is very misleading;[31] the book concerns functions of the body; it does not deal with parts (limbs and organs), but with what we would call tissues. At the beginning of it, three kinds of composition are recognized, the first being the purely physical kind, the second, homogeneous parts or tissues, the third, heterogeneous parts or organs. Six kinds of "tissues" are referred to: blood, fat, marrow, brain, flesh, bone. The *De partibus animalium* is the earliest treatise on animal physiology in any language. *De incessu animalium* also deals with physiology in the same way; it explains how the bodies of animals are made to serve their purposes. We must remember that every living being is made up of matter and form (soul). The two books just named deal with the matter; *De anima*, on the other hand, deals with the form; it is a treatise on psychology.

The two other treatises, *De motu animalium* and *De generatione animalium*, plus the smaller treatises grouped under the general title *Parva naturalia*, discuss functions common to matter and form (body and soul) and various peculiarities of behavior. If we realize that physics (as we understand it) hardly existed in Aristotle's time, and chemistry not at all, we can hardly expect his physiology to be other than rudimentary. Indeed, in all fairness, we must consider it as a kind of protophysiology. Yet it is surprising how much of the truth Aristotle had already managed to catch a glimpse of. He had no understanding of respiration, but he had some general idea of nutrition. He conceived it as the transformation of the ingested food into nutriment which was carried by the blood to the parts of the body. Not bad at all. How could he have imagined, without any chemical knowledge, the infinitely complex chemical reactions that are involved? He realized the existence of excretory organs and the meaning of excreta such as bile, urine, and sweat. His account of the gall bladder was remarkably correct (with due regard to the limitations unavoidable in his time), but he thought[32] that some

[31] In the *De generatione animalium*, 782A 21, Aristotle calls his other book "The causes of the parts of the animals," which is better than the title with which we are familiar.

[32] *History of animals*, 506A, 22; *Parts of animals*, 676B, 27.

viviparous quadrupeds lacked it; in this he was wrong, for all the mammals have it.

We must return for a moment to teleology, which we have already described as an essential part of Aristotelian thought. In order to understand its application to life, or rather to living beings, it will be helpful to reconsider the Aristotelian notions of cause and of soul. Though both words have a general meaning, there are various kinds of causes and various kinds of souls.

As to causes, we must distinguish (1) the final cause or rational purpose, something that pulls from ahead, (2) the motive or efficient cause, (3) the formal cause, (4) the material cause. More simply, one might group the first three as formal causes, as against the material cause. Both (1) and (3) are occasionally represented by the same word, *logos*. Yet we must often draw the line between the final cause and the formal cause *stricto sensu* as we do between future and present.

The most general definition of the soul is given in *De anima*: "The soul is the first grade of actuality of a natural body having life potentially in it. The body so described is a body which is organized." [33] All living beings have a nutritive soul (a soul that guides their nutrition and their material life); in addition, all animals have a sentient soul, which enables them to feel; in addition, some higher animals have an appetitive and locomotive soul; in addition, men have a rational soul.[34] All these souls are parts (or faculties) of the soul. To put it otherwise, we might say that the soul of a living being becomes more and more complex as we go up the scale of relative perfection, the acme being reached in the highest being, man. In any case, the soul belongs to the body and cannot be detached from it (as the Pythagoreans thought); it is not separate from it, but is the form or realization (*entelecheia*, actuality) of it. Every living body is animated (*empsychos* as opposed to *apsychos*); that is, every living body is made up of matter and form.[35]

Aristotle's teleology was of the limited kind called by Bergson the "doctrine of internal finality"; in the case of each individual all the parts are united for the greatest good of its wholeness and are intelligently organized in view of that end, but without regard for other individuals. That doctrine was generally accepted until Darwin elaborated his theory of natural selection (1859). Teleology could then be extended ("doctrine of external finality") from the individual or from the species, to all the individuals or to all the species, which constitute a greater whole, the whole of life.[36]

Aristotelian teleology is expressed by the formula "Nature never makes anything that is superfluous";[37] hence, it did not consider vestigial or rudimentary organs which could be interpreted only in terms of "evolution," that is, in terms not of

[33] *De anima*, 412A, 28; translation by J. A. Smith in the Oxford *Aristotle*. Scholars should refer to the Greek text, which cannot be translated adequately. It is a good specimen of Aristotelian prose made heavy by excessive condensation.

[34] More briefly, there are three kinds of soul: (1) vegetative or nutritive in all living beings, (2) animal or sensitive in all the animals, (3) rational in men only. (Thus men have the three kinds.) That classification was generally accepted until modern times. Note that the Aristotelian idea of soul or psyche is not differentiated from mind. Vital spirit, soul, mind are all one.

[35] Compare Genesis 2:7: "And the Lord God formed man of the dust of the ground and breathed into his nostrils the breath of life; and man became a living soul."

[36] These remarks on internal *versus* external finality are suggested by Francis Hugh Adam Marshall in his foreword to the Loeb edition of the *Parts of animals* (1937).

[37] *Parts of animals*, 691B, 4.

any single individual but of a long series of them. Nature does nothing without a purpose. But what is the purpose of each individual? His purpose is revealed to us by his activities, and chiefly by their best and ultimate fruits.

These views have been developed by many biologists and are accepted to this day with technical qualifications by many of them, who are called vitalists.[38]

Aristotle's classification of souls of increasing complexity as one goes up the scale of nature implies a belief in the existence of such a scale, and that belief was very clearly expressed by him in the *Historia animalium.*

Nature proceeds little by little from things lifeless to animal life in such a way that it is impossible to determine the exact line of demarcation, nor on which side thereof an intermediate form should lie. Thus, next after lifeless things in the upward scale comes the plant, and of plants one will differ from another as to its amount of apparent vitality; and, in a word, the whole genus of plants, whilst it is devoid of life as compared with an animal, is endowed with life as compared with other corporeal entities. Indeed, as we just remarked, there is observed in plants a continuous scale of ascent towards the animal. So, in the sea, there are certain objects concerning which one would be at a loss to determine whether they be animal or vegetable. For instance, certain of these objects are fairly rooted, and in several cases perish if detached; thus the pinna is rooted to a particular spot, and the solen (or razor-shell) cannot survive withdrawal from its burrow. Indeed, broadly speaking, the entire genus of testaceans have a resemblance to vegetables, if they be contrasted with such animals as are capable of progression.

In regard to sensibility, some animals give no indication whatsoever of it, whilst others indicate it but indistinctly. Further, the substance of some of these intermediate creatures is fleshlike, as is the case with the so-called tethya (or ascidians) and the acalephae (or sea-anemones); but the sponge is in every respect like a vegetable. And so throughout the entire animal scale there is a graduated differentiation in amount of vitality and in capacity for motion.

A similar statement holds good with regard to habits of life. Thus of plants that spring from seed the one function seems to be the reproduction of their own particular species, and the sphere of action with certain animals is similarly limited. The faculty of reproduction, then, is common to all alike. If sensibility be superadded, then their lives will differ from one another in respect to sexual intercourse through the varying amount of pleasure derived therefrom, and also in regard to modes of parturition and ways of rearing their young. Some animals, like plants, simply procreate their own species at definite seasons; other animals busy themselves also in procuring food for their young, and after they are reared quit them and have no further dealings with them; other animals are more intelligent and endowed with memory, and they live with their offspring for a longer period and on a more social footing.

The life of animals, then, may be divided into two acts – procreation and feeding; for on these two acts all their interests and life concentrate. Their food depends chiefly on the substance of which they are severally constituted; for the source of their growth in all cases will be this substance. And whatsoever is in conformity with nature is pleasant, and all animals pursue pleasure in keeping with their nature.[39]

Note that the Aristotelian *scala naturae* does not necessarily imply evolution, for that *scala* might be conceived as static and the idea of the fixity of species is not

[38] Hans Driesch (1867–1941), *The history and theory of vitalism* (347 pp.; London, 1914) [*Isis* 3, 439–440 (1920–21)]; *Mind and body* (London, 1927).

[39] *History of animals,* 588B, 4; quoted from D'Arcy W. Thompson's translation in the Oxford *Aristotle.* I extend the quotation beyond my immediate need to 589A, 9, because it illustrates the richness of Aristotle's thought.

incompatible with it.[40] The *scala* appealed to medieval imagination, especially in the Muslim world. Arabic men of science often spoke of it, and those who were mystically minded liked to think of a continual scale or chain of being leading from the minerals to plants, from plants to animals, from animals to men, and from men to God.[41] The *scala naturae* was a means of illustrating the fundamental unity and order of nature. It implied a classification, but Aristotle did not stop there. He had recognized some 540 species of animals, a number that may seem ridiculously small to modern taxonomists but was immense in his age. Many of these animals have such obvious relations that they group themselves, as it were; yet a complete classification involves great difficulties. Aristotle faced those difficulties and solved many of them; for example, the fishlike Cetacea did not fool him, and he recognized their mammalian nature. In spite of the availability of his zoölogical books to the Latin West from the thirteenth century on, medieval scholars overlooked his just conception of the cetaceans, which was forgotten until Pierre Belon rediscovered it and published (in 1551) a description of the cetacean placenta. It is clear that Aristotle had devoted considerable thought to problems of classification; he showed that one had to be very careful and not mistake apparent resemblances, such as are due to homology (bone and fish spine, scales and feather, nail and hoof), with real ones, due to the "more or less" (excess or defect) of this or that part. He had certainly a table of classification in mind, and it is highly probable that he put it down either in words or in synoptic form; such a table has not come down to us, but it is not difficult to reconstruct it.

Aristotle objected to excessive dichotomy, but his classification began with a fundamental one, the division of the animal kingdom into two very different parts, which he called the bloody animals and the bloodless (this fundamental division is preserved under the names vertebrates and invertebrates). We cannot discuss the details of his classification; it will suffice to publish its reconstruction. Thanks to the kindness of Charles Singer, we reproduce here two forms of it (Figs. 97 and 98).[42]

That scheme included many errors and imperfections, but if one bears in mind the number of facts available to Aristotle, most of which had been collected under his direction, and the poverty of his observational means, one cannot help admire the results he obtained.

Comparative anatomy and physiology. Most of the anatomic notes are to be found in the *Historia animalium*, but they are mixed with physiologic remarks. The other books are more physiologic. The distinction between anatomy and physiology was not as clear-cut as it is today. Aristotle's main purpose was to describe the animals, and it was hardly possible to speak of the organs without speaking of their functions; from the Aristotelian point of view the function created the organ rather than the opposite. To give a full account of Aristotle's anatomy and physiology would be endless. It will suffice to give a few examples of good and bad.

Because Aristotle was a zoölogist, his anatomy naturally took the form of com-

[40] Harry Beal Torrey and Frances Felin, "Was Aristotle an evolutionist?" *Quart. Rev. Biol.* 12, 1–18 (1937). After reviewing all the evidence they cannot answer yes or no.

[41] For details on Arabic, Persian, and Turkish documents on this subject see *Introduction*, vol. 3, pp. 211–213, 1170.

[42] As given by him in his article on "Greek biology" in *Studies in the history and method of science* (Oxford, 1921), vol. 2, pp. 1–101; see pp. 16, 21.

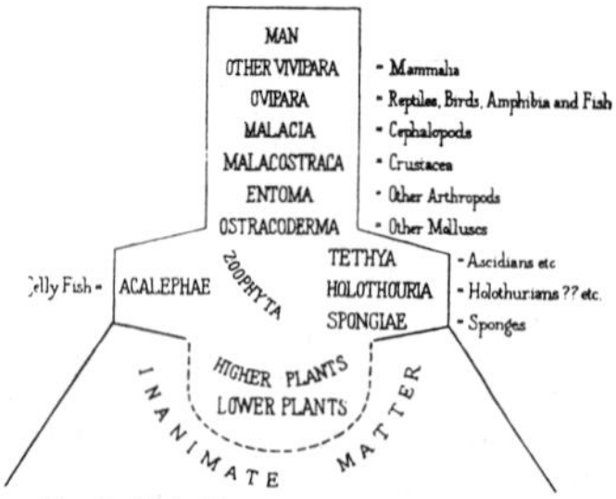

The order of living things, put together from the descriptions of Aristotle.

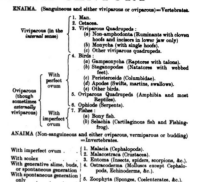

Fig. 97. The Aristotelian classification of animals as reconstructed from the *Historia animalium*. [Reproduced from *Studies in the history and method of science* (Oxford, 1921), vol. 2, p. 16, with kind permission of Dr. Charles Singer and of the Clarendon Press.]

ENAIMA. (Sanguineous and either viviparous or oviparous)=Vertebrates.

Viviparous (in the *internal* sense)
- 1. Man.
- 2. Cetacea.
- 3. Viviparous Quadrupeds:
  - (*a*) Non-amphodonta (Ruminants with cloven hoofs and incisors in lower jaw only)
  - (*b*) Monycha (with single hoofs).
  - (*c*) Other viviparous quadrupeds.

Oviparous (though sometimes *externally* viviparous)

With perfect ovum
- 4. Birds:
  - (*a*) Gampsonycha (Raptores with talons).
  - (*b*) Steganopodes (Natatores with webbed feet).
  - (*c*) Peristeroeide (Columbidae).
  - (*d*) Apodes (Swifts, martins, swallows).
  - (*e*) Other birds.
- 5. Oviparous Quadrupeds (Amphibia and most Reptiles).
- 6. Ophiode (Serpents).

With imperfect ovum
- 7. Fishes:
  - (*a*) Bony fish.
  - (*b*) Selachia (Cartilaginous fish and Fishing-frog).

ANAIMA (Non-sanguineous and either oviparous, vermiparous or budding) =Invertebrates.

With imperfect ovum . . . 1. Malacia (Cephalopods). 2. Malacostraca (Crustacea).
With scolex . . . . 3. Entoma (Insects, spiders, scorpions, &c.).
With generative slime, buds, or spontaneous generation 4. Ostracoderma (Molluscs except Cephalopods, Echinoderms, &c.).
With spontaneous generation only . . . . 5. Zoophyta (Sponges, Coelenterates, &c.).

Fig. 98. Diagram reproducing the Aristotelian idea of *scala naturae* implicit in the *Historia animalium*. [Reproduced from *Studies in the history and method of science*, vol. 2, p. 21, with kind permission of the author and publisher.]

parative anatomy, and his classification was based, as it ought to be, on anatomic evidence. For example, he studied the stomach of the ruminants and gave a correct account of the four chambers.

In spite of his prudence he was sometimes led into dangerous comparisons. Here is a good example of bad Aristotle, which I need not discuss. It combines a whole series of unrelated topics.

Of all animals human beings are the ones which go bald most noticeably; but still baldness is a general and widespread condition. Thus, although some plants are evergreen, others shed their leaves, and birds which hibernate shed their feathers. Baldness, in those human beings whom it affects, is a comparable condition to these. Of course, a partial and gradual shedding of leaves takes place in all plants, and of feathers and hair in those animals that have them; but it is when the shedding affects the whole of the hair, feathers, etc., at once that the condition is described by the terms already mentioned (baldness, moulting, etc.). The cause of this condition is a deficiency of hot fluid, the chief hot fluid being greasy fluid, and that is why greasy plants tend more to be evergreen than others. However, we shall have to deal with the cause of this condition so far as plants are concerned in another treatise, since in their case there are other contributory causes of it. Now in plants this condition occurs in winter: this seasonal change overrides in importance the change in the time of life. The same is true of the hibernating animals; they too are in their nature less fluid and less hot than human beings. For human beings, however, it is the seasons of life which play the part of summer and winter; and that is why no one goes bald before the time of sexual intercourse, and also why that is the time when those who are naturally prone to intercourse go bald. The reason is that the effect of sexual intercourse is to cool, as it is the excretion of some of the pure, natural heat, and the brain is by its nature the coldest part of the body; thus, as we should expect, it is the first part to feel the effect: anything that is weak and poorly needs only a slight cause, a slight momentum, to make it react . . . And it is owing to the same cause that it is on the front part of the head only that human beings go bald, and that they are the only animals which do so at all; i.e., they go bald in front because the brain is there, and they alone do so, because they have by far the largest brain of all and the most fluid. Women do not go bald because their nature is similar to that of children: both are incapable of producing seminal secretion. Eunuchs, too, do not go bald, because of their transition into the female state, and the hair that comes at a later stage they fail to grow at all, or if they already have it, they lose it, except for the pubic hair: similarly, women do not have the later hair, though they do grow the pubic hair. This deformity constitutes a change from the male state to the female.[43]

These statements are foolish, but they are not contemptible. They are not uncritical folklore, but premature generalizations based upon too small a number of facts, facts that had not been observed with sufficient care and had been brought together too hastily. Some of the topics dealt with in such a casual way are extremely difficult.[44]

What is much worse, Aristotle had very wrong ideas concerning the brain and the heart, in spite of the fact that the main function of the brain had been recognized almost two centuries before by Alcmaion of Croton. Aristotle considered the heart the seat of intelligence, the function of the brain being then simply to cool the heart by the secretion of phlegm and to prevent its overheating. How could this experienced and wise man reach such preposterous conclusions? The insensitivity of the exposed brain to handling and wounding is striking, and even more so the sensitiveness of the heart to emotions; the brain seems comparatively bloodless; and so on.[45] At any rate, Aristotle's position was clear: the brain may serve the mind indirectly (by its action upon the heart), but it is not the seat of reason. It is very curious that Aristotle, son of a physician, was less interested in medicine than in science and philosophy, and was apparently unacquainted with the Hippocratic writings.[46] One cannot help being shocked, however, by his wrongness on one of the most fundamental points of human life.

Habits of animals. The *Historia animalium* is full of observations concerning the curious habits of animals. Much of that had been known to observant farmers or fishermen long before Aristotle's time, but it required scientific curiosity and persistence such as the master possessed to criticize those stories and put them together in scientific language. His criticism varies greatly from one case to another; sometimes we are astonished by its depth and sureness; sometimes we shake our head and wonder how he could be so careless. The answer to that is, of course, that genius even at its best is never continuous. *Aliquando bonus dormitat . . . Aristoteles!* This remark must be made, because the good examples that I am

[43] *De generatione animalium*, 783B, 9, quoted from the Loeb translation by A. L. Peck (1943).

[44] For example, hibernation, about which see M. A. Herzog, *Aristoteles Anschauungen über die Lehre vom Winterschlaf* (Festschrift für Zschokke, No. 41, 28 pp.; Basel, 1920) [*Isis 4*, 128 (1921–22)]. Francis G. Benedict and Robert C. Lee, *Hibernation and marmot physiology* (250 pp., 2 pls., 11 figs.; Washington: Carnegie Institution, 1938) [*Isis 30*, 398 (1939)]. For the latest views see Charles P. Lyman and Paul O. Chatfield, "Hibernation," *Scientific American* (December 1950). The mechanism of hibernation is better understood, but the fundamental process is still a mystery.

[45] Sir Charles Sherrington, *Man on his nature* (Cambridge: University Press, 1940), p. 238 [*Isis 33*, 544–545 (1941–42); *34*, 48 (1942–43)]. "How came it," asks Sir Charles, "that Aristotle, the father of psychology, missed the localization of the mind in the brain?"

[46] There are many references in Aristotle to the mathematician Hippocrates of Chios, but only one, insignificant, to the physician Hippocrates (*Politica*, 1326A, 15). Aristotle's relative lack of interest in medicine is not so curious after all, but rather normal. The mathematical and the medical minds are, if not antagonistic, at least very different and sometimes poles asunder.

going to offer should not give one a false idea of the *Historia animalium.* The critics have devoted their attention mainly to the good pieces. It would be interesting to make a statistical analysis of the whole work and determine how often the master is right, how often wrong, and to evaluate his degree of correctness in every case.

His description of the shock produced by the torpedo fish[47] is perhaps not surprising, for many fishermen must have experienced that phenomenon; yet Aristotle's description is significant, because it is sober and objective -- the description of a man who could not yet know the electric nature of that shock, who had no knowledge of electricity whatsoever, yet was not swept off his feet and did not speak of marvels, but described simply what he had observed.

Now read his account of the breeding habits of the catfish:

> The catfish deposits its eggs in shallow water, generally close to roots or close to reeds. The eggs are sticky and adhere to the roots. "The female catfish, having laid her eggs, goes away. The male stays on and watches over the eggs, keeping off all other little fishes that might steal the eggs or fry. He thus continues for forty or fifty days, till the young are sufficiently grown to escape from the other fishes for themselves. Fishermen can tell where he is on guard, for, in warding off the little fishes, he sometimes makes a rush in the water and gives utterance to a kind of muttering noise. Knowing his earnestness in parental duty the fishermen drag into a shallow place the roots of water plants to which the eggs are attached, and there the male fish, still keeping by the young, is caught by the hook when snapping at the little fish that come by. Even if he perceive the hook, he will still keep by his charge, and will even bite the hook in pieces with his teeth." [48]

Aristotle's story of the catfish was disbelieved, because the catfish of Western Europe do not look after their young in such a fashion, but Louis Agassiz discovered that the American catfish confirmed the story. Some of the catfish of the Achelous river (running into the Gulf of Corinth) were sent in 1856 to Agassiz, who could verify Aristotle's account and calls them *Parasilurus Aristotelis.* It was only in 1906, however, that those facts became generally known to men of science.

Aristotle also remarked[49] that the catfish and some other fishes produce sounds by rubbing their gills (more exactly the gill covers); it is not true then to say that all fishes are silent.[50]

The Greeks were familiar with bees, to which they owed honey – a substance incredibly precious when no other sugar was available – and it is natural enough to find many references to them in the *Historia animalium.* The account that Aristotle gave of them is admirable, except that he did not realize unambiguously that the ruler of the hive is a female, a queen, not a king.

Aristotle's descriptions are the more startling when we recall the extreme paucity of his means; not only did he lack the instruments (magnifying glasses, etc.)

[47] *Historia animalium,* 620B, 18–29.

[48] *Ibid.,* 568A. The account is not quoted verbatim but as abbreviated by Charles Singer in his *Story of living things* (New York: Harper, 1931), p. 18. Compare with similar observations made by the poet Henry David Thoreau in 1858, in his *Journal* (1906), vol. 10, pp. 483–484. Thoreau made his observations just as Aristotle had made them twenty-two centuries earlier, without any new equipment.

[49] *Historia animalium,* 535B, 13.

[50] For modern investigations on the sound-producing organs of fishes see Bashford Dean and Eugene Willis Gudger, *Bibliography of fishes* (New York, 1923), vol. 3, p. 594 [*Isis 6,* 456–459 (1924)]. During World War II it was discovered in the operation of underwater sound gear that many kinds of fish produce sounds. See, for instance, Donald P. Love and Don A. Proudfoot, "Underwater noise due to marine life," *J. Acoustical Soc. Am. 18,* 446–449 (1946).

Fig. 99. Lewes's book of 1864 was a brilliant, if imperfect, synthesis of Aristotle's scientific thought. It was the first elaborate study of Aristotelian science and the first instalment of his projected history of science. He was one of the pioneer historians of science, and is now unfairly decried, especially by people, men of letters or men of science, who have but little knowledge of the subject. Wrote Lewes in his Preface, "I have been for many years preparing myself to attempt a sketch of the Embryology of Science, so to speak, — an exposition of the great *momenta* in scientific development; and the present volume is the first portion of such an exposition. . ." [From the copy in the Harvard College Library.]

ARISTOTLE:

A CHAPTER FROM

THE HISTORY OF SCIENCE,

INCLUDING

ANALYSES OF ARISTOTLE'S SCIENTIFIC WRITINGS.

BY

GEORGE HENRY LEWES.

LONDON:
SMITH, ELDER AND CO., 65, CORNHILL.
M.DCCC.LXIV.

[*The right of Translation is reserved.*]

and drugs with which our naturalists are always armed, but he had no reference books and dictionaries which enable us to check and recheck our own conclusions in a moment. There was probably a library at the Lyceum, but it was rudimentary, especially with regard to scientific subjects. The language, without which ideas cannot be communicated, did not even exist. The marvelous tool created by the poets and historians lacked the technical terms without which a brief and clear description is impossible. Aristotle had to create many of the necessary terms as he needed them. But even the most highly developed technical language is insufficient in biologic description without drawings. It is certain that Aristotle (or his collaborators) added drawings, though we have no means of estimating their number and their value. For example, speaking of the womb, he says, "As to the appearance of this organ I must refer the reader to diagrams in my *Anatomy*."[51] As to the bladder and penis, he remarks a little further in the same book, "All these descriptive particulars may be seen in the accompanying diagram,"[52] and he refers to various parts of it by means of letters, just as was done in the geometric figures. In another book, he remarks, "All this should be studied with the help of the illustrative diagrams given in the *Dissections* and *Researches*."[53]

[51] *Historia animalium*, 497A, 32.

[52] *Ibid.*, 510A, 30.

[53] *Generatio animalium*, 746A, 14. We must remember, however, that the relative scarcity and expensiveness of papyrus did not enable Aristotle's contemporaries to use it as recklessly as we use paper. There was, therefore, a tendency to avoid drawings and graphs rather than to multiply them as we do. Even when drawings were provided in the author's original manuscript, it was so difficult and bothersome to reproduce them exactly that they were likely to be dropped by the copyists. No Aristotelian drawing has come down to us. The technical terms used by Aristotle for his drawings are *schēmata*, *diagraphē*, *paradeigmata*.

Embryology. One of the earliest elaborate studies in Aristotle's scientific thought was written by George Henry Lewes in 1864 (Fig. 99).[54] Lewes was not by any means an uncritical admirer of Aristotle, but when he came to the biologic writings which, being himself a naturalist, he could fully appreciate, he could no longer restrain his admiration. Here is the way he spoke of *De generatione animalium*:

> It is an extraordinary production. No ancient, and few modern works, equal it in comprehensiveness of detail and profound speculative insight. We there find some of the obscurest problems of Biology treated with a mastery which, when we consider the condition of science at that day, is truly astounding. That there are many errors, many deficiencies, and not a little carelessness in the admission of facts, may be readily imagined; nevertheless, at times the work is frequently on a level with, and occasionally even rises above, the speculations of many advanced embryologists. At least so it appears to me; and the reader knows how little I am disposed to discover in ancient texts the fuller meanings of modern science, and how anxiously I strive to represent what Aristotle actually thought. It is difficult to disengage ancient texts from the suggestions of modern thought; but I should not be candid were I to conceal the impression which the study of this work left on my mind, that the labors of the last two centuries from Harvey to Kölliker have furnished the anatomical data to confirm many of the views of this prescient genius. Indeed, I know no better eulogy to pass on Aristotle than to compare his work with the "Exercitations concerning Generation" of our immortal Harvey. The founder of modern physiology was a man of keen insight, of patient research, of eminently scientific mind. His work is superior to that of Aristotle in some few anatomical details; but it is so inferior to it in philosophy, that at the present day it is much more antiquated, much less accordant with our views.[55]

The English critic did not hesitate to place the *De generatione animalium* of Aristotle above that of his illustrious countryman, published in 1651, some two thousands years later!

As this is a subject remote from my own investigations, it will be better to withdraw for a moment and allow one of my friends, who is a very distinguished embryologist, to judge his forerunner:

> His outstanding contributions to embryology may be put in the following way:
>
> 1. He carried to their logical conclusion the principles of the observation of facts suggested by the unknown Hippocratic embryologist, and added to them a discipline of classification and correlation of facts which gave embryology a quite new coherence.
>
> 2. He introduced the comparative method into embryology, and by studying a multitude of living forms was able to lay the foundation for future science of the various ways in which embryonic growth can take place. Thus he knew of oviparity, ovoviviparity, and viviparity, and one of his distinctions is substantially the same as that known to modern embryology between holoblastic and meroblastic yolks.
>
> 3. He distinguished between primary and secondary sexual characteristics.
>
> 4. He pushed back the origin of sex determination to the very beginning of embryonic development.
>
> 5. He associated the phenomena of regeneration with the embryonic state.
>
> 6. He realized that the previous speculations on the formation of the embryo could be absorbed into the definite antithesis of preformation and epigenesis, and he decided that the latter alternative was the true one.
>
> 7. He put forward a conception of the unfertilized egg as a complicated machine,

[54] George Henry Lewes (1817–1878), *Aristotle. A chapter from the history of science, including analyses of Aristotle's scientific writings* (414 pp.; London, 1864). Lewes is best known to most people as the devoted "husband" of George Eliot from 1854 to his death in 1878. R. E. Ockenden, "George Henry Lewes," *Isis* 32, 70–86 (1947–49), with portrait.

[55] Lewes, *Aristotle*, p. 325.

the wheels of which would move and perform their appointed function in due course when once the master lever had been released.

8. He foreshadowed the theory of recapitulation in his speculations on the order in which the souls came to inhabit the embryo during its growth, and in his observation that universal characteristics precede particular characteristics in embryogeny.

9. He foreshadowed the theory of axial gradients by his observations on the greater and more rapid development of the cephalic end in the embryo.

10. He allotted the correct functions to the placenta and the umbilical cord.

11. He gave a description of embryonic development involving comparison with the action of rennet and yeast, foreshadowing thus our knowledge of organic catalysts in embryogeny.

But there was another side to the picture. Aristotle made three big mistakes, and here I do not refer to any matters of detail, in which it would not have been humanly possible to be more than very often right, but rather to general notions, such as the eleven correct ones.

They were as follows:

1. He was incorrect in his view that the male supplies nothing tangible to the female in the process of fertilization. To say that the semen gave the "form" to the inchoate "matter" of the menstrual blood was equivalent to saying that the seminal fluid carried nothing in it but simply an immaterial breath along with it. Aristotle did not, of course, envisage the existence of spermatozoa.

2. He was entirely wrong in his teaching about the scolex. The caterpillar is not, as he supposed, an egg laid too soon, but has already passed through the embryonic state.

3. He was misled by some observations on castrated animals and so did not ascribe to the testis its true function.[56]

Let us now give four concrete illustrations of Aristotle's genius as an embryologist; they deal with the chick, the placental shark, the cephalopods, and the belone.[57]

The case of the chick is the simplest, for it was easy enough (provided one had the idea of doing it) to break and examine eggs of known age (freshly laid, 1, 2, 3, . . . days old). Aristotle observed the first indication of an embryo just after 3 days (a little earlier, he noted, in smaller birds, a little later in larger ones). He saw the heart beating, a speck of blood which later writers called *punctum saliens* (the heart primordium). It may be that this observation, the appearance of the heart ahead of all the other organs, confirmed his opinion that the heart was the seat of the soul or of the mind. Observing older eggs, he described the growth of the embryo, the absorption of the yolk, the shriveling of the membranes, and so forth. It was a magnificent beginning of scientific embryology, which remained unsurpassed until Harvey's time or even until later (if we accept Lewes's judgment quoted above).

Aristotle knew that most fishes bring out their young in the potential form of eggs, but that a group of these, which he called *selache*, brought them out fully shaped and actively alive. In one of the group, the resemblance to mammals goes even further.

The so-called smooth shark has its eggs in betwixt the wombs like the dog-fish; these eggs shift into each of the two horns of the womb and descend, and the young develop with the navel-string attached to the womb, so that, as the egg-substance

[56] Joseph Needham, *History of embryology* (Cambridge: University Press, (1934) pp. 36–37 [*Isis* 27, 98-102 (1937)]. Another set of conclusions on Aristotle's embryology was included by A. L. Peck in his introduction to the *Generation of animals* (Loeb Classical Library, 1943); these conclusions have been reprinted in *Isis* 35, 181 (1944).

[57] A part of this would be classified under "breeding habits" rather than "embryology," but this does not matter, my main purpose being to illustrate Aristotle's genius as a naturalist.

gets used up, the embryo is sustained to all appearance just as in the case of quadrupeds. The navel-string is long and adheres to the under part of the womb (each navel-string being attached as it were by a sucker), and also to the center of the embryo in the place where the liver is situated. If the embryo be cut open, even though it has the egg-substance no longer, the food inside is egg-like in appearance. Each embryo, as in the case of quadrupeds, is provided with a chorion and separate membranes.[58]

This was so extraordinary a phenomenon that it was almost disregarded until modern times. Pierre Belon (1553) and Guillaume Rondelet (1554) were aware of the connection between the embryo and the maternal oviduct or uterus, and Niels Stensen (Steno) realized a century later (1673) that this was for the nourishment of the embryo, and was in short a true functional placenta. Yet for all that Aristotle's early discovery was overlooked until Johannes Müller reëxplained it (1839–1842).[59] One must admit that there is something almost uncanny in the anticipation by Aristotle, unequipped with tools and books, of a discovery remade a century ago by one of the leading physiologists of the nineteenth century.

The sexual congress of cephalopods such as the octopus, the sepia, and the calamary was imperfectly described by Aristotle[60] and involved him in self-contradiction, yet he adumbrated the discovery of the process known as hectocotylization, which was not correctly described until the nineteenth century. It is not expedient to quote Aristotle's own words because that would entail too many qualifications, nor the modern descriptions, for the best anatomists of the last century did not find the truth except after many detours. Suffice it to say that hectocotylus is the name given to one arm of the male of most cephalopods, which is specially modified to accomplish the fertilization of the eggs. In the *Argonauta* (such as the paper nautilus),[61] the hectocotylus after receiving the spermatophores is detached from the male and attaches itself to the female. When the detached hectocotylus was first discovered, it was mistaken (even by such a man as Cuvier) for a kind of worm parasitic on the female. The mystery was first solved by Albrecht von Kölliker in 1842 (1847), but many more investigations were required to elucidate it and some details remain unexplained to this day.[62]

Aristotle's description of the belone is ambiguous; part of it applies to one kind of fish, another part to another kind. The part that applies to the pipefish or needlefish (*Syngnathus acus*) describes with remarkable precision the singular mode of

[58] *Historia animalium*, 565B, 2, D'Arcy W. Thompson's translation in the Oxford *Aristotle*.

[59] For details and illustrations, see D'Arcy Thompson and Singer. Wilhelm Haberling, "Der glatte Hai des Aristoteles. Briefe Johannes Müller über seine Wiederauffindung an Wilhelm Karl Hartwig Peters 1839–40," *Arch. Geschichte Math. Wiss. 10*, 166–184 (1927).

[60] *Historia animalium*, 541B, 1; *Generatio animalium*, 720B, 25.

[61] It must be added, to Aristotle's discredit, that he helped to spread a legend concerning the paper nautilus (*Argonauta argo*). "It rises up from deep water and swims on the surface; it rises with its shell down-turned in order that it may rise the more easily and swim with it empty, but after reaching the surface it shifts the position of the shell. In between its feelers it has a certain amount of web-growth, resembling the substance between the toes of web-footed birds; only that with these latter the substance is thick, while with the nautilus it is thin and like a spider's web. It uses this structure, when a breeze is blowing, for a sail, and lets down some of its feelers alongside as rudder-oars. If it be frightened, it fills its shell with water and sinks" (*Historia animalium*, 622B, 5–15). That amiable legend of the nautilus using its membrane as a sail and its arms as oars has been spread by later writings and by illustrations (for example, Belon's, 1551).

[62] For details, illustrations, and bibliography, see Singer, "Greek biology," *Studies*, vol. 2, pp. 39–46.

reproduction of that very small and needlelike fish. According to his statements in various places:

> Fishes then in general produce their young by copulation, and lay their eggs; but the pipefish, as some call it, when the time of parturition arrives, bursts in two, and the eggs escape out. For the fish has a diaphysis or cloven growth under the belly and abdomen (like the blind snakes), and, after it has spawned by the splitting of this diaphysis, the sides of the split grow together again.[63]
>
> There are some fishes, such as the one known as belone, which burst asunder owing to the size of the eggs, the fetations of this fish being large instead of numerous; here Nature has taken away from their number and added to their size.[64]
>
> The so-called needlefish (or pipefish) is late in spawning, and the greater portion of them are burst asunder by the eggs before spawning; and the eggs are not so many in number as large in size. The young fish cluster around the parent like so many young spiders, for the fish spawns onto herself; and, if any one touch the young, they swim away.[65]

So far so good, but Aristotle did not see that the pouch is developed under the belly of the male fish, that the eggs are deposited into it by the female, and that it is the male that continues to nurse and look after the young. Aristotle's discovery was completed only in 1784 by John Walcott of Teignmouth, and published half a century later by William Yarrell.[66] Later investigations (in our own century) established that the marsupium of the male of such fishes with its epithelial lining and its capillaries and lymph vessels is a functional uterus placenta.[67] One could not expect Aristotle to discover all that; it was materially impossible for him to do so; but is it not astounding that he came so close to the edge of the mystery, and – we must always insist on this – spoke of it in a sensible and quiet way, even as any zoölogist of our own times?

Geographic distribution of living beings. The Greeks were a roving kind of people, restless amphibians[68] sailing across the Mediterranean or caravaning across foreign lands in search of business or of knowledge. They were intelligent and alert, good observers, and Aristotle must have enjoyed many opportunities of interviewing travelers. His own journeys were not extensive; yet they included a good diversity of landscapes and climates and the travelers whom he met in Macedonia, in the Troad, or in Athens were able to give him some idea of the other climates. Above all, Alexander could bring many novelties to his knowledge; we may imagine that the *mot d'ordre* had been given to his scientific entourage, that whatever Aristotle asked for should be granted and that every novelty should be communicated to him. Hence the richness of Aristotle's biologic point of view, and his keen realization of the geographic distribution of plants and animals. Plants were fixed to the land of their birth, but animals could move and did move when the

[63] *Historia animalium*, 567B, 22.

[64] *Generatio animalium*, 755A, 33.

[65] *Historia animalium*, 571A, 3.

[66] William Yarrell (1784–1856), "Note on the foetal pouch of the male needle pipe-fish," *Proc. Zool. Soc.* (1835), pp. 3, 183; *History of British fishes* (2 vols.; London, 1836). Eugene Willis Gudger, "The breeding habits and the segmentation of the egg of the pipefish, *Siphostoma Floridae*," *Proc. U. S. National Museum* 29, 447–499 (1906), 11 pls., including historical outline of our knowledge of the reproduction of *Lophobranchii* (pp. 449–462). D. W. Thompson, *Greek fishes*, pp. 29–31.

[67] See Gudger, note 66.

[68] As Strabon (I–2 B.C.) called them; see *Osiris* 2, 411 (1936).

climate became unsuitable to them or unpleasant. Listen to this, the earliest text concerning one of the most mysterious of all biologic subjects, the migration of animals:

> The habits of animals are all connected with either breeding and the rearing of young, or with the procuring a due supply of food; and these habits are modified so as to suit cold and heat and the variations of the seasons. For all animals have an instinctive perception of the changes of temperature, and, just as men seek shelter in houses in winter, or as men of great possessions spend their summer in cool places and their winter in sunny ones, so also all animals that can do so shift their habitat at various seasons. Some creatures can make provision against change without stirring from their ordinary haunts; others migrate, quitting Pontos and the cold countries after the autumnal equinox to avoid the approaching winter, and after the spring equinox migrating from warm lands to cool lands to avoid the coming heat. In some cases they migrate from places near at hand, in others they may be said to come from the ends of the world, as in the case of the crane; for these birds migrate from the steppes of Scythia to the marshlands south of Egypt where the Nile has its source. And it is here, by the way, that they are said to fight with the pygmies; and the story is not fabulous, but there is in reality a race of dwarfish men, and the horses are little in proportion, and the men live in caves underground. Pelicans also migrate, and fly from the Strymon to the Ister, and breed on the banks of this river. They depart in flocks, and the birds in front wait for those in the rear, owing to the fact that when the flock is passing over the intervening mountain range, the birds in the rear lose sight of their companions in the van. Fishes also in a similar manner shift their habitat now out of the Euxine and now into it. In winter they move from the outer sea in towards land in quest of heat; in summer they shift from shallow waters to the deep sea to escape the heat. Weakly birds in winter and in frosty weather come down to the plains for warmth, and in summer migrate to the hills for coolness. The more weakly an animal is the greater hurry will it be in to migrate on account of extremes of temperature, either hot or cold; thus the mackerel migrates in advance of the tunnies, and the quail in advance of the cranes. The former migrates in the month of Boēdromiōn [22 August–22 September], and the latter in the month of Maimactēriōn [22 October–22 November]. All creatures are fatter in migrating from cold to heat than in migrating from heat to cold; thus the quail is fatter when he emigrates in autumn than when he arrives in spring. The migration from cold countries is contemporaneous with the close of the hot season. Animals are in better trim for breeding purposes in spring-time, when they change from hot to cool lands. . . .[69]

Not only was Aristotle acquainted with what we would call today geographic biology or biological geography; he had some definite knowledge of ecology, the relation not only between living beings and their physical environment, but also between living beings and their biologic environment. How is each animal affected by other animals or by the plants living near it? Some of the other animals prey on it, or it preys on them. Some animals are rivals, others are collaborators. But this subject brings us so close to sociology that it is better to save our examples of Aristotelian ecology until the next chapter.

Our enumeration of his biologic knowledge could be extended considerably; yet enough has been said to illustrate the magnitude of his biologic genius. He was not only the first great one in his field, somewhat like Hippocrates in medicine, but he remained the greatest for two thousand years.

After a period of anti-Aristotelian reaction and oblivion, Aristotle the biologist was fully rehabilitated and vindicated by the end of last century. This could be

[69] *Historia animalium*, 596B, 20.

proved in many ways, but I shall restrict myself to a single document, the letter that Charles Darwin wrote to Dr. William Ogle to acknowledge receipt of the latter's translation of the *Parts of animals*.[70] A part of that letter has often been quoted, but I reproduce it completely, because it is typical of Darwin's kindness and honesty.

Down, February 22, 1882
My Dear Dr. Ogle:

You must let me thank you for the pleasure which the introduction to the Aristotle book has given me. I have rarely read anything which has interested me more, though I have not read as yet more than a quarter of the book proper.

From quotations which I had seen, I had a high notion of Aristotle's merits, but I had not the most remote notion what a wonderful man he was. Linnaeus and Cuvier have been my two gods, though in very different ways, but they were mere schoolboys to old Aristotle. How very curious, also, his ignorance on some points, as on muscles as the means of movement. I am glad that you have explained in so probable a manner some of the grossest mistakes attributed to him. I never realized, before reading your book, to what an enormous summation of labour we owe even our common knowledge. I wish old Aristotle could know what a grand Defender of the Faith he had found in you. Believe me, my dear Dr. Ogle,

Yours very sincerely,
Ch. Darwin[71]

What greater testimony could one obtain than the one freely given by the master of general biology in the second half of the last century? If Hippocrates deserves to some extent to be called the father of medicine, Aristotle deserves more fully to be called the father of biology.

BOTANY

THE RHIZOTOMISTS

When we tried to explain the Greek background of Hippocratic medicine, we spoke of the herb gatherers, thanks to whom an enormous amount of plant lore had been patiently accumulated. How long the accumulation lasted we cannot say; one might speak of millennia as well as of centuries. People learned to know gradually, very slowly, by a process of trial and error that had to be endlessly repeated, for its results were never properly recorded, that certain plants were useful and others dangerous; some were foods, palatable and nutritious, others were refreshing; some were sweet and others bitter; and so on. The main discoveries concerned the pharmacologic qualities of herbs and roots, which might be laxative, emetic, sedative, diuretic, emmenagogic, analgesic, antipyretic, etc.; it was observed that the best results were obtained with a definite dose and that if the dose were excessive death might ensue. In other words, the people of Greece, like the people of other countries, discovered foods, drugs, and poisons. In the course of ages, there developed among them a special profession of herb gatherers or herbalists; as the virtues of plants were often concentrated in the roots, the familiar Greek name for them was rhizotomists. These men were indispensable and rendered great services; it is probable that the folklore transmitted by them concerned not only drugs but also poisons and magic potions. Judging from allusions to them in Greek

[70] Ogle's translation was first published in 1882. A revised edition of it is included in the Oxford *Aristotle*, vol. 5 (1911). For the copious biological notes it is necessary to refer to the original edition.

[71] Francis Darwin, *The life and letters of Charles Darwin* (ed. 2, London, 1887), vol. 3, p. 251.

literature, the rhizotomists did not enjoy a good reputation; they were accused of being magicians, witches, poisoners; they certainly knew dark secrets and were ready to use them, and even to abuse them. There was no ethical code to restrain their activities, but their manners and customs were permeated with superstitious rites.[72]

ARISTOTLE THE BOTANIST

That immense botanic folklore was available to the men of science as well as to the common people and it was up to the former to investigate it, to verify the claims that were made for each plant, and to incorporate some items into their scientific publications. Thus, we find some 300 plants mentioned in the Hippocratic corpus;[73] the mention, being restricted to medical use, takes for granted that the plant is known to the reader and makes it impossible for him to identify it if he is not already familiar with it.

It is certain that botanic topics were discussed in the Academy and the Lyceum. Aristotle and his pupils were not interested only in the practical value of plants, they were anxious to define them and to discuss their form and growth.[74] Unfortunately, we can give no precision to that idea, because Aristotle's botanic writings, if any, have disappeared. The *De plantis* included in the *Opuscula* is certainly apocryphal; it is generally ascribed to Herod's friend, Nicholas of Damascus (I–2 B.C.), and its tradition is so crooked that we may digress a moment in order to summarize it. It is a good example of the precariousness and capriciousness of literary tradition.

The original Greek text of the *De plantis* was translated at least once into Arabic, by Isḥāq ibn Ḥunain (IX–2). The Arabic text was translated into Latin by the Englishman Alfred of Sareshel (XIII–1), and into Hebrew by the Provençal Qalonymos ben Qalonymos (XIV–1). The Greek and Arabic texts are lost. The Greek text in the Bekker edition[75] is a retranslation of the Latin into Greek! In this case, it is thus better to refer to the Latin, one step closer to the lost original than to the Greek, which is thrice removed from it.[76] Though the *De plantis* is certainly not Aristotelian, it contains many passages that are parallel to various writings of Aristotle and Theoprastos.[77] Its general structure is typical of Peripatetic thought.

Book I is divided into seven chapters: 1. The nature of plant life. 2. Sex in plants. 3. The parts of plants. 4. Structure and classification of plants. 5. Composition and products of plants. 6. Methods of propagation and fertilization. 7. Changes and

[72] See p. 333. The best study is A. Delatte, *Herbarius* (Paris: Académie royale de Belgique, 1936) [*Isis 27*, 531 (1937)]; ed. 2 (1938) [*Isis 30*, 295 (1939)]. Greek superstitions relating to herbs were continued in Roman days; examples of them can be found in the Latin writings as well as in the Greek ones, for example, in the *Apologia* of Apuleius (II–2) or in that Vergilian cento the *Medea* of Hosidius Geta, who flourished in Apuleius' time or soon afterwards. Joseph J. Mooney, *Hosidius Geta's tragedy Medea, text and metrical translation, with an outline of ancient Roman magic* (96 pp.; Birmingham, 1919). Superstition is of necessity more conservative than science, because it is uncorrectible and unprogressive.

[73] Some 63 in Homer.

[74] Agnes Arber, *The natural philosophy of plant form* (first pages; Cambridge: University Press, 1950) [*Isis 41*, 322–323 (1950)].

[75] Pages 815A–829B.

[76] The Latin text was edited by E. H. F. Meyer (Leipzig, 1846). It contains many Arabicisms. The Arabic text may still be found; if it were, its edition would enable one to clear various obscurities.

[77] Friedrich Wimmer began but did not complete a collection of Aristotle's botanical fragments, *Phytologiae Aristotelicae fragmenta* (Breslau, 1838); not seen.

variations in plants. Book II is divided into ten chapters: 1. Origins of plant life; "concoction." 2. Digression on "concoction" in the earth and sea. 3. The material of plants; effects of outward conditions and climate. 4. Water plants. 5. Rock plants. 6. Other effects of locality on plants; parasitism. 7. Production of fruit and leaves. 8, 9. Colors and shapes of plants. 10. Fruits and their flavor.[78]

We need not worry too much about the botanic knowledge of Aristotle; he was probably like many other naturalists of all ages who know botany, and may even have a fair knowledge of it, yet who are more interested in animals. Moreover, Aristotle had an immense amount of work to do, for he had cut out for himself the encyclopedic survey of all knowledge. When a master is thus overburdened by a gigantic labor and finds an intelligent pupil who is ready to do a part of it, he may be willing enough to abandon that part to him. This is exactly what happened; his best disciple, Theophrastos, took a special interest in botany and Aristotle abandoned botany to him. Who was Theophrastos and how did he meet Aristotle and become his best collaborator and his successor?

THEOPHRASTOS OF ERESOS

We have already taken our readers to Lesbos (main city, Mytilene), the largest island along the Asiatic coast in the Aegean Sea, the native region of the Aeolian school of lyric poetry. During the seventh century it gave birth to four illustrious poets: Terpandros, Arion, Alcaios, and, greatest of all, sweet Sappho.[79] The word Lesbian has stupidly acquired a bad connotation; to me it suggests lyric poetry and beauty. During the same century Lesbos gave to Greece one of her Seven Wise Men, Pittacos; in the fifth century, one of the earliest historians, Hellanicos; and finally, in the fourth century it gave her two philosophers and to Aristotle two disciples, Phanias and Theophrastos.

Theophrastos, son of Melantas, a fuller, was born at Eresos *c.* 372, and he died in very old age, *c.* 288. He came to Athens to sit at Plato's feet, and during that period must have become acquainted with Aristotle. Their acquaintance was renewed and their friendship established when Aristotle lived in Assos, Aterneus, and Lesbos. It was probably during that period that both men engaged in the study of natural history, in the island, along its shores, or sailing on the blue sea. They belonged to the same generation, for Theophrastos was only a dozen years younger than Aristotle. They flourished together at the Lyceum and when Aristotle was obliged to leave Athens in 323/22, he appointed him his successor [80] and bequeathed to him his library and the manuscripts of his own works. Theophrastos continued the master's tradition in great style and may be called the second founder of the Lyceum; he was head of the school for thirty-five years (thrice as long as Aristotle); [81] he reorganized the school and enlarged it. His rich pupil, Demetrios of

[78] The contents of *De plantis*, taken from the Oxford *Aristotle* (vol. 6), are quoted for the sake of comparison with the botanic books of Theophrastos, below.

[79] Terpandros belongs to the first half of the seventh century; Arion and Alcaios flourished respectively in 625 and 613; Sappho was born *c.* 612. Her name should be pronounced Sap-phó.

[80] We have told above the story of Aristotle's hesitation between Eudemos of Rhodes and Theophrastos and how he finally preferred the Lesbian wine to the Rhodian. According to another story, the original name of Aristotle's favorite disciple was Tyrtamos, which he changed to Theophrastos (divine speaker). *Se non è vero . . .* Yet a poor woman selling herbs in the market of Athens recognized at once from his accent that the "the divine speaker" was a provincial.

[81] Theophrastos' stay in Athens was interrupted only for a short while in 318, when he was exiled because of an edict of Demetrios Poliorcetes, king of Macedonia, against schools of philosophy.

Phalerôn, enabled him to buy an adjoining estate and to increase the garden of the Lyceum. His fame as a lecturer was so great that he gathered around him some two thousand disciples;[82] this is a very large number, but it refers presumably to the whole of Theophrastos' career; it would mean a yearly average of fewer than sixty pupils – still a large number in the Athens of those days, but acceptable. He was at least eighty-five years old when he died, and, like every great man who had the privilege of remaining intelligent and lucid until the end, he complained that life is so short that a man must go just when he is beginning to understand its mysteries.

He continued Aristotle's encyclopedic purpose and his activity was prodigious. Diogenes Laërtios ascribes 227 treatises to him, dealing with religion, politics, ethics, education, rhetoric, mathematics, astronomy, logic, meteorology, natural history, etc. The main works that have come down to us are the two on plants and the one on stones that will be discussed below. Fragments exist of his treatises *De sensu et sensibilibus*, *De igne*, *De odoribus*, *De ventis*, *De signis tempestatum* (*pluviarum, ventorum, tempestatis et serenitatis*), *De lassitudine*, *De vertigine*, *De sudore*, *De animi defectione* (*lipopsychia*), *De nervorum resolutione* (paralysis), *Metaphysica*, etc.

The most convenient edition of the *opera omnia* is the Greco-Latin by Friedrich Wimmer (Paris, 1866), with *indices nominum, graecitatis et rerum, plantarum.* The text covers 462 pp., of which 319 are botanical; it does not include the *Characters.*

On stones, Greek and English edition by Sir John Hill[83] (234 pp.; London, 1746; 2nd ed., London, 1774).

On winds and *On weather signs*, translation by James George Wood (97 pp.; London, 1894).

Enquiry into plants, *On odours* and *On weathersigns*, Greek-English by Sir Arthur Hort (2 vols.; Loeb Classical Library, 1916) [*Isis* 3, 92 (1920–21)].

On the senses, Greek-English by George Malcolm Stratton (London, 1917).

We have not yet mentioned the most popular of all of Theophrastos' writings, the *Characters* (*Ēthicoi charactēres*), a series of thirty sketches of typical weaknesses, such as arrogance, backbiting, boorishness, buffoonery, written in 319. Their genuineness has been questioned, but they have never been ascribed to another author. They were not discovered together but gradually and thus the date of the editio princeps varies according to the number of characters included.

The first edition, by Willibald Pirckheimer (Nuremberg, 1527), included only *Characters* I to XV; *Characters* XVI to XXIII were first published by Giambattista Camozzi (Venice, 1552), *Characters* XXIV to XXVIII by Isaac Casaubon (in his second edition of the *Characters*, Leyden, 1599; the first edition had appeared in 1592), *Characters* XXIX to XXX by Giovanni Cristoforo Amaduzzi (Parma, 1786). The first edition of the thirty *Characters* was published, singularly enough, by the English dilettante, John Wilkes (London, 1790) (Fig. 100). There is a very convenient Greek-English edition by John Maxwell Edmonds in Loeb Classical Library (1929).

We reproduce the full text of *Character* XVI, *Superstitiousness*,[84] for two reasons:

[82] Including Menandros (342–291), the leading poet of the New Comedy. Menandros was the disciple and friend of both Theophrastos and Epicuros.

[83] Clark Emery, " 'Sir' John Hill versus the Royal Society," *Isis 34*, 16–20 (1942–43). John Hill (1716?–1775), apothecary, quack, botanist, historian, was a queer individual. He called himself "Sir," because he had been awarded the Swedish order of Vasa; *Dictionary of National Biography*, vol. 26, pp. 397–401.

[84] Theophrastos, Loeb Classical Library. The word used by Theophrastos is *deisidaimonia*, which means fear of gods and has a good sense (piety) and a bad one (superstition).

Fig. 100. First page of the first edition of the *Thirty characters*, by John Wilkes (1727–1797), politician, lord mayor of London in 1774; edition de luxe, 103 copies printed (84 pp.; 21 cm; London, 1790). The author's dilettantism is illustrated by the fact that the text is printed without breathings and without accents! [From the copy in the Harvard College Library.]

ΘΕΟΦΡΑΣΤΟΥ

ΧΑΡΑΚΤΗΡΕΣ

ΗΘΙΚΟΙ,

ΠΡΟΟΙΜΙΟΝ.

ΗΔΗ μεν και προτερον πολλακις επιστησας την διανοιαν, εθαυμασα, ισως δε ου παυσομαι θαυμαζων, τι δηποτε, της Ελλαδος υπο τον αυτον αερα κειμενης, και παντων των Ελληνων ομοιως παιδευομενων, συμβεβηκεν

A 2 ημιν

Superstitiousness, I need hardly say, would seem to be a sort of cowardice with respect to the divine; and your Superstitious man such as will not sally forth for the day till he have washed his hands and sprinkled himself at the Nine Springs, and put a bit of bayleaf from a temple in his mouth. And if a cat cross his path he will not proceed on his way till someone else be gone by, or he have cast three stones across the street. Should he espy a snake in his house, if it be one of the red sort he will call upon Sabazios, if of the sacred, build a shrine then and there.
When he passes one of the smooth stones set up at crossroads he anoints it with oil from his flask, and will not go his ways till he have knelt down and worshiped it. If a mouse gnaw a bag of his meal, he will off to the wizard's and ask what he must do, and if the answer be "send it to the cobbler's to be patched," he neglects the advice and frees himself of the ill by rites of aversion. He is forever purifying his house on the plea that Hecate has been drawn thither. Should owls hoot when he is abroad, he is much put about, and will not on his way till he have cried "Athena forfend!" Set foot on a tomb he will not, nor come nigh a dead body nor a woman in childbed; he must keep himself unpolluted. On the fourth and seventh days of every month he has wine mulled for his household, and goes out to buy myrtle boughs, frankincense, and a holy picture, and then returning spends the livelong day doing sacrifice to the Hermaphrodites and putting garlands about them. He never has a dream but he flies to a diviner, or a soothsayer, or an interpreter of visions, to ask what God or Goddess he should appease; and when he is about to be initiated into the holy orders of Orpheus, he visits the priests every month and his wife with him, or if she have not the time, the nurse and children. He would seem to be one of those who are forever going to the seaside to besprinkle themselves; and if ever he see one of the figures of Hecate at the crossroads wreathed with garlic, he is off home to wash his head and summon priestesses whom he bids purify him with the carrying around him of a squill or a puppy dog. If he catch sight of a madman or an epilept, he shudders and spits in his bosom.

We reprint this text in the first place, because it is a good description of the darker side of Greek thought in its golden age. There were superstitious people in Athens close to the Academy and even to the Lyceum, even as there are superstitious people today in the shadow of our own academies and colleges. Our second reason is that this sketch renders Theophrastos' authorship very plausible. Indeed, we would expect a man of science to deride superstitions in that very manner. Assuming his authorship, Theophrastos was in his early fifties when he composed them; they show that he was not a pedant (*scholasticos, micrologos*); he was a philosopher and had a good sense of humor.

Such characteristic portraits were not invented by him; we find some in Herodotos, Plato, Aristotle, not to mention Aristophanes and Menandros; but Theophrastos was the first to publish a gallery of them, and by so doing he created a new literary genre. La Bruyère's French translation, to which he added a series of sketches characterizing the manners and customs of his own century, was published in Paris in 1688 [85] and became one of the classics of French literature (Fig. 101). More than two millennia (2008 years) separate the two books, composed respectively in the golden age of Athens and the *grand siècle* of France, but they are very close, except that Theophrastos was primarily a man of science, and La Bruyère, a man of letters.

My last statement should not be misunderstood. Theophrastos at his best wrote simply but well; he wrote like a man of science who appreciates literary values yet must subordinate them to scientific purposes. Truth first, beauty next. He realized the danger of superfluous words from the scientific point of view, but also from the artistic one. "It is better not to say everything at great length but to leave some things for the reader to guess at and to find for himself. The reader who has guessed what was left unsaid becomes a collaborator and a friend. Should you try to say everything to him as one would to a fool, he would sense your distrust of his intelligence." [86]

The characters of Theophrastos are more concrete than the sketches that Aristotle introduced in his *Rhetoric* to illustrate various passions, but they are less individual than those of La Bruyère.

To return for a moment to Theophrastos' nonbotanic writings, the following remarks will suffice.

One of his most important small writings is the *De signis tempestatum* (On weather signs) which was used by Aratos of Soli (III–1 B.C.); as Aratos' poem was commented upon by Hipparchos (II–2 B.C.), Theophrastos helped to initiate a great astronomic tradition.

His treatise on odors, good and bad, perfumes and disagreeable smells, is very curious. It illustrates the Peripatetics' eagerness to explain everything and their insatiable curiosity. Theophrastos discusses the various smells of plants and animals, for example, the smells of the latter during the breeding season. We cannot expect him to throw much light on a subject that is still very obscure today, yet we cannot help admiring the boldness of his initiative.

He apparently [87] replaced the seat of intelligence in the brain, instead of putting it in the heart as Aristotle did. He knew that some animals living in northern regions have a white coat of fur in the winter.

[85] Jean de La Bruyère (Paris, 1645–1696), *Les Caractères de Théophraste avec les Caractères ou les moeurs de ce siècle* (Paris, 1688).

[86] Free translation of fragment 96 in Friedrich Wimmer's edition, p. 440. Theophrastos refers to the hearer (*acroatēs*), not to the reader, because in his time more people listened to readings than read themselves.

[87] The word "apparently" is added because the matter is not quite clear to me. In his *De sensibus* Theophrastos discusses the views of Alcmaion, Anaxagoras, Democritos, and Diogenes of Apollonia, but does not express his own views with sufficient unambiguity.

LES

CARACTERES

DE THEOPHRASTE

TRADVITS DV GREC.

AVEC

LES CARACTERES

OU

LES MOEURS

DE CE SIECLE.

A PARIS,

Chez ESTIENNE MICHALLET, premier Imprimeur du Roy, ruë S. Jacques, à l'Image saint Paul.

M. DC. LXXXVIII.

Avec Privilege de Sa Majesté.

Fig. 101. Title page of the first edition of the *Caractères* of La Bruyère (Paris, 1688), a small volume (15.5 cm) containing La Bruyère's discourse on Theophrastos, then his French translation of Theophrastos' *Characters* (97 pp.), and his own *Characters* (210 pp.). [From the copy in the Harvard College Library.]

Among his lost works was the *Physicōn doxai*, the opinions of the physicists, which is one of our best indirect sources for the history of Greek philosophy and science.[88]

THE FATHER OF BOTANY

We may now approach the botanic works of Theophrastos, which have survived in their integrity and are the earliest books of this kind in world literature. Our readers already know that he was not by any means the first botanist, for they will assume that the most intelligent of the rhizotomists did not simply gather roots and herbs but thought about them; yet his books are the earliest and, at their best, they are excellent. He fully deserves to be called the father of botany.[89]

Theophrastos wrote two large botanic works entitled respectively *Historia de plantis* (History of plants, or Enquiry into plants) and *De causis plantarum* (The causes of plants) (Figs. 102 and 103). The first is very largely descriptive; Theophrastos tries to distinguish the different parts of plants, and the differences obtaining between plants (*tōn phytōn tas diaphoras*). As its title suggests, the second is more philosophic, or more physiologic. Given those differences between plants or between their organs, how shall we account for them, in Aristotelian (teleologic)

[88] Fragments edited by Hermann Diels in his *Doxographi Graeci* (Berlin, 1879). The treatise on senses edited in Greek and English by G. M. Stratton (1917) is the largest of those fragments and does not give one a high idea of Theophrastos' impartiality as a historian of thought. Such impartiality, or better, the ability to judge the views of other people in their integrity and in relation to their social background, was hardly understood until modern times; it has been achieved only by very few scholars.

[89] *Celeberrimus autem omnium, verus rei herbariae parens Theophrastus fuit Eresius.* K. P. J. Sprengel (1766–1833), *Historia nei herbariae* (Amsterdam, 1807), vol. 1, p. 66.

ΘΕΟΦΡΑΣΤΟΥ ΠΕΡΙ ΦΥΤΩΝ ΙΣΤΟΡΙΑΣ ΤΟ Α

ΤΩΝ φυτῶν τὰς διαφορὰς καὶ τὴν ἄλλην
φύσιν ληπτέον, κατά τε τὰ μέρη, καὶ τὰ πάθη, καὶ τὰς γενέσεις, καὶ τοὺς βίους· ἤθη γὰρ καὶ
πράξεις οὐκ ἔχουσιν ὥσπερ τὰ ζῷα. εἰσὶ δ' αἱ μὲν
κατὰ τὴν γένεσιν καὶ τὰ πάθη καὶ τοὺς βί-
ους, εὐθεωρητότεραι καὶ ῥᾴους· αἱ δὲ κατὰ τὰ
μέρη, πλείους ἔχουσι ποικιλίας. αὐτὸ γὰρ
τοῦτο πρῶτον οὐχ ἱκανῶς ἀφώρισται, τὰ ποῖα δεῖ μέρη, καὶ μὴ μέρη καλεῖν, ἀλλ' ἔχει τινὰ ἀπορίαν. τὸ μὲν γὰρ μέρος, ἅτε ἐκ τῆς ἰδίας
φύσεως ὄν, ἀεὶ δοκεῖ διαμένειν ἢ ἁπλῶς ἢ ὅταν γένηται· καθάπερ ἐν τοῖς ζῴοις τὰ ὕστερον γενησόμενα· πλὴν εἴ τι διὰ νόσον ἢ γῆρας ἢ
πήρωσιν ἀπεβάλετο. τῶν δ' ἐν τοῖς φυτοῖς ἔνια τοιαῦτ' ἐστίν, ὥστε ἐπέ-
τειον ἔχειν τὴν οὐσίαν· οἷον ἄνθος, βρύον, φύλλον, καρπόν· ἁπλῶς
ὅσα πρὸ τῶν καρπῶν ἢ ἅμα γίνεται τοῖς καρποῖς· ἔτι δὲ αὐτὸς
ὁ βλαστός· ἀεὶ γὰρ ἐπίφυσιν λαμβάνει τὰ δένδρα κατ' ἐνιαυτὸν, ὁμοίως ἔν τε τοῖς ἄνω καὶ ἐν τοῖς περὶ τὰς ῥίζας. ὥστε εἰ μέν τις ταῦτα θήσει μέρη, τό, τε πλῆθος ἀόριστον ἔσται, καὶ οὐδέποτε τὸ αὐτὸ
τῶν μορίων· εἰ δ' αὖ μὴ μέρη, συμβήσεται δι' ὧν τέλεια γίνεται καὶ φαίνεται, ταῦτα μὴ εἶναι μέρη. βλαστάνοντα γὰρ καὶ θάλλοντα καὶ καρ-
πὸν ἔχοντα, πάντα καλλίω καὶ τελειότερα καὶ δοκεῖ καὶ ἔστιν.
αἱ μὲν οὖν ἀπορίαι σχεδὸν εἰσὶν αὗται. τάχα δὲ οὐχ ὁμοίως

aaa ααα

Fig. 102. First page of the *princeps* edition of Theophrastos' *Historia plantarum*, forming vol. 4 of the Greek *princeps* of Aristotle (5 vols., folio, 30.5 cm; Venice: Aldus Manutius, 1495–1498; Klebs, 83). Volume 4, printed in June 1497, contains the Greek text of the *Historia plantarum* and also of *De causis plantarum*. [From the copy in the Harvard College Library.]

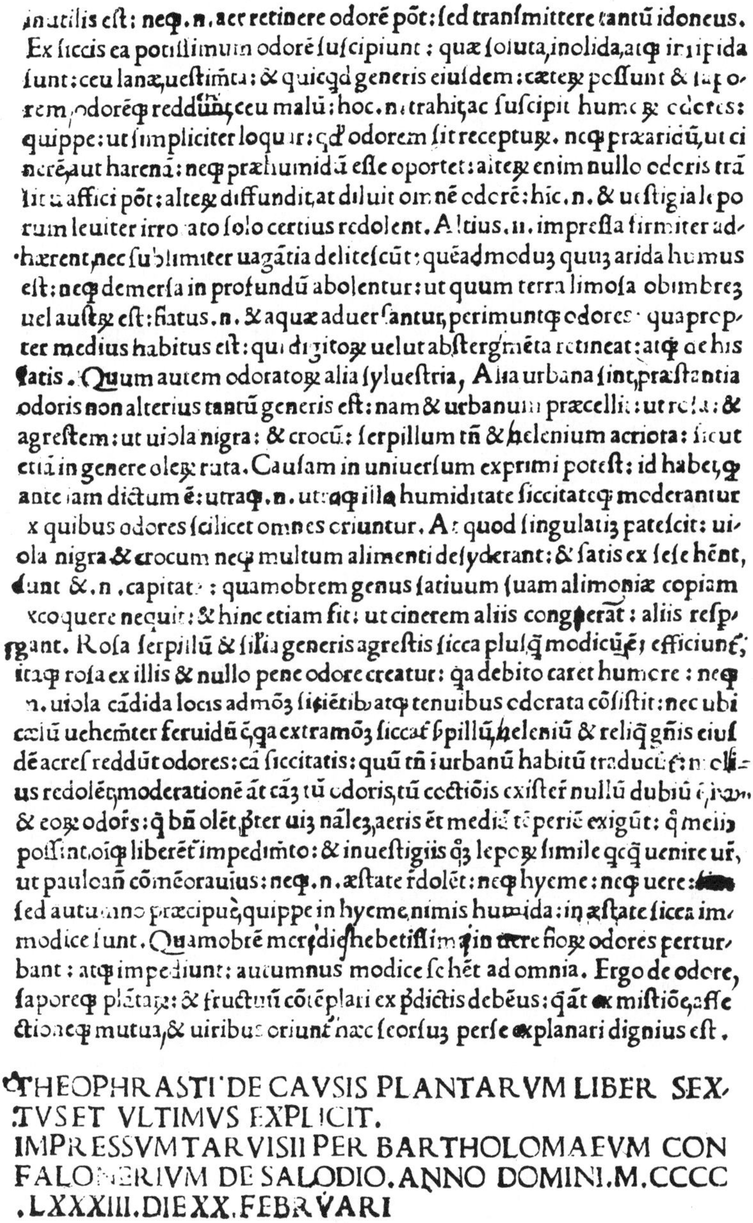

inutilis est: neq̃.n.aer retinere odorẽ põt: sed transmittere tantũ idoneus. Ex siccis ea potissimum odorẽ suscipiunt: quæ soluta, inolida, atq̃ insipida sunt: ceu lanæ, uestimta: & quicq̃d generis eiusdem: cæteꝝ possunt & sapo- rem, odorẽq̃ reddũt, ceu malũ: hoc.n. trahit, ac suscipit humoꝝ odores: quippe: ut simpliciter loquar: q̃d odorem sit receptuꝝ. neq̃ præaridũ, ut ci nerẽ, aut harenã: neq̃ præhumidũ esse oportet: alteꝝ enim nullo odoris trã situ affici põt: alteꝝ diffundit, at diluit omnẽ odorẽ: hic.n. & uestigia lepo rum leuiter irrorato solo certius redolent. Altius.n. impressa firmiter ad- hærent, nec sublimiter uagãtia delitescũt: queadmodũ quum arida humus est: neq̃ demersa in profundũ abolentur: ut quum terra limosa obimbrẽ uel austꝝ est: flatus.n. & aquæ aduersantur, perimuntq̃ odores: quapropter medius habitus est: qui digitoꝝ uelut abstergmẽta retineat: atq̃ de his satis. Quum autem odoratoꝝ alia syluestria, Alia urbana sint, præstantia odoris non alterius tantũ generis est: nam & urbanum præcellit: ut rosa: & agrestem: ut uiola nigra: & crocũ: serpillum tñ & helenium acriora: sicut etiã in genere oleꝝ ruta. Causam in uniuersum exprimi potest: id habet, q̃ ante iam dictum ẽ: utraq̃.n. utroq̃ illa humiditate siccitateq̃ moderantur ex quibus odores scilicet omnes oriuntur. At quod singulatiʒ patescit: ui- ola nigra & crocum neq̃ multum alimenti desyderant: & satis ex sese hñt, sunt &.n. capitata: quamobrem genus satiuum suam alimoniæ copiam excoquere nequit: & hinc etiam fit: ut cinerem aliis congerat: aliis resp- rgant. Rosa serpillũ & silia generis agrestis sicca plusq̃ modicũ efficiunt; itaq̃ rosa ex illis & nullo pene odore creatur: q̃a debito caret humore: neq̃ .n. uiola cãdida locis admõʒ sitiẽtib. atq̃ tenuibus odorata cõsistit: nec ubi cæliũ uehemẽter feruidũ ẽ q̃a extramõʒ siccat. ßpillũ, heleniũ & reliq̃ gñis eius dẽ acres reddũt odores: cã siccitatis: quũ tñ i urbanũ habitũ traducũt: n: ... us redolẽt, moderationẽ ãt cãʒ tũ odoris, tũ coctiõis existẽr nullũ dubiũ ẽ, iam & eoꝝ odoꝝs: q̃ bñ olẽt, ꝑter uiʒ nãleʒ, aeris ẽt mediã tẽperiẽ exigũt: q̃ melius possint, oĩq̃ liberẽt impedimto: & inuestigiis q̃ʒ lepoꝝ simile qcq̃ uenire ur, ut paulo añ cõmẽorauius: neq̃.n. æstate rdolẽt: neq̃ hyeme: neq̃ uere: ... sed autumno præcipuẽ, quippe in hyeme, nimis humida: in æstate sicca im- modice sunt. Quamobrẽ meridie hebetissim ... in ... fioꝝ odores pertur- bant: atq̃ impediunt: autumnus modice se hẽt ad omnia. Ergo de odore, saporeq̃ plãtaꝝ: & fructuũ cõtẽplari ex ꝑdictis debẽus: q̃ãt ex mistiõe, affe ctioneq̃ mutua, & uiribus oriunẽ hæc seorsũ perse explanari dignius est.

THEOPHRASTI DE CAVSIS PLANTARVM LIBER SEX- TVS ET VLTIMVS EXPLICIT.
IMPRESSVM TARVISII PER BARTHOLOMAEVM CON FALONERIVM DE SALODIO. ANNO DOMINI. M.CCCC .LXXXIII. DIE XX. FEBRVARI

Fig. 103. Colophon of Theophrastos' *De causis plantarum*; first Latin edition (Treviso: Confalonerius, 1483; Klebs, 958). This folio volume (30.5 cm) contains the Latin text of both *De historia* and *De causis plantarum*, as translated by Theodoros Gaza (*c.* 1400–1475), who added a long preface. [From the copy in the Harvard College Library.]

terms? What are the intentions of Nature, which does nothing in vain? How do the plants live, grow, and multiply? In spite of its being less descriptive than the first work, it is full of facts. Theophrastos' accumulation of botanic knowledge is as wonderful as Aristotle's accumulation of zoölogic knowledge; both are almost unbelievable. We must admit the same partial explanation in both cases: Theoprastos (as well as Aristotle) wrote the synthesis and did most of the work, but his facts were gathered not only by himself but by many other men. Among his two thousand disciples, he must have obtained the collaboration of many; though Alexander died before Theophrastos' accession to the mastership, we may be sure that his staff sent botanic specimens as well as zoölogic ones to the Lyceum, and Theophrastos' knowledge of foreign plants (for example, those of India) was partly due to Alexander's generous patronage.

Let us first see how these two treatises are built. The *History of plants* is divided into nine books, dealing roughly with the following topics: 1. Parts of plants and their nature, classification. 2. Propagation, especially of trees. 3. Wild trees. 4. Trees and plants special to particular districts and situations (geographic botany). 5. Timber of various trees and its uses. 6. Undershrubs. 7. Herbaceous plants, other than coronary; potherbs and similar wild herbs. 8. Herbaceous plants: cereals, pulses, and "summer crops." 9. Juices of plants and the medicinal properties of herbs.

The *Causes of plants* is divided into fewer books but is almost as long as the former work:[90] 1. Generation and propagation of plants, fructification and maturity of fruits. 2. Things which help most the increase of plants; horticulture and sylviculture. 3. Plantation of shrubs and preparation of the soil; viticulture. 4. Goodness of seeds and their degeneration; culture of legumes. 5. Diseases and other causes of failure. 6. Savors and odors of plants.

Theophrastos deals with about 500–550 species and varieties of plants, most of them cultivated; the wild plants, he adds, are very largely unknown and unnamed, yet he often refers to such. He assumed that certain wild plants could not be domesticated; this implies that attempts at acclimatization had already been made, and we are not surprised that some had failed.

What is most striking in both works is their methodic nature, in the best Aristotelian tradition. Some curious and idle facts are mentioned here and there, because the author found them too interesting to be thrown out, but in general there is a clear purpose of explanation, of differentiation and classification. Theophrastos (like Aristotle before him) was handicapped by insufficient terminology, but he introduced a few technical terms, the most needful, such as *carpos* for fruit, *pericarpion* for seed vessel, *metra* or matrix for the central core of stems.

He distinguished various modes of reproduction of plants — spontaneous,[91] from a seed, from a root, or from other parts. What is more remarkable, he had observed the behavior of germinating seeds and had seen the fundamental difference that

[90] In Wimmer's Greek-Latin edition (Paris, 1866), 155 pp. against 163. The first work is readily available in the Loeb Classical Library, two volumes translated by Sir Arthur Hort. Wimmer's and Hort's editions include glossaries of plant names, but Wimmer's glossary covers both works.

[91] It is hardly necessary to recall that the idea of spontaneous generation continued to be accepted (for lower and lower forms of life) until the time of Pasteur, 1861, less than a century ago.

we express by the words monocotyledon and dicotyledon.[92] His explanation was insufficient, yet it held the ground until it was corrected and completed in the second half of the seventeenth century by Marcello Malpighi (1628–1694).

The urge for botanic knowledge was at the beginning nothing but the urge for food and drugs. Theophrastos had long passed that stage and was interested in botany for its own sake, the understanding of plant life in all its forms, yet he did not lose interest in the many applications of botany to human purposes. The ninth book of the *History of plants* is largely medical. We find in it a good account of the superstitious rites of the rhizotomists in the gathering of roots and herbs.[93] Another illustration of his scientific spirit is given in the same work when he describes the "spontaneous changes in the character of trees and certain marvels" and remarks that "soothsayers call such changes portents." [94] He is not able to indicate the cause of each, but assumes that there is a cause; the changes are not miraculous but natural.

Book IX will appeal to students of economics and sociology, as well as to students of botany and pharmacy, for its chapters describe the methods of collecting resin and pitch, of making pitch in Macedonia and Syria, of collecting frankincense and myrrh in Arabia, and so on. The products and methods are described with some detail, though they often refer to countries that Theophrastos had not visited, and this proves again that much of his information was obtained from other people.

There are even references to Indian plants.[95] The first is a fig (*Ficus bengalensis*, banyan) and he noticed the ability of its branches to reach to the ground and become roots, the second a reed, the third a plant having a powerful aphrodisiac virtue.[96] Theophrastos must have received such information either from Hindu merchants who came to Athens, from members of Alexander's expedition, or perhaps from old students of his who had traveled to India.

The *De causis plantarum* is less known than the other work, but my examination of it suggests that it would deserve to be investigated more deeply and to be translated into English. Let us sample it. There is an account of mistletoe and its habit of refusing to sprout except upon the bark of living oaks.[97]

We have discussed above Herodotos' confused account of fructification of palms and of the caprification of fig trees; Theophrastos' account is much better, and that is what we would expect, for not only did he come a century later but he was a professional botanist while Herodotos was an amateur. Theophrastos' account of caprification is still very imperfect (there is a confusion in his mind between caprification and the formation of galls by insects), but let me quote his way of explaining the fecundation of palm trees:

With dates it is helpful to bring the male to the female; for it is the male which causes the fruit to persist and ripen, and this process some call, by analogy, "the use of the wild fruit." The process is thus performed: when the male palm is in flower, they at once cut off the spathe on which the flower is, just as it is, and shake

[92] Plants beginning with one seed leaf or with two. Theophrastos' distinction between these two groups is well indicated by Singer, *Story of living things* p. 50. The word *cotylēdones* occurs in Theophrastos, History of plants, IX, 13, 6, but with the meaning of suckers.

[93] *History of plants*, IX, 8.

[94] *Ibid.*, II, 3.

[95] *Ibid.*, I, 7, 3; IV, 11, 13; IX, 18, 9.

[96] The whole end of that chapter (IX, 18) was omitted by Hort in the Loeb edition. Such prudishness in a scientific book is truly shocking.

[97] *Causes of plants*, II, 17.

the bloom with the flower and the dust over the fruit of the female, and, if this is done to it, it retains the fruit and does not shed it. In the case both of the fig and of the date it appears that the "male" renders aid to the "female," – for the fruit-bearing tree is called "female" – but while in the latter case there is a union of the two sexes, in the former the result is brought about somewhat differently.[98]

Is this clear-cut introduction of plant sexuality not amazing, especially if one considers that it was almost completely forgotten until its reintroduction more than two millennia later?

The amount of detailed information included in the two works (and both must be considered if we wish to judge fairly the extent of Theophrastos' botanic knowledge) is so vast that he must have had fairly continual access to a number of plants. The garden of the Lyceum was to some extent a botanic garden; it may be that the estate added to the original property thanks to the munificence of Demetrios of Phaleron was partly devoted to that purpose. In his will (preserved by Diogenes Laërtios), Theophrastos requests that he be buried in the garden and hopes that Pamphylos "who dwells in it shall keep it and everything else in the same condition as it has been in hitherto." This does not prove, of course, that the garden was a botanic garden, but when does a garden become botanic? Or, to put it otherwise, does not every garden become botanic when a botanist uses it for his own scientific needs? The garden of the Lyceum was probably a botanic garden of that simple kind. It could not be botanic in the sense of a later age when taxonomy was given supreme importance and gardens were arranged for the main purpose of teaching it.[99]

There is also a fair amount of phytopathology in both works,[100] and why not? Phytopathology is a learned word which the Greeks never knew, but every Greek husbandsman could not help being aware of the degeneration and untimely destruction of some of his crops. These were terrible facts which hurt him and might even ruin him, and there was no means of forgetting them. Greek farmers would discuss such events in their respective families or with other farmers. Writers on horticultural subjects like Theophrastos did not need to invent anything new when they spoke of various pests; they simply dealt with the obvious.

As for pests, – radish is attacked by fleas, cabbage by caterpillars and grubs, while in lettuce, leek, and many other herbs occur "leek-cutters." These are destroyed by collecting green fodder, or when they have been caught somewhere in a mass of dung, the pest being fond of dung emerges, and having entered the heap, remains dormant there; wherefore it is then easy to catch, which otherwise it is not. To protect radishes against fleas it is of use to sow vetch among the crop; to prevent the fleas from being engendered they say that there is no specific.[101]

Other passages of the same kind occur in the *History of plants.*[102] The insects referred to can sometimes be identified by modern entomologists.

[98] *History of plants,* end of II.

[99] If we accept the Lyceum garden as the first botanic garden, we have to wait four hundred years for the second, to wit, the garden created in Rome by Antonius Castor. Pliny (I-2) visited that garden in which Antonius, though he had passed his hundredth year "cultivated vast numbers of plants with the greatest care"; Pliny, *Natural history,* XX, 100; XXV, 5.

[100] For example, *History of plants,* VII, 5; VIII, 10; *Causes of plants,* IV.

[101] *History of plants,* VII, 5.

[102] *History of plants,* VIII, 10; VIII, 11; IV, 14; V, 4; etc.

The fleas on radish were flea-beetles; the caterpillars on cabbage were cabbage butterflies; the "horned-worm," a cerambycid beetle; the grubs engendered in seeds, pea-weevils; the cobweb of olive, red spider; the worm of fruits, codling-moth; the *teredo* of timber in sea-water, the ship worm.[103]

Theophrastos' phytopathology was restricted to the damage caused by insects and worms; he could not yet know of those caused by vegetable parasites. Yet his was a good beginning.

The best summary of Theophrastos' botanic achievements has been given by Greene, and we reproduce it. That summary is a little deceptive in that it includes, for the sake of clearness and brevity, some technical terms (for example, petal, corolla, androecium) that were unknown to Theophrastos, and thus his knowledge seems sharper than it really was.

1. [Theophrastos] distinguished the external organs of plants, naming and discussing them in regular sequence from root to fruit; the naturalness of which sequence was afterwards pointedly denied; but in modern botany it stands everywhere approved.

2. He classified such organs as (*a*) permanent, and (*b*) transient; a division of them which may yet be shown more scientific than the modern distinguishing of them as (*a*) vegetative, and (*b*) reproductive.

3. The existence of aërial roots, as being of the nature of roots, and thus different from tendrils and other prehensile organs, was discovered by him and has never since been disputed.

4. He remarked upon the inconsistency of retaining in the category of roots certain enlarged, solidified, jointed, and otherwise peculiar underground parts; a suggestion which lay unheeded during two thousand years of botanical history, and has only recently led to the open recognition of the category of subterranean stems.

5. He recognized, by differences of size, solidity, and other particulars of structure, three classes of stems: the trunk, stalk, and culm.

6. By never speaking of calyx and corolla as peculiar and separate organs, but always referring to their parts as leaves merely, it is evident he regarded the flower but as a metamorphosed leafy branch; to which forgotten Theophrastan philosophy of the flower neither Goethe nor Linnaeus had but returned, when each supposed himself the discoverer of a new anthogeny.

7. He divided the plant world into the two subkingdoms of the flowering and flowerless.

8. The subkingdom of the flowering he again saw to be made up of plants leafy-flowered and capillary-flowered; really the distinction between the petaliferous and the apetalous; one the deep import of which was first realized and taken advantage of by the systematists of some two centuries ago.

9. He indicated the still more important differences of the hypogynous, perigynous, and epigynous insertion of corolla and androecium.

10. He distinguished between the centripetal and centrifugal in inflorescences.

11. He was first to use the term fruit in the technical sense, as applying to every form and phase of seed encasement, seed included; and gave to carpology the term pericarp.

12. He classified all seed plants as (*a*) angiospermous and (*b*) gymnospermous.

13. Respecting the texture and duration of their parts he classified all plants as tree, shrub, half-shrub, and herb; also noted that herbs were of perennial, biennial, or annual duration.

14. He indicated with clearness several of those differences in the structure of stems, leaves, and seeds by which the botany of later times separates plants monocotyledonous and dicotyledonous.

15. He described the differences between the excrescent and deliquescent in tree development.

16. He knew how the annual rings in the

[103] The previous quotation and these identifications are borrowed from Melville H. Hatch, "Theophrastos as economic entomologist," *J. New York Entomological Soc.*, *46*, 223–227 (1938). More identifications are suggested by F. S. Bodenheimer, *Materialien zur Geschichte der Entomologie* (Berlin, 1928), vol. 1, pp. 70–76 [*Isis* 3, 388–392 (1920–21)].

stems or trunks of certain woody growths were formed.

17. Theophrastos, with natural vision unaided by so much as the simplest lens, and without having seen a vegetable cell, yet distinguished clearly between parenchymatous and prosenchymatous tissues; even correctly relating the distribution of each to the fabrics of pith, bark, wood, leaves, flowers, and fruits.[104]

It is very strange that so much botanic knowledge should have been accumulated by the end of the fourth century, and that so little, if anything, was added to it in ancient times. Theophrastos is not only the first botanic writer but also the greatest until the Renaissance of the sixteenth century in Germany. His Greek followers, Nicandros of Colophon (III-1 B.C.), Cratevas (I-1 B.C.) and the latter's royal employer, Mithridates Eupator (I-1 B.C.), Dioscorides of Anazarbos (I-2), enriched the Greek herbal and Cratevas illustrated it, but I am not aware of any material contribution of theirs to botanic science. As to the Romans, Cato the Censor (II-1 B.C.), Varro (I-2 B.C.), Columella of Gades (I-2), their main contributions were agricultural. Pliny (I-2) put together all the knowledge available in his time but added nothing. The botany of Theophrastos and the zoölogy of Aristotle represent the climax of natural history in antiquity.

GEOLOGY AND MINERALOGY

EARLY KNOWLEDGE

Much knowledge of geology and mineralogy had been collected for centuries because of mining undertakings in Egypt, Greece, and elsewhere. The quest for metallic ores and for gems began very early. Many curious geologic phenomena could be witnessed in the Near East, such as earthquakes, volcanic explosions, hot and mineral springs, caverns and underground waters, strange mountain shapes, narrow canyons. People who were attentive enough and reflective, as many Greeks were, could not help cogitating on those mysteries. Why did they happen? And how? The first explanations were mythologic and could not satisfy very long those gifted men, wise in their generation. The Pythagoreans postulated the existence of a central fire within the earth, an idea that could not be disproved and has continued to be entertained almost to our own day, coalescing with the conception of an underground hell.[105] In our note on Xenophanes of Colophon above, he was called with justice the earliest geologist as well as the earliest paleontologist. Herodotos explained the formation of Lower Egypt by alluvia. The extraordinary behavior of the Nile had exercised the curiosity of Greek travelers from early days, and they speculated on the causes of the annual flood. Even the most intelligent admitted the possibility of interchanges between earth and water; it was possible, they thought, for earth to appear where water was, and vice versa. Xenophanes' ideas concerning fossils were accepted by Xanthos of Sardis,[106] by Herodotos,

[104] Edward Lee Greene (1843-1915), *Landmarks of botanical history prior to 1562* (Washington, 1909), pp. 140-142. The most recent studies of Theophrastos are due to the Swiss botanist, Gustav Senn-Bernoulli (1875-1945), *Die Pflanzensystematik bei Theophrast* (Bern, 1922) [*Isis* 6, 139 (1923-24)]; *Die Entwicklung der biologischen Forschungsmethode in der Antike und ihre grundsätzliche Förderung durch Theophrast* (262 pp.; Aarau: Sauerländer, 1933) [*Isis* 27, 68-69 (1937)].

[105] For example, Dante's *Inferno*; see *Introduction*, vol. 3, p. 487, Fig. 8. The ideas should not be confused with modern ideas on the constitution of the earth's nucleus or on the foci of earthquakes, for the modern scientific ideas are absolutely independent of the ancient or medieval fancies.

[106] The Lydian Xanthos son of Candaules flourished in the time of the first Artaxerxes (ruled 464-424); he concerned himself with geology and botany.

Eudoxos of Cnidos, Aristotle, Theophrastos; and such ideas would have remained in circulation had they not been driven down and out by the Jewish-Christian dogma of creation.

Precious stones were collected from the earliest times for the adornment of women or for ceremonial purposes.[107] There was an immemorial knowledge of them, comparable to the immemorial knowledge of animals and plants. The three kingdoms of nature were equally familiar to prehistoric men. The novelty in the age of Aristotle was not so much in the knowledge as in the scientific form given to it, in its partial disentanglement from folklore and superstition.

Various geologic topics are discussed in the *Meteorologica* ascribed to Aristotle.[108] It is significant that in ancient and medieval times the two fields, meteorology and geology, were closely interlocked. For Aristotle and for all the men of science of classical antiquity, earthquakes and volcanic explosions were interrelated. They continued to accept the idea of a central fire, and Aristotle tried to explain it by his assumption of underground winds which would be heated up by friction and shock; that might lead to explosions, even submarine explosions, such as occurred in one of the Lipari Islands. The idea of underground winds was itself very ancient; [109] it was symbolized by the myth of Aiolos; Aiolos was supposed to dwell in or under the Aiolian Islands (that is, the Lipari Islands, where volcanic explosions were not unfrequent). It was thus natural enough to pass from the consideration of winds above ground (meteorology) to winds underground (seismology, geology). The genesis of stones, metals, and minerals was expressed in terms of winds or exhalations, some of them giving birth to minerals and insoluble stones, others to metals which are fusible or ductile.

Aristotle's explanation of earthquakes is interesting in itself, and in addition it includes an account of the earlier views of Anaximenes, Anaxagoras, and Democritos. That subject had forced itself upon the attention of Greek philosophers; it is not even necessary to be a philosopher in order to be aware of a volcanic explosion or of an earthquake, and, according to one's temperament and education, such a sharp warning will start one dreading and praying, or wondering, imagining, and cogitating. Some Greeks invented appropriate myths and incantations; others, the natural philosophers, tried to find explanations and started a new branch of science, seismology.

THEOPHRASTOS THE MINERALOGIST

It so happens that the earliest scientific book on stones (minerals and gems) was written by Theophrastos. It is as if Aristotle and he had shared the three kingdoms of nature between them: the first two were dealt with by Theophrastos, the third by Aristotle himself.[110]

The *De lapidibus* is counted as a fragment, but it is fairly long (some ten closely printed pages in the Didot edition), and it would be better to call it a treatise, even if the whole of it has not come down to us. It deals with stones in the

[107] See, for example, the description of the gems on Aaron's breastplate in Exodus 28.

[108] *Isis 6*, 138 (1924).

[109] The conception of winds imprisoned in underground caverns is not yet completely abandoned today. The fancy subsists in Persia; see anecdote told by E. G. Browne, *A year among the Persians* (Cambridge, ed. 2, 1926), p. 257.

[110] Aristotle's geologic explanations have been already referred to, but his main work as a naturalist was done in zoölogy.

broadest sense; one might call it a treatise on petrography, the very first, of course. It describes the characteristics of various rocks and minerals and indicates their provenience and their uses. Theophrastos' ideas on fossils were not explained in this book, however, but in another one on fossil fishes,[111] wherein he refers to the remains of fishes found in the rocks of the region south of the Black Sea.

He thought that these fossils were developed from fish spawn left in the earth, or that fishes had wandered from neighboring waters and had finally been turned into stone. He also expressed the idea that a plastic force is inherent in the earth whereby bones and other organic bodies are imitated.[112]

To return to the rocks, Theophrastos described their several kinds and tried to classify them, according to the action of fire upon them. Part of that is naturally chemical, for a mineralogic analysis, however crude, leads to the consideration of chemical reactions or chemical applications. For example, Theophrastos describes the preparation of white lead:

A piece of lead the size of a brick is placed over vinegar in an earthenware vessel. When the lead has acquired a [rustlike] layer, which usually happens in ten days, they open the vessel and scrape off the decayed part. The process is then repeated again and again until the lead is entirely consumed. They take what has been scraped off and keep pulverizing it in a mortar and filtering it. What finally settles to the bottom is white lead.[113]

Theophrastos, continuing Aristotle's meditations, tried to account for the genesis in inorganic nature of two kinds of things that were extremely different, stones and metals. The stones, he suggested, were of earthy origin (stones disintegrate into earth), the metals of watery origin. Among the stones he attached special importance to those wonders of the lifeless world, the precious stones, the gems. A good part (one quarter) of his treatise deals with gems and it was that part which interested posterity most. In his description of gems he recognized many physical peculiarities, such as weight, color, transparency, luster, fracture, fusibility, hardness. He indicated the locality where some gems could be found, and the high prices that they fetched. His descriptions are sufficient to identify a number of stones: alabaster, amber, amethyst, emerald, garnet, lapis lazuli, jasper, agate, onyx, carnelian, rock crystal, prase, chrysocolla, malachite, magnetite, and hematite. About many others we cannot be sure or we are entirely in the dark. For example, *adamas*, which is not injured by fire. What is that? Diamond? It is impossible to say. His information came from almost every quarter of the world known to the Greeks, from the three continents centered upon the Mediterranean. Much of it was very ancient, Babylonian or Egyptian perhaps, immemorial, prehistoric folklore. Hence, we must not be surprised when we come across some

[111] In a long fragment (frag. 171) entitled *De piscibus in sicco degentibus* (Didot Greek-Latin edition, p. 455–58), On the fishes remaining in a dry condition – really on fossil fishes. This fragment is long enough to be called the first treatise on paleontology. Theophrastos was "first" in many fields.

[112] Sir Archibald Geikie, *The founders of geology* (London: Macmillan, ed. 2, 1905), p. 16.

[113] In par. 56; translation as given in M. R. Cohen and I. E. Drabkin, *Source book in Greek Science* (600 pp.; New York: McGraw-Hill, 1948), p. 359. As Drabkin remarks in a footnote, the end product of that reaction is not necessarily lead carbonate (white lead, ceruse) but lead acetate; it would require plenty of carbonic acid to change the acetate into a carbonate.

irrational sayings; yet the book as a whole is remarkably rational, or call it scientific. Some of his conclusions were correct. He knew that pearls were secreted by oysters (of course, pearls were always found in oysters and nowhere else), that corals grow in the sea; he was aware of the existence of fossil ivory. Theophrastos' *De lapidibus* was the main source of Book XXXVII of Pliny's *Natural history*[114] and through Pliny it influenced all the more scientific lapidaries until modern times. A comparison between Theophrastos and Pliny is entirely to the advantage of the former. In spite of his coming no less than four centuries later, Pliny is far less scientific than Theophrastos; he knew incomparably more, but his knowledge was definitely of a poorer quality. This illustrates the abyss that was formed between Hellenic science and Roman science, the latter being at best a very imperfect offspring of the former.

MEDICINE

ARISTOTLE THE PHYSICIAN

In our account of Aristotle's life, we remarked that he perhaps owed his scientific proclivities to his father, who was a physician. Yet Aristotle did not become a physician, and there are very few medical references in his writings. The few references in *Topica* and *Politica* are insignificant; it is true that a whole book of *Problemata*, the first one, discusses "problems connected with medicine," but we cannot draw anything from that as the *Problemata* is certainly apocryphal and probably very late; some critics would place it as late as the fifth or sixth century.[115] The Peripatetic character of that book is admitted, but it can tell us nothing concerning Aristotle's own ideas.

It is curious, on the other hand, that his anatomic and physiologic observations were so often right in the case of animals, yet wrong in the case of men. He drew a distinction between the sutures of the skull in men and women, he stated that there are eight ribs, and only three chambers in the heart (he overlooked the interauricular septum). It is clear that he did not make human dissections but was satisfied to accept statements on human anatomy without attempting to verify them. The case of Aristotle is not as abnormal as it seems to be at first sight. Many sons of physicians have inherited from their fathers a love of science and a strong dislike of medicine; the two feelings are not by any means incompatible.

Aristotle was not interested in medicine, but some physicians took a deep interest in his philosophy and in his scientific methods, and therefore he exerted a definite influence upon the progress of medicine, as is proved by the emergence of the Dogmatic school.

THE DOGMATIC SCHOOL. DIOCLES OF CARYSTOS

The history of the Dogmatic school has been badly outlined by historians of medicine because of a fundamental error. The founder of the school, Diocles of Carystos, was supposed to have preceded Aristotle and influenced him. It has been

[114] A recent English version of that book by Sydney H. Ball, *A Roman book on precious stones* (Los Angeles: Gemological Institute, 1950 [*Isis 42*, 52 (1951)] is valuable because the author had a practical knowledge of the gems themselves.

[115] *Isis 11*, 155 (1928).

proved by Jaeger [116] that Diocles was, on the contrary, a younger contemporary of Aristotle, and that his medical theories were formed under the influence of the Lyceum.

What happened to medicine in the second half of the fourth century does not surprise the historian, for similar phenomena have happened many times. Athenian and Greek education was dominated by two illustrious schools, the Academy and the Lyceum, which gave to ambitious young men a new style of research, discussion, and exposition. A group led by Diocles realized the need of reshaping medical doctrines in accordance with academic usage and of reëxplaining them in philologic language.[117] There have always been physicians who loved learning, who were learned themselves, or who liked to be taken for learned men, and used the most high-brow jargon of their age. Diocles did that very well and by so doing founded a new school of medicine, the Dogmatic school; he was called by the Athenians "another Hippocrates."

It is very significant that he was the first physician to write in the Attic dialect instead of the Ionian, and that linguistic change is, perhaps, the best symbol of the intellectual revolution that occurred under his direction. Up to that time, Hippocrates' dialect had been accepted as the medical dialect par excellence; it was now superseded by the language that had been stabilized by Plato and Aristotle. This marks a new epoch in medical thought. Diocles was also the first to refer to a Hippocratic collection, and this shows that for him Hippocrates was still the main guide; he did not necessarily object to Hippocratic knowledge but he believed – rightly – that scientific knowledge should be expressed in the best logical order and in the most elegant language. He was well acquainted also with the physiologic theories of the Sicilian school, as set forth by Philistion of Locroi, and combined them with the traditional views of the school of Cos.

Though Diocles is called the founder of the Dogmatic school, the foundation had been gradually prepared by other men. This was a natural development of the old Hippocratic teaching. The teaching of a man of genius is generally informal, but it cannot be preserved and continued except in a more systematic manner; this was realized unconsciously by Hippocrates' followers, to begin with, his son Thessalos, his son-in-law Polybos, his grandsons, as well as his immediate disciples, Apollonios of Cos and Dexippos of Cos, and finally Diocles. The name given to them later (by Galen and others) was *logicoi*, logicians. The translation of that term that I have just given and the traditional one, Dogmatists, are equally imperfect; the word *logicos* means many things, such as "intellectual," "dialectical," "argumentative"; it is clear that Galen used it to distinguish logical and philosophic methods of exposition from simpler ones. To put it as briefly as possible, the Dogmatists gave to the medicine of Aristotle's time its speculative coloring.

As far as can be judged from fragments (for none of his many writings has been preserved in its integrity) and from the early commentators, Diocles was not simply an able writer who put the medical knowledge of his day in logical order; he

[116] Werner Jaeger, *Diokles von Karystos. Die griechische Medizin und die Schule des Aristoteles* (244 pp.; Berlin: Walter de Gruyter, 1938) [*Isis* 33, 86 (1941–42)]; "Vergessene Fragmente des Peripatetikers Diokles, nebst zwei Anhängen zur Chronologie der dogmatischen Ärzteschule," *Abhandl. Preuss. Akad., Phil. hist. Kl.*, No. 3 (46 pp.; 1938).

[117] The same thing happened in the late thirteenth century and in the fourteenth when Italian physicians, hypnotized by the expository methods of theologians and jurists, wrote medical books in a similar style; *Introduction*, vol. 2, p. 70; vol. 3, pp. 264, 1222.

increased that knowledge by his own observations. He made embryologic, gynecologic, and obstetric studies and carried on animal dissections (for example, he dissected the womb of a mule). He described the cotyledonous placenta of ruminants and early human embryos. He held that both man and woman contribute "seed" toward the creation of their children. It is said that he wrote the first textbooks on anatomy and on medical botany.[118]

Diocles' successor as leader of the Dogmatists was Praxagoras of Cos, who was the first to make a clear distinction between veins and arteries, holding that the former carry blood while the others are filled with air.[119] His study of the blood vessels led him to that of the pulse, strangely neglected in the Hippocratic corpus. Three of Praxagoras' pupils are known – Philotimos, who paid special attention to gymnastics and diet, Mnesitheos of Athens, who made anatomic investigations (on the bodies of animals) and tried to classify diseases, and the illustrious Herophilos. If we accept (as we do) Jaeger's new dating of Diocles, he ended his life in the first quarter of the third century and was thus already a witness of the Hellenistic period; a fortiori, Praxagoras and Mnesitheos, *fin de siècle* children of Hellenism, belong to the new period; they are contemporaries of Herophilos of Chalcedon (III-1 B.C.), and we would have been justified in leaving them out of this volume.

We know the ideas of the Dogmatists only in a fragmentary way, but their evolution from Polybos to Mnesitheos suggests that their dogmatism was tempered with genuine observations and sound criticism. The Dogmatic school was the necessary transition from Hippocratism to the new anatomy and physiology; it constructed one of the bridges leading from Cos to Alexandria.

MENON

It is with some diffidence that we bring this chapter to a close with a brief account of the mysterious Menon. According to Galen, if one wishes to know the ideas of the ancient physicians, one should read the historical outline ascribed to Aristotle but written by his disciple, Menon, and therefore called *Menoneia*.[120] If Menon was Aristotle's disciple, then, of course, his place is here, but Galen's assumption is vague; Menon might be a distant disciple, instead of an immediate one.

The tradition of Menon's outline is curious. The British Museum acquired in 1891 a medical papyrus of considerable size,[121] the importance of which was soon appreciated and advertised by Sir Frederick Kenyon.[122] The text was written at

[118] His *Rhizotomicon* may perhaps be called a treatise on botany, and it may be prior to Theophrastos. He and Theophrastos were close contemporaries, Diocles being probably a little younger. This does not exclude the possibility that Theophrastos used the botany of his younger colleague. Diocles (this Diocles?) is mentioned but once in Theophrastos' writings, not in the botanical ones but in the book on stones (28), apropos of *lyngurion* (amber or tourmaline?).

[119] The error was excusable, because the elastic arteries empty themselves when the heart ceases to beat. It was accepted for centuries and was one of the causes of the long delay in the discovery of circulation, the whole circulation (Harvey, 1628).

[120] K. G. Kühn, *Galeni opera omnia* (Leipzig, 1821–1833), vol. 15, p. 25, "Galeni in Hippocratem de natura hominis commentarius."

[121] The papyrus called Anonymus Londinensis is 12 ft long and includes 39 columns or portions of columns, each about 3 in. wide; in all, about 1900 lines. The text is acephalous. Paleographic considerations tally with a date such as (II-1).

[122] F. G. Kenyon, "A medical papyrus in the British Museum," *Classical Rev.* 6, 237–240 (1892). The text was entirely transcribed by Kenyon but first edited by Hermann Diels, *Supplementum Aristotelicum* (Berlin, 1893), vol. 3, part 1; new edition by W. H. S. Jones, *The medical writings of Anonymus Londinensis* (176 pp.; Cambridge: University Press, 1947) [*Isis* 39, 73 (1948)].

the beginning of the Christian era, before Galen's time, perhaps just before that time, in the first half of the second century. The first half of it is a historical outline derived from Menon's. The outline ends with the second half of the fourth century B.C., and this would confirm the hypothesis that Menon flourished in that period or very soon afterward.

The fact that a disciple of Aristotle thought it necessary to write a history of ancient medicine is significant. It will not surprise the readers who have followed in this volume our own very brief outline of it. By the end of the fourth century, medicine was not only an art and a profession of immemorial antiquity; it was also a science with many centuries of experience, and it was, or tried to be, a philosophy. The learned physician who was flourishing at the turn of the fourth century in Athens was a very sophisticated man. If he was wise enough, he recognized his ignorance of many things and the urgent need of deeper investigation, especially in the basic fields of anatomy and physiology. Hellenic medicine was ending in a dignified philosophic atmosphere, with a magnificent record of achievements to its credit; it had gone as far as was possible with its own methods. More investigations were needed to justify new theories. The last Hellenic physicians were preparing the way for the Hellenistic anatomists.

XXII

ARISTOTELIAN HUMANITIES AND HISTORIOGRAPHY IN THE SECOND HALF OF THE FOURTH CENTURY B.C.

ECOLOGY

Aristotle was primarily a man of science, looking at everything from a rational angle, but he was also a philosopher, even a metaphysician, and was deeply interested in all the humanities. It is highly typical of him that we must introduce his political and sociologic studies with ecologic considerations.

What is ecology? The word is Greek, of course, but is no part of the ancient Greek vocabulary. Its English form was first (more correctly) oecology, and the earliest example of "oecology" mentioned in the *Oxford English Dictionary* is dated as late as 1873 (Haeckel); the earliest example of "ecology" in the *Supplement* of the *Dictionary* is dated 1896.[1] The *Oxford Dictionary* defines it as "the science of the economy of animals and plants; that branch of biology which deals with the relations of living organisms to their surroundings, their habits and modes of life, etc."

The name "ecology" is very recent, but the science itself is ancient; as old as Aristotle himself. Every intelligent naturalist has dealt occasionally with ecologic problems, just as the *bourgeois gentilhomme* of Molière *"faisait de la prose sans le savoir."* We may be sure that even before Aristotle's time bright farmers, hunters, or fishermen had had opportunities of observing ecologic phenomena. Aristotle, however, was the first to write about them, and thus to introduce ecologic conceptions into scientific literature.

Let me adduce two examples. The first is that of the pinna.[2] The pinna [3] is a bivalve mollusk whose comfort is ensured by the presence of a small crab living in its mantle cavity and helping it to obtain its food. This little crab is called *pinotērēs* or *pinophylax* (guardian of the pinna). Says Aristotle, "If the pinna be deprived of

[1] The original meanings of the words "ecology" and "economy" are almost equivalent. It would be silly to insist on the spelling "oecology," and not to spell the other word "oeconomy." The words *nomos* and *logos* are frequently interchanged in our terminology; we call one science "geology" and another "astronomy," while the word "astrology" is used to denote a set of superstitions. Every language is a mixture of reason and caprice.

[2] *Historia animalium*, 547*b*–548*a*.

[3] The old Greek spelling is *pina* or *pinē*, with one *n*; we write *pinna* according to English usage, but *pinotērēs* and *pinophylax* to follow the best Greek one. For pinna folklore, see *Isis* 33, 569 (1941–42).

its guard it soon dies." It is highly probable that fishermen had observed that strange case of commensalism long before Aristotle, and that the name *pinotērēs* (or *pinophylax*) is a popular word rather than a scientific term. Popular knowledge on that subject is proved by the use of the word *pinotērēs* to designate human parasites! We may be sure that the first people to give to a sycophant that ingenious nickname had found it not in the *Historia animalium* but in the living language.

The other example is even more curious. I quote the whole text, though the end of it is irrelevant; it is a good example of Aristotelian zoölogic description. The discussion following the quotation is restricted to the problem of population, which Aristotle was first to raise.

> The phenomena of generation in regard to the mouse are the most astonishing both for the number of the young and for the rapidity of recurrence in the births. On one occasion a she-mouse in a state of pregnancy was shut up by accident in a jar containing millet-seed, and after a little while the lid of the jar was removed and upwards of one hundred and twenty mice were found inside it.
>
> The rate of propagation of field mice in country places and the destruction that they cause, are beyond all telling. In many places their number is so incalculable that but very little of the corn-crop is left to the farmer; and so rapid is their mode of proceeding that sometimes a small farmer will one day observe that it is time for reaping, and on the following morning, when he takes his reapers afield, he finds his entire crop devoured. Their disappearance is unaccountable: in a few days not a mouse will there be to be seen. And yet in the time before these few days men fail to keep down their numbers by fumigating and unearthing them, or by regularly hunting them and turning in swine upon them; for pigs, by the way, turn up the mouse-holes by rooting with their snouts. Foxes also hunt them, and the wild ferrets in particular destroy them, but they make no way against the prolific qualities of the animal and the rapidity of its breeding. When they are superabundant, nothing succeeds in thinning them down except the rain; but after heavy rains they disappear rapidly.
>
> In a certain district of Persia when a female mouse is dissected the female embryos appear to be pregnant. Some people assert, and positively assert, that a female mouse by licking salt can become pregnant without the intervention of the male.
>
> Mice in Egypt are covered with bristles like the hedgehog. There is also a different breed of mice that walk on their two hind-legs; their front legs are small and their hind-legs long;[4] the breed is exceedingly numerous. There are many other breeds of mice than are here referred to.[5]

Aristotle has well noticed the sudden and quick increase in the population of an animal species, followed by its decrease or complete disappearance. A very recent writer on the subject remarked that:

> Aristotle's measured and balanced description of the rise and fall of a mouse population might be taken for a text for the present book. For it contains most components of the problem of natural fluctuations.[6]

We need not be surprised that Aristotle did not see the bottom of that riddle, for it is very deep indeed, and the essential part of it was not seen until our own day (1925–1935). Says Elton:

> The general idea that animal communities simply by their structure and organization have the ability to generate fluctuations was not explicitly discussed by anyone at all (except Spencer) until about 1925. In this year Lotka, an American mathematical

[4] This is a reference to the jerboa (*Dipus aegyptiacus*).

[5] *Historia animalium*, 580b10.

[6] Charles Elton, *Voles, mice and lemmings. Problems in population dynamics* (Oxford: Clarendon Press, 1942), p. 3 [*Isis* 35, 82 (1944)].

expert on human population dynamics, published his remarkable analysis of the world as an ecosystem; and about the same time Volterra, a pure mathematician working in Italy, arrived at somewhat similar ideas about fluctuations.

The great difference between their theories and those of ecologists like myself was that I had thought of external disturbances such as climate as the primary generating force in causing populations to oscillate, the other factors such as epidemics and predator population changes being a secondary result. But Lotka and Volterra believed that they could prove by rigid mathematical arguments that groups of ecologically linked species must fluctuate, so that climate and other external influences would merely tend to interfere with the natural rhythms, producing very complex consequences. There is very little doubt that their conclusions are broadly true. It is remarkable that such an important concept should have originated independently in the minds of two mathematicians living four thousand miles apart, one officially studying human vital statistics and the other not directly connected with biology at all.[7]

This quotation has taken us far away from Aristotle; it serves to illustrate the incredible resonance of genuine scientific problems. Superstitions turn in circles and lead nowhere, but the rational questions that were asked twenty-three centuries ago by men of science such as Aristotle and Theophrastos are still exercising and fertilizing the minds of men today.

ETHICS

Founder of logic, founder of many branches of the natural sciences, Aristotle was also a founder of ethics. The ethical treatises ascribed to him are indeed the earliest formal treatises of their kind.[8]

Four ethical treatises are included in the Aristotelian corpus.[9] The first and largest, called *Nicomachean ethics*, is almost certainly genuine. The three others are (2) the *Eudemian ethics*, a briefer recasting of the same subject probably by another person;[10] (3) *Magna moralia*, a later work partly derived from the two preceding; (4) a very brief treatise on *Virtues and vices*, which is later still, perhaps much later. The first treatise is larger than the three others put together.[11] When one studies Aristotelian ethics, with the accent on Aristotle's personal contribution to it, it is sufficient to consider the *Nicomachean ethics*. For a deeper study of Peripatetic morals, it would be necessary to examine as well the *Eudemian ethics* and the *Magna moralia* and to discuss the interrelations of these three treatises, something comparable to the interrelations of three synoptic Gospels.

The *Nicomachean ethics* is so called because Aristotle wrote it for one Nicomachos, probably his son by his second wife, Herpyllis of Stageira. It has been argued, however, that Aristotle wrote it not for his son but for his father; according

[7] *Ibid.*, p. 158.

[8] Plato's dialogues belong to a different category of books.

[9] 1094*a*–1251*b*.

[10] There are many resemblances between the *Nicomachean* and the *Eudemian ethics*, and Books IV, V, VI of the latter are equivalent to Books V, VI, VII of the former. It has been argued that these three books originally belonged to the *Eudemian ethics* and were later interpolated into the *Nicomachean*. Both treatises might have been written by the same person, though if Aristotle was that person, one does not see why, busy as he was, he should have rewritten one of his own works. The *Magna moralia* is certainly by another person, as is proved by differences in vocabulary and grammar; there are no less than 40 words in it that do not occur in the two others.

[11] The *Nicomachean ethics* covers 176 columns in the Bekker edition, the three others 144 (72, 66, 6).

to a third theory, the treatise was not dedicated to his son but edited by him. The first hypothesis is the one most commonly accepted today.

Aristotle's aim was to discover what scheme of life is the most desirable and the best, or, to put it otherwise, to determine man's highest good, which being determined it is his duty to pursue. The highest good is the fulfillment of one's human mission, the development of the virtues of which the human soul is capable, and the attainment of happiness (true happiness, not the popular conception of it). The possession of external goods is helpful but not essential. Virtue is praiseworthy but happiness is above praise. There are two great classes of virtues – moral (such as courage, temperance, magnanimity, justice) and intellectual (wisdom, the contemplation of truth). The highest good is to be found in the life of contemplation (*theōria*).

The *Nicomachean ethics* is divided into ten books: 1. The good for man. 2–5. Moral virtues. 6. Intellectual virtue. 7. Continence and incontinence, Pleasure. 8–9. Friendship. 10. Pleasure and happiness.

Aristotle explained that virtue is neither innate, nor a result of knowledge (as Plato believed); it is a habit of the soul which can be acquired and perfected. The highest habit is the exercise of the divine part of our soul, our reason. The development of the divine in us brings us closer to God. The *Nicomachean ethics* is not only the earliest treatise on ethics; it is the earliest treatise on rational ethics, and on many questions it has not yet been improved. One cannot help envying the students and auditors of the Lyceum who were admitted to such a lofty discussion, a discussion that was conspicuous by its reasonableness and moderation, with a minimal appeal to emotions and enthusiasm.

The title of the *Eudemian ethics* is explained in the same way as that of the Nicomachean and has caused the same ambiguity. The treatise is called Eudemian either because Aristotle dedicated it to one Eudemos, or because Eudemos actually wrote or edited it. In both cases, the source would remain the same, Aristotle, and there can be no doubt about that on account of the many similarities between the *Eudemian* and the *Nicomachean ethics*.

As to Eudemos, one can think of only one, the mathematician Eudemos of Rhodes, who was one of Aristotle's favorite disciples, and who might have succeeded him as master of the Lyceum. Theophrastos was finally chosen not because he was a more faithful disciple than Eudemos, but rather because of the greater suavity of his manners. Aristotle, who knew so many things, had already realized that a headmaster must have many qualities, and that the purely intellectual qualities are perhaps not the most important. How much did the master appreciate the value of "feelings" *versus* intelligence, of the heart *versus* the brain?[12] It is impossible to say, almost as impossible as to answer the query: Did the architect of the Parthenon have a good heart, was he a kindly and generous man?

If the *Magna moralia* could be definitely ascribed to Aristotle, we would be on safer ground. It has about the same length as the *Eudemian ethics* (66 columns to 72), and is a summary of both the Nicomachean and the Eudemian treatises, but with a startling novelty:

[12] Strictly speaking, this last expression is wrong when applied to Aristotle, who placed the seat of reason in the heart itself, not in the brain. I use it for the sake of clearness.

Speaking generally, it is not the case, as the rest of the world think, that reason is the principle and guide to virtue, but rather the feelings. For there must first be produced in us (as indeed is the case) an irrational impulse to the right, and then later on reason must put the question to the vote and decide it.[13]

Another passage of the *Magna moralia* is equally significant.

We ought to speak about the soul in which [virtue] resides, not to say what the soul is (for to speak about that is another matter), but to divide it in outline. Now the soul is, as we say, divided into two parts, the rational and the irrational. In the rational part, then, there reside wisdom, readiness of wit, philosophy, aptitude to learn, memory, and so on; but in the irrational those which are called the virtues – temperance, justice, courage, and such other moral states as are held to be praiseworthy. For it is in respect of these that we are called praiseworthy; but no one is praised for the virtues of the rational part. For no one is praised for being philosophical nor for being wise, nor generally on the ground of anything of that sort. Nor indeed is the irrational part praised, except in so far as it is capable of subserving or actually subserves the rational part.[14]

The author of the *Magna moralia* (was he Aristotle, or did he simply repeat the master's sayings?) combined reason with emotivity and did so without losing his intellectual balance. Feelings can never be separated from intelligence in human nature; it is the essence of wisdom not to separate them in one's philosophy.

POLITICS

The passage from ethics to politics is natural enough; both concern the same field, except that ethics is more personal. Politics is concerned with the well-being of the whole community, ethics with that of the individual, but the well-being of the community and that of the individuals who constitute it are so closely interrelated that it is almost impossible to abstract the one from the other. In many cases, the line is impossible to draw; it is ethics if you look from one side and politics if you look from the other.

Economics is in some respects a transition between ethics and politics, but the work bearing that title in the Aristotelian corpus[15] is certainly apocryphal. It is divided into two (or three) books. The first is derived from Aristotle and Xenophon, and may be a product of the end of their century; the second, dividing economics into four kinds (royal, satrapic, political, personal) and covering that curious field in an anecdotic and chaotic way, was probably written by a Greek of the Hellenistic age living in Egypt or Asia; the third (existing only in Latin), concerned only with a wife's position and duties, is even more distant from the Aristotelian source.[16]

It would be going too far to claim that Aristotle's *Politics* was entirely derived from biologic considerations, but there is no doubt that such considerations helped to guide his thought. When he discussed the many forms of government he made a comparison with the different species of animals. Each animal is a combination of organs; different organs or different combinations of them will naturally produce different species. In the same way, any society is constituted by the mutual subordination to one another of many kinds of men, who exercise various functions, such as husbandmen, mechanics, traders, laborers, soldiers, judges, councilors. Moreover, some are rich while the majority are poor. The final product may be one

[13] P. 1206*b*19.

[14] P. 1185*b*.

[15] Pp. 1343-1353.

[16] The Bekker edition and the Oxford translation include only the first two books (pp. 1343–1353). For the third book, see Franz Susemihl, *Aristotelis quae feruntur Oeconomica* (Leipzig, 1877).

of many things.[17] It is clear that in Aristotle the politician could not be separated from the physiologist and the biologist; the same is true of the philosopher in him; it is significant that his *Metaphysics* itself begins with zoölogic comparisons.

Not only was Aristotle the first to compare the state with an organism, the body politic with a single man's body, but he pursued his political investigations in the same way as those dealing with natural history. Just as he compared different species of fish in order better to understand what a fish is, so he made a comparative study of some two hundred constitutions of Greek cities. Unfortunately, only one of these constitutional histories has come down to us, but it happens to be the most important.[18] Aristotle was not satisfied to describe the constitution of Athens as it existed in his day; he introduced that description with an account of the development of the Athenian government down to that time; we must know the past evolution of an organism in order to appreciate clearly its present condition. He did in the second half of the fourth century B.C. what Herbert Spencer undertook in the second half of the nineteenth century, and Spencer's *Descriptive sociology*, in spite of a more elaborate and systematic analysis, is not superior, as a synthesis, to Aristotle's *Constitution of Athens*.

Aristotle fully realized the value of political history from the point of view of sociologic case studies, and Book II of his *Politics* was devoted to the description of the actual political communities as well as the ideal commonwealths invented by Plato, Phaleas of Chalcedon,[19] and Hippodamos of Miletos.

The historical survey was preceded, however, by an examination of the basic facts of any government, an examination that could be safely accomplished without reference to the past. Therefore, Book I of *Politics* deals with the definition and structure of the state. "The state is a creation of nature and man is by nature a political animal." [20] There are various stages of social organization, the family, the village, the city (the Greek city, *hē polis*, corresponded somewhat to the modern state). The fundamental bonds thanks to which the body politic is kept together are the bonds between master and slave, man and wife, father and children.[21] It is necessary to consider these bonds and their implications before trying to understand the organization of the whole state.

Inasmuch as we have indicated the contents of the first two books of *Politics*, we might complete this with a rapid analysis of the other books: Book III. The citizen, civic virtue, and the civic body; classification of constitutions, democracy and oligarchy, kingship; forms of monarchy. Book IV.[22] Variations of the main types of constitutions; the best state in general and under special circumstances; how to proceed in framing a constitution (deliberative, executive, and judicial functions).

[17] *Politics*, 1290*b*21–1291*b*13.

[18] As explained above, the Greek text of the *Athenensium Respublica* was discovered only in 1891 by Frederic G. Kenyon. English translation by Sir Frederic in vol. 10 of the Oxford *Aristotle* (1920). See the edition by Sir John Edwin Sandys (London, first, 1893; second, 1912). All editions and translations are divided into chapters numbered 1 to 69 as by Kenyon and no reference is made to pages in the Bekker style, for this treatise was not included in the Bekker edition.

[19] "Phaleas of Chalcedon, who was the first to affirm that the citizens of a state ought to have equal possessions," *Politics*, 1266*a*40, 1274*b*9.

[20] *Ibid.*, 1253*a*2.

[21] It would be interesting to compare Aristotle's concepts of those fundamental bonds, and more generally his politics and sociology, with the Chinese concepts developed by Confucius (VI B.C.), Mo Ti (V B.C.), and Mencius (IV-2 B.C.), but that would lead us too far astray. Mencius (372–289) was a younger contemporary of Aristotle.

[22] Books were numbered differently in various manuscripts and editions. Books IV to VIII were numbered also VI, VIII, VII, IV, V.

Book v. Revolutions and their general causes; revolutions in particular states and how they may be avoided. Book vi. Organization of democracies and oligarchies. Book vii. *Summum bonum* for individuals and states; picture of the ideal state; educational system of the ideal state, its aim and early stages. Book viii. The ideal education continued; music and gymnastic.

So many topics and problems are reviewed that the briefest enumeration of them would take too much space. Perhaps the best example of Aristotle's political wisdom are given in Book v, which might be called a natural history of revolutions. Aristotle asked himself what were the causes, symptoms, and remedies of revolutions in the same spirit as a physician would consider the diagnosis and treatment of a disease. Why do revolutions happen? They are caused by social inequalities, by conflicts between political views, by passions; one must distinguish between the causes, which may be very deep and chronic, and the provocative accidents, which may start a revolution as in the drawing of a trigger. How is it possible to prevent the occurrence of such calamities? One should avoid illegality and frauds upon the unprivileged, maintain good feelings between the rulers and the people, keep a close watch on subversive agencies, alter property qualifications from time to time, let no individual or class become too powerful, prevent the corruption of magistracies, practice moderation in everything. Anyone who reads the whole book [23] will realize once more the comprehensiveness of Aristotle's thinking and its up-to-dateness. His *Politics* might still be used as a textbook in a school of government and administration.

The two final books, which remained unfinished, describe and discuss the ideal republic. They remind us of Plato, who is frequently referred to and criticized, but what a difference between the blind dogmatism of Plato and Aristotle's sweet reasonableness! We would not say that Aristotle was never dogmatic, or that he lacked superstitions. Like every other great man, he had his blind spots, but we must bear in mind as always that those blind spots were largely of social origin; no man, however original and however great, can completely escape the limitations of his own time and space.

One of his limitations arose from the smallness of the Greek state, generally restricted to the city and its surroundings. A kind of democratic government was possible which at its best was like those of the New England town meetings or of the Swiss cantons. There was no need of delegation of power and Aristotle did not have to discuss the very difficult problems of representative government.

The worst blind spot concerned slavery, which he considered "natural." Consider these sayings of his:

> It is clear, then, that some men are by nature free, and others slaves, and that for these latter slavery is both expedient and right.[24]
>
> It must be admitted that some are slaves everywhere, others nowhere. The same principle applies to nobility. Hellenes regard themselves as noble everywhere, and not only in their own country, but they deem the barbarians noble only when at home, thereby implying that there are two sorts of nobility and freedom, the one absolute, the other relative.[25]

[23] Pp. 1301–1316.

[24] *Politics,* 1255a1.

[25] *Ibid.,* 1255a31. So many slaves had proved their distinction and magnanimity that one could not claim that slaves were essentially different from other men. There was a loophole, however, as there always is; one could claim that those "good" slaves were not "real" or "natural"

Aristotle was so convinced of this that he gave his blessing to the kind of war that our grandparents would have called a "colonial" war. He remarked:

> Now if nature makes nothing incomplete, and nothing in vain, the inference must be that she has made all animals for the sake of man. And so, in one point of view, the art of war is a natural art of acquisition, for the art of acquisition includes hunting, an art which we ought to practice against wild beasts, and against men who, though intended by nature to be governed, will not submit; for war of such a kind is naturally just.[26]

Atrocious, is it not? But is our own thinking about war and peace so blameless that we can indulge in the denunciation of others?

After this, it is not necessary to investigate Aristotle's views on war and peace any longer; warring being assimilated to hunting, those views were vitiated at the very source. For once, a biologic analogy had led him completely astray. Remember, however, how many centuries had to pass, how many horrors and crimes had to be perpetrated, before men became able to face the injustice and inhumanity of war and to condemn it. Remember again that in the seventeenth century, almost two thousand years after Aristotle, a fine gentleman like Descartes considered it right to enlist in the Dutch army and to take part in a war that did not concern him in the remotest way: it was good exercise, good sport, and naught else.

There remains in us, however, a legitimate cause of anxiety. How could a philosopher, a man as wise and great as Aristotle, ever say such things about other men – slaves? Slavery had existed from time immemorial, and it was so well organized in Athens that it was considered a part of the order of nature. From that point of view, one may say that Athens was never a popular democracy, but an oligarchy dominating and exploiting a large mass of inarticulate slaves. Remember that the great Catholic philosopher, St. Thomas Aquinas (XIII-2), who flourished more than sixteen centuries after Aristotle, still considered slavery justified. Non-Catholics will hasten to object that St. Thomas was a product of the Middle Ages, the "Dark Ages" – what would you expect? Well, let us forget St. Thomas, and the Middle Ages. These were followed by the Renaissance, the Reformation, the Enlightenment, the American and the French Revolutions, and after all that, there were still Christian gentlemen, less than a century ago, who believed that the slavery of black people was justified and natural. Less than a century ago! Can you blame Aristotle for his inability to realize the inhumanity of deeds, the guilt of which is still weighing on our own consciences?

Aristotle's views on commerce seem equally primitive, but again we do not have to go very far into the past to come across fine gentlemen who considered taking part in business a kind of misdemeanor and disgrace and who looked down upon "tradesmen" as people of an inferior breed.

> Enough has been said about the theory of wealth-getting; we will now proceed to the practical part. The discussion of such matters is not unworthy of philosophy, but to be engaged in them practically is illiberal and irksome.[27]

We have already quoted above (p. 173) the story of Thales' financial speculation. Aristotle told another story of the same kind in the same context:

---

slaves, that they were free men who had become slaves by accident. Aristotle granted that if a slave had the soul of a freeman, he should be freed.

[26] *Ibid.*, 1256*b*20; see also 1255*b*39, 1333*b*38.

[27] *Ibid.*, 1258*b*8.

There was a man of Sicily, who, having money deposited with him, bought up all the iron from the iron mines; afterwards, when the merchants from their various markets came to buy, he was the only seller, and without much increasing the price he gained 200 percent. Which when Dionysios[28] heard, he told him that he might take away his money, but that he must not remain at Syracuse, for he thought that the man had discovered a way of making money which was injurious to his own interests. He made the same discovery as Thales; they both contrived to create a monopoly for themselves. And statesmen as well ought to know these things; for a state is often as much in want of money and of such devices for obtaining it as a household, or even more so; hence some public men devote themselves entirely to finance.[29]

It is curious that the loaning of money is hardly referred to, though there were plenty of moneylenders, bankers, and financiers in Aristotle's time. Usury is mentioned without any explanation as one of the means of wealth-getting.[30] The prejudice against the taking of interest on loans was increased and nourished by the Jewish and Christian religions and as a consequence we find it condemned by St. Thomas. It took many more centuries to establish a distinction between the taking of moderate and of excessive interest, between legitimate business and usury proper.[31] It is clear that Aristotle was not an economist, and that it was not as natural to him to understand economic questions as it was to realize the existence of political or sociologic problems. This raises a curious puzzle: economic facts are as old as society itself; how is it that it took so long to integrate them in science and philosophy?

It is clear that Aristotle's politics were not right, but they were not fundamentally wrong like Plato's. They were redeemed by the master's willingness to compromise; they were far from perfect, but they were perfectible. Aristotle had examined all the kinds of government that had been tried in or before his time, and had concluded that democracy was full of risks. The solution that appealed most to his mind was a compromise between Platonic aristocracy, balanced feudalism, and some democratic ideas. All the citizens should have a chance to participate in the government. The working classes should not rule; the ruling classes should not work, nor should they earn money. Rulers would have to be educated in the proper way, like gentlemen. Philosophers should not be rulers, but teachers; philosophy is an essential part of a gentleman's education. The Aristotelian city was not to be a kind of military monastery like the Platonic, but a moderate republic, the virtues of which would be derived from the virtues of the individual families. Aristotle recognized the fact that no form of government is absolutely good; each form may be good relatively to certain kinds of people or to certain conditions.

His good sense appears in his discussion of communism;[32] communism could hardly be enforced upon the people, but they would progress toward it naturally as their dispositions became more benevolent. Aristotle's conclusion on that subject is still valid today. The community of material goods is a magnificent ideal,

[28] This Dionysios was the tyrant of Syracuse, either the father, Dionysios the Elder (430–367), or the son, Dionysios the Younger, who succeeded his father in 367 and died obscurely in Corinth after 343. Both had befriended Plato.

[29] *Politics*, 1259*a*23.

[30] *Ibid.*, 1258*b*25.

[31] For the history of usury, see *Encyclopedia of Religion and Ethics*, vol. 12 (1922), pp. 548–558. Benjamin N. Nelson, *The idea of usury. From tribal to universal brotherhood* (280 pp.; Princeton: Princeton University Press, 1949) [*Isis 41*, 406 (1950)].

[32] *Politics*, 1263.

but we are not yet worthy of it, and therefore it is better not to realize it, except gradually in proportion as we are prepared for it and deserve it.

The publication of the *Politics* before the end of the fourth century is as astounding as the very greatest achievements of the artists, mathematicians, and men of science of that golden age. To measure its greatness it will suffice to realize that there was nothing comparable to it until modern times. There was nothing remotely comparable to it in antiquity or the Middle Ages. Even after the Flemish Dominican Willem of Moerbeke (XIII–2) had translated the *Politics* from Greek into Latin at St. Thomas's request in 1260, it did not make the impression that one might have expected and did not change the political atmosphere of the age. St. Thomas used it for the development of his own ideas, and while he preserved some of Aristotle's prejudices, he definitely improved the master's teaching in the democratic sense.[33] Practical politics remained as little influenced by St. Thomas as by Aristotle. The rational politics that Aristotle had so brilliantly begun in the fourth century B.C. is still in an embryonic stage today. The problems that Aristotle and St. Thomas discussed are still worrying us, and very few people have yet understood the need of approaching them without any other passion than the love of truth and of justice.

HISTORIOGRAPHY

At the very beginning of the "historical library" that Diodoros of Sicily (I–2 B.C) completed in Rome *c.* 30 B.C., he remarks:

> It is fitting that all men should ever accord great gratitude to those writers who have composed universal histories,[34] since they have aspired to help by their individual labors human society as a whole . . . For just as Providence, having brought the orderly arrangement of the visible stars and the natures of men together into one common relationship, continually directs their courses through all eternity, apportioning to each that which falls to it by the direction of fate, so likewise the historians, in recording the common affairs of the inhabited world as though they were those of a single state, have made of their treatises a single reckoning of past events and a common clearinghouse of knowledge concerning them . . . For this reason one may hold that the acquisition of a knowledge of history is of the greatest utility for every conceivable circumstance of life. For it endows the young with the wisdom of the aged, while for the old it multiplies the experience which they already possess; citizens in private station it qualifies for leadership, and the leaders it incites, through the immortality of the glory which it confers, to undertake the noblest deeds.

Of whom was Diodoros thinking? He was acquainted with Hecataios, Herodotos, Thucydides, Xenophon, and others, but the accent he put on "universal histories" suggests that he was thinking first of all of the ambitious historical efforts that began in this very age, the age of Aristotle, and culminated with those of Polybios (II–1 B.C.). It is true that Herodotos was, in his own naïve and charming way, a "universal" historian, but many things had happened since his time,

[33] He insisted that the state exists for the good of the citizen and not the citizen for the good of the state. This was one of the first declarations of the rights of man; *Introduction*, vol. 2, p. 915.

[34] *Tois tas coinas historias pragmateusamenois.* Translation of C. H. Oldfather in the Loeb Classical Library (1933).

and the days of innocence were definitely over. Histories in Herodotos' manner had become impossible, and so, for different reasons, had monographs like that of Thucydides. The Greek world that those two giants had known was gone forever. When the Greeks were united, they had been able to defeat the Persian empire; weakened by internecine jealousies, they were at the mercy of their northern neighbor. Greece, or, we may say, Athens, was defeated and superseded by Macedonia. On the spiritual level the fight was led on the one side by Isocrates, on the other by Demosthenes. Isocrates had finally won because Philip had won. His triumph was not only political but literary. Indeed, Isocrates the Athenian (436–338) was above all a great man of letters who helped to bring the Greek language to its formal perfection; he was also a politician, a publicist, an orator (one of the "ten Attic orators"); in spite of his being the chief of "collaborators," one cannot say that he lacked patriotism. He saw the need of internal peace for the salvation of Hellas, and he realized that internal peace was impossible without outside (Macedonian) pressure. His influence upon Greek letters (and even, through Cicero, upon Latin letters) was enormous. It was a literary influence, not a philosophic one, and was thus on a much lower level than Aristotle's; it lasted a shorter time, but while it lasted, it dominated the ancient humanities. Aristotle's teaching was restricted to advanced students of philosophy or science; Isocrates could influence all the young men who loved their language and were ambitious to handle it as elegantly as possible. When freedom is lost, education becomes rhetorical, and Isocrates was the supreme rhetorician.

His orations were often historical, for it was natural to praise the glory of Greece, and especially that of Athens, and such praise dealt with the past, not with the present. Two of his disciples, Ephoros and Theopompos, were historians, the most distinguished ones of this age. These two men had many qualities in common, but they were very different in temperament. According to Suidas, Isocrates used to say that Theopompos needed the curb and Ephoros the spur. Time dealt harshly with them, for their works are lost. As far as we can judge from the fragments, they were very inferior to the giants of the preceding century, Herodotos and Thucydides; yet we should make an effort to win their acquaintance. In a time of national disillusion, they set a new emphasis on international history, and also on the geographic background of human events.

Ephoros of Cyme.[35] Cyme, where Ephoros was born (*c.* 405), was the largest of the Aiolian cities of Asia Minor, a city of old Hellenic traditions.[36] He left it to obtain a better education in Athens and became one of Isocrates' favorite pupils. We do not know exactly when he died; it was probably during Alexander's lifetime, say *c.* 330. He wrote a universal history from the return of the Heracleidai and the Dorian settlements in the Peloponnesos at the end of the eleventh century (which he thought were the earliest tangible deeds) down to 341. It was divided into thirty books, the last of which was completed by his son, Demophilos. His purpose is revealed by the title *Historia coinōn praxeōn*,[37] which might be trans-

[35] Godfrey Louis Barber, *The historian Ephoros* (202 pp.; Cambridge: University Press, 1935) [*Isis* 26, 157–158 (1936)].

[36] Hesiod's father had emigrated from Cyme to Boeotia. Cyme faces the open sea between Lesbos and Chios. Modern Turkish name, Sandakli.

[37] Compare the words at the beginning of Diodoros' history quoted in footnote 34.

lated "History of (or Inquiry into) the common affairs of men"; it would be expressed in modern language by "comparative history"; the idea was to investigate what happens to men placed in various geographic and political circumstances. Some eighty-six fragments of his work have come to us and some glimpses of him are given in the writings of later historians such as Polybios, Diodoros, Strabon, Plutarchos. Polybios said of him that he was "the first and only man to write a universal history." [38] We should not take this too literally. Ephoros' universalism was certainly focused on Greece; how else could it be? Even the universal historians of our own time, who have access to a great variety of sources, can never overcome completely their national prepossessions. Ephoros tried to avoid myths and to give rational explanations, for example, to account for the deeds of nations by geographic necessities.

Theopompos of Chios. Theopompos came from the same part of the Greek world as Ephoros, for it did not take very long to sail from the island of Chios to the bay of Cyme. He was born *c.* 380 and a few years later his father Damasistratos was banished from the island for political reasons, perhaps for Laconism. The boy was educated in Athens; he became one of Isocrates' pupils, and eventually, like the latter, a famous orator. His first great success was the award of a prize by Queen Artemisia for his panegyric on her late brother and husband Mausolos; as Mausolos died in 353, that must have happened soon afterward.[39] He traveled considerably throughout Greece, delivering lectures, teaching, and obtaining the favor of rulers such as the kings of Macedonia. Alexander the Great brought him back to Chios, but, after the conqueror's death, Theopompos was exiled for the second time from his native island. He took refuge in Ephesos and later in Egypt, where he was received by the first Ptolemy (king from 323 to 285) and where he probably died.

Among his voluminous writings were a continuation of Thucydides' history from 410 to 398, and the *Philippica* in fifty-eight books, a history of Greece from the battle of Mantinaia in 362 (where Xenophon's *Hellenica* ended) to the death of Philip in 336. The works of Theopompos are lost, but we have some 383 fragments, chiefly taken from the *Philippica*; a longish text (about 30 pages) found in an Oxyrhynchos papyrus in 1911 is ascribed to him. Some of his qualities were the same as those of Ephoros, and that is natural enough, for they were schoolmates in the school of Isocrates, and they were fruits of their age, an age of disillusion. Both appreciated the value of geographic factors and the need of an international outlook. After the victories of Philip, and even more so after those of Alexander, Greek parochialism had become impossible; the intellectual leaders could not lead any more unless they were able to look far beyond the prostrate body of Greece.

The unique quality of Theopompos' writing lies in his psychologic tendencies. Events may be accounted for in geographic and political terms, but the main motives are to be found in the minds of great men. He was very learned and critical, incredibly conceited, a keen politician, and an outstanding psychologist,

[38] Polybios, *Histories,* v, 33.

[39] We have already come across Mausolos. He was satrap of Caria from 377 to his death in 353, and had achieved almost complete independence from Persian rule. His palace and later his "mausoleum" were in Halicarnassos.

a forerunner of Sallustius (I–2 B.C.)[40] and even of Tacitus (I–2). He wrote fearlessly and made many enemies. He did not spare the very ruler whom he admired most and his description of Philip's behavior was as black as could be. Was he malicious or truthful, evil-minded or simply clear-sighted? He was certainly bitter, sarcastic, cynical. He has been accused of being pro-Spartan, like his father; that is possible; he found more to criticize in Athens than in Sparta, but he did not spare the latter. He was a satirist, always ready to denounce evil wherever he saw it, or believed he saw it; it was not a matter of courage but rather an irrepressible instinct. It is highly probable that his maliciousness was increased by his rhetorical habits and his literary skill; men of his kidney often say biting things, because they cannot resist the lure of an incisive sentence or of striking and cruel image.

Gilbert Murray has thrown much light on Theopompos' character. As to his conceit, he wittily said:

> Critics speak severely of his lapses in this respect. But we must remember that a modern writer need never praise himself. He simply arranges with his publisher that such-and-such a sum is to be paid for advertisement, and thus having secured the blowing of a large and expensive bugle, can afford in his own preface to be modest as a violet. Theopompos had not these advantages.[41]

Ephoros tried to avoid myths; Theopompos, on the contrary, seems to have liked them. He looked at them with the eyes not of a mediocre rationalist but rather of a philosopher, just as Plato did. Virtue was disappearing, truth was evasive. Perhaps one would discover it among the myths, the "things which never happened but always are."[42] Theopompos was a cynic, not only in the common meaning of that word (we would expect an intelligent man living in a cockeyed world, his own nation utterly defeated, to assume a cynical attitude) but also in the technical sense. The only philosopher whom he praised was Antisthenes, founder of the Cynic school. The cynical reaction was natural and to some extent healthy; it was a revolt of the free spirit against overpowering circumstances. The world was going to pieces, everything was vanity except a man's soul. Theopompos was probably not as thoroughgoing a cynic as Antisthenes or Diogenes, but he understood and appreciated their message.

In times as dark as they were for the Greeks who had been permitted to live under the Macedonian yoke, there were two extreme kinds of reaction, the cynical and skeptical, as exemplified by Theopompos, and the superstitious, which was perhaps more common among the uneducated people yet by no means exclusive to them. We may be sure that the magicians, the soothsayers, the wonderworkers, and the priests in charge of temples, grottoes, holy springs, and oracles did a thriving business. Men and women can stand misfortune up to a certain point;

[40] Sallustius suggests Thucydides. To call Theopompos the founder of psychologic history, as I did (*Introd.*, 1, 147), is perhaps unfair to Thucydides, who would seem to deserve that title.

[41] Three lectures delivered in Cambridge in 1928 on "Paracharaxis or the restamping of conventional coins." Reprinted in his *Greek studies* (Oxford: Clarendon Press, 1946), pp. 149–170 [*Isis* 38, 3–11 (1947–48)].

[42] *Tauta de egeneto men udepote, esti de aei.* Salustios put it thus in his book concerning the gods and the universe. This Salustios was acquainted with Neoplatonism in the form given to it by Iamblichos (IV–1) and he was probably a friend of Julian the Apostate (IV–2). His book was probably written shortly before Julian's death (363) and published sub rosa. See edition and translation of it by Arthur Darby Nock (Cambridge, 1926), p. 8.

when that point is reached, they must defend themselves with sarcasms and other forms of rebellion, or else they will prostrate themselves before the unavoidable, humiliate their intelligence, and outrage reason.

HISTORIANS OF SCIENCE

These two kinds of reaction were extreme forms to the left and right; we must assume that the wisest men did not lose their balance in those ways but continued their work with as much equanimity as they were able to achieve. They suffered as deeply as the other people, perhaps more, but they tried and managed not to show it. This was the case not only for a great master like Aristotle but also for smaller men, who might lack creative ability, but had enough prudence and forbearance.

Among those quieter people, I would like to give a place of honor to the men who were our own spiritual ancestors, the first historians of science. We have already met three of them, all of whom lived in Aristotle's age – Eudemos of Rhodes and Theophrastos of Eresos, who wrote histories of arithmetic, geometry, and astronomy, and, on a much lower level, Menon, who described the vicissitudes of medicine.

The labor of these men is encouraging for two reasons. First, it proves that science had already become so rich and complex that historical surveys and philosophic cogitations had become necessary. By the end of the fourth century, men of science and physicians had gone so far beyond primitive experiments and naïve adumbrations, that it was exciting to ask oneself: "Where did we come from? Where have we been roaming? How did we reach the present situation?", and even more exciting to ask another question: "Whither are we going?"

This is perhaps easier to understand now than it would have been in a quieter period, say in the middle of the Victorian age. As to political and economic matters, we are as disillusioned as the Athenians of twenty-three centuries ago, and at the same time we are even more flabbergasted than they could be by the prodigious advance in knowledge and technology.

In the second place, these early historians of science were like ourselves, defenders of reason against irrationalism and of freedom against superstitions and spiritual thralldom.

RHETORIC

Aristotle was not only a master of science and philosophy, but also a master of the humanities. He composed one or two treatises on rhetoric, and one on poetry.

Who wants to study rhetoric today, except simpletons? The reader may even ask: "What is rhetoric?" Such a question would have been unnecessary some fifty years ago, but at present the subject is almost completely neglected in our colleges (except theological schools), or it is taught only by implication. Rhetoric is the art of expressive and persuasive speech. Aristotle's main treatise on the subject is divided into three books. There is no point in analyzing them, for the subject – being a part of the humanities – is extremely complex, but we shall offer a few general remarks.

The first book is largely devoted to definitions of rhetoric in general and of its varieties. The rhetorician, or let us call him the orator, must try to explain his message and persuade his audience that that message is true and worth while. There are three kinds of rhetoric (or oratory) which may be called political, forensic, and academic. Political orators must learn to debate political questions in the popular assemblies; forensic orators are like lawyers pleading before a court; academic orators are like professors discussing life and letters, philosophy, or art before an audience of colleagues or students. These three kinds of oratory are obviously different and require different techniques, which Aristotle describes. It was not necessary to give detailed explanations, for every student at the Lyceum, every educated Athenian, had already a practical acquaintance with those matters and all that was needed was to clarify the essential points. In fact, Athenians were so familiar with every form of oratory, almost from the time of their boyhood, that one cannot help wondering why Aristotle included the art of rhetoric in his teaching. He did so, perhaps, because, however familiar, it was extremely important; things may become so familiar that it is the more necessary to reconsider them from a new, unfamiliar point of view.

Oratory implies passions, the speaker's passions and those of his audience. It is a conflict of passions, and the orator's art consist in shaping and orienting the passions of the people who listen to him in the manner that he considers to be proper and fair. Therefore, Book II analyzes many passions, such as calmness and anger, friendship and enmity, fear and confidence, shame and shamelessness, kindness and unkindness, pity, indignation, envy, emulation, the passions that are characteristic of various age levels, those that accompany the use (or the lack) of wealth and power. This might be called a little treatise on practical psychology. The orator must be a keen psychologist; it is not enough for him to know his own mind, he must know the mind, the qualities, and the weaknesses of the people whom it is his business to convince and to convert. This part of *Rhetoric* exerted a very strong influence on medieval thought, as is witnessed by innumerable books discussing human passions, either from the rhetorical point of view, or from the point of view of morality and religious salvation. One of the many digressions of that second book is devoted to the use of maxims (or proverbs); the popular sayings are an epitome of the people's experience and of their ancestral wisdom. The orator must learn to use maxims as vehicles of his own arguments; the better the people already know these maxims, the better will they help him to inculcate what he wishes to be understood and remembered.

Book III, which may have been a separate work, but is as genuine as the two others, deals more particularly with style and language. Much of this hardly concerns the modern reader, unless he wishes to know more deeply the Greek language; for example, ancient orators (Roman as well as Greek) attached much importance to the musical characteristics of their speeches, such as prose rhythm and periodic style. The discussion of good language, fully appropriate to the aim that it must serve, implies problems that we would call grammatical.

It is hard to realize that in Aristotle's time, when most of the masterpieces of Greek literature had already been composed, formal grammar (such as we find in schoolbooks) hardly existed. Few of the grammatical categories that we learn painfully in our childhood were recognized as such. The first formal Greek grammar was composed only much later, by Crates of Mallos (II-1 B.C.), but that grammar is lost; the earliest Greek grammar extant is that of Dionysios Thrax (II–2 B.C.); Apollonios Dyscolos (II–1), who flourished in Alexandria, much later, has been called the founder of scientific grammar, the inventor of syntax. Apollonios' date is difficult to fix, but assuming that he flourished about the middle of Hadrian's rule (*c.* A.D. 127), that was four and a half centuries after Aristotle's death![43]

The authors most often quoted in the *Rhetoric* are, in order of decreasing number of references: Homer, Euripides, Sophocles, Isocrates, Plato, Gorgias, Socrates, Theodectes.[44] Demosthenes is hardly mentioned, Thucydides not at all.

The three books of the *Rhetoric* are not as sharply divided as my all too brief analysis would suggest; the order is a little capricious and some topics are discussed many times. For example, the use of maxims is taken up again in Book III.

Endless remarks might be made about separate statements. Let me restrict myself to a single quotation:

> Now the style of oratory addressed to public assemblies is really just like scene-painting. The bigger the throng, the more distant is the point of view: so that, in the one and the other, high finish in detail is superfluous and seems better away. The forensic style is more highly finished; still more so is the style of language addressed to a single judge, with whom there is very little room for rhetorical artifices, since he can take the whole thing in better, and judge of what is to the point and what is not; the struggle is less intense and so the judgement is undisturbed. This is why the same speakers do not distinguish themselves in all these branches at once; high finish is wanted least where dramatic delivery is wanted most, and here the speaker must have a good voice, and above all, a strong one. It is ceremonial oratory that is most literary, for it is meant to be read; and next to it forensic oratory.[45]

Note the first comparison, that of public speaking before a large audience with scene painting. Aristotle wrote that before 322 B.C., and many public speakers have not yet understood it in 1952, twenty-two centuries later. Pedantic speakers persist in painting miniatures when they should paint large frescoes, and they bore their audiences to death. Boring them does not matter so much, perhaps, but what is worse, they fail to convey whatever message they intended to convey. Why do they speak at all? Aristotle knew better.

The other *Rhetoric* is much shorter than the first (54 columns in Bekker against 134). It is generally designated by the title *De rhetorica ad Alexandrum*. It begins with the words "Aristotle to Alexander. Salutation . . ." which are followed by a dedication covering more than three columns, wherein the author explains why a king should know rhetoric. Erasmus considered that dedication to be a forgery, but I am not convinced. It sounds Aristotelian; it is a bit tedious, but very dignified, in great contrast with the obsequious and sycophantic prefaces that Renaissance authors did not blush to address to their patrons, and which were actually printed, to the perpetual shame of patron and author. Not only the preface but

[43] One might object that some grammatical ideas were discovered before Aristotle. Protagoras (V B.C.) has been called the first grammarian, but there was a very long distance between the beginning of grammatical consciousness and the establishment of the first rudimentary grammar. The interval between Protagoras and Crates is about two and a half centuries.

[44] All these names are already familiar to the reader except the last one. Theodectes (*c.* 375–334) of Phaselis (Lycia) flourished mainly in Athens, studied under Plato, Isocrates, and Aristotle, and became famous as an orator and playwright. His monument at Phaselis was honored by Alexander the Great.

[45] *Rhetoric*, 1414*a*. Translation by W. D. Ross in the Oxford *Aristotle*.

the whole work has been suspected of being apocryphal. Some scholars would ascribe it to Anaximenes of Lampsacos (*c.* 380–320), who was Aristotle's contemporary and like him a tutor to Alexander; others regard it as a later work, though not much later. Many fragments of it have been identified in a papyrus found by Grenfell and Hunt [46] at Hibeh and published by them in 1906. The assumptions that the treatise was written for Alexander the Great and by Aristotle seem plausible to me but are not susceptible of proof. If it was not written by Aristotle, it was very probably written not long after his death, before the end of the century. The student of Aristotle's longer *Rhetoric* will find few novelties in the smaller one.

POETICS

The treatise on poetics that has come down to us is fairly short, less than 30 columns, and it is incomplete; we have only one book out of two or more. Did Aristotle fail to complete it? Or was part of his work a victim of time? The first alternative seems more credible, because one would think that the manuscript copies of such a work would have been treasured with especial care, and because Aristotle's *Poetics* (as well as his *Rhetoric*) was composed toward the end of his life. A man's last work is more likely than the others to remain unfinished.

Poetics, as Aristotle understood it, is something much broader than our modern conception of it. It is the literature of imagination, as against scientific (or objective) literature. Aristotle begins thus:

> Our subject being Poetry, I propose to speak not only of the art in general but also of its species and their respective capacities; of the structure of plot required for a good poem; of the number and nature of the constituent parts of a poem; and likewise of any other matters in the same line of inquiry. Let us follow the natural order and begin with the primary facts.
>
> Epic poetry and Tragedy, as also Comedy, Dithyrambic poetry, and most flute-playing and lyre-playing, are all, viewed as a whole, modes of imitation. But at the same time they differ from one another in three ways, either by a difference of kind in their means, or by differences in the objects, or in the manner of their imitations.[47]

(The text as we have it deals only with tragedy; the part dealing with comedy and music is lost or remained unwritten.)

Aristotle's definition of poetry is very well put in his chapter 9:

> From what we have said it will be seen that the poet's function is to describe, not the thing that has happened, but a kind of thing that might happen, i.e. what is possible as being probable or necessary. The distinction between historian and poet is not in the one writing prose and the other verse — you might put the work of Herodotus into verse, and it would still be a species of history; it consists really in this, that the one describes the thing that has been, and the other a kind of thing that might be. Hence poetry is something more philosophic and of graver import than history, since its statements are of the nature rather of universals, whereas those of history are singulars.[48]

The comparison with history is highly significant. It is curious that Aristotle referred many times to Herodotos but never to Thucydides. This is the more aston-

[46] Bernard Pyne Grenfell (1869-1926) and Arthur Surridge Hunt (1871-1934), famous English papyrologists.

[47] This and other quotations from the *Poetics* are taken from Ingram Bywater, *Aristotle on the art of poetry, Greek and English* (434 pp.; Oxford, 1909); the translation was reprinted in the Oxford *Aristotle* (1924).

[48] *Poetics,* 1451 end of *a.*

ishing because in the *Politics* he discussed the Peloponnesian War. How could Thucydides be unknown in Athens? How could Aristotle not know of him? And if he read his *History*, how is it that he never referred to it? This puzzles me deeply; the very man who was best able to appreciate Thucydides' objectivity ignored him, it would seem, deliberately. Such happenings are sad but not uncommon; the history of science gives many examples of them. The men of science who seem to be closer to one another than other men fail to come together; their paths come so near that one would expect them to cross, but they do not.

The part of the *Poetics* with which most people are familiar is the one wherein tragedy is likened to a purgation (*catharsis*). This occurs in Aristotle's definition of tragedy:

> A tragedy, then, is the imitation of an action that is serious and also, as having magnitude, complete in itself; in language with pleasurable accessories, each kind brought in separately in the parts of the work; in a dramatic, not in a narrative form; with incidents arousing pity and fear, wherewith to accomplish its *catharsis* of such emotions. Here by "language with pleasurable accessories" I mean that with rhythm and harmony or song superadded; and by "the kinds separately" I mean that some portions are worked out with verse only, and others in turn with song.[49]

The definition refers also to what we might call the unity of action: the tragedy must be "complete in itself"; and a little further on he speaks more definitely of the "unity of plot."[50] There is a brief reference to the "unity of time,"[51] but none to the unity of place. The theory of the three unities, which the writers of the classical age in France (Corneille, Racine, Boileau) accepted as a kind of literary dogma, was not an ancient dogma but a new one, not clearly formulated until 1636 (Le Cid).[52]

It would be easy enough to object that Aristotle's *Poetics* does not really deal with the magical art of poetry. No poet will ever wish to read it, or if he did read it, he would find no inspiration in it. The *Poetics* was not written for poets but for critics and philosophers; it was not written for seers but for men of science. We may criticize it, but we should not criticize it on false grounds.

CONCLUSION

Some of my readers may say that I should not have spoken, except in the briefest fashion, of the *Rhetoric* and *Poetics*, because they are outside my field, the history of science. The reason why I spoke of them, and had to speak of them, is to illustrate the comprehensive scope of Aristotelianism. We are dealing in this book with ancient science, not with modern science; we have to discuss Aristotelian science in the light of his own conception of it and not of our own. His idea was to analyze the whole of knowledge in scientific terms; rhetoric and poetry were not parts of science even in his own eyes, but they came very close to it, and the man of science must be acquainted with them. If so, his acquaintance must be a scientific acquaintance.

49 *Ibid.*, 1449*b*.

50 *Ibid.*, 1451*a*16.

51 "Tragedy endeavors to keep as far as possible within a single circuit of the sun, or something near that" (1449*b*13).

52 The "rule of three unities" continued to be accepted in France as a kind of dramatic ideal until the angry challenge of Victor Hugo in the preface to his *Cromwell* (Paris, December 1827), the manifesto of the Romantic school.

The man of science must be a humanist. Aristotle did almost the opposite of what Plato had done. Plato had reduced science, philosophy, sociology to fantastic metaphysical conceptions; he had driven the poets and artists out of the city. Aristotle tried to embrace in his philosophy the whole of knowledge and the whole of life. He accepted art, but he tried to explain it and mixed science with it. He was in that sense the forerunner of the historians of art and the historians of poetry of our own day. Artists and poets often object to the scientific analysis of their achievements, but they are wrong in that as long as such study is removed from pedantry, does not try to regulate those achievements, but is willing to accept them in the same spirit as it accepts the creations of nature.

Nevertheless, one understands how Aristotle can easily become (and did become) the bête noire of the men who dislike and distrust science, of the would-be poets and artists, and how he became on the other hand the patron saint of the men of science and of the lovers of objective truth.

XXIII

OTHER THEORIES OF LIFE AND OF KNOWLEDGE THE GARDEN AND THE PORCH

As the ancient world, the old Hellenic culture, came to an end, there were many thinkers who were not satisfied with the conclusions that found acceptance in the Academy or the Lyceum. In the midst of political and economic anxieties, the Greek mind continued to assert its originality and its independence. It was perhaps a consolation for the Greeks in their spiritual agony to believe that the most important thing in the world was not by any means to wield power but rather to know the truth and to practice virtue, and therefore they were ready to declare that one must give top priority to fundamental questions: What are the origin, nature, and purpose of the universe, and particularly of ourselves? When did the universe begin, if it ever began? Is it material or spiritual? What are we? Whence did we come and whither are we going? What is truth? Is it possible to know it? And if it is, how shall we know that we know? Can we understand the world and our place in it? What is virtue? Is it possible to reach it? . . . We have already considered the answers that some philosophers, especially Plato and Aristotle, gave to these anxious questions, but other philosophers suggested other answers which we shall examine presently. The main point to keep in mind always is that these questions were not academic or idle. We might perhaps consider them that way, but that would be only because we ourselves have lost all sense of value, and are like sailors whose compass is lost or broken and who find that their ship no longer answers to the helm.

For the Greeks these were not academic questions but vital ones which are more urgent than such other questions as who is the king or the boss, how shall we pay the rent next month, do we ourselves deserve to be happy or not? Let us interrogate those earnest men. They belong to the following schools or sects: Cynics, Skeptics, Euhemerists, Epicureans, and Stoics.

THE CYNICS

The Cynic school is much older than Aristotle's age; it can be traced back to Socrates (there were indeed cynical tendencies in his outlook and in his behavior), and Antisthenes, who was one of Socrates' immediate pupils, is generally considered the founder of the sect. His father was an Athenian, but his mother was a Thracian. Therefore, he was educated at the Cynosarges, a gymnasium outside of Athens, sacred to Heracles, and reserved for those who were not of pure Athenian descent; he also taught in that school, and it has been suggested that the name of his sect was derived from Cynosarges. That is possible; it is more probable that the word cynic is derived from one of the roots of Cynosarges (*cyon*, *cynos*= dog) and thus originally meant doglike, because Antisthenes accentuated Socrates' tendency to live in the simplest fashion and to disregard many of the social conventions and amenities.

Antisthenes' dates are unknown; as he was a pupil of Gorgias and Socrates, he was still a youngish man at the end of the fifth century. His most famous disciple was Diogenes of Sinope,[1] whose excesses of austerity have become proverbial. Diogenes' father had been in charge of the mint of Sinope and had got into trouble, being accused of falsifying the coinage (*paracharattein to nomisma*). Whether his guilt was personal or political, he was obliged to leave Sinope.[2] He and his son Diogenes lived in great poverty; Antisthenes' teaching was very welcome, at least to the younger man, who realized that poverty should not be considered a punishment but rather an accomplishment, the reward of exceptional virtue. Diogenes proclaimed the necessity of self-sufficiency (*autarceia*), austerity (*ascēsis*), and shamelessness (*anaideia*), and made an aggressive display of his contempt of conventions. He did not add anything new to Antisthenes' teaching but dramatized and advertised it. We have already told the (legendary) story of his rebuking the Master of the World, a story that is very much to Alexander's credit.

His main disciple was Crates, son of Ascondas of Thebes (*c.* 365–285),[3] who renounced a large fortune for the sake of philosophy and reduced his needs to the strict minimum; he converted the children of a distinguished Thracian family, Hipparchia and her brother Metrocles of Maroneia, and married the girl; they lived together like the poorest missionaries, like two beggars; he was somewhat of a poet and both of them seem to have been very lovable.

Let us name one more of Diogenes' disciples, Onesicritos of Astypalaia (one of the Sporades). He was a seaman who had accompanied Alexander to Asia; he was chief pilot of the fleet built on the Hydaspes and remained in charge throughout the voyage down the Indus and up the Persian gulf. He was one of Alexander's historians, of questionable veracity. Being a cynic he made of Alexander a cynic hero. He may have been right in this; it is highly probable that Alexander had acquired cynical tendencies; a successful dictator cannot help becoming cynical.

Of these four men – Antisthenes, Diogenes, Crates, and Onesicritos – only the first was a philosopher in the technical sense. Diogenes, Crates, and his wife Hipparchia were comparable to many other saints and ascetics, such as have flourished in almost every country, chiefly in the East. Crates especially was like a Hindu

[1] Diogenes was born *c.* 412–400 at Sinope (about the middle of the south shore of the Black Sea); he died at Corinth in very old age, *c.* 325–323.

[2] My Harvard colleague, George H. Chase, kindly wrote to me (13 February 1951) that the translation "to falsify the coinage" seems the best one to him; *paracharattein* means to engrave falsely. "So I suspect that Diogenes' father got into trouble by stamping coins of Sinope with other than the officially accepted design, rather than by restamping them." The coinage might be considered "falsified," however, by one party and not by another.

[3] It is said that Crates was a pupil of Bryson before following Diogenes. Yes, but this was Bryson of Achaïa, not the mathematician Bryson of Heraclea.

faqīr, a Muslim darwīsh, and many a Christian hermit. There is a touch, or more, of cynicism in every saint. One wonders whether Diogenes or Crates had been influenced by Indian examples? That is possible but is not necessary to explain their behavior. Onesicritos must have seen fuqarā' in India, but again he did not need, nor did Alexander need, such examples in order to advertise their contempt of the trappings and the vanities of life.

Cynicism was never a formal school. Antisthenes, it is true, had been explaining what might be called a cynical doctrine: happiness is based on virtue, virtue on knowledge; knowledge can be taught, hence virtue and happiness can be obtained, and the happiness thus obtained cannot be lost. His followers accepted that teaching, but their cynicism was a matter of behavior rather than of theory. They were more like missionaries and salvationists than theologians. Cynicism is a temperamental state of mind, independent of doctrine. Every philosophy or religion may produce its own cynics, its own saints.

THE SKEPTICS

While Onesicritos was trying to interpret life in cynical terms, another Greco-Indian, Pyrrhon, was developing a new doctrine which was, or might become, equally upsetting. Pyrrhon (*c.* 360–*c.* 270), son of Pleistarchos, hailed from Elis (northwest Peloponnesos); his parents being poor, he was obliged to learn a trade and became a painter. He was deeply interested in philosophy, however, and sat at the feet, first of Bryson, son of Stilpon,[4] and later of Anaxarchos of Abdera, of the school of Democritos. It is said that both Anaxarchos and Pyrrhon accompanied Alexander into Asia (it is interesting to find so many philosophers and men of science associated with the conqueror; so did Bonaparte select many men of science for his expedition to Egypt).[5] After his return, Pyrrhon settled in his native city Elis, where he lived in retirement, with great simplicity. He wrote nothing, except a poem addressed to Alexander, but he was immortalized by a faithful disciple, Timon of Phlius (*c.* 320–230),[6] who extolled his master's wisdom and virtue.

It cannot be said of Pyrrhon, as of most prophets, that he was without honor in his own country. On the contrary, his fellow citizens made him their high priest and erected a monument to his memory soon after his death. While other philosophers questioned the reality of matter (or the reality of no-matter), he was bolder still in that he doubted the possibility of knowledge. How can we be sure of any-

[4] This Bryson is different from the two Brysons mentioned in the previous footnote. The name Bryson was not uncommon. In his *Life of Pythagoras* (par. 104), Iamblichos (IV–1) speaks of an early disciple bearing it. A treatise on economics is ascribed to one Bryson; the author was a Neo-Pythagorean, who flourished in Alexandria or Rome in the first or second century A.D.; that treatise was edited by Martin Plessner in 1928 [*Isis 13*, 529 (1929–30)]. To return to the present Bryson, son of Stilpon, we wonder whether his father was the famous Stilpon, third head of the Megarian school? This Stilpon (*c.* 380–300) had been influenced by Diogenes of Sinope as well as by Eucleides of Megara; under his mastership the school of Megara obtained considerable popularity, but that was also the end of it.

[5] F. Charles-Roux, *Bonaparte, gouverneur d'Egypte* (Paris: Plon, 1935) [*Isis 26*, 465–470 (1936)].

[6] Timon, son of Timarchos of Phlius (northeast Peloponnesos), came also from a poor family and started life as a dancer. He studied under Stilpon at Megara, then under Pyrrhon, who converted him. Having been obliged to leave Elis, he exercised the profession of sophist in the country around the Hellespont and the Propontis, then retired *fortune faite* to Athens, where he lived to a very old age. He is remembered chiefly because of his individual type of satirical poems (*silloi*).

thing? In particular, how could we know the nature of things? Do we not continually observe the contradictions of our sensual perceptions, of opinions, of customs? These contradictions prove the impossibility of knowledge. Therefore, if we are honest, we shall say not "This is so" but "This might be so," nor "This is true," but "This might be true."[7] This suspension of judgment (*acatalēpsia*, *epochē*) created impassiveness (*ataraxia*), that is, a complete repose of the soul, freedom from passion (*apatheia*), indifference (*adiaphoria*) to outward things, to pleasure and pain. Pyrrhonism was a kind of quietism.

Pyrrhon did not create a regular school but he had admirers like Timon and he influenced a few other men, like Arcesilaos[8] (*c.* 315–240) founder of the Middle Academy, Carneades (*c.* 213–129),[9] founder of the New Academy, Ainesidemos[10] in the time of Cicero (I–1 B.C.) or later, and Sextos Empiricos (II–2). Pyrrhonism, like Cynicism, is a state of mind rather than a philosophic system. There are always and everywhere some people who are skeptically minded, yet skepticism, in the Pyrrhonian sense or otherwise, is always limited and relative; nobody ever doubts everything or believes everything. The Pyrrhonian spirit is more or less illustrated by Montaigne's motto, "*Que sais-je?*" and by Lagrange's favorite answer, "*Je ne sais pas.*" A man of science cannot do good work if his imagination is not continually restrained by skepticism or agnosticism.

EUHEMERISM

Another set of opinions was crystallized at about this time by the Sicilian Euhemeros of Messina, who flourished at the court of Cassandros.[11] He was said to have sailed down the Red Sea and across the Arabian Sea and to have reached an Indian island called Panchaia, where he found sacred inscriptions. Whether his travels and discoveries are real or imaginary, he wrote a description of them entitled *Hiera anagraphē* (Sacred history), wherein he emphasized the historical origin of myths. It was an attempt to rationalize mythology, that is, Greek religion.

This was hardly a novelty, though Euhemeros' book (of which only fragments remain) may have been the first publication, or the first popular publication, of these views. He may have been impressed by the Egyptian usage, imitated by the Greeks, of deification or apotheosis of mortals. Thus, the Egyptian physician Imhotep had become a hero and later a god, and the same thing had happened to the Greek physician Asclepios. There were intermediary beings between men and gods, namely, heroes; the lines between men and heroes on the one hand, and between heroes and gods on the other, were not sharply drawn. It was possible to pass from one group to the other, and if so, was it not natural enough to postulate human origins or relationships for all the gods? Was not Greek mythology exceedingly anthropomorphic? How could one believe that the gods were

[7] According to an old tradition, after Pyrrhon's death he was asked "Are you dead, Pyrrhon?", and he answered, "I don't know."

[8] Arcesilaos of Pitane (Aiolis), pupil of the mathematician Autolycos of Pitane; then went to Athens, where he sat at the feet of Theophrastos, Polemon, and Crantor; succeeded Crates as master of the Academy.

[9] Carneades of Cyrene introduced skepticism into Rome in 155, and Cato asked the Senate to send this dangerous seducer of Roman youth back to his own Athens.

[10] Ainesidemos of Cnossos, whose lost work is one of the main sources of Sextos Empiricos (II-2).

[11] Cassandros was regent of Macedonia from 316 to 306, king from 306 to 297. He was the founder of Thessalonica (Salonica).

essentially different from men, when every story told about them illustrated human characteristics and weaknesses? We may safely assume that long before Euhemeros' time almost every man of science had become accustomed to consider mythology as a kind of poetry which it sufficed to love; none of them expected one to believe it. The reality of religion was to be found, not in the myths, but rather in the rites and festivals, in the celebration of which the Greeks satisfied their love of beauty and sublimity, uttered their consciousness of divine mysteries, and expressed their spiritual brotherhood. Unfortunately, the celebration of those festivals encouraged clerical impostures which were bound to evoke as much criticism as the very myths.

The same kind of anticlerical[12] criticism was taught by the Cyrenaic school, founded by one of Socrates' pupils, Aristippos of Cyrene. His philosophy was hedonistic and rationalist. The teaching of it was continued by his daughter Arete, her son Aristippos the Younger (*ho mētrodidactos*, he who received his training from his mother), and a few others, Antipater of Cyrene, Theodoros the Atheist, Hegesias, and Anniceris the younger. Euhemeros may have been influenced by the Cyrenaic school, but there is no means of proving that and no need of postulating it. Rationalism was as congenial to a few Greeks as superstition was natural to many others.

Euhemerism was reëxplained in Latin by Ennius (II–1 B.C.) and in Greek by Diodoros of Sicily (I–2 B.C.); it was exploited by the early Christians in their anti-Pagan propaganda. It is one aspect, out of many, of the eternal war between reason and superstition.

THE GARDEN OF EPICUROS[13]

EPICUROS OF SAMOS

We have tried to give our readers some idea of the greatness of Democritos of Abdera (pp. 251–256), one of the purest glories of the second half of the fifth century. There was in Greece such an exuberance of genius that some of it was lost and forgotten. Democritos was overlooked during the best part of the fourth century. Plato never mentioned him; Aristotle referred to him often but only for the sake of adverse criticism. Happily, his philosophy, if not his personality, was resurrected in the last quarter of the century by a new prophet, Epicuros.

Epicuros (341–270) was a scion of a noble Athenian family, but his father Neocles had moved to Samos, and Epicuros was probably born, and certainly educated, in that island. He was a precocious boy, who took to the study of philosophy at the age of fourteen and was already well educated when he went to Athens four years later, no doubt in order to pass the civic examination (*doci-*

[12] The word anticlerical is used advisedly; it designates a reaction that is bound to occur in every country where the clerics (the ministers of any religion) tend to abuse their power and their privileges. The priests of innumerable temples and sanctuaries all over the Greek world wielded a great deal of power and, being human, they wanted more power and more wealth; they had vested interests to protect and to extend, and by so doing they could not help creating enemies.

[13] Diogenes Laërtios (Book x). Cyril Bailey, *Epicurus, the extant remains* (Greek and English, 432 pp.; Oxford, 1926); *The Greek atomists and Epicurus* (630 pp.; Oxford, 1928) [*Isis* 13, 123–125 (1929–30)].

Marie Jean Guyau (1854–88), *La morale d'Epicure et ses rapports avec les doctrines contemporaines* (285 pp.; Paris, 1878; ed. 7, 1927). Benjamin Farrington, *Science and politics in the ancient world* (244 pp.; New York: Oxford University Press, 1940) [*Isis* 33, 270–273 (1941–42)], a book written to Epicuros' glory.

masia) that would entitle him to be enrolled among the ephēboi of his ancestral deme. At the time of his visit (323), Perdiccas, guardian of Alexander's sons, the general who tyrannized the city, obliged the Athenian colonists established in Samos to leave the island. Thus, Epicuros did not return to Samos but roamed with his family along the Asiatic coast, remaining for a short time in various places, chiefly the Ionian cities Colophon and Teos (try to visualize a group of uprooted Athenians, refugees, "D.P.'s," moving from place to place). In Teos, he received some instruction from Nausiphanes,[14] who explained the philosophy of Democritos. At the age of 30 (in 311) he settled down in Mitylene and began his own career as an independent philosopher. His influence even then must have been considerable because his three brothers[15] were among his disciples; this extraordinary circumstance is a credit not only to his persuasiveness but to his fundamental goodness. After a while, the new school was moved to Lampsacos, on the Asiatic side of the Dardanelles, and there Epicuros gained the adherence of more disciples, such as Metrodoros, Colotes, Polyainos, Idomeneus, and Leonteus and his wife Themista.[16]

The success thus far obtained determined Epicuros to move his school to Athens, for there only could the influence of a new philosophic school be completely established. He returned to his native country in 307, during the tyranny of Demetrios Poliorcetes (king of Macedonia), and bought a house and garden[17] in Melita (between the city and the harbor Peiraieus). The rest of his life was spent there, some thirty-seven years. He could begin in good style, like a recognized master, for many of his disciples, including his own family, had come with him, and soon new disciples were attracted, among them Hermarchos of Mitylene, who was to be his successor, Pythocles, Timocrates, brother of Metrodoros. Slaves were admitted, such as Mys, whom Epicuros manumitted, and women, even courtesans, such as Leontion, who became Metrodoros' wife.

Teaching in the "garden of Epicuros" was informal and life very simple and brotherly. Yet the presence of women was soon a pretext for gossip, and the success of the school, a cause of jealousy. Some of the adversaries affected to be scandalized and the ill repute that still sticks to the name "Epicurean" was attached to it in Melita before the end of the century.

These slanders increased the devotion of the disciples to their master, and life continued to be friendly and simple for many years. At the age of seventy Epicuros died; the house and garden had been bequeathed to Hermarchos for the use of the school, and provision was made for the celebration of festivals and for the care of the son and daughter of Metrodoros, who had died before Epicuros.

Epicuros' writings were numerous, filling three hundred rolls; most of them are lost, but we have extracts from many in Greek or Latin. The most important were the *Canon*, said to have been derived from the *Tripod* of Nausiphanes of Teos, and the treatise on *Nature* (in 37 books) which contained the most elaborate account of his scientific views. Diogenes Laërtios has transmitted to us a collection

[14] Nausiphanes of Teos had been trained by Pyrrhon of Elis, perhaps while they were taking part together in Alexander's Asiatic campaign. Later, he became an atomist, but he differed from Democritos in that he insisted that the scholar should take part in public life.

[15] Neocles, Jr., Chairedemos, Aristobulos. I know of no other philosopher who counted three of his own brothers among his disciples.

[16] All of them natives or residents of Lampsacos.

[17] Or orchard (*ho cēpos*).

of forty *Sovran maxims* (*cyriai doxai*) and the letters of Epicuros to three of his disciples, Herodotos, Pythocles, and Menoiceus. Another collection of eighty maxims was discovered in a Vatican manuscript and published in 1888. In addition to these writings and to the fragments embedded in classical literature, we must still mention two unusual sources that have enriched our knowledge of Epicuros and of the Epicurean tradition. First, papyrus rolls found in the excavations of Herculaneum gave us writings of the Epicurean Philodemos of Gadara (Palestine), a contemporary of Cicero (I–1 B.C.); second, a stone inscription found at Oinoanda in Lycia in 1884 preserved the Epicurean catechism written by a certain Diogenes.[18] This devoted Epicurean had caused the inscription to be engraved for the admonition of the passers-by. The best source of Epicurean doctrine, however, is the *De rerum natura* which Lucretius wrote two centuries after the master's death, the most remarkable monument ever built to the memory of a great philosopher.

EPICUREAN PHYSICS AND PHILOSOPHY

The main physical theory of Epicuros was the atomic theory, which had been explained by Leucippos and Democritos, but of which he modified various details. Everything, whether material or spiritual, is made out of atoms. These atoms, of many shapes, are scattered everywhere; they are not necessarily close together; they are in a vacuum, so that it is possible for them to move from place to place and to collide. When a man dies, the atoms that constitute his soul are disengaged and distributed[19] just like those of his body. The gods themselves are made of atoms; they exist in a kind of intermediate paradise (*ta metacosmia*), empty spaces between the integrated worlds. Mind (*nus*) is a concentration of very fine atoms, while the vital spirit (*psychē*) is composed of subtle atoms distributed all over the body. Spiritual entities (such as gods, souls, minds) differ from the material ones only in the finer and subtler nature of their atoms. Thus, everything is materialized, and it is not incorrect to speak of Epicurean atomism as materialism.

And yet Epicuros qualified this materialism and determinism in two ways. He admitted that the soul included some nameless (*acatonomastos*) element. For Epicuros fire (heat), wind (breath), air were elements added to the atoms and ubiquitous, but the soul and the mind implied the existence of a fourth element, subtler than the three others, which was as it were the soul of the soul.[20] The other qualification was the conception of atomic swerve (*parenclisis tōn atomōn, clinamen*), that is, the assumption of an irreducible amount of spontaneity and capriciousness in atomic motions.

These two qualifications are extraordinary; they illustrate Epicuros' poetic genius but also the impossibility of driving spirituality completely out, even of the most thoroughgoing materialism. Throw the spirit out of the window and it comes back through invisible holes in the wall. That is what happened to Epi-

[18] This Diogenes is called Diogenes of Oinoanda, date unknown. Oinoanda was a Cabalia, a district north of Lycia in southwest Asia Minor. His inscription was edited in the Teubner Library by Johannes William, *Diogenis Oenoandensis fragmenta* (151 pp.; Leipzig, 1907).

[19] The word distribute is here given the connotation familiar to the old printers; they separated the type that had been used to print a text and distributed it into the boxes of the type cases in order that it might be used to set up another text.

[20] This is a very obscure subject, which I do not claim to understand. See Bailey, *The Greek atomists and Epicurus*, Appendix V, pp. 580–587, on the relation of the "nameless" element and the "mind."

curos and to every materialist after him. He was a rationalist, yet his "nameless" element of the soul opened the door to occultism.

Epicureanism was much more, however, than atomism; atomism, we might say, was the physical core of Epicurean philosophy, and one that the master had modified in order to diminish friction and to leave a minimum of clearance and freedom.

One of his main ideas was that pleasure is the only good, but his conception of pleasure was very remote from coarse sensualism; the kind of pleasure he had in mind could be attained only by the exercise of many virtues, such as prudence and justice and the extirpation of many desires; it implied moderation, if not asceticism. Epicuros gave a new meaning to the old Greek maxim, *mēden agan*, nothing too much (*ne quid nimis*).

Another idea of his that has been frequently misunderstood may be called sensationalism. In reaction to Pythagorean and Platonic fantasies he claimed that all our knowledge is derived from our senses. Experimental science hardly existed in his time; otherwise he might have said that our knowledge must have an experimental basis. He could not go as far as that, but he claimed that one must have some kind of sensual evidence; our words must correspond to tangible things. Of course, his atomism went far beyond the possibility of verification; it was not even a workable theory in the modern sense; Epicuros was a philosopher, not a man of science.

He was before everything else a moralist, trying to cut out a new way to virtue and happiness. Virtue implies freedom, and the freedom of the human spirit was so essential to him that in order to make it possible he was obliged to modify the basic atomic doctrine. The "swerving" of the atoms established chance and freedom within the most material objects; that element of chance and freedom increased as matter was more and more spiritualized and reached its climax in the soul of man.

Happiness should be attained by the practice of forbearance and abstinence, that is, in a negative manner. The master of the garden advised his disciples not to marry, not to beget children, not to attract public attention to themselves. Epicurean hedonism was ill-judged because its enemies represented it as a search for pleasure (chiefly sensual pleasures, for they themselves could conceive no other), whereas it was rather an attempt to free oneself from pain and trouble; the Epicureans tried to cast off fears, such as the fear of death or poverty, and to attain imperturbability (*ataraxia*); they tended to withdraw from life, and one might accuse them of defeatism; their general attitude lacked heroism, but it was not immoral. They might seem to be selfish, but we should not forget that they lived in dangerous times, when arbitrariness was more common than justice, when everything was more precarious than ever, and when it was wiser to hide one's life than to invite jealousy and violence.[21]

EPICUROS' STRUGGLE AGAINST CLERICALISM AND SUPERSTITION

The main feature of Epicuros' philosophy of life, and the one that created for him and his teaching many irreconcilable enemies, was his struggle against

[21] A few centuries later the poet Ovid (43 B.C.–A.D. 18) could still repeat *Bene qui latuit bene vixit* (*Tristium* lib. III, el. IV, 1, 25). This is still good advice today, but it was more urgent in the fourth century or in the first than it is today, at least in civilized countries.

superstition. We have already indicated many times that superstitions were rampant in the Greek world; the love of magic and miracles which had existed from the earliest times (witness the ancient mysteries, the myths, the healing shrines) was exacerbated by the miseries of war and by political and economic insecurity. The miseries that had grown during the civil wars had reached a new climax after Alexander's death and the dissolution of his empire; their abundance and ubiquity strengthened the power of the guardians of the temples, of the priests and the soothsayers.

Epicuros was animated by at least one passion, the hatred of superstition. It has been observed that the passions which dominate a man's activity are frequently the result of personal experience, especially such as have stamped a man's soul during the most impressionable years of his life. Diogenes Laërtios reports [22] that young Epicuros used "to go round with his mother Chairestrate to cottages and to read charms and assist his father in his school for a pitiful fee." This conjures up the vision of a family struggling to keep the wolf from the door, the father being an underpaid schoolteacher, the mother helping out by acting as a kind of sham priestess or magician. If the precocious boy was thus obliged to witness his mother's spiritual prostitution, one can easily conceive the growth of his disgust and of his lifelong anger. He had seen early what incantations meant for the insider; he had been obliged to help his mother deceive their neighbors. Could any experience be more awful?

At any rate, he had realized that poor men were simply the victims of circumstance, and he hated not so much the so-called popular superstitions, the fantastic folklore of illiterate and overcredulous people, as the pious lies disseminated by clerics and the "noble lies" so beautifully expressed by the Platonists. The distinction between popular and learned superstition is not always easy to make, because so many interests were vested in folklore that there was a tendency to assimilate it to the learned nonsense. The query whether superstitions are of popular origin or not is an academic and insoluble one. The ultraconservatives who believed that "religion is good for the people" knew well enough that any kind of superstition fostered other superstitions and hence was useful.[23] They were like sellers of whisky who would foster the love of alcohol (in general) rather than discourage it. Let the common people have all the superstitions they want, Plato and his disciples would have said, they are too stupid to contemplate the truth; they prefer lies.

That may be true, but the immense difference between Plato and Epicuros consists in this very fact, that the former was ready to exploit popular ignorance and credulity, while the latter did his best to eradicate them. For example, Epicuros did not hesitate to reject the whole of divination, and divination was big business. All the sects, except the Epicurean, accepted the reality of magic.

Epicuros was definitely anticlerical, but he was not antireligious. He claimed that gods exist; one must look for them, however, not in the stars but in the hearts of men. This is put beyond any doubt in his admirable letter to Menoiceus.

[22] Diogenes Laërtios, x, 1.

[23] In his *Journal* under date of 21 March 1906, André Gide remarks, "Certainement le but secret de la mythologie était d'empêcher le développement de la science." That is an exaggeration of the truth; the purpose of deceiving the people and of sidetracking them was more unconscious than deliberate. It is Epicuros' main glory to have detected and fought the purpose.

> Those things which without ceasing I have declared unto thee, those do, and exercise thyself therein, holding them to be the elements of right life. First believe that God is a living being immortal and blessed, according to the notion of a god indicated by the common sense of mankind; and so believing, thou shalt not affirm of him aught that is foreign to his immortality or that agrees not with blessedness; but shalt believe about him whatever may uphold both his blessedness and his immortality. For verily there are gods, and the knowledge of them is manifest; but they are not such as the multitude believe, seeing that men do not steadfastly maintain the notions they form respecting them. Not the man who denies the gods worshiped by the multitude, but he who affirms of the gods what the multitude believes about them is truly impious. For the utterances of the multitude about the gods are not true preconceptions but false assumptions; hence it is that the greatest evils happen to the wicked and the greatest blessings happen to the good from the hand of the gods, seeing that they are always favorable to their own good qualities and take pleasure in men like unto themselves, but reject as alien whatever is not of their kind.[24]

God's existence is proved by the goodness in man (this is still, I believe, the best proof). Epicuros had no quarrel with pure religion, but he hated the religion fostered by Platonists and by aristocrats, the kind of religion that was encouraged by the "best people" for the good of the lower classes, and that was mixed not only with base superstitions, but with police power, spying, and persecutions. He rejected the idea of Providence (*pronoia*) – dear to the Stoa; he even rejected the idea of creation, or at any rate of continuous creation. God created the world, then withdrew from it and left it to its own evolution. The laws of nature are not to be disturbed by any kind of arbitrariness.

Epicuros was the first to proclaim the social danger of superstition and the primary need of fighting it. The people must not be lied to according to the Platonic method; they must be told the truth; if they are not sufficiently educated for that, then they must be educated; the truth will make them free, naught else.[25]

He represents liberalism and rationalism against Plato's conservatism and deliberate obscurantism. His rationalism was not absolute, but limited. What rationalism is not?

Epicureanism is full of inconsistencies: its atomism is mitigated by atomic caprices, its materialism by the recognition of soul and gods; but the greatest inconsistency was its idea of a crusade against superstition, for that did not tally at all with the purpose of keeping free from pain and trouble. If the purpose had been to cause more trouble for themselves, Epicureans could not have discovered anything better than to fight against social lies and superstitions. Their choice of the most troublesome and dangerous cause to which to devote themselves proves their radical inconsistency and their moral greatness.

Epicuros was not antireligious. It is equally untrue that he was an enemy of science. He was more interested in ethics than in the pursuit of pure knowledge, yet he realized that our first duty is to know the truth, or rather that we must

[24] This very long letter is quoted *in extenso* by Diogenes Laërtios, x, 122–135; it is a very good summary of Epicurean ethics. We quote only the beginning, concerning the gods; after that he deals with the groundless fear of death, good and bad desires, pleasure, etc. Translation by R. D. Hicks in Loeb Classical Library, vol. 2 (1925).

[25] There is no evidence that the Epicureans did much to educate the poor and illiterate people, but nobody bothered about them in antiquity. Public education could be organized only by the state or by powerful bodies. The Epicureans understood the need of education but could not and did not implement it. The main weakness of their doctrines was the indifference and passivity that they induced. They lacked energy.

know the truth in order to do our duty. His opposition to what might have been called "pure science" was caused by the many falsifications of its purity; he despised logic because of the aberrations of the dialecticians; he distrusted mathematics because of Pythagorean numerology and Platonic geometry; above all, he rejected the astral theology that was debasing astronomy as well as religion. The tendency to confuse pure science with Platonic magic completely justified Epicuros in his rejection of both. His fight against superstition and irrationalism became unavoidably a fight against false sciences as well as against false religions.

This being said, it must be admitted that Epicuros had no scientific curiosity; there was in him no urge to discover the truth. This explains why Aristotle did not appeal to him; he would probably have regarded as idle all the stories collected by Aristotle in his zoölogic books. He would have said, What do we care about the breeding of fishes or the copulation of snails? Let us devote our attention to matters of human concern. We repeat: he was primarily a moralist, not a man of science.

He was a moralist and a politician concerned with the education of men, of all men and women, their education and their happiness. It is amusing to put together the epigrammatic descriptions of him made by two English philologists. Said Gilbert Murray, "The Epicureans were in a sense the Tolstoyans of antiquity," and Benjamin Farrington, "The Epicureans were a sort of Society of Friends with a system of Natural Philosophy as its intellectual core." [26] These two sayings taken *grosso modo*, as they should be taken, do not contradict each other. The second is fuller in that it recognized Epicuros' scientific interest. Indeed, it would be paradoxical to deny that interest altogether in the man who passed the torch of atomism from Democritos to Lucretius.

THE SCHOOL

The Epicurean school was fairly well established by the master himself. Epicuros had one of the essential qualities needed for that purpose; he was able to kindle the enthusiasm of his auditors and to ensure their loyalty. Already in Lampsacos he had managed to gather around himself many men of promise. The greatest of these early disciples was Metrodoros, who died many years before Epicuros, in 277, at the age of fifty-three. Other early disciples, such as Polyainos, Colotes, Idomeneus, have been mentioned. Polyainos was a mathematician who forsook mathematics after his Epicurean conversion. This has been used as a proof that Epicuros was antagonistic to science, but the proof is very insufficient. In the first place, Epicuros' objections to Pythagorean arithmetic and to Platonic geometry could be fully justified on scientific grounds, and in the second place many men have passed from mathematics to philosophy or to religion.[27]

The continuity of the school was ensured by the master's will bequeathing the leadership and the garden itself to Hermarchos of Mytilene. That will is so moving a document that we quote it verbatim:

[26] Murray in his *Greek studies* (Oxford: Clarendon Press, 1946), p. 85; Farrington in *Science and politics in the ancient world*, p. 159.

[27] Think of Pascal! Why do these men abandon mathematics? Is it because philosophy or religion appeals more to them, or because their mathematical work is done? They do not abandon mathematics, one might suggest, it is mathematics that abandons them.

On this wise I give and bequeath all my property to Amynomachos, son of Philocrates of Bate, and Timocrates, son of Demetrios of Potamos, to each severally according to the items of the deed of gift laid up in the Mētrōon, on condition that they shall place the garden and all that pertains to it at the disposal of Hermachos, son of Agemortos, of Mitylene, and the members of his society, and those whom Hermarchos may leave as his successors, to live and study in. And I entrust to my School in perpetuity the task of aiding Amynomachos and Timocrates and their heirs to preserve to the best of their power the common life in the garden in whatever way is best, and that these also (the heirs of the trustees) may help to maintain the garden in the same way as those to whom our successors in the School may bequeath it. And let Amynomachos and Timocrates permit Hermarchos and his fellow members to live in the house in Melite for the lifetime of Hermarchos.

And from the revenues made over by me to Amynomachos and Timocrates let them to the best of their power in consultation with Hermarchos make separate provision (1) for the funeral offerings to my father, mother, and brothers, and (2) for the customary celebration of my birthday on the tenth day of Gamēliōn in each year, and for the meeting of all my School held every month on the twentieth day to commemorate Metrodoros and myself according to the rules now in force. Let them also join in celebrating the day in Poseideōn which commemorates my brothers, and likewise the day in Metageitniōn which commemorates Polyainos, as I have done hitherto.[28]

And let Amynomachos and Timocrates take care of Epicuros, the son of Metrodoros, and of the son of Polyainos, so long as they study and live with Hermarchos. Let them likewise provide for the maintenance of Metrodoros's daughter, so long as she is well ordered and obedient to Hermarchos; and, when she comes of age, give her in marriage to a husband selected by Hermarchos from among the members of the School; and out of the revenues accruing to me let Amynomachos and Timocrates in consultation with Hermarchos give to them as much as they think proper for their maintenance year by year.

Let them make Hermarchos trustee of the funds along with themselves, in order that everything may be done in concert with him, who has grown old with me in philosophy and is left at the head of the School. And when the girl comes of age, let Amynomachos and Timocrates pay for her dowry, taking from the property as much as circumstances allow, subject to the approval of Hermarchos. Let them provide for Nicanor as I have hitherto done, so that none of those members of the School who have rendered service to me in private life and have shown me kindness in every way and have chosen to grow old with me in the School should, so far as my means go, lack the necessaries of life.

All my books to be given to Hermarchos.

And if anything should happen to Hermarchos before the children of Metrodoros grow up, Amynomachos and Timocrates shall give from the funds bequeathed by me, so far as possible, enough for their several needs, as long as they are well ordered. And let them provide for the rest according to my arrangements; that everything may be carried out, so far as it lies in their power. Of my slaves I manumit Mys, Nicias, Lycon, and I also give Phaidrion her liberty.[29]

Hermarchos succeeded Epicuros in 270; and he was succeeded himself by Polystratos, then by Dionysios, and he by Basileides. Apollodoros, known as the tyrant of the garden, who wrote over four hundred books, is also famous; and the two Ptolemaioi of Alexandria, the one black and the other white;[30] and Zenon of Sidon, the pupil of Apollodoros, a voluminous author; and Demetrios, who was called the Laconian; and Diogenes of Tarsos, who compiled the select lectures; and Orion, and others whom the genuine Epicureans call Sophists.[31]

28 The Greek months mentioned in this paragraph correspond nearly: Gamēliōn to January, Poseideōn to December, Metageitniōn to August.

29 Diogenes Laërtios, x, 16–21, as translated by R. D. Hicks (Loeb Classical Library, 1925).

30 The two Ptolemaioi, "the one black and the other white" (*ho te melas cai ho leucos*). If the word black is to be taken literally, the Black Ptolemaios was the first Negro philosopher (second century B.C.). That is quite plausible. The Epicureans were exceedingly human.

31 Diogenes Laërtios, x, 25–26.

These names are mentioned to illustrate the continuity and vitality of the Epicurean school. Zenon of Sidon brings us already into the first century, for Cicero heard him in Athens; that must have been in 79 B.C., but Cicero had been initiated into Epicureanism before he went to Greece, for he had heard Phaidros (140–70) lecture in Rome before 88.[32] Another Epicurean of Cicero's time was Philodemos of Gadara (in Palestine). The greatest of all was Lucretius (I-1 B.C.), of whom we need not say anything more at present. For Lucretius, Epicuros was almost a god (see the opening of *De rerum natura*[33]). This appreciation was not popular afterward, however, though it was shared by such exceptional persons as Lucian of Samosata and his friend Celsos,[34] both of whom considered Epicuros a divine hero, a benefactor of mankind.

This appreciation *could* not be popular. The glory of Epicuros and later of Lucretius was their fight against superstition. Such a struggle has never brought and will never bring popularity to anybody. Even when superstitions were finally overcome, it was only because they were replaced by other superstitions, just as the weeds of our gardens when we pluck them out make room for other weeds. In spite of Epicurean efforts, the pagan superstitions did not decrease; the lack of political and economic stability tended, on the contrary, to increase them. The best of ancient religion was being gradually debased, corrupted; its poetry was lost. The philosophic (non-Epicurean!) elite replaced it with a new astrologic religion too difficult for the people to grasp and too abstract to warm their hearts. There remained only rituals, processions, pilgrimages, and superstitions of every kind. The religious vacuum was filled with fantastic ideas borrowed from Egypt and other parts of the Near East. The growth of superstition implied the growth of clerical bumptiousness and intolerance. The plain people were so deeply afflicted, their miseries were so many and so complex, that they abandoned rational efforts toward improvement and thought only of "salvation" – a kind of mystical salvation in another world.[35]

The Epicureans had against them also the philosophers of the other sects, chiefly the Stoics. For example, the astronomer Cleomedes[36] expressed his contempt of Epicuros for using a vulgar language such as was current "among the harlots, the women who celebrated the Ceres festivals, the beggars, etc." The roots of Cleomedes' anger were deeper; what irritated him was not so much

[32] Phaidros the Epicurean (140–70) was head of the Epicurean school in Rome. One of his books inspired Cicero's *De natura deorum*; fragments of it were discovered in Herculaneum and edited by Christian Petersen (52 p.; Hamburg, 1833).

[33] See opening of Book v of *De rerum natura*:

. . . deus ille fuit, deus, inclyte Memmi,
qui princeps vitae rationem invenit . . .

(He was a god indeed, illustrious Memmius, he who discovered that rule of life . . .) C. Memmius Gaius, to whom Lucretius dedicated his poem, was a Roman statesman and orator (fl. 66–49).

[34] Probably but not certainly identical with Celsos (II-2), who flourished in the Near East (Egypt?) and wrote the *True word* (*Alēthēs logos*) the first systematic criticism of Christianity, a work known only through the refutation of Origen (III-1).

[35] For abundant details see Franz Cumont (1868–1947), *Les religions orientales dans le paganism romain* (Paris, 1929) [*Isis 15*, 271 (1931)].

[36] Cleomedes was placed too early (I-1 B.C.) in my *Introduction*, vol. 1, p. 211. His date is very uncertain; he may be as early as the end of the first century B.C., and as late as the third century of our era. For Cleomedes' reaction to Epicuros see Saul Lieberman, *Hellenism in Jewish Palestine* (New York: Jewish Theological Seminary, 1950) [*Isis 42*, 266 (1951)].

Epicuros' language as his rejection of the astrologic religion and his friendliness to the plain people.

The Epicurean hatred of superstition irritated everybody from the Stoics down to the soothsayers and to the demagogues who confused it with hatred of religion. That is an old trick which is still played today. A rationalist is generally accused of trying to pervert the young and to repudiate the gods. It was easy to exploit against Epicuros not only his anticlericalism but also his hedonism, which was shamelessly traduced. There is nothing strange in that. Could the Greeks of that time, whose minds were frustrated and demoralized by defeat and misery, be expected to give a welcome to those premature Quakers and to those Tolstoyans "avant la lettre"?

The opposition to Epicureanism was even greater among the religious communities, particularly the Jewish ones. Epicuros was a rebel and an infidel. It was relatively easy to represent his disciples as sordid materialists, lovers of pleasure, doubters, and liars. Both Philon (I-1) and Joseph Flavius (I-2) called him an atheist. "Epicurean" became an abusive term in Hebrew and has remained so to this day.[37]

All this concerns the historian of science directly, because it affected the fate of atomistic ideas. These, being mixed up with Epicurean philosophy, were considered themselves subversive. Atomism was driven underground; it was not killed (one can hardly destroy an idea) but continued a secret life and re-emerged sometimes with strange associates.[38] To the superstitious and unthinking people, atomism was simply a rebellion, a kind of satanic rebellion, as if the wicked atomists were trying to pulverize their very faith. In the Christian West it was not rehabilitated until the seventeenth century, first by Pierre Gassendi (1592–1655), later by Robert Boyle (1627–1691),[39] and it was not presented in a form acceptable to men of science before the beginning of the nineteenth century, by John Dalton (1766–1844).

The further vicissitudes of scientific atomism would take us too far away from our field, but the reader may permit me to introduce the following remarks. It took almost the whole of the nineteenth century to establish atomism on a sound experimental basis and this required an immense amount of chemical investigation. When success was finally in sight, a number of men of science and philosophers who tried to reach a deeper understanding of things rejected atomism as a kind of illusion. Antiatomic views were published by such men as Ernst Mach (1838–1916),[40] Pierre Duhem (1861–1916), even by a practical chemist like Wilhelm Ostwald (1853–1932); these men were fighting a rearguard action at the very

[37] "Apiqoros" or "Epiqoros" has been used since Mishnaic times to mean "freethinker, unbeliever, a man who makes fun of the rabbis and does not believe in the World Beyond." See article by Bernard Heller, *Encyclopaedia Judaica,* vol. 6 (1930), pp. 686–688. My friend Gandz writes to me (15 February 1951) that in Hebrew literature the word "Epicurean" does not mean a *bon vivant* and sensualist, but just an unbeliever and an infidel. See also his remarks in *Isis 43,* 58, 1952.

[38] For example, with Ismā'īlī doctrines in the Muslim East; *Introduction,* vol. 3, p. 149. The history of atomism, open and secret, is made exceedingly complex because the fundamental ideas were not simply Greek, but also Jaina and Buddhist. Moreover, their very secrecy and deliberate elusiveness discourages the investigators and, what is worse, sidetracks them.

[39] G. Sarton, "Boyle and Bayle. The sceptical chemist and the sceptical historian," *Chymia* 3, 155–189 (Philadelphia: University of Pennsylvania Press, 1950).

[40] With regard to Mach, see Einstein's statement in Isaac Benrubi, *Les sources et les courants de la philosophie contemporaine* (Paris: Alcan, 1933), p. 416, n. 3.

time when atomism had ceased to be a hypothesis, when atoms could be counted and weighed, yet ceased to be atoms in the literal sense, for they were reduced to other elements incredibly smaller than themselves.

To return to Epicuros, we should repeat that the rejection of atomism by Ostwald and others was infinitely more scientific than his own blind acceptance of it. Epicuros' discovery or rediscovery of atomism was *not* a scientific achievement. The historian of science will give him more credit for his general philosophy, and especially for his struggle against superstition. Science cannot flourish in the darkness; in order to make its growth possible one must be ready to fight magic and superstition at every step, and Epicuros did that or tried to do it.

EPICUROS' CHARACTER. HIS DEATH

The best way to end this chapter is to give an idea of the personality of Epicuros. It is good to be able to do that, especially when we remember that we know practically nothing of the personalities of most of the great men of science of antiquity. Most of them are like abstractions, but Epicuros is alive.

It is pleasant to see him walking with his disciples in the garden of Melita, talking and discussing with them. He found time to write considerably, but apparently he did not deliver set lectures. He was not a lecturer but a genuine teacher, deeply concerned with his students. What he founded was not simply a school but a brotherhood. Not only men, but also women and children gathered around him. Here is a letter from him to one of the children:

> We have arrived at Lampsacos safe and sound, Pythocles and Hermarchos and Ctesippos and I, and there we found Themista and our other friends all well. I hope you too are well and your mamma, and that you are always obedient to papa and Matro, as you used to be. Let me tell you that the reason that I and all the rest of us love you is that you are always obedient to them.[41]

This document is unique in ancient literature. Other letters of his include similar proofs of his kindliness to his parents, brothers, disciples, even to his slaves. Far from being the devil and debauchee that his enemies represented him to be, he was a simple and friendly creature, loving life and loving men. His manner of living was moderate, but he had realized the need of occasional feasts to break the monotony of days and accentuate their succession. The twentieth day of each month was set apart for a feast, which after his death became a memorial to himself and to Metrodoros. Unfortunately, we do not know how one was admitted into the Epicurean brotherhood. To be permitted to enter the garden and talk with the brothers and sisters must have been a blessing – a blessing without nonsense added to it, just love and reason.

The only unpleasant feature in Epicuros' personality (and it displeases me very much) was his very ungrateful judgment of his teachers and of other philosophers. His own teacher, Nausiphanes, he called the jellyfish.[42] He used other nicknames, equally nasty, to designate Heracleitos (the muddler), Democritos (nonsense), Aristotle (the profligate); Leucippos he refused to consider at all.

[41] Transmitted to us in the Herculaneum papyrus 176; translation by Cyril Bailey, *The Greek atomists and Epicurus*, p. 225.

[42] *Pleumōn* or *pneumōn*, the very word used by Pytheas (sea lungs). The word is far from clear, but the abusive intention is unmistakable.

A very original man may deny his teachers because he does not realize how much he owes to them, or in his own ardor he may have forgotten them; he may be sincere, yet such a lack of recognition of others is a lack of grace. This puzzles me very much in Epicuros, for the disregard of others and the belittling of greatness in them is almost always a symptom of mediocrity. Yet Epicuros was a very great man. How could he be so blind to the greatness of his predecessors, and to the merit of his teachers?

Just as we know Epicuros' life much better than the life of other Greek philosophers, we know better also the circumstances of his death. Of course, we know well enough the circumstances of Socrates' death, because that death was a public execution, but about the other philosophers who died a natural death we are less well informed. As to Epicuros' last illness and death, Diogenes Laërtios gives us definite information.

> He died in the second year of the 127th Olympiad [= 271–270], in the archonship of Pytharatos, at the age of seventy-two; and Hermarchos the son of Agemortos, a Mitylenaean, took over the School. Epicuros died of renal calculus after an illness which lasted a fortnight: so Hermarchos tells us in his letters. Hermippos relates that he entered a bronze bath of lukewarm water and asked for unmixed wine, which he swallowed, and then, having bidden his friends remember his doctrines, breathed his last.

At the very end of his life Epicuros wrote a letter to his friend Idomeneus, which contains another account of his pains and a final, unforgettable, image of his kindness.

> On this blissful day, which is also the last of my life, I write this to you. My continual sufferings from strangury and dysentery are so great that nothing could augment them; but over against them all I set gladness of mind at the remembrance of our past conversations. But I would have you, as becomes your lifelong attitude to me and to philosophy, watch over the children of Metrodoros.[43]

THE STOA

The birth of Stoicism cannot be determined exactly, because we do not know when its founder, Zenon, was born. If it was as late as 336, then Stoicism could hardly be a product of this century, or it would belong to its very last years, but his birth has been placed as early as 348 and even 356. Zenon would then be an older contemporary of Epicuros. There is another reason, more fundamental, why Stoicism must be dealt with in this chapter: no matter when it matured, it is a fruit of Alexander's age.

ZENON OF CITION

Zenon son of Mnaseas was born at Cition. It has been claimed that he was of Phoenician race, and that is not impossible, because Cition was or had been one of the Phoenician settlements in Cypros, probably the oldest in the island.[44] He

[43] These two extracts are taken from Diogenes Laërtios, x, 15; x, 22, as translated by R. D. Hicks (Loeb Classical Library, 1925).

[44] Cition was on the site of Larnaca, the main harbor of Cypros, on the southeast coast. The Phoenician settlement was prehistoric. Even if

was almost certainly subjected to Phoenician influences. He came to Athens at the age of 22 or 30, and his Athenian studies lasted over 20 years, presumably before the foundation of his own school; he was the head of that school for 58 years and died at 98 (or 72?).[45]

The circumstances of his arrival at Athens deserve to be recorded. Says Diogenes Laërtios:

> He was shipwrecked on a voyage from Phoenicia to Peiraieus with a cargo of purple. He went up into Athens and sat down in a bookseller's shop, being then a man of thirty. As he went on reading the second book of Xenophon's *Memorabilia*, he was so pleased that he inquired where men like Socrates were to be found. Crates passed by in the nick of time, so the bookseller pointed to him and said, "Follow yonder man." From that day he became Crates' pupil, showing in other respects a strong bent for philosophy, though with too much native modesty to assimilate Cynic shamelessness. Hence Crates, desirous of curing this defect in him, gave him a potful of lentil soup to carry through the Ceramicos; and when he saw that he was ashamed and tried to keep it out of sight, with a blow of his staff he broke the pot. As Zenon took to flght with the lentil soup flowing down his legs, "Why run away, my little Phoenician?" quoth Crates, "nothing terrible has befallen you."[46]

This account is suggestive in many ways. It was because of an accident reducing him to poverty that Zenon became a philosopher, and he remarked later, "I made a prosperous voyage when I suffered shipwreck."[47] That is plausible enough. Second, Crates' calling him "little Phoenician" (*Phoinicidion*) confirms the story of Zenon's "Phoenician" origin. The main point is that Zenon was a disciple of Crates the Cynic. According to old traditions, Zenon's teaching was connected with that of Socrates via Antisthenes, Diogenes, Crates and thus the early history of both Stoicism and Cynicism was confused. There can be no doubt, however, about the Cynic roots of Stoicism: Cynic traces can be detected in all the Stoic writings, even in the reminiscences of Marcus Aurelius.

Athens at the end of the fourth century had many things to offer to an ambitious man, as Zenon was, and though he attached himself mainly to Crates of Thebes (who lived until 285), he listened to other teachers at the Academy and elsewhere. Among his teachers were mentioned Xenocrates and Polemon of the Academy, Stilpon and Diodoros Cronos of the school of Megara.[48] Polemon teased him, saying, "You slip in by the garden door, you pilfer my ideas and give them a Phoenician appearance."[49] What matters most, however, is not the philosophers whom he frequented in Athens, but the definite orientation of his own mind, and there can be no more doubt in his case than in that of Epicuros that his way of thinking was a reaction against the Academy and the Lyceum. There was an immense difference between Epicuros and him, a difference cutting down

---

Zenon had no Phoenician chromosomes in his cells, he may easily have been influenced by Phoenician (Semitic) examples during his youth. Yet to build an argument on the Semitic origin of Zenon and of Stoicism is unwarranted and foolish.

[45] In my *Introduction*, vol. 1, p. 137, I gave for Zenon's birth and death the years *c.* 336, *c.* 264, thus assuming that he died at 72. By making various selections of the figures given by Diogenes Laërtios, vii, 25, and others, one can obtain many different dates of almost equal probability. It is safe to conclude that Stoicism was a *fin de siècle* product.

[46] Diogenes Laërtios, vii, 2.

[47] The Greek is terser: "*nyn euploēca, ote nenauagēca*"; Diogenes Laërtios, vii, 4.

[48] If he sat at the feet of Xenocrates he must have arrived in Athens before 315/14, because Xenocrates died in that year. Stilpon taught mainly in Megara and Diodoros Cronos of Iasos (Caria) in Alexandria under Ptolemaios Soter. Zenon may have met them in Athens, however.

[49] Diogenes Laërtios, vii, 25.

to their youth, in that while Epicuros was harking back to Democritos, Zenon was under the influence of Heracleitos; Democritos meant rationalism, while Heracleitos was an occultist. Those influences, going back to the fifth century, justify my inclusion of both Epicuros and Zenon in this volume. Both philosophies were hatched and born before the end of the fourth century.

Diogenes Laërtios relates many anecdotes concerning Zenon, and yet we do not see him as clearly as we see Epicuros. Some of the traits mentioned by Diogenes are striking, however. For example, we are told he had a wry neck, was lean, fairly tall, and swarthy, that he was fond of green figs and of sunbaths.[50] It is clear enough, however, that Zenon was popular in Athens, and that the Athenians loved him; witness the two decrees that they voted in his honor and his official burial in the Ceramicos.

The manner of his death was as follows. As he was leaving the school he tripped and fell, breaking a toe. Striking the ground with his fist, he quoted the line from the *Niobe*:

"I come, I come,
why dost thou call for me?"

and died on the spot through holding his breath.[51]

STOIC SCIENCE AND PHILOSOPHY

Zenon began his teaching in Athens in a hall or portico which was called the painted hall or stoa (*hē stoa hē poicilē*) because it had been decorated about the middle of the fifth century by Polygnotos of Thasos, "the inventor of painting." That hall had been used as a meeting place by poets, and it was probably open to all who chose to gather there. The use that was now made of it by Zenon caused his school to be called the Stoa, and his followers were called Stoics.

It is sometimes difficult to separate in the Stoic doctrines the elements that must be ascribed to the founder from those that were added later by Cleanthes and others.[52] My impression is that Zenon had already explained the essentials and that he was undoubtedly the creator of Stoic philosophy; in the course of centuries many changes were made in the doctrines but those changes were of little importance. The sayings of Marcus Aurelius can be generally illustrated with references taken from Zenon's fragments.

Zenon divided philosophy into three main sections; physics, ethics, logic. Physics is the foundation of knowledge, logic the instrument, ethics the end.

[50] All this should be read in Greek, for the original terms are amusing; I must resist the temptation of putting too much Greek in this volume, however; nor is this necessary, as it is easy enough to read Diogenes Laërtios in the Loeb edition (Book vii, 1–160). One of Diogenes' remarks (vii, 32) puzzles me: "They say that Zenon was in the habit of swearing by capers just as Socrates used to swear by the dog." Capers is the Mediterranean shrub bearing in Latin its original Greek name Capparis. That is a curious bit of folklore. Did the Greeks appreciate the capers' buds?

My friend, A. Delatte, kindly wrote in answer to my enquiry (Liége, 26 March 1951) that Zenon, like Socrates and the Pythagoreans, did not like to swear by the gods (to use their names in vain); he preferred then to swear by something insignificant, the more insignificant the better.

[51] Diogenes Laërtios, vii, 28, Hicks' translation. *Niobe* was written by the famous Athenian poet and musician Timotheos of Miletos (446–357), who increased the number of strings of the cithara. The line quoted from Niobe reads: *erchomai; ti m'aueis?*

[52] A. C. Pearson, *The fragments of Zeno and Cleanthes* (352 pp.; London, 1891), in Greek or Latin with English commentary. Zenon covers 181 pp.; Cleanthes, 95 pp. There are 202 fragments of Zenon, 114 of Cleanthes. Very convenient Greek glossary referring immediately to Zenon or Cleanthes.

His logic was derived from Antisthenes and Diodoros Cronos, that is, from Cynic and Megarian examples, yet it developed independently in various directions. For example, it led to a deeper grammatical consciousness and Greek grammar may be said to be largely a Stoic creation. The grammatical work of Zenon was continued by Chrysippos and completed by Diogenes the Babylonian and Crates of Mallos.[53] Other branches of logic were rhetoric and dialectic. The epistemology of the Stoics was also original. Knowledge, they held, is obtained from sense impressions, yet one should consider them prudently and not allow oneself to be carried away by "fantasies." [54]

Stoic physics was a combination of materialism and pantheism. The Stoics conceived the existence everywhere of forces or tensions, coextensive with matter; these tensions cause the flux and reflux of the universe. They were involved in the same contradictions or ambiguities as the Epicureans, for they admitted the existence of souls, but these souls were made of matter, a subtler kind of matter than that of the more tangible bodies; these souls were corporeal, not spiritual.

Their main interest was ethical. The Socratic idea that virtue is knowledge was developed by them; true goodness consists in living according to reason or to nature, but this implies a sufficient knowledge of nature (physics, theology). Their purely scientific knowledge was derived from Plato rather than from Aristotle; therefore it lacked clearness and was somewhat impure. For example, the Platonic parallelism between macrocosm and microcosm misled them into attaching much importance to divination. In this they followed old Greek traditions, but proved themselves very inferior to the Epicureans.

They rejected atomism, but the substance of their universe was not less material for that. Everything is made up of the four elements, in order of increasing subtleness: earth, water, air, and fire. God himself is material, and so is reason, the cosmic reason or the individual reason, which is like "a fragment detached from God." [55] That reason is like a kind of hot breath. The souls are made of fire and at the end of a cosmic period a universal conflagration (*ecpyrōsis*) will bring them all back into the divine fire, after which there may be a new creation (*palingenesia*).[56]

These are later sophistications, however, and we must not anticipate. The main point from the time of Zenon on is that the world is made up of matter and reason, yet matter and reason are but two aspects of the same reality. There is no reason without matter, no matter without reason. To put it otherwise, God is the single all pervading force, yet that force cannot be separated from the rest. Understand that if you can. In short, Stoicism was not less materialist than Epicureanism, but it was less rational.

Ethics are the climax and the eternal glory of Stoicism. The chief good is virtue,

[53] Note that all these men had some knowledge of foreign languages. Zenon came from Cypros (not to say Phoenicia), Chrysippos came from Cilicia, Diogenes flourished for a time in Rome, and Crates was the head of the library of Pergamon. Grammatical awakening is much facilitated by the comparison of one's own language with others.

[54] For Stoic logic in general, see Antoinette Virieux-Reymond, *La logique et l'épistémologie des Stoïciens, leurs rapports avec la logique d'Aristote, la logistique et la pensée contemporaine* (338 pp.; Chambéry: Lire, 1949) [*Isis 41*, 316 (1950)].

[55] This is a later expression, *apospasma tu theu* (Epictetos, I, 14, 6; II, 8, 11), but the idea is as old as Zenon.

[56] A new form of the old myth of eternal return or eternal recurrence which was probably of Oriental origin but was popularized by Pythagoras and Plato and reappears periodically in the writings of philosophers and apocalyptic historians.

and virtue is simply to live according to nature or to reason (*homologumenōs physei zēn*). To be virtuous is the only good, not to be virtuous is the only evil; everything else, including poverty, disease, pain, death, is indifferent. The good man, who cannot be deprived of his virtue, is invulnerable. When he has properly withdrawn into himself, and has realized that most miseries are matters of opinion, his virtue gives him self-sufficiency (*autarceia*) and impassiveness, freedom from pain (*ataraxia*). This quietism was similar to the Epicurean, yet less passive, more virile (or it became so in Roman times). It is not enough for a man to bear and to forbear, he should be brave.

One consequence of Stoicism was the obligation of the wise man to obtain the available knowledge, for in order to live according to nature one must understand the cosmos. Unfortunately, most Stoics were satisfied with a very imperfect knowledge of nature. They lacked scientific curiosity. Stoicism lifted the heart, it did not sharpen the mind.

The Stoics accepted the idea of providence (*pronoia*) and thought the ways of providence could be discovered by means of divination (*manteia*) – two good examples of inconsistency, caused by their lack of scientific rigor and by the lack of vigor against traditional feelings.

The most often quoted of Zenon's lost writings was his treatise on government (*Politeia*); according to Plutarch, it was an answer to Plato's *Republic*. At any rate, the Stoics were interested in politics. In this respect, they were superior to the Epicureans, whose quietism led them to political aloofness; the Stoics felt that it was part of a man's duty to assume his full share of the political burden. This explains the success of Stoicism in the frame of Roman law and administration.

The most original and pleasant feature of Stoic ethics and politics was their feeling of communion (*coinōnia*) or participation, not only with the people of their own deme or country, but with those of the whole universe. Under the influence of the tremendous revolution caused by Alexander's conquest of the world they escaped one of the oldest and strongest Greek traditions, the city-centered or provincial spirit of the Hellenic age; they were cosmopolitans, the first in history. Plutarch said that behind Zenon's dream lay Alexander's reality. That is not quite correct. Zenon was inspired not so much by Alexander's empire (which was crumbling to pieces) as by Alexander's conception of the unity of mankind (*homonoia*); he made of that individual conception a philosophic doctrine.[57]

The doctrine of *homonoia* (or *concordia*, consensus of mankind) was one of the sources of Roman law, of the *jus gentium*, the law of all nations, the law of nature.[58] On the other hand, that idea might (and did) justify widespread preju-

[57] Excellent discussion of this by William Woodthorpe Tarn, "Alexander the Great and the unity of mankind," *Proc. British Acad.* 9, 46 pp. (1933). Tarn has shown, conclusively in my opinion, that Alexander's idea of *homonoia* is prior to Stoicism and not a projection backward from Stoicism into the Alexandrian tradition. Tarn has reaffirmed these views in his recent book, *Alexander the Great* (Cambridge: University Press, 1948) [*Isis* 40, 357 (1949)].

[58] In English, "law of nature" or "natural law" means generally scientific laws (as distinguished from human laws). That is so at least since the creation of the Royal Society (Oxford English Dictionary, vol. 6, p. 115), or even since 1609, when Bacon wrote the *Advancement of learning*. According to French usage of about the same time (Pascal), "loi naturelle" meant the moral principles and ideas of justice which are independent of the written law and prior to it. The Greek conception of *homonoia* was of necessity closer to the French meaning of "natural law" than to the English, because the Greeks were more concerned with "moral laws" than with "scientific laws," and knew no clear example of the latter.

dices. If all men believed in divination, was it not wiser, less dangerous, to share their belief? The political value of cosmopolitanism appealed to the Romans, but it could easily take a subversive appearance. The idea that all men are brethren might be considered a dangerous doctrine; that idea was strengthened later by the early Christians and was one of the causes of the persecutions that they suffered.

For us who look from a great distance, we realize that the Stoic ethics in general and its cosmopolitanism in particular constituted an immense progress, so immense that whatever of it was realized was destroyed or jeopardized over and over again. We appreciate this more vividly than ever because of the terrific experiences, calamities, and passions of our own time.[59]

Unfortunately, the Stoics had accepted too lightly all kinds of Pythagorean, Heracleitean, and Platonic fantasies; the benefits of Stoic morality were weakened, because it was combined wtih a poor cosmology and with the astral religion. In spite of the charity that informed it, it was too abstract, too scientific, to satisfy the uneducated people, who were the vast majority. Stoicism became a creed, but a creed without rituals and without wonders, which left the eyes dry and the heart cold; it could not compete with the ritualistic and miraculous religions which gave comfort in spite of endless misery and promised salvation in the midst of terror. Such as it was, Stoic ethics, combined with bad science and cold religion, was the last barrier of paganism against Christianity; we are not surprised at its failure, but rather at its relative popularity.

BRIEF HISTORY OF THE SCHOOL

The whole of Stoic philosophy was already developed in Zenon's time, and even before the end of the century, but we must tell briefly its later evolution, for one cannot appreciate the seed before one has observed its germination and watched the buds, the flowers, and the fruits.

Zenon was succeeded by his disciple Cleanthes of Assos (III-1 B.C.), who was head of the Stoa from 264 to 232.[60] The following headmasters were Chrysippos of Soloi (III-2 B.C.), Zenon of Tarsos (c. 208–180), Diogenes of Seleucia (II-1 B.C.), who carried Stoicism to Rome in 156–155,[61] Antipatros of Tarsos, Panaitios of Rhodes (II-2 B.C.). Panaitios was thus the seventh headmaster; he lived for a time with Polybios (II-1 B.C.) in Rome and completed the Stoicization of the Roman elite that Diogenes had begun. His main disciple, Poseidonios of Apamea (I-1 B.C.), settled in Rhodes, where Cicero attended his lectures in 78.

These men were headmasters (*prostatai*) and philosophers; they did not modify

[59] In order to illustrate the present-day fundamental conflict on that subject, consider on the one hand the ideal explained by Wendell Willkie in *One world* (New York: Simon and Schuster, 1943) and on the other the fact that the word "cosmopolitan" has become a term of abuse in the Russian language. From the point of view of intransigent orthodoxy, tolerance is nothing but infidelity; from the Soviet point of view, cosmopolitanism is treason.

[60] Two other immediate disciples of Zenon must be named, Ariston of Chios and Herillos of Carthage. Ariston was a more thoroughgoing Cynic than his master and despised all forms of culture. He was one of the first to exaggerate ethics (in comparison with logic and physics); that exaggeration became typical of the whole school. On the contrary, Herillos attached much importance to knowledge (*epistēmē*). About the middle of the third century, Ariston and Arcesilaos of the Academy were the outstanding philosophers in Athens.

[61] This Diogenes hailed from Seleucia on the Tigris. It was during his mastership that Crates of Mallos (II-1 B.C.) wrote the first Greek grammar (lost). Crates was the first director, the founder of the library of Pergamon.

Stoic doctrine in any essential way but each of them continued investigations of his own. Cleanthes was a poet; Chrysippos was a logician and grammarian (his own contributions to Stoic doctrine were so considerable that it used to be said, "Without Chrysippos no Stoa"),[62] Diogenes the Babylonian was interested in grammar, archaeology, divination; Panaitios was chiefly a moralist; Poseidonios was a geographer and astronomer.

Note that all those early Stoics came from Western Asia:[63] the founder, Zenon, came from Cypros, three others came from Cilicia[64] (Chrysippos of Soloi and Zenon and Antipatros of Tarsos), Poseidonios came from Apamea on the Orontes, and Diogenes from Seleucia on the Tigris; three others were closer to the Aegean Sea and the Greek world proper, Cleanthes of Assos (close to Lesbos), Ariston of Chios, and Panaitios of Rhodes. The Stoic teachings were born in Asia, found their form in Athens, and attained their maturity and popularity in Rome.

While Epicureanism was brought to a climax, and almost to an end, by Lucretius (I-1 B.C.), the development of Stoicism was slower and continued longer. Later Stoicism is represented by three giants, Seneca of Cordova (I-2), Epictetos (II-1), and Marcus Aurelius Antoninus (II-2).[65] It is interesting to note that the great emperor created in 176 in Athens four chairs of philosophy, to represent four schools, the Stoic, the Epicurean, the Academic, and the Peripatetic. This illustrates the generosity and tolerance of Marcus Aurelius, and the survival of these four schools and of no others in Athens at the end of the second century.[66] Thus did Plato, Aristotle, Epicuros, and Zenon live until the end of paganism; the triumph of Christianity drove them underground for centuries; yet they are still very much alive today.

[62] This statement should be taken, I think, in a material rather than in a spiritual sense. Thanks to his abundant writings and his logical power, he was the main defender of the Stoa (against the Academy) and its organizer. He strengthened the Stoa in the same way that Theophrastos strengthened the Lyceum. The greatest headmasters are not so much the innovators as those who help to clarify and explain the new teachings.

[63] With the apparent exception of Herillos of Carthage. We do not know whence he came; his birthplace might be Carthage, but as he was an immediate disciple of Zenon of Cition, it is more probable that he came either from Greece or from Western Asia, like the others.

[64] Cilicia was the closest neighbor to the island of Cypros. It was much easier for Cilicians to sail to Cypros than to travel to most places inland, for these could not be reached without crossing the Tauros range. Cypros, the Cilician and the north Syrian coasts formed a geographic unit. Hence, we can say that the two Stoics, Zenon and Chrysippos, and even Poseidonios, came from the same region.

[65] This confirms the statement that Stoicism reached its maturity in Rome, not only in the Roman world but in the city of Rome. Marcus Aurelius was a son of that city; the Spaniard Seneca and the Phrygian Epictetos flourished in it.

[66] At that time Athens had become a provincial city, but it had remained a center of learning and of pagan wisdom. Rome was the metropolis of the empire, Athens the outstanding sanctuary.

EPILOGUE

XXIV

THE END OF A CYCLE

Looking backward, either from the year 300 B.C. or from the more enlightened (?) year of grace A.D. 1950, the greatest achievement, the climax, of the enormously long period that this book has covered seems to be the Aristotelian synthesis. The greatness and wisdom of that synthesis appear equally well whether one considers it against the background of the Greek past, brilliant and adventurous, artistic, lyrical, scientific, or from the point of view of the many-sided discussions that agitated Greek minds during the short twilight of Hellenism.

Aristotle had put in good order the knowledge then available in astronomy, physics, zoölogy, ethics, politics, but in addition he had built up a philosophy that was well documented, rational, and moderate. He established the *via media* that can be traced after him across the ages down to our own day, the *via media* that was followed in the course of time by many Muslim and Jewish philosophers, by St. Thomas, the neo-Thomists, and many Jesuits, as well as by the majority of the men of science. The history of that middle road includes a large part of the history of philosophy and of the history of science; to put it otherwise, when one contemplates the history of science in its wholeness, one can see very distinctly that road passing through it, right in the middle of it, from the fourth century B.C. to the twentieth after Christ.

The very mention of a middle road suggests the existence of many other roads around it, which may converge or diverge but remain different. Such roads existed, and they were followed by such men as the Cynics, the Skeptics, the Epicureans, and the Stoics. The *via media* was very broad, however; it attracted not only Aristotle's own disciples but also the latest alumni of the Academy who had jettisoned the theory of ideas and the political fantasies of Plato. There was more and more concern for ethics and for common-sense politics, and the *via media* would have been even more popular than it was but for the terrible vicissitudes of those hard days. The ancient world was going to pieces — but is not the world always disintegrating? Death is the condition of life, war is the condition of peace, suffering is the condition of happiness. Every coin has its reverse; everything, however beautiful, has its wrong side. The old world was dying in order that a new world might be born.

The twilight of Hellenism may be said to begin in the twenties of the fourth century. Alexander the Great died in 323, Aristotle in 322. The Greek world had lost its independence a few years earlier, in 338. The dissolution of the Alexandrian empire introduced the complexities of the Hellenistic age, and prepared a little later the "new deal" of Roman culture. The death of Aristotle coincided with a kind of philosophic recrudescence, as if all the problems of life and knowledge had to be settled before the night set in. The Lyceum and the Academy were still the main schools, but newer schools were trying to crowd them out, chiefly the Epicurean and the Stoic.

These two schools had come into being largely as revulsions against the Academy and even against the Lyceum (the new schools are always of necessity reactions against the old ones; that is a law of life and death). The Garden of Epicuros and the Porch of Zenon had much in common, aside from their distrust of the Academy, and, judging from the writings that have come down to us, many students must have walked from the Porch to the Garden or vice versa. Later writers, like Seneca and Marcus Aurelius, mixed Epicurean and Stoic teachings and were not always able to decide between the two.

The post-Alexandrian philosophies had unavoidably in common a sense of disillusion.[1] Philosophies as well as religions develop because men in the midst of their recurrent miseries require spiritual comfort; the bodies are trembling and the hearts need solace. The Epicureans and the Stoics realized that need and agreed that man can find comfort in himself and nowhere else; they were able equally to please rational men and to offend and irritate the irrational ones. It is true that Stoic physics included various fantasies, but one might be a good Stoic without bothering about these; Stoic morality was eminently acceptable and comforting. No philosophy has ever done more to reconcile man to his fate.

The Stoics and the Epicureans had little interest in science; their supreme concern was ethics, the conduct of life. To that extent one might say that they agreed in discouraging scientific research; yet there was in that respect an essential difference between them. The Epicureans neglected science but did no harm to it; on the contrary, in so far as they were fighting superstition, they helped to clear the ground for the quest of truth. The Stoics indulged in occultism; they favored divination, their acceptance and fosterage of astral religion was a real betrayal of truth (as men of science understand it). The paradoxical consequence of this was that while the Stoics devoted far more attention to science than the Epicureans they jeopardized its progress.

Aside from their physical theories, the main differences between Stoics and Epicureans concerned the life after death and Providence. According to the Stoics, the dead body returned to the "seminal reason" of the cosmos; according to the Epicureans, it was scattered into atoms. The difference was not essential, for none of them believed in personal immortality,[2] but it was obscured by commentators and controversialists, who jumbled together two different sets of alternatives —

[1] The same remark applies, of course, to Greek literature. The "New Comedy" of Menandros (*c.* 343–*c.* 291) is as typical of this age as the Old Comedy of Aristophanes was of the end of the fifth century. Menandros was a friend of Epicuros and his influence upon the Hellenistic and Roman stage and letters was very great.

[2] Marcus Aurelius could hesitate between the two alternatives. See his autobiography. For example, "Death reduced to the same condition Alexander the Macedonian and his muleteer, for either they were taken back into the same Seminal Reason of the Universe or scattered alike into the atoms" (IV, 24). Marcus preferred the first alternative, but he was not dogmatic. The best discussion of Epicurean and Stoic ideas on the hereafter will be found in Franz Cumont, *Lux perpetua* (Paris: Geuthner, 1949), pp. 109–156 [*Isis 41*, 371 (1950)].

atomism versus nonatomism, Providence versus no-Providence – and dealt with them as if the real alternatives were atomism versus Providence.

The Epicureans combined atomism with no-Providence, and the Stoics, Providence with a denial of atomism. However, these two choices were not exhaustive; one might very well believe in atomism and Providence; this was discovered by Muslim philosophers and again by modern men of science beginning with Gassendi.

By the end of the fourth century, the main branches of science (except physics and chemistry) were established, many fundamental problems had been formulated, and nearly every philosophic attitude had been prefigured.

The various philosophic tendencies were interwoven. When one investigates the life of any philosopher, one generally discovers that he sat at the feet of many masters. This is not surprising, because the opportunities existed, especially in Athens, where it was impossible not to know of the competitive theories that were advocated at the same time, and an honest man in search of the truth would shop a long while before making his choice.

The variety was increased by the size of the Greek world and its ramifications into Asia, Africa, and various parts of Europe outside of the Greek peninsula. That vast Greek world was fairly homogenous, yet local differences were abundant. Although Athens was the main center of attraction, where every philosopher, man of science, or artist would spend at least a part of his life, they used to travel considerably from one end of their homoglot territory to another. Sensitive people living near the boundaries could not help being aware of the feelings and ideas that obtained currency beyond them, and thus exotic ideas, especially religious ideas, could and did permeate inside. We must never forget that to the Greek knowledge, experience, and wisdom were added the superstitions that would come naturally to any people, and little by little the Oriental religions which satisfied more completely their hopes and longings.

During the twilight of Hellenism, thinking men had been offered all the possible alternatives, rationalism versus superstition, cynicism, agnosticism, mysticism, defeatism in all its forms. We may assume that the majority of them had chosen the *via media* of the Peripatos, or the ethical quietism of the Epicureans or of the Stoics.

The main issue, then as now, was not between materialism and spiritualism, but between rationalism and irrationalism. It is amazing to discover that in that early time almost all the Greek philosophers had already realized that. No system of theirs, not even the Epicurean, was purely materialistic; none, not even the Platonic, was purely spiritualistic. They all understood that one needs some kind of matter even for thinking, and that one cannot refute spiritualism except with some kind of mind or spirit. In addition, they had asked all the great questions that we are still trying to answer today.

Hellenism was going down in unique splendor, or rather, it was moving out; one can hardly speak of going down, for it was not a real decadence but the end of an incubation, the preparation for a metamorphosis.

The Greek peoples had been weakened by military and political disasters, by wars and revolutions. It is possible that they had been weakened also (and more deeply) by an infectious disease. During the fourth century malaria became endemic throughout a large part of the Greek world.[3] Malarial conditions may help to account for the fact that the new culture was begun not in Greece proper – Greece was exhausted – but in a Greek colony in Egypt, Alexandria.[4]

The end of the fourth century witnessed the end of a cycle and the emergence of a new one. The Greek spirit was not dead, not by any means; it is immortal; it was resurrected in the following centuries in Alexandria, in Pergamon, in Rhodes, in Rome, and in other places scattered around the Mediterranean Sea. We shall tell the history of that resurrection in the next volume.

[3] William Henry Samuel Jones, *Malaria and Greek history*, with an appendix by Edward Theodore Withington (186 pp.; Manchester, 1909) [*Isis* 6, 47 (1923–24)].

[4] The Greeks called it Alexandria *near* Egypt (*Alexandreia hē pros Aigyptō, Alexandria ad Aegyptum*).

GENERAL BIBLIOGRAPHY

INDEX

GENERAL BIBLIOGRAPHY

Cohen, Morris Raphael (1880–1947), and I. E. Drabkin, *Source book in Greek science* (600 pp.; New York: McGraw-Hill, 1948) [*Isis 40*, 277 (1949)].

Diels, Hermann (1848–1922), *Doxographi graeci. Collegit recensuit prolegomenis indicibusque instruxit* (Berlin, 1879 *Editio iterata*, 864 pp.; Berlin, 1929).

– *Die Fragmente der Vorsokratiker* (612 pp.; Berlin, 1903; ed. 2, 2 vols. in 3, 1906–1910; ed. 3, 3 vols., 1912–1922; ed. 4, 3 vols., Berlin, 1922; ed. 5, by Walther Kranz, 3 vols., Berlin: Weidmann, 1934–35). See Freeman, below.

Freeman, Kathleen, *The pre-Socratic philosophers. A companion to Diels' Fragmente* (500 pp.; Oxford: Blackwell, 1946, reprinted 1949).

This work is seldom if ever mentioned by me, because I did not know of its existence until the end of my own labor. It is listed here because it will be of very great use to scholars who do not read Greek.

Heath, Sir Thomas Little (1861–1940), *History of Greek mathematics* (2 vols.; Oxford, 1921) [*Isis 4*, 532–535 (1921–22)].

– *Manual of Greek mathematics* (568 pp.; Oxford: Clarendon Press, 1931) [*Isis 16*, 450–451 (1931)].

– *Greek astronomy* (250 pp.; London: Dent, 1932) [*Isis 22*, 585 (1934–35)].

Isis. International review devoted to the history of science and civilization. Official journal of the History of Science Society. Founded and edited by George Sarton (43 vols., 1913–1952).

The many references to *Isis* in this volume are generally made to complete in the briefest manner the information given about this or that book or memoir. From them the reader may obtain quickly, if he wishes, either a critical review of the book or some other additional information the development of which would take too much space.

Osiris. Commentationes de scientiarum et eruditionis historia rationeque. Edidit Georgius Sarton (10 vols. Bruges 1936–51).

Oxford classical dictionary (998 pp.; Oxford: Clarendon Press, 1949).

Pauly-Wissowa, *Real-Encyclopädie der classischen Altertumswissenschaft* (Stuttgart, 1894 ff.).

Sarton, George, *Introduction to the history of science* (3 vols. in 5; Baltimore: Williams and Wilkins, 1927–1948).

Tannery, Paul (1843–1904), *Mémoires scientifiques* (17 vols.; Paris, 1912–1950) see *Introduction*, vol. 3, p. 1906.

INDEX

The index is meant to make possible the finding of information of definite persons and subjects; for broad topics such as "mathematics" or "humanities," the Table of Contents will provide better guidance.

References to papyri will be found under the heading "papyrus" (Ebers, Rhind, Smith, etc.); references to definite numbers under the heading "number" (one, two, . . . , sixty).

The spelling of Greek names is explained in the preface; we use the Greek vowels (not the Latin), and we write *u* for *ou*.

The index was compiled under my direction by Frances Siegel; Miss Siegel also typed it and read the proofs.

Summer Solstice 1952
Cambridge, Massachusetts

G. S.

A CATALOG OF SELECTED
DOVER BOOKS
IN ALL FIELDS OF INTEREST

A CATALOG OF SELECTED

DOVER BOOKS

IN ALL FIELDS OF INTEREST

SUNDIALS, Albert Waugh. Far and away the best, most thorough coverage of ideas, mathematics concerned, types, construction, adjusting anywhere. Over 100 illustrations. 230pp. 5⅜ × 8½. 22947-5 Pa. $5.95

PICTURE HISTORY OF THE NORMANDIE: With 190 Illustrations, Frank O. Braynard. Full story of legendary French ocean liner: Art Deco interiors, design innovations, furnishings, celebrities, maiden voyage, tragic fire, much more. Extensive text. 144pp. 8⅞ × 11¾. 25257-4 Pa. $10.95

THE FIRST AMERICAN COOKBOOK: A Facsimile of "American Cookery," 1796, Amelia Simmons. Facsimile of the first American-written cookbook published in the United States contains authentic recipes for colonial favorites—pumpkin pudding, winter squash pudding, spruce beer, Indian slapjacks, and more. Introductory Essay and Glossary of colonial cooking terms. 80pp. 5⅜ × 8½. 24710-4 Pa. $3.50

101 PUZZLES IN THOUGHT AND LOGIC, C. R. Wylie, Jr. Solve murders and robberies, find out which fishermen are liars, how a blind man could possibly identify a color—purely by your own reasoning! 107pp. 5⅜ × 8½. 20367-0 Pa. $2.95

ANCIENT EGYPTIAN MYTHS AND LEGENDS, Lewis Spence. Examines animism, totemism, fetishism, creation myths, deities, alchemy, art and magic, other topics. Over 50 illustrations. 432pp. 5⅜ × 8½. 26525-0 Pa. $8.95

ANTHROPOLOGY AND MODERN LIFE, Franz Boas. Great anthropologist's classic treatise on race and culture. Introduction by Ruth Bunzel. Only inexpensive paperback edition. 255pp. 5⅜ × 8½. 25245-0 Pa. $7.95

THE TALE OF PETER RABBIT, Beatrix Potter. The inimitable Peter's terrifying adventure in Mr. McGregor's garden, with all 27 wonderful, full-color Potter illustrations. 55pp. 4¼ × 5½. (Available in U.S. only) 22827-4 Pa. $1.75

THREE PROPHETIC SCIENCE FICTION NOVELS, H. G. Wells. *When the Sleeper Wakes, A Story of the Days to Come* and *The Time Machine* (full version). 335pp. 5⅜ × 8½. (Available in U.S. only) 20605-X Pa. $8.95

APICIUS COOKERY AND DINING IN IMPERIAL ROME, edited and translated by Joseph Dommers Vehling. Oldest known cookbook in existence offers readers a clear picture of what foods Romans ate, how they prepared them, etc. 49 illustrations. 301pp. 6⅛ × 9¼. 23563-7 Pa. $7.95

SHAKESPEARE LEXICON AND QUOTATION DICTIONARY, Alexander Schmidt. Full definitions, locations, shades of meaning of every word in plays and poems. More than 50,000 exact quotations. 1,485pp. 6½ × 9¼. 22726-X, 22727-8 Pa., Two-vol. set $31.90

THE WORLD'S GREAT SPEECHES, edited by Lewis Copeland and Lawrence W. Lamm. Vast collection of 278 speeches from Greeks to 1970. Powerful and effective models; unique look at history. 842pp. 5⅜ × 8½. 20468-5 Pa. $12.95

A CONCISE HISTORY OF PHOTOGRAPHY: Third Revised Edition, Helmut Gernsheim. Best one-volume history—camera obscura, photochemistry, daguerreotypes, evolution of cameras, film, more. Also artistic aspects—landscape, portraits, fine art, etc. 281 black-and-white photographs. 26 in color. 176pp. 8⅜ × 11¼.
25128-4 Pa. $14.95

THE DORÉ BIBLE ILLUSTRATIONS, Gustave Doré. 241 detailed plates from the Bible: the Creation scenes, Adam and Eve, Flood, Babylon, battle sequences, life of Jesus, etc. Each plate is accompanied by the verses from the King James version of the Bible. 241pp. 9 × 12. 23004-X Pa. $9.95

WANDERINGS IN WEST AFRICA, Richard F. Burton. Great Victorian scholar/adventurer's invaluable descriptions of African tribal rituals, fetishism, culture, art, much more. Fascinating 19th-century account. 624pp. 5⅜ × 8½. 26890-X Pa. $12.95

HISTORIC HOMES OF THE AMERICAN PRESIDENTS, Second Revised Edition, Irvin Haas. Guide to homes occupied by every president from Washington to Bush. Visiting hours, travel routes, more. 175 photos. 160pp. 8¼ × 11.
26751-2 Pa. $9.95

THE HISTORY OF THE LEWIS AND CLARK EXPEDITION, Meriwether Lewis and William Clark, edited by Elliott Coues. Classic edition of Lewis and Clark's day-by-day journals that later became the basis for U.S. claims to Oregon and the West. Accurate and invaluable geographical, botanical, biological, meteorological and anthropological material. Total of 1,508pp. 5⅜ × 8½.
21268-8, 21269-6, 21270-X Pa., Three-vol. set $29.85

LANGUAGE, TRUTH AND LOGIC, Alfred J. Ayer. Famous, clear introduction to Vienna, Cambridge schools of Logical Positivism. Role of philosophy, elimination of metaphysics, nature of analysis, etc. 160pp. 5⅜ × 8½. (Available in U.S. and Canada only) 20010-8 Pa. $3.95

MATHEMATICS FOR THE NONMATHEMATICIAN, Morris Kline. Detailed, college-level treatment of mathematics in cultural and historical context, with numerous exercises. For liberal arts students. Preface. Recommended Reading Lists. Tables. Index. Numerous black-and-white figures. xvi + 641pp. 5⅜ × 8½.
24823-2 Pa. $11.95

HANDBOOK OF PICTORIAL SYMBOLS, Rudolph Modley. 3,250 signs and symbols, many systems in full; official or heavy commercial use. Arranged by subject. Most in Pictorial Archive series. 143pp. 8⅛ × 11. 23357-X Pa. $7.95

INCIDENTS OF TRAVEL IN YUCATAN, John L. Stephens. Classic (1843) exploration of jungles of Yucatan, looking for evidences of Maya civilization. Travel adventures, Mexican and Indian culture, etc. Total of 669pp. 5⅜ × 8½.
20926-1, 20927-X Pa., Two-vol. set $13.90

DEGAS: An Intimate Portrait, Ambroise Vollard. Charming, anecdotal memoir by famous art dealer of one of the greatest 19th-century French painters. 14 black-and-white illustrations. Introduction by Harold L. Van Doren. 96pp. 5⅜ × 8½.
25131-4 Pa. $4.95

PERSONAL NARRATIVE OF A PILGRIMAGE TO AL-MADINAH AND MECCAH, Richard F. Burton. Great travel classic by remarkably colorful personality. Burton, disguised as a Moroccan, visited sacred shrines of Islam, narrowly escaping death. 47 illustrations. 959pp. 5⅜ × 8½.
21217-3, 21218-1 Pa., Two-vol. set $19.90

PHRASE AND WORD ORIGINS, A. H. Holt. Entertaining, reliable, modern study of more than 1,200 colorful words, phrases, origins and histories. Much unexpected information. 254pp. 5⅜ × 8½. 20758-7 Pa. $5.95

THE RED THUMB MARK, R. Austin Freeman. In this first Dr. Thorndyke case, the great scientific detective draws fascinating conclusions from the nature of a single fingerprint. Exciting story, authentic science. 320pp. 5⅜ × 8½. (Available in U.S. only) 25210-8 Pa. $6.95

AN EGYPTIAN HIEROGLYPHIC DICTIONARY, E. A. Wallis Budge. Monumental work containing about 25,000 words or terms that occur in texts ranging from 3000 B.C. to 600 A.D. Each entry consists of a transliteration of the word, the word in hieroglyphs, and the meaning in English. 1,314pp. 6⅜ × 10.
23615-3, 23616-1 Pa., Two-vol. set $35.90

THE COMPLEAT STRATEGYST: Being a Primer on the Theory of Games of Strategy, J. D. Williams. Highly entertaining classic describes, with many illustrated examples, how to select best strategies in conflict situations. Prefaces. Appendices. xvi + 268pp. 5⅜ × 8½. 25101-2 Pa. $6.95

THE ROAD TO OZ, L. Frank Baum. Dorothy meets the Shaggy Man, little Button-Bright and the Rainbow's beautiful daughter in this delightful trip to the magical Land of Oz. 272pp. 5⅜ × 8. 25208-6 Pa. $5.95

POINT AND LINE TO PLANE, Wassily Kandinsky. Seminal exposition of role of point, line, other elements in nonobjective painting. Essential to understanding 20th-century art. 127 illustrations. 192pp. 6½ × 9¼. 23808-3 Pa. $5.95

LADY ANNA, Anthony Trollope. Moving chronicle of Countess Lovel's bitter struggle to win for herself and daughter Anna their rightful rank and fortune—perhaps at cost of sanity itself. 384pp. 5⅜ × 8½. 24669-8 Pa. $8.95

EGYPTIAN MAGIC, E. A. Wallis Budge. Sums up all that is known about magic in Ancient Egypt: the role of magic in controlling the gods, powerful amulets that warded off evil spirits, scarabs of immortality, use of wax images, formulas and spells, the secret name, much more. 253pp. 5⅜ × 8½. 22681-6 Pa. $4.95

THE DANCE OF SIVA, Ananda Coomaraswamy. Preeminent authority unfolds the vast metaphysic of India: the revelation of her art, conception of the universe, social organization, etc. 27 reproductions of art masterpieces. 192pp. 5⅜ × 8½.
24817-8 Pa. $6.95

CHRISTMAS CUSTOMS AND TRADITIONS, Clement A. Miles. Origin, evolution, significance of religious, secular practices. Caroling, gifts, yule logs, much more. Full, scholarly yet fascinating; non-sectarian. 400pp. 5⅜ × 8½.
23354-5 Pa. $7.95

THE HUMAN FIGURE IN MOTION, Eadweard Muybridge. More than 4,500 stopped-action photos, in action series, showing undraped men, women, children jumping, lying down, throwing, sitting, wrestling, carrying, etc. 390pp. 7⅞ × 10⅝.
20204-6 Cloth. $24.95

THE MAN WHO WAS THURSDAY, Gilbert Keith Chesterton. Witty, fast-paced novel about a club of anarchists in turn-of-the-century London. Brilliant social, religious, philosophical speculations. 128pp. 5⅜ × 8½. 25121-7 Pa. $3.95

A CÉZANNE SKETCHBOOK: Figures, Portraits, Landscapes and Still Lifes, Paul Cézanne. Great artist experiments with tonal effects, light, mass, other qualities in over 100 drawings. A revealing view of developing master painter, precursor of Cubism. 102 black-and-white illustrations. 144pp. 8¾ × 6⅜. 24790-2 Pa. $6.95

AN ENCYCLOPEDIA OF BATTLES: Accounts of Over 1,560 Battles from 1479 B.C. to the Present, David Eggenberger. Presents essential details of every major battle in recorded history, from the first battle of Megiddo in 1479 B.C. to Grenada in 1984. List of Battle Maps. New Appendix covering the years 1967–1984. Index. 99 illustrations. 544pp. 6½ × 9¼. 24913-1 Pa. $14.95

AN ETYMOLOGICAL DICTIONARY OF MODERN ENGLISH, Ernest Weekley. Richest, fullest work, by foremost British lexicographer. Detailed word histories. Inexhaustible. Total of 856pp. 6½ × 9¼.
21873-2, 21874-0 Pa., Two-vol. set $19.90

WEBSTER'S AMERICAN MILITARY BIOGRAPHIES, edited by Robert McHenry. Over 1,000 figures who shaped 3 centuries of American military history. Detailed biographies of Nathan Hale, Douglas MacArthur, Mary Hallaren, others. Chronologies of engagements, more. Introduction. Addenda. 1,033 entries in alphabetical order. xi + 548pp. 6½ × 9¼. (Available in U.S. only)
24758-9 Pa. $13.95

LIFE IN ANCIENT EGYPT, Adolf Erman. Detailed older account, with much not in more recent books: domestic life, religion, magic, medicine, commerce, and whatever else needed for complete picture. Many illustrations. 597pp. 5⅜ × 8½.
22632-8 Pa. $9.95

HISTORIC COSTUME IN PICTURES, Braun & Schneider. Over 1,450 costumed figures shown, covering a wide variety of peoples: kings, emperors, nobles, priests, servants, soldiers, scholars, townsfolk, peasants, merchants, courtiers, cavaliers, and more. 256pp. 8⅜ × 11¼. 23150-X Pa. $9.95

THE NOTEBOOKS OF LEONARDO DA VINCI, edited by J. P. Richter. Extracts from manuscripts reveal great genius; on painting, sculpture, anatomy, sciences, geography, etc. Both Italian and English. 186 ms. pages reproduced, plus 500 additional drawings, including studies for *Last Supper*, *Sforza* monument, etc. 860pp. 7⅞ × 10¾. (Available in U.S. only) 22572-0, 22573-9 Pa., Two-vol. set $35.90

THE ART NOUVEAU STYLE BOOK OF ALPHONSE MUCHA: All 72 Plates from "Documents Decoratifs" in Original Color, Alphonse Mucha. Rare copyright-free design portfolio by high priest of Art Nouveau. Jewelry, wallpaper, stained glass, furniture, figure studies, plant and animal motifs, etc. Only complete one-volume edition. 80pp. 9⅝ × 12¼. 24044-4 Pa. $9.95

ANIMALS: 1,419 COPYRIGHT-FREE ILLUSTRATIONS OF MAMMALS, BIRDS, FISH, INSECTS, ETC., edited by Jim Harter. Clear wood engravings present, in extremely lifelike poses, over 1,000 species of animals. One of the most extensive pictorial sourcebooks of its kind. Captions. Index. 284pp. 9 × 12. 23766-4 Pa. $9.95

OBELISTS FLY HIGH, C. Daly King. Masterpiece of American detective fiction, long out of print, involves murder on a 1935 transcontinental flight—"a very thrilling story"—NY Times. Unabridged and unaltered republication of the edition published by William Collins Sons & Co. Ltd., London, 1935. 288pp. 5⅜ × 8½. (Available in U.S. only) 25036-9 Pa. $5.95

VICTORIAN AND EDWARDIAN FASHION: A Photographic Survey, Alison Gernsheim. First fashion history completely illustrated by contemporary photographs. Full text plus 235 photos, 1840–1914, in which many celebrities appear. 240pp. 6½ × 9¼. 24205-6 Pa. $8.95

THE ART OF THE FRENCH ILLUSTRATED BOOK, 1700–1914, Gordon N. Ray. Over 630 superb book illustrations by Fragonard, Delacroix, Daumier, Doré, Grandville, Manet, Mucha, Steinlen, Toulouse-Lautrec and many others. Preface. Introduction. 633 halftones. Indices of artists, authors & titles, binders and provenances. Appendices. Bibliography. 608pp. 8⅜ × 11¼. 25086-5 Pa. $24.95

THE WONDERFUL WIZARD OF OZ, L. Frank Baum. Facsimile in full color of America's finest children's classic. 143 illustrations by W. W. Denslow. 267pp. 5⅜ × 8½. 20691-2 Pa. $7.95

FOLLOWING THE EQUATOR: A Journey Around the World, Mark Twain. Great writer's 1897 account of circumnavigating the globe by steamship. Ironic humor, keen observations, vivid and fascinating descriptions of exotic places. 197 illustrations. 720pp. 5⅜ × 8½. 26113-1 Pa. $15.95

THE FRIENDLY STARS, Martha Evans Martin & Donald Howard Menzel. Classic text marshalls the stars together in an engaging, non-technical survey, presenting them as sources of beauty in night sky. 23 illustrations. Foreword. 2 star charts. Index. 147pp. 5⅜ × 8½. 21099-5 Pa. $3.95

FADS AND FALLACIES IN THE NAME OF SCIENCE, Martin Gardner. Fair, witty appraisal of cranks, quacks, and quackeries of science and pseudoscience: hollow earth, Velikovsky, orgone energy, Dianetics, flying saucers, Bridey Murphy, food and medical fads, etc. Revised, expanded In the Name of Science. "A very able and even-tempered presentation."—The New Yorker. 363pp. 5⅜ × 8. 20394-8 Pa. $6.95

ANCIENT EGYPT: ITS CULTURE AND HISTORY, J. E Manchip White. From pre-dynastics through Ptolemies: society, history, political structure, religion, daily life, literature, cultural heritage. 48 plates. 217pp. 5⅜ × 8½. 22548-8 Pa. $5.95